Rotifera VIII: A Comparative Approach

Developments in Hydrobiology 134

Series editor
H. J. Dumont

Rotifera VIII: A Comparative Approach

Proceedings of the VIIIth International Rotifer Symposium,
held in Collegeville, Minn., U.S.A., 22–27 June 1997

Edited by

E. Wurdak, R. Wallace and H. Segers

Reprinted from Hydrobiologia, volumes 387/388 (1998)

Springer-Science+Business Media, B.V.

Library of Congress Cataloging-in-Publication Data

International Rotifer Symposium (8th : 1997 : Collegeville, Minn.)
    Rotifera VIII : a comparative approach / edited by E. Wurdak, R.
Wallace, and H. Segers.
        p.   cm. -- (Developments in hydrobiology ; 134)
    ISBN 978-94-010-6009-7   ISBN 978-94-011-4782-8 (eBook)
    DOI 10.1007/978-94-011-4782-8
    1. Rotifera--Congresses.   I. Wurdak, E. (Elizabeth) II. Wallace,
R. (Robert), A.M.   III. Segers, Hendrik. IV. Title. V. Series.
QL391.R8I57 1997
592'.52--dc21   98-52160

ISBN 978-94-010-6009-7

Printed on acid-free paper

Hydrobiologia **387/388:** v–ix, 1998.
E. Wurdak, R. Wallace & H. Segers (eds), Rotifera VIII: A Comparative Approach.

Contents

Part VI. Ecosystem Ecology: Lotic Systems

Part VII. Chemical Ecology

Part VIII. Autecology and Population Ecology

Part IX. Population Genetics

Part X. Trophic Interactions

Part XI. Applied Studies and Aquaculture

Hydrobiologia **387/388:** xi–xii, 1998.
E. Wurdak, R. Wallace & H. Segers (eds), Rotifera VIII: A Comparative Approach.

Preface

The VIIIth International Rotifer Symposium was organized by Elizabeth S. Wurdak at St. John's University in Collegeville, Minnesota, USA during the week of June 22, 1997. It was the first rotifer symposium held outside Europe. Ninety-seven participants from 22 countries distributed over 4 continents attended.

Following the tradition of earlier symposia, the conference was divided into oral sessions, poster sessions and workshops. Contributions were initially grouped into ecological studies, population structure, genetics, mating, phylogeny, biogeography, systematics, habitat, life history, biotechnology, culture, and cell biology. Morning and afternoon sessions began with an invited review followed by individual presentations of recent research findings. The presentations were punctuated by discussions over coffee. Home movies of rotifer behavior were screened in the evening amidst lively talk and amusement over the antics of male brachionids.

At the mid-week banquet the assembled rotiferologists paid their respect to two international-known scientists for their long-term commitment to research on rotifers: Professors Ludmillia Kutikova (Russia) and Birger Pejler (Sweden). Dr Kutikova was present and accepted, with much graciousness, a certificate of achievement and the brief speech made by Bob Wallace on her behalf. The family of rotiferologists provided a hardy round of applause. A similar tribute to Dr Pejler, who because of a sudden illness had to cancel his plans to attend the meetings, follows this preface.

During breaks and after formal sessions participants were free to explore the lakes, woodlands and prairie which surround St. John's University and to enjoy the hospitality offered by Br. Willie's Pub, the SJU summer events staff and the City of St. Cloud. Highlights of the meeting, other than the presentations, were the Mississippi River cruise on board the *Anne Bonnie*, picnic at the Stearns County Park along the River, banquet and dance at the College of Saint Benedict, post-conference excursions to Lake Itasca State Park and the Twin Cities of Minneapolis and Saint Paul.

The success of the conference was a result of the hard work of the organizers, the exemplary cooperation of the participants, financial support from private donors, industry (i.e., Aquaculture Systems®, Chlorella Industries®) and assistance from the CSB/SJU Biology Department and the Science Hall Secretarial Staff.

Many things have changed since the previous rotifer meeting. Rotifers have gone into space and, unlike earlier symposia, the VIIIth Rotifer Symposium was somewhat unique in that it had a unifying theme, the comparative approach, which can be traced through the contributions included in this volume. Following we highlight some examples of this theme within specific papers.

In an invited review, microplankton responses to habitat disturbance in arctic lakes are compared to their responses in temperate lakes for which more extensive data are available. In another contribution the effects of disturbance on the rotifer community within an experimental lake are documented and the progress of recovery after habitat restoration is followed. Manipulated and unmanipulated lakes in the same geographical area are contrasted. The effect of colonization history and nutrient levels on zooplankton assemblages are explored in separate reports. Several studies examine spatial and temporal variability in rotifer abundance, and community structure in relation to variations in biotic and abiotic factors along gradients in natural settings, others approach these questions in an experimental way. Population responses to diet and chemical signals are tested in the field and through manipulation of laboratory cultures. The reproductive strategy of different rotifer taxa is investigated, by attempts to correlate egg size with life history parameters in order to evaluate the adaptive merits of large *versus* small eggs. The status of various rotifer taxa and species boundaries are reexamined in the light of results obtained by the most recent biochemical and molecular approaches, and other modern techniques (scanning electron microscopy, and cladistics and numerical techniques). The broader issue of the relationships of the Rotifera to other invertebrate taxa is taken up in a number of provocative papers. The distribution of Rotifera in remote geographical regions is explored by several authors. All these papers reflect the wide diversity of approaches, methods of analysis and conclusions that have been the hallmark of the rotifer symposia since their inception.

Kluwer Academic Publishers and Dr H. J. Dumont have graciously accepted the publication of the symposium proceedings in a special volume in the series Developments in Hydrobiology. The manuscripts published in this volume have undergone a careful review and revision process. Portions were further amended by the editors for the sake of clarity and shortened to fit the allotted number of pages. The final product is a result of the efforts of the authors, reviewers and editors.

The scientific committee and the organizers were gratified to note the participation of amateur colleagues at this and the previous symposia. Their enthusiasm invigorates the professional researchers and enlivens debate. Even as we finish the task of editing the *Proceedings of the VIIIth Rotifer Symposium*, we are looking forward to seeing new and old faces at the IXth Rotifer Symposium scheduled for Thailand in the year 2000.

The Editors
ELIZABETH WURDAK
ROBERT WALLACE
HENDRIK SEGERS

Hydrobiologia **387/388:** xiii, 1998.
E. Wurdak, R. Wallace & H. Segers (eds), Rotifera VIII: A Comparative Approach.

Dedication

This volume is dedicated to Professor Birger Pejler in recognition of his lifelong contribution to the advancement of our understanding of the biology of the Rotifera.

Birger Pejler was born on February 3, 1924, in Lindesberg, a small town outside Orebro in central Sweden. At that time, young school children were required to collect plants (hopefully, not rare species!). Though Birger was ambivalent about this assignment at first, his enthusiasm, fostered by the influence of his mother, grew. He also took an interest in birds. He attributes his choice of biology as a career to the natural beauty of the surroundings of his home town and the richness of its flora and fauna.

Birger enrolled at Uppsala University for his doctoral work. His first research project, under the direction of Professor Sven Ekman, was a study of the highest lakes in northern Sweden. Birger completed his dissertation on planktonic rotifers in 1957. His ideas on introgressive hybridization stimulated much discussion. Though his thesis was in zoology, he was employed as professor in the Department of Limnology which was under the directorship of Wilhelm Rhode. His original duties were teaching and administration. These were followed by the training of PhD students. In his teaching Birger placed a heavy emphasis on ecology, a field which had been largely neglected in Sweden up until that point.

The central theme of Birger's research has been the analysis of biotic and abiotic factors that play a role in rotifer distribution. His observations were based largely on his own collections. However, after the early death of Bruno Berzins, Birger took over the large body of accumulated data and based some of his scientific work on it.

We regret that poor health prevented Professor Pejler from joining us in Minnesota for the VIIIth International Rotifer Symposium. He had a perfect record of attendance at previous symposia and was looking forward to visiting with the descendants of relatives from Sweden who had emigrated to America. A report on the history of rotifer research in northern Europe by Birger Pejler is included in this volume.

ELIZABETH WURDAK
on behalf of the family of Rotiferologists

Participants in the VIIIth International Rotifer Symposium

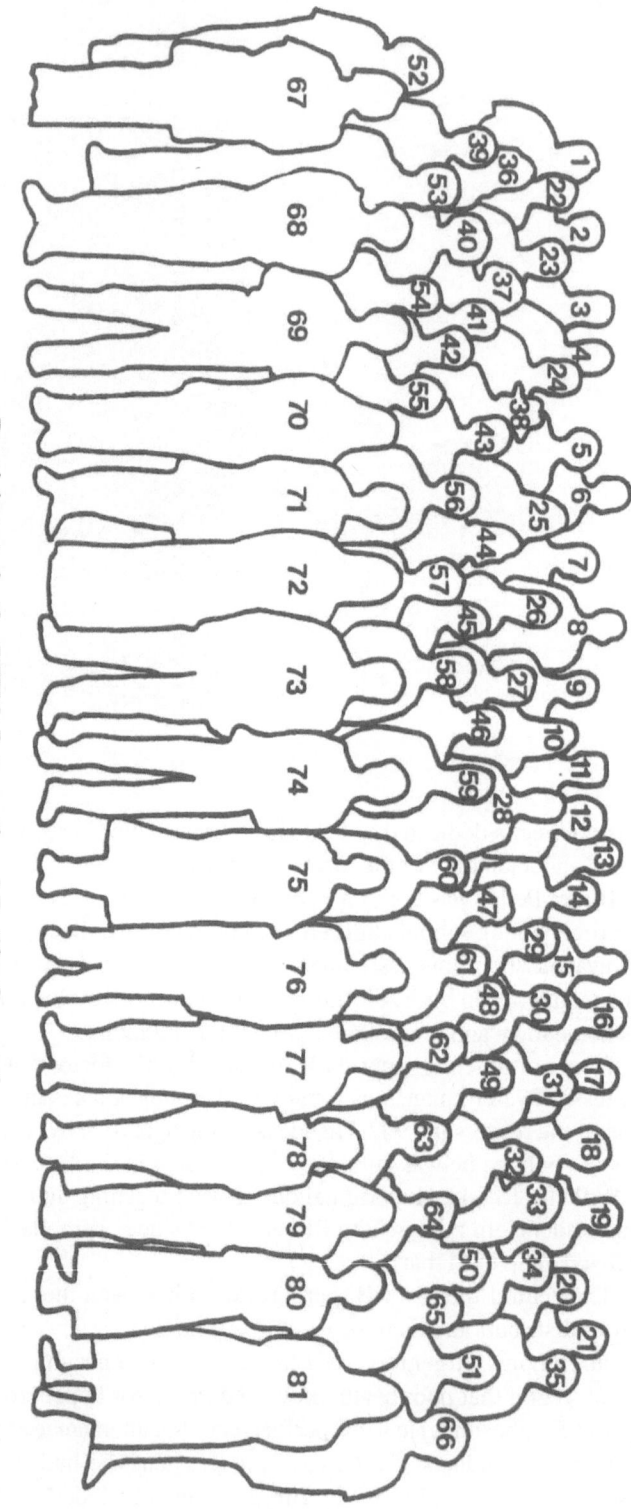

1. Tony Saunders-Davies
2. Christian Jersabek
3. Willem De Smet
4. Charles King
5. Jolanta Ejsmont-Karabin
6. Howard Taylor
7. Manuel Yúfera
8. David Mark Welch
9. Yves Mameffe
10. Jessica Mark Welch
11. Walter Kleinow
12. Beatriz Modenutti
13. Esteban Balseiro
14. Maria Diéguez
15. Jim Garey
16. Rainer Deneke
17. Tom Schroder
18. Tomonari Kotani
19. John Gilbert
20. Taavi Virro
21. Jim Green

22. Veronica De La Riva
23. Marcelo Silva-Briano
24. Jose M. Conde-Porcuna
25. Sharon Dornhoff
26. Veronique Gosselain
27. Tom Nogrady
28. Giulio Melone
29. Brian Dingmann
30. Atsushi Hagiwara
31. Tom Frost
32. Diethelm Ronneberger
33. Irena Tetesh
34. Juta Haberman
35. Eduardo Aparici
36. Liz Walsh
37. Lois Bateman
38. Joseph Boxhorn
39. Yan Zhuge
40. Maria Miracle
41. Terry Snell

42. Hachiro Hirata
43. Manuel Serra
44. Simona Orsenico
45. Manuela Caprioli
46. Bob Wallace
47. Eric Hollowday
48. Isao Maruyama
49. Norbert Walz
50. Ludmilla Kutikova
51. Tony Baumert
52. Liane Cochran-Stafira
53. Ignacio Pérez-Legaspi
54. Tania De Wolf
55. Amina Pollard
56. Claus-Peter Stelzer
57. Rebeca Rueda
58. Manuel Elías-Gutierrez
59. Paul Turner
60. Claudia Ricci
61. B.K. Sharma

62. Ian Duggan
63. Miloslav Devetter
64. Elena Kotikova
65. Russ Shiel
66. Martin Boraas
67. Erica Estrada
68. Pomsilp Pholpuntin
69. Hendrik Segers
70. La-orsri Sanoamuang
71. África Gómez
72. Jenny Schmid-Araya
73. Roberto Rico-Martinez
74. Ramesh Gulati
75. Elizabeth Wurdak
76. K.R. Rao
77. Aydin Orstan
78. Maria González
79. Stephanie Hampton
80. Linda May
81. Diane Seale

Participants in the VIIIth International Rotifer Symposium

Hydrobiologia **387/388**: 1–8, 1998.
E. Wurdak, R. Wallace & H. Segers (eds), Rotifera VIII: A Comparative Approach.
© *1998 Kluwer Academic Publishers.*

History of rotifer research in northern Europe

Birger Pejler
Institute of Limnology, Norbyvägen 20, 752 36 Uppsala, Sweden

Key words: rotifera, history of research, northern Europe

Abstract

The research of rotifers in northern European countries is recorded and discussed, beginning with Carolus Linnaeus and finishing in recent times. The general development of rotifer research is reflected as well, ever since the first time, dominated by morphological taxonomy, and up till now, when also modern fields, like biochemistry, ecotoxicology and aquaculture, are included.

Introduction

At two previous rotifer symposia historical surveys were published, one about western European research (Koste & Hollowday, 1993) and one about Polish rotiferology (Hillbricht-Ilkowska, 1995). As the first-mentioned paper includes Denmark now only the remaining nordic countries are treated, i.e. Finland, Iceland, Norway and Sweden. If all short notes and faunistic reports would be mentioned, a very long list would result; hence the present publication is restricted to the contributions which have been considered most important, and some marine species lists were excluded.

The pioneers

According to Koste & Hollowday (1993) the first proper description of a rotifer was made by the inventor of the microscope, van Leeuwenhoek in Holland. In Scandinavia another biological pioneer, Carolus von Linnaeus, included rotifers in his wide interests, and in his fundamental work, Systema Naturae (1758), he mentions three species, using his binomial nomenclature, i.e. *Tubipora urceus*, *Serpula ringens* and *Hydra socialis*. It is, however, characteristic of the taxonomic development that all these species have changed names one or several times. They are now named *Brachionus urceolaris*, *Floscularia ringens* and *Sinantherina socialis*, respectively. At the time of Linnaeus, Rotifera was not recognized as a group of its own, and though the three above mentioned spe-

cies have all been regarded as belonging to the class 'Vermes', they were included in different subgroups in Linnaeus' system: *Brachionus* in 'Lithophyta', *Floscularia* in 'Testacea' and *Sinantherina* in Zoophyta'. In this way the rotifers were brought together with other completely unrelated animals.

It is difficult to find out in which way Linnaeus obtained his results, e.g. how he collected his material, and I have found no indication that he used microscopes for this purpose. However, it is quite evident that he had a very extensive correspondence. Thus, it is conceivable that the original findings may have been made by other investigators.

The taxonomic-faunistic era

After Linnaeus it took a rather long time until any noticeable contribution from the Nordic countries can be found, except for the Dane, Otto Friedrich Müller (treated in detail by Koste & Hollowday). However, during the last part of the nineteenth century a large number of zoologists showed interest in rotifer biology. These researchers worked with different animal groups, both marine and limnic. A typical exponent of this category was Nils von Hofsten at Uppsala University in Sweden. He wrote his thesis about turbellarians but later studied a number of invertebrates, as well as diverse general disciplines, e.g. zoogeography, genetics, evolution and history of biology. For a time he was also the Vice-Chancellor (President) of Uppsala University and in that capacity he played an important political role. His rotifer papers were published

2

Plate 1. Carolus Linnaeus (Carl von Linné) 1707–1778. Inga Borg & Fibben Hald: I naturens riken. 1979. Litteraturfrämjandet.

relatively early in his life (1909, 1912, 1923) and constitute only a small part of his production. However, they are of high quality and represent the most serious Swedish contributions to rotifer biology produced to that date.

The publications of this time are mostly of a taxonomic and morphological nature. From Sweden, Bergendal (1892), Runnström (1909, 1925), Olofsson (1917, 1918) and Lang (1928) may be mentioned; from Finland, Stenroos (1898) and Levander (1894a, 1898b); and from Norway, Lie-Pettersen (1905, 1909, 1911) and Huitfeldt-Kaas (1906). While a higher activity could be observed in central Europe (see Koste and Hollowday, *op. cit.*), studies in northern Europe

contributed considerably to the general taxonomic knowledge amassed during this time.

Since some of these papers deal with arctic areas it might have been expected to find cold-stenotherm endemics. This, however, was not the case. Olofsson described several new species from Spitsbergen, but only three of them have later been shown to be restricted to the arctic region, viz. *Notholca latistyla*, *Lecane rotundata* and *Trichocerca longistyla*. However, due to the low number of species a principally new result was obtained: it was found that the first generation of *Polyarthra* has a deviating apperance, the aptera form. Because in Spitsbergen only one species of this genus is present, *P. dolichoptera*, it was easier to draw this

Plate 2. Nils von Hofsten 1881–1967. Zool. Bidr. Uppsala 25 (festskrift) 1947.

conclusion here than in other areas housing several *Polyarthra* species with different seasonal strategies. For Olofsson the result was merely a supposition, but proof for these findings was obtained later by Nipkow (1952).

Advancement of taxonomy

The early taxonomy was based upon subjective descriptions. However, introduction of a quantitative approach resulted in a major breakthrough. The honour for this goes to the Swede Börje Carlin. His first investigations were in a more traditional style, but in his thesis (1943) biometric methods were introduced. This was made possible by the availability of a very large collection of planktic rotifers. The combination

with ecological data was also essential. In this way it was possible to detect species which were previously 'hidden' within muddled form complexes. E.g., Carlin was able to separate the cold-water species *Keratella hiemalis* within the 'Anuraea aculeata-group' which had caused so much trouble for previous authors. Another good example is shown by the reorganization of the genus *Polyarthra*. After his important thesis Carlin devoted the rest of his life to fishery biology, and for many years he was the director of the Swedish Salmon Research Institute. The more modern taxonomic work was continued by Pejler, who also found evidence of introgressive hybridization (1956) and analysed the background of variation in rotifers (1957a, 1962c, 1980). The material investigated by Pejler was collec-

4

Plate 3. Börje Carlin 1910–1971. Courtesy Nils Johansson.

ted in Swedish Lapland (1957b, 1962b), and central Sweden (1957c). The variation in *Keratella cochlearis* was studied by Eloranta (1982) as well. Godske Eriksen (1968, 1969) and Godske Björklund (1972; same person) instead devoted herself mainly to the marine rotifer fauna in Finland and Norway.

The advance of microscopy has given rise to new morphological results, though this has been more evident in other parts of the world: I have found only one Nordic paper based upon electron microscopy, viz. Hendelberg et al., 1979.

Ecology

Carlin was also a pioneer regarding ecological factors, and in his thesis he distinguished several ecological groups based on temperature requirements. The relation of rotifer distribution to temperature was studied also by Pejler (1957b, 1961, 1962c), Nauwerck (1963), Hakkari (1969) and Bérziņš & Pejler (1989a),

and the importance of oxygen was discussed by Pejler (1961), Larsson (1971) and Bérziņš & Pejler (1989b). Extremely cold water lakes can be found in northern Europe. The highest situated lake in Scandinavia was investigated by Blakar & Jacobsen (1979), whereby 14 rotifer species were found. Still more extreme environments were studied by Amrén (1964a–c) in Spitsbergen. The rotifer fauna here was very poor, favouring the discernment of generations (introduced by Olofsson; see above), and Amrén could trace this phenomenon for *Keratella quadrata*. As regards this species an internal rhythm could also be suspected, because the formation of mictic forms was seemingly independent of environmental factors.

The dependence of rotifers on pH has become more and more in focus in light of the ongoing acidification in large parts of Scandinavia and Finland (see Bérziņš & Pejler, 1987; Morling & Pejler, 1990 and references herein). Similarly, the role of trophic degree was discussed by Pejler (1957c, 1965 and 1981),

Hakkari (1972) and Bérziņš & Pejler (1989c). The effect on zooplankton by the water regulation of a lake in northern Sweden was investigated by Axelson (1961), who found that the impoundment caused a considerable increase in the number of organisms as well as the total zooplankton biomass. As regards the choice of food an important investigation was made by Nauwerck (1963), who stressed the importance of detritus and bacteria for zooplankton. Among the algae, chrysomonads and cryptomonads were shown to be especially important. The preference for the cryptomonad *Rhodomonas minuta* by rotifers was pointed out by Lindström & Pejler (1975), Pejler (1977b) and Lindström (1983). Little is known about the predation upon rotifers (by other invertebrates as well as by fish). However, Stenson (1982, 1983) reported some interesting data regarding direct and indirect effects of predation. Also, competition between different rotifer species may be an essential factor for their distribution, as thoroughly studied in a field experiment performed by Grundström (1987).

Recent research is often characterized by an approach involving population ecology, often connected with studies of other organisms in the same lake. Such investigations were made in northern Sweden by Persson (1978), in a Norwegian mountain region by Larsson (1978), in Iceland by Adalsteinsson (1979) and Antonsson (1992), in Finnish lakes by Hakkari (1978) and Cajander (1983), and in the Baltic sea by Johansson (1983).

Because the research has been more and more orientated towards quantitative approaches and population ecology, the main focus during recent years has been on planktic species. However, the benthic and periphytic rotifers have not been entirely neglected in the past. In an article published 1962 (Pejler, 1962b) the taxonomic and ecological problems as regards non-planktic species were discussed. Moreover, in a recent paper (Pejler, 1995), the choice of habitat by non-planktic species was reviewed. In addition, a periphytic species in the Baltic, viz. *Proales reinhardti*, was analysed more in detail by Jansson (1969).

Bruno Bérziņš

In the context of nordic rotifer research Bruno Bérziņš should be especially remembered. He came to Sweden as a refugee from Latvia during the second World War (see further Pejler, 1987, where also a photo is included) and performed most of his rotifer work in our country. A long row of papers were published, with several different approaches. His nice study of the Collothecacean Rotatoria (1951) ought especially to be mentioned, containing, amongst other things, the description of several new species. Furthermore, sponges were detected as a new biotope for rotifers (1950). The investigation of the small Lake Skärshultsjön (1958) is also remarkable, in which both horizontal and vertical distributions of rotifers were simultaneously registered. Bérziņš was engaged as a plankton specialist in many limnological investigations. A large zooplankton material was hereby collected, and was subsequently analysed and partly published, together with me (see above and the references). Also marine rotifers were part of his interest (see e.g., 1952). Bérziņš received rotifer samples from diverse parts of the world, giving rise to a large number of taxonomic descriptions and revisions (see the reference list in the determination guide by Koste, 1978). As a result of this work many new species were described, and even some new genera, viz. *Pleurotrochopsis* and *Paracephalodella* from Sweden (1973 and 1976, respectively), and *Repauliana* and *Veltae* from Madagascar (1960 and 1982, respectively). In this way Bérziņš made an important contribution to the present faunistic and taxonomic knowledge of Nordic rotifers, and at present no less than 724 species have been registered from these countries (according to Pejler, 1993). Although the importance of Bruno Bérziņš in rotifer biology is appreciated, it should be emphasized that some of his conclusions have recently been questioned (Segers, 1995).

Biogeography

During his taxonomic studies Bérziņš often also discussed biogeographical problems, e.g. in connection with the revisions of the *Keratella valga* group (1955) and *Anuraeopsis* (1962). Thomasson included rotifers in his comprehensive investigations of lakes in Brazil (1953, 1971), Argentine (1959) and Chile (1963), whereby *Lecane melini, Lecane armata, Lecane brundinii* and *Notholca haueri* were described. The probable immigration of *Kellicottia bostoniensis* from America was stated by Carlin (1943) and Arnemo et al. (1968) in Sweden and by Eloranta (1988) in Finland. Finally, Pejler (1977a) wrote a survey about the global distribution of the family Brachionidae.

Aquaculture and biochemistry

During the last decades marine aquaculture has largely been connected with mass cultivation of *Brachionus plicatilis*. Within northern Europe only Norway has practiced this industry at a larger scale. In this context several investigations have been performed, with the purpose of improving the techniques and analysing adjacent theoretical problems (see Korstad et al., 1989a, b; Olsen et al., 1989; Olsen et al., 1993; Øie & Olsen, 1993; Skjermo & Vadstein, 1993; Korstad et al., 1995).

Final comments

It is apparent from the above that the dominating directions of Nordic rotifer research have varied strongly during the course of the past two centuries, and has been connected with great theoretical and technical progress. The development of Nordic rotifer research is largely similar to the development that has occurred in other geographic areas (cf. Koste & Hollowday, 1993). It reflects the basic development of all fields of biology, occurring parallelly in the whole scientific world.

Acknowledgements

I am grateful to Gunnar Pejler for helping me in the final stages of the preparation of this manuscript.

References

Adalsteinsson, H., 1979. Zooplankton and its relation to available food in Lake Myvatn. Oikos 32: 162–194.

Amrén, H., 1964a. Ecological studies of zooplankton populations in some ponds on Spitsbergen. Zool. Bidr. Uppsala 36: 161–191.

Amrén, H., 1964b. Temporal variation of the rotifer *Keratella quadrata* (Müll.) in some ponds on Spitsbergen. Zool. Bidr. Uppsala 36: 193–208.

Amrén, H., 1964c. Ecological and taxonomical studies on zooplankton from Spitsbergen. Zool. Bidr. Uppsala, 36: 209–276.

Antonsson, U., 1992. The structure and function of zooplankton in Thingvallavatn, Iceland. Oikos 64: 188–221.

Arnemo, R., B. Bérziņš, B. Grönberg & I. Mellgren, 1968. The dispersal in Swedish waters of *Kellicottia bostoniensis* (Rousselet) (Rotatoria). Oikos 19: 351–358.

Axelson, J., 1961. Zooplankton and impoundment of two lakes in Northern Sweden (Ransaren and Kultsjön). Rept. Inst. Freshw. Res. Drottningholm 42: 84–168.

Bergendal, D., 1892. Beiträge zur Fauna Grönlands. Ergebnisse einer im Jahre 1890 in Grönland vorgenommenen Reise. I. Zur Rotatorienfauna Grönlands. K. Fysiogr. Sällsk. Lund. Handl. N.F. Vol. 3. 180 pp.

Bérziņš, B., 1950. Observations on rotifers on sponges. Trans. Amer. Microsc. Soc. 59: 189–193.

Bérziņš, B., 1951. On the Collothecacean Rotatoria. Ark. Zool., ser. 2, bd 1, nr 37: 565–592.

Bérziņš, B., 1952. Contributions to the knowledge of the marine Rotatoria of Norway. Univ. Bergen, Årbok 1951, Nat.vit. rekke 6. 11 pp.

Bérziņš, B., 1955. Taxonomie und Verbreitung von *Keratella valga* und verwandten Formen. Ark. Zool. ser. 2, bd 8., nr 7: 549–559.

Bérziņš, B., 1958. Ein planktologisches Querprofil. Rept. Inst. Freshw. Res. Drottningholm 39: 5–22.

Bérziņš, B., 1960. Neue Rotatorienarten aus Madagaskar. Mém. Inst. Sci. Madagascar, sér. A, tom 14. 6 pp.

Bérziņš, B., 1962. Revision der Gattung *Anuraeopsis* Lauterborn (Rotatoria). Kungl. Fysiogr. Sällsk. Lund förhandl. bd 32. nr 5: 33–47.

Bérziņš, B., 1973. *Pleurotrochopsis anebodica* n.g., n.sp. (Rotatoria) aus Schweden. Hydrobiologia 41: 449–451.

Bérziņš, B., 1976. Notes on the Rotifera from Aneboda, Sweden. AV-centralen, Lund. 16 pp.

Bérziņš, B., 1982. Zur Kenntnis der Rotatorienfauna von Madagaskar. Limnol. Inst., Lund. 24 pp.

Bérziņš, B. & B. Pejler, 1987. Rotifer occurrence in relation to pH. Hydrobiologia 147: 107–116.

Bérziņš, B. & B. Pejler, 1989a. Rotifer occurrence in relation to temperature. Hydrobiologia 175: 223–231.

Bérziņš, B. & B. Pejler, 1989b. Rotifer occurrence in relation to oxygen content. Hydrobiologia 183: 165–172.

Bérziņš, B. & B. Pejler, 1989c. Rotifer occurrence and trophic degree. Hydrobiologia 182: 171–180.

Blakar, I. A. & O. J. Jacobsen, 1979. Zooplankton distribution and abundance in seven lakes from Jotunheimen, a Norwegian high mountain area. Arch. Hydrobiol. 85: 277–290.

Cajander, V.-R., 1983. Production of planktonic Rotatoria in Ormajärvi, an eutrophicated lake in southern Finland. Hydrobiologia 104: 329–333.

Carlin, B., 1943. Die Planktonrotatorien des Motalaström. Zur Taxonomie und Ökologie der Planktonrotatorien. Medd. Lunds Univ. limnol. Instn, 5. 256 pp.

Eloranta, P., 1982. Notes on the morphological variation of the rotifer species *Keratella cochlearis* (Gosse) s.l. in one eutrophic pond. J. Plankton Res. 4: 299–312.

Eloranta, P., 1988. *Kellicottia bostoniensis* (Rousselet), a planktonic rotifer species new to Finland. Ann. Zool. Fennici 25: 249–252.

Godske Björklund, B., 1972. The rotifer fauna of rock-pools in the Tvärminne Archipelago, southern Finland. Acta zool. fenn. 135. 30 pp.

Godske Eriksen, B., 1968. Marine rotifers found in Norway, with descriptions of two new and one little known species. Sarsia 33: 23–34.

Godske Eriksen, B., 1969. Rotifers from two tarns in southern Finland, with a description of a new species, and a list of rotifers previously found in Finland. Acta zool. fenn. 125. 36 pp.

Grundström, R., 1987. Changes in the population dynamics of *Keratella cochlearis* (Gosse), *Kellicottia longispina* (Gosse) and *Polyarthra vulgaris* (Carlin) in a fertilised enclosure. Hydrobiologia 147: 215–219.

Hakkari, L., 1969. Zooplankton studies in the lake Längelmävesi, south Finland. Ann. Zool. Fennici 6: 313–326.

Hakkari, L., 1972. Zooplankton species as indicators of environment. Aqua fenn. 1972: 46–54.

Hakkari, L., 1978. On the productivity and ecology of zooplankton and its role as food for fish in some lakes in Central Finland. Biol. Res. Rep. Univ. Jyväskyklä 4: 3–87.

Hendelberg, M., G. Morling & B. Pejler, 1979. The ultrastructure of the lorica of the rotifer *Keratella serrulata* (Ehrbg). Zoon 7: 49–54.

Hillbricht-Ilkowska, A., 1995. One hundred years of Polish rotiferology – scientists and their work. Hydrobiologia 313/314: 1–14.

Hofsten, N. von, 1909. Rotatorien aus dem Mästermyr (Gottland) und einigen andern schwedischen Binnengewassern. Ark. zool. bd 6, no 1. 125 pp.

Hofsten, N. von, 1912. Marine, litorale Rotatorien der skandinavischen Westküste. Zool. Bidr. Uppsala 1: 163–228.

Hofsten, N. von, 1923. Rotatorien der nordschwedischen Hochgebirge. Naturw. Untersuch. Sarekgeb. 4: 829–894.

Huitfeldt-Kaas, H., 1906. Planktonundersogelser i norske vande. 199 pp. Christiania, Nationaltrykkeriet. (Norwegian, with German summary.)

Jansson, A. M., 1969. Competition within an algal community. Limnologica 7: 113–117.

Johansson, S., 1983. Annual dynamics and production of rotifers in an eutrophication gradient in the Baltic Sea. Hydrobiologia 104: 335–340.

Korstad, J., Y. Olsen & O. Vadstein, 1989a. Life history characteristics of *Brachionus plicatilis* (Rotifera) fed different algae. Hydrobiologia 186/187: 43–50.

Korstad, J., O. Vadstein & Y. Olsen, 1989b. Feeding kinetics of *Brachionus plicatilis* fed *Isochrysis galbana*. Hydrobiologia 186/187: 51–57.

Korstad, J., A. Neyts, T. Danielsen, I. Overrein & Y. Olsen, 1995. Use of swimming speed and egg ratio as predictors of the status of rotifer cultures in aquaculture. Hydrobiologia 313/314: 395–398.

Koste, W., 1978. Rotatoria. Die Rädertiere Mitteleuropas. Ein Bestimmungswerk begr. von Max Voigt. Überordnung Monogononta. Vol. 1–2. 673 pp. + 234 pl.

Koste, W. & E. D. Hollowday, 1993. A short history of western European rotifer research. Hydrobiologia 255/256: 557–572.

Lang, K., 1928. Beiträge zur Kenntnis der Süsswasserrotatorien Schwedens. Lunds univ. årsskr., N.F. Avd. 2, Bd 23, Nr 13. 36 pp.

Larsson, P., 1971. Vertical distribution of planktonic rotifers in a meromictic lake: Blankvatn near Oslo, Norway. Norw. J. Zool. 19: 47–75.

Larsson, P., 1978. The life cycle dynamics and production of the zooplankton in Ovre Heimdalsvatn. Holarct. Ecol. 1: 162–218.

Levander, K.M., 1894a. Beiträage zur Kenntniss der *Pedalion*-Arten. Acta soc. pro fauna et flora fenn., bd 11, no 1. 33 pp.

Levander, K. M., 1894b. Materialen zur Kenntniss der Wasserfauna in der Umgebung von Helsingfors, mit besonderer Berücksichtigung der Meeresfauna. II. Rotatoria. Acta soc. pro fauna et flora fenn., bd 12, no. 3. 72 pp.

Lie-Pettersen, O. J., 1905. Beiträge zur Kenntnis der marinen Rädertier-Fauna Norwegens. Bergens Museums Aarbog 1905, no. 10. 44 pp.

Lie-Pettersen, O. J., 1909. Zur Kenntnis der Süsswasser-Rädertier-Fauna Norwegens. Bergens Museums Aarbog 1909, no. 15. 99 pp.

Lie-Pettersen, O. J., 1911. Rotatoriefaunaen paa Tromso. Tromso museums aarshefte 33: 41–73. (Norwegian only.)

Lindström, K., 1983. Changes in growth and size of *Keratella cochlearis* (Gosse) in relation to some environmental factors in cultures. Hydrobiologia 104: 325–328.

Lindström, K. & B. Pejler 1975. Experimental studies on the seasonal variation of the rotifer *Keratella cochlearis* (Gosse). Hydrobiologia 46: 191–197.

Linnaeus, C. von, 1758. Systema naturae, per regna tria naturae. Tomus 1. Editio decima, reformata. Holmiae 1758. 824 pp.

Morling, G. & B. Pejler, 1990. Acidification and zooplankton development in some west-Swedish lakes 1966–1983. Limnologica 20: 307–318.

Nauwerck, A., 1963. Die Beziehungen zwischen Zooplankton und Phytoplankton im See Erken. Symb. bot. upsaliens., vol. 17, no. 5. 163 pp.

Nipkow, F., 1952. Die Gattung *Polyarthra* Ehrenberg im Plankton des Zürichsees und einiger anderer Schweizer Seen. Schweiz. Z. Hydrol. 14: 135–181.

Olofsson, O., 1917. Süsswasser-Entomostraken und -Rotatorien von der Murmanküste und aus dem nördlichsten Norwegen. Zool. Bidr. Uppsala 5: 959–294.

Olofsson, O., 1918. Studien über die Süsswasserfauna Spitzbergens. Beitrag zur Systematik, Biologie und Tiergeographie der Crustaceen und Rotatorien. Zool. Bidr. Uppsala 6: 183–646.

Olsen, Y., J. Rodriguez Rainuzzo, O. Vadstein & A. Jensen, 1989. Kinetics of n-3 fatty acids in *Brachionus plicatilis* and changes in the food supply. Hydrobiologia 186/187: 409–413.

Olsen, Y., K. I. Reitan & O. Vadstein, 1993. Dependence of temperature on loss rates of rotifers, lipids, and fatty acids in starved *Brachionus plicatilis* cultures. Hydrobiologia 255/256: 13–20.

Øie, G. & Y. Olsen, 1993. Influence of rapid changes in salinity and temperature on the mobility of the rotifer *Brachionus plicatilis*. Hydrobiologia 255/256: 81–86.

Pejler, B., 1956. Introgression in planktonic Rotatoria, with some points of view on its causes and conceivable results. Evolution 10: 246–261.

Pejler, B., 1957a. On variation and evolution in planktonic Rotatoria. Zool. Bidr. Uppsala 32: 1–66.

Pejler, B., 1957b. Taxonomical and ecological studies on planktonic Rotatoria from northern Swedish Lapland. K. svenska Vetensk Akad. Handl., ser. 4, bd 6, no 5. 68 pp.

Pejler, B., 1957c. Taxonomical and ecological studies on planktonic Rotatoria from Central Sweden. K. svenska Vetensk Akad. Handl., ser. 4, bd 6, no. 7. 52 pp.

Pejler, B., 1961. The zooplankton of Ösbysjön, Djursholm. I. Seasonal and vertical distribution of the species. Oikos 12: 225–248.

Pejler, B., 1962a. The zooplankton of Ösbysjön. II. Further ecological aspects. Oikos 13: 216–231.

Pejler, B., 1962b. On the taxonomy and ecology of benthic and periphytic Rotatoria. Investigations in northern Swedish Lapland. Zool. Bidr. Uppsala 33: 327–422.

Pejler, B., 1962c. On the variation of the rotifer *Keratella cochlearis* (Gosse). Zool. Bidr. Uppsala 35: 1–17.

Pejler, B., 1965. Regional-ecological studies of Swedish fresh-water zooplankton. Zool. Bidr. Uppsala 36: 407–515.

Pejler, B., 1977a. On the global distribution of the family Brachionidae (Rotatoria). Arch. Hydrobiol., Suppl. 53: 255–306.

Pejler, B., 1977b. Experience with rotifer cultures based on *Rhodomonas*. Arch. Hydrobiol., Beih. 8: 264–266.

Pejler, B., 1980. Variation in the genus *Keratella*. Hydrobiologia 73: 207–213.

Pejler, B., 1981. On the use of zooplankters as environmental indicators. In: M. Sudzuki (ed.), Some Approaches to Saprobiological Problems. Sanseido, Tokyo, pp. 9–12.

Pejler, B., 1987. Bruno Bérziņš in memoriam, 1909–1985. Hydrobiologia 147: 1–2.

8

Pejler, B., 1993. Nordic Rotifera. Term Lists for Environmental Data Standardization. Nordic Code Centre & Swedish Mus. Nat. Hist., 124 pp.

Pejler, B., 1995. Relation to habitat in rotifers. Hydrobiologia 313/314: 267–278.

Persson, G., 1978. Experimental lake fertilization in the Kuokkel area, northern Sweden: The response by the planktonic rotifer community. Verh. int. Ver. Limnol. 20: 875–880.

Runnström, J., 1909. Beiträge zur Kenntnis der Rotatorienfauna Schwedens. Zool. Anz. 34: 263–279.

Runnström, J., 1925. *Synchaeta neapolitana* Rousselet und *S. littoralis* Rousselet von der schwedischen Westküste. Ark. zool. bd 18 A, no. 17. 5 pp.

Segers, H., 1996. The biogeography of littoral *Lecane* Rotifera. Hydrobiologia 323: 169–197.

Skjermo, J. & O. Vadstein, 1993. Characterization of the bacterial flora of mass cultivated *Brachionus plicatilis*. Hydrobiologia 255/256: 185–191.

Stenroos, K. E., 1898. Das Thierleben im Nurmijärvi-See. Eine faunistisch-biologische Studie. Acta Soc. Fauna Flora fenn. 17, no. 1. 259 pp.

Stenson, J. A. E., 1982. Fish impact on rotifer community structure. Hydrobiologia 87: 57–64.

Stenson, J. A. E., 1983. Changes in the relative abundance of *Polyarthra vulgaris* and *P. dolichoptera*, following the elimination of fish. Hydrobiologia 104: 269–273.

Thomasson, K., 1953. Studien über das südamerikanische Süsswasserplankton. 2. Zur Kenntnis des südamerikankschen Zooplanktons. Ark. zool., ser. 2, bd 6: 189–194.

Thomasson, K., 1959. Nahuel Huapi. Plankton of some lakes in an Argentine National Park. Acta phytogeogr. suec. 42: 1–83.

Thomasson, K., 1963. Araucanian lakes. Plankton studies in North Patagonia. Acta phytogeogr. suec. 47: 1–139.

Thomasson, K., 1971. Amazonian Algae. Inst. roy. sci. natur. Belg. 86. 57 pp.

Hydrobiologia **387/388**: 9–14, 1998.
E. Wurdak, R. Wallace & H. Segers (eds), Rotifera VIII: A Comparative Approach.
© *1998 Kluwer Academic Publishers.*

An analysis of taxonomic studies on Rotifera: a case study

Hendrik Segers

Laboratory of Animal Ecology, Zoogeography and Nature Conservation, Department M.S.E., University of Ghent, K.L. Ledeganckstraat 35, B-9000 Gent, Belgium

Key words: taxonomy, evaluation, Rotifera

Abstract

Results are presented of a historical analysis of taxonomic research on Rotifera, as reflected by the case of α-taxonomy of Lecanidae and Dicranophoridae. The number of available names established, as well as the fraction presently considered valid are counted per decade. Two peak periods in taxonomic research are revealed, viz. a minor one in the last decades of the 19th century, and a major one in the 1920s–1930s. Especially work published during the second period contains a high proportion of names that are currently considered valid. The second half of the 20th century witnessed a decrease in quantity, but also in quality of taxonomic research. The basic cause for this is probably the typological approach to a group exhibiting high intraspecific morphological variability, but also poor taxonomic education, as reflected by a high incidence of insufficient descriptions, and poor knowledge of the rules governing zoological nomenclature, are of incisive importance.

Introduction

Many major works on Rotifera (e.g., Koste, 1978; Nogrady et al., 1993; Ruttner-Kolisko, 1974), but also works dealing with general problems of taxonomy or zoogeography (Dumont, 1980, 1983; Koste & Shiel, 1989; Pejler, 1977a,b; Ruttner-Kolisko, 1989, 1993; Segers & Dumont, 1993; Shiel & Sanoamuang, 1993), complain about contemporary rotifer taxonomy. The most illustrative statement in this respect is that of Koste (*in* Dumont, 1980): "we are today witnessing the stone age of rotifer taxonomy". On the other hand, rotifer taxonomy has remained reasonably stable for the last 30 years or so, with minor shifts at higher taxonomic levels (Nogrady et al., 1993). But this stability is misleading, as illustrated by the work of Markevich (Markevich, 1989, 1990; Markevich & Kutikova, 1989), who suggested a radically different scheme for rotifer systematics. De Smet & Pourriot (1997), Segers et al. (1993) and Segers (1995b), also recently proposed taxonomic changes at the genus or family rank.

These criticisms reflect the general experience of researchers. Support for them is given by Ruttner-Kolisko (1989) and Snell (1989), who comment on the problems inherent to the work on this parthenogenetic-

ally reproducing, highly variable, yet morphologically simple group. However, few attempts at a critical analysis, highlighting the actual state of taxonomic research on Rotifera have been made, although some studies dealing with the development of rotifer research (e.g., Hussey, 1980; Koste & Hollowday, 1993; Nogrady et al., 1993; Sarma, 1988b) provide elements, such as rotifer studies on a local basis. Here, I attempt such an analysis by evaluating α-taxonomy of Lecanidae Bartos, 1959 and Dicranophoridae Harring, 1913.

Material and methods

The analysis is based on the revisions of Lecanidae by Segers (1995a) and of Dicranophoridae by De Smet & Pourriot (1997), supplemented by recently published additions or corrections (Table 1). Counts were made of all available names (in the sense of the International Code of Zoological Nomenclature, hence *nomina nuda* and unavailable names of infrasubspecific taxa are excluded), established for taxa at and below the species level, per complete decade starting from 1770. Distinction is made between valid names – names now considered to denote valid (sub)species – and other

Table 1. Additions and corections to the revisions of Lecanidae and Dicranophoridae

---

Species names now considered valid:
> *Lecane mitis* Harring & Myers, 1926
> *L. fadeevi* (Neiswestnowa-Shadina, 1935)

Valid names added: *Lecane boliviana* Segers, 1994
> *L. broaensis* Segers & Dumont, 1995
> *L. chinesensis* Zhuge & Koste, 1996
> *L. difficilis* Segers & Pourriot, 1997
> *L. enowi* Segers & Mertens, 1997
> *L. namatai* Segers & Mertens, 1997
> *L. pluto* Segers & Mertens, 1997
> *L. segersi* Sanoamuang, 1996
> *L. tanganyikae* Segers & Baribwegure, 1996

Invalid name added: *Lecane fadeewi* (Wiszniewski, 1954)

Corrections: *Encentrum grande* (Western, 1891)
> *E. simillimum* Remane, 1929

---

names (*species inquirendae*; junior synonyms including names of invalid subspecies and available names, presently considered to denote taxa of infrasubspecific rank). The proportion of valid names among all names established per decade is one measure of the quality of taxonomic work published (Figures 1 and 2). This pro-

portion has not been calculated for the current decade (1990–2000), as this period is not closed yet, and as the names established during this decade are too young to have been subject to a critical revision. To compare the quality of species descriptions, the ratio of *species inquirendae* (insufficiently described species; *species inquirendae* and *nomina dubia* in De Smet & Pourriot, 1997) to the total number of names, established during two periods of 40 years, one from 1911 to 1950 and one from 1951 to 1990, also was calculated.

Results and discussion

Results of this analysis are presented in Figures 1 and 2 for Lecanidae and Dicranophoridae, respectively. The number of names established per decade as well as their cumulative number are indicated in the upper part of the figures, while the percentage of valid names established per decade is reported in the lower box.

The past: 1770–1950

Taxonomic research on Lecanidae and Dicranophoridae started in the last decades of the 18th century (Figure 1). The first names established in Lecanidae were *Cercaria luna* Müller, 1776 and *Trichoda cornuta* Müller, 1786. In Dicranophoridae (Figure 2), the first-established names are *Vorticella felis* Müller,

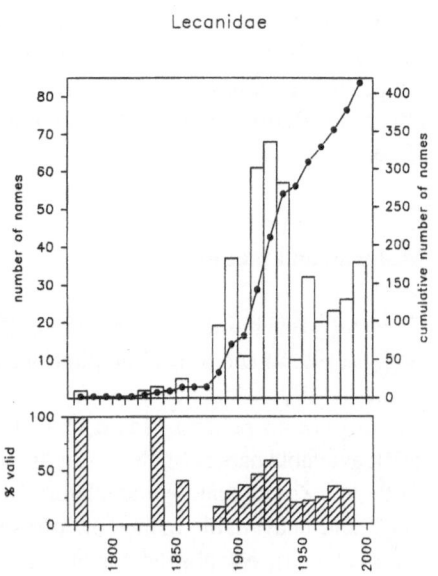
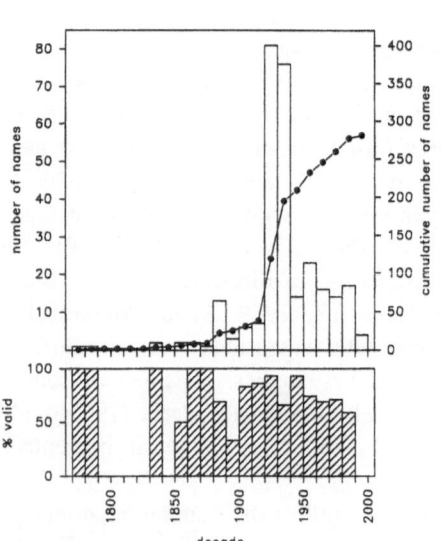

Figures 1 and 2. Growth in naming Lecanidae (1) and Dicranophoridae (2). Upper box: number (bars) and cumulative number (line) of names established per decade. Lower box: fraction of valid names established per decade.

1773 (now *Encentrum felis*) and *Cercaria forcipata* Müller, 1786 (now *Dicranophorus forcipatus*). These descriptions were published after the first flourishing period of research on Rotifera, the mid-18th century (Hussey, 1980; Sarma, 1988b). This is most likely because both Lecanidae and Dicranophoridae are littoral–benthic animals, whereas research first focused on the more accessible pelagic habitat. That these names are still in use, now in combination with other generic names, is by tradition rather than because of the accuracy of their description. Illustrative for this is Hauer's (1929) treatment on *Lecane cornuta* (Müller, 1786). In that study Hauer makes it clear that the name had previously been used to denote both 'real' *L. cornuta* and *L. closterocerca* (Schmarda), two taxa that are only superficially similar.

A first upsurge in number of named taxa occurred during the last twenty years of the 19th century (Figures 1 and 2), when D. Bryce (Bryce, 1891, 1892), E. Daday (Daday, 1897, 1898) and, especially, C.T. Hudson and P.H. Gosse (Hudson, 1885; Hudson & Gosse, 1886, 1889; Gosse, 1887a–c) published their work. Unfortunately, many of the descriptions contained therein are unrecognizable, perhaps due to the low-quality optics available at that time. The period from 1910 to 1940 was prolific regarding species descriptions, with as most noticeable contributors J. Murray (Murray, 1913a–c), J. Hauer (Hauer, 1924, 1925, 1929, 1931, 1935a,b, 1936a,b, 1937, 1938, 1940), and, particularly, H.K. Harring and F.J. Myers (Harring, 1913, 1914, 1921; Harring & Myers, 1922, 1926, 1928; Myers, 1933a,b, 1936a–c, 1937, 1938). This 'golden age' not only refers to quantity, but also to quality of published work. For Lecanidae, a peak of over 50% of names that are still considered valid was reached in the 1920s, the decade during which Harring & Myers' (1926) revision of the group was published. This high proportion of valid names probably resulted from the fortuitous combination of early work (high probability of encountering unnamed taxa), with the availability of adequate optics, enabling accurate descriptions and recognizable drawings.

The 'present': 1950–1997

After an interval of reduced activity during the 1940s, the number of names established per decade more or less remained at a constant level from 1950 onwards. Not only more authors than before contributed to the taxonomy of Rotifera (e.g., E. Bartoš, M. De Ridder, J. Hauer, W. Koste (and collaborators), L. A. Kutikova,

L. Rudescu, K. Wulfert and the following), but also authors living in nearly unexplored regions, as far as rotifers are concerned (e.g., B.K. Sharma (India); R.J. Shiel (Australia); M. Sudzuki and K. Yamamoto (Japan). When the proportion of valid names established is considered, however, we see a decrease to 20% valid names established during 1940–1950, with, subsequently, a slow increase to 28% during 1980–1990 for Lecanidae. Generally, the fraction of valid names established during 1951–1990 is 28%. In Dicranophoridae, the proportion of valid names established is noticeably higher than in Lecanidae, and a similar decrease is not evident. The fraction of valid names established during 1951–1990 is 69%. A possible explanation for the difference between the proportion of valid names in Lecanidae and Dicranophoridae may be that taxonomy of the latter is a field more restricted to experienced researchers, partly because it is based on morphology of a soft body and trophi, whereas Lecanidae taxonomy has historically been based on morphology of the stiff lorica only.

Several characters of Rotifera have been inferred to account for the confused taxonomy of the group (see Koste & Shiel, 1989; Pejler, 1977a; Snell, 1989; for reviews). However, there are more than these reasons that account for the taxonomic difficulties. The fraction of valid names established during 1911–1950 is noticeably higher for both groups (48% in Lecanidae, 81% in Dicranophoridae) than during 1951–1990 (28 and 69%, respectively). This may be a result of the proliferation and scatter of relevant literature causing duplications. This is well-illustrated by Lecanidae, that contains a single, speciose genus with over 400 available names. Moreover, the typological approach to a group exhibiting large intraspecific variability may have played a role. That this is so, is illustrated by the fact that quite a few of the invalid names established after 1950 were originally given to taxa of subspecific or infrasubspecific rank (see also Hussey, 1980).

Another probable cause is illustrated by the occurrence of *species inquirenda* amongst the descriptions, viz. names accompanied by a description that is too poor to permit recognition of the examined taxon. Of the Lecanidae names established during the period 1911–1950, 10.3% have to be considered insufficiently described. For the period 1951–1990, these amount up to 17.8% of the names established. The proportions for the same periods are 3.4 and 20.3%, respectively, for Dicranophoridae. So, relatively more unrecognizable descriptions were published during the second than during the first part of the 20th century. Hence, the quality of taxonomic research as exemplified by the quality of species descriptions, was lower during the second part of the 20th century, notwithstanding the technical progress that, amongst other things, resulted in the availability of better microscopes.

Although the correct application of the rules governing zoological nomenclature has been advocated by many researchers (e.g., Koste & Shiel, 1989), it is obvious that rotifer taxonomists appear not to be fully aware of these rules. Errors of gender (e.g., *Brachionus bidentatus* Anderson, 1889 still commonly being referred to as *B. bidentata*; *Scaridium 'grandis'* Segers, 1995; see Segers, 1996b), misuse of brackets for author citations (e.g., brackets wrongly applied in the case where a name is elevated or lowered in rank), use of the term '*nomen novum*' in other cases than the replacement of a junior homonym, *inter alia*, are common errors, even in recent publications. Senior synonyms have been ignored, homonymy overlooked, names inappropriately replaced, etc. (e.g., the history of use of the name '*Lecane amazonica*', as outlined by Segers, 1993). Moreover, fundamental misconceptions exist regarding some key elements basic to the structure of the Code, and of zoological nomenclature generally. The misconception that the Code would recommend only one category below species, and that infrasubspecific descriptors should take subspecific rank (see Koste & Shiel, 1989 for one example), has seriously complicated rotifer nomenclature regarding names at infrasub-, sub- and specific level.

The future

Clearly, the above-mentioned causes support the suggestion by Nogrady et al. (1993), that taxonomic education has been neglected for a long period of time. This may be an unfortunate side effect of the development of new fields in biological science. The confused taxonomy now jeopardizes the use of Rotifera in applied studies, and the resulting low reliability of records is in turn largely responsible for reports of cosmopolitanism in Rotifera (Segers, 1996a). It is urgently needed that steps are taken to address the situation, and a start has been made by the preparation of new taxonomic works or identification guides treating higher taxa on a world-wide scale (e.g., De Smet, 1996; De Smet & Pourriot, 1997; Nogrady et al., 1993, 1995; Segers, 1995a), a work that has be-

nefited greatly from the pioneer works of Voigt (1957), Kutikova (1970) and Koste (1978).

From the evolution of cumulative numbers in Figures 1 and 2 follows that we are far from reaching the end of naming in Rotifera. Illustrated records of unnamed taxa from various regions exist (Australia, Figure 12.1 in Koste & Shiel, 1990; Bolivia, Segers et al., 1994; Nigeria, Segers et al., 1993b; Norway (Bjørnøya), *Lecane* sp., De Smet, 1988; Thailand, Segers & Sanoamuang, 1994). New species are still being described from well-studied regions, such as Europe (De Manuel, 1994; Galindo et al., 1994; Jersabek, 1994; Segers et al., 1996) or north-east U.S.A. (Segers, 1997). The availability of new techniques (see present volume), and the new research effort on littoral habitats in the tropics and subtropics are responsible for this. The striking discrepancy between the increase in numbers of valid versus all names illustrates the need for more caution when naming taxa.

Conclusions

The evolution of numbers of names established per decade shows no trend towards stabilization. This, combined with reports of unnamed taxa, makes it likely that many more mophospecies await discovery.

The proportion of valid names established during the last decades is lower that that of the first part of the 20th century. Scattered literature, the application of a typological methodology to a group exhibiting a wide morphological variability, and poor taxonomic education are inferred as probable causes for this. More diligence is urged when naming taxa, not only in the groups included in this analysis, but especially in genera that contain variable taxa.

References

Bryce, D., 1891. Remarks on *Distyla*, with descriptions of three new rotifers. Sci. gossip 27: 204–207.

Bryce, D., 1892. On some moss-dwelling Cathypnidae; with descriptions of five new species. Sci. gossip 28: 271–275.

Daday, E., 1897. Új-Guineai Rotatoriák. (Rotatoria novae Guineae). Math. Termész. Értesitö15: 131–148.

Daday, E., 1898. Mikroscopische Süsswasserthiere aus Ceylon. Termész. Füzetek, Budapest, vol. 21, Anhangsheft, 123 pp.

De Manuel, J., 1994. Taxonomic and zoogeographic considerations on Lecanidae (Rotifera: Monogononta) of the Balearic archipelago, with the description of *Lecane margalefi* n. sp. Hydrobiologia 288: 97–105.

De Smet, W. H., 1988. Rotifers from Bjørnøya (Svalbard), with the description of *Cephalodella evabroedi* n. sp. and *Synchaeta lakowitziana arctica* n. subsp. Fauna norv. Ser. A 9: 1–18.

De Smet, W. H., 1996. Rotifera 4: The Proalidae (Monogononta). In Dumont, H. J. & T. Nogrady (eds), Guides to the Identification of the Microinvertebrates of the Continental Waters of the World 9. SPB Academic Publishing, The Hague, The Netherlands: 102 pp.

De Smet, W. H. & R. Pourriot, 1997. Rotifera 5: The Dicranophoridae (Monogonta) and The Ituridae (Monogononta). In Dumont, H. J. & T. Nogrady (eds), Guides to the Identification of the Microinvertebrates of the Continental Waters of the World 12. SPB Academic Publishing, The Hague, The Netherlands: 344 pp.

Dumont, H. J., 1980. Workshop on taxonomy and biogeography. Hydrobiologia 73: 205–206.

Dumont, H. J., 1983. Biogeography of rotifers. Hydrobiologia 104: 19–30.

Galindo, M. D., L. Serrano, H. Segers & N. Mazuelos, 1994. *Lecane donyanaensis* n. sp. (Rotifera: Monogononta, Lecanidae), from the Doñana National Park (Spain). Hydrobiologia 284: 235–239.

Gosse, P. H., 1887a. Twenty-four new species of Rotifera. J. r. Mircrosc. Soc. 1–7.

Gosse, P. H., 1887b. Twelve new species of Rotifera. J. r. Microsc. Soc. 361–367.

Gosse, P. H., 1887c. Twenty-four more new species of Rotifera. J. r. microsc. Soc. 861–871.

Harring, H. K., 1913. A list of the Rotatoria of Washington and vicinity, with descriptions of a new genus and ten new species. Proc. U.S. Nat. Museum 46: 387–405.

Harring, H. K., 1914. Report on Rotifera from Panama with descriptions of new species. Proc. U.S. Nat. Museum 47: 525–564.

Harring, H. K., 1921. Rotatoria. Rep. Canadian Arctic Exped. 1913–1918, Ottawa, 8: 1–23.

Harring, H. K. & F. J. Myers, 1922. The rotifer Fauna of Wisconsin. Trans. Wisconsin Acad. Sci. Arts Lett. 20: 553–662, pl. 51–61.

Harring, H. K. & F. J. Myers, 1926. The Rotifer Fauna of Wisconsin. III. A revision of thegenera *Lecane* and *Monostyla*. Trans. Wisconsin Acad. Sci. Arts Lett. 22: 315–423.

Harring, H. K. & F. J. Myers, 1928. The rotifer Fauna of Wisconsin. IV. The Dicranophorinae. Trans. Wisconsin Acad. Sci. Arts Lett. 23: 667–808, pl. 23–49.

Hauer, J., 1924. *Lecane lauterborni* n. sp. und einige für die deutsche Fauna neue *Lecane*- und *Monostyla*-Arten. Zool. Anz. 61: 145–149.

Hauer, J., 1925. Rotatorien aus des Salzgewässern von Oldesloe (Holstein). Mitt. Geogr.Gesell. nat. Hist. Museum Lübeck, II Reiche 30: 152–195.

Hauer, J., 1929. Zur Kenntnis der Rotatoriengenera *Lecane* und *Monostyla*. Zool. Anz. 83: 143–164.

Hauer, J., 1931. Zur Rotatorienfauna Deutschlands (II). Zool. Anz. 93: 7–13.

Hauer, J., 1935a. Zur Rotatorienfauna Deutschlands (IV). Zool. Anz. 110: 260–264.

Hauer, J., 1935b. Rotatorien aus dem Schluchseemoor und seiner Umgebung. Ein Beitrag zur Kenntnis der Rotatorienfauna der Schwarzwaldhochmoore. Verh. Naturwiss. ver. Karlsruhe 31: 47–130.

Hauer, J., 1936a. Zur Rotatorienfauna Deutschlands (V). Zool. Anz. 113: 154–157.

Hauer, J., 1936b. Neue Rotatorienarten aus Indien. Zool. Anz. 116: 77–80.

Hauer, J., 1937. Die Rotatorien von Sumatra, Java und Bali nach den Ergebnissen der Deutschen Limnologischen Sunda-Expedition. Teil I. Arch. Hydrobiol., suppl. Bd.XV (2), 296–384.

Hauer, J., 1938. Die Rotatorien von Sumatra, Java und Bali nach den Ergebnissen der Deutschen Limnologischen Sunda-Expedition. Teil II. Arch. Hydrobiol., suppl. Bd.XV (3), 507–602.

Hauer, J., 1940. Beitrag zur Kenntnis der Rotatorien warmer Quellen Deutschlands. Zool. Anz. 130: 156–158.

Hudson, C. T., 1885. On four new species of the genus *Floscularia*, and five other newspecies of Rotifera. J. r. microsc. Soc. London: 608–614.

Hudson, C. T. & P. H. Gosse, 1886. The Rotifera or Wheel-Animalcules, both British and Foreign. Longmans, Green and Co., London. Vol. 1: VI+128 pp., Vol. 2: 144 pp.

Hudson, C. T. & P. H. Gosse, 1889. The Rotifera or Wheel-Animalcules, both British and Foreign. Supplement. Longmans, Green, and Co., London., 64 pp.

Hussey, C. G., 1980. A historical survey of the collection and study of rotifer in Britain. Hydrobiologia 73: 237–240.

Jersabek, C. D., 1994. *Encentrum (Parencentrum) walterkostei* n.sp., a new dicranophorid rotifer (Rotatoria: Monogononta) from the high alpine zone of the Central Alps (Austria). Hydrobiologia 281: 51–61.

Koste, W., 1978. Rotatoria. Die Rädertiere Mitteleuropas. Borntraeger, Berlin, 2 vols: 673pp., 234 plates.

Koste, W. & E. D. Hollowday, 1993. A short history of western European rotifer research. Hydrobiologia 255/256: 557–572.

Koste, W. & R. J. Shiel, 1989. Classical taxonomy and modern methodology. Hydrobiologia 186/187: 279–284.

Koste, W. & R. J. Shiel, 1990. Rotifera from Australian inland waters. V. *Lecanidae* (Rotifera: Monogononta). Trans. r. Soc. S. Aust. 114: 1–36.

Kutikova, L. A., 1970. Kolovratki Fauna SSSR. Fauna SSSR 104, Academia Nauk. Moscow, 744pp.

Markevich, G. I., 1989. Morphology and principal organization of the sclerotized system ofthe rotifer mastax. In Biologiya, Sistematika i Funksionalnaya Morfologiya Presnovodick Zhivotnick. Inst. Biol. Vnutrenny Vod. Acad. Nauk SSSR. Trudy 56: 27–82 (in Russian).

Markevich, G. I., 1990. A historic reconstruction of phylogenesis of rotifers as a basis for their macrosystem. Rotifera. Proc. of the third All-Union rotifer symposium. Zool. Inst. Acad. Nauk Leningrad: 140–156 (in Russian).

Markevich, G. I. & L. A. Kutikova, 1989. Mastax morphology under SEM and its usefulness in reconstructiong rotifer phylogeny and systematics. Hydrobiologia 186/187: 285–289.

Murray, J., 1913a. South American Rotifera. Part II. J. r. microsc. Soc. 341–362.

Murray, J., 1913b. Australasian Rotifera. J. r. microsc. Soc. 455–461

Murray, J., 1913c. Notes on the Family *Cathypnidae*. J. r. microsc. Soc. 545–564.

Myers, F. J., 1933a. A new genus of rotifers (*Dorria*). With observations on *Cephalodella crassipes* (Lord): *Cephalodella crassipes* (Lord) and *Dorria dalecarlica* Gen.n., Sp.n. J. r. microsc. Soc. 53 (ser. 3): 118–121.

Myers, F. J., 1993b. The distribution of Rotifera on Mount Desert Island. III. New Notommatidae of the genera *Pleurotrocha, Lindia, Eothina, Proalinopsis,* and *Encentrum*. Am. Muz. Nov. 660: 1–18.

Myers, F. J., 1936a. Psammolittoral rotifers of Lenape and Union Lakes, New Jersey. Am. Mus. Novitates 830: 22 pp.

Myers, F. J., 1936b. Three new brackish water and one new marine species of Rotatoria. Trans. Am. Microsc. Soc. 55: 428–432

Myers, F. J., 1936c. Rotifers from the Laurentides National Park with descriptions of two new species. Can. Field Natur. 50: 82–84.

Myers, F. J., 1937. Rotifera from the Adirondack region of New York. Am. Mus. Novitates 903: 17 pp.

Myers, F. J., 1938. New species of Rotifera from the collection of the American Museum of Natural History. Am. Mus. Novitates 1011: 17 pp.

Nogrady, T., R. L. Wallace & T. W. Snell, 1993. Rotifera 1: Biology, Ecology and Systematics. In Guides to the identification of the Microinvertebrates of the Continental Waters of the World 4. SPB Academic Publishing, The Hague, The Netherlands, 142 pp.

Nogrady, T., R. Pourriot & H. Segers, 1995. Rotifera 3: The Notommatidae and: The Scaridiidae. In Dumont, H. J. & T. Nogrady (eds), Guides to the Identification of the Microinvertebrates of the Continental Waters of the World 8. SPB Academic Publishing, The Hague, The Netherlands: 248 pp.

Pejler, B., 1977a. General problems on rotifer taxonomy and global distribution. Arch.Hydrobiol. Beih. 8: 212–220.

Pejler, B., 1977b. On the global distribution of the family Brachionidae (Rotatoria). Arch. Hydrobiol./Suppl. 53: 255–306.

Ruttner-Kolisko, A., 1974. Planktonic rotifers: biology and taxonomy. Binnengewässer (Suppl.) 26: 1–146.

Ruttner-Kolisko, A., 1989. Problems in the taxonomy of rotifers, exemplified by the *Filinia longiseta–terminalis* complex. Hydrobiologia 186/187: 291–298.

Ruttner-Kolisko, A., 1993. Taxonomic problems with the species *Keratella hiemalis*.Hydrobiologia 255/256: 441–443

Sarma, S. S. S., 1988. World trends in Rotifer research. Biol. Educ. 5: 240–243.

Segers, H., 1993. Rotifera of some lakes in the floodplain of the River Niger (Imo State, Nigeria). I. New species and other taxonomic considerations. Hydrobiologia 250: 39–61.

Segers, H., 1995a. Rotifera 2: The Lecanidae (Monogononta). In Dumont, H. J. & T. Nogrady (eds), Guides to the Identification of the Microinvertebrates of the Continental Waters of the World 6. SPB Academic Publishing, The Hague, The Netherlands: 226 pp.

Segers, H., 1995b. A reappraisal of the Scaridiidae (Rotifera: Monogononta). Zool. Scr. 24: 91–100.

Segers, H., 1996a. The biogeography of littoral *Lecane* Rotifera. Hydrobiologia 323: 169–197.

Segers, H., 1996b. On a new *Scaridium* (Rotifera: Monogononta: Scaridiidae) from Brazil. Belg. J. Zool. 126: 57–63.

Segers, H., 1997. Rotifera from the collection of the Academy of Natural Sciences of Philadelphia. Proc. Acad. natl. Sci. Philadelphia, 148: 147–156.

Segers, H. & H. J. Dumont, 1993. Rotifera from Arabia, with descriptions of two new species.Fauna of Saudi-Arabia 13: 3–26.

Segers, H. & L. Sanoamuang, 1994. Two more new species of *Lecane* (Rotifera: Monogononta), from Thailand. Belg. J. Zool. 124: 39–46.

Segers, H., G. Murugan & H. J. Dumont, 1993a. On the taxonomy of the Brachionidae: description of *Plationus* n. gen. (Rotifera, Monogononta). Hydrobiologia 268: 1–8.

Segers, H., C. S. Nwadiaro & H. J. Dumont, 1993b. Rotifera of some lakes on the floodplain of the River Niger (Imo State, Nigeria). II. faunal composition and diversity. Hydrobiologia 250: 63–71.

Segers, H., L. Meneses & M. Del Castillo, 1994. Rotifera (Monogononta) from Lake Kothia,a high-altitude lake in the Bolivian Andes. Arch. Hydrobiol. 132: 227–236.

Segers, H., D. Bonte & W. H. De Smet, 1996. Description of *Lepadella deridderae deridderae* n.sp., n.subsp. and *L. deridderae alaskae* n.sp., n.subsp. (Rotifera: Monogononta, Colurellidae). Belg. J. Zool. 126: 117–122.

Shiel, R. J. & L. Sanoamuang, 1993. Trans-Tasman variation in Australasian *Filinia*-populations. Hydrobiologia 255/256: 455–462.

Snell, T., 1989. Systematics, reproductive isolation and species boundaries in monogonont rotifers. Hydrobiologia 136/187: 299–310.

Voigt, M., 1957. Rotatoria. Die Rädertiere Mitteleuropas. Nikolassee, Berlin: 508 pp., 115 plates.

Hydrobiologia **387/388**: 15–21, 1998.
E. Wurdak, R. Wallace & H. Segers (eds), Rotifera VIII: A Comparative Approach.
© *1998 Kluwer Academic Publishers.*

Zooplankton may not disperse readily in wind, rain, or waterfowl

David G. Jenkins & Marilyn O. Underwood
Department of Biology, University of Illinois at Springfield, P.O. Box 19243, Springfield, IL 62794-9243, U.S.A.
E-mail: jenkins.david@uis.edu

Key words: Rotifera, zooplankton, dispersal, colonization, wind, rain, waterfowl

Abstract

Zooplankton, and especially rotifers, have long been thought to be readily dispersed by wind, rain and animals (especially waterfowl). Given that premise, local processes (tolerance to abiotic conditions, biotic interactions) have been the main focus of ecological studies. We tested the premise of high dispersal rates by incubating particulates collected with windsocks and rain samplers at two sites over 1 year. The sites were 80 km apart and differed in proximity to water and surrounding terrain. We also incubated fecal material of wild ducks. Pond sediments were identically incubated as a test of incubation method. Only bdelloid rotifers were collected in wind samples, and only four rotifer species were collected in rain samples: *Lecane leontina, Lecane closterocerca, Keratella cochlearis,* and a bdelloid. No metazoans were found in incubated duck feces, yet incubated pond sediments yielded 11 rotifer, one copepod, four cladoceran, and three ostracod species. Our results do not support the premise of readily dispersed zooplankton. If zooplankton dispersal is infrequent and limited to few species, a series of other questions should be addressed on processes regulating zooplankton population dynamics and community composition.

Introduction

Questions of dispersal are central to our perceptions of zooplankton community organization. If sites are saturated with regionally available species, then local processes should regulate community composition (Ricklefs, 1987; Roughgarden, 1989). However, if dispersal processes do not deliver species to a site, local biotic and abiotic processes will act on a subset of potential species and lead to a different community structure: consider the potential differences between a community with a dominant competitor and one without that competitor. The presence or absence of some species may be predictably attributed to local conditions (e.g., Pejler, 1995; Pontin & Langley, 1993), but an alternative hypothesis exists that should be tested before local conditions are assigned general primacy: dispersal processes may also regulate community composition. More explicitly, dispersal could constrain community composition if dispersal is generally limited, as in isolated sites (Jenkins & Buikema, 1998), or dispersal could affect community composition by residual effects of colonization se-

quence (Robinson & Dickerson, 1987). In reality, both local and regional processes surely determine zooplankton community composition, but most of our understanding of zooplankton communities is based on studies of local processes.

Zooplankton dispersal processes have been underevaluated because it is commonly assumed that zooplankton (especially rotifers) disperse readily, as evidenced by 'cosmopolitan' distributions, dessication-resistant dormant stages, small size, and parthenogenetic reproduction (Brown & Gibson, 1983; Gislen, 1947; Hutchinson, 1967; King, 1980; Lampert & Sommer, 1997; McAtee, 1917; Maguire, 1963; Pejler, 1995; Pennak, 1989; Wetzel, 1983). Widespread distributions of some taxa are testimony to the eventual dispersal of those taxa overland (e.g., Chaplin & Ayre, 1997). However, many zooplankton species do not, in fact, have cosmopolitan distributions (Carter et al., 1980; Frey, 1986; Hebert & Hann, 1986; Hebert & Wilson, 1994; Stemberger, 1995), and gene flow among populations can be limited (Berg & Garton, 1994; Boileau & Hebert, 1991; Thier, 1994; Weider, 1989). Therefore, one should not infer that potential dispersal necessarily translates to actual dispersal and cosmopolitan distributions. In addition, zooplankton species vary in vagility (Jenkins, 1995), indicating that more detailed analyses are needed in place of generalizations.

Zooplankton are potentially dispersed overland via several possible vectors, including vertebrates, insects, wind, and rain (Lowndes, 1930; Maguire, 1959, 1963; Malone, 1965; Proctor, 1964; Proctor & Malone, 1965; Proctor et al., 1967; Schlichting & Milliger, 1969; Sides, 1973; Stewart & Schlichting, 1966). However, most studies of dispersal vectors have been conducted under artificial conditions: the results speak to the potential for dispersal, but leave actual dispersal events largely unstudied.

Maguire (1963) examined the passive dispersal of aquatic microorganisms near ponds. He concluded that wind, rain, insects and vertebrates all play roles in the dispersal of small aquatic organisms, but did not attempt to distinguish the relative importances of each vector in the transport process. In addition, Maguire's local-scale study did not address long-range dispersal; we know of no studies that have empirically assessed long-range zooplankton dispersal frequency by wind and rain or the relative importances of wind and rain dispersal.

It is clear that rotifers and other zooplankton *can* disperse passively overland. However, it is not clear if zooplankton actually disperse at sufficient frequency to saturate a habitat with regionally available species, and thereby elevate the importance of local processes. The assumption that zooplankton disperse readily needs to be evaluated if the relative importances of regional and local processes are to be understood. In addition, zooplankton species do not disperse equally well (Jenkins, 1995), and generalities about zooplankton dispersal may be inadequate for developing an understanding of the role of dispersal dynamics in zooplankton community structure.

The purpose of this study was to examine the importance of wind, rain and waterfowl as zooplankton dispersal vectors. Wind and rain dispersal of zooplankton were evaluated by collecting and incubating airborne particles and rain to identify organisms carried by each vector and the frequency of dispersal events. In addition, water fowl feces were collected and incubated to determine if they contained viable disseminules of zooplankton. Incubation methods were tested with pond sediments.

Materials and methods

Wind and rain samples were collected at two sites, 80 km apart, for over 1 year (October 1994 through December 1995). Samples were collected at various intervals, ranging from biweekly to 4 months, depending on precipitation patterns and season (see Results below for sample dates). Samples were collected continuously, with the exception of wind samples at the UIS site: the windsock was destroyed in a storm in late April 1995 and replaced in July 1995.

Wind samples were collected using nylon windsocks (one per site) lined with 10-μm plankton netting. The windsock mouth (9.5 cm diameter) was shaped like a Wisconsin-style plankton net to reduce backflow and was constructed of stiff plastic to keep it open. The windsock pivoted 360° in the wind atop a 2-m pole.

Rain samplers were designed to collect rain distinct from wind samples. Samplers were modified Buchner funnels; four such samplers were used at each site. The first 100 ml of rain entering a funnel flowed into a small bottle: this volume was sufficient to wash the funnel and remove aerially deposited materials. A floating styrofoam ball sealed the small bottle when it was full, and the rest of the rainwater was diverted through 10-μm Nitex mesh. Collected particles were retained on the 10-μm mesh above water level of the collected rainwater so that any zooplankton propagules would not hatch in the sampler. The mesh was removed and washed with dechlorinated tap water to collect a sample, and rainwater volume was measured in a graduated cylinder. Four rain samplers were placed at each site.

Wind and rainwater samplers were placed at two sites: the University of Illinois at Springfield (UIS) and near Shick Shack Pond (SS). The UIS site was selected to be remote from potential sources of zooplankton propagules, by virtue of its elevation and distance from upwind water bodies. The SS site was near a natural pond, closer to ground level, and was considered more likely to collect propagules than the UIS site. However, the SS sample collectors were more remote from the pond than Maguire's (1963) samplers.

The UIS samplers were placed on the roof of a two-story building, with rain samplers and the windsock approximately 9 and 11 m above ground level, respectively. Land around the UIS site is very flat, and primarily agricultural land use. Samplers were near the west edge of the building, so as to collect particles from the prevailing wind direction (westerly) unaffected by rooftop structures. The UIS samplers were 0.4 km southwest of the campus pond, which was the closest body of water to the sample site. The Campus Pond is relatively young (25 years) with some submerged vegetation at the margins and a muddy bottom. Because prevailing winds were westerly and the Campus Pond was northeast of the samplers, we considered the pond an unlikely source of propagules. The closest water upwind of the samplers (other than roadside ditches) was Lake Springfield, a reservoir approximately 1 km away. No other upwind source of propagules was nearby: any organisms cultured from this site probably travelled some distance in the wind or rain.

Shick Shack Pond (Cass Co.) is approximately 80 km west–northwest of Springfield and one of the Illinois Natural Preserves (Karnes & McFall, 1995). Shick Shack Pond is old (ca 10 000 years) and completely surrounded by shrubs and trees. Surrounding land is hilly and mixed pasture and woods. Samplers were placed on a small hill approximately 150 m northeast of the pond. Rain samplers were placed on the ground and the windsock was 2.5 m above ground. Since the pond was upwind (for prevailing winds), relatively nearby, and samplers were at or near ground level, the pond was a probable source of propagules. In addition, no other water bodies were nearby.

Wind samples were removed from windsocks by transferring collected dry particles to a bottle which had been sterilized by rinsing with 70% alcohol and dechlorinated tap water and exposure to a germicidal UV bulb for 35–60 min. Dechlorinated tap water was used to rinse any remaining material out of the windsock into the bottle.

Rain samples were collected by removing the plankton netting from each apparatus and washing it thoroughly into a sterile bottle using dechlorinated tap water.

Duck feces were collected from mallard ducks at the Washington Park pond (Springfield, IL) in October 1995 and March 1996. The collection procedure was identical on both dates, except that March samples were carefully broken apart and observed under a dissecting scope for ephippia or adult zooplankton: none were observed. Ducks were observed feeding in the water and on shore. Feces were collected within minutes of deposition with a sterile metal spatula. Only the upper three-quarters of a fecal pellet was collected to avoid that part of the feces contaminated by the ground. Feces were placed in sterile bottles

transported to the lab. The spatula was rinsed in 70% isopropyl alcohol between each collection.

In the laboratory, fecal material was rinsed with dechlorinated tap water through a 0.45-mm sieve to remove coarse particulate matter. The rinsate was used to culture zooplankton. Rinsed samples were refrigerated for 7 days prior to culturing in an attempt to break dormancy of copepods and cladocerans (Marcus, 1979; Schwartz & Hebert, 1987).

Samples were divided and incubated under one of two conditions: 20°C±2°C with constant fluorescent light (cool white bulbs) or 20°C ±2°C with UV light (Hagiwara, 1995). Samples were observed weekly under a dissecting scope and organisms were preserved with formalin or 70% alcohol. Samples were observed until no organisms were observed for 3 consecutive weeks.

Sediment core samples from Shick Shack Pond and Campus Pond were incubated as a test of incubation technique. Surface sediment was mixed and divided into several parts, diluted with sterile dechlorinated tap water in sterile containers, and incubated as above. Organisms were preserved with formalin or 70% isopropyl alcohol.

Results

Only four species were collected in wind and rain samples: bdelloids, *Lecane leontina*, *Lecane closterocerca* and *Keratella cochlearis* (Table 1). No crustacean species were observed. Bdelloid rotifers were the most common organism found in any of the samples (Table 1). Bdelloids were the only organisms found in the windsocks at both the campus and Shick Shack sites, but sites differed in the frequency of bdelloid occurrence. Bdelloid rotifers were found in only one of 11 (9%) of UIS windsock samples, and in seven of 11 (63.6%) SS windsock samples. Similarly, bdelloid rotifers occurred in 10% of UIS rain samples and 20% of SS rain samples (Table 1). Interestingly, *Keratella cochlearis* ccurred in one rain sample at each site, and on the same sample date. *Lecane closterocerca* and *Lecane leontina* also occurred in one of 40 (2.5%) of SS rain samples.

No metazoans were collected from incubated duck fecal matter (ciliates were present). However, rotifers, copepods, ostracods, and cladocerans were obtained by our incubation of both Campus Pond and Shick Shack Pond sediments (Table 2).

Table 1. Wind and rain dispersal results

Sample date	Sampling interval (days)	UIS Wind	UIS Rain	SS Wind	SS Rain
22 Oct 94	14	0/1	0/4	Bdell 1/1	Bdell 1/1
					Lleon 1/4
12 Nov 94	21	0/1	0/4	Bdell 1/1	0/4
19 Nov 94	7	0/1	0/4	Bdell 1/1	Bdell 1/4
14 Mar 95	115	0/1	Bdell 1/4	Bdell 1/1	0/4
28 Mar 95	14	0/1	0/4	Bdell 1/1	0/4
12 Apr 95	15	0/1	0/4	0/1	0/4
24 Apr 95	12	0/1	0/4	0/1	0/4
14 Aug 95	112	Bdell 1/1	Bdell 3/3	0/1	Bdell 1/3
			Kcoch 1/3		Kcoch 1/3
					Lclos 1/3
29 Aug 95	15	0/1	0/3	Bdell 1/1	Bdell 2/3
18 Sep 95	20	0/1	0/3	0/1	Bdell 2/3
4 Dec 95	77	0/1	0/3	Bdell 1/1	Bdell 1/3
TOTAL	422	Bdell 1/11	Bdell 4/40	Bdell 7/11	Bdell 8/40
			Kcoch 1/40		Kcoch 1/40
					Lleon 1/40
					Lclos 1/40

UIS site was on rooftop at University of Illinois at Springfield campus, SS site was near Shick Shack Pond, 80 km away from UIS. Bdell, bdelloid rotifer; Lleon, *Lecane leontina*; Lclos, *Lecane closterocerca*; Kcoch, *Keratella cochlearis*. Data represent the fraction of samples collected that contained each species. One windsock and four rain samplers were located at each site through 24 April 95, after which three rain samplers were used.

Table 2. Organisms incubated from sediments by same procedures used for wind, rain, and waterfowl fecal samples

	UIS Campus Pond	Shick Shack Pond
Rotifera	Bdelloid	Bdelloid
	Lecane leontina	*Lecane copeis*
	Lecane sp.	*Lecane closterocerca*
	Lepadella ovalis	*Lecane leontina*
	Lepadella patella	*Lepadella triptera*
	Lepadella rhomboides	
	Platyias patulus	
	Platyias quadricornis	
	Testudinella sp.	
Copepoda	*Cyclops bicuspidatus thomasi*	
	nauplii	
Cladocera	*Alona guttata*	
	Ceriodaphnia quadrangula	
	Kurzia latissima	
	Pleuroxus denticulatus	
Ostracoda	Two unidentified species	*Candona biangulata*

Discussion

Only four species were detected as dispersing by wind and rain during one year of sampling at two sites 80 km apart. All four species were rotifers: no crustaceans were observed in cultured wind or rain samples, nor were ephippia or diapaused copepods observed.

The two sample sites were intended to sample dispersal differently. Although both sites collected very limited sets of dispersing zooplankton, the site intended to be more remote (UIS) collected propagules less frequently than the site intended to be near a potential source pool (SS). This result may be due to several factors: (1) proximity of samplers to water (150 m at SS versus 1 km at UIS; (2) elevation (up to 2.5 m at SS versus 16 m at UIS); and/or (3) position of samplers relative to the prevailing wind direction and the pond. Although spatial pattern of dispersal canot be adequately analyzed with two sample sites, our results suggest that distances on the order of a kilometer severely restrict wind and rain dispersal of zooplankton.

Keratella cochlearis was observed in rain samples collected on the same day but 80 km apart. Rain samples were collected after precipitation events, and it is possible that *K. cochlearis* were dispersed by one storm front to both sites. However, this was the only such occurrence during the year.

No zooplankton were incubated from duck fecal material. Previous studies of zooplankton passing through waterfowl (Malone, 1965; Mellors, 1975; Proctor, 1964; Proctor & Malone, 1965; Proctor et al., 1967) were conducted in laboratories, with zooplankton eggs or adults fed to birds in the lab and feces then collected and incubated. None of the studies involved collection of feces from birds feeding in the wild. Clearly, birds can *potentially* transport zooplankton internally, but our results indicate that such events may not be naturally common. In addition, natural dispersal events would depend on compound probabilities: ingestion of viable propagules, survival of propagules in the gut, transport to a site within the gut passage time, and deposition in the site.

Limited species diversity in incubated samples was not due to poor incubation conditions, as evidenced by species collected from identically incubated pond sediments. It is possible that some propagules were present that did not hatch or break diapause in our experimental conditions. May (1986) used three temperatures to incubate sediments, with great success. We used one temperature, but the diversity of species from pond sediments and the paucity of species in wind, rain, and fecal samples indicates that incubation conditions did not cause low diversity.

Based on the results of this study, zooplankton are *not* readily dispersed by wind, rain, and duck feces. Wind and rain may play an occasional role in the dispersal of bdelloid rotifers but do not play a significant role in the dispersal of monogonant rotifers or other zooplankton. Aquatic organisms may be transported externally on waterfowl and other animals (Swanson, 1984), but actual rates, distances, and species involved are not clear.

Note that we do not say wind, rain, and internal transport by waterfowl does not happen, only that it appears to happen infrequently and for a few species. Our results are consistent with genetic and biogeographic studies that indicate species ranges change at geologic time scales (Boileau & Hebert, 1991; Carter et al., 1980; Stemberger, 1995). Obviously, more studies of zooplankton dispersal are warranted, but our results have important implications for the forces that shape zooplankton communities:

Abiotic conditions

Limited dispersal may impede our ability to use rotifers and other zooplankton as indicators of water quality (Pejler, 1995; Pontin & Langley, 1993). Instead, community composition, or the presence/absence of certain species may be confounded by chance dispersal events. Of course, the older the system and the more interconnections with other water bodies, the more likely it is that regional species would have been dispersed to a site. The problem is that so little empirical data exists on natural dispersal rates that the magnitude of this confounding factor remains largely unknown.

Biotic interactions

Competition and predation can be significant forces in community composition and seasonal successions of zooplankton communities. Dispersal may also be important, by virtue of its function as a 'rate-limiting step.' Ricklefs (1987) argued that unsaturated communities are shaped more by regional processes (speciation and dispersal) than by local processes (e.g., competition and predation). Models and empirical studies of rocky intertidal communities have indicated that settlement rate is the controlling parameter in determining those community dynamics

(Roughgarden, 1989; Roughgarden et al., 1987; Underwood et al., 1983). When settlement rates are high, post-settlement processes (e.g., competition, predation) determine community composition (Connell, 1961; Paine, 1974); but when settlement rates are low, community composition is strongly influenced by settlement rate. This 'supply-side ecology' (Lewin, 1986) may be a common theme among different ecosystems.

Invasions

If zooplankton communities are not saturated by regionally available species, then local abiotic and biotic processes act on but a subset of potential community members. The addition of a new member by a rare dispersal event could have major consequences for community composition, especially if that species has strong interactions with existing species. Therefore, changes in community structure and function may occur that could rival or exceed changes that occur due to local processes. This is essentially the problem with invading species (Drake et al., 1991), although exotic species transferred among continents by humans have typically received most attention. Our results suggest that similar 'invasions' could occur intra-continentally by species native to a region, although the results may go unrecognized if zooplankton are presumed to have cosmopolitan distributions.

Succession

Robinson & Dickerson (1987) experimentally manipulated inoculation sequence and found sequence was important to resulting community structure, especially at low arrival rates. If many zooplankton species rarely disperse overland, the sequence of colonization will have lasting priority effects on subsequent community structure, especially given the ability of many zooplankton to develop large populations quickly and produce dormant life stages. Variation among zooplankton communities of regional, even closely-spaced ponds (Fryer, 1985) may be due to such effects.

Population genetics

Dispersal is significant to the maintenance of regional metapopulations (Taylor, 1990). Given low dispersal rates among water bodies, populations founded by single or few propagules may exhibit lasting founder effects (Berg & Garton, 1994; Boileau & Hebert, 1991; Thier, 1994). Therefore, egg banks (DeStasio,

1987) may store little genetic variation, and populations may be subject to inbreeding or outbreeding depression (Brown, 1991). Zooplankton populations among isolated water bodies may not operate as metapopulations.

Disturbance

Without minimal dispersal to provide a 'rescue effect' (Gotelli, 1991), local extinctions may occur, potentially changing community dynamics. Although local extinction risk is mitigated by an egg bank, the genetic bottleneck involved in colonization by one of few propagules may render some populations susceptible to disturbances that would be relatively innocuous to other, more diverse populations. Different populations may then respond to disturbance differently.

In summary, we did not find zooplankton to be readily dispersed by wind, rain, and waterfowl. Bdelloid rotifers alone were wind dispersed, but infrequently at distances of 1 km from a water body. Limited dispersal has important ramifications for common perceptions about processes regulating zooplankton community structure in freshwaters.

References

Berg, D. J. & D. W. Garton, 1994. Genetic differentiation in North American and European populations of the cladoceran Bythotrephes. Limnol. Oceanogr. 39: 1503–1516.

Boileau, M. G. & P. D. N. Hebert, 1991. Genetic consequences of passive dispersal in pond dwelling copepods. Evolution 45: 721–733.

Brown, J. H. & A. C. Gibson, 1983. Biogeography. C. V. Mosby Company, St.Louis, MO, USA.

Brown, A. F, 1991. Outbreeding depression as a cost of dispersal in the harpacticoid copepod, Tigriopus californicus. Biol. Bull. 181: 123–126.

Carter, J. C. H., M. J. Dadswell, J. C. Roff & W. G. Sprules, 1980. Distribution and zoogeography of planktonic crustaceans and dipterans in glaciated eastern North America. Can. J. Zool. 58: 1355–1387.

Chaplin, J. A. & D. J. Ayre, 1997. Genetic evidence of widespread dispersal in a parthenogenetic freshwater ostracod. Heredity 78: 57–67.

Connell, J. H., 1961. The influence of interspecific competition and other factors on the distribution of the barnacle Chthamalus stellatus. Ecology 42: 710–723.

De Stasio, B. T. Jr., 1991. The seed bank of a freshwater crustacean: copepodology for the plant ecologist. Ecology 70: 1377–1389.

Drake, J. A., H. A. Mooney, F. di Castri et al. (eds), 1989. Biological Invasions. A Global Perspective. SCOPE 37. John Wiley & Sons, New York, NY, U.S.A.

Frey, D. G., 1986. The non-cosmopolitanism of chydorid Cladocera: implications for biogeography and evolution. In: Gore, R. H. & K. L. Heck (eds), Crustacean Biogeography. Balkema, Rotterdam: 237–256.

Fryer, G., 1985. Crustacean diversity in relation to the size of water bodies: some facts and problems. Freshwat. Biol. 15: 347–361.

Gislen, T., 1947. Aerial plankton and its conditions of life. Biol. Rev. 23: 109–126.

Gotelli, N. J., 1991. Metapopulation models: the rescue effect, the propagule rain, and the core-satellite hypothesis. Am. Nat. 138: 768–776.

Hagiwara, A., 1995. Resting eggs of the marine rotifer Brachionus plicatilis Muller; development, and effect of irradiation on hatching. Hydrobiologia 313/314: 223–229.

Hebert, P. D. N. & B. J. Hann, 1986. Patterns in the composition of arctic tundra pond microcrustacean communities. Can. J. Fish. aquat. Sci. 43: 1416–1425.

Hebert, P. D. N. & C. Wilson, 1994. Provincialism in plankton: endemism and allopatric speciation in Australian Daphnia. Evolution 48: 1333–1349.

Hutchinson, G. E., 1967. A Treatise on Limnology. Volume II. Introduction to Lake Biology and the Limnoplankton. John Wiley and Sons, New York, NY, U.S.A.

Jenkins, D. G. & A. L. Buikema, Jr., 1998. Do similar communities develop in similar sites? A test with zooplankton community structure and function in new ponds. Ecol. Monogr. 68: 421–443.

Jenkins, D. G., 1995. Dispersal-limited zooplankton distribution and community composition in new ponds. Hydrobiologia 313/314: 15–20.

Karnes, J. & D. McFall (eds), 1995. A directory of Illinois Nature Preserves. Vol. 2 — Northwest, Central, and Southern Illinois. Illinois Dept. of Natural Resources, Springfield, IL, U.S.A.: 321 pp.

King, C. E., 1980. The genetic structure of zooplankton populations. In W. C. Kerfoot (ed.), Evolution and Ecology of Zooplankton Communities. University Press of New England, Hanover, NH, U.S.A.: 315–328.

Lampert, W. & U. Sommer, 1997. Limnoecology. The Ecology of Lakes and Streams. Oxford University Press, New York, NY, U.S.A.: 156.

Lewin, R., 1986. Supply-side ecology. Science 234: 25–27.

Lowndes, A. G., 1930. Living ostracods in the rectum of a frog. Nature 126: 958.

Maguire, B., 1959. Passive overland transport of small aquatic organisms. Ecology 40: 312.

Maguire, B., 1963. The passive dispersal of small aquatic organisms and their colonization of isolated bodies of water. Ecol. Monogr. 33: 161–185.

Malone, C. R., 1965. Dispersal of plankton: rate of food passage in mallard ducks. J. Wildlife Management 29: 529–533.

Marcus, N., 1979. The population biology and nature of diapause of Labidocera aestiva (Copepoda: Calanoida). Biol. Bull. 157: 297–305.

May, L., 1986. Rotifer sampling – a complete species list from one visit? Hydrobiologia 134: 117–120.

McAtee, W. L., 1917. Showers of organic matter. Monthly Weather Rev. May: 217–224.

Mellors, W. K., 1975. Selective predation of ephippial Daphnia and the resistance of ephippial eggs to digestion. Ecology 56: 974–980.

Paine, R. T., 1974. Intertidal community structure. Experimental studies on the relationship between a dominant competitor and its principal predator. Oecologia 15: 93–120.

Pennak, R. W., 1989. Freshwater invertebrates of the United States, 3rd edn. John Wiley and Sons, New York, NY, U.S.A.: 628 pp.

Proctor, V. W., 1964. Viability of crustacean eggs recovered from ducks. Ecology 45: 656–658.

Proctor, V. W. & C. R. Malone, 1965. Further evidence of the passive dispersal of small aquatic organisms via the intestinal tract of birds. Ecology 46: 728–729.

Proctor, V. W., C. R. Malone & V. L. DeVlaming, 1967. Dispersal of aquatic organisms: viability of disseminules recovered from the intestinal tract of captive killdeer. Ecology 48: 672–676.

Ricklefs, R. E., 1987. Community diversity: relative roles of local and regional processes. Science 235: 167–171.

Robinson, J. V. & J. E. Dickerson, 1987. Does invasion sequence affect community structure? Ecology 68: 587–595.

Roughgarden, J., 1989. The structure and assembly of communities. In Roughgarden, J., R. M. May & S. A. Levin (eds), Perspectives in Ecological Theory. Princeton University Press, Princeton, NJ, U.S.A.: 203–226.

Roughgarden, J., S. D. Gaines & S. W. Pacala, 1987. Supply side ecology: the role of physical transport processes. In Gee, J. H. R. & P. S. Giller (eds), 27th Symposium of British Ecological Society. Blackwell, Boston, MA, U.S.A.: 491–518.

Schlichting, H. E. & L. E. Milliger, 1969. The dispersal of microorganisms by a hemipteran, *Lethocerus uhleri* (Montandon). Trans. am. Microsc. Soc. 88: 452–454.

Schwartz, S. S. & P. D. Hebert, 1987. Methods for the activation of the resting eggs of *Daphnia*. Freshwat. Biol. 171: 173–179.

Sides, S. L., 1973. Observation on dispersal of algae and protozoa by the cabbage butterfly. Trans. am. Microsc. Soc. 9: 96–97.

Stemberger, R. S., 1995. Pleistocene refuge areas and postglacial dispersal of copepods of the northeastern United States. Can. J. Fish. aquat. Sci. 52: 2197–2210.

Stewart, K. W. & H. E. Schlichting, 1966. Dispersal of algae and protozoa by selected aquatic insects. J. Ecol. 54: 551–562.

Swanson, G. A., 1984. Dissemination of amphipods by waterfowl. J. Wildl. Manage. 48: 988–991.

Taylor, A. D., 1990. Metapopulations, dispersal, and predator-prey dynamics: an overview. Ecology 71: 429–433.

Thier, E., 1994. Allozyme variation among natural populations of *Holopedium gibberum* (Crustacea; Cladocera). Freshwat. Biol. 31: 87–96.

Underwood, A. J. & E. J. Denley, 1984. Paradigms, explanations and generalizations in models for the structure of intertidal communities on rocky shores. In Strong, D. R. Jr., D. Simberloff, L. G. Abele & A. B. Thistle (eds), Ecological Communities: Conceptual Issues and the Evidence. Princeton University Press, Princeton, NJ, U.S.A.: 151–180.

Wetzel, R. G., 1983. Limnology, 2nd edn. Saunders, Philadelphia: 767 pp.

Weider, L. J., 1989. Spatial heterogeneity and clonal structure in arctic populations of apomictic *Daphnia*. Evolution 70: 1405–1413.

Hydrobiologia **387/388**: 23–26, 1998.
E. Wurdak, R. Wallace & H. Segers (eds), Rotifera VIII: A Comparative Approach.
© *1998 Kluwer Academic Publishers.*

Freshwater Rotifera of the genus *Lecane* from Songkhla Province, southern Thailand

Pornsilp Pholpunthin & Supenya Chittapun
Department of Biology, Faculty of Science, Prince of Songkhla University, Hat-Yai, Songkhla 90112, Thailand

Key words: Rotifera, *Lecane*, Songkhla province, Thailand

Abstract

Eighteen freshwater bodies in Songkhla Province, southern Thailand were investigated for rotifers of the genus *Lecane*. A total of 23 species were identified. The majority of species found were cosmopolitan (43%) or tropico-politan (39%). The rest were oriental (9%) and palaeotropical (9%). The most common species was *L. bulla* (61% of the plankton samples taken), while *L. aculeata*, *L. arcula*, *L. blachei*, *L. stenroosi* and *L. tenuiseta* were rare (only found once). The greatest species diversity was found in Khlong-Hla reservoir (14 species).

Introduction

Although there have been several contributions dealing with the Thai rotifer fauna (Boonsom, 1984; Sanoamuang et al., 1995; Segers & Pholpunthin, 1997; Segers & Sanoamuang, 1994), most of these studies were conducted on plankton samples from several localities in north or northeast Thailand. Information on the freshwater rotifers from the southern part of Thailand is restricted to Thale-Noi Lake (Segers & Pholpunthin, 1997). The aim of the present study, reporting on the diversity of the genus *Lecane* in Songkhla province, was to inform on the rotifer fauna in the south.

Materials and methods

Plankton samples were collected from 18 freshwater habitats, ranging from dams, reservoirs, ponds, and roadside canals to rivers of Songkhla province (Figure 1). They were collected quantitatively by several oblique hauls, using a 26-μm mesh plankton net during September and October 1996. All samples were preserved in 4% formaldehyde. Specimens were examined and photographed using an Olympus VM dissection, and a Nikon OPTIPHOT-2 microscope with differential interference contrast equipment. Nomenclature follows Segers (1995).

All measurements are in μm.

Results and discussion

To date, 54 species of Thai rotifer fauna of the genus *Lecane* were recorded (Segers & Pholpunthin, 1997; Sanoamuang et al., 1995), 26 of which appeared in both the northeastern and southern part of Thailand. Twenty-one species were only reported from the northeast and seven species were exclusively found in the south. In the present study, a total of 23 species were identified as shown in Table 1. Of these, 20 species have been previously recorded in Thale-Noi Lake (Segers & Pholpunthin, 1997) and three species (*Lecane blachei*, *L. stenroosi* and *L. thailandensis*) were found for the first time in the south of Thailand. The finding of *L. blachei* (Figures 2 and 3), which has been reported from several northeastern localities (Sanoamuang et al., 1995), is an extension of its distribution to the south. *L. stenroosi* (Figure 4) is considered to be a cosmopolitan species (Segers, 1996). *L. thailandensis* (Figure 5) was previously recorded from Nam Pung reservoir (Sakon Nakhon province, northeast Thailand) and Donqian Lake (Zhejiang province, China) (Segers & Sanoamuang, 1994). The present finding is the third record for the species. Its general morphology agrees with the original description by Segers & Sanoamuang (1994).

The greatest number of species was found in Khlong-Hla reservoir (14 species). In contrast, the genus *Lecane* was not represented in Rawa dam, Ban-Sauntoon reservoir and PSU reservoir. The most

24

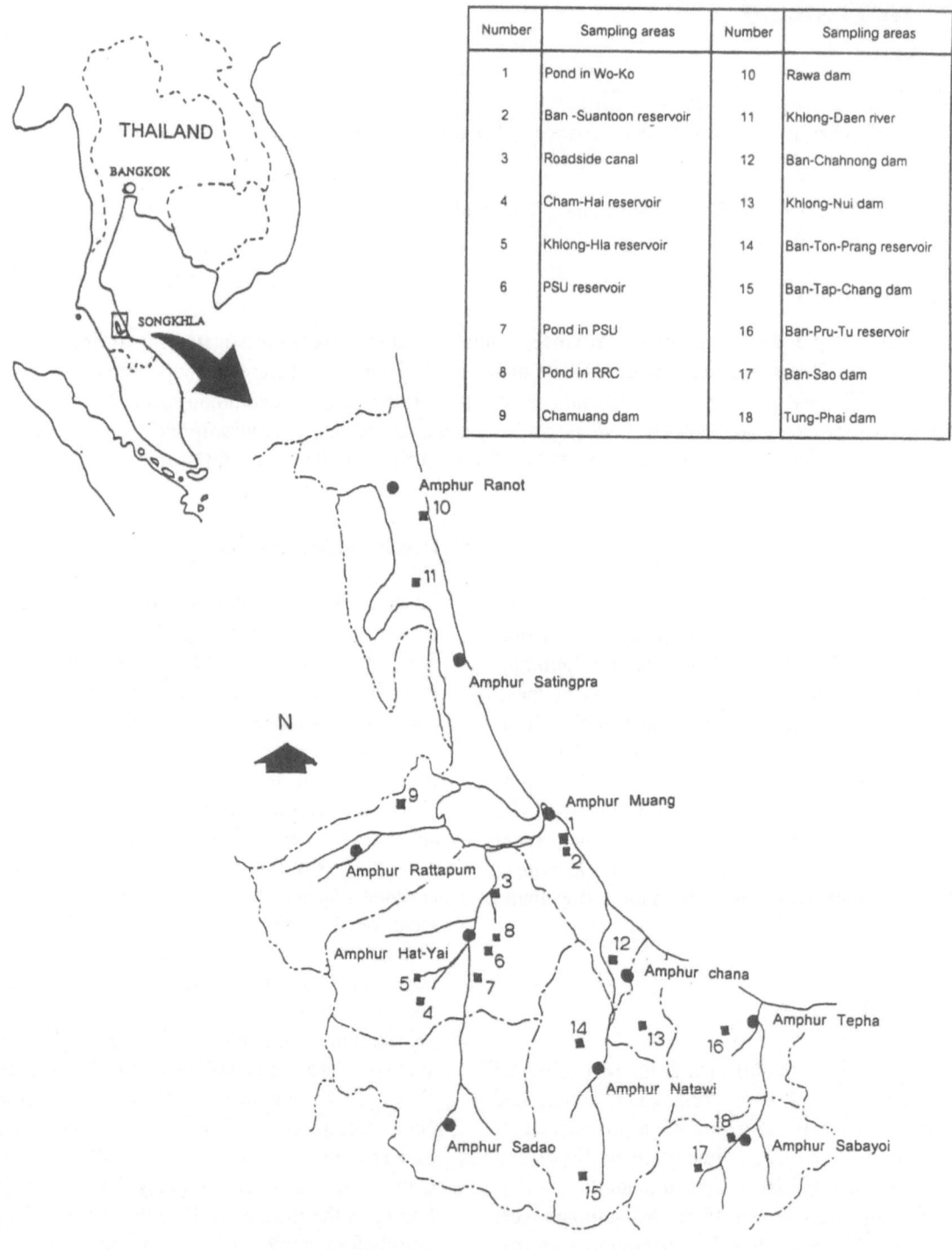

Number	Sampling areas	Number	Sampling areas
1	Pond in Wo-Ko	10	Rawa dam
2	Ban -Suantoon reservoir	11	Khlong-Daen river
3	Roadside canal	12	Ban-Chahnong dam
4	Cham-Hai reservoir	13	Khlong-Nui dam
5	Khlong-Hla reservoir	14	Ban-Ton-Prang reservoir
6	PSU reservoir	15	Ban-Tap-Chang dam
7	Pond in PSU	16	Ban-Pru-Tu reservoir
8	Pond in RRC	17	Ban-Sao dam
9	Chamuang dam	18	Tung-Phai dam

Figure 1. Situation of Songkhla province. Inset: sampling stations of freshwater habitats in Songkhla Province.

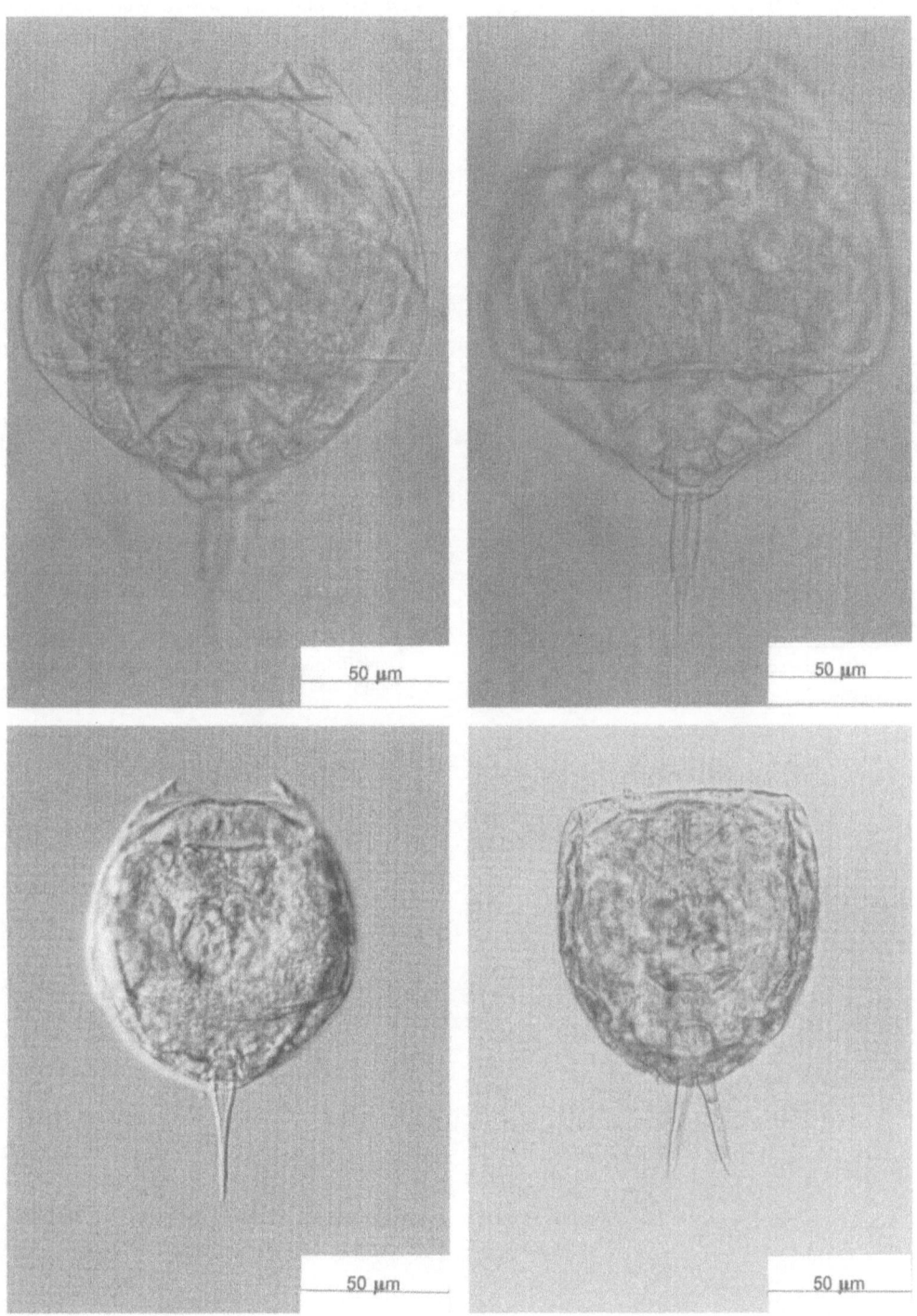

Figures 2–5. **Figures 2 and 3.** *Lecane blachei.* **Figure 4.** *Lecane stenroosi.* **Figure 5.** *Lecane thailandensis.*

26

Table 1. List of Rotifera of the genus *Lecane* from Songkhla Province, Thailand

Lecane aculeata (Jakubski, 1912): 5
L. arcula (Harring, 1914): 5
*L. blachei (Berzins, 1973): 13
L. bulla (Gosse, 1851): 1, 3, 5, 7, 9, 11, 13, 14, 17
L. curvicornis (Murray, 1913): 3, 8, 11, 12, 14, 16
L. furcata (Murray, 1913): 4, 5, 8, 11
L. hamata (Stokes, 1896): 4, 5, 7, 8, 11, 14, 15, 17, 18
L. hornemanni (Ehrenberg, 1834): 1, 5, 13, 14, 16
L. lateralis (Sharma, 1978): 11, 17
L. leontina (Turner, 1892): 3, 4, 5, 7, 8, 9, 13, 14
L. ludwigii (Eckstein, 1883): 3, 5
L. luna (O.F. Müller, 1883): 4, 5, 7, 13, 14, 16
L. lunaris (Ehrenberg, 1832): 7, 8, 9, 13, 16
L. obtusa (Murray, 1913): 4, 13
L. papuana (Murray, 1913): 1, 3, 5, 8, 11, 17
L. quadridentata (Ehrenberg, 1832): 3, 5, 11, 13, 14
L. rhenana (Hauer, 1964): 5, 11
L. signifera (Jennings, 1896): 1, 3, 4, 5, 8, 17
*L. stenroosi (Meissner, 1908): 11
L. tenuiseta (Harring, 1914): 11
*L. thailandensis (Segers & Sanoamuang, 1994): 5, 16
L. unguitata (Fadeev, 1925): 3, 4, 7, 8, 9, 13, 17
L. ungulata (Gosse, 1887): 3, 8, 9

The numbers refer to the sampling station. *New to southern part of Thailand.

common species was *L. bulla* (61% of the plankton samples taken), followed by *L. hamata* (50%) and *L. leontina* (45%). *L. aculeata*, *L. arcula*, *L. blachei*, *L. stenroosi* and *L. tenuiseta* were rare and only found at one sampling station each. Among the 23 species found in the present study, 10 (43%) are cosmopolitan and nine (39%) are tropicopolitan.

References

Boonsom, J., 1984. The freshwater zooplankton of Thailand (Rotifera and Crustacea). Hydrobiologia 113: 233–229.

Sanoamuang, L., H. Segers & H. J. Dumont, 1995. Additions to the rotifer fauna of south-east Asia: new and rare species from northeast Thailand. Hydrobiologia 313/314 (Dev. Hydrobiol. 109): 35–45.

Segers, H., 1995. Rotifera. Volume 2: The Lecanidae (Monogononta). Guides to the Identification of the Microinvertebrates of the Continental Waters of the World. SPB Academic Publishing, The Netherlands: 226 pp.

Segers, H., 1996. The biogeography of littoral *Lecane* Rotifera. Hydrobiologia 323: 169–197.

Segers, H. & P. Pholpunthin, 1997. New and rare Rotifera from Thale-Noi Lake, Pattalung Province, Thailand, with a note on the taxonomy of *Cephalodella* (Notommatidae). Ann. Limnol. 33: 13–21.

Segers, H. & L. Sanoamuang, 1994. Two more new species of *Lecane* (Rotifera, Monogononta) from Thailand. Belg. J. Zool. 124: 39–46.

Hydrobiologia **387/388**: 27–33, 1998.
E. Wurdak, R. Wallace & H. Segers (eds), Rotifera VIII: A Comparative Approach.
© *1998 Kluwer Academic Publishers.*

Rotifera of some freshwater habitats in the floodplain of the River Nan, northern Thailand

La-orsri Sanoamuang
Department of Biology, Faculty of Science, Khon Kaen University, Khon Kaen 40002, Thailand

Key words: Rotifera, biodiversity, floodplain, taxonomy, Thailand

Abstract

A survey of 11 freshwater habitats in the floodplain of the River Nan, northern Thailand was carried out during April and September 1996. The rotifer samples were collected qualitatively from paddy fields, ponds, canals and reservoirs, using a 60 μm mesh net. One hundred and eighteen species were identified, four (*Lepadella quinquecostata* (Lucks), *Macrochaetus danneeli* Koste & Shiel, *Testudinella ahlstromi* Hauer and *T. greeni* Koste) of which are new to Thailand and one (*L. quinquecostata*) is new to Asia. The numbers of species found in two localities are relatively high, with 86 and 73 rotifer taxa. Most of the species recorded are common, cosmopolitan or pantropical and warm-stenotherms. The occurrence of a species previously considered endemic to Australia, *M. danneeli* provides more evidence illustrating a relation between the rotifer faunas of southeast Asia and Australia. Comments are presented on some insufficiently known taxa in particular on the new records for Thailand.

Introduction

Thailand is situated in a tropical and humid climatic zone and supports a variety of freshwater ecosystems. In the last five years, attempts have been made to study the species composition of the Thai rotifer fauna. In 1994, Segers & Sanoamuang described two new rotifers, *Lecane shieli* and *Lecane thailandensis* from Nam Pung reservoir, north-east Thailand. Later, Sanoamuang et al. (1995) described *Brachionus niwati* and identified 200 species from 93 localities in the northeast. They also documented 120 species new to Thailand, bringing the rotifer records to 251. Recently, Sanoamuang (1996) described *Lecane segersi* from a swamp in the north-east and added *L. braumi* Koste to the Thai checklist. Three more newly described species are *Cephalodella songkhlaensis* Segers & Pholpunthin, 1997, *Trichocerca siamensis* Segers & Pholpunthin, 1997 and *Lecane superaculeata* Sanoamuang & Segers, 1997. Fifteen and one (*L. eswari* Dhanapathi) new Thai records were recently published by Segers & Pholpunthin (1997) and Sanoamuang & Segers (1997), respectively. As a result, 272 rotifer species are recorded from the

country. However, very little is known about the rotifer communities in northern Thailand. The purpose of this contribution is to document the rotifer community composition in a range of habitats in the floodplain of the River Nan, northern Thailand.

Study area

The majority of water in northern Thailand is supplied by four main rivers (Ping, Wang, Yom and Nan). The River Nan originates in the Luang Prabang mountain ranges in the north-east of the northern area, and stretches over about 740 kilometers passing through the Sirikit Dam and some major cities. It flows southward and meets the Rivers Yom and Ping at Phijit and Nakhon Sawan provinces, respectively (Figure 1). During the south-west and north-east monsoon period (May to November), the high precipitation usually causes temporarily flooding of vast areas. The samples for this study were collected from 11 habitats, ranging from paddy fields, ponds and canals to reservoirs.

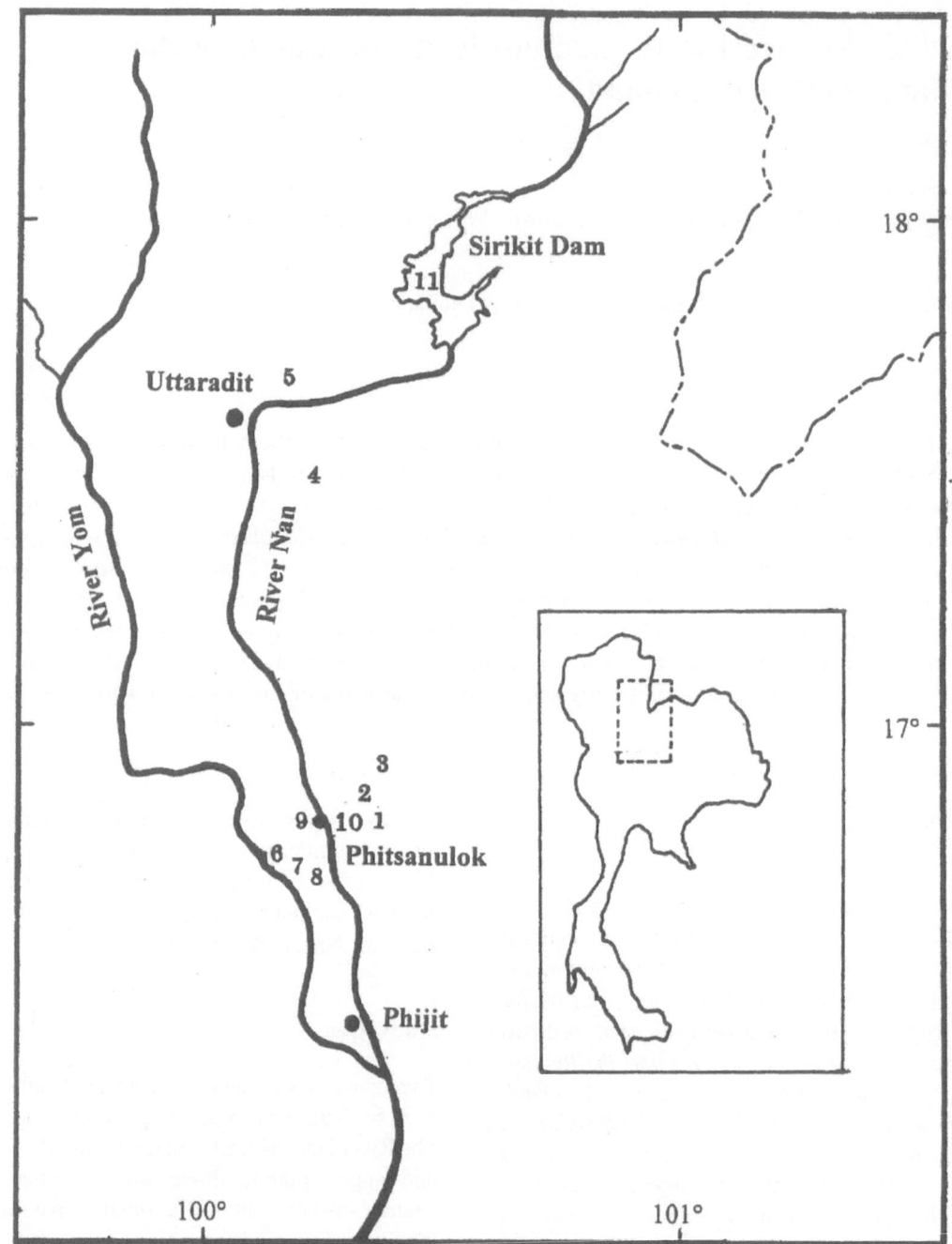

Figure 1. Map of northern Thailand to show sampling stations.

Methods

Qualitative samples were collected in 11 habitats using a standard plankton net with 60 μm mesh size, during April and September 1996. Table 1 lists the sampling localities and some environmental variables. The rotifers were preserved in 4% formaldehyde. Specimens were searched under a dissection microscope, and examined using an Olympus CHD microscope. Drawings were done using a camera lucida. Scanning

Table 1. List of localities sampled during this study with dates and some environmental variables

Locality	Co-ordinates	Sampling dates	Temp. (°C)	pH	Cond. (μS cm^{-1})
1. Paddy field I, Phitsanulok	16° 49' N–100° 23' E	27-09-96	30	7.6	55
2. Paddy field II, Phitsanulok	16° 51' N–100° 20' E	27-09-96	30	7.2	90
3. Roadside pond at Phitsanulok	16° 54' N–100° 24' E	27-09-96	30	6.8	65
4. Irrigation canal at Uttaradit	17° 27' N–100° 15' E	27-09-96	33	7.1	60
5. Roadside canal at Phitsanulok	17° 39' N–100° 10' E	28-09-96	29	7.2	130
6. Pond in Narasuan Univ., Phitsanulok	16° 45' N–100° 12' E	29-09-96	29	7.2	180
7. Roadside canal, Phitsanulok	16° 45' N–100° 11' E	29-09-96	29	7.5	160
8. Lotus Pond in Narasuan Univ., Phitsanulok	16° 44' N–100° 12' E	29-09-96	30	7.3	185
9. Pond in Dept. of Irrigation, Phitsanulok	16° 8' N–100° 15' E	29-09-96	29	7.8	140
10. Canal at Phitsanulok	16° 8' N–100° 18' E	29-09-96	30	7.5	40
11. Sirikit reservoir, Uttaradit		19-04-96	34	7.2	140

electron microscopy (SEM) was performed using a Hitachi S-320ON microscope on critical-point dried specimens.

Results and discussion

A list of the Rotifera recorded from the samples examined is presented in Table 2. 118 taxa were identified, four of which are new records for Thailand. The material also included a recently described species, *Lecane superaculeata* Sanoamuang & Segers (1997) (Figure 2). One species, *Lepadella quinquecostata* (Lucks), had not been recorded from Asia before. The samples furthermore yielded a species hitherto known from Australia only, *Macrochaetus danneeli* Koste & Shiel. The rotifer species record of Thailand now stands at 276 species. Most numerous are representatives of the genus *Lecane*, with 29.7% of the species listed, followed by species of the genera *Lepadella* (8.5%), *Brachionus* (8.5%) and *Trichocerca* (6.8%).

Eight of the rotifers recorded (6.8%) are restricted to the tropical and subtropical regions of the Old World. These are *Brachionus forficula* Wierzejski, *Lecane lateralis* Sharma, *L. unguitata* (Fadeev), *Lepadella discoidea* Segers, *L. vandenbrandei* Gillard, *Scaridium grandis* Segers, *Testudinella brevicaudata* Yamamoto and *Trochosphaera aequatorialis* (Semper). Three species (2.5%), *Brachionus donneri* Brehm, *Lecane blachei* Berzins and *L. superaculeata* Sanoamuang & Segers, are Oriental endemics. Additionally, 3 taxa (2.5%) *Brachionus kostei* Shiel,

Brachionus dichotomus Shephard f. *reductus* Koste & Shiel and *Macrochaetus danneeli* Koste & Shiel, are Australasian species.

Remarkably rich rotifer faunas were represented in samples from locality 10 and 2, where 86 and 73 taxa were found, respectively. These numbers do not include a large number of unidentifiable Bdelloidea and other illoricate rotifers. An explanation for this high species richness is probably the recent flood that washed resting eggs from higher areas into these localities. Comparable rotifer species diversity were recorded in Lakes Iyi-Efi and Oguta in the floodplain of the River Niger, Nigeria (with 136 and 124 species, Segers et al., 1993), and billabongs in the River Murray floodplain, Australia (with 71 species, Shiel, 1990). Most of the rotifers recorded are common, cosmopolitan, probably warm-stenothermic and have already been recorded from the north-east (Sanoamuang et al., 1995) and the south (Segers & Pholpunthin, 1997) of Thailand.

The occurrence of several species which were previously considered endemic to Australia, e.g., *B. dichotomus* f. *reductus*, *B. kostei*, *B. lyratus*, *L. batillifer* (see Sanoamuang et al., 1995) and *M. danneeli* (present study) documents more evidence illustrating a relation between the rotifer faunas of the tropical region of south Asia and Australia.

Comments on some rare or poorly known species are as follows.

Table 2. Rotifer species recorded from some habitats in the flood-plain of the River Nan, northern Thailand. Numbers refer to sample localities (Table 1). *: New record for Thailand

Anuraeopsis fissa (Gosse): 2, 11
Ascomorpha ecaudis (Perty): 4
Asplanchna priodonta Gosse: 4, 6
A. sieboldi (Leydig): 1, 7, 8
Brachionus angularis Gosse: 2, 5, 6, 8, 10
B. calyciflorus Pallas: 6, 9, 10
B. caudatus Barrois & Daday: 10
B. dichotomus Shephard f. *reductus* (Koste & Shiel): 2, 4, 10
B. diversicornis (Daday): 6
B. donneri Brehm: 4, 10
B. falcatus Zacharias: 2, 4, 5, 6, 8, 10
B. forficula Wierzejski: 4, 6, 11
B. kostei Shiel: 10
B. quadridentatus Hermann: 2, 10
Cephalodella sp.: 1
Collotheca sp.: 4, 11
Colurella colurus (Ehrenberg): 2
C. uncinata Muller 1, 2, 10
Conochilus dossuarius (Hudson): 2, 4
Dicranophoroides caudatus (Ehrenberg): 10
D. grandis (Ehrenberg): 10
Dipleuchlanis propatula (Gosse): 1, 2, 3, 10
Epiphanes clavulata (Ehrenberg): 2, 3, 5, 10
Euchlanis dilatata Ehrenberg: 1, 3, 10
E. incisa Carlin: 2, 3
Filinia camasecla Myers: 2, 4, 10
F. longiseta (Ehrenberg): 2, 4, 8, 10
F. opoliensis (Zacharias): 2, 4, 8, 10
F. pejleri Hutchinson: 2, 4, 6, 10
F. saltator (Gosse): 2
Hexarthra intermedia Wiszniewski: 4, 6, 7, 8
H. mira (Hudson): 2, 10
Keratella cochlearis (Gosse): 1, 2, 4, 10, 11
K. lenzi Hauer: 2, 10
K. tropica (Apstein): 2, 8, 10, 11
Lecane aculeata (Jakubski): 10
L. aeganea Harring: 2
L. arcula Harring: 2
L. aspasia Myers: 2, 10
L. blachei Berzins: 10
L. bulla (Gosse): 1, 2, 3, 5, 7, 10
L. closterocerca (Schmarda): 10
L. crepida Harring: 1, 5, 10
L. curvicornis (Murray): 1, 2, 3, 4, 10
L. doryssa Harring: 2, 10
L. elegans Harring: 10
L. furcata (Murray): 2, 10
L. haliclysta Harring & Myers: 2, 10
L. hamata (Stokes): 1, 2, 10
L. hastata (Murray): 5
L. hornemanni (Ehrenberg): 2, 10
L. inopinata Harring & Myers: 2
L. lateralis Sharma 1, 2, 5, 10
L. leontina (Turner): 1, 2, 5, 10
L. ludwigii (Eckstein): 2, 10
L. luna (Muller): 1, 2, 3, 5, 10
L. lunaris (Ehrenberg): 2, 10
L. obtusa (Murray): 10

Table 2. Continued.

L. papuana (Murray): 1, 2, 3, 5, 10
L. pertica Harring & Myers: 10
L. pusilla Harring: 2
L. quadridentata (Ehrenberg): 2, 10
L. rhenana Hauer: 2
L. rhytida Harring & Myers: 2, 10
L. signifera (Jennings): 2, 10
L. stenroosi (Meissner): 2, 5, 10
L. superaculeata Sanoamuang & Segers: 10
L. undulata Hauer: 10
L. unguitata (Fadeev): 1, 2, 5, 10
L. ungulata (Gosse): 2, 10
Lepadella costatoides Segers: 2, 10
L. discoidea Segers: 1, 2, 3, 10
L. ehrenbergi (Perty): 10
L. latusinus (Hilgendorf): 1, 2, 10
L. ovalis (Muller): 10
L. patella (Muller): 2
L. quadricarinata (Stenroos): I
* *L. quinquecostata* Lucks: 10
L. rhomboides (Gosse): 1, 2, 3, 10
L. vandenbrandei Gillard: 10
Lophocharis salpina (Ehrenberg): 2, 5, 10
Macrochaetus collinsi (Gosse): 1, 2, 5, 10
* *M. danneeli* Koste & Shiel: 2
M. longipes Myers: 1, 2, 10
Manfredium eudactylotum (Gosse): 1, 2, 10
Monommata sp.: 10
Mytilina acanthophora Hauer: 2, 10
M. bisulcata (Lucks): 2, 10
M. unguipes (Lucks): 10
M. ventralis (Ehrenberg): 2, 3, 10
Notommata pachyura (Gosse): 2, 10
Plationus patulus (Muller): 1, 2, 3, 4, 5, 10
Platyias quadricornis (Ehrenberg): 1, 2, 3, 10
Polyarthra major Burckhardt: I
P. vulgaris Carlin: 2, 3, 4, 5, 6, 7, 8, 9, 10, 11
Pompholyx complanata Gosse: 10
Scaridium bostjani Daems & Dumont: 10
S. grandis Segers: 10
S. longicaudum (Muller): 2
Sinantherina semibullata (Thorpe): 2, 10
Synchaeta pectinata Ehrenberg: 7
Synchaeta sp.: 4, 6
* *Testudinella ahlstromi* Hauer: 10
T. brevicaudata Yamamoto: 2
* *T. greeni* Koste: 2
T. patina (Hermann): 1, 2, 3, 5, 10
T. tridentata Smirnov: 10
Trichocerca bicristata (Gosse): 2
T. braziliensis (Murray): 2, 10
T. capucina Wierzejski & Zacharias: 4, 5
T. flagellata Hauer: 10
T. insignis (Herrick): 2
T. pusilla (Lauterborn): 1, 2, 3, 10
T. similis (Wierzej ski): 2, 3, 4, 10
T. tenuior Gosse: 2
Trichotria tetractis (Ehrenberg): 1, 2, 10
Tripleuchlanis plicata (Levander): 10
Trochosphaera aequatorialis (Semper): 10

31

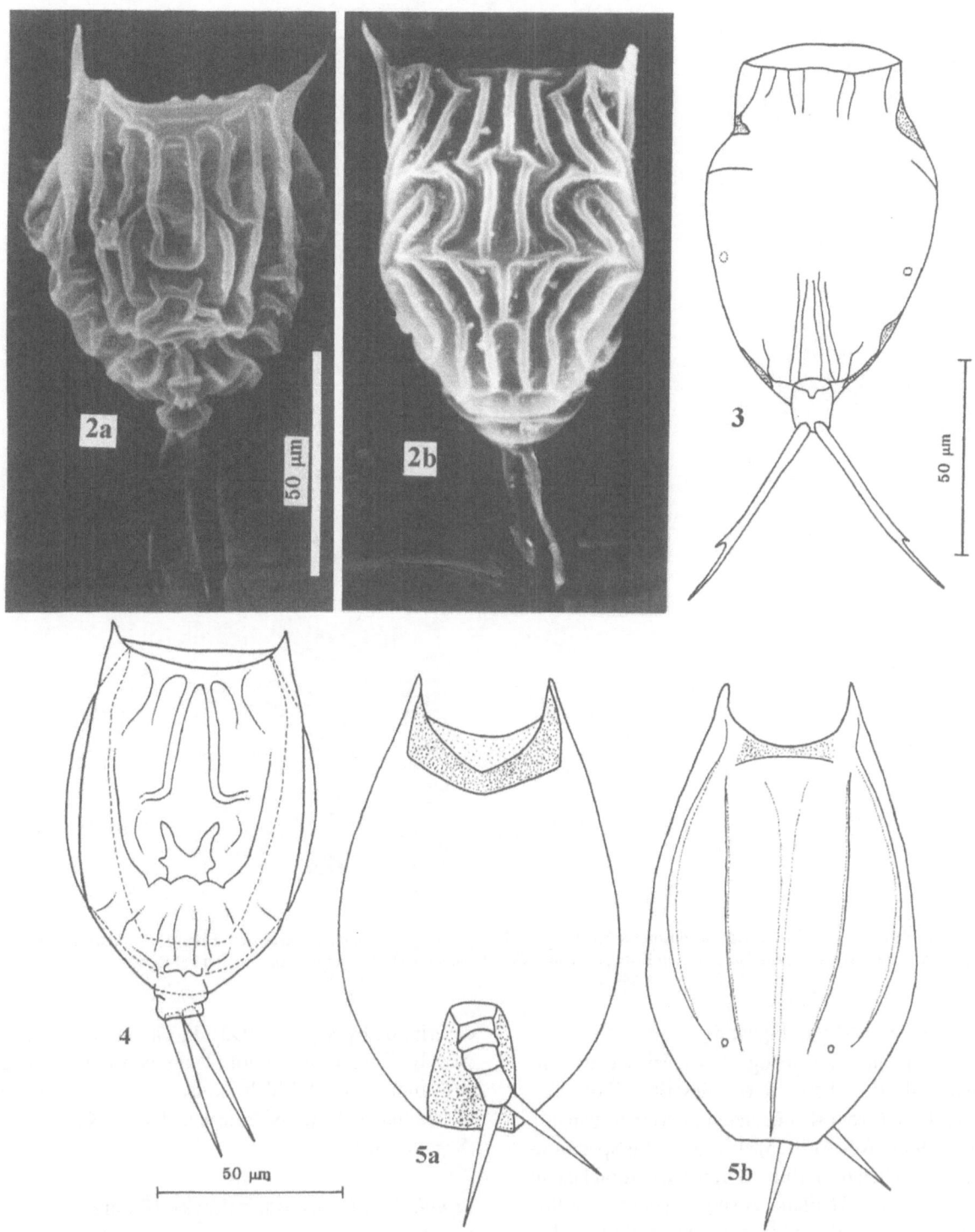

Figures 2–5. **Figure 2.** *Lecane superaculeata* Sanoamuang & Segers, a: ventral view, b: dorsal view (SEM photomicrograph). **Figure 3.** *Lecane elegans* Harring, ventral view. **Figure 4.** *Lecane rhytida* Harring & Myers, ventral view. **Figure 5.** *Lepadella quinquecostata* (Lucks), a: ventral view, b: dorsal view.

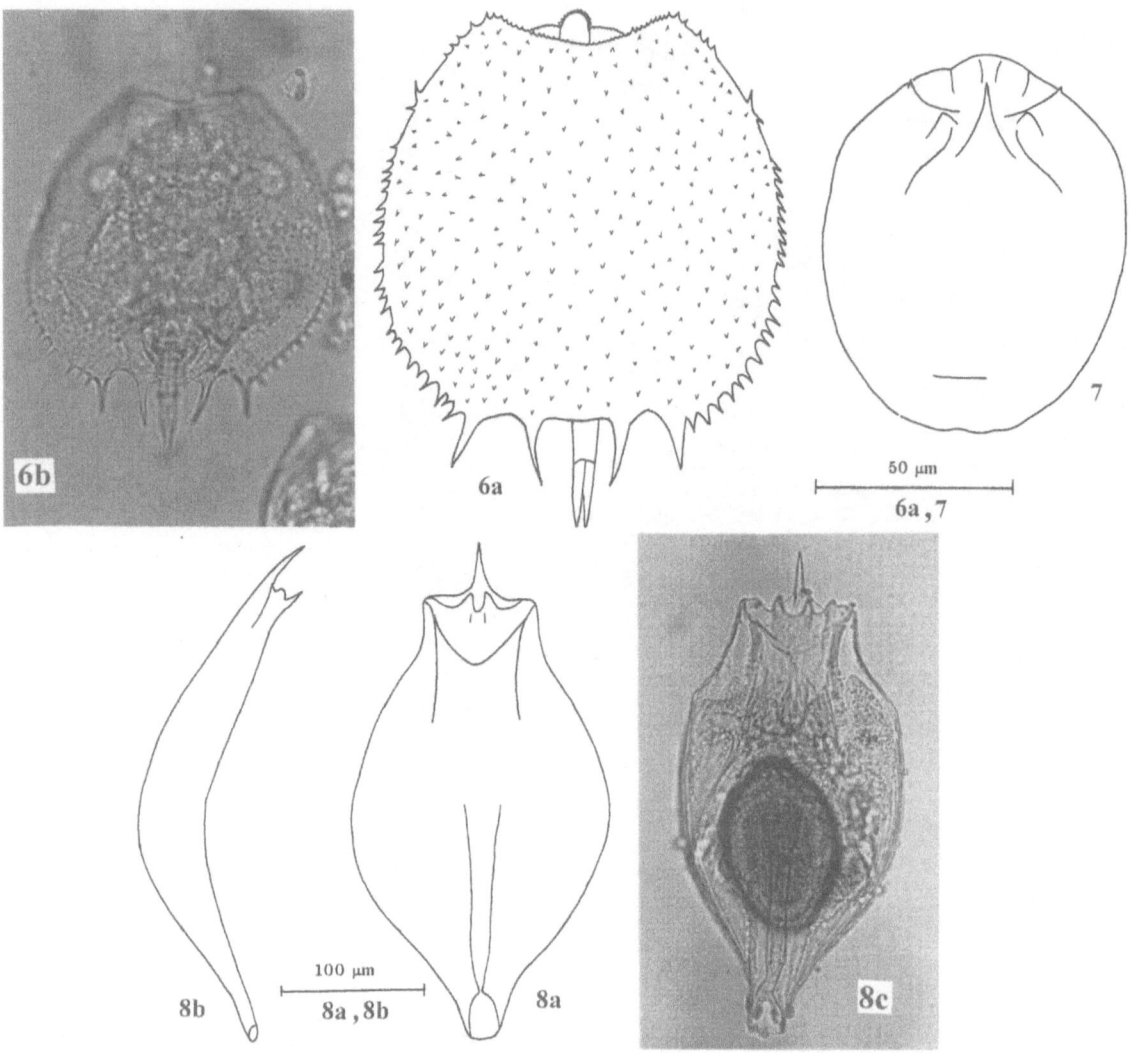

Figures 6–8. **Figure 6.** *Macrochaetus danneeli* Koste & Shiel, a: dorsal view, b: photomicrograph, dorsal view. **Figure 7.** *Testudinella ah!stromi* Hauer, ventral view. **Figure 8.** *Testudinella greeni* Koste, a: ventral view, b: lateral view, c: photomicrograph, ventral view.

Lecane elegans Harring (Figure 3)

L. elegans is an easily recognized species with soft, elongate, illoricate body. It was described from material collected from Rio Grande reservoir in Panama (Segers, 1995). Koste (1975) also found this species in the samples taken from Bung Borapet, a natural lake in the central part of Thailand. A single specimen of this rare species was collected from a canal in Phitsanulok (locality 10).

Lecane rhytida Harring & Myers (Figure 4)

Numerous specimens were found in an inundated paddy field (locality 2) and a canal at Phitsanulok (locality 10). The general characteristics follow the description by Segers (1995). In Thailand, it was previously found in the south (Thale-Noi Lake) (Segers & Pholpunthin, 1997). It is also known from Brazil, Nicaragua, U.S.A., Nigeria and Papua New Guinea (Segers, 1995).

Lepadella quinquecostata (Lucks) (Figure 5)

L. quinquecostata is characterized by having five sharp longitudinal carinas. The general shape of lorica and habitat of the Thai specimens conforms with figures of Nigerian specimens (see Figs 10a–e in Segers et al., 1992). It has been recorded from Australia (Koste & Shiel, 1989), Argentina (Jose de Paggi, 1990), Brazil (Segers & Sarma, 1993) and Nigeria (Segers et al.,

1993). The present record of this rare species from Thailand is the first for Asia.

Macrochaetus danneeli Koste & Shiel (Figure 6)

This remarkable species was described from a single locality, Buffalo billabong in Northern Territory, Australia (Koste & Shiel, 1983). It is distinguished from other species of *Macrochaetus* by the absence of large dorsal spines. Only *M. aspinus* Segers & Sarma from Brazil also lacks these spines, but *M. danneeli* is armed with 2 pairs of caudal spines as in *M. americanus* Segers & Sarma (Segers & Sarma, 1993). *M. danneeli* is only known from Australia and the south of India (unpublished data, cited by Segers & Sarma, 1993). Thus, the present finding is the first confirmed record of this species from Asia.

Testudinella ahlstromi Hauer (Figure 7)

T. ahlstromi has been recorded from several localities in Tasmania (Koste et al., 1988), Argentina (Jose de Paggi, 1990) and Malaysia (Green, 1995). The general morphology of the Thai material is in good agreement with the description the species as in Koste (1978).

Testudinella greeni Koste (Figure 8)

This rare taxon was originally described as a variety of *T. tridentata* (Wulfert, 1965, cited by Koste, 1978). However, Koste (1981) considered that it was sufficiently different from *T. tridentata* to be treated as as a separate species, under the name *T. greeni*. It is known from Nigeria (Wulfert, 1965), tropical Australia (Koste, 1981), Malaysia (Fernando & Zankai, 1981), and now from Thailand.

Acknowledgements

The author appreciates the grant supported by the Ministry of University Affairs to attend the VIII International Rotifer Symposium. The author also thanks Mr Pipatphong Kanla for his efforts on SEM photography and Dr Hendrik Segers for providing some interesting papers. This work was supported by the TRF/BIOTEC Special Program for Biodiversity Research and Training grant BRT 140028.

References

Fernando, C.H. & N.P. Zankai, 1981. The Rotifera of Malaysia and Singapore with remarks on some species. Hydrobiologia 78: 205–219.

Green, J., 1995. Associations of planktonic and periphytic rotifers in a Malaysian estuary and two nearby ponds. Hydrobiologia 313/314: 47–56.

Jose de Paggi, S. J., 1990. Ecological and biogeographical remarks on the rotifer fauna of Argentina. Rev. Hydrobiol. trop. 23: 297–311.

Koste, W., 1975. Über den Rotatorienbestand einer Mikrobiozonose in einem tropischen aquatischen Saumbiotop, der *Eichhornia-crassipes*-Zone im Littoral des Bung-Borapet, einem Stausee in Zentralthailand. Gewass. Abwass. 57/58: 43–58.

Koste, W., 1978. Rotatoria. Die Rädertiere Mitteleuropas. Ein Bestimmungswerk begr. Von Max Voigt. Überordnung Monogononta Vol. 1–2. 673 pp. + 234 pl.

Koste, W., 1981. Zur Morphologie, Systematik und Ökologie von neuen momogononten Rädertieren (Rotatoria) aus dem Überschwemmungsgebiet des Magela Creek in der Alligator-River-Region Australiens, N.T. Tiel 1. Osnabrucker naturwiss. Mitt. 8: 97–126.

Koste, W. & R. J. Shiel, 1983. Morphology, systematics and ecology of new monogonont Rotifera (Rotatoria) from the Alligator Rivers region, Northern Territory. Trans. R. Soc. S. Aust. 107: 109–121.

Koste, W. & R. J. Shiel, 1989. Rotifera from Australian inland waters. IV. Colurellidae (Rotifera: Monogononta). Trans. R. Soc. S. Aust. 113: 119–143.

Koste, W., R. J. Shiel & L. W. Tan, 1988. New rotifers (Rotifera) from Tasmania. Trans. R. Soc. S. Aust. 112: 119–131.

Sanoamuang, L. 1996. *Lecane segersi* n. sp. (Rotifera, Lecanidae) from Thailand. Hydrobiologia 339: 23–25.

Sanoamuang, L. & H. Segers, 1997. Additions to the *Lecane* fauna (Rotifera: Monogononta) of Thailand. Int. Rev. ges. Hydrobiol. 82: 525–530.

Sanoamuang, L., H. Segers & H. J. Dumont, 1995. Additions to the rotifer fauna of south-east Asia: new and rare species from north-east Thailand. Hydrobiologia 313/314: 35–45.

Segers, H., 1995. Rotifera. Vol 2: The Lecanidae (Monogononta). Guides to the identification of the Microinvertebrates of the Continental Waters of the World 6. SPB Academic Publishing, The Hague, The Netherlands, 226 pp.

Segers, H., 1996. The biogeography of littoral *Lecane* Rotifera. Hydrobiologia 323: 169–197.

Segers, H., N. Emir & J. Mertens. 1992 Rotifera from north and northeast Anatolia (Turkey). Hydrobiologia 245: 179–189.

Segers, H., C. S. Nwadiaro & H. J. Dumont, 1993. Rotifera of some lakes in the floodplain of the River Niger (Imo State, Nigeria), II Faunal composition and diversity. Hydrobiologia 250: 63–71.

Segers, H. & P. Pholpunthin, 1997. New and rare rotifera from Thale-Noi Lake, Pattalung province, Thailand, with a note on the taxonomy of *Cephalodella* (Notommatidae). Annals Limnol. 33: 13–21.

Segers, H. & L. Sanoamuang, 1994. Two more new species of *Lecane* (Rotifera, Monogononta) from Thailand. Belg. J. Zool. 124: 39–46.

Segers, H. & S. S. S. Sarma, 1993. Notes on some new or little known Rotifera from Brazil. Rev. Hydrobiol. Trop. 26: 175–185.

Shiel, R. J., 1990. Zooplankton. In N. Mackay & D. Eastburn (eds), The Murray. Murray Darling Basin Commission, Canberra: 275–284.

Wulfert, K., 1965. Die Radertiere saurer Gewasser der Dubener Heide. Arch. Hydrobiol. 58: 72–102.

Hydrobiologia **387/388**: 35–37, 1998.
E. Wurdak, R. Wallace & H. Segers (eds), Rotifera VIII: A Comparative Approach.
© *1998 Kluwer Academic Publishers.*

On a new species of *Keratella* (Rotifera: Monogononta: Brachionidae)

Yan Zhuge & Xiangfei Huang
Institute of Hydrobiology, Chinese Academy of Sciences, Wuhan, Hubei 430072, P.R. China

Key words: Rotifera, *Keratella trapezoida* n. sp., taxonomy

Abstract

A new species of planktonic rotifer, *Keratella trapezoida* n. sp. is described from the Yangtze River, P.R. China. The new morphospecies is characterized by its four enclosed dorsal median facets, nearly trapezoid shape of the first median facet on dorsal plate, and the caudal median facet with parallel margins and being open posteriorly.

Introduction

In the past 40 years, Rotifera of Chinese inland waters have received some attention from rotiferologists. According to the literature (Wang, 1958, 1961; Gong, 1983; Huang & Hu, 1985; Zhuge et al., 1998), about 500 rotifer taxa occur in China, including 13 species and subspecies of the genus *Keratella* Bory de St. Vincent. Rotifera of the genus *Keratella* are characterized by a rigid lorica featuring pattern of facets. This species-specific foundation pattern is the most important feature in the taxonomy and phylogeny of *Keratella* (Segers & Wang, 1997). In the present work, a hitherto unknown species of this speciose genus, with distinctive foundation pattern is described on material from the Yangtze River, China.

Keratella trapezoidea new species (Figures 1–3)
Material examined

Holotype(ROT-019) and two paratypes(ROT-020) deposited in the Invertebrate specimen room of the Institute of Hydrobiology, Chinese Academy of Sciences; additional material preserved in 5% formalin in the Donghu Ecosystem Research Station of Institute of Hydrobiology, Chinese Academy of Sciences.

Differential diagnosis

Keratella trapezoida n. sp. has a distinctive pattern of facets on its lorica. It can be distinguished from all other *Keratella* species by its four enclosed median dorsal facets, nearly trapezoid shape of the first median dorsal facet and the parallel-sided caudal median facet open posteriorly.

Description

Lorica stout, rounded bilaterally; widest in posterior third; lorica with a very thick line on each side; median frontal facet trapezoid, with very thin posterior line; its frontal arcs bent passing to antero-median spines; second, third and forth median facets nearly rectangular, conspicuous lateral ridges arising from lateral angle of second and third median facets, bifurcate terminally; the last facet open posteriorly with parallel-sided margins; anterior margin of dorsal plate with three pairs of spines, with acutely pointed tips; ventral plate narrower than dorsal plate, surface smooth; its anterior margin bilobate; posterior margin of ventral plate trilobate; caudal spines short, tips acutely pointed and curving inwards. Male unknown.

Measurements (in μm): length of lorica 140–150, maximum width 110–120, apical width 81–89; length of anterior spines: median 25–30, intermediate 8–11, lateral 11–13; length of caudal spines: 20–25.

Distribution and ecology

The sample containing the new species was collected from the Yangtze River near Dongting Lake, on March 3, 1996. The environmental characteristics are: temperature = 10.7 °C; pH = 7.4; dissolved oxygen 12.41 mg l^{-1}. The co-occurring rotifer fauna is listed in Table 1.

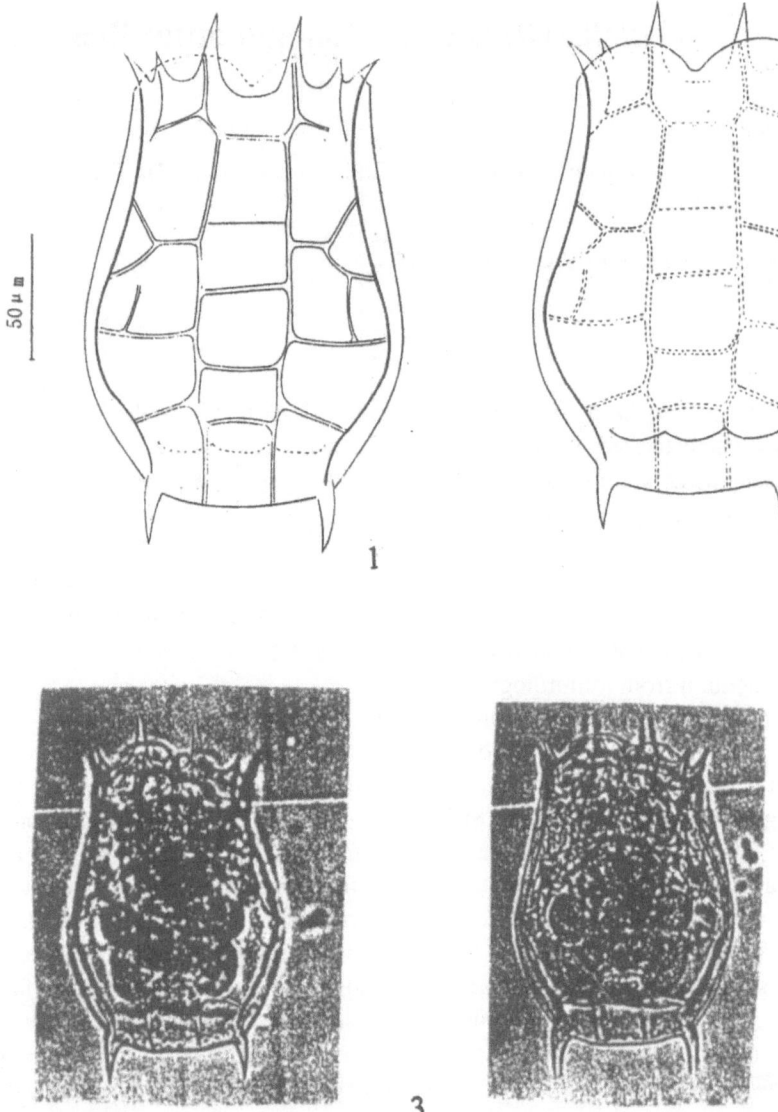

Figures 1–3. Keratella trapezoida n. sp. 1, dorsal; 2, ventral; 3, photomicrographs.

Table 1. Rotifera recorded from the type locality

Brachionus calyciflorus Pallas	*Keratella cochlearis* Gosse
Brachionus angularis (Gosse)	*Keratella quadrata* (O. F. Müller)
Brachionus quadridentatus Hermann	*Keratella valga* (Ehrenberg)
Brachionus leydigi Cohn	*Notholca labis* (Gosse)
Brachionus urceolaris (O. F. Müller)	*Polyarthra dolichoptera* Idelson
Cephalodella gibba (Ehrenberg)	*Synchaeta oblonga* Ehrenberg
Conochilus unicornis Rousselet	*Synchaeta pectinata* Ehrenberg
Euchlanis defiexa (Gosse)	*Trichotria curta* (Skorikov)
Filinia longiseta (Ehrenberg)	*Trichocerca pusilla* (Lauterborn)

Etymology

The name of the new species is derived from the shape of the first median facet of the dorsal plate.

Acknowledgments

This work is supported by the key project 'Formation and Sustainability of Ecosystem Productivities', Chinese Academy of Sciences and the 'Funds of Taxonomy of Chinese Fauna', Chinese Academy of Sciences.

References

Gong, X. J., 1983. The Rotifera from the high plateau of Tibet. In Jiang X. Z. & Y. F. Shen (eds), Freshwater Invertebrates from Tibet. Science Press of China, Beijing, 335–442 (in Chinese).

Huang, X. F. & C. Y. Hu, 1985. Rotifera from Donghu Lake, Wuhan. Acta. Hydrobiol. Sinica 3: 345–358 (in Chinese).

Segers, H. & Q. X. Wang, 1997. On a new species of *Keratella* (Rotifera: Monogononta: Brachionidae). Hydrobiologia 344: 163–167.

Wang, J. J., 1958. The ecological distribution of freshwater Rotifera of China. Acta. Hydrobiol. Sinica (1): 26–40 (in Chinese).

Wang, J. J., 1961. Fauna of Freshwater Rotifera of China. Science press of China, Beijing, 288 pp. +252 Figures (in Chinese).

Zhuge, Y., X. F. Huang & W. Koste, 1998. Rotifera recorded from China, 1893–1997, with remarks on their composition and distribution. Internat. Rev. Hydrobiol. 83(3): 217–232.

Hydrobiologia **387/388**: 39–46, 1998.
E. Wurdak, R. Wallace & H. Segers (eds), Rotifera VIII: A Comparative Approach.
© *1998 Kluwer Academic Publishers.*

Floodplain biodiversity: why are there so many species?

Russell J. Shiel[1], John D. Green[2] & Daryl L. Nielsen[1]
1. *Co-operative Research Centre for Freshwater Ecology, MDFRC, P.O. Box 921, Albury, NSW 2640, Australia*
[2]*Dept of Biological Sciences, University of Waikato, P.B. 3105, Hamilton, New Zealand*

Key words: Rotifera, floodplain, ephemeral waters, species diversity, habitat partitioning, opportunism, food webs, predation, Copepoda

Abstract

Spring surveys of 112 temporary floodplain waters on River Murray tributaries demonstrated a heterogeneous habitat series, with ca. 500 species of microfauna encountered. Rotifers comprised the most diverse group (>250 taxa), however mean diversity was low (10.93 ± 7.5), in part reflecting predation by copepods and macroinvertebrates. Notably, only 10 rotifer species could be considered widespread in the study area. Ephemeral pool microfaunal communities were distinct from those of adjacent permanent billabongs; their community variability is seen as a function of, or response to, habitat heterogeneity. The significance of high species diversity in ephemeral waters is considered in the context of age of the Murray-Darling Basin, which has persisted in its present location since the breakup of Gondwana, >65 MY BP.

Introduction

The floodplains of the Murray-Darling Basin, south-eastern Australia, are distinguished by low gradients, meandering rivers, and a characteristic profusion of cut-off meanders (ox-bows or *billabongs*), abandoned channels and swales, ephemerally-filled depressions, *inter alia* (Mackay & Eastburn, 1990). Preliminary studies on the ecology of billabongs established that highly diverse biotic communities (in the context of species richness) were present (Boon et al., 1990). Communities also were extremely heterogeneous between billabongs – even billabongs which could be connected in periods of high flow shared few species (Hillman, 1986; Hillman & Shiel, 1991). Subsequent monitoring of microfaunal responses to flooding in a billabong demonstrated the rapidity of community change (Tan & Shiel, 1993), the importance of habitat partitioning by vegetation (Pontin & Shiel, 1995), and (experimentally) the opportunistic nature of microbiota responses (Nielsen, unpubl. thesis). The microfauna of floodplain water bodies elsewhere also has come under recent scrutiny, with comparable recognition of high biotic diversity and rapid community dynamics, e.g. in Bolivia (Segers et al., in press), Brazil (Dabés, 1995; Bonecker & Lansac-Tôha, 1996;

Bonecker et al., 1994, 1998; Heckman, 1998), Nigeria (Segers et al., 1993), Venezuela (Vásquez & Rey, 1994).

Floodplain waters other than the large and distinctive billabongs in the R. Murray catchment remained unrecognized as a source of microfaunal biodiversity, until a visit to the upper R. Murray area by Dr Walter Koste in 1990. He sampled a range of previously dry ephemeral pools after rain, identifying a diverse suite of rotifers therein, in one case (Ryan's #3 (Figure 2), cf. Pontin & Shiel, 1995) over 100 species. Rotifers of these habitats in Australia were otherwise unknown – Morton & Bayly (1977) reported on the microcrustacea of 53 temporary ponds in Victoria, noting only that rotifers occurred. Microcrustacean and macroinvertebrate succession in a (non-floodplain) temporary pond subsequently was detailed by Lake et al. (1989). The appearance of rotifers from dry R. Murray floodplain sediments flooded experimentally in the laboratory was first documented by Boulton & Lloyd (1992). They were investigating the effects of altered flooding frequency on the floodplain 'seedbank' – the resting stages, cysts, ephippia, etc, of micro- and macroinvertebrates. Protists and rotifers commonly were the first to emerge from flooded sediments. The potential for high biotic diversity in Australian ephemeral pools

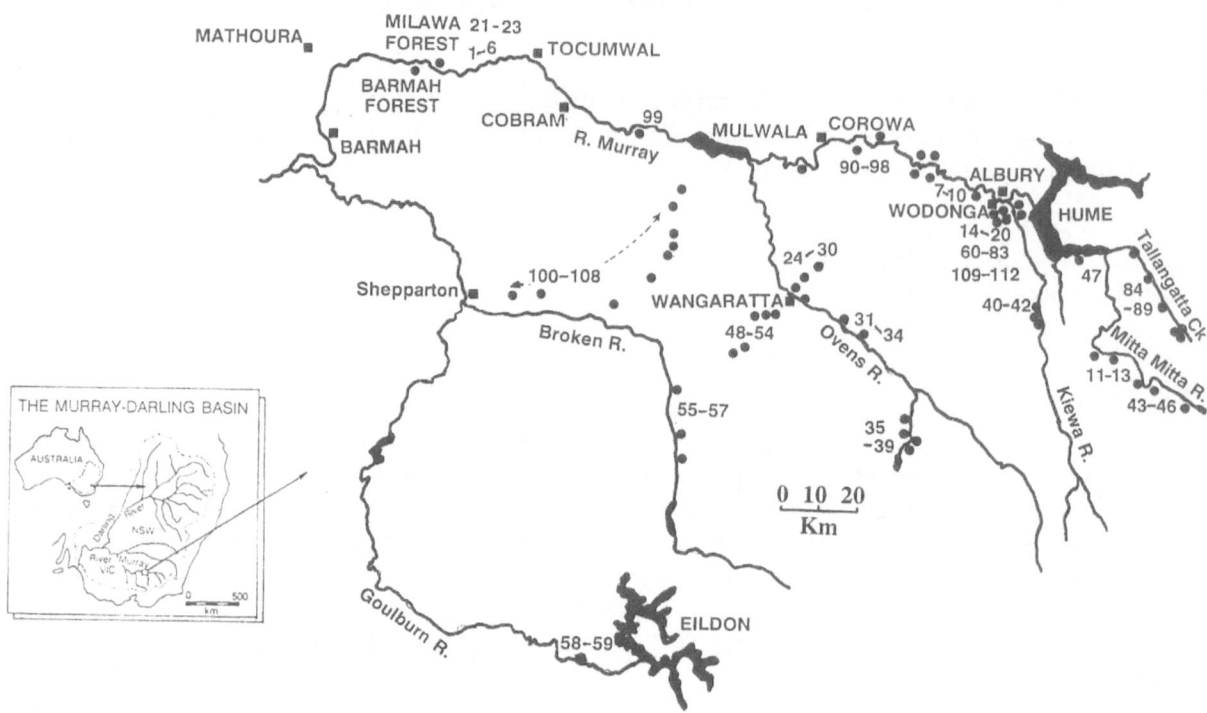

Figure 1. Sampling sites on floodplains of Upper Murray tributaries.

Figure 2. Ryan's #3, autumn (site 14, Figure 1).

was discussed by Williams (1985). He later argued that temporary waters may be important evolutionary loci (Williams 1988).

With the evidence of high microfaunal diversity provided by Dr Koste's 1990 net tows from roadside marginal pools, we undertook to survey a wide variety of non-billabong floodplain ephemeral habitats to determine the extent of variability in their microfaunal communities. We were interested to determine if catchment specificity in microfaunal assemblages could be identified. Concurrently, taking advantage of the second author's sabbatical year at MDFRC, a single site (Ryan's #3) was to be monitored quantitatively at least on a weekly basis for the duration of wetting. The first part of this project, the spatial component, is reported here. The temporal component will be reported elsewhere – Ryan's #3 was flooded for over 12 months in the atypical wet year following commencement of sampling, after which we elected to resample Ryan's #3 during a more 'typical' El Niño dry year.

Study area

In Spring (Aug.–Sept.) 1991, 112 ephemeral pools across the R. Murray and tributaries upstream of the Murray-Goulburn junction (Figure 1) were sampled once. All were associated with a floodplain, but were sufficiently high on the floodplain not to receive water from the parent river or nearby billabongs (when present). At the time of sampling all sites had received winter rainfall as local runoff. All could be considered 'pools', with considerable variability in physical and chemical parameters, presence or absence of aquatic vegetation, access to stock and so on. Their ephemerality depends on local conditions, waterlogging of floodplain, local rainfall, etc., however a characteristic of all was that they dry out, but not necessarily each year. As noted above, Ryan's #3 (Figure 2), for example, a shallow (<1 m at the time of sampling) roadside pool, was inundated for over 12 months then dry for almost 2 years subsequently.

Materials and methods

Sampling of all sites during the initial survey was qualitative, by means of a 37 μm-mesh plankton net or 53 μm-mesh Birge cone net where sufficient water depth was present to use them. A long-handled

37 μm Frey net (with 4 mm stainless steel mesh over the net opening to minimize ingress of debris) was used where shallow water or emergent vegetation precluded use of the tow nets. In some very shallow sites, samples were collected in a 1 litre plastic beaker. All samples were concentrated to 50–60 ml on site, and preserved with 75% ethanol or 4% formalin. To minimize cross-contamination, nets were washed in alcohol or detergent (Decon 90, Selby Biolab Scientific Ltd) and rinsed in filtered water between sites. Basic water chemistry (temp., pH, conductivity) was taken on site with portable field meters.

Microfauna were sorted in the laboratory, generally from 20 ml aliquots of the samples, on a Zeiss SV-8 dark-field dissecting microscope, with identifications performed on appropriately treated material on an Olympus BH-2 compound microscope (Nomarski optics). Rotifers in the samples were identified using the keys in Koste (1978), Segers (1995), Shiel (1995), De Smet (1996, 1997).

Rotifer abundance data from all sites was analysed by PRIMER. Non-metric multidimensional scaling (MDS) of presence and absence data, using a Bray-Curtis similarity matrix, was used to produce MDS ordinations (Clarke, 1993; Clarke & Warwick, 1994). Rotifer species numbers in the presence and absence of the predatory copepods *Boeckella major* Searle, 1938 and *Australocyclops australis* (Sars, 1896) were compared by t-test.

Results and discussion

Most of the temporary waters sampled were recently filled (Winter 1991) fresh waters, at the time of sampling in the temperature range 9.0–17.5 °C, pH 6.1–9.1, conductivity 27–250 μS cm^{-1}, with most clear or lightly coloured water, probably due to tannins derived from eucalypt leaf fall. The predominant canopy vegetation in the study area is *Eucalyptus* spp. Most recently flooded sites had submerged terrestrial grasses, longer duration wetted sites had some aquatic vegetation, while longest flooded sites had a diverse suite of submerged, emergent and floating aquatic plants (cf. Boon et al. 1990).

Some 500 taxa in the groups Rhizopoda (17%), Rotifera (51%), Microcrustacea (22%) and various small macroinvertebrates, primarily insect juveniles or nymphs (9%), were recorded. Rotifers were the most abundant single group, with at least 252 spp. recognized. It is likely that more species remain to be recorded in the samples, as only 20 ml aliquots were examined. Of the Rotifera, the most common family represented was Lecanidae (30 taxa) followed by Trichocercidae (25), Brachionidae (23) and No- tommatidae (22). A further 17 rotifer families were represented by 1–12 taxa each. A full species list can be obtained on request from the senior author.

Of the 252 taxa, only three (one species each of *Cephalodella, Dicranophorus* and *Lecane*) apparently are undescribed. They will be described elsewhere. A proalid very close to, if not conspecific with, *Proales sigmoidea* (Skorikov, 1896) occurred in one roadside pool (Site 83, Figure 1), and is the first record of this taxon from Australia and apparently also from the southern hemisphere (De Smet, 1996). The remaining taxa were mostly widely distributed or cosmopolitan, with the exception of brachionids (*Brachionus lyr- atus* Shephard, *B. kostei* Shiel, *B. reductus* Koste & Shiel, *Keratella australis* Berzins, *K. slacki* Berzins) previously thought to be Australian endemics, but recently recorded more widely in the Australasian re- gion (Sanoamuang et al., 1995; Segers & De Meester, 1994; Shiel & Green, 1996; Sudzuki, 1992).

Table 1 lists the occurrence of the recorded spe- cies. 109 taxa (>43%) were encountered only once. Only 11 taxa occurred at more than 20% of sites, and could be considered relatively common: in decreasing order of frequency, *Keratella procurva* (Thorpe, 1891) (54 sites, cf. Table 1), *Trichocerca bidens* (Lucks, 1912), *Euchlanis dilatata* (Ehrenberg, 1832), *Lecane luna* (Müller, 1776), *Lepadella ovalis* (Müller, 1786), *Trichocerca rattus* (Müller, 1776), *Rotaria* (Scopoli) sp. A, *Lecane hamata* (Stokes, 1896), *L. closterocerca* (Schmarda, 1896) and *Testudinella patina* (Hermann, 1783).

No sites in the 1991 survey were as speciose as the >100 rotifer taxa noted from Ryan's #3 in November 1990 (W. Koste, pers. comm.), even though Ryan's #3 was included in this sample series the following year (18.VIII.91): that is, only 1 rotifer species found! (see below). Interannual variation in rotifer diversity in this particular habitat was described by Pontin & Shiel (1995). Only three sites had more than 30 rotifer species in their microfaunal communities (Figure 3), otherwise from 0–25 spp. were recorded from each sample. Notably, sites with no rotifers recorded, or low rotifer diversity, were dominated by calanoid and cyclopoid copepods, including the genera *Boeckella, Hemiboeckella, Australocyclops, Diacyclops, Meso- cyclops* and *Microcyclops*, with up to 10 copepod species co-occurring in some sites.

Table 1. Subset of sites at which specified number of ro- tifer species occurred, e.g. 109 species (43%) occurred at only a single site.

Occurrence	No. of species	% of total number of species	Cumulative %
1	109	43.25	43.25
2	40	15.87	59.13
3	24	9.25	68.65
4	15	5.95	74.60
5	6	2.38	76.98
6	8	3.17	80.16
7	10	3.97	84.13
8	3	1.19	85.32
9	3	1.19	86.51
10	6	2.38	88.89
11	1	0.40	89.29
12	5	1.98	91.27
13	1	0.40	91.67
14	1	0.40	92.06
17	3	1.19	93.25
18	2	0.79	94.05
20	1	0.40	94.44
21	2	0.79	95.24
22	1	0.40	95.63
23	3	1.19	96.83
24	2	0.79	97.62
28	2	0.79	98.41
35	1	0.40	98.81
50	1	0.40	99.21
52	1	0.40	99.60
54	1	0.40	100.00
Total	252		

We expected some of these, at least, to be sig- nificant predators of rotifers (cf. Williamson, 1991). Rotifer species numbers in the presence (9 sites) and absence (21 sites) of the large calanoid *Boeckella major,* for example, when tested by t-test proved sig- nificantly different ($t = -2.618$; DF = 16.6; $P<0.05$), prompting an examination of gut contents of several calanoid species (Green & Shiel, 1995). *Boeckella major* is a significant predator of rotifers and mi- crocrustacea in temporary habitats, as, to a lesser extent, are other species of omnivorous *Boeckella* and *Hemiboeckella* (Green et al., in press). Although only a single species of rotifer, *Brachionus quadridentatus* Hermann, was collected from Ryan's #3 during this survey, a diverse array of rotifer remains was found

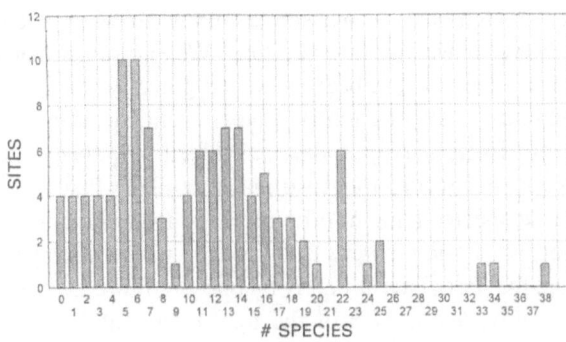

Figure 3. Cumulative plot of numbers of rotifer species x sites.

Table 2.

Code	F	F	C	X	C	X	F	A	C	A	O	A	B	A	A	M	M	G	B	O	G	O	C	T	T	K	T	A	A	T
Site	106	103	93	1	98	3	102	96	78	32	60	57	63	61	45	59	52	43	39	92	86	34	41	87	80	74	85			
Trichotria	X			X	X			X													X	X		X						
Mytilina	X		X	X		X		X				X			X					X		X								
Lacinularia	X		X	X		X		X									X			X			X							
Scaridium			X	X	X		X									X														
Asplanchnopus			X	X	X	X					X																			
Collotheca	X	X		X			X									X														
Trichocerca	X	X		X	X		X	X	X	X	X	X	X	X	X	X		X	X	X					X	X			X	X
Lecane	X	X		X	X	X	X	X	X	X	X	X		X	X		X	X		X	X	X			X	X	X	X	X	X
Lepadella			X	X		X	X	X	X					X	X			X	X					X	X	X	X	X
Testudinella	X		X			X	X	X		X	X							X		X	X			X	X	X
Keratella	X	X	X	X	X			X	X	X		X	X	X	X	X		X	X	X	X	X	X	X			X	X	X
Euchlanis	X	X		X	X	X	X	X	X	X			X	X	X	X									X	X		X	X	X
Colurella	X	X		X	X		X	X	X	X				X	X					X	X	X			X	X	X
Bdelloid	X	X				X	X		X	X	X									X	X				X	X	X
Cephalodella	X			X			X					X				X	X	X	X			X	X			X	X
Gastropus	X			X		X		X	X	X	X	X					X										X
Polyarthra	X		X	X	X		X	X	X	X	X				X					X				X		X	X
Brachionus	X		X		X	X	X					X	X		X											X	X
Squatinella	X			X			X									X					X						
Eosphora	X			X												X					X						
Lindia	X			X			X			X	X	X															
Lophocharis	X		X	X			X		X		X	X	X			X				X			X				
Monommata	X		X		X		X																X				
Encentrum			X	X		X		X															X				
Manfredium	X		X													X											
Entroplea	X		X																								
Proales	X		X			X			X			X			X			X									
Itura	X		X	X										X													
Epiphanes				X	X							X	X	X	X												
Notommata			X		X	X		X		X									X								
Eothinia			X										X														
Rotaria							X	X		X	X																
Proalides			X																								
Filinia			X						X	X			X	X		X	X										
Conochilus			X						X	X																	
Asplanchna			X		X				X	X	X			X													
Platyias	X	X						X			X				X						X	X					
Synchaeta	X					X			X			X															X
Unidentified	X			X			X									X											X
Dicranophoroides												X	X		X											X	X
Resticula	X												X													X	X
?Dipleuchlanis																											X
Aspelta																											X
Ptygura			X								X																
indet. notommatid	X		X						X			X									X	X			X	X	X
Macrochaetus																	X										
Ascomorpha																	X										
Flosculariid					X											X											
Notholca							X																				
Habrotrocha							X								X												

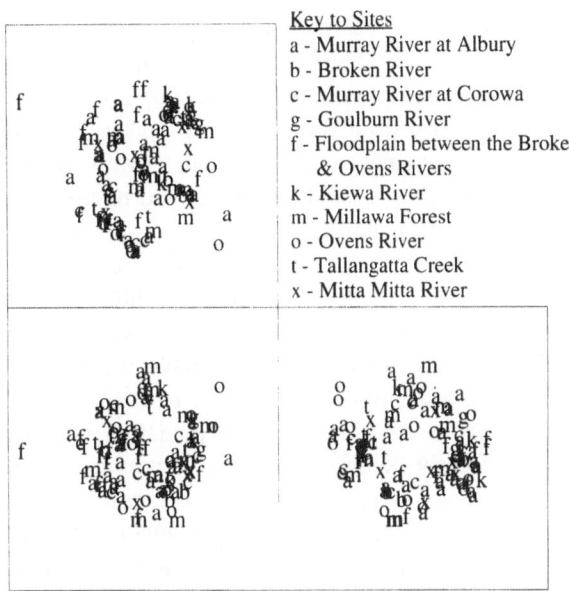

Key to Sites
a - Murray River at Albury
b - Broken River
c - Murray River at Corowa
g - Goulburn River
f - Floodplain between the Broken
    & Ovens Rivers
k - Kiewa River
m - Millawa Forest
o - Ovens River
t - Tallangatta Creek
x - Mitta Mitta River

Figure 4. 3-dimensional ordination derived from non-metric multi-dimensional scaling from a Bray-Curtis similarity matrix of presence/absence data. Stress = 0.21.

The shaded region indicates the 11 genera which occurred at more than ten sites, including the common species noted above. Additional relatively common genera in which two or more species occurred, albeit in less than 20% of sites overall (*Colurella, Cephalodella, Gastropus, Polyarthra*), were discriminated by the matrix. Cluster analysis (not shown) and a 3-D MDS ordination generated from the species-level dataset (Figure 4) provided similar information: there was little community similarity between the rotifer assemblages present at any site, even those in close proximity on the same floodplain. The community heterogenity evident in the Rotifera of larger/more permanent floodplain habitats clearly extends to the inhabitants of ephemeral waters.

Despite the absence of other ephemeral pond rotifer community studies with which to compare our data, it is clear from the microcrustacean community composition that similar diverse microcrustacean assemblages to those reported by Morton & Bayly (1977) and Lake et al. (1989) occurred across our sample series. For example, the rapid appearance of *Boeckella major* and *Boeckella pseudochelae* and a suite of other copepods after flooding is characteristic of such habitats. The larger calanoids are not found in permanent billabongs, where their size would render them subject to predation by fish and macroinvertebrates. Smaller calanoids *Boeckella triarticulata, B. fluvialis*, and species of *Calamoecia* are typical of per-

within the predominant planker, *B. major*! Rotifer diversity in the presence and absence of the cyclopoid predator(s) *Australocyclops* spp. was not detectably different in this sample series, however it is likely that predaceous cyclopoids also exert a selective pressure on temporary pond rotifer communities.

Rotifer presence-absence data were analysed using PRIMER as noted above. Thirty arbitrarily selected sites from across the sample area were used to generate a shade matrix derived from a Bray-Curtis similarity matrix using generic level discrimination (Table 2).

manent billabongs in the region. In ephemeral waters, a good option would be to emerge from the sediments immediately after the pool fills, develop rapidly to a large adult size (to avoid predators), and to prey on smaller microfaunal species (particularly rotifers) that are able to utilise available bacteria and micro-algae unavailable to calanoids.

This strategy suggests that selective pressure by copepods, for example, would be intermittent in such habitats, and may account for the temporal disparity noted in Ryan's #3 rotifer assemblages. Early in the season after initial wetting, the emerging rotifer community is subject to intense predation pressure from the early colonizing *B. major*, among others. The calanoids complete their life cycles in one to three weeks, produce resting stages which drop into the sediments, and disappear. Later hatching rotifers are not subject to the same predation pressure (from this particular predator at least), by which time there is the possibility of more shelter or habitat partitioning provided by growth of aquatic vegetation. Hence, our sample series in August–September, shortly after flooding of most sites, was relatively early in the wetting cycle, and coincided with the intense predation pressure of early colonizing copepods. The November 1990 samples collected by Dr Koste apparently were later in the cycle, from microspatially more diverse habitats, where a more diverse rotifer community was able to develop.

The observed heterogeneity of rotifers in these ephemeral habitats suggests that they are very successfully adapted to unpredictability, as are the complex food-web interactions. Perhaps this is not surprising given the long evolutionary association with a flood-drought regime, as operates in the Murray-Darling Basin. As part of ancient Gondwana, the ancestral Murray occupied the Basin over 60 million years ago, and its present course for several million years. Billabongs and temporary floodplain waters are the rule rather than the exception in terms of standing waters over much of the continent. There are few natural freshwater lakes, most that now exist are man-made in the last 200 years. The limited evidence of rotifer speciation in Australia supports Williams' (1988) contention that temporary waters in Australia have been important evolutionary loci: most of ca. 40 rotifers described from Australia in the last 20 years have been reported only from floodplain waters, with a few from the pelagic of central plateau lakes or coastal dune lakes in Tasmania.

On the basis of our single survey, to what extent can we answer our title question? We know how many rotifer species have been described globally (ca. 2000) and how many of them are known from Australia (ca. 640). Every survey to date has added new species to the rotifer record for Australia. It is likely that many more species remain to be recorded from the continent; as yet only a very small area has been surveyed, and that only intermittently by one or a few collectors. Notably, about 47% of the total continental record, 300 rotifer species, have been found in a single 1 km long billabong on the R. Murray (Ryan's #1) over the course of 25 years collecting. It is unlikely, given latitudinal disparities in species assemblages, that all the rotifers likely to be found in Australia are present in resting stages in the sediments of the Murray-Darling floodplain, but it seems evident that many of them are. We can only speculate that this seedbank provides the colonizers for each flood event, each newly-filled pool; dispersal of rotifer resting eggs by flood events is yet to be demonstrated.

Circumstantial evidence from the sediment-flooding experiments of Boulton & Lloyd (1992) indicates that rotifer resting eggs can survive for many years between flood events. Rotifer emergence from intermittently flooded sediments on the Darling River, NSW, also was reported by Jenkins & Briggs (1997), who speculated on their survival in a dry lakebed for >100 years. They pointed out that wind dispersal needed to be investigated. Regardless of the dispersal mechanism, the repository of resting stages of rotifers clearly is the source of colonization of our ephemeral pond series – who emerges to colonize seems to depend on the chance deposition of resting eggs at that site – who survives to reproduce clearly is dependent on surviving predation by other early colonizers, which in turn determines whose resting eggs are deposited in the sediment.

The heterogeneity of rotifer dispersion reflects the heterogeneity of habitats. All of the temporary ponds were visibly and measurably 'different' at the time of sampling, reflecting local differences in floodplain substrata, vegetation, time of first wetting, amount of runoff, i.e. catchment, depth, water quality and so on. The cues to emergence in one habitat may occur in another, but not necessarily at the same time, therefore emergence of a particular species, even if present in the sediments of different sites, is not necessarily coincident. Eurytolerant generalists who can cope with a wide range of habitat conditions, food preferences, etc, are likely to be found more widely, those with more stringent cues or needs, less so. The potential for a predator to drive a rotifer species or even community to extinction in a given pool is suggested by the dynamics of *Boeckella major*. The microfaunal ecology of ephemeral pools is comparable to that reported for island biogeography – the pools are effectively (most of the time at least) isolated 'islands' in the floodplain.

The importance of such habitats on floodplains in terms of biodiversity remains largely unappreciated and uninvestigated in Australia, and probably elsewhere. There can be no doubt that extensive river regulation and subsequent modification of river flow regimes has had deleterious effects on the floodplain communities of all Australian rivers (Shiel, 1996). Profound declines have been reported in most native aquatic macrobiota (e.g. Gehrke et al., 1995), yet the potential loss of species diversity represented by the floodplain seedbank – the cysts, resting eggs, ephippia, diapausing stages deposited in the multitude of floodplain standing waters, or transported over the floodplain proper in times of high flows – has hardly been addressed.

Acknowledgements

Garth Watson provided technical assistance in both the field and laboratory. Fare support to attend the VIIIth Int. Rotifer Symposium from the Co-operative Research Centre for Freshwater Ecology is gratefully acknowledged (by RJS). JDG is grateful to the MD-FRC for providing facilities and a congenial work environment during his study leave there. His visit to MDFRC was funded by a University of Waikato Study Leave grant. The constructive criticism of the volume editors also is appreciated.

References

Bonecker, C. C. & F. A. Lansac-Tôha, 1996. Community structure of rotifers in two environments of the upper River Paraná (MS) Brazil. Hydrobiologia 325: 137–150.

Bonecker, C. C., F. A. Lansac-Tôha & A. Staub, 1994. Qualitative study of rotifers in different environments of the High Parana River floodplain (MS) – Brazil. Rev. UNIMAR 16 (Suppl.): 1–16.

Bonecker, C. C., F. A. Lansac-Tôha & L. M. Bini, 1998. Composition of zooplankton in different environments of the Mato Grosso Pananal, Mato Grosso, Brazil. Ann. VIII Sem. Reg. Ecol. 3: 1123–1135.

Boon, P. I., J. Frankenberg, T. J. Hillman, R. L. Oliver & R. J. Shiel, 1990. Billabongs. In N. Mackay & D. Eastburn (eds), The Murray. Murray-Darling Basin Commission, Canberra: 183–198.

Boulton, A. J. & L. N. Lloyd, 1992. Mean flood recurrence frequency and invertebrate emergence from dry sediments of the Chowilla floodplain, River Murray, Australia. Reg. Riv. Res. Managem. 7: 137–151.

Clarke, K. R., 1993. Non-parametric multivariate analyses of changes in community structure. Aust. J. Ecol. 18: 117–143.

Clarke, K. R. & R. M. Warwick, 1994. Changes in marine communities: an approach to statistical analysis and interpretation. Nat. Env. Res. Council, UK: 144 pp.

Dabés, M. G. B. S., 1995. Composicão e descrição do zooplâncton de 5 (cinco) lagoas marginais do Rio São Francisco, Pirapora/ Três Marias/ Minas Gerais, Brasil. Rev. Brasil. Biol. 55: 831–845.

De Smet, W. H., 1996. Rotifera. Vol. 4. The Proalidae (Monogononta). Guides to the Identification of the microinvertebrates of the Continental Waters of the World 9: 1–102.

De Smet, W. H., 1997. Rotifera Vol 5. The Dicranophoridae (Monogononta). Guides to the Microinvertebrates of the Continental Waters of the World 12: 1–325.

Gehrke, P. C., P. Brown, C. B. Schiller, D. B. Moffatt & A. M. Bruce, 1995. River regulation and fish communities in the Murray-Darling River System, Australia. Reg. Riv. Res. Managem. 11: 363–375.

Green, J. D. & R. J. Shiel, 1995. Calanoid copepods as rotifer taxonomists. Quek. J. Microscopy 37: 491–492.

Green, J. D. & R. J. Shiel, in press. Predation by the centro-pagid calanoid, *Boeckella major*, structuring microinvertebrate communities in the absence of fish. Verh. Int. Ver. Limnol.

Heckman, C. W., 1998. The seasonal succession of biotic communities in wetlands of the tropical wet-and-dry climatic zone: V. Aquatic invertebrate communities in the Pantanal of the Mato Grosso, Brazil. Int. Rev. Hydrobiol. 83: 31–63.

Hillman, T. J., 1986. Billabongs. In P. De Deckker & W. D. Williams (eds), Limnology in Australia. CSIRO/Junk B.V., Melbourne/Dordrecht: 457–470.

Hillman, T. J. & R. J. Shiel, 1991. Macro- and microinvertebrates in Australian billabongs. Verh. Internat. Verein. Limnol. 24: 1581–1587.

Jenkins, K. & S. Briggs, 1997. Wetland invertebrates and flood frequency on lakes along Teryawenya Creek. NSW Nat. Parks & Wildlife Service Rept., July 1997: 40 pp.

Koste, W., 1978. Rotatoria – Die Rädertiere Mitteleuropas (Uberordnung Monogononta). Revision after M. Voigt (1956/7) 2 Vols. Borntraeger, Stuttgart. Text 673 pp.

Lake, P. S., I. A. E. Bayly & D. W. Morton, 1989. The phenology of a temporary pond in western Victoria, Australia, with special reference to invertebrate succession. Arch. Hydrobiol. 115: 171–202.

Mackay, N. & D. Eastburn (eds), 1990. The Murray. Murray-Darling Basin Commission, Canberra: 363 pp.

Morton, D. W. & I. A. E. Bayly, 1977. Studies on the ecology of some temporary freshwater pools in Victoria with specieal reference to microcrustaceans. Aust. J. Mar. Freshwat. Res. 28: 439–454.

Pontin, R. M. & R. J. Shiel, 1995. Periphytic rotifer communities of an Australian seasonal floodplain pool. Hydrobiologia 313/314: 63–67.

Sanoamuang, L., H. Segers & H. J. Dumont, 1995. Additions to the rotifer fauna of south-east Asia: new and rare species from north-east Thailand. Hydrobiologia 313/314: 35–45.

Segers, H., 1995. Rotifera Vol. 2. The Lecanidae (Monogononta). Guides to the microinvertebrates of the continental waters of the world 6: 1–226.

Segers, H. & L. De Meester, 1994. Rotifera of Papua New Guinea, with the description of a new *Scaridium* Ehrenberg, 1830. Arch. Hydrobiol. 131: 111–125.

Segers, H., C. S. Nwadiaro & H. J. Dumont, 1993. Rotifera of some lakes in the floodplain of the River Niger (Imo State, Nigeria). II. Faunal composition and diversity. Hydrobiologia 250: 63–71.

Segers, H., L. Ferrufino & L. De Meester, in press. Diversity and zoogeography of Rotifera (Monogononta) in a flood plain lake of the Ichilo river, Bolivia, with notes on little-known species. Int. Rev. Hydrobiol.

Shiel, R. J., 1995. A guide to identification of rotifers, cladocerans and copepods from Australian inland waters. CRCFE Ident. Guide 3: 1–144.

Shiel, R. J., 1996. Human population growth and over-utilization of the biotic resources of the Murray-Darling River system, Australia. GeoJournal 40: 101–113.

Shiel, R. J. & J. D. Green, 1996. Rotifera recorded from New Zealand, 1859–1995, with comments on zoogeography. N.Z. J. Zool. 23: 193–209.

Sudzuki, M., 1992. New Rotifera from southwestern Islands of Japan. Proc. Japan. Soc. Syst. Zool. 46: 17–28.

Tan, L. W. & R. J. Shiel, 1993. Responses of billabong rotifer communities to inundation. Hydrobiologia 255/256: 361–369.

Vásquez, E. & J. Rey, 1994. Rotifers and cladoceran zooplankton assemblages in lakes on the Orinoco River floodplain (Venezuela). Verh. Internat. Verein. Limnol. 25: 912–917.

Williams, W. D., 1985. Biotic adaptations in temporary lentic waters, with special reference to those in semi-arid and arid regions. Hydrobiologia 125: 85–110.

Williams, W. D., 1988. Limnological imbalances: an Antpodean viewpoint. Freshwat Biol. 20: 407–420.

Williamson, C. E., 1991. Copepoda. In J. H. Thorp & A. P. Covich Ecology and Classification of North American Freshwater Invertebrates. Academic Press, N.Y.: 787–822.

Hydrobiologia **387/388**: 47–54, 1998.
E. Wurdak, R. Wallace & H. Segers (eds), Rotifera VIII: A Comparative Approach.
© *1998 Kluwer Academic Publishers.*

47

Rotifer diversity in a central Mexican pond

S. S. S. Sarma & Elías-Gutiérrez Manuel
National Autonomous University of Mexico Campus Iztacala, AP 314 CP 54000 Los Reyes, Iztacala, Tlalnepantla Edo. de México, México

Key words: Mexico, Rotifera, new record, taxonomy

Abstract

A survey of rotifers from a small pond (less than 2 ha in area and 3 m deep), located at Kilometer 28 in the federal highway Ixtlahuaca-Jilotepec (19° 49′ 13″ N, 99° 42′ 22″ W) at an altitude of 2503 m above sea level, resulted in a total of 78 species. From these, 20 are new records for Mexico. This study confirms the presence of some of the rotifer species listed only in earlier studies. Comments on some species are made from a zoogeographical point of view.

Introduction

Studies on freshwater zooplankton of Mexico yield a high degree of diversity and, in some cases, new species have been described (Cladocera: Ciros-Perez et al., 1995; Copepoda: Suárez & Campos, 1994; Rotifera: Rico-Martinez & Silva-Briano, 1993; Kutikova & Silva-Briano, 1995; Örstan, 1995). However, taxonomic studies, particularly on Rotifera have been sporadic and limited to analyses based on planktonic samples and filtered pond water. Such samples generally yield low species diversity. Rotifers have been recognized as indicators of saprobic status of Mexican water bodies (Vilaclara & Sladecek, 1989). They are also currently being cultured for rearing fish larvae in Mexico (Ramirez-Sevilla et al., 1991). In spite of these important developments, taxonomic works on Mexican rotifers should be considered as insufficient. Nogrady et al. (1993), after considering world developments in rotifer systematics, concluded that research on rotifer systematics in Mexico is poor. The present study is aimed to add information on rotifer species diversity of Mexico.

Material and methods

Rotifer samples were collected from a perennial pond located at Kilometer 28 on the federal highway Ixtlahuaca-Jilotepec (19° 49′ 13″ N, 99° 42′ 22″ W;

altitude 2503 m above sea level) on 3rd June and 7th July, 1994. At the time of the surveys, physical and chemical variables of the water body were: temperature 20–23 °C; Secchi transparency 0.3–0.4 m; maximum depth 2.82 m; dissolved oxygen 7–11 mg/l; pH 7.8–9.0 and conductivity 150–240 μS.

Collections were made using a 50 μm plankton mesh, which collected planktonic and littoral rotifers. Samples were fixed in 10% formalin in the field. Rotifers from the samples were identified in the laboratory by using a stereomicroscope at a magnification of 40× and later, with a compound microscope at different magnifications (100–1000×). For identification of rotifers, we mainly followed Koste (1978). When available, more recent literature (e.g., Koste and Shiel, 1987, 1989, 1990, 1991; Segers, 1995) was consulted. The classification followed here is after Koste (1978).

Rotifer figures were drawn under a compound microscope using camera lucida (Nikon Labophot-2 model), calibrated with a stage micrometer. Trophi preparations when needed, were made by dissolving rotifers with dilute a sodium hypochlorite solution.

Results

Analysis of rotifer samples from the study pond yielded 78 rotifer species. These belong to 19 families and 33 genera.

48

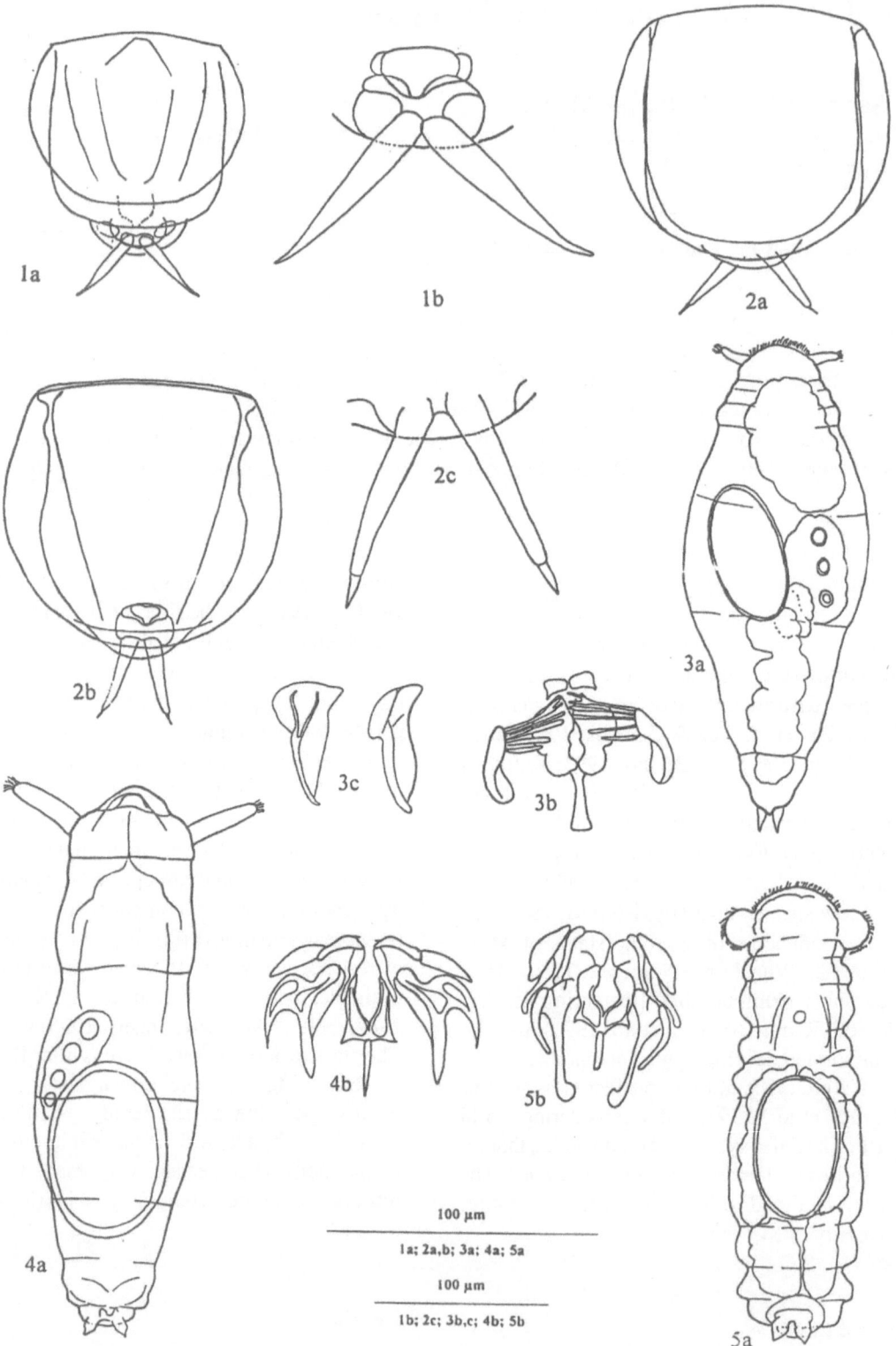

Figures 1–5. **Figure 1.** *Lecane hornemanni.* a: dorsal, b: toes. **Figure 2.** *L. latissima.* a: dorsal, b: ventral, c: toes. **Figure 3.** *Proales fallaciosa.* a: dorsal, b: trophi ventral, c: manubria. **Figure 4.** *Lindia torulosa.* a: dorsal, b: trophi. **Figure 5.** *L. truncata.* a: dorsal, b: trophi.

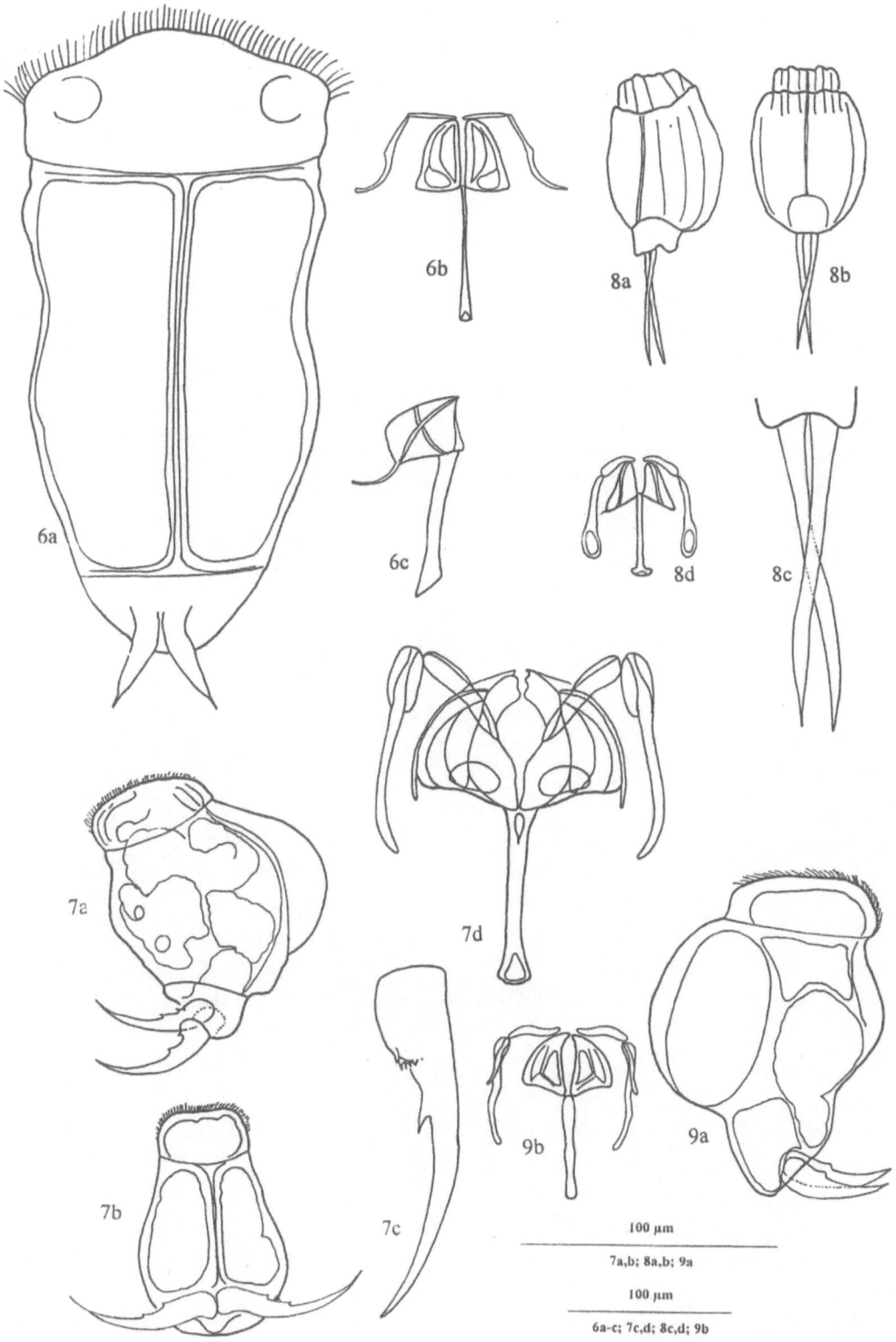

Figures 6–9. **Figure 6.** *Cephalodella exigua.* a: dorsal, b: trophi (dorsal), c: trophi (lateral). **Figure 7.** *C. forficula.* a: lateral, b: ventral, c: trophi (dorsal). **Figure 8.** *C. misgurnus.* a: dorsal (contracted), b: lateral (contracted), c: trophi (dorsal), d: toes. **Figure 9.** *C. stenroosi.* a: lateral, b: trophi (dorsal).

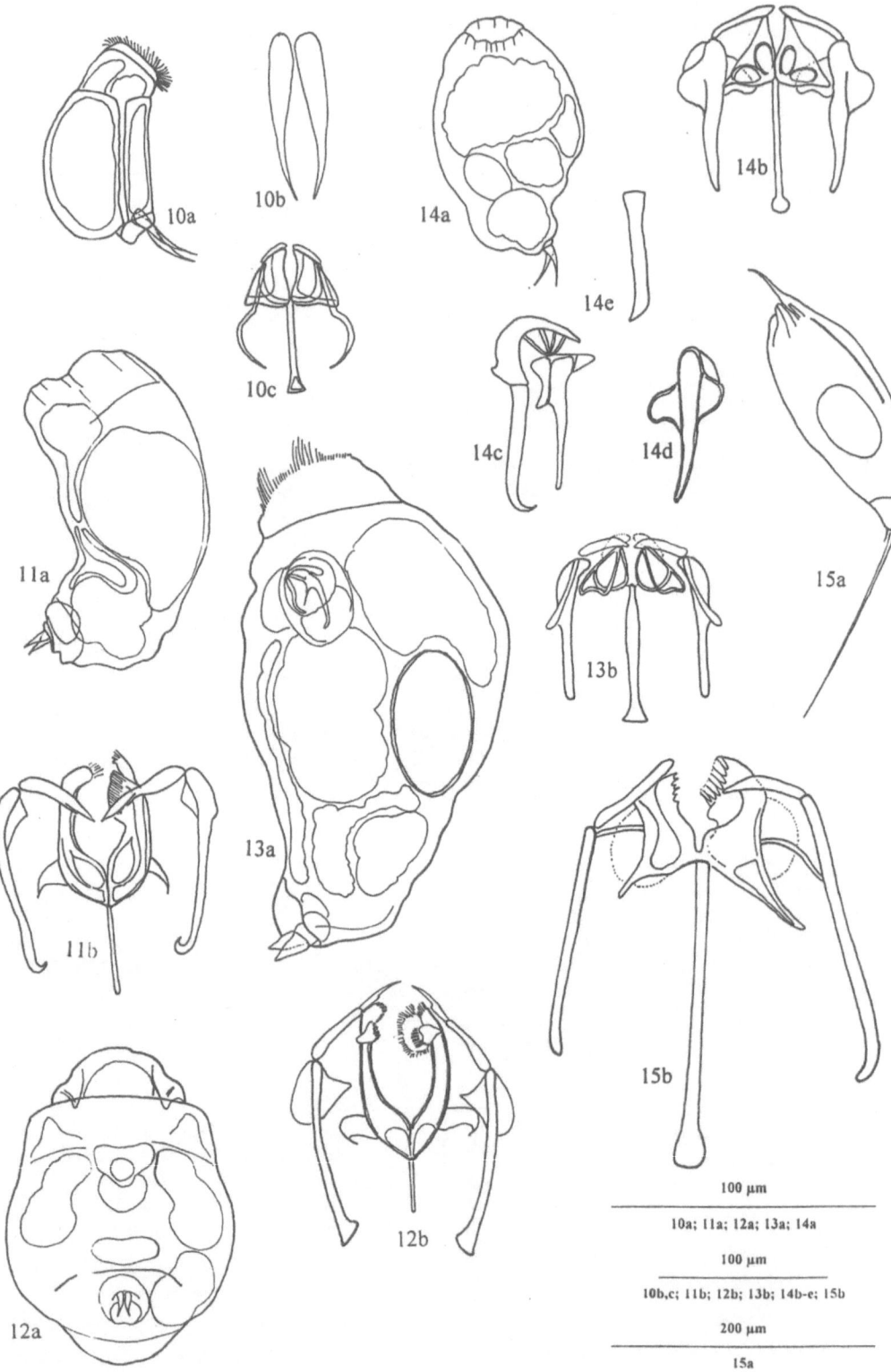

Figures 10–15. **Figure 10.** *C. ventripes.* a: lateral, b: toes, c: trophi (dorsal). **Figure 11.** *Itura aurita.* a: lateral (contracted), b: trophi. **Figure 12.** *I. myersi.* a: dorsal (contracted), b: trophi. **Figure 13.** *Pleurotrocha petromyzon.* a: lateral (contracted), b: trophi (dorsal), c: trophi (lateral). **Figure 14.** *Resticula melandocus.* a: lateral (contracted), b: trophi, c: fulcrum, d: manubrium. **Figure 15.** *Trichocerca cylindrica.* a: lorica (lateral), b: trophi.

51

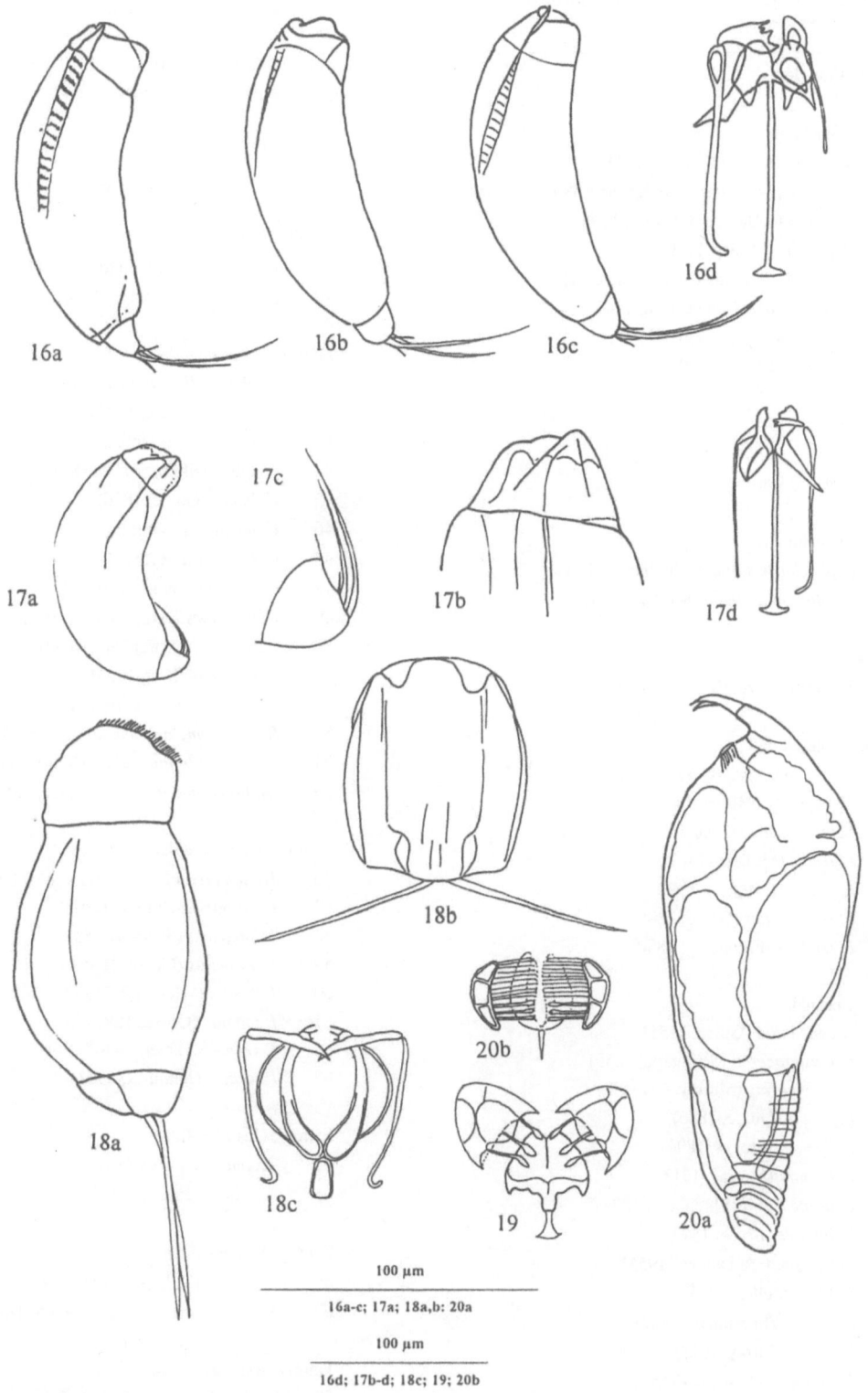

Figures 16–19. **Figure 16.** *T. tenuior.* a–c: habitus in different positions, d: trophi. **Figure 17.** *T. vernalis.* a: lateral, b: toes, c: anterior end, d: trophi. **Figure 18.** *Dicranophoroides caudatus.* a: habitus (lateral), b: dorsal (contracted), c: trophi (ventral). **Figure 19.** *Octotrocha speciosa* trophi (ventral). **Figure 20.** *Ptygura furcillata.* a: lateral (contracted), b: trophi.

Table 1. Systematic list of Rotifera recorded. *:
species new to Mexico

Order: Ploimida

Family: Brachionidae
1.　*Anuraeopsis fissa* (Gosse, 1851)
2.　*Brachionus bidentatus* Anderson, 1889
3.　*B. caudatus* Barrois & Daday, 1894
4.　*B. patulus* (Müller, 1786)
5.　*Kellicottia bostoniensis* (Rousselet, 1908)
6.　*Keratella americana* Carlin, 1943
7.　*K. cochlearis* (Gosse, 1851)
8.　*K. tropica* (Apstein, 1907)

Family: Euchlanidae
9.　*Euchlanis dilatata* Ehrenberg, 1832
10.　*E. incisa* Carlin, 1939

Family: Mytilinidae
11.　*Lophocharis salpina* (Ehrenberg, 1834)
12.　*Mytilina mucronata* (Müller, 1773)

Family: Trichotriidae
13.　*Trichotria tetractis* (Ehrenberg, 1830)

Family: Colurellidae
14.　*Colurella obtusa* (Gosse, 1886)
15.　*C. uncinata* (Müller, 1773)
16.　*Lepadella acuminata* (Ehrenberg, 1834)
17.　*L. ovalis* (Müller, 1786)
18.　*L. patella* (Müller, 1786)
19.　*L. rhomboides* (Gosse, 1886)
20.　*L. triptera* (Ehrenberg, 1830)

Famil: Lecanidae
21.　*Lecane bulla* (Gosse, 1851)
22.　*L. closterocerca* (Schmarda, 1859)
23.　*L. curvicornis* (Murray, 1913)
24.　*L. flexilis* (Gosse, 1886)
25.　*L. hamata* (Stokes, 1896)
26.　*L. hastata* (Murray, 1913)
27.　*L. hornemanni* (Ehrenberg, 1834)*
28.　*L. inermis* (Bryce, 1892)
29.　*L. latissima* Yamamoto, 1955*
30.　*L. luna* (Müller, 1776)
31.　*L. lunaris* (Ehrenberg, 1832)
32.　*L. nana* (Murray, 1913)
33.　*L. papuana* (Murray, 1913)
34.　*L. pyriformis* (Daday, 1905)

Table 1. Continued.

35.　*L. stenroosi* (Meissner, 1908)
36.　*L tenuiseta* Harring, 1914
37.　*L. unguitata* (Fadeev, 1925)

Family: Proalidae
38.　*Proales fallaciosa* Wulfert, 1937*

Family: Lindiidae
39.　*Lindia torulosa* Dujardin, 1841*
40.　*L. truncata* (Jennings, 1894)*

Family: Notommatidae
41.　*Cephalodella catellina* (Müller, 1786)
42.　*C. exigua* (Gosse, 1886)*
43.　*C. forficula* (Ehrenberg, 1832)*
44.　*C. gibba* (Ehrenberg, 1838)
45.　*C. hoodi* (Gosse, 1896)
46.　*C. misgurnus* Wulfert, 1937*
47.　*C. panarista* Myers, 1924
48.　*C. stenroosi* Wulfert, 1937*
49.　*C. ventripes* Dixon-Nuttall, 1901*
50.　*Itura aurita* (Ehrenberg, 1830)*
51.　*I. myersi* Wulfert, 1935*
52.　*Pleurotrocha petromyzon* Ehrenberg, 1830*
53.　*Resticula melandocus* (Gosse, 1887)*
54.　*Scaridium longicaudum* (Müller, 1786)
55.　*Taphrocampa annulosa* Gosse, 1851

Family: Trichocercidae
56.　*Trichocerca bicristata* (Gosse, 1887)
57.　*T. cylindrica* (Imhof, 1891)*
58.　*T. longiseta* (Schrank, 1802)
59.　*T. porcellus* (Gosse, 1886)
60.　*T. similis* (Wierzejski, 1893)
61.　*T. tenuior* (Gosse, 1886)*
62.　*T. vernalis* Hauer, 1936*
63.　*T. weberi* (Jennings, 1903)

Family: Gastropodidae
64.　*Ascomorpha ecaudis* (Perty, 1850)
65.　*A. saltans* Bartsch, 1870

Family: Synchaetidae
66.　*Polyarthra vulgaris* Carlin, 1943
67.　*Synchaeta pectinata* Ehrenberg, 1832

Family: Asplanchnidae
68.　*Asplanchnopus multiceps* (Schrank, 1793)

Table 1. Continued.

Family: Dicranophoridae
69. *Dicranophoroides caudatus* (Ehrenberg, 1834)*
70. *Dicranophorus forcipatus* (Müller, 1786)

Order: Gnesiotrocha

Family: Testudinellidae
71. *Pompholyx sulcata* (Hudson, 1885)
72. *Testudinella patina* (Hermann, 1783)

Family: Flosculariidae
73. *Octotrocha speciosa* Thorpe, 1893*
74. *Ptygura furcillata* (Kellicott, 1889)*

Family: Conochilidae
75. *Conochilus natans* (Selgo, 1900)

Family: Hexarthridae
76. *Hexarthra intermedia* Wiszniewski, 1929

Family: Filiniidae
77. *Filinia longiseta* (Ehrenberg, 1834)
78. *F. opoliensis* (Zacharias, 1898)

Bdelloidea
79. *Philodina* sp.
80. *Rotaria* sp.

Table 1 lists species (* new records) encountered during this study. Comments on selected rotifer taxa are as follows:

T. vernalis (Figure 17). Anterior margin of left lorica with a rounded plate. Dorsal keel indistinct. Anterior end of lorica with folds. Two toes, long and unequal. Trophi virgate. Recorded from North America and Europe (Koste, 1978).

Dicranophoroides caudatus (Figure 18). Loricate species with long toes. Body when contracted becoming short and stout with parallel foldings. Trophi forcipate with short fulcrum. Rami with lamellar extensions. Manubria curved outwards at distal ends. Uncus long and pointed. Cosmopolitan species (Koste, 1978).

Octotrocha speciosa (Figure 19). Illoricate floscularid rotifer. Contracted specimens not identifiable. Trophi uncinate. Common in the American continent (Koste, 1978).

Ptygura furcillata (Figure 20). Soft-bodied floscularid rotifer. Contracted specimens with characteristic anterior projections. Trophi malleoramate.

Discussion

Some tropical systems yield a high number of species (e.g. Segers et al., 1993; Segers & Dumont, 1995). Mexico is located in the transitional zone between nearctics and neotropics, and the high altitude of the surveyed pond allows a mixture of cold and warm water species (Rico-Martinez & Silva-Briano, 1993). Probably due to this reason we encountered some species like *L. latissima* and *C. stenroosi* known from cold systems (Koste, 1978; Segers, 1995). The high diversity of the water body studied here is not restricted to rotifers; we also found a complex species composition of Cladocera (Ciros-Pérez & Elías-Gutiérrez, 1996) and copepoda (Grimaldo et al., 1997).

The present study adds four genera new to Mexico: *Itura*, *Pleurotrocha*, *Octotrocha* and *Ptygura*. This study confirms the presence of rotifer species listed only as list from Mexican waters. Most taxa recorded here have a wide distribution. Data on soft-bodied rotifers of Flosculariidae and Collothecidae, in addition to Bdelloidea, from Mexico are needed to get a more complete picture for understanding the geographical distribution of rotifers in the North American continent.

Acknowledgements

This study was supported by CONABIO (no. H-116) granted to the authors. SSSS is also thankful to CONACYT for Catedra Patrimonial (no. 940481) and SNI (18723) at Iztacala campus of UNAM.

References

Ciros-Pérez, J., M. Silva-Briano & M. Elías-Gutiérrez, 1995. A new species of *Macrothix* (Anomopoda: Macrothricidae) from Central Mexico. Hydrobiologia 319: 1–8.

Ciros-Pérez, J. & M. Elías-Gutiérrez, 1996. Nuevos registros de cladoceros (Crustacea: Anomopoda) en México. Rev. Biol. Trop. 44: 297–304.

Grimaldo O. D., M. Elías-Gutiérrez, M. Camacho-Lemus & J. Ciros-Perez, 1997. Additions to Mexican freshwater copepods with the description of the female *Leptodiaptomus mexicanus* (Marsh). Proc. VI Int. Copepod Conf. (In press).

Koste, W., 1978. Rotatoria. Die Rädertiere Mitteleuropas. Borntraeger, Berlin, Stuttgart, vol. 1: 673 pp., vol. 2: 234 pp.

Koste, W. & R. J. Shiel, 1987. Rotifera from Australian inland waters. 2. Epiphanidae and Brachionidae (Rotifera: Monogononta). Invertebr. Taxon. 7: 949–1021.

Koste, W. & R. J. Shiel, 1989. Rotifera from Australian inland waters. 4. Colurellidae (Rotifera: Monogononta). Trans. r. Soc. S. Aust. 113: 119–143.

Koste, W. & R. J. Shiel, 1990. Rotifera from Australian inland waters. 5. Lecanidae (Rotifera: Monogononta). Trans. r. Soc. S. Aust. 114: 1–36.

Koste, W. & R. J. Shiel, 1991. Rotifera from Australian inland waters. 7. Notommatidae (Rotifera: Monogononta). Trans. r. Soc. Aust. 115: 111–159.

Kutikova, L. A. & M. Silva-Briano, 1995. *Keratella mexicana* sp. nov., a new rotifer from Aguascalientes, Mexico. Hydrobiologia 310: 119–122.

Nogrady, T., R. L. Wallace & T. W. Snell, 1993. Rotifera 1. Biology, ecology and systematics. In H. J. Dumont & T. Nogrady (eds), Guides to the Identification of the Microinvertebrates of the Continental Waters of the World. 6. SPB Academic Publishers, The Hague, The Netherlands, 142 pp.

Örstan, A., 1995. A new species of bdelloid rotifer from Sonora, Mexico. Southwest. Nat. 40: 255–258.

Ramirez-Sevilla, R., R. Rueda-Jasso, J. L. Ortiz-Galindo & Gonzalez-Acosta, 1991. Metodologia para el cultivo experimental del rotifero *Brachionus plicatilis*. Inv. Mar. CICIMAR 6: 287–290.

Rico-Martinez, M. & M. Silva-Briano, 1993. Contribution to the knowledge of the Rotifera of Mexico. Hydrobiologia 255/256: 467–474.

Segers, H., 1995. Rotifera 2. The Lecanidae (Monogononta). In H. J. Dumont & T. Nogrady (eds), Guides to the Identification of the Microinvertebrates of the Continental Waters of the World. 6. SPB Academic Publishers, The Hague, The Netherlands, 226 pp.

Segers, H., C. S. Nwadiaro & H. J. Dumont, 1993. Rotifera of some lakes in the floodplain of the river Niger (Imo State, Nigeria) II. Faunal composition and the diversity. Hydrobiologia 250: 63–71.

Segers, H. & H. J. Dumont, 1995. 102+ rotifer species (Rotifera: Monogononta) in Broa reservoir (SP., Brazil) on 26 August, 1994, with description of three new species. Hydrobiologia 316: 183–197.

Suárez, M. E. & H. A. Campos, 1994. Copépodos pelágicos del Golfo de México y Mar Caribe. 1. Biología y Sistematica. CIQRO, Mexico.

Vilaclara, G., & V. Sladecek, 1989. Mexican rotifers as indicators of water quality with description of *Collotheca riverai*, new species. Arch. Hydrobiol. 115: 257–264.

Hydrobiologia **387/388**: 55–62, 1998.
E. Wurdak, R. Wallace & H. Segers (eds), Rotifera VIII: A Comparative Approach.
© *1998 Kluwer Academic Publishers.*

Rotifers new to Florida, U.S.A.

Paul N. Turner[1] & Howard L. Taylor[2]
[1] *1003 Sutters Rim, San Antonio, TX 78258, U.S.A.*
[2] *1812 Wood Hollow Court, Sarasota, FL 34235, U.S.A.*

Key words: Rotifera, distribution, Florida, new records

Abstract

Two-hundred thirty-four rotifers were identified over a 14-year period from 30 sites in south and south-west Florida. Ninety-six species are newly recorded from the state, bringing the total number of rotifers now known from Florida to 419. Four rotifers not recorded from North America prior to this study are noted with respect to their geographical distribution and specific status: *Cephalodella pentaplax* Wulfert (1943), *Cephalodella pachydactyla* Wulfert (1937), *Lecane gwileti* Tarnogradski (1930) and *Lepadella tenella* Wulfert. *Lecane curvicornis* (Murray (1913) f. *miamiensis* Myers (1933) is discussed concerning its variability and distribution.

Introduction

Little work has been done on the rotifer fauna of Florida since Ahlstrom (1934). In the 62 years since his report, only 11 papers recorded rotifers from Florida (Fry & Osborne, 1980; King & Snell, 1983; Nogrady, 1983; Snell & Hawkinson, 1983; Snell et. al., 1983; Snell & Carrillo, 1984; Billets & Osborne, 1985; Bienert, 1986; Scheda & Cowell, 1988; Turner, 1990, 1993). Although 7 papers were ecologically oriented and not species compilations, together they added 48 known species to the 275 valid species Ahlstrom listed in 1934, for a total of 323 currently valid species, subspecies and forms (Ahlstrom's paper cited 279 rotifers, yet listed only 278). Of the 11 papers cited above, only two papers added rotifers new to science: Myers (1941) added *Lecane curvicornis* v. *miamiensis* as a new variety, and Bienert (1986) added *Lecane ordwayi* as a new species. Also, despite the fact that Florida has the longest saltwater coastline in the US, only Turner (1990, 1993) dealt with rotifers in brackish or salt coastal waters. Finally, as is common in most rotifer research efforts, the Bdelloidea are greatly understated, with only three listed by Ahlstrom (1934), one *Rotaria* sp. listed by Turner (1993), and one in this study. The present study adds 96 new freshwater records to the Floridian rotifer fauna, bringing the total rotifers now known from Florida to 419 taxa.

Description of localities

A qualitative study of rotifers in southern Florida, including 24 different sample sites, was carried out by the second author from 1986–1996. Five additional sites were sampled from 1983–1989 by the first author (Figure 1).

The Everglades National Park sites (2–8, 25–27), located at the southern tip of Florida, receive their water from the large Everglades swamp which flows south, 80 km wide and about 15 cm deep, from Lake Okeechobee (27-00 N latitude). The terrain elevation drops only 4.6 meters on its entire 167 kilometer length to Florida Bay.

Myakka River State Park sites (14–17, 24) are located approximately 24 km south-east of Sarasota. The River in the Park receive their water from rain, except for the water from Clay Gully which is spring fed. Myakka Upper Lake has a maximum depth of about one meter. (See Table 1 for localities and sample sites.)

Data on water quality and aquatic vegetation of the studied sites are summarized in Tables 2 and 3, respectively.

Methods

Collections were taken using a combination of plankton net throws and bulk sampling. In samples from

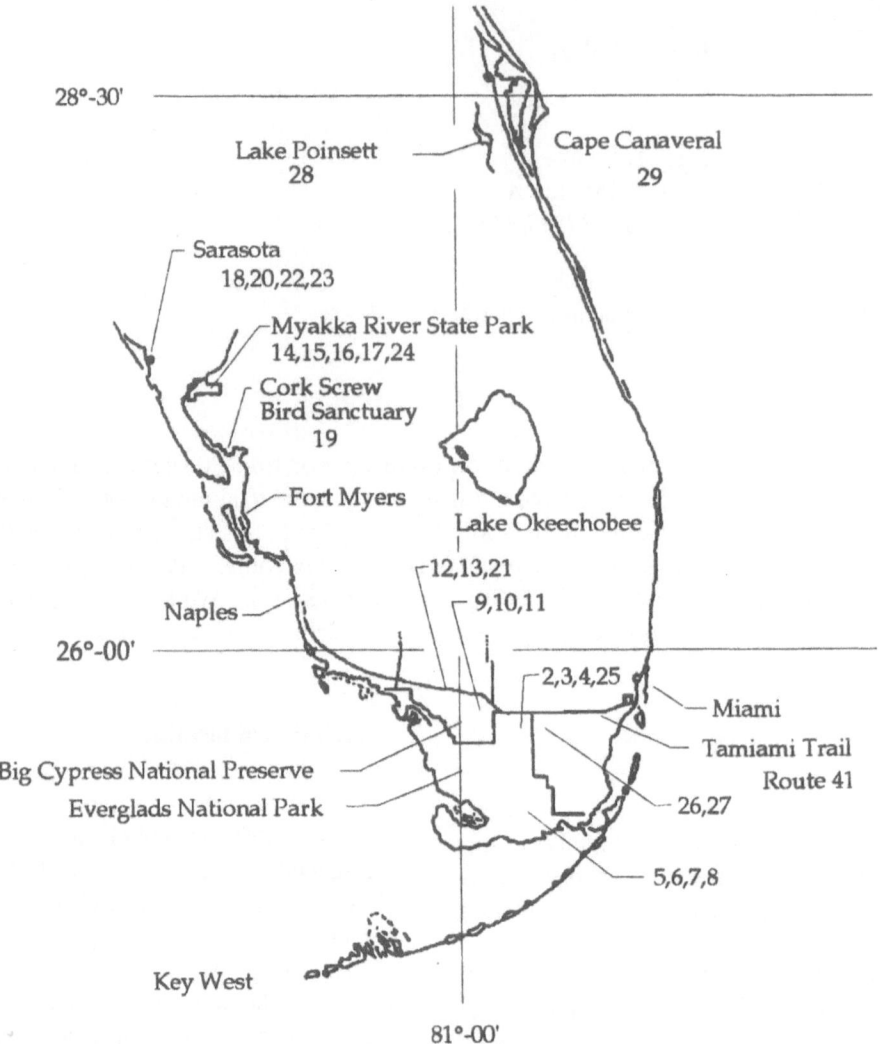

Figure 1. Map of the study area showing sampled regions.

littoral sites 1, 5–9, 17–20, 22 and 23, a small sample of each aquatic plant was floated into the collection jar, and the plankton net was then drawn through each area of aquatics at the exact collection site. In the laboratory, these collections were examined for free-swimming species and the aquatic plants were examined for sessile rotifers. All samples were preserved on the spot and taken to the lab for identification.

Results

Table 4 lists the 96 rotifers found new to Florida. Following each rotifer name is a bracketed numerical identifier of the site at which the rotifer was found. Because some sites were sampled many times at different times of the year, the sampled months are shown with a numerical month identifier following the site found, and outside the bracket. The rotifer species listed in Table 4 for sites 1, 5–9, 17–20, 22 and 23 were found in, around or directly on the submerged macrophytes listed for the same site in Table 3.

As shown in Table 4, eight rotifers were found at more than 6 different sites. *Euchlanis dilatata* f. *lucksiana* Hauer (1930) was found at 10 sites, *Polyarthra dolichoptera* Idelson (1925) and *Mytilina ventralis* (Ehrenberg 1832) f. *macracantha* (Gosse 1886) at 9 sites, *Trichocerca elongata* (Murray 1913) at 8 sites, *Lecane arcuata* (Bryce 1891), *Monommata dentata* Wulfert (1940) at 7 sites, *Dicrano-*

Table 1. List of Florida, USA sampled during this study

Site	Name of location
1	Shady Oaks Pond, North Miami
2	Everglades national Park, Shark Rivver, at Bobccat Trail
3, 4	Everglades, Spillways S-12A and S-12B on U.S. 41
5	Everglades National Park, Royal Palm Slough
6	Everglades National Park, Long Pine Key
7	Everglades National Park, Pine Glades Lake
8	Everglades National Park, Pa-hay-okee at Shark River Slough
9, 10, 11	Big Cypress National Preserve, Old Loop Road
12, 13	Big Cypress National Preservve, U.S. 41, 12 and 14 miles west of Shark River
14	Myakka River State Park, Main Drive bridge
15	Myakka River State Park, Myakka River bridge on Route 780
16	Myakka River State Park, Clay Gully bridge on Route 780
17	Myakka River State Park, Bird Walk
18, 19, 20	Sarasota environs, retaining ponds
21	Big Cypress National Preserve, Turner River at U.S. 41
22, 23	Sarasota environs, retaining ponds
24	Myakka River State Park, Howard Creek dam
25	Everglades National Park, Station no. 6, 2 miles south-east of Shark River Tower
26	Everglades, Station no. 23, Everglades National Park research 8.2 miles west of State Road 27 and 6 miles south of U.S. 41
27	Everglades, Station no. 50, Everglades National Park research, 5.2 miles west of State Road 27 and 2.8 miles South of U.S. 41
28	Lake Poinsett
29	Banana River, Cape Canaveral area

phorus epicharis Harring & Myers (1928) and *Limnias melicerta* Weisse (1848) at 6 sites.

Species discussion

Four species found were new to Florida, the United States and North America, having only been found in Europe prior to these records.

Cephalodella pentaplax Wulfert (1943) was originally found in a saltwater pool in Belarus. The animal's hinged lorica, large foot, and a dorsal, double-keeled 'tail' are indicative. Two frontal eyes and short, laterally separated and stippled toes are also characteristic. Trophi is an asymmetrical Type B. Geolocation: Germany (Wulfert, 1943), and now Florida, USA.

Cephalodella pachydactyla Wulfert (1937) was found originally in a ditch in Zeitz, Germany. The animal has three semi-rigid lorica plates connected by flexible sulci. With a short foot, two frontal eyes, and a similar dorsal 'tail' to *C. pentaplax*, it can easily be confused. Trophi is symmetrical Type B. Geolocation:

Germany (Wulfert, 1960; Koch-Althaus, 1963; Koste & Poltz, 1987), Finland (Eriksen, 1969; Bjorklund, 1972), Sweden (Pejler & Berzins, 1993) and now Florida, USA.

Lecane gwileti (Tarnogradski, 1930) was originally found in Ossetia, Caucasus mountains. This rotifer has a dorsal lorica plate wider than ventral plate, except anteriorly where the margins are nearly coincident. Ventral plate longer than dorsal plate, with margin longer and covering foot joint. One toe terminates in incompletely separated pseudoclaws. This rotifer strongly resembles *Lecane furcata*, except that *L. furcata* toe claws are clearly and distinctly separated from toe. Geolocation: Spain (Fores, E., Menendez, M. & Comin, F. A., 1986), Germany (Hauer, 1931; Wulfert, 1960), Belgium (De Ridder, 1989, 1992), and now Florida, USA.

Lepadella tenella Wulfert (1942) is closely related to *L. triptera*, distinguished largely by the much smaller height and scale of the dorsal lorica ridge, as well as the anterior margin profile which is concave on

Table 2. Water quality data from selected sampled locations

Loc.	Temp. Co	D.O. mg/1	Conductivity μmhos/cm	Tot.Dis.Mg. μg/1	Ortho PO_4^{2-} μg/1	NO4 μg/1(N)	pH
1	–	–	–	–	–	–	–
2	30	–	–	–	–	7.4	
3	–	1.8–8.23	–	197–767	–	.004–.069	6.6–8.4
4	–	1.8–8.23	– 197–767	–	.004–.069	6.6–8.4	
5	15–25	–	–	207–881	–	–	8.2
6	25	–	–	207–881	–	–	7.3–7.4
7	25	–	–	207–881	–	–	8.2
8	24	–	–	207–881	–	–	8.2
9	21–22	–	–	–	–	–	7.1–7.2
10	21–22	–	–	–	–	–	7.1–7.2
11	21–22	–	–	–	–	–	7.1–7.2
12	21	–	–	–	–	–	7.2
13	21	–	–	–	–	–	7.2
14	19–31	0.5–6.7	191–473	–	0.13–0.35	0.02–0.18	7.2
15	19–31	0.5–6.7	191–473	–	0.13–0.35	0.02–0.18	7.2
16	18–31	3.1–9.8	240–468	–	0.12–0.33	0.04–0.18	6.9–7.2
17	31	–	–	–	–	–	6.9
18	32	–	–	–	–	–	7.0–7.2
19	32	–	–	–	–	–	7.0–7.2
20	32	–	–	–	–	–	7.0–7.2
21	30	–	–	–	–	–	7.0–7.2
22	33	–	–	–	–	–	7.7–8.2
23	33	–	–	–	–	–	7.7–8.2
24	19–29	3.1–7.4	–	–	0.27–0.87	0.0–0.87	7.2
28	18	8.4	–	–	–	–	6.5
29	20	–	–	–	–	–	7.0

*The Everglades National Park receives wind-blown contaminants such as mercury from incinerators along the east coast of the state.

Table 3. Aquatic plants found at sampled sites

Species	Sites	Species	Sites
Alternathera philoxeroides	23,24	*Nuphar luteum* Sibth. and Smith	17,22,23,24
Bacopa caroliniana (Q Walt.) Robins	5	*Nymphaea mexicana* Zucc.	22,23
Brachiara purpurabscebce Radd	14	*Nymphaea odorate* Aiton	2,4,9,10,17,19,22,23,24
Brasenia schreveri Bmelin	14,17,24	*Paspalum repens* Berg	14,15,16,17
Ceratphyllum demersium L.	16,23	*Pistia stratiosis* L.	3,24
Eichornia ccrassipes (Mart.) Solms	12,13	*Pontenderia cordata* L.	22,23
Eleocharis baldwinii Michx.	14	*Salvia rotundifolia* Wild	16,24
Fontinalis spp.	16,18,19,20	*Sagittaria graminea* Michx	18,19,20
Hydrilla verticillata Royle	14,15,16,24	*Typoha spp.*	1,22,23
Lemna minor L	19	*Utricularia gibba* L.	2,3,6,7,8
Ludwigia repens Forst	12,13,14,15	*Spirogyra sup.*	1,5,14,15,17,18,19,20

Table 4. Rotifers new to Florida found in 29 different sites from 1983 to 1996. Bracketed numbers denote sampled location, non-bracketed number indicates month sampled

Asplanchna brightwelli (Gosse, 1850) [28] 2

A. girodi (De Guerne, 1888) [23,24] 2,12

A. sieboldi (Leydig, 1854) [3,4] 11,12

Asplanchnopus hyalinus Harring, 1913 [18] 7

Brachionus caudatus Barrois & Daday 1894 f. *apsteini* Fadeew 1925 [28] 2

Brachionus falcatus Zacharias 1898 [4] 12

Brachionus quadridentatus (Hermann, 1783) f. *cluniorbicularis* (Skorikov, 1894) [23,25,28] 2,7

Brachionus quadridentatus mirabilis (Daday, 1897) [23] 8

Brachionus quadridentatus (Hermann, 1783) f. *rhenanus* (Lauterborn, 1893) [16,23] 2,9

Cephalodella catellina volvocicola (Zavadovsky, 1916) [9,14] 3,12

Cephalodella doryphora Myers 1934 [9] 1

Cephalodella gracilis (Ehrenberg 1832) [29] 2

Cephalodella pachydactyla (Wulfert 1937) [28] 2

Cephalodella pentaplax Wulfert 1937 [18] 6

Cephalodella physalis Myers,1924 [16] 9

Cephalodella rotunda Wulfert, 1937 [6] 12

Cephalodella ventripes (Dixon-Nutall, 1901) [4,13,15,23] 1,2,5,12

Collotheca ornata (Ehrenberg, 1932) f. *cornuta* (Dobie, 1849) [9] 12

Colurella hindenburgi Steinecke, 1917 [5] 12

Conochilus natans (Selligo, 1900) [18] 12

Cupelopagis vorax (Leidy, 1857) [28] 2

Dicranophorus caudatus (Ehrenberg, 1834) [16] 9

Dicranophorus epicharis Harring & Myers, 1928 [3,4,9,14,15,16,28] 1,2,7,9,12

Dicranophorus mesotis Harring & Myers, 1928 [9,14] 1,9,12

Dicranophorus tegillus Harring & Myers, 1928 [5] 12

Dicranophoroides claviger (Hauer, 1965) [16] 8

Dipleuchlanis propatula (Gosse, 1886) [25,26,27] 7

Dissotrocha aculeata [28] 2

Eosphora thoides Wulfert, 1935 [14] 9

Epiphanes senta (Müller, 1773) [3,6,15] 5,11,12

Euchlanis dilatata Ehrenberg, 1832 f. *lucksiana* (Hauer, 1930) [4,5,8,9,14,19,21,23,25,27] 1,3,5,6,7,8,9,11,12

Filinia opoliensis (Zacharias, 1898) [18] 6

Filinia pejleri Hutchinson, 1964 [18] 3,5

Filinia terminalis (Plate, 1886) [16,18] 6,8

Floscularia ringens (LinnÈ, 1758) f. *conifera* (Hudson, 1886) [2,3] 7,11,12

Horaella thomassoni Koste, 1973 [18] 5

Itura chamadis Harring & Myers, 1928 [8,18] 4,12

Itura myersi (Wulfert, 1935) [18] 5

Keratella lenzi (Hauer, 1953) [3,4] 11,12

Keratella thomassoni Hauer, 1958 [14] 8

Keratella valga (Ehrenberg, 1934) f. *monospina* (Klausener, 1908) [3,4] 1

Lacinularia flosculosa (Müller, 1758) [2,3] 7,12

Lecane aculeata (Jakubski, 1912) [9] 3

Lecane aegana Harring, 1914 [5,15,16] 9,12

Lecane arcuata (Bryce, 1891) [1,3,4,6,7,9,13] 6,7,11,12

Lecane arcula Harring, 1914 [25,26] 7

Lecane elongata Harring & Myers, 1926 [7,15,28] 2,5,12

Lecane gwileti (Tarnogradski, 1913) [2] 12

Continued on p. 60

Table 4. Continued

Lecane haliclysta (Harring & Myers, 1926) [25] 7
Lecane levistyla (Olofsson, 1917) [5,26,27] 7,12
Lecane pusilla Harring, 1914 [15] 7
Lecane rhenana Hauer, 1929 [27] 7
Lecane rhopalura (Harring & Myers, 1926) [2] 7
Lecane spinulifera Edmondson 1935 [25] 7
Lecane stenroosi (Meissner, 1908) [25] 7
Lecane thienemanni (Hauer, 1938) [8] 12
Lecane sp. A. (to be described by Segers, 1997) [27] 7
Lepadella pteroigoides (Dunlop, 1897) [25,26] 7
Lepadella pyriformis Myers, 1938 [14] 3
Lepadella tenella Wulfert, 1942 [19] 5
Lepadella triptera Ehrenbert, 1830 f. *rhomboidula* (Bryce, 1890) [2,3] 7
Lepadella dactyliseta (Stenroos, 1898) [25,27] 7
Limnias melicerta Weisse, 1848 [9,13,15,16,17,18] 3,7,8,9,11,12
Lophocharis salpina (Ehrenberg, 1834) [15] 2 *Manfredium eudactylota* (Gosse, 1886) [3,15] 7
Monommata dentata Wulfert, 1940 [2,3,4,8,9,17,21] 1,7,9,11,12
Monommata phoxa Myers, 1930 [3,15] 5,7
Mytilina bisulcata (Lucks, 1912) [25] 7
Mytilina ventralis (Ehrenberg, 1832) f. *macracantha* (Gosse, 1886) [2,9,13,114,15,23] 1,2,3,7,12
Notholca acuminata (Ehrenberg, 1832) [28] 2
Notommata codonella Harring & Myers, 1924 [3,13] 11,12
Notommata haueri Wulfert, 1939 [25] 7
Notommata prodota Myers, 1923 [5] 12
Notommata tripus Ehrenberg, 1838 [8,15,16] 2,3,9,12
Platonius polyacanthus (Ehrenberg, 1834) [2,25,27] 1,5,7,11,12
Platyias quadricornis brevispinus Daday, 1905 [25,27,28] 7,2
Polyarthra dolichoptera Idelson, 1925 [1,2,3,4,13,14,15,16,18] 3,5,6,7,9,11,12
Polyarthra remata Skorikov 1896 [25,28] 7,2
Proales fallaciosa Wulfert, 1937 [9,14,15] 3,5,12
Pseudoharringia similis Fadeew, 1925 [15] 7
Ptygura brachiata (Hudson, 1886) [14] 9
Ptygura pectinifera (Murray, 1913) [4] 1
Synchaeta tremula (Müller, 1786) [3,9,13,21] 6,11,12
Testudinella amphora Hauer, 1938 [25] 7
Testudinella discoidea Ahlstrom 1934 [25] 7
Testudinella mucronata (Gosse, 1886) [5] 12
Testudinella parva semiparva (Hauer, 1938) [25,26] 7
Testudinella patina dendradena (DeBeauchamp, 1955) [14] 12
Testudinella patina (Hermann, 1783) f. *intermedia* (Anderson, 1889) [9] 3,12
Trichocerca brachyura (Gosse, 1851) [28] 2
Trichocerca elongata braziliensis (Murray, 1913) [2,3,4,7,14,15,16,17,18] 1,3,7,9,12
Trichocerca relicta (Donner, 1950) [15,28] 2,3
Trichocerca similis (Wierzejski, 1893) [14,15,16,18] 1,3,9
Trichocerca uncinata (Voigt, 1902) [1] 1
Trichocerca flagellata Hauer, 1937 [8,9] 3,12
Trichocerca rattus (Müller, 1776) [18] 7
Trichocerca vernalis Hauer, 1936 [28] 2

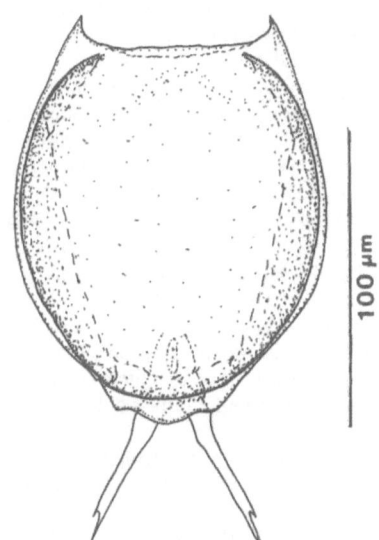

Figure 2. Lecane curvicornis (Murray 1913); Dorsal view of female form *miamiensis* Myers 1941.

both dorsal and ventral plates. Geolocation: Germany (Wulfert, 1942), and now Florida, USA.

Lecane curvicornis Murray (1913) is an extremely variable species, with form *'miamiensis'* being potentially common to the tropics and semi-tropics. The variability of the s. str. rotifer is documented by Martinez & Paggi (1988), who conclude that the subtle modifications in shape of the dorsal and ventral posterior region of the lorica, as well as superficial sculpture of the lorica, are of little specific value, and reflect only different forms. They did not include *'miamiensis'* forms in their discussion, except to say they had no specimens, and therfore no opinion.

The apparent type and degree of variability in posterior lorica projection in *L. curvicornis* Murray (1913) form *'miamiensis'* Myers (1941) (Figure 2) is similar to that of *Lecane leontina* (Turner, 1892), *L. ludwigii* (Eckstein, 1883), and to some extent *L. ligona* (Dunlop, 1901). These four rotifers show variability, particularly in the posterio-ventral lorica; with *L. leontina* and *L. curvicornis* ranging from no projection at all to significant, lobed, two-spined and even three spined posterior projection. Although almost certainly environmentally affected, there is currently no data to account for the cause of this variability in any of the species. The *'miamiensis'* form is usually larger than the s. str. with shorter overall toes and significant shoulders to the base of the toes. Segers (1995) synonymizes all forms and re-

gards the taxon as cosmopolitan. Geolocation: Florida (Myers, 1941; Biernert, 1986; and this record), Sri Lanka (Chengalath and Fernando, 1973), Philippines (Mamaril & Fernando, 1978), India (Sharma, 1980), and Thailand (Sanoamuang, pers. comm.).

Acknowledgements

Thanks go to Roxanne Conrow for making available material from the difficult to access sites 25, 26 and 27 in the Everglades. We thank biologist Belinda Perry, Myakka River State Park, for help with the identification of aquatic plants in the Park, and biologist Christopher Becker, Oscar Sherer State Park, for data on locations in Myakka River State Park. Thanks to Angela Chong, South Florida Water Management District, for data on Everglades National Park.

References

Ahlstrom, E. H., 1934. Rotatoria of Florida. Trans. Amer. Micros. Soc. 53: 251–266.

Bienert, R. W., 1986. A new species of *Lecane* (Rotifera: Lecanidae) from subtropical Florida. Hydrobiologia 141: 175–177.

Billets, B. D. & J. A. Osborne, 1985. Zooplankton abundance and diversity in Spring Lake, Florida. Florida Scientist 48(3): 129–139.

De Ridder, M., 1989. De huidige stand van het raderdieronderzoek in Belgie. Verhandelingen van het Synposium 'Invertebvraten van Belgie': 31–41.

De Ridder, M., 1992. Distribution of Belgian Rotifera. Proc. 8th Int. Coll. European Invert. Surv., Brussels, 9–10 September 1991: 199–212

Pejler, B. & B. Berzins, 1993. On the ecology of *Cephalodella*. Hydrobiologia 259: 125–128.

Bjřklund, B. G., 1972. Taxonomic and ecological studies of species of *Notholca* (Rotatoria) found in sea- and brackish water, with description of a new species. Sarsia 51: 25–66.

Chengalath, R. & C. H. Fernando, 1973. Rotifera from Sri Lanka (Ceylon) I. The genus *Lecane* with descriptions of two new species. Bull. Fish. Res. Stn., Sri Lanka (Ceylon) 24: 13–27.

Eriksen, Brit Godske, 1969. Rotifers from two tarns in Southern Finland, with a description of a new species, and a list of Rotifers previously found in Finland. Acta Zool. Fenn. 125: 1–35.

Fry, D. L. & J. A. Osborne, 1980. Zooplankton abundance and diversity in central Florida grass carp ponds. Hydrobiologia 68: 145–155.

Hauer, J., 1931. Zur Rotatorienfauna Deutschlands (II). Zool. Anz. 93: 7–13.

King, C. E. & T. W. Snell., 1983. Density-dependent sexual reproduction in natural populations of the rotifer *Asplanchna girodi*. Hydrobiologia 73: 149–152.

Koch-Althaus, B., 1963. Systematische und okologische Studien an Rotatorien des Stechlinsees. Limnologica (Berlin) 1: 375–456.

Koste, W. & J. Poltz, 1987. Über die Rädertiere (Rotatoria, Phylum Aschelminthes) des Alfsees, eines Hochwasser-Rückhaltebeckens der Hase, NW Deutschland, FRG. Osnabrücker naturwiss. Mitt. 13: 185–220.

Mamaril, A. C. & C. H. Fernando, 1978. Freshwater zooplankton of the Philippines (Rotifera, Cladocera, and Copepoda). Nat. & Appl. sci. bull. 30: 169–221.

Martinez, C. C. & S. J. De Paggi, 1988. Especies de *Lecane* Nitzch (Rotifera, Monogononta) en ambientes acuáticos del Chaco Oriental y del valle aluvial del río Paraná (Argentina). Rev. Hydrobiol. trop. 21: 279–295.

Myers, F. J., 1933. The Distribution of Rotifera on Mt Desert Island, II. New Notommatidae of the Genus *Notommata* and *Proales*. Amer. Mus. Nov.: 1–26.

Myers, F. J., 1941. *Lecane curvicornis* var. *miamiensis*, new variety of Rotatoria, with observations on the feeding habits of rotifers. Notul. Nat. No 75, 1–8.

Murray, J., 1913. South American Rotifera. Part II. XI. J. Royal Micro. Soc. XI: 341–362.

Nogrady, T., 1983. Some new and rare warmwater rotifers. Hydrobiologia 106: 107–114.

Scheda, A. M. & B. C. Cowell, 1988. Rotifer grazers and phytoplankton: seasonal experiments on natural communities. Arch. Hydrobiol. 114(1): 31–44.

Segers, H., 1995. Rotifera 2: The Lecanidae. In H. J. Dumont & T. Nogrady (eds), Guides to the Identification of the Microinvertebrates of the Continental Waters of the World. SPB Academic Publishing, 226 pp.

Sharma, B. K., 1980. Synopsis of taxonomic studies on Indian Rotatoria. Hydrobiologia 73: 229–236.

Snell, T. W. & C. A. Hawkinson, 1983. Behavioral reproductive isolation among populations of the rotifer *Brachionus plicatilis*. Evolution 37(6): 1294–1305.

Snell, T. W. & K. Carrillo, 1984. Body size variation among strains of the rotifer *Brachionus plicatilis*. Aquaculture 37: 359–367.

Snell, T.W., B. E. Burke & S. D. Messur, 1983. Size and distribution of resting eggs in a natural population of the rotifer *Brachionus plicatilis*. Gulf Research Reports 7(3): 285–287.

Tarnogradski, D., 1930. [The Rotiferfauna of the Northern Caukasus belonging to the genera *Lecane, Monostyla* and *Colurella*]. Raboty Sev. -Kaukazskoj Gidrobiol. Stancii 2: 111–144 (in Russian).

Turner, P. N., 1990. Some Interstitial Rotifera from a Florida, U.S.A., Beach. Trans. Am. Microsc. Soc. 109: 417–421.

Turner, P. N., 1993. Distribution of rotifers in a Floridian saltwater beach, with a note on rotifer dispersal. Hydrobiologia 255/256: 435–439.

Wulfert, K., 1937. Beiträge zur Kenntniss der Rädertierfauna Deutschlands. Teil 3. *Cephalodella*. Arch. Hydrobiol. 31: 592–636.

Wulfert, K., 1942. Neue Rotatorienarten aus deutschen Mineralquellen. Zool. Anz., 137: 187–200.

Wulfert, K., 1943. Rädertiere aus dem Salzwasser von Hermannsbad. Zool. Anz. 143: 164–172.

Wulfert, K., 1960. Die Rädertiere saurer Gewässer der Dubener Heide. II. Die Rotatorien des Krebsscherentumpels bei Winkelmuhle. Arch. f. Hydrobiol. 56: 311–333.

Wulfert, K., 1966. Rotatorien aus dem Stausee Ajwa und der Trinkwasser-Aufbereitung der Stadt Baroda (Indien). Limnologica (Berlin) 4: 53–93.

Hydrobiologia **387/388**: 63–77, 1998.
E. Wurdak, R. Wallace & H. Segers (eds), Rotifera VIII: A Comparative Approach.
© *1998 Kluwer Academic Publishers.*

Dicranophoridae (Rotifera) from the Alps

Christian D. Jersabek

University of Salzburg, Institute of Zoology, Hellbrunnerstr. 34, A-5020 Salzburg, Austria
E-mail: Christian.Jersabek@sbg.ac.at

Key words: Dicranophoridae, Austria, alpine water bodies, taxonomy, biogeography, ecology

Abstract

Rotifera of the family Dicranophoridae Harring, 1913 were recorded from mountainous altitudes of the Austrian Alps. Here, their morphology, distribution and ecology is detailed. The description of *Encentrum walterkostei* Jersabek is amended by observations on living animals. Of 19 species encountered, all but four are new to the alpine region, nine species are first records for the biogeographic region 'Alps'. Four species are new to science and will be published elsewhere. The majority are cosmopolites or widely distributed taxa, but also species with a more limited range, possibly endemics, seem to exist. Most species can be characterized as being more commonly found in cold environments, some of them are known to be psammobiontic or psammophilic.

Introduction

Benthic rotifers of alpine regions are sparsely investigated. For many of Europe's mountain regions there is little or no information as to which species are present, and local patterns of distribution and abundance are also little known. This is especially striking with illoricate forms. So, in previously published species lists from other palaearctic mountain ranges, like the Caucasus (Tarnogradski, 1959, 1961), the Pyrenees (Margalef, 1948; De Ridder, 1964; Miracle, 1978), the Atlas Mountains (Coussement & Dumont, 1980), and the Sierra Nevada (Morales-Baquero, 1987; Morales-Baquero et al., 1989) not even a single dicranophorid species was mentioned. Single reports of dicranophorids exist from the Tatra Mountains (Koniar, 1955, 1957; Fott, 1960), the Carpathians (Godeanu, 1963a, b; 1970) and the Massif Central in France (Francez, 1981). The low number of species previously reported from the Alps reflects the paucity of rotiferologists in that geographic area, although Central Europe was a stronghold of rotifer taxonomy since decades. An early compilation of rotifers occurring in alpine lakes, mainly of Switzerland, was published by Zschokke (1900). Further sketchy notes on littoral forms are available from a number of studies based upon plankton samples. However, in these non-taxonomic studies illoricate rotifers were considered at most on generic level. More detailed species lists for subalpine and alpine environments of the Austrian Alps were provided by Donner (1954, 1970) and Jersabek (1995), focusing mainly on soil samples, river habitats and stagnant waters, respectively. From these studies a total of about 88, mainly edaphic, bdelloid rotifers and 169 monogonont taxa are listed for alpine sites > 1000 m a.s.l. In a compilation of Rotifera for the 'Limnofauna Europaea', Bērziņš (1978) mentioned 23 dicranophorid species to be known from the zoogeographic region 'Alps', which, however, includes also low altitude regions.

As part of a continuing study on alpine Rotifera, this contribution illustrates the occurrence of dicranophorids in high altitude lakes and pools of the Austrian Alps. The study was accompanied by examining drawings and records by Dr Josef Donner in his original notebooks.

Methods

Field methods and laboratory procedures were applied as described by Jersabek (1995). Drawings and morphometric measurements were made at 500 to 1250 × magnification, using a Leitz-Laborlux D microscope fitted with a camera lucida.

The study focuses on records of dicranophorids at altitudes > 1000 m a.s.l. This somewhat arbitrary borderline includes altitudes well below the alpine vegetation zone, which is found between timberline and snowline. The timberline lies at ca. 1500 m a.s.l. in the Northeastern Limestone Alps, and fluctuates between 1600 (northern flank) and 1700 m a.s.l. (southern flank) in the Central Alps. Species found above this climatically determined borderline are regarded as 'alpine *s.str.*', whereas the remainder would be 'alpine *s.l.*'. The boundaries of the biogeographic region 'Alps' are according to Illies (1978). A table listing short descriptions of the sampled water bodies and environmental characteristics can be obtained from the author on request.

Results

Data on occurrence, abundance and habitat characteristics for the dicranophorids collected during this survey are summarized in Table 1. In all, 19 species were found, four of which are new to science (Jersabek, in press).

Comments on recorded species

Aspelta lestes H. & M., 1928 (Figures 1–8)

Material examined: Two specimens each from two alpine lakes, one specimen from a subalpine pool.

Foot and toes retracted within body in contracted specimens. Toes slightly decurved ventrally and forceps-shaped in ventral view (Figure 1). Posterior 1/3 swollen, abruptly tapering to acutely pointed tip (Figure 2). Cuticle highly transparent. Trophi asymmetric, with acute hook-shaped alula on right ramus (Figure 3). Manubria with minute notch terminally (Figure 4), not crutched as described by Harring and Myers (1928); this was also observed in Czechian specimens examined by Donner (1954). Unci highly characteristic (Figures 6,7).

Dimensions (μm): n=3; total length 98–109 (contracted); toes 29–33; trophus 30; rami d/s 13/13.5–14.5/15; fulcrum 7–9.5; unci d/s 10.5/8–11.5/9; manubria d/s 18.5/19.

Occurrence and ecology: Encountered on sand bottom and among macrophytes (*Ranunculus* sp.) of oligotrophic alpine lakes (2174–2220 m a.s.l.), as well as on sludge of a permanent pool in subalpine pasture land (1360 m a.s.l.). Previous records restricted to muddy substrates and mosses (*Fontinalis*); Palaearctis, Nearctis.

Previous records from the Alps: First record for the Alps.

Records from other palaearctic mountain ranges: Massif du Canigou, Pyrénées Orientales, France (De Smet & Pourriot, 1997).

Dicranophorus forcipatus (O.F.M., 1786) (Figures 9–11)

Material examined: Two specimens from a limestone pool.

Conspicuous by its large size and sabre-shaped toes with basal septum, and 2 small, black eyespots (Figure 10, 11). Highly transparent. Trophi normal, measurements within the range given by previous authors, but unci very long, reaching > 2/3 ramus length. Shearing teeth blunt elliptic, on ventral side of rami. Apical rami teeth with 2 tips (Figure 9).

Dimensions (μm): n=2; body length 183–189 (contracted); toes 85; trophus 72; ramus 47; fulcrum 20; uncus 36; manubrium 55–57.5; epipharynx 34; distance between eyes 17; number of shearing teeth 8/9 (d/s).

Occurrence and ecology: Northeastern Limestone Alps (Sengsengebirge, northern flank); on sludge in productive pool below timber line (1370 m a.s.l.), slightly auxotrophicated by game. Probably the most common *Dicranophorus*, previous findings in limnosaprobic environments of fresh and brackish waters, but also recorded as psammoxenic species from the hygropsammon of lakes; cosmopolite.

Previous records from the Alps: Tyrolean Alps (Austria), without exact location (Dalla-Torre, 1889); Swiss Alps (Unterer Murgsee, 1673 m a.s.l.) (Zschokke, 1900).

Records from other palaearctic mountain ranges: Reported from two localities (ca. 1147–1280 m a.s.l.) in the Massif Central, France (Francez, 1981) and from the Massif du Canigou, Pyrénées Orientales, France (De Smet & Pourriot, 1997).

Dicranophorus luetkeni (Bergendal, 1892) (Figures 12–15)

Material examined: Several specimens from seven localities in the Central Alps, and from two bog ponds in the Northeastern Limestone Alps.

Conspicuous by long toes terminating in movable, blunt claws of 1/3 toe length (Figure 15). Trophus with acute elliptic shearing teeth, posteriorly decurved

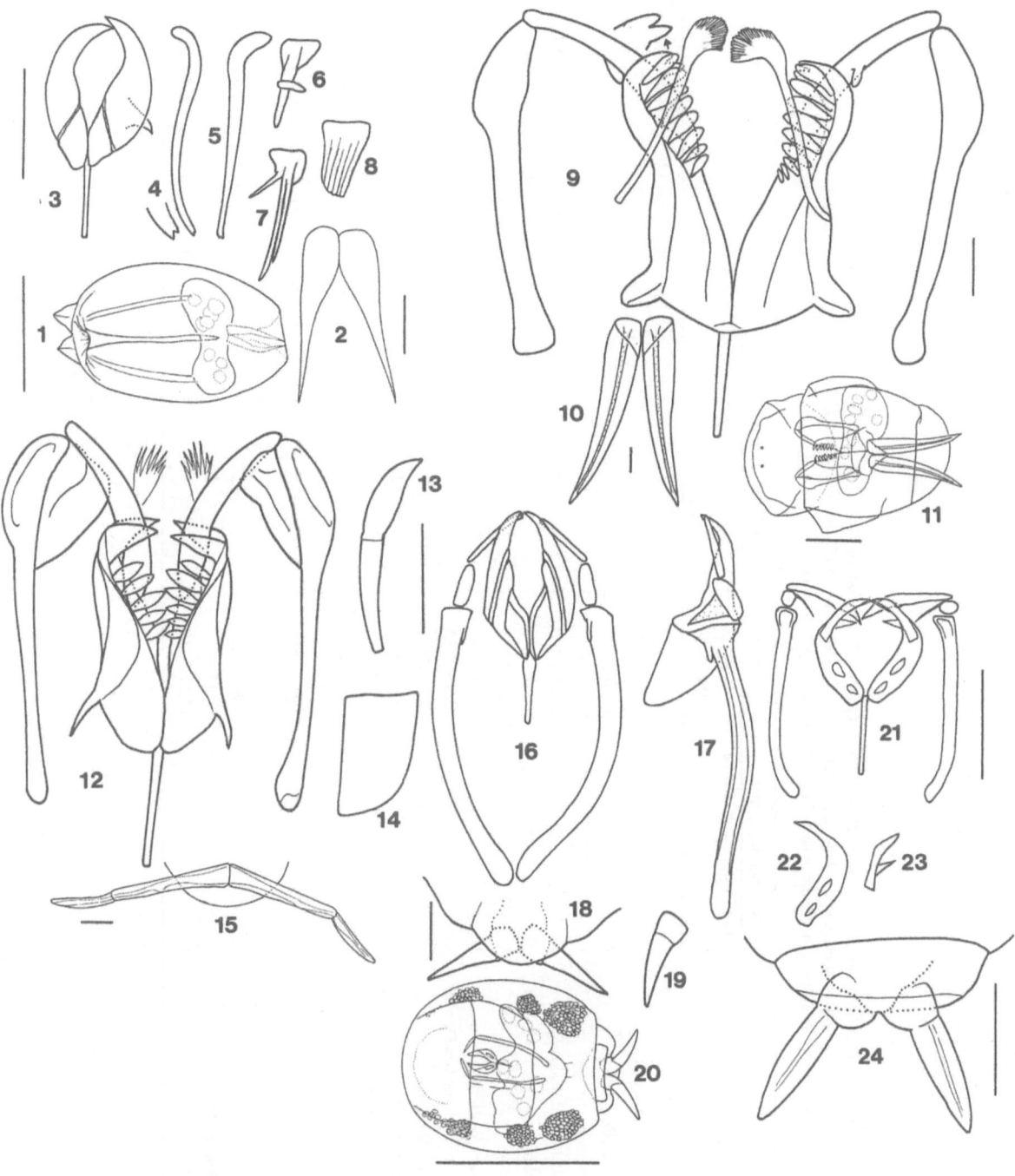

Figures 1–24. Aspelta lestes (H. & M.) – 1: contracted specimen, ventral view; 2: toes, ventral view, 3: incus, dorsal view; 4, 5: left and right manubrium, lateral views; distal end of left manubrium enlarged; 6, 7: left and right uncus, lateral views; 8: fulcrum, lateral view. *Figures 9–11. Dicranophorus forcipatus* (O.F.M.) – 9: trophus, ventral view; arrow indicates apical ramus tooth in dorsal view; 10: toes, ventral view; 11: contracted specimen, ventral view. *Figures 12–15. Dicranophorus luetkeni* (Bergendal) – 12: trophus, ventral view; 13: uncus, lateral view; 14: fulcrum, lateral view; 15: toes, ventral view. *Figures 16–20. Encentrum diglandula* (Zavadowski) – 16: trophus, ventral view; 17: ditto, lateral view; 18: posterior end of contracted specimen, dorsal view; 19: toe, lateral view; 20: contracted specimen, ventral view. *Figures 21–24. Encentrum incisum* Wulfert – 21: trophus, ventral view; 22: ramus, ventral view; 23: precuncinal teeth; 24: foot with toes, ventral view. Scale bars: habitus 50 μm; trophi, toes 10 μm.

Table 1. Occurrence of dicranophorids in high-altitude sites of the Central Alps (CA) and Northeastern Limestone Alps (LA) (Austria); wb – waterbodies. Substrate types: min – mineral bottom (fine sand, coarse gravel, rocks), uca – dense aggregations of unicellular algae (*Cylindrocystis* sp.), slu – sludge, mos – mosses, mac – macrophytes. Abundance values are highest observed densities: 1 single encounter, 2 rare, 3 sporadically, 4 abundant

Species	Number of records		Altitudinal range (m a.s.l.)	Substrate, abundance					pH	cond (µS)	T (°C)
	CA (43 wb)	LA (18 wb)		min	uca	slu	mos	mac			
Aspelta lestes (H. & M.)	2	1	1360–2220	2		1		2	6.9–8.2	11–140	8.3–24.2
Dicranophorus forcipatus (O.F.M.)	1		1370			2			8.2	204	17.1
Dicranophorus luekeni (Bergendal)	7	2	1290–2505	2		3	2		4.3–7.2	5–24	8.2–28.6
Dicranophoridae Gen. et sp. nov.	1		2547	2					8.3	24–28	3.3–6.0
Encentrum diglandula (Zavadowski)	1	2	1410–2270	3		2		2	5.6–6.4	9–14	12.7–22.1
Encentrum incisum Wulfert	1		1370			2			8.2	204	17.1
Encentrum lutra Wulfert	6	1	1643–2753	3	3		3		4.9–8.3	7–240	3.2–16.8
Encentrum mucronatum Wulfert	3		2420–2640	2	1				4.9–6.8	6–10	7.8–10.5
Encentrum cf. *putorius* var. *armatum* Donner	3		2560–2753	2				2	4.9–8.1	7–21	3.2–7.8
Encentrum saundersiae (Hudson)		1	1410			2			5.6	11	12.7
Encentrum uncinatum (Milne)	1	1	1360–2630	2		2			7.9–8.2	12–140	6.4–24.2
Encentrum walterkostei Jersabek	1		2543	4					7.5	16	8.0
Encentrum sp. nov. A	1		2547	2					8.3	26	4.3
Encentrum sp. nov. B	1		2543	3					7.5	16	6.0
Encentrum sp. nov. C	8		2220–2753	3			3		4.9–8.3	7–26	3.2–11.9
Encentrum sp. 1	1		2753	1					5.4	8	3.2
Encentrum sp. 2	2		1890–2420	2		2			5.3–6.4	5–10	8.5–10.5
Parencentrum plicatum (Eyferth)		1	1500			4			6.4	–	23.3
Wierzejskiella velox (Wiszniewski)	3		2505–2630	3					6.8–7.9	6–16	6.4–8.2

alulae, and 2 stout apical rami teeth. Epipharynx 2 soft pinnate plates (Figure 12). Triangular prominences on the antero-lateral rami edges, as observed by Wulfert (1960), Donner (1954; *D. lütkeni* var.?), and De Smet (1997), were not seen in the alpine specimens, which in this respect agree with the drawings of *D. luetkeni*-trophi by Harring and Myers (1928), Wiszniewski (1934), Hauer (1958), Koste (1965), and Sharma (1979). Number of shearing teeth varying within and between populations; the following tooth formulae were found: 5/6, 6/6, 7/6, 7/7 (d/s). A thin lamellar rib extends along the inner rami margins to the apical rami teeth. Measurements of toe length and trophi dimensions in the upper range of hitherto reported values.

Dimensions (μm): $n=7$; body length 95–125 (contracted); toes 57–63; claw 20–22; trophus 33–42; ramus 19–22; alulae 3–4.5; fulcrum 11–12; uncus 16–18; manubrium 32–33.

Occurrence and ecology: Widespread in the Central Alps, occurring on mineral substrate in ultra-oligotrophic lakes and pools, on sand and sludge in eutrophic pools in alpine pasture land, as well as on sludge and among *Sphagnum* in acid bog ponds (Hohe Tauern, 1893–2505 m a.s.l.). In the Northeastern Limestone Alps restricted to acid bog ponds (1290 m a.s.l.). Seems to prefer acid to neutral soft water. Feeds on diatoms and rotifers (*Lecane*, *Encentrum*, Bdelloidea). Eurytopic cosmopolite.

Previous records from the Alps: First record in the alpine region, but previously found among macrophytes (*Menyanthes trifoliata*, *Nymphaea alba*) in an inner-alpine lake (Piburger See, 915 m a.s.l., Austria) (Gabriel, 1981).

Records from other palaearctic mountain ranges: Reported from two localities (ca. 1200 m a.s.l.) in the Massif Central, France (Francez, 1981) and from the Massif du Canigou, Pyrénées Orientales, France (De Smet & Pourriot, 1997).

Encentrum diglandula (Zavadowski, 1926) (Figures 16–20)

(*E. felis* (O.F.M.) after Jersabek (1995))

Material examined: Specimens from three ± eutrophic pools in subalpine and alpine pasture land from Central Alps and Limestone Alps.

The species complex *E. diglandula-felis-glaucum* still confronts rotiferologists with problems, as transitional trophi characters seem to exist between these closely related species. In the present material the manubria are slightly expanded anteriorly (Figure 17), as it is typical of *E. felis* (O.F.M.), which, however, has posteriorly crutched manubria. From this species it further differs by its relatively longer fulcrum and intramallei. Drawings of *E. diglandula*-manubria with expanded heads were also found in Donner's notebooks. Koste & Shiel (1986) reported a Tasmanian *E.* cf. *diglandula* with posteriorly crutched manubria. Another character that seems to be variable is the dorsal view of the fulcrum, which may show a proximal expansion, as found in several alpine specimens (Figure 16) (but also normal rod-shaped ones occurred) and was figured for *E. diglandula* by Wiszniewski (1934) and Donner (1978; *E. felis-diglandula*), as well as by Althaus (1957) and Koste (1976) for *E. glaucum* Wulfert and by Harring & Myers (1928) for the closely related marine species *E. villosum* H. & M. Structures which have not been reported to date are small rod-shaped processes ventrally on the head of the manubria, and very soft supramanubria inserted ventrally on the intramallei (Figure 17). Digestive tract filled with zoochlorellae. Toes slightly bulbous at their base and curved ventrally (Figure 19). They may appear straight (Figure 18) or slightly curved outwardly (Figure 20) in dorsal/ventral view.

Dimensions (μm): $n=6$; total length 66–91 (contracted); toes 15–17.5; trophus 31–33; ramus 13.5–14.5; fulcrum 6.5–8; uncus 5.5–6; manubrium 23–25; intramalleus 4; supramanubrium 4–4.5.

Occurrence and ecology: Central Alps (Hohe Tauern, northern flank): on sand in pool (2270 m a.s.l.) in alpine pasture land; Northeastern Limestone Alps (Sengsengebirge, northern flank): on sludge and among macrophytes (*Callitriche* sp.) of two pools (1410 m a.s.l.) in subalpine pasture land. Known from diverse lotic and lentic habitats from the Palaearctis, Orientalis and Notogaea.

Previous records from the Alps: First record in the Alps.

Records from other palaearctic mountain ranges: none.

Encentrum incisum Wulfert, 1936 (Figures 21–24)

Material examined: Two specimens from a subalpine limestone pool.

Characteristic by form of toes, which are bulbous in their proximal part, continue parallel-sided and taper into blunt tips in their distal 1/3 (Figure 24). Deviations from the nominal species were noted by

68

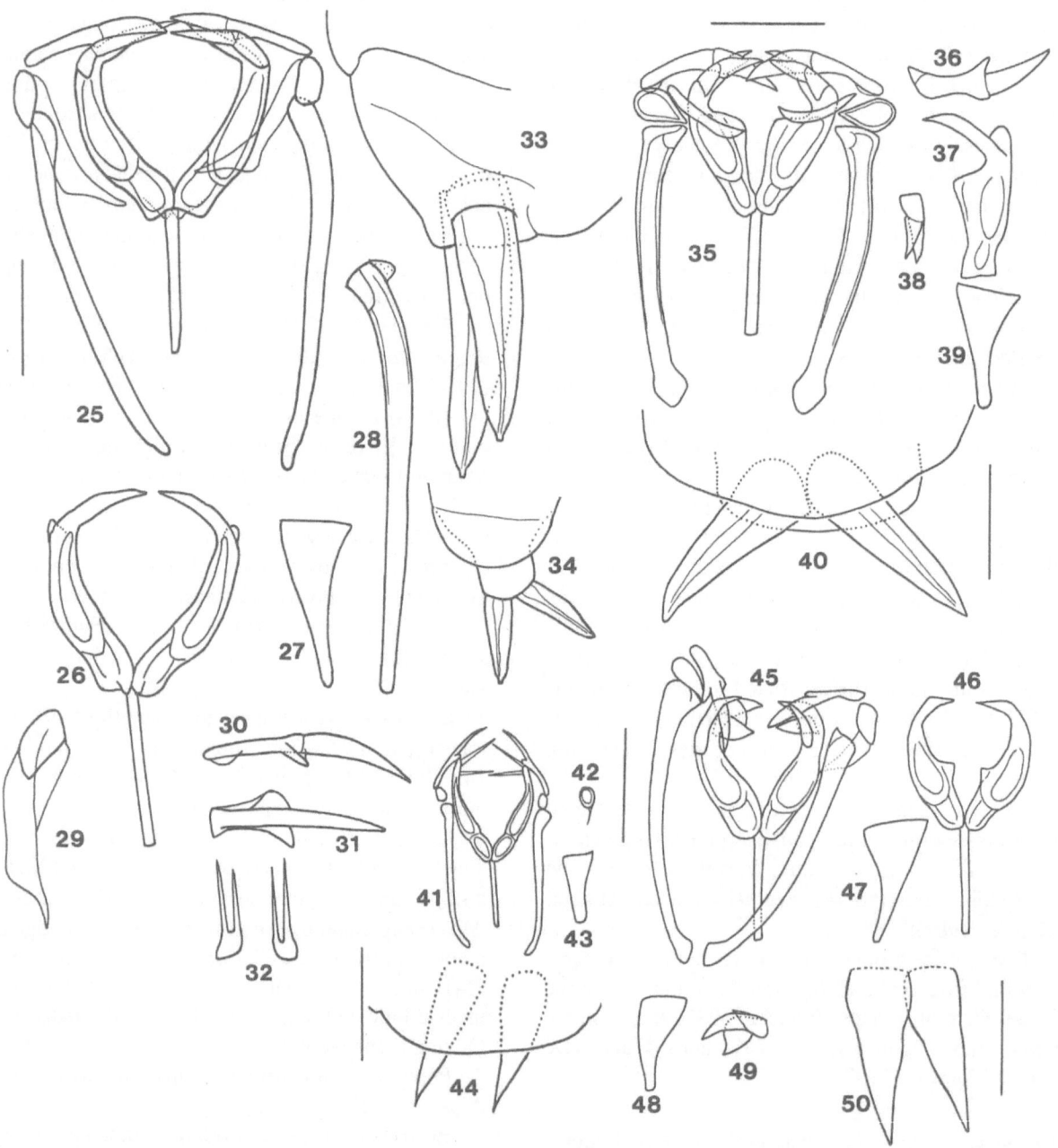

Figures 25–34. Encentrum lutra Wulfert – 25: trophus, ventral view; 26: incus of (25), dorsal view; 27: fulcrum, lateral view; 28: manubrium, lateral view; 29: intramalleus with supramanubrium; 30: uncus, lateral view; 31: uncus, apical view; 32: preuncinal teeth, apical view; 33: posterior end of contracted specimen with toes, lateral view; 34: foot with toes, dorsal view. *Figures 35–40. Encentrum mucronatum* Wulfert – 35: trophus, ventral view; 36: uncus, apical view; 37: ramus, ventral view; 38: preuncinal teeth; 39: fulcrum, lateral view; 40: posterior end of contracted specimen with toes, dorsal view. *Figures 41–44. Encentrum* sp. 1 – 41: trophus, ventral view; 42: intramalleus with supramanubrium, apical view; 43: fulcrum, lateral view; 44: posterior end of contracted specimen with toes, dorsal view. *Figures 45–50. Encentrum* cf. *putorius* var. *armatum* Donner – 45: trophus, ventral view; 46: incus of other specimen, ventral view; 47: fulcrum of (46), lateral view; 48: fulcrum of (45), lateral view; 49: preuncinal teeth, ventral view; 50: toes, ventral view. Scale bars: 10 μm.

the presence of a second pair of preuncinal teeth (Figure 23). However, these structures were frequently overlooked in *Encentrum* by elder rotiferologists. A second pair of preuncinal teeth was also reported for *E. incisum* from the river Salzach, Austria (Donner, 1970) and for Belgian animals (De Smet & Pourriot, 1997). *Encentrum oxyodon* Wulfert has similar trophi with 2 preuncinal teeth, but clearly differs by the more slender rami and the different form of its conical, shorter and acutely pointed toes.

Dimensions (μm): *n*=2; total length 142–150 ± contracted); toes 13.5–14; trophus 18–18.5; ramus 7; fulcrum 6.5; uncus 7.5; manubrium 17–18; intramalleus 1.3; supramanubrium 2.5.

Occurrence and ecology: Northeastern Limestone Alps (Sengsengebirge, northern flank); on sludge in productive pool below timber line (1370 m a.s.l.), slightly auxotrophicated by game. Known from diverse aquatic and semi-aquatic environments in Central Europe.

Previous records in the Alps: First record in the alpine region, but reported from the same geographic area at lower altitudes in the river Salzach, province of Salzburg (Donner, 1970).

Records from other palaearctic mountain ranges: none.

Encentrum lutra Wulfert, 1936 (Figures 25–34)

(Encentrum putorius Wulfert after Jersabek (1995))
Material examined: Several specimens from 7 subalpine and alpine localities.

Differs from the original description by the form of its toes, which are not acutely pointed, but terminate in blunt, retractable (?), short tubules (Figure 33). In this respect resembling *E. putorius* Wulfert. From this species, however, it is quite distinct by its trophi characteristics and the narrow foot segment (Figure 34).

Dimensions (μm): *n*=8; total length 160–225 (contracted) to 280 (± stretched); toes 20.5–29; trophus 32–45; ramus 14–17.5; fulcrum 12–15; uncus 12–17.5; manubrium 29–38; intramalleus 4–5; supramanubrium 12–16.5; outer/inner preuncinal teeth 9/8.

Occurrence and ecology: Central Alps (Hohe Tauern, southern flank): fairly common on mineral bottom sediments of glacier margin lakes, ultra-oligotrophic lakes and pools above timber line. Also encountered among submersed mosses of temporary puddle (2520 m a.s.l.), and among aggregations of unicellular desmids (*Cylindrocystis* sp.) of permanent

pools (2640 to 2753 m a.s.l.). Northeastern Limestone Alps (Totes Gebirge): among macrophytes (*Chara contraria, Ranunculus eradicatus*) in productive karst lake. Frequently observed to feed on diatoms. Known from diverse aquatic and semi-aquatic environments; Palaearctis, Neotropis.

Previous records in the Alps: First record in the alpine region, but known from the same geographic area at lower altitudes (Donner, 1970).

Records from other palaearctic mountain ranges: none.

Encentrum mucronatum Wulfert, 1936 (Figures 35–40)

Material examined: Two specimens from ultra-oligotrophic lakes, three specimens from ultra-oligotrophic pool.

Trophi morphology in close agreement with that of arctic (De Smet & Pourriot, 1997) and alpine (Donner, 1970) animals.

Dimensions (μm): *n*=3; total length 165–175 (contracted); toes 19; trophus 26; ramus 15; fulcrum 9.5; uncus 12.5; manubrium 24; intramalleus 5; supramanubrium 7.

Occurrence and ecology: Central Alps (Hohe Tauern); very low densities on fine sand in ultra-oligotrophic lakes on southern (2505 m a.s.l.) and northern flank (2420 m a.s.l.), and among aggregations of unicellular desmids (*Cylindrocystis* sp.) in permanent, ultra-oligotrophic pool (2640 m a.s.l.) on southern flank. Known from diverse aquatic and semi-aquatic environments; cosmopolite.

Previous records in the Alps: River Salzach, 1425 m a.s.l., in littoral moss, northern flank of Central Alps, Austria (Donner, 1970).

Records from other palaearctic mountain ranges: none.

Encentrum cf. *putorius* var. *armatum* Donner, 1943 (Figures 45–50)

Material examined: Five specimens from three localities.

Trophi similar to that of an ambiguous relative of *Encentrum putorius* Wulfert, described as var. *armatum* by Donner (1943), with one preuncinal tooth strongly broadened, blade shaped (l/w 4/2.5 μm) (Figure 49). Fulcrum not hook-shaped terminally. This character seems to be variable in *E. putorius* var. *armatum* (cf. Donner, 1964). One specimen with triangular expansion on the posterior part of inner rami

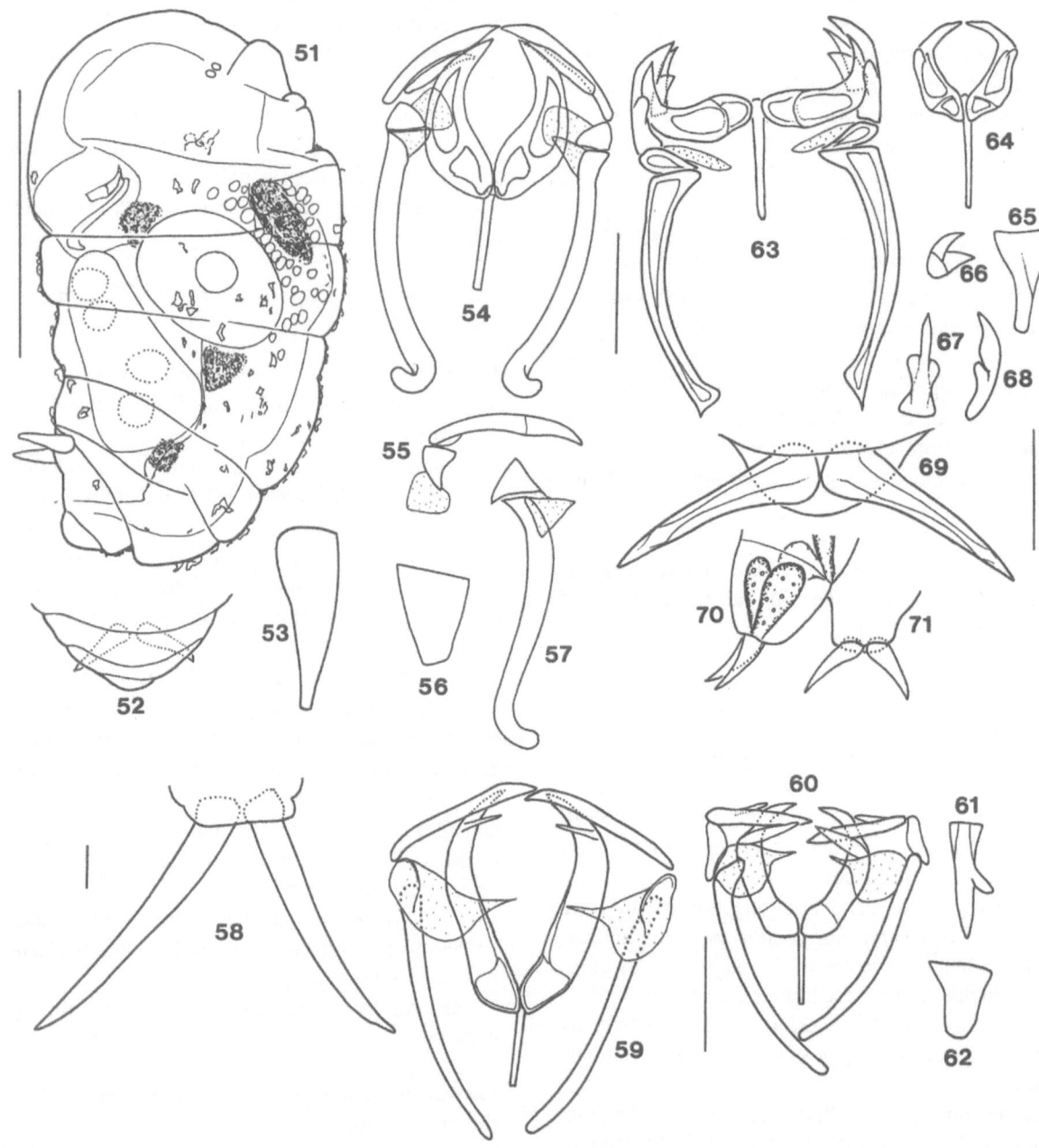

Figures 51–57. Encentrum saundersiae (Hudson) – 51: contracted specimen, lateral view; 52: contracted hind body and toes, dorsal view; 53: toe, lateral view; 54: trophus, ventral view; 55: uncus, intramalleus and supramanubrium; 56: fulcrum, lateral view; 57: manubriun and intramalleus, lateral view. *Figures 58–59. Encentrum uncinatum* (Milne) – 58: toes, ventral view; 59: trophus, ventral view. *Figures 60–62. Encentrum* cf. *uncinatum* – 60: trophus, ventral view; 61: uncus, apical view; 62: fulcrum, lateral view (58–60, and 61,62 from different localities). *Figures 63–71. Encentrum* sp. 2 – 63: opened trophus, dorsal view; 64: incus, ventral view; 65: fulcrum, lateral view; 66: preuncinal teeth; 67, 68: uncus, lateral and apical view, resp.; 69: toes, ventral view; 70: contracted foot and toes, lateral view; 71: ditto, dorsal view; (63,65,71) and (64,66,69,70) from different lake specimens; (67,68) from bog specimen. Scale bars: habitus 50 μm; trophi, toes 10 μm.

margins, similar to *E. parvum* Donner (Figure 46), but in any other respect agreeing with trophi of other specimens. Toes acutely pointed, without indications of retractable tips (Figure 50). Dimensions distinctly smaller than in nominal variety.

Dimensions (μm): *n*=5; total length 88–130 (contracted); toes 15–21; trophus 27–28; ramus 11.5–12; fulcrum 8–10; uncus 9; manubrium 22.5–25; intramalleus 4; supramanubrium 5; preuncinal teeth 4–5.5.

Occurrence and ecology: Central Alps (Hohe Tauern); on mineral bottom of three ultra-oligotrophic pools on southern flank (2560–2753 m a.s.l.). *Encentrum putorius* var. *armatum* was previously known only from organic substrates of running waters in Central Europe.

Previous records in the Alps: First record for the alpine region. The nominal variety was hitherto recorded only from running waters at lower altitudes (Donner, 1964; 1970).

Records from other palaearctic mountain ranges: none.

Encentrum saundersiae (Hudson, 1885) (Figures 51–57)

Material examined: Two specimens from a subalpine pool.

Very conspicuous even in preserved samples (Figure 51). Trunk strongly plicated by 5 transversal folds, tail prominent. Two light refracting eyespots without pigment. Integument sticky, covered with detritus and sand grains. Digestive tract with zoochlorellae and brown defecation pellets. Vitellarium with 4 nuclei. Trophi normal. Preuncinal teeth only loosely attached to ventral side of ramus (in Figure 54 slipped aside with unci). Left uncus and preuncinal tooth slightly shorter than right ones (Figure 54).

Dimensions (μm): *n*=2; total length 98–124 (contracted) to > 200 (stretched); toes 18; trophus 28–31; ramus 14–15; fulcrum 8; uncus 11–13; manubrium 21–22; intramalleus 2.5; supramanubrium 3.5–4; preuncinal teeth d/s 6.5/7.5

Occurrence and ecology: Northeastern Limestone Alps (Sengsengebirge, northern flank); sporadically on sludge of permanent pool in subalpine pasture land (1410 m a.s.l.); cosmopolite.

Previous records in the Alps: First record for the Alps.

Records from other palaearctic mountain ranges: Reported from the Massif Central (1206 m a.s.l.),

France (Francez, 1981), and from the Pyrenées Orientales, France (De Smet, pers. comm.).

Encentrum uncinatum (Milne, 1886) + *E.* cf. *uncinatum* (Figures 58–62)

Material examined: Three specimens each from a high altitude lake and a pool in subalpine pasture land. Toes very long, slightly curved outwardly in ventral view, with tips not or only indistinctly offset (Figure 58). Ciliary tufts on both sides of broad rostrum. Vitellarium with 8 nuclei. No eyes. Trophi in general agreement with previously described specimens, but apical ramus tooth relatively stout. L:W ratio of closed rami outline ca. 1.6 (Figure 59). Intramallei form a close unit with the supramanubria, from which they are difficult to separate. In the specimens from the alpine lake and in one specimen from the subalpine pool, however, considerable deviations occurred (Figures 60–62). Here, the rami were less elongate (l:w ratio of closed rami outline ca. 1.2), the preuncinal teeth stronger, and the unci shorter, with the tooth being distinctly longer than the shaft (Figure 60), or of similar length (Figure 61). The intramallei were distinct and separable from the supramanubria, which had drawn-out, sharply pointed tips as in the typical form. Except smaller dimensions no differences between the deviating and the typical morph were observed with respect to the external appearance.

Donner (1978) mentioned a similarly high variability in trophi morphology in a population of *E. uncinatum* in cut-off waters of the Austrian River Danube. Here, in one specimen the preuncinal teeth were even more pronounced and strongly broadened, the l:w ratio of the outline of the closed, stout rami was ca. 1.2. A somewhat intermediate case was reported for *E. uncinatum* from the South Shetlands (Antarctica) by José de Paggi (1982). In her specimen figured the l:w ratio of the rami outline was ca. 1.4. In addition to the shorter rami the antarctic animals differed by the form of the fulcrum and the absence of the drawn-out tips on the supramanubria. It is questionable, whether these 'deviating morphs' belong to *E. uncinatum*.

Dimensions (μm) *n*=3 (values in brackets refer to dimensions of deviating morph; *n*=2): body length 95 (77–129) (contracted); toes 68 (54–59.5); trophus 31 (24–26); ramus 19 (11.5–16); fulcrum 6.5 (6–6.5); uncus 14 (9–9.5); manubrium 23 (18.5–20); intramalleus 5; supramanubria 10 (7.5).

Occurrence and ecology: Central Alps (Hohe Tauern, southern flank): on mineral bottom of ultra-

72

oligotrophic lake (2630 m a.s.l.). Northeastern Limestone Alps (Sengsengebirge, northern flank): on sludge of permanent pool in subalpine pasture land (1340 m a.s.l.). Observed to feed on diatoms. Common species in diverse lentic and lotic environments; cosmopolite.

Previous records in the Alps: First record in the alpine region, but reported from the same geographic area at lower altitudes (Donner, 1954; 1964; 1970; Schmid-Araya, 1993).

Records from other palaearctic mountain ranges: none.

Encentrum walterkostei Jersabek, 1994 (Figures 72–82)

Material examined: Several specimens from the type locality.

The species is conspicuous by its long sword-shaped toes, a caudal papilla between the toes and the strongly plicated body, as well as only 4 nuclei in the vitellarium (Figures 72, 73). The description of Jersabek (1994) is amended by observations on living animals: Head and trunk of almost the same width. Head with transverse fold, tilted ventrally in swimming animal. Corona strongly oblique, almost ventral. Rostrum prominent, broadly rounded anteriorly, not decurved ventrally. Dorsal antenna on posterior of brain, opening behind neck fold. Trunk with distinct longitudinal folds extending obliquely from the anterior ventral side to the posterior dorsal side. Posterior part of trunk pseudosegmented by 2 transversal folds. Head organs without pigment or eyespots. A pair of small, rounded salivary glands terminally on mastax. Gastric glands big ellipsoid, unstalked. Tail small. Foot-pseudosegment short-conical, distinctly narrower than trunk. Toes permanently kept in horizontal position in swimming animals (Figure 73).

Dimensions (μm): n=20; body length 51–77 (contracted) to 102 (swimming); toes 22.5–25; caudal papilla 3.5–4; trophus 15–15.5; ramus 7–7.5; fulcrum 5.5; uncus 5.5–6; manubrium 14–14.5; intramalleus 2.5; supramanubrium 5.

Occurrence and ecology: In the psammolittoral of 'Unterer Schwarzhornsee' (2543 m a.s.l.), Hohe Tauern, Central Alps.

Previous records in the Alps: Known from the type locality only.

Encentrum sp. 1 (Figures 41–44)

Material examined: Single specimen from ultra-oligotrophic pool.

The minute trophus superficially resembles that of *Encentrum mustela* (Milne), but differs in several respects. The rami are more slender and elongate, resembling closely those of *E. acrodon* Wulfert. From this species, however, it is quite distinct by the presence of apical rami teeth and the position of the teeth on the inner rami margins (preuncinal teeth?). These teeth are inserted subapically of the blunt rami tips in the present specimen (Figure 41). They point directly inwards with their tips, almost touching in the closed trophus. Apical rami teeth, preuncinal teeth and unci teeth are of about the same length, as are uncus-tooth and uncus-shaft. This is in contrast to the findings of Wulfert (1936), Donner (1978), Dartnall & Hollowday (1985), De Smet (1993), Dartnall (1995) and De Smet & Pourriot (1997) for *E. mustela*, in which the uncus-tooth was found to be distinctly longer than the uncus-shaft, as well as the rami teeth. However, they were shorter in Austrian specimens from soil samples (Donner, 1952). The supramanubria are particularly delicate in the alpine specimen and were not seen in ventral view; they were also not reported for *E. mustela* by Wulfert (1936), Donner (1978), Dartnall & Hollowday (1985) and De Smet (1993), but were found to be very conspicuous in Dutch and Canadian animals (De Smet & Pourriot, 1997), in specimens from sub-Antarctic Heard Island (Dartnall, 1995), as well as in Donner's soil sample (op.cit.). Further differences are found in the more slender fulcrum (lateral view) (Figure 43), and the posterior end of the manubria, which are slightly crutched in *E. mustela*, and the smaller trophi dimensions. Intestine filled with large amounts of small globules of 1×1.5 μm (zoochlorellae?). A similar trophus was described for *E. insolitum* Myers, which, however, differs markedly by its shorter fulcrum, shorter rami teeth, longer uncus tooth, as well as the different form of toes (Figure 44). Most likely, the alpine specimen is a still undescribed species.

Dimensions (μm): n=1; total length 120 (contracted); toes 12; trophus 17; ramus 12 (incl. apical tooth); fulcrum 6; uncus 7; manubrium 12; intramalleus 1.5; supramanubrium 1.3.

Occurrence and ecology: Central Alps (Hohe Tauern); on fine sand in ultra-oligotrophic pool on southern flank (2753 m a.s.l.). Feeds on small diatoms.

73

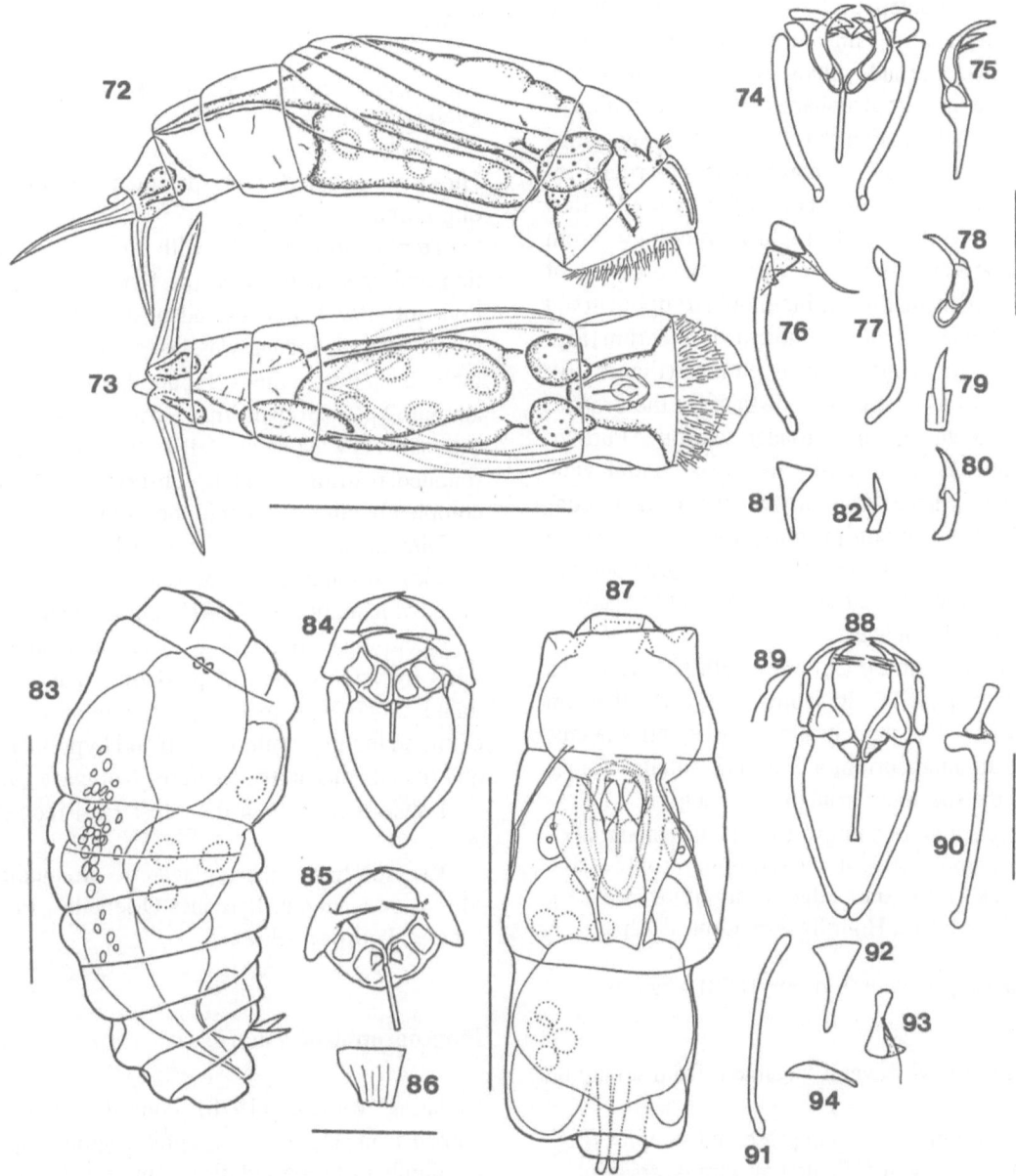

Figures 72–91. Encentrum walterkostei Jersabek – 72: swimming animal, lateral view; 73: ditto, ventral view; 74: trophus, dorsal view; 75: incus, lateral view; 76: manubrium, intramalleus and supramanubrium; 77: manubrium, lateral view; 78: ramus, ventral view; 79, 80: uncus, apical and lateral view resp.; 81: fulcrum, lateral view; 82: preuncinal-teeth. *Figures 83–86. Parencentrum plicatum* (Eyferth) – 83: contracted specimen, lateral view; 84: trophus, dorsal view; 85: incus, ventral view; 86: fulcrum, lateral view. *Figures 87–94. Wierzejskiella velox* (Wiszniewski) – 87: contracted specimen, ventral view; 88: trophus, ventral view; 89: apical ramus tooth; 90: manubrium and intramalleus, lateral view; 91: manubrium, dorsal view; 92: fulcrum, lateral view; 93: intramalleus with bristle-like appendage; 94: uncus, lateral view. Figures 74–82 from Jersabek (1994). Scale bars: habitus 50 μm; trophi 10 μm.

Encentrum sp. 2 (Figures 63–71)

Material examined: Three specimens from ultra-oligotrophic lake, one specimen from bog pond. This rare species could not be related to any known congener. It is characteristic by relatively robust trophi with a ± hexagonal rami outline (Figures 63, 64), 2 pairs of very broad, almost triangular preuncinal teeth (Figure 66), droplet-shaped intramallei with narrow supramanubria (Figure 63), as well as the shape of its manubria. Thus, it shows some superficial resemblance to *E. martoides* Fott and *E. sorex* Wulfert with respect to its overall trophi morphology. From the first species it differs, however, by the presence of 2 pairs of preuncinal teeth, the more strongly curved manubria, the lateral view of the broader fulcrum (Figure 65) and the ventrally decurved toes (Figure 70). From *E. sorex* it seems to be distinct by the broader preuncinal teeth and the broader fulcrum. Further, the toes are curved inwardly in dorsal/ventral view (Figure 69), whereas they are outcurved in its congener, which is also conspicuous by a large tail and a dorsal papilla between the toes. The measurements are within or slightly above the values given for *E. sorex*. Vitellarium with 4 nuclei.

Dimensions (μm): *n*=3; body length 104–123; toes 12.5–17.5; trophus 20–30.5; incus 13.5–18; fulcrum 7–8.5; uncus 7–8.5; manubrium 14.5–19; intramalleus 3.5–4.5; supramanubrium 4.5–5. The smaller values belong to the specimen from the peat bog.

Occurrence and ecology: In very low abundance in the psammolittoral of an ultra-oligotrophic lake (2543 m a.s.l.) and on sludge of an alpine peat-bog pond (1890 m a.s.l.), Hohe Tauern, Central Alps.

Parencentrum plicatum (Eyferth, 1878) (Figures 83–86)

Material examined: Several specimens from subalpine pool.

Very conspicuous even in preserved samples (Figure 83). Trunk strongly plicated by 7 transversal folds. Tail prominent. Two light refracting eyespots without pigment. Digestive tract filled with zoochlorellae. Vitellarium with 4 nuclei. Trophi normal, very small.

Dimensions (μm): *n*=2; total length 104 (contracted); toes 12; trophus 20–20.5; ramus 6; fulcrum 4; uncus 9.5; subuncus 4.5; manubrium 13.

Occurrence and ecology: Northeastern Limestone Alps (Koppenalm, Sengsengebirge, southern flank); abundant on loamy sludge of a temporary pool in subalpine pasture land (1500 m a.s.l.); pH 6.4, T

23.3 °C. Eurytopic species, widespread throughout the Holarctis.

Previous records in the Alps: first record for the Alps.

Records from other palaearctic mountain ranges: none.

Wierzejskiella velox (Wiszniewski, 1932) (Figures 87–94)

Material examined: Five specimens from three ultra-oligotrophic lakes.

Trophi agree largely with the original description and specimens from the Pyrenees (De Smet & Pourriot, 1997), but instead of a supramanubrium a bristle-like appendage was present on the outer margin of each intramalleus (Figure 93). In the contracted animals 2 pairs of fusiform gastric glands and 1 pair of oval salivary glands, as well as the retracted, broadly rounded rostrum could be observed. Foot and toes completely retracted within the trunk (Figure 87).

Dimensions (μm): *n*=5; total length 85–90 (contracted); trophus 20–22.5; ramus 9–10; fulcrum 7; uncus 5; manubrium 14.5–15; intramalleus 5.

Occurrence and ecology: In littoral sand of three lakes from the Central Alps (Hohe Tauern, southern flank; 2505–2630 m a.s.l.). Psammobiontic species common in the psammolittoral and hyporheal of lakes and rivers; Palaearctis, Nearctis, Notogaea, Aethiopis.

Previous records in the Alps: First record for the Alps.

Records from other palaearctic mountain ranges: Massif du Canigou, Pyrenées Orientales, France (De Smet & Pourriot, 1997).

Biogeographical remarks

Updating Bērziņš (1978) compilation of rotifers known from the zoogeographic region 'Alps' raises the number of reported dicranophorids from 23 to 44. Out of these, however, only 21 are presently known to occur in mountainous regions above 1000 m a.s.l., and 15 were reported above timber line. Only the latter may be regarded as alpine fauna *sensu stricto*.

From the data and references available to date, it seems that the rotifer fauna from high altitudes is composed largely of widely distributed species with generalist habitat requirements and a good ability for dispersal. Yet, the biogeography of the dicranophorids found in the Alps provides partial support for this idea, as several species are cosmopolites (*Dicranophorus forcipatus*, *D. luetkeni*, *Encentrum saundersiae*, *E. uncinatum*), or show a wider range than the Palae- or Holarctis (*Dicranophorus hercules*, *Encentrum diglandula*, *E. lutra*, *E. mucronatum*, *E. putorius*, *Wierzejskiella velox*). A holarctic range of occurrence is known for *Aspelta lestes* and *Parencentrum plicatum*, whereas a more restricted distribution may be suspected in only two species, with *Encentrum incisum* and *E. walterkostei* being presently known only from Central Europe and the Alps, respectively. On the other hand, considering the fact that five new species were recently discovered in high altitude lakes of the Central Alps (Jersabek, 1994; in press), the question whether endemicity occurs in the Alps may be picked up. All the more so, since Central Europe belongs to the best investigated regions in the world, and highly distinctive local biota occur in the Alps. But more information on distribution and ecology of alpine Rotifera is required to seriously discuss the matter of alpine endemicity.

Although there are single records of some widespread species, the species-rich genus *Encentrum* is almost absent on faunal lists of the Neotropis (Koste & José de Paggi, 1982; José de Paggi & Koste, 1995) and the Aethiopis (De Ridder, 1986, 1991), but it seems that surveying comparable habitats, e.g. waters from the alpine region of tropical mountains (cf. De Smet & Bafort, 1990; Jersabek, unpubl.) will considerably extend the known ranges of these predominantly cold adapted rotifers.

Still, the interpretation of the chorology of the Dicranophoridae should be done with caution, as in some cases the limited distribution of a species is likely to reflect rather that it belongs to a difficult, and hence frequently neglected, taxonomic group, than its rarity. Conversely, as not all reports are sufficiently documented, and hence checkable, insufficient taxonomic resolution may be another 'source of dispersal' of species with an actually more restricted distribution.

Ecological remarks

A common feature of species encountered in alpine waters is, that they either show a clear preference of low temperatures, or have a wide range of thermal tolerance. Notably the genus *Encentrum* is common in cold habitats, where it inhabits even sea ice (Chengalath, 1985) and depressions in the surface of glaciers (cryoconite holes) (De Smet & Van Rompu,

1994). Eurythermic species which occur in alpine and tropical waters as well are *Dicranophorus forcipatus*, *D. luetkeni*, *Encentrum saundersiae* and *E. uncinatum*.

Several species are particularly adapted to exist in the interstitial of mineral sediments, which is a frequent habitat in ultra-oligotrophic waters in alpine areas. Of the species recorded for the Alps several are common in the hygropsammon of lakes, and were regarded as psammobiontic (*Dicranophorus hercules*, *Encentrum diglandula*, *Wierzejskiella velox*), as psammophilic (*D. luetkeni*), or as psammoxenic (*D. forcipatus*) by Wiszniewski (1934). Species recorded from stagnant alpine waters (this paper) and the hyporheic interstitial of an Austrian mountain gravel stream as well (Schmid-Araya, 1993), are *D. luetkeni*, *E. incisum*, *E. mucronatum*, and *E. uncinatum*. Probably confined to the hygropsammon of alpine lakes is *E. walterkostei*. However, in a recent study on the habitat choice of dicranophorids, Pejler and Bērziņš (1993) included seven of the above mentioned species, but only one of them, the eurytopic *D. luetkeni*, was found on mineral bottoms as well. Generally, all species were found on a variety of periphytic substrates and the authors concluded, that the low substrate specifity, even of species previously regarded as 'psammobionts', might be a consequence of the high agility of these raptorial predators. The population densities of species inhabiting fine-grained mineral sediments, such as glacial flour and very fine sand may, however, remain extremely low throughout the year (Jersabek, unpubl.), due to the low carrying capacity of such habitats. Such species may easily be missing even in thouroughly prepared inventories.

The severity of the physical environment, the rather young age of water bodies originating from Pleistocene glaciations, along with a decreasing habitat complexity, cause a decrease of species-richness with increasing altitude (Jersabek, 1995). Analogously, the rather low species richness and diversity in arctic waters may be explained (De Smet, 1993; De Smet & Beyens, 1995). But not all taxocoenoses are equally affected by the changing conditions in the alpine region, which normally results in distinct shifts in the relative composition of rotifer communities. Considering the family Dicranophoridae, an increase of its relative importance with increasing altitude is obvious in the present material, as exemplified most clearly in the genus *Encentrum*, which even increased its species number from the subalpine to the alpine region.

76

Conclusions

Although the species richness of more than 180 aquatic rotifers hitherto recorded from mountainous regions of the Alps surpasses what was to be expected from earlier studies, I still refuse to consider it representative of this area. Without doubt our present knowledge on alpine rotifers is incomplete in several ways: (1) with respect to its extremely diversified climatic and geological conditions, only a negligible part of the Alps has been subjected to systematic study of its aquatic biota; (2) some environments, notably those in glaciated areas, have hitherto been almost totally neglected; (3) results of local surveys may be scattered through inaccessible reports and collections; (4) further fieldwork will produce additional records and very probably even new species. Moreover, it is difficult to determine boundaries. Thus, we have to expect that some of the taxa currently listed only from alpine sites will finally prove to be more widely distributed.

Acknowledgements

I thank Drs Walter Koste (Quakenbrück) and Willem De Smet (Antwerp) for valuable comments on earlier versions of this work. Dr Hendrik Segers (Ghent) is acknowledged for providing difficult accessible literature. Primary financial support was provided by the 'Forschungsinstitut Gastein-Tauernregion', while secondary funds came from the 'Stiftungs- und Förderungsgesellschaft der Paris-Lodron-Universität Salzburg' and the 'Bundesministerium für Wissenschaft und Verkehr'.

References

Althaus, B., 1957. Faunistisch-ökologische Studien an Rotatorien salzhaltiger Gewässer Mitteldeutschlands. Wiss. Z. Univ. Halle, Math.-Nat. VI/1: 117–158.

Bērziņš, B., 1978. Rotatoria. In J. Illies (ed.), Limnofauna europaea. G. Fischer Verlag, Stuttgart: 54–91.

Chengalath, R., 1985. The Rotifera of the Canadian Arctic sea ice, with description of a new species. Can. J. Zool. 63: 2212–2218.

Coussement, M. & H. J. Dumont, 1980. Some peculiar elements in the rotifer fauna of the atlantic Sahara and of the Atlas Mountains. Hydrobiologia 73: 249–254.

Dalla-Torre, K.W., 1889. Studien über die mikroskopische Thierwelt Tirols, I. Theil: Rotatoria. Zeitschrift Ferdinandeum, Innsbruck 33: 239–252.

Dartnall, H. J. G., 1995. The rotifers of Heard Island: Preliminary survey, with notes on other freshwater groups. Pap. Proc. R. Soc. Tasm. 129: 7–15.

Dartnall, H. J. G. & E. D. Hollowday, 1985. Antarctic rotifers. Brit. Antarctic Surv. Sci. Rep. 100: 1–46.

De Ridder, M., 1964. Enkele raderdiertjes uit de Spaanse Pyreneen. Biol. Jb. Dodonaea 32: 185–201.

De Ridder, M., 1986. Annotated checklist of non-marine rotifers (Rotifera) from african inland waters. Koninklijk Museum voor Midden-Afrika Tervuren, Belgie. Zoologische Dokumentatie, 123 pp.

De Ridder, M., 1991. Additions to the 'Annotated checklist of non-marine rotifers from african inland waters'. Rev. Hydrobiol. trop. 24: 25–46.

De Smet, W. H., 1993. Report on rotifers from Barentsøya, Svalbard (78°30′ N). Fauna norv. Ser. A 14: 1–26.

De Smet, W. H. & J. M. Bafort, 1990. Rotifers from the Kilimanjaro. Biol. Jb. Dodonaea 58: 120–130.

De Smet, W. H. & L. Beyens, 1995. Rotifers from the Canadian High Arctic (Devon Island, Northwest Territories). Hydrobiologia 313/314: 29–34.

De Smet, W. H. & R. Pourriot, 1997. Rotifera 5. Dicranophoridae and Ituridae (Monogononta). Guides to the identification of the microinvertebrates of the continental waters of the world 12. SPB Academic Publishing, The Hague, The Netherlands, 344 pp.

De Smet, W. H. & E. A. Van Rompu, 1994. Rotifera and Tardigrada from some cryoconite holes on a Spitsbergen (Svalbard) glacier. Belg. J. Zool. 124: 27–37.

Donner, J., 1943. Zur Rotatorienfauna Südmährens. II. Zool. Anz. 143: 63–75.

Donner, J., 1952. Bodenrotatorien im Winter. Mikrokosmos 42: 29–33.

Donner, J., 1954. Rotatoria. In: Franz H.: Die Nordostalpen im Spiegel ihrer Landtierwelt. Univ. Verlag Wagner, Innsbruck I. Bd., 136–157.

Donner, J., 1964. Die Rotatorien-Synusien submerser Makrophyten der Donau bei Wien und mehrerer Alpenbäche. Arch. Hydrobiol./Suppl. 27: 227–324.

Donner, J., 1970. Die Rädertierbestände submerser Moose der Salzach und anderer Wasser-Biotope des Flußgebietes. Arch. Hydrobiol./Suppl. 36: 109–254.

Donner, J., 1978. Material zur saprobiologischen Beurteilung mehrerer Gewässer des Donau-Systems bei Wallsee und in der Lobau, Österreich, mit besonderer Berücksichtigung der litoralen Rotatorien. Arch. Hydrobiol., Suppl. 52: 117–228.

Fott, J., 1960. Zwei Encentrum-Arten (Rotatoria) aus der Hohen Tatra mit Beschreibung von Encentrum martoides n.sp. Vest. Ceskosl. zool. spolec. 24: 175–182.

Francez, A. J., 1981. Rotiferes de quelques tourbieres d'Auvergne. Annls. Stn. Biol. Besse-en-Chandesse 15: 276–287.

Gabriel, A., 1981. Die Rotatorienfauna der Seggen-, Fieberklee- und Schilfbestände des Piburger Sees (Ötztal, Tirol). Master Thesis, Univ. Innsbruck, Austria, 148 pp.

Godeanu, S., 1963a. Neue und bemerkenswerte Rädertiere aus dem Bucegi-Gebirge (Südkarpathen) Rumäniens. Zool. Anz. 170: 374–380.

Godeanu, S., 1963b. Contributii la studiul rotiferilor din unele ape ale mutilor bucegi (I.). Studii cerc. Biol., Ser. Biol. Animala 15: 365–389.

Godeanu, S., 1970. Fauna de Rotifere a tinovului laptici. Studii cerc. Biol., Ser. Zool. 22: 157–165.

Harring, H. K. & F. J. Myers, 1928. The rotifer fauna of Wisconsin. IV. The Dicranophorinae. Trans. Wisconsin Acad. Sci., Arts and letters 23: 667–808, pl. 23–49.

Hauer, J., 1958. Rädertiere aus dem Sumpfe 'Große Seewiese' bei Kist. Nachr. Naturwiss. Mus. Aschaffenburg 60: 1–52.

Jersabek, C. D., 1994. *Encentrum (Parencentrum) walterkostei* n.sp., a new dicranophorid rotifer (Rotatoria: Monogononta) from the high alpine zone of the Central Alps (Austria). Hydrobiologia 281: 51–56.

Jersabek, C. D., 1995. Distribution and ecology of rotifer communities from high-altitude alpine sites – a multivariate approach. Hydrobiologia 313/314: 75–89.

Jersabek, C. D., New dicranophorids (Rotifera, Monogononta) from the Austrian Alps, including a new genus. J. nat. Hist., in press.

José de Paggi, S., 1982. *Notholca walterkostei* sp.nov. y otros rotiferos dulceacuicolas de la Peninsula Potter, Isla 25 de Mayo (Shetland del Sur, Antartida). Rev. Asoc. Cienc. Nat. Litoral 13: 81–95.

José de Paggi, S. & W. Koste, 1995. Additions to the checklist of rotifers of the superorder Monogononta recorded from the Neotropis. Int. Rev. ges. Hydrobiol. 80: 133–140.

Koniar, P., 1955. Beitrag zur Kenntnis der Rädertierfauna (Rotatoria) der Moose der Hohen Tatra. Biológia (Bratislava) 10: 449–463 (in Slovakian, German summary).

Koniar, P., 1957. Zoozönose der Moose in den Wasserfällen und Bächen der Hohen Tatra (in Slovakian, German summary). Acta Fac. Rer. Nat. Univ. Comen., Zool.2: 87–109.

Koste, W., 1965. Die Rotatorien des Naturdenkmals 'Engelbergs Moor' in Druchhorn, Kreis Bersenbrück. Veröff. Naturw. Ver. Osnabrück 31: 49–82.

Koste, W., 1976. Über die Rädertierbestände (Rotatoria) der oberen und mittleren Hase in den Jahren 1966–1969. Osnabrücker Naturw. Mitt. 4: 191–263.

Koste, W. & S. José de Paggi, 1982. Rotifera of the superorder Monogononta recorded from the Neotropis. Gewässer und Abwässer 68/69: 71–102.

Koste, W. & R. Shiel, 1986. New Rotifera (Aschelminthes) from Tasmania. Trans. R. Soc. S. Australia 109: 93–109.

Margalef, R., 1948. Flora, fauna y comunidades bióticas de las aguas dulces del Pirineo de la Cerdana. Monogr. Inst. Est. Pirenaicos 11: 1–226.

Miracle, M. R., 1978. Composición específica de las comunidades zooplanctónicas de 153 lagos de los Pirineos y su interés biogeográfico. Oecol. aquatica 3: 167–191.

Morales-Baquero, R., 1987. Rotifer fauna of lakes and ponds over 2500 m above sea level in the Sierra Nevada, Spain, with description of a new subspecies. Hydrobiologia 147: 97–101.

Morales-Baquero, R., Cruz-Pizarro, L. & P. Carillo, 1989. Patterns in the composition of the rotifer communities from high mountain lakes and ponds in Sierra Nevada (Spain). Hydrobiologia 186/187: 215–221.

Pejler, B. & B. Bērziņš, 1993. On the ecology of Dicranophoridae. Hydrobiologia 259: 129–131.

Schmid-Araya, J. M., 1993. Benthic Rotifera inhabiting the bed sediments of a mountain gravel stream. Jber. Biol. Stn. Lunz 14: 75–101.

Sharma, B. K., 1979. Rotifers from West Bengal. IV. Further contributions to the Eurotatoria. Hydrobiologia 65: 39–47.

Tarnogradski, D. A., 1959. [Microflora and microfauna of peats in the Caucasus. 8.] Raboty Sev.-Kaukazskoj Gidrobiol. Stancii 6: 1–91 (in Russian).

Tarnogradski, D. A., 1961. [Microflora and microfauna of peats in the Caucasus. 5. Sphagnetums of Maharsky gonge (Karatscheavo – Circassian Republic)]. Raboty Sev.-Kaukazskoj Gidrobiol. Stancii 22: 3–32 (in Russian).

Wiszniewski, J., 1934. Wrotki psammonowe. Les rotifères psammiques. Ann. Mus. Zool. Polon. 10: 339–399, pl. 58–63.

Wulfert, K., 1936. Beiträge zur Kenntnis der Rädertierfauna Deutschlands. Teil II. Arch. Hydrobiol. 30: 401–437.

Wulfert, K., 1960. Die Rädertiere saurer Gewässer der Dübener Heide. I. Die Rotatorien des Zadlitzmoores und des Wildenhainer Bruchs. Arch. Hydrobiol. 56: 261–298.

Zschokke, F., 1900. Die Tierwelt der Hochgebirgsseen. Denkschr. allg. schweiz. Ges. Naturw. 37: 1–400.

Hydrobiologia **387/388**: 79–82, 1998.
E. Wurdak, R. Wallace & H. Segers (eds), Rotifera VIII: A Comparative Approach.
© *1998 Kluwer Academic Publishers.*

Remarks on the rotifer fauna of north and northwestern Russia

L. A. Kutikova
The Zoological Institute RAS, Saint-Petersburg, 199034, Russia

Key words: Rotifera, north and northwest Russia, biogeography

Abstract

The rotifer fauna (about 460 species) of different waterbodies of the north and northwest of Russia is documented in the present thorough review of published and unpublished data. The biogeography of this fauna is discussed.

Introduction

The beginning of investigations on the Russian rotifer fauna dates back from the first decades of the 19th century, when Ehrenberg examined the fauna of some Russian fresh water bodies (Kutorga, 1839). The first information on rotifers of the northwest of Russia was obtained from waterbodies around St. Petersburg and the Karelia, Pskov and Novgorod regions. In the beginning of the 20th century, there was growing interest in zooplankton studies in the regions of the northwest and northern regions, where more than 200 rotifer species from 65 genera were recorded by N. V. Voronkov by the 1920s (archive data). Later, more intensive studies were performed in the 1950s through 1980s.

In this work I present a review of published (over 420 publications involving 220+ authors) and unpublished information on the rotifer species composition from waterbodies in north and northwest Russia in an attempt to prepare a data base of rotifer records. The region considered here is that between Chudskoe Lake to the Ural mountains (approximately between 25° E and 60° E) and from Novaya Zemlya to central Russia (approximately 77° N and 55° N).

Results

All waterbodies were classified into four types: lakes (about 100), rivers and their basins (70+), reservoirs (10) and small waterbodies and marshes. The considered region is situated in the tundra and taiga climate zones. Forest tundra occurs on a small territory on Kola Peninsula and Bolshezemelskaya tundra,

while mixed forest is found near lakes Chudskoe and Ilmen. A total of 461 rotifer species (not including intraspecific taxa) belonging to 82 genera have been identified (see Table 1). Most of the rotifers have cosmopolitan or palaearctic distributions. The frequency of species occurrence permits a characterization of the ecology and latitudinal distribution of some species (Kutikova, 1970; Green, 1972; Pejler; 1977; Dumont, 1983). The first advocate of the latitudinal distribution of rotifers was Voronkov (1925, 1927). He associated the distribution of planktonic rotifers with zones, more or less corresponding to climatic zones. Voronkov (1925, 1927) subdivided the European part of Russia into four zones: northern, northern temperate, more southern and true southern. To the habitants of the northern zone he referred *Polyarthra platyptera* Ehrb., *Asplanchna priodonta* Gosse, *Keratella cochlearis* (Gosse), *K. quadrata* (Müll.), *Kellicottia longispina* (Kell.), *Conochilus unicornis* Rouss. and *Filinia longiseta* (Ehrb). Later Gerd (1946) and Kutikova (1975) identified a complex of species, characteristic of the plankton northern latitudes. These are *Kellicottia longispina, Keratella cochlearis, K. hiemalis* Carl., *Polyarthra major* Burck., *P. remata* Skor., *Conochilus unicornis, C. hippocrepis* Rouss., *Asplanchna priodonta, Ploesoma hudsoni* (Imh.), *Gastropus stylifer* Imh., *Collotheca mutabilis* (Huds.), *Filinia terminalis* (Plate) and *F. major* (Cold.). At the present time, *Polyarthra dolichoptera* Idel., *Synchaeta stylata* Wier., *S. verrucosa* Nip. and *S. lakowitziana* Lucks can be included in this complex. The latitudinal complex of the temperate zone includes *Trichocerca cylindrica* (Imh.), *T. rousseleti* (Voigt), *T. capucina* (Wierz. et Zach.), *Asplanchna priodonta, Conochiloides natans*

Table 1. List of genera and number of recorded species in every genus

Acyclus	1	Limnias	4
Adineta	4	Lindia	5
Albertia	2	Lophocharis	3
Anuraeopsis	1	Macrochaetus	1
Ascomorpha	5	Macrotrachela	6
Ascomorphella	1	Manfredium	1
Aspelta	3	Microcodon	1
Asplanchna	7	Microcodides	3
Asplanchnopus	1	Mniobia	4
Beauchampia	1	Monommata	8
Brachionus	14	Myersinella	1
Cephalodella	29	Mytilina	8
Collotheca	15	Notholca	9
Colurella	8	Notommata	16
Conochiloides	2	Philodina	7
Conochilus	2	Platyias	2
Cupelopagis	1	Pleuretra	3
Cyrtonia	1	Pleurotrocha	1
Dicranophorus	12	Ploesoma	4
Dipleuchlanis	1	Polyarthra	9
Diplois	1	Pompholyx	2
Dissotrocha	4	Proales	13
Drilophaga	1	Proalinopsis	1
Encentrum	6	Pseudoharringia	1
Enteroplea	1	Ptygura	3
Eosphora	2	Resticula	2
Eothinia	2	Rotaria	6
Epiphanes	5	Scaridium	1
Euchlanis	14	Sinantherina	1
Filinia	5	Squatinella	6
Floscularia	3	Stephanoceros	1
Gastropus	3	Synchaeta	15
Habrotrocha	11	Taphrocampa	1
Harringia	1	Testudinella	10
Hexarthra	1	Tetrasiphon	1
Itura	3	Trichocerca	15
Kellicottia	1	Trichotria	5
Keratella	11	Tylotrocha	1
Lacinularia	2	Zelinkiella	1
Lecane	64	Wierzejskiella	2
Lepadella	21	Wolga	1

(Sel.), *C. dossuarius* (Hud.). They are the intermediate group between the northern and more southern complex (*Brachionus angularis* Gosse, *B. calyciflorus* Pal., *B. diversicornis* (Dad.), *Ascomorpha ovalis* (Berg.), *Hexarthra mira* (Hud.), *Trichocerca stylata* (Gosse)). The group inhabiting southern latitudes is represented

by *Brachionus falcatus* Zach., *B. forficula* Wier., *B. caudatus* Barr. et Dad., *Filinia opoliensis* Zach. and some other thermophilic species, which are also common in tropical waterbodies. Boundaries between the zones are indistinct and do not permit neat separation of definite zoogeographical regions, because the distribution of rotifers is connected to physical, chemical and biological (Voronkov, 1927). For instance, Pejler (1962) regards *Notholca caudata* Carl., usually not inhabiting regions below 58° N as a glacial relict. The common *Keratella quadrata platei* (Jar.) and *K. cochlearis baltica* (Sok.) from the plankton of the Gulf of Finland can be considered local subspecies. *Notholca cornuta* Carl., found in Sweden and the Rybinsk reservoir, and *N. triarthroides* are rare species, the latter has a disjunct distribution, being also noted in Ladoga Lake and Baikal. In the northern waterbodies, species of the northern circumpolar complex naturally dominate; however, populations of thermophilic species (i.e. *Brachionus budapestinensis* in the Leningrad region) may develop occasionally, i.e. in the estuaries of northern rivers like Severnaya Dvina (*Brachionus angularis*, *B. calyciflorus*) and Pechora (*Asplanchna brightwelli*; see Deeksbach, 1926).

The frequency of occurrence of species permit us to distinguish the dominant group of rotifers in the studied regions (more than 10% of the total species number) (Table 2).

Conclusion

The list of Rotifera recorded in waterbodies of the north and northwestern part of European Russia should not be regarded as exhaustive because it is mainly based on data collected different methods. The rotifer fauna of the St. Petersburg region and Karelia proved to be most completely known, however, the vast territories of the tundra and taiga regions remain insufficiently studied. Comparing the qualitative composition of the rotifer fauna in Russian waterbodies with the more completely known ones in Fennoscandia, Denmark and adjacent areas (Pejler, 1993) it can be noted that they have 68.4% of species and 95% of genera in common. The species complex typical of the plankton of lakes appears quite adequately known. Many species from this complex are cosmopolitans. When compared to the rotifer fauna of northern latitudes in North America, we see that, apart of a similar complex of planktonic species, there is a group of endemics, especially in the genera *Keratella, Lecane*

Table 2. Frequency of occurrence of dominant rotifers in northern and northwestern Russia

Dominant species	No. of reports	Lakes	Rivers	Reservoirs	Small water bodies	% of Total
Keratella cochlearis Gosse	205	107	52	28	18	41.0
Kellicottia longispina (Kellicott)	201	92	38	23	48	40.6
Asplanchna priodonta Gosse	194	83	39	25	47	39.2
Keratella quadrata (Müller)	164	55	41	30	35	32.5
Filinia longiseta (Ehrenberg)	123	58	26	10	29	24.8
Conochilus unicornis Rousselet	121	60	22	12	27	24.4
Lecane luna (Müller)	105	33	28	10	44	21.2
Euchlanis dilatata Ehrenberg	93	38	32	6	17	18.8
Lecane lunaris (Ehrenberg)	82	25	27	4	26	15.8
Gastropus stylifer Imhof	68	29	12	3	24	13.3
Brachionus quadridentatus Hermann	68	12	25	2	29	13.7
Brachionus angularis Gosse	65	18	13	10	24	13.1
Conochilus hippocrepis (Schrank)	65	28	13	11	13	13.1
Brachionus calyciflorus Pallas	64	12	22	7	23	12.9
Ploesoma hudsoni (Imhof)	63	28	23	3	9	12.7
Lecane luna (Gosse)	56	13	17	1	25	11.1
Lepadella ovalis (Willer)	57	9	13	2	33	11.1
Brachionus urceolaris (Linnaeus)	46	2	15	2	27	9.3

and *Trichocerca* (Stemberger, 1990). The geographic situation of the latidunal zones is different for northern Europe and North America, as a result of the warm Gulf Stream that prevents strong cooling of the outlying area in north Europe. Unfortunately, a rather incomplete rotifer species record is available for the arctic regions (Novaja Zemlya, Franz Josef Land) and mainly loricate rotifers are recorded (Idelson, 1925; Retovsky, 1935). Because of this, it is hard to conduct a comparison with the intensively studied regions of Canada (Chengalath & Koste, 1983; 1987; 1989; De Smet & Beyens, 1995).

Littoral taxa, as Segers (1996) has shown with reference to the genus *Lecane*, are more promising in the elucidation of questions of rotifers biogeography. Of the 64 species of this genus noted in the regions considered, 20 (31.3%) are cosmopolitans and have been found in all 6 zoogeographical regions indicated by Segers (1996). Nine species (14%) are known only from the Palaearctic and Nearctic. Probably, littoral taxa are similar to planktonic ones regarding their zonal distribution.

Acknowledgements

Thanks are due to the St. Johns University and College of St. Benedict for the grant of financial assistance to attend the VIIIth International Rotifer Symposium in USA. I wish to thank I. Nikolaeva for her help in collection of published data. The author is deeply indebted to anonymous reviewers for their valuable comments to my manuscript.

References

Chengalath, R. & W. Koste, 1983. Rotifera from northeastern Quebec, Newfoundland and Labrador, Canada. Hydrobiologia 104: 49–56.

Chengalath, R. & W. Koste, 1987. Rotifera from Northwestern Canada. Hydrobiologia 147: 49–56.

Chengalath, R. & W. Koste, 1989. Composition and distributional patterns in arctic rotifers. Hydrobiologia 186/187: 191–200.

De Smet, W. H. & L. Beyens, 1995 Rotifers from the Canadian High Arctic (Devon Island, Northwest Territories). Hydrobiologia 313/314: 29–34.

Decksbach, H. I., 1926. Studien über das Zooplankton des Petsjora-Beckens und der südlichen Nebenflusse der Dwina. Int. Rev. Hydrobiol. 14: 323–338.

Dumont, H. J., 1983. Biogeography of rotifers. Hydrobiologia 104: 19–30.

82

Gerd, S. V., 1946. Planktonic complexes of large lakes of Karelia and summer migrations of shallow-water cisco. Uchen. zap. Kar.-fin. univ. 1: 305–340.

Green, J., 1972. Latitudinal variation in associations of planktonic Rotifera. J. Zool. Lond. 167: 31–39.

Kutikova, L. A., 1970. Kolovratki Fauna SSSR. (The rotifer fauna of USSR). Key of fauna USSR 104, Leningrad. 744 pp.

Kutikova, L. A., 1975. Planktonic rotifers. Biological productivity of northern lakes. Part 1. Lakes Krivoye and Krugloye. Trudy Zool. Inst. 56: 67–76.

Kutorga, S. S., 1839. Natural history (microbes) compiled mostly on the basis of observations by Ehrenberg. St. Peterburg, Ed. E. Praca: 16 pp.

Pejler, B., 1962. *Notholca caudata* Carlin (Rotatoria) a new presumed glacial relict. Zool. Bidr. Uppsala 33: 307–319.

Pejler, B., 1977. General problems on rotifer taxonomy and global distribution. Arch. Hydrobiol. Beih. 8: 212–220.

Pejler, B., 1993. Nordic Rotifera. Term Lists for Environm. Data Standardization. Term List RF. Version 931123/KS.

Retowski, L. O., 1925. Süsswasserfauna von Nowaja Semlja und Franz Joseph Land. Trans. Arctic. Inst. 14: 3–72.

Segers, H., 1996. The biogeography of littoral Lecane Rotifera. Hydrobiologia 323: 169–197.

Stemberger, R. S., 1990. A inventory of rotifer diversity of north Michigan inland lakes. Arch. Hydrobiol. 118(3): 283–302.

Voronkov, N. V., 1925. On the geographic distribution of rotifers, in particular Russia. N 1. Krasnojarsk: 1–111, 19 pp.

Voronkov, N. V., 1927. On the geographic distribution of rotifers, in particular Russia. N 2. Krasnojarsk: 32 pp.

Hydrobiologia **387/388**: 83–91, 1998.
E. Wurdak, R. Wallace & H. Segers (eds), Rotifera VIII: A Comparative Approach.
© *1998 Kluwer Academic Publishers.*

Review paper

The evolutionary relationships of rotifers and acanthocephalans

James R. Garey[1,*], Andreas Schmidt-Rhaesa[1], Thomas J. Near[2] & Steven A. Nadler[3]
[1]*Department of Biology, University of South Florida, 4202 East Fowler Avenue, SCA 110 Tampa, FL 33620-5150, U.S.A. (*author for correspondence)*
[2]*Center for Biodiversity, Illinois Natural History Survey, Champaign, IL 61820, U.S.A.*
[3]*Department of Nematology, University of California, Davis, CA 95616-8668, U.S.A.*

Key words: phylogeny, acanthocephala, rotifera, bilateria, evolution, 18S rRNA gene

Abstract

Advances in morphological and molecular studies of metazoan evolution have led to a better understanding of the relationships among Rotifera (Monogononta, Bdelloidea, Seisonidea) and Acanthocephala, and their relationships to other bilateral animals. The most accepted morphological analysis places Acanthocephala as a sister group to Rotifera, although other studies have placed Acanthocephala as a sister taxon to Bdellodea or Seisonidea. Molecular analyses using nuclear 18S rRNA and mitochondrial 16S rRNA genes support Acanthocephala as a sister taxon to Bdelloidea, although no molecular data is available for Seisonidea. Combining molecular and morphological analyses of Bilateria leads to a tree with Platyhelminthes, Rotifera, Acanthocephala and Gnathostomulida (and probably Gastrotricha) as a sister group to the annelid-mollusc lineage of the Spiralia (Lophotrochozoa).

Introduction

The phylogenetic position of rotifers and acanthocephalans among metazoans has been a major problem in evolutionary studies for many years. Traditionally, both rotifers and acanthocephalans have been included within the Aschelminthes, and a close association between the two groups has been suspected since Haffner (1950), although not generally accepted until recently. The purpose of this article is to (1) review the morphological and molecular evidence for the relationships among the three major rotifer groups (Bdelloidea, Monogonontea, Seisonidea) and acanthocephalans, and their evolutionary relationships to other metazoans, and (2) to suggest areas of future studies. Two important advances since the last Rotifer Symposium have been new ultrastructural studies of *Seison* and the use of molecular phylogenetic analyses.

Morphological evidence for evolutionary relationships among the Rotifera

Phylum Rotifera consists of three groups, the classes Bdelloidea, Monogononta, and Seisonidea. A three

taxon tree has only three rooted solutions, and each has been proposed at various times for the rotifers. These are illustrated in Figure 1. Tree A (Figure 1) is probably the most accepted, because it unites Bdelloidea and Monogononta with a number of characters that are most certainly synapomorphic for the two taxa such as clefts but no pores in the terminal organ of the protonephridia, unpaired retrocerebral glands, salivary glands integrated into the mastax (Ahlrichs, 1995, 1997) and the presence of a vitellarium (Wallace & Colburn, 1989). In this tree, Seisonidea is the most basal group. Wallace and Colburn (1989) suggested that Bdelloidea + Monogononta be united as the Eurotatoria, and that all three classes make up the phylum Rotifera, while Ahlrichs (1997) only applies the name Rotifera to Bdelloidea + Monogononta. Tree B (Figure 1) has been suggested by Pennak (1989) with Seisonidea and Bdelloidea united as the digonont rotifers (paired female gonads), forming a sister group to Monogononta (unpaired female gonads). Paired gonads are most likely the plesiomorphic condition within Bilateria, and would not unite Seisonidea with Bdelloidea. Tree C (Figure 1) has Bdelloidea as the most basal rotifer with Seisonidea and Monogononta united

84

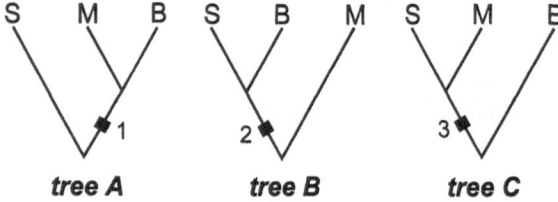

Figure 1. Possible relationships between Seisonidea (S), Monogononta (M) and Bdelloidea (B). 1: Clefts but no pores in terminal organ of the protonephridia; rotatory organ; unpaired retrocerebral glands; salivary glands integrated into the mastax (Ahlrichs, 1997); vitellarium (Wallace & Colburn, 1989). 2: Paired ovaries, ramate mastax, absence of secreted tube (Pennak, 1989). 3: Males present, no bladder, cellular stomach with microvilli (Ricci et al., 1993), similarities of internal layer in their syncytial integument (Clement, 1993).

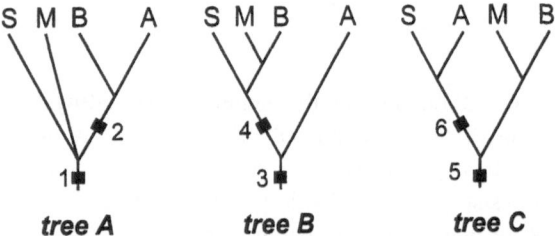

Figure 2. Proposed relationships between Seisonidea (S), Monogononta (M), Bdelloidea (B) and Acanthocephala (A). 1: Internal layer of syncytial epidermis. 2: Lemnisci and proboscis present (Lorenzen, 1985). 3: Pseudocoel present, syncytial epidermis, monociliated pit absent, hermaphorditism absent, acrosome present, anteriorly inserting flagellum on sperm (Wallace et al., 1996), internal layer in the syncytial epidermis (Nielsen, 1995). 4: Parthenogenesis, hypodermic impregnation, collagen absent (Wallace et al., 1996), toes with adhesive glands (Nielsen 1995). 5: Internal layer in the syncytial epidermis, anteriorly inserted flagellum on sperm cell, outer epidermal cell membrane intrusions with bulbs. 6: Dense bodies within spermatozoa, epidermis with filament bundles (Ahlrichs, 1997).

based on males being present, no bladder, and cellular stomach with microvilli (Ricci et al., 1993). However, these characters are most likely plesiomorphic because they are found in outgroup taxa such as Platyhelminthes. Another character, similarities of the internal layer in the syncytial integument (Clement, 1993) has been discussed by Ahlrichs (1997).

Morphological evidence for the evolutionary relationship between Rotifera and Acanthocephala

Although rotifers and acanthocephalans have historically been included among the Aschelminthes (Ruppert & Barnes, 1994), it is clear that Aschelminthes is a polyphyletic (Lorenzen, 1985; Malakhov, 1994; Neuhaus, 1994; Winnepenninckx et al., 1995; Ehlers

et al., 1996; Wallace et al., 1996) or paraphyletic (Nielsen, 1996) assemblage and that the pseudocoelom evolved independently in several aschelminth phyla (Remane, 1963; Ruppert, 1991; Nielsen, 1995). Despite this, a close affinity between Rotifera and Acanthocephala was suspected by Haffner (1950) based on common characters such as a cloaca, protonephridia, egg segmentation, and muscles that retract the anterior region of the body (Remane, 1963). Figure 2 shows three trees that have been proposed for the relationship between Rotifera and Acanthocephala based on morphological data.

Lorenzen (1985) suggested that rotifers and acanthocephalans can be united based on the internal layer of the syncytial epidermis found in both (Storch & Welsch, 1969) and the testis attached to a reduced intestine in monogononts comparable to the ligament cord found in acanthocephalans (Haffner, 1950). Lorenzen's analysis did not resolve the relationship between seisonid and monogonont rotifers, but he united Bdelloidea + Acanthocephala based on the presence of lemnisci and similarities of the proboscis in both taxa (Figure 2, tree A). These two characters have been rejected by Clement (1993) and Nielsen (1995) as synapomorphies for Bdelloidea + Acanthocephala because the 'proboscis' of acanthocephalans develops from different regions in the embryo than the comparable structure in bdelloid rotifers. The lemnisci are sac-like structures with a high number of lacunes and a still not completely understood function (Miller & Dunagan, 1985; Dunagan & Miller, 1991), while the structures in bdelloids are most likely thickened regions of the epidermis that carry the rotatory organ (Ahlrichs, pers. comm.). However, ultrastructural investigations of this region are still lacking.

Nielsen (1995) and Wallace et al. (1996) have both proposed a sister relationship between Rotifera and Acanthocephala, leaving each phylum monophyletic (Figure 2, tree B). The characters used to group all three classes of rotifers separately from acanthocephalans are parthenogenesis, hypodermic impregnation, absence of collagen (Wallace et al., 1996) and toes with adhesive glands (Nielsen, 1995). However, many of those characters may not be autapomorphies for Rotifera. Seisonidea reproduce exclusively by sexual reproduction (Clement & Wurdak, 1991), so parthenogenesis is not an autapomorphy for Rotifera. Apparently, copulation has never been observed in *Seison*, which, unlike other rotifers, lacks a penis but has a spermatophore-like structure (Ricci et al., 1993; Ahlrichs, 1995). Free sperm cells have been observed only in the reproductive tract of female *Seison*, and it is likely that sperm enter through the cloaca, so hypodermic impregnation is not likely to be an autoapomorphy for Rotifera. We are not aware of any studies that conclusively demonstrate that collagen is absent from *Seison*. The presence of toes with adhesive glands as an autoapomorphy of Rotifera has come under question because the cement glands of acanthocephalans may be homologous to the adhesive glands of rotifers (Near et al., 1998).

A novel scheme has recently been proposed (Ahlrichs, 1997) that most closely relates Acanthocephala with Seisonidea (Figure 2, tree C) using dense bodies within the spermatozoa and bundles of filaments within the epidermis as synapomorphies. These characters have not before been used for phylogenetic studies and so their significance remains to be confirmed. Ahlrichs retains a monophyletic Rotifera as Bdelloidea + Monogononta, and uses the taxon name Syndermata for Rotifera + Seisonidea + Acanthocephala based on the presence of a syncytial epidermis with an internal layer, outer epidermal cell membrane intrusions with bulbs and an anterior insertion of the flagellum on sperm cells.

Molecular studies of rotifers and acanthocephalans

Molecular studies of phylogeny are based on aligning the DNA sequences of orthologous genes, and deducing trees by one of three common methods (reviewed in Li, 1997). In distance methods, a matrix of evolutionary distances between all pairs of sequences are calculated, and a tree is deduced from the distance matrix most commonly by the Neighbor-Joining (NJ) method. A number of algorithms can be used to calculate distances from the alignment which correct for multiple substitutions at the same site and/or correct for different nucleotide substitution rates at different sites (site to site variation). NJ trees can be calculated very quickly and their polarity is determined by an outgroup. In Maximum Parsimony (MP) trees, the alignment is used to choose the tree with the shortest path that accounts for the nucleotide changes. Considering that there are over 34 million possible topologies for even a 10 taxon tree, MP trees can take a lot of computation time. MP analysis generally does not correct for multiple substitutions at the same site or site to site variation. In Maximum Likelihood (ML) trees, a maximum likelihood value for character state

configurations among the sequences are calculated for each possible tree and the tree with the largest value chosen. This method can accommodate corrections for multiple substitutions at the same site and for site to site variation. The ML method is usually the slowest of the three kinds of analyses.

Confidence in molecular trees is most often determined by bootstrap analysis (Felsenstein, 1988; Hillis & Bull, 1993) in which new datasets are constructed from the original alignment by selecting sites from the original alignment randomly with replacement. Trees are made from each bootstrapped dataset and the percent of bootstrapped trees that support each branch is reported when greater than 50%, and the closer a value is to 100%, the more confidence one has in that region of the tree. Bootstrap analysis can be carried out on any type of tree, although ML bootstrap analysis is usually impractical because of long computation times. Other statistical methods include Confidence Probability (CP) values for NJ trees in which the confidence that a given branch is greater than zero is calculated (Kumar et al., 1994). The closer a CP value for a given branch is to 100, the higher the confidence one has in that branch of the tree. Decay analysis is used in MP trees, and refers to the number of steps that a tree can be lengthened and still retain a particular clade (Donoghue et al., 1992). The higher the number, the more probable the clade, although computation time often limits the number of steps that can be tried. Although various statistical analyses are the most widely used determinant of confidence in a tree, it is possible to have statistical support for an incorrect tree (Hillis et al., 1994).

Molecular studies have contributed to the evidence that rotifers and acanthocephalans are closely related. The complete 18S rRNA gene of the archiacanthocephalan *Moniliformis moniliformis* was published in 1993 (Telford & Holland) in a study of chaetognath affinites, but no rotifer sequence was included. The first mention of an association between rotifers and acanthocephalans (Raff et al., 1994) was a reference to an unpublished study in a review article on animal phylogeny, but no statistical support for the association was given, and a subsequent paper describing the analysis was not published. The first rigorous molecular study of aschelminth phylogeny (Winnepenninckx et al., 1995) included nearly complete 18S rRNA gene sequences from the acanthocephalan *M. moniliformis*, the monogonont rotifer *Brachionus plicatilis*, numerous nematodes, a gastrotrich, a nematomorph, and a priapulid. The study showed that aschelminths are

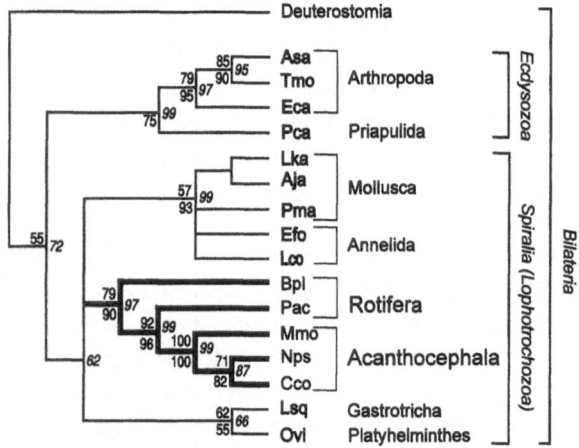

Figure 3. Molecular phylogeny of Bilateria based on the 18S rRNA gene. The tree shown is a strict consensus of NJ, MP, and ML analyses from Garey et al. (1996a). Numbers above and below each fork represent the percentage of 1,000 bootstrap replicates that support the branch in the MP and NJ trees, respectively. Numbers to the right of each fork are CP values from the NJ tree. Values are shown only when greater than 50. The Rotifera + Acanthocephala clade, Bdellodea + Acanthocephala clade, and the Acanthocephala clade were all supported by decay indices greater than 20. Taxon abbreviations: *Artemia salina*, **Asa**; *Tenebrio molitor*, **Tmo**; *Eurypelma californica*, **Eca**; *Priapulus caudatus*, **Pca**; *Limicolaria kambeul*, **Lka**; *Acanthopleura japonica*, **Aja**;*Placopecten magellanicus*, **Pma**; *Eisenia foetida*, **Efo**; *Lanice conchilega*, **Lco**; *Brachionus plicatilis*, **Bpl**;*Philodina acuticornis*, **Pac**;*Moniliformis moniliformis*, **Mmo**; *Neoechinorhynchus pseudemydis*, **Nps**; *Centrorhynchus conspectus*, **Cco**; *Lepidodermella squammata*, **Lsq**; Platyhelminthes: *Opisthorchis viverrini*, **Ovi**. See Garey et al. (1996a) for Genbank accession numbers and other details of the analysis.

polyphyletic, but supported a rotifer + acanthocephalan clade with a weak bootstrap value of 52%, and a CP value of 86 in an NJ tree. The MP tree revealed the rotifer + acanthocephalan clade but with a bootstrap value below 50%.

A more recent study (Garey et al., 1996a) contributed new 18S rRNA gene sequences from the bdelloid rotifer *Philodina acuticornis*, the palaeacanthocephalan *Centrorhynchus conspectus* and the eoacanthocephalan *Neoechinorhynchus pseudemydis*. To date, the monogonont *B. plicatilis* and the bdelloid *P. acuticornis* are the only rotifers for which 18S rRNA sequences have been published. Therefore, the presently available molecular data cannot discriminate between any of the trees in Figure 1 concerning the relationships among the three rotifer classes.

The presently available molecular evidence (Garey et al., 1996a) overwhelmingly supports a sister relationship between Bdelloidea and Acanthocephala (Figure 3), favoring tree A in Figure 2, based on

the hypothesis of Lorenzen (1985), contradicting the idea of Acanthocephala as a sister taxon to monophyletic Rotifera (tree B, Figure 2), or a monophyletic Bdelloidea + Monogononta (tree C, Figure 2). In the analyses, NJ, ML, and MP trees were found to be congruent in regard to the relationship between rotifers and acanthocephalans with remarkably strong statistical support. Bootstrap support for Rotifera + Acanthocephala ranged from 79 to 90%, and was 92 to 96% for Bdelloidea + Acanthocephala. CP values were similarly high, and decay analyses (not shown) indicated that even 20 steps were insufficient to decay the two clades.

One problem with the study is that the rotifer *P. acuticornis* and the acanthocephalan *C. conspectus* have 18S rRNA genes that evolve much more rapidly than other metazoans, and it is possible that unequal rate effects could cause an incorrect tree. It has been shown that unequal rates and other problems can cause all tree making methods (NJ, MP and ML) to produce identical but incorrect trees (Hillis et al., 1994). In Figure 4, the NJ tree from Garey et al. (1996a) is shown with branches drawn to scale. It can be seen that the long branches leading to *C. conspectus* and *P. acuticornis* are not directly adjacent to one another and they are not in a basal part of the tree, both which would be expected if unequal rate effects were a factor (see Figure 1 in Aguinaldo et al., 1997). The tree in Figure 4 is only a portion of the tree produced by the analysis, which also included nematode 18S rRNA genes with very long branches that appeared incorrectly as basal to the bilateria. If unequal rate effects were a factor, one would expect that the *C. conspectus* and *P. acuticornis* branches would have been attracted toward the long branches of the nematode genes and appeared more basal. In additional analyses, the sequences from *C. conspectus* and *P. acuticornis* were removed from the tree separately and together, with no change in topology of the tree (Garey et al., 1996a), further evidence that unequal rates are not a factor.

The mitochondrial genome contains a number of highly conserved genes that are useful for phylogenetic analysis. One of the most conserved is that of the mitochondrial 16S rRNA gene which is inherited independently from the nuclear rRNA genes. We sequenced a 600 bp fragment of the 16S rRNA gene from *B. plicatilis*, *P. acuticornis*, and *M. moniliformis* and present the analysis in Figure 5. Although the 16S rRNA gene is less conserved that the nuclear 18S rRNA, there is sufficient signal to indicate relationships between closely related taxa. For example, the

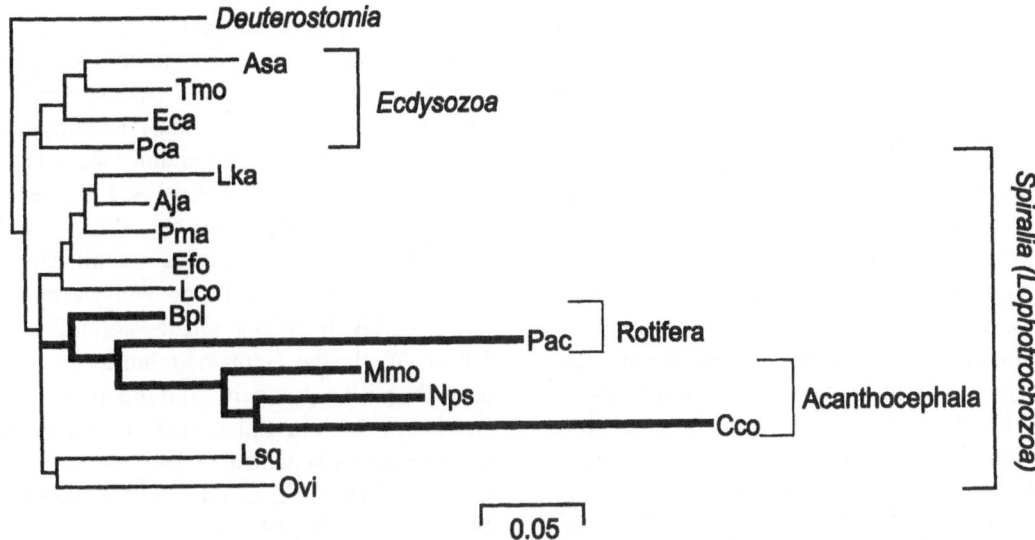

Figure 4. The tree from Figure 3 drawn with branch lengths proportional to evolutionary distance to illustrate the unequal evolutionary rates of rotifers and acanthocephalans. The rotifer *P. acuticornis* and the acanthocephalan *C. conspectus* are evolving at a rate approximately 5 times as fast as most other taxa in the tree. When the fastest evolving rotifer sequence (*P. acuticornis*) was removed from the analysis, the acanthocephalans remained as a sister taxon of the rotifers. When the fastest acanthocephalan sequence (*C. conspectus*) was removed, the other acanthocephalans remained within the rotifer clade, demonstrating that the position of acanthocephalans as a sister taxon to bdelloid rotifers is not likely to be an artifact due to unequal rate effects (from Garey et al., 1996a). Taxon labels are defined in Figure 3.

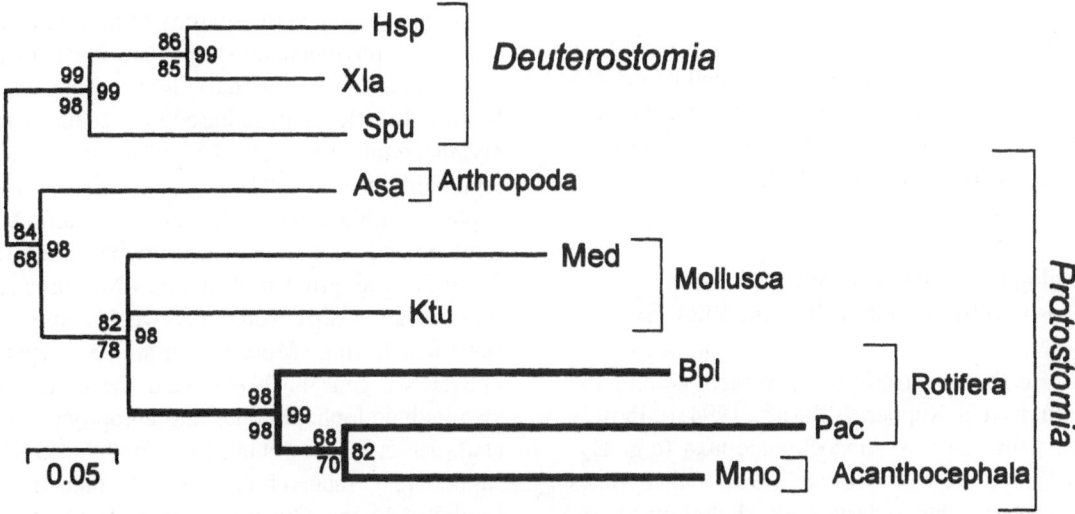

Figure 5. Molecular phylogeny of Bilateria based on a 600 bp alignment of the mitochondrial 16S rRNA gene. The tree shown is a NJ tree. Bootstrap values for Kimura distances with gamma correction (a = 0.72) are shown above the forks, values for Tamura & Nei distances are below and numbers to the right are CP values for Kimura distances. See Kumar et al. (1995) for details. The same topology was recovered with all NJ analyses and with ML analysis with multiple rate categories but not with MP or ML analysis without multiple rate categories (see text). Taxon abbreviations and Genbank accession numbers: *Artemia salina*, **Asa**, M21833; *Brachionus plicatilis*, **Bpl**, AF108106; *Homo sapiens*, **Hsa**, D38112; *Katherina tunicata*, **Ktu**, U09810; *Moniliformis moniliformis*, **Mmo**, AF108107; *Mytilus edulis*, **Med**, M83756; *Philodina acuticornis*, **Pac**, AF108108; *Strongylocentrotus purpuratus*, **Spu**, X12631; *Xenopus laevis*, **Xla**, X01601. Portions of mitochondrial 16S rRNA genes corresponding to a sea urchin 16S rRNA gene (Genbank Accession 12825) from nucleotides 814–833 and 1406–1425 were PCR amplified from cellular DNA isolated from *P. acuticornis*, *B. plicatilis*, and *M. moniliformis*, and resulted in a fragment about 450 nucleotides in length. Primers were 16S-RNA1:16S-RNA1 CCGGAATTCCGCCTGTTTATCAAAAACAT, and 16S-RNA2: CCCAAGCTTCTCCGGTTTGAACT-CAGATC, which have EcoRI and HindIII site tails, respectively. PCR products were cloned into M13 and sequenced in both directions. All sequences were aligned according to a secondary structure model (De Rijk & De Wachter, 1993) and trees produced using MEGA (Kumar et al., 1994) for NJ trees and Phylip (Felsenstein 1993) for ML and MP trees. Sites with gaps were not used in the analyses.

tree in Figure 5 properly groups deuterostomes together, and indicates the same relationship between Rotifera and Acanthocephala as the 18S rRNA study in Figures 3 and 4. The NJ tree in Figure 5 was calculated using a variety of distance-correction methods which correct for multiple substitutions at the same site and site to site rate variation, all resulting in the topology shown in Figure 5 with similar bootstrap values. It can be seen that branch lengths are more consistent among taxa than in the 18S rRNA gene analyses, so rate effects are unlikely to be significant.

The importance of correcting for site to site variation and multiple substitutions at the same site is important for the fast evolving mitochondrial 16S gene. MP analysis does not carry out those corrections, and produced a tree similar to tree B in Figure 2, with Acanthocephala as a sister taxon to Rotifera. Similarly, when ML analysis was carried out without correcting for site to site variation, Acanthocephala appeared as the sister taxon to Rotifera, but when the analysis was repeated with multiple rate categories (four categories: 10% of sites with no variation, 20% each with rates of 1, 5, and 20) the topology shown in Figure 5 (Bdelloida and Acanthocephala as sister taxa) was recovered with a higher likelihood (ln likelihood = −2651) than without the correction (ln likelihood = −2754). It is well established that rRNA genes demonstrate site to site variation of evolutionary rates as some sites are completely conserved and others are not (Hillis & Dixon, 1991).

Morphological evidence for the position of Rotifera-Acanthocephala within the Bilateria

In most textbooks, rotifers are placed among the aschelminths (e.g. Ruppert & Barnes, 1994) or loosely grouped with other pseudocoelomate taxa (e.g. Hyman, 1951; Brusca & Brusca, 1990). Other views such as a relationship to derived platyhelminth groups (Markevich, 1993) are rare. Some cladistic studies of the entire Metazoa consider the pseudocoelom an important character which can result in a monophyletic aschelminth clade (e.g. Schram, 1991; Eernisse et al., 1992), but the pseudocoelom has been shown to be a doubtful phylogenetic character (Ruppert, 1991). It is clear that the aschelminths are polyphyletic, but the more rigorous treatments of aschelminth taxa often fail to include mainstream metazoan phyla such as arthropods, annelids and molluscs (e.g. Lorenzen, 1985; Wallace et al., 1996), although a few

recent studies have (Ehlers et al., 1996; Nielsen et al., 1996). Recent morphological analyses from a number of laboratories seem to be converging on the concept of two 'aschelminth' clades, one (Nemathelminthes) containing Priapulida + Kinorhyncha + Loricifera + Nematoda + Nematomorpha + Gastrotricha (Nebelsick, 1993; Neuhaus, 1994; Ehlers et al., 1996; Nielsen et al., 1996, Wallace et al., 1996), the other clade (Syndermata) containing Acanthocephala + Rotifera (Nielsen et al., 1996; Wallace et al., 1996) and possibly including Gnathostomulida (Ahlrichs, 1997). While the Nemathelminthes are most probably the sister group of Spiralia (Lophotrochozoa) within Protostomia (Ehlers et al., 1996), Syndermata + Gnathostomulida (named Gnathifera) have been hypothesized as the sister taxon of Platyhelminthes within Spiralia (Ahlrichs, 1995).

Molecular evidence for the position of Rotifera-Acanthocephala within the Bilateria

A major difficulty in relating minute animals like rotifers to larger animals such as annelids, molluscs, and arthropods has been the scarcity of uniting characters. Molecular phylogenetic studies are ideal for relating more distant taxa that have few uniting characters. Several 18S rRNA gene based metazoan phylogenies (Winnepenninckx et al., 1995; Garey et al., 1996b; Aguinaldo et al., 1997) placed Rotifera + Acanthocephala within a clade loosely including Platyhelminthes and sometimes Gastrotricha which together formed a sister group to Annelida + Mollusca although these relationships were only weakly supported by statistical testing. More recent studies based on the 18S rRNA gene have extended the entire clade to also include lophophorates and entoprocts (Halanych et al., 1995; Mackey et al., 1996) with better statistical support and Halanych et al. (1995) named the clade Lophotrochozoa. Our mitochondrial 16S rRNA gene analysis is consistent with the 18S rRNA gene findings (Figure 5).

In most 18S rRNA gene studies, Priapulida appeared in another protostome clade as a sister group to Arthropoda. Nematoda + Nematomorpha usually appeared basal to the bilateria, an artifact now recognized as caused by unequal rate effects (Aguinaldo et al., 1997). Another 18S rRNA study (Garey et al., 1996b) extended the clade of Arthropoda + Priapulida to include Tardigrada. Recently, Aguinaldo et al. (1997) solved the problem of the placement of Nematoda within Bilateria by finding a nematode (*Trichinella spiralis*) with a slow evolving 18s rRNA gene. With careful attention to unequal rate effects, they provided evidence that the protostomes consist of two clades: the Ecdysozoa includes all molting animals (e.g. nematodes and arthropods), and is the sister taxon to Spiralia (Lophotrochozoa). Since ecdysozoans generally lack spiral cleavage which is present in spiralians, we prefer to use the term Spiralia instead of Lophotrochozoa (see Malakhov, 1994 and Nielsen, 1995 for descriptions of cleavage in nematodes). The Ecdysozoa/Spiralia (Lophotrochozoa) structure of protostomes appears consistent with morphological characters. For example, the developmental pattern of growth by molting under the control of the steroid hormone ecdysone has been confirmed among Arthropoda, Nematoda, and Tardigrada (see Gupta, 1990; Davies & Fisher, 1994).

Conclusions and future directions

The molecular and morphological evidence is overwhelmingly in favor of a close relationship between Rotifera and Acanthocephala. Analyses of nuclear 18S rRNA and mitochondrial 16S rRNA genes strongly favor a sister relationship between Bdelloidea and Acanthocephala, one of three possible relationships argued by morphological studies, but the morphological support for the sister relationship of Bdelloidea and Acanthocephala appears weak and is very controversial. In this regard, the time is ripe for a series of rigorous ultrastructural comparisons of the epidermis underlying the rotatory organ of bdelloid rotifers to the lemnisci of acanthocephalans. Similarly, ultrastructural studies should be carried out to compare rotifer adhesive glands and acanthocephalan cement glands.

The molecular data supporting the sister group relationship between Bdelloidea and Acanthocephala appears very strong, but it is based on only two genes from two species of rotifers, and artifacts due to unequal rate effects cannot be completely ruled out. The sequences of more genes from more rotifer taxa should be analyzed, particularly from *Seison*. Studies of rotifer 18S rRNA genes from a large number of rotifer taxa are underway (Walsh, pers. comm.). Other suitable genes would include those of elongation factor-1α, heat shock proteins, triose phosphate isomerase, and some of the more conserved mitochondrial protein genes such as those from cytochrome b and cytochrome oxidase subunit I. We are currently

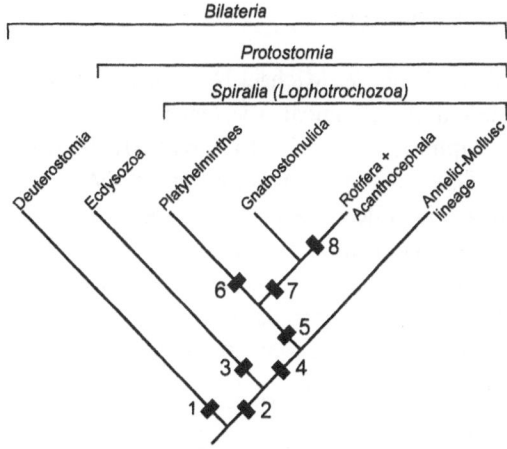

Figure 6. Proposed position of Rotifera within the Bilateria based on morphological and molecular data. The annelid-mollusc lineage refers to the bulk of the non-ecdysozoan protostomes, but not necessarily all of them. Only a few key characters are given. 1: Blastopore becomes the anus. 2: Ventral lateral nerve chord (Ahlrichs, 1995). 3: Molting by ecdysis (Aguinaldo et al., 1997). 4: Spiral cleavage. 5: Filiform sperm without accessory centriole (Ahlrichs, 1995). 6. Biciliary terminal cell in the protonephridia (Ax, 1996). 7: Jaws composed of rods imbedded in a cuticular matrix (Ahlrichs, 1997). 8: Internal layer in the syncytial epidermis (Storch & Welsch, 1969).

sequencing the complete mitochondrial genome of *B. plicatilis* (Li and Garey, unpublished) to make it easier to PCR amplify the mitochondrial genes of other rotifers. It is important for new molecular analyses to be carried out rigorously, with special attention paid to alignments, unequal rate effects, site to site variation, and multiple substitutions at the same site.

The position of Rotifera is becoming clearer as morphological and molecular evidence are considered together. We propose a scheme (Figure 6) that places Rotifera among the Bilateria and appears to be consistent with many of the more recent molecular and morphological studies. In this phylogeny, Rotifera + Acanthocephala is considered a sister group to Gnathostomulida because they all have jaws composed of rods imbedded in a cuticular matrix (Rieger & Tyler, 1995; Ahlrichs, 1997). Rotifera + Acanthocephala + Gnathostomulida are then placed as a sister group to Platyhelminthes with filiform sperm cells without accessory centrioles as a possible synapomorphy (Ahlrichs, 1995). Rotifera + Acanthocephala + Gnathostomulida + Platyhelminthes are considered a sister group to the annelid-mollusc lineage of Spiralia, which in turn is the sister group to Ecdysozoa.

Acknowledgments

We thank Qing Ye, Michael Nonnemacher and Ryan Landefeld for technical assistance. M.N. and R.L. were supported by NSF REU grant BIR-9322152 to JRG. This work was supported by USDA grant 95-37302-2359 to JRG. AS-R was supported by Deutsche Forschungsgemeinschaft (DFG). We thank W. Ahlrichs for helpful discussions and B. Nickol for providing frozen *M. moniliformis* specimens. We also thank Liz Wurdak and four anonymous reviewers for helpful comments.

References

Aguinaldo, A. M. A., J. M. Turbeville, L. S. Linford, M. C. Rivera, J. R. Garey, R. A. Raff & J. A. Lake, 1997. Evidence for a clade of nematodes, arthropods and other moulting animals. Nature 387: 489–493.

Ahlrichs, W., 1995. Zur Ultrastruktur und Phylogenie von *Seison nebaliae* Grube, 1859 und *Seison annulatus* Claus, 1876 – Hypothesen zu phylogenetischen Verwandtschaftsverhältnissen innerhalb der Bilateria. Cuvillier, Göttingen, 310 pp.

Ahlrichs, W., 1997. Epidermal ultrastructure of *Seison nebaliae* and *Seison annulatus*, and a comparison of epidermal structures within the Gnathifera. Zoomorphology 117: 41–48.

Ax, P., 1996. Multicellular Animals, a new approach to the phylogenetic order in nature, Vol. I. Springer, Berlin, 225 pp.

Brusca, R. C. & G. J. Brusca, 1990. Invertebrates. Sinauer, Sunderland, 922 pp.

Clement, P. & E. Wurdak, 1991. Rotifera. In F. W. Harrison & E. E. Ruppert (eds), Microscopic Anatomy of Invertebrates, Vol. 4. Wiley-Liss, Inc., New York: 219–297.

Clement, P., 1993. The phylogeny of rotifers: molecular, ultrastructural and behavioural data. Hydrobiologia 255/256: 527–544.

Davies, K. A. & J. M. Fisher, 1994. On hormonal control of moulting in *Aphelenchus avenae* (Nematoda: Aphelenchida). Int. J. Parasitol. 24: 649–655.

De Rijk, P. & R. De Wachter, 1993. DCSE, an interactive tool for sequence alignment and secondary struture research. Comput. Appl. Biosci. 9: 735–740.

Donoghue, M. J., R. G. Olmstead, F. F. Smith & J. D. Palmer, 1992. Phylogenetic relationships of Dipsacaleson RbcL sequences. Ann. MO Bot. Gard. 79: 333–345.

Dunagan, T. T. & D. M. Miller, 1991. Acanthocephala. In F. W. Harrison & E. E. Ruppert (eds), Microscopic Anatomy of Invertebrates, Vol. 4. Wiley-Liss, Inc., New York: 299–332.

Eernisse, D. J., J. S. Albert, & F. E. Anderson, 1992. Annelida and Arthropoda are not sister taxa: a phylogenetic analysis of spiralian metazoan morphology. Syst. Biol. 41: 305–330.

Ehlers, U., W. Ahlrichs, C. Lemburg & A. Schmidt-Rhaesa, 1996. Phylogenetic systematization of the Nemathelminthes (Aschelminthes). Verh. Dtsch. Zool. Ges. 89.1: 8.

Felsenstein, J., 1988. Phylogenies from molecular sequences: inference and reliability. Annu. Rev. Genet. 22: 521–565.

Felsenstein, J., 1993. PHYLIP: phylogeny inference package, version 3.5, University of Washington, Seattle.

Felsenstein, J. & G. A. Churchill, 1996. A hidden Markov model approach to variation among sites in rate of evolution. Mol. Biol. Evol. 13: 930–104.

Garey J. R., T. J. Near, M. R. Nonnemacher & S. A. Nadler, 1996a. Molecular evidence for Acanthocephala as a sub-taxon of Rotifera. J. Mol. Evol. 43: 287–292.

Garey, J. R., M. Krotec, D. R. Nelson & J. Brooks, 1996b. Molecular analysis supports a Tardigrade-Arthropod association. Invert. Biol. 115: 79–88.

Gupta, A. P., 1990. Morphogenetic hormones and their glands in arthropods: evolutionary aspects. In: Gupta (ed.), Morphogenetic Hormones of Arthropods. New Brunswick: Rutgers University Press: 1–34.

Haffner, K. von, 1950. Organisation und systematische Stellung der Acanthocephalen. Zool. Anz. 145: (suppl.) 243–274.

Halanych, K. M., J. D. Bacheller, A. A. Aguinaldo, S. M. Liva, D. M. Hillis & J. A. Lake, 1995. Evidence of 18S ribosomal DNA that the Lophophorates are Protostome Animals. Science, 267: 1641–1643.

Hillis, D. M., J. P. Huelsenbeck & C. W. Cunningham, 1994. Application and accuracy of molecular phylogenies. Science 264: 671–677.

Hillis, D. M. & M. T. Dixon, 1991. Ribosomal DNA: molecular evolution and phylogenetic inference. Quart. Rev. Biol. 66: 411–446.

Hillis, D. M. & J. J. Bull, 1993. An empirical test of bootstrapping as a method for assessing confidence in phylogenetic analysis. Syst. Biol. 42: 182–192.

Hyman, L.,1951. The invertebrates: Acanthocephala, Aschelminthes, and Entoprocta. McGraw Hill, New York, 572 pp.

Kumar S, K. Tamura & M. Nei, 1994. MEGA: molecular evolutionary genetics analysis software for microcomputers. Comput. Appl. Biosci. 10: 189–191.

Li, W.-H., 1997. Molecular Evolution. Sinauer Associates, Inc., Sunderland, MA, 487 pp.

Lorenzen, S., 1985. Phylogenetic aspects of pseudocoelomate evolution. In S. Conway-Morris, J. D. George, R. Gibson & H. M. Platt (eds), The Origin and Relationships of Lower Invertebrates. Claredon Press, Oxford: 210-2-23.

Mackey, L. Y., B. Winnepennickx, T. Backeljau, R. De Wachter, P. Emschermann & J. R. Garey, 1996. 18S rRNA suggests that Entoprocta are Protostomes, unrelated to Ectoprocta. J. Mol. Evol. 42: 552–559.

Malakhov, V. V., 1994. Nematodes. Smithsonian Institution Press, Washington, 286 pp.

Markevich, F. I., 1993. Phylogenetic relationships of Rotifera to other vermiform taxa. Hydrobiologia 255/256: 521–526.

Miller, D. M. & T. T. Dunagan, 1985. Functional morphology. In D. W. T. Crompton & B. B. Nickol (eds), Biology of the Acanthocephala. Cambridge University Press, Cambridge: 73–123.

Near, T. J., J. R. Garey & S. A. Nadler, 1997. Phylogenetic relationships of the acanthocephala inferred from 18S ribosomal DNA sequences. Mol. Phyl. Evol., in press.

Nebelsick, M., 1993. Introvert, mouth cone, and nervous system of *Echinoderes capitatus* (Kinorhyncha, Cyclorhagida) and implications for the phylogenetic relationships of Kinorhyncha. Zoomorphology 113: 211–232.

Neuhaus, B., 1994. Ultrastructure of alimentary canal and body cavity, ground pattern, and phylogenetic relationships of the Kinorhyncha. Microfauna Marina 9: 61–156.

Nielsen, C., 1995. Animal evolution. Interrelationships of the living phyla. Oxford University Press, Oxford, 467 pp.

Nielsen, C., N. Scharff & D. Eibye-Jacobsen, 1996. Cladistic analysis of the animal kingdom. Biol. J. linn. Soc. 57: 385–410.

Pennak, R. W., 1989. Fresh-water invertebrates of the United States, Protozoa to Mollusca, 3rd Ed. John Wiley and Sons, New York, 628 pp.

Raff, R. A., C. R. Marshall & J. M. Turbeville, 1994. Using DNA sequences to unravel the Cambrian radiation of the animal phyla, Annu. Rev. Ecol. Syst. 25: 351–374.

Remane, A., 1963. The systematic position and phylogeny of the pseudocoelomates. In E. C. Dougherty (ed.), The Lower Metazoa. Univ. of Calif. Press, Berkeley: 247–255.

Ricci, C., G. Melone & C. Sotgia, 1993. Old and new data on Seisonidea (Rotifera). Hydrobiologia 255/256: 495–511.

Rieger, R. M. & S. Tyler, 1995. Sister-group relationship of Gnathostomulida and Rotifera-Acanthocephala. Invertebrate Biology 114: 186–188.

Ruppert, E. E. (1991) Introduction to the aschelminth phyla: a consideration of mesoderm, body cavities, and cuticle. In F. W. Harrison & E. E. Ruppert (eds), Microscopic Anatomy of Invertebrates, Vol. 4: Aschelminthes. Wiley-Liss, New York: 1–17.

Ruppert, E. E. & R. D. Barnes, 1994. Invertebrate Zoology, 6th Ed., Saunders College Publ., Fort Worth, 1056 pp.

Schram, F. R., 1991. Cladistic analysis of metazoan phyla and the placement of fossil problematica. In A. Simonetta & S. Conway Morris (eds), The Early Evolution of Metazoa and the Significance of Problematic Taxa. Cambridge University Press, Cambridge: 35–46.

Storch, V. & U. Welsch, 1969. Uber den Aufbau des Rotatorienintegumentes. Z. Zellforsch. 95: 405–414.

Telford, M. J. & P. W. H. Holland, 1993. The phylogenetic affinities of the chaetognaths: A molecular analysis. Mol. Biol. Evol. 10: 660–676.

Wallace, R. L., C. Ricci & G. Melone, 1996. A cladistic analysis of pseudocoelomate (aschelminth) morphology. Invertebrate Biol. 115: 104–112.

Wallace, R. L. & R. A. Colburn, 1989. Phylogenetic relationships within phylum Rotifera: orders and genus Notholca. Hydrobiologia 186/187: 311–318.

Winnepenninckx, B., T. Backeljau, L. Y. Mackey, J. M. Brooks, R. De Wachter, S. Kumar & J. R. Garey, 1995. 18S rRNA data indicate that the aschelminthes are polyphyletic and consist of at least three distinct clades. Molec. Biol. Evol. 12: 1132–1137.

Hydrobiologia **387/388**: 93–96, 1998.
E. Wurdak, R. Wallace & H. Segers (eds), Rotifera VIII: A Comparative Approach.
© *1998 Kluwer Academic Publishers.*

Opinion paper

Are lemnisci and proboscis present in the Bdelloidea?

Claudia Ricci
Department of Biology, Milan State University, Milan, Italy

Phylum Rotifera is typically defined by the presence of a (a) primary body cavity, (b) specialized ciliary apparatus called the corona located at the anterior end, and (c) jaw apparatus of hard parts (trophi) (e.g., de Beauchamp, 1965). To this list of characters Nielsen (1995) added (d) an intracytoplasmic skeletal lamina, (e) a retrocerebral apparatus, and (f) a foot with adhesive glands that open at the toes. On the basis of ultrastructural studies other distinctive characters for the phylum have been proposed (Clément & Wurdak, 1991): (g) absence of collagen, (h) a glia-free centralized nervous system composed of a dorsal main ganglion (brain), (i) multiciliated flame cells. Other general characters have been added to the list (Nogrady et al., 1993): (l) a tiny, unsegmented body divided into three typical regions (head, trunk and a post-cloacal foot), and (k) sensory antennae present in variable numbers on the animal's surface.

Although all these characters are present in rotifers, not all are autapomorphic. For example, the primary body cavity is common to almost all of the so-called aschelminths, a jaw apparatus is present in the Gnathostomulida and in Kristensen's (1995) 'new taxon No. 1' (not yet described), and the multiciliated flame cells are shared by several pseudocoelomate phyla.

Even though the taxonomic position of the rotifers has been addressed by several authors, there is no consensus as to the ancestral form. In fact, very different origins have been proposed: i.e., from a turbellarian-like ancestor (e.g., de Beauchamp, 1907; Clément, 1985; Kutikova, 1983) to a primitive or, at least, much simplified annelid worm (e.g., Remane, 1963). Since the general rejection of the 'Aschelminths' as a taxon (e.g., Willmer, 1990), the taxonomic position of the Rotifers has received much attention (Clément, 1993; Haszprunar, 1996). Thus, while rotifers have been linked to several unique taxa, including Monogenea

(Markevich, 1993), Chaetognatha (Nielsen, 1995), and Gnathostomulida (Rieger & Tyler, 1995), there is a general consensus that they have a close phylogenetic tie to phylum Acanthocephala (Lorenzen, 1985; Wallace et al., 1996; Winnepenninckx et al., 1995). This affinity was first proposed by Wright & Lumsden (1969) on the basis of the ultrastructure of the syncytial epidermis, and later on the basis of other characters, such as sperm and muscle ultrastructure (Whitfield, 1971). Actually the strongest morphological evidence of a close relationship between the two taxa rests on the (1) presence of a skeletal lamina inside the syncytial epidermis (Ruppert, 1991; Storch, 1979), (2) arrangement of the blastomeres during cleavage (Nielsen, 1995), and (3) fine structure of the spermatozoon (Melone & Ferraguti, 1994).

Phylum Rotifera consists of three classes: marine dioecious Seisonidea (one genus with two species), Bdelloidea, exclusively parthenogenetic and able to undergo anhydrobiosis and to creep with leech-like movements (three orders with about 370 species), and Monogononta, representing the largest group of rotifers, characterized by the well-known heterogonic cycle (three orders with about 1600 species). Although the systematic position of the rotifers was an open question and the relationships among the different rotifer groups were being debated, no one questioned the monophyly of the phylum until Lorenzen (1985) proposed Acanthocephala to be highly specialized bdelloids due to presence of two 'holapomorphic' (=synapomorphic) characters: proboscis and paired lemnisci. Unfortunately, Lorenzen's hypothesis has been used as external support for a phylogeny based on molecular characters (Garey et al., 1996) and has even become established within main-stream biological knowledge with its appearance in a recent zoology textbook (Westheide & Rieger, 1996). This is indeed unfortunate because Lorenzen's hypothesis has been rejected by both students of Rotifera (e.g., Clément, 1993; Markevich, 1993; Wallace & Colburn, 1989) and by systematicists (e.g., Nielsen, 1995). More importantly, Lorenzen's conclusions have been accepted without a critical examination of the arguments supporting his hypothesis.

So while the molecular support for a close relationship between acanthocephalans and rotifers is strong, even to the point of placing Acanthocephala within the Rotifera clade (Garey et al., 1996, 1998; Mark-Welch, pers. comm.), morphologically based analyses do not support this conclusion (Melone et al., 1998). Thus, if we are to understand the phylogenetic relationship of these two taxa, both molecularly based and morphologically based data sets need to be inspected carefully. Here I examine Lorenzen's conclusions that the characters lemnisci and proboscis also are present in the Bdelloidea.

Character analysis

Lemnisci

Always present in the Acanthocephala from the older acanthella stage onwards, lemnisci are usually paired, elongated structures, located between the body wall and the retractor muscles in the neck. Their surface is covered by a layer of collagenous connective tissue. Lemnisci develop by ingrowth of the syncytial tegument (Dunagan & Miller, 1991) and thus are of ectodermal origin. Their function is unclear, but they contain deposits of glycogen and fatty material.

Lorenzen identified structures in the drawings of two bdelloid species, *Mniobia* (=*Callidina*) *symbiotica* and *Zelinkiella* (=*Discopus*) *synaptae* (Zelinka, 1886, 1888), to be lemnisci. Both Zelinka (1886, 1888) and Remane (1928–33) described such structures as hypodermic cushions made of cells of the rotatory apparatus, common to both bdelloid and monogonont Rotifers. The relationship between hypodermic cushions and the rotatory apparatus is made clear by the observation that when the corona is made by multiple parts, multiple cushions are present, and that none are found in species that do not possess a well developed corona (e.g., Bdelloidea: Adinetida). For example, Seisonidea do not have a well-developed rotatory apparatus, nor do they possess hypodermic cushions. Ultrastructural studies revealed that the rotatory apparatus of both monogonont and bdelloid rotifers has hypodermal cushions that accomodate the

long, striated rootlets of the cilia (Clément & Wurdak, 1991). Thus, the rotifer hypodermic cushions and the acanthocephalan lemnisci share only an ectodermal derivation; the former are ciliated cell bodies and the latter are an ingrowth of the syncythial tegument. On the basis of this analysis, I argue that there is no evidence for structural homology between these two structures.

Proboscis

In Acanthocephala the proboscis is invaginable and its wall contains fluid-filled lacunar channels, similar to, but independent from, the lacunar system of the trunk. When invaginated, the proboscis is accomodated in a receptacle distinct from the body cavity and when extended, it anchors the worm to the intestine wall of its host. A cerebral ganglion that innervates the apical sense organ at the tip of the proboscis is found in the proboscis receptacle. Radially symmetrical, the proboscis of the Acanthocephala cannot be considered an introvert because of the absence of the mouth. The proboscis develops in the acanthella from the most anterior group of nuclei of the peculiar 'inner nuclear mass,' originating from some micromeres (Crompton, 1989).

Among rotifers, only bdelloids possess an apical rostrum, that is bilaterally symmetrical. Generally withdrawn when the corona is extended, it is used when the animal is crawling on a surface or is creeping with leech-like movements. Typically the rostrum is made of a dorsal lamella, a sensitive region and a ciliated area (Melone & Ricci, 1995). Apically, the rostrum bears the outlet of the retrocerebral apparatus ducts. It does not possess its own ganglion, but is innervated by the brain. The bdelloid rostrum is derived embryologically from the dorsal medial part of the apical field; this is the same region that gives rise to the rotatory apparatus (Zelinka, 1891). Notwithstanding that both acanthocephalan proboscis and bdelloid rostrum are involved in the animal's movement, the former to impede dislodgement, the latter to facilitate locomotion, it appears premature to consider these two structures homologous.

Phylogeny

While the absence of reliable morphologically based, homologous structures does not offer support for the hypothesis that Bdelloidea and Acanthocephala are closely tied, it does not negate the possibility that they are related. In fact, a relationship between Rotifera and Acanthocephala is clearly established (e.g., Garey et al., 1996; Garey et al., 1998; Melone et al., 1998).

Bdelloids, because of their obligate parthenogenesis and capacity of surviving desiccation, may have separated early from the rest of the phylum Rotifera; however, the dioecious Seisonidea may be a better candidate for the most primitive rotifer. On the other hand, Monogononta appear more evolved, especially those belonging to order Ploima. All three taxa comprise a single taxon, Rotifera, because they share all of the characters discussed above, in particular corona, trophi, retrocerebral apparatus, foot with adhesive glands, three body regions, lack of collagen, and intracytoplasmic skeletal lamina (Melone et al., this volume). All of these characters, with the exception of the intracytoplasmic skeletal lamina, are absent in acanthocephalans. The latter taxon is comprised of highly specialized endoparasites that do not have free-living stages, and can be assumed to possess reductions typical to adaptation to such a life style: loss of the digestive system, simplification of nervous system, amplification of the reproductive system. The absence of structures like the gut and probably others could be interpreted as a secondary loss

However, while the close relationship Bdelloidea + Acanthocephala is not supported by morphological data, which is probably influenced by the very different life histories of both groups (Melone et al., this volume), it is strongly supported by molecular data. After the comparison of 18S rRNA sequences of a highly evolved monogonont species (*Brachionus plicatilis*) and of a bdelloid species (*Philodina acuticornis*) with sequences of three acanthocephalan species, a tree was obtained that links Bdelloidea and Acanthocephala in the superclass Lemniscea, established on the basis of the supposed common lemnisci (Garey et al., 1996). Close resemblance between the two taxa came from studies of mitochondrial 16S rRNA (Garey et al., 1998) and from hsp82 (Mark-Welch, pers. comm.), lending support to a close relationship between Acanthocephala and Bdelloidea.

The close resemblance of the Bdelloid and Acanthocephala sequences is remarkable, and the nature of the supporting evidence is convincing. Such results are very important and represent a stimulating challenge for further morphological studies. However, establishing new taxa may still be premature: I suggest that some taxa like *Seison* or Flosculariacea are added to the list of taxa to be investigated using molecular

techniques. They offer several characters which are either held in common with Acanthocephala or are considered primitive within Rotifera, and are prime candidates to lie near the basal node of the Rotifera. It is clear that additional morphological, biochemical, and behavioral studies ought to be conducted to bring molecularly and morphologically based trees into greater harmony.

Acknowledgements

I wish to thank David Mark-Welch for the communication of his very recent and important results on the comparison between Acanthocephala and Bdelloidea *hsp*82 sequences. A number of colleagues helped in clarifying my ideas. I am grateful for critical reading of manuscript draft by M. Ferraguti, G. Melone, P. Menozzi, A. Minelli, H. Segers, R. Wallace, C. Sotgia and A. Zullini.

References

Bereiter-Hahn, J., A. G. Matoltsky, & K. S. Richards, 1984. Biology of the Integument. 1. Invertebrates. Springer-Verlag, Berlin.

Clément, P., 1985. The relationships of rotifers. In Conway Morris, S., J. D. George, R. Gibson & H. M. Platt (eds), The Origins and Relationships of Lower Invertebrates Claredon Press, Oxford: 224–247.

Clément, P., 1993. The phylogeny of rotifers: molecular, ultrastructural and behavioural data. Hydrobiologia 255/256: 527–544.

Clément, P. & E. Wurdak, 1991. Rotifera. In Harrison, F. W. & E. E. Ruppert (eds), Microscopic Anatomy of Invertebrates, Vol. 4: Aschelminthes. Wiley-Liss, New York: 219–297

Crompton, D. W. T., 1989. Acanthocephala. In Adijodi, K. G. & R. Adijodi (eds), Reproductive Biology of Invertebrates, vol. IV. Fertilization, Development and Parental Care (Part A). John Wiley and Sons, New York: 215–258.

de Beauchamp, P., 1907. Morphologie et variations de l'appareil rotateur des Rotifères. Arch. Zool. exp. gén. 6: 1–29.

de Beauchamp, P., 1965. Classe des Rotifères. In P. P. Grassé, Traite de Zoologie. Tome IV Masson et C, Paris, 1225–1379.

Dunagan, T. T. & D. M. Miller. 1991. Acanthocephala. In F. W. Harrison & E. E. Ruppert (eds), Microscopic Anatomy of Invertebrates. Vol. 4: Aschelminthes. Wiley-Liss, New York: 299–332.

Garey, J. R., J. N. Near, M. R. Nonnemacher & S. A. Nadler, 1996. Molecular evidence for Acantocephala as a subtaxon of Rotifera. J. Mol. Evol. 43: 287–292.

Garey, J. R., A. Schmidt-Rhaesa, T. J. Near & S. A. Nadler, 1998. The evolutionary relationships of rotifers and acanthocephalans. Hydrobiologia 387/388 (Dev. Hydrobiol. 134): 83–91.

Haszprunar, G., 1996. Plathelminthes and Plathelminthomorpha—paraphyletic taxa. J. Zool. Syst. Evol. Res. 34: 41–48.

Kristensen, R. M., 1995. Are Aschelminthes pseudocoelomate or acoelomate? In Lanzavecchia, G., R. Valvassori & M. D. Candia Carnevali (eds), Body Cavities: Function and Phylogeny. Mucchi, Modena: 41–43.

Kutikova, L. A., 1983. Parallelism in the evolution of rotifers. Hydrobiologia 104: 3–7.

Lorenzen, S., 1985. Phylogenetic aspects of pseudocoelomate evolution. In Conway Morris, S., J. D. George, R. Gibson & H. M. Platt (eds), The Origins and Relationships of Lower Invertebrates. Claredon Press, Oxford: 210–223.

Markevich, G. I., 1993. Phylogenetic relationships of Rotifera to other vermiform taxa. Hydrobiologia 255/256: 521–526.

Melone, G. & M. Ferraguti, 1994. The spermatozoon of *Brachionus plicatilis* (Rotifera, Monogononta) with some notes on sperm ultrastructure in Rotifera. Acta Zool. 75: 81–88.

Melone, G. & C. Ricci, 1995. Rotatory apparatus in Bdelloids. Hydrobiologia 313/314: 91–98.

Melone, G., C. Ricci, H. Segers & R. L. Wallace, 1998. Phylogenetic relationships of phylum Rotifera with emphasis on the families of Bdelloidea. Hydrobiologia 387/388 (Dev. Hydrobiol. 134): 101–107.

Nielsen, C., 1995. Animal Evolution. Oxford University Press, Oxford.

Nogrady, T., R. L. Wallace & T. W. Snell, 1993. Rotifera. SPB Academic Publishing, The Hague, The Netherlands.

Remane, A., 1929/33. Rotatorien. Bronn's Klassen und Ordnungen des Tierreichs. Akademische Verlagsgesellschaft, Leipzig.

Remane, A., 1963. The systematic position and phylogeny of the Pseudocoelomata. In Dougherty, E. C. (ed.), The Lower Invertebrates. University of California Press, Berkeley: 247–255.

Rieger, R. M. & S. Tyler, 1995. Sister-group relationship of Gnathostomulida and Rotifera–Acanthocephala. Invertebr. Biol. 114: 186–189.

Ruppert, E. E., 1991. Introduction to the Aschelminth phyla: a consideration of mesoderm, body cavities, and cuticle. In Harrison, F. W. & E. E. Ruppert (eds), Microscopic Anatomy of Invertebrates, Vol. 4: Aschelmintes. Wiley-Liss, New York: 1–17.

Storch, V., 1979. Contributions of comparative ultrastructural research to problems of invertebrate evolution. Am. Zool. 19: 637–645.

Wallace, R. L. & R. A. Colburn, 1989. Phylogenetic relationships within phylum Rotifera: orders and genus *Notholca*. Hydrobiologia 186/187: 311–318.

Wallace, R. L., C. Ricci & G. Melone, 1996. A cladistic analysis of pseudocoelomate (aschelminth) morphology. Invertebr. Biol. 115: 104–112.

Westheide, W. & R. Rieger (eds), 1996. Spezielle Zoologie. Fischer Verlag, Stuttgardd, Jena, New York.

Whitfield, P. J., 1971. Phylogenetic affinities of Acanthocephala: an assessment of ultrastructural evidence. Parasitology 63: 49–58.

Willmer, P., 1990. Invertebrate Relationships: Patterns in Animal Evolution. Cambridge University Press, Cambridge.

Winnepenninckx, B., T. Backeljau, L. Y. Mackey, J. M. Brooks, R. De Wachter, S. Kumar & J. R. Garey, 1995. 18S rRNA data indicate that Aschelminthes are polyphyletic in origin and consist of at least three distinct clades. Mol. Biol. Evol. 12: 1132–1137.

Wright, R. D. & R. D. Lumsden, 1970. The acanthor tegument of *Moniliformis dubius*. J. Parasitol. 56: 727–735.

Zelinka, C., 1886. Studien uber Rädertiere. I. Uber die Symbiose und Anatomie von Rotatorien aus dem Genus *Callidina*. Z. wiss. Zool. 44: 396–506.

Zelinka, C., 1888. Studien uber Rädertiere. II. Der Raumparasitismus und die Anatomie von *Discopus synaptae* n.g. nov. sp. Z. wiss. Zool. 47: 353–458

Zelinka, C., 1891. Studien uber Rädertiere. III. Zur Entwicklunsgeschichte der Radertiere nebst Bemerkungen über ihre Anatomie und Biologie. Z. wiss. Zool. 53: 1–159.

Hydrobiologia **387/388**: 97–100, 1998.
E. Wurdak, R. Wallace & H. Segers (eds), Rotifera VIII: A Comparative Approach.
© *1998 Kluwer Academic Publishers.*

Numerical phenetic taxonomy and its heuristic aspects

Thomas Nogrady

Department of Biology, Queen's University, Kingston, Ontario K7L 3N6, Canada

Abstract

The principal multivariate methods used in numerical taxonomy and their graphic representations are outlined. Advantages of, and problems with, numerical taxonomy and the definition of phenetic taxonomic characters, recent use of ultrastructural, biochemical and genetic tools and their application to taxonomy, and elucidation of non-cosmopolitanism and speciation is discussed. The application of a combination of molecular and numerical methodologies is recommended. The capability of numerical taxonomy to reveal new knowledge is explored. It could lead to recognition of multiple correlations hidden in the data, to ordination of populations, and the use of phenetic descriptors in defining multidimensional niches.

Introduction

Numerical taxonomy is the grouping of taxonomic units by numerical methods into taxa on the basis of their character states (Sneath & Sokal, 1973). Multivariate techniques have no inherent biological meaning; discriminant functions are simply able to distinguish among many overlapping data sets with added information, i.e. more data lead to better differentiation. The grouping or clusters obtained are based on phenetic resemblance only and have no phyletic connotations. There are many excellent textbooks on the topic. The most widely used ones are Sneath & Sokal (1973), Romesburg (1984), Abbot et al. (1985), Everitt & Dunn (1992), Minelli (1993), just to mention a few. An eminently practical collection of computer programs is that of NTSYS-pc v. 2.0 by Rohlf (1997), now available for the Windows 95 platform.

The most commonly used multivariate tools are similarity and dissimilarity programs which then are used in clustering (Romesburg, 1984), most often by the SAHN (sequential, agglomerative, hierarchical and nested clustering, Sneath & Sokal, 1973) method, or in constructing minimal spanning trees. The ordination methods allow the interpretation of the differences among objects through their display in space of whose dimensions are defined by the characters of the operational taxonomic units (OTUs, 'descriptors', 'characters'). Correspondence analysis, principal coordinates analysis, and multidimensional scaling are the most widely used ordination tech-

niques. Their graphic representation can be as scatter diagrams and two- or three-dimensional plots. Clustering methods assign OTUs to groups forming hierarchical non-overlapping groups, known as two-dimensional dendrograms (trees). Examples of these can be seen in Nogrady & Wallace (1995). Many of these techniques are also widely applied in ecological studies.

Taxonomic characters

The prevalent view of phenetic classification is probabilistic: in a polythetic taxon membership is based on the greatest number of shared character states; no single state is either sufficient or essential for inclusion in the group. Instead of using the classical dichotomous step-by-step keys, multivariate methods use data matrices containing all OTUs for all taxa under consideration. The proper selection of OTUs is crucial for arriving at a stable and convenient classification. However, preconceived ideas held by 'experienced taxonomists' as to which OTUs to use may lead to 'expected' but certainly dogmatic classifications, and the heuristic advantage of the numerical method is lost. Equally problematic are the questions of the requisite number and weighting of OTUs. There are no valid answers concerning the number of characters used, but an increase in numbers usually results in an asymptotic decrease in the change of phenetic similarity allowing to arrive at a practical and optimal number of OTUs.

Weighting characters is undesirable as it may be subjective. If used at all, *a priori* weighting is the only acceptable method, and strict rules must be defined beforehand (Neff, 1986).

Numerical taxonomy applied to known systems may reveal the insufficiency of available taxonomic characters, either by leading to unsatisfactory groupings or showing a lack of fine definitions (Benzie, 1986b; Nogrady & Wallace, 1995). This can, on occasion, be remedied by mathematical manipulation: Benzie (1988) could account for successful isolation and resolution of cyclomorphotic female forms of *Daphnia carinata* after log-transformation of data, where clustering, principal component analysis and discriminant methods were not capable to separate them from other crested forms. All these problems indicate that the exclusive use of classical microscopic OTUs is insufficient in arriving at classifications required in solving increasingly sophisticated taxonomic problems.

The use of taxonomic characters that go beyond classical α-taxonomy by utilizing recent advances in ultrastructural microscopy and biochemistry has been championed by Koste & Shiel as early as 1989. The book by Avise (1991) deals with tools and methodology in molecular phylogenetics, speciation and clones. In the context of rotifers, methods are discussed in other presentations of this workshop (Gomez, Rico-Martinez, Watson & Russel and Walsh) but a brief review and leading references are still offered. SEM pictures (cf. Claugher, 1990) of trophi are already widely used as taxonomic characters in recent rotifer publications, e.g., by De Smet (this volume); the easily accessible allozyme electrophoresis has been pioneered in rotifer taxonomy by King (1977), Snell (1989), and Serra and co-workers (this volume) and used on cladocerans by Benzie (1986). Other leading references to these topics are those of Buth (1984), and Sokal (1986). Among molecular methods nucleic acid sequencing is the most promising and easily accessible by recent techniques, notably polymerase chain reaction (see Starkweather, 1993; Walsh & Starkweather, 1993; communications of Garey et al., Walsh et al., this workshop). Amplification of minute amounts of nucleotide sequences is equally useful in systematic, cladistic and biogeographic research. Restriction fragment mapping and DNA hybridization are also valuable methods. For example, it has been shown (Hughes, 1992) that avian species could be described on the basis of DNA only. All these data are eminently suitable as OTUs for

numerical taxonomic evaluation in addition to classical characters or, even better, by themselves, as they may shed light on the shortcomings and dead-ends of results obtained by α-taxonomic methodology.

While progress has been reported in many fields, Sneath (1995), a leading figure in numerical taxonomy, still faults the taxonomist of having become sidetracked by cladistic controversies and not having moved fast enough on molecular taxonomy. Taxonomic problems can be anticipated in genera comprising hermaphroditic species or parthenogenetic species of hybrid or clonal origin, notably bdelloid rotifers (Walker, 1986).

Heuristic aspects of numerical taxonomy

New knowledge, as mentioned above, can be obtained by applying numerical methods to existing orthodox taxonomy. Numerical taxonomy may either confirm a classical arrangement or, frequently, suggest major revisions by uncovering confused systems or others refractory to orthodox methodology. Insufficiency of available characters for taxonomic purposes is also often revealed (see Nogrady & Wallace, 1995).

One of the most interesting consequences of utilizing numerical methodology was the corroboration that 'cosmopolitanism' of many species is an erroneous assumption based on the eurocentricity and insufficient techniques of 19th century zoological research. Frey (1987) discusses this at length in his seminal paper on cladoceran biogeography, the existence of discrete local species in spite of apparent phenetic similarity, and the fallacious assumptions of facile transport of resting eggs (to the latter topic see Jenkins, this volume). Different species carrying the same name have been separated using electrophoresis, time of the year for gamogenesis, and morphology of populations (not individuals) by regressing measurements against length of body or length of postabdomen (Frey, 1987). The non-cosmopolitan nature of rotifer species has been repeatedly emphasised by researchers working in Australia and Southeast Asia, notably Russel Shiel in numerous publications, as well as Dumont & Segers (1996), but needs to be followed up in detail. It is surprising how little exchange of genetic material occurs even between contiguous areas like the northern and southern regions of the United States (Frey, 1987). Untangling these groups of morphologically similar, but well-defined, species is a major challenge in rotifer taxonomy, solvable almost exclusively by biochemical and numerical tools only, as well as working with populations instead of individuals. Hebert (e.g., 1994 and numerous other papers) has pioneered the molecular treatment of such problems with cladocerans and dealt with species boundaries, phenotypic plasticity, hybridization, morphologically distinct but genetically similar species and other such problems, pushing such knowledge way ahead of rotifers.

Another major advance based on the use of numerical methodology is the recognition of the niche as a multi-dimensional hyperspace resolvable into niche width, dimensionality and shape. Species within genera or populations within a species reflect variability; the genetic continuum may result in continua in phenetic space. These, as well as discontinua reflect on ecological niches and even phylogenetic processes. Phenetic spaces are also reflections on habitat diversity (Hutchinson, 1957, 1968; Sokal, 1986). Multiple correlations jointly comparing morphology, genetics and geography can partition taxonomic relationships by using a modification of the Mantel test (a significance test) so that space maps and habitats become a matrix correlation and seem to be more restrictive than previously thought (see Sokal, 1986).

Conclusions

The recent review of Quieroz & Good (1997) provides an appraisal of the pros and cons of the use and practice of numerical taxonomy, is eminently relevant, deserving a brief summary in this paper.

Phenetics was a reform movement to remedy the subjective and non-quantitative nature of the taxonomy of the 1950s. Its limitations have been known for a long time, and it has not become as successful a taxonomic tool as hoped, at least not with the prevalent methodology. The fact that it has produced so many methods without being able to offer a clear preference attests to that.

Phenetic clustering rests on questionable assumptions on nested hierarchical structures, and correspondence between similarity and degree of genetic continuity, thus its usefulness is limited in defining species limits (genetic continuity) and geographic variations. Nevertheless, cluster analysis is still a convenient, easy and visual method, even though it is well known that any set of similarity data will produce a dendrogram. Thus caution is advisable.

Ordination methods are preferable to clustering when summarizing non-hierarchical patterns, but

do not form explicit groups. Principal component and principal coordinate analyses as well as multi-dimensional scaling summarize similarity data in as few as three dimensions without imposing nested hierarchies.

The quality and relevance of OTUs seems to have a profound influence on achieving a rational and heuristically productive taxonomy, thus the use of molecular methodology may help to initiate progress in a potentially profitable field.

References

Abbott, L. A., F. A. Bisby & D. J. Rogers, 1985. Taxonomic Analysis in Biology. Columbia University Press: 336 pp.

Avise, J., 1991. Molecular Markers in Natural History and Evolution. Chapman & Hall, New York: 511 pp.

Benzie, J. H. A., 1986a. Phylogenetic relationships within the genus Daphnia (Cladocera: Daphniidae) in Australia, determined by electrophoretically detectable protein variation. Aust. J. Freshwat. Res. 37: 251–260.

Benzie, J. H. A., 1986b. Phenetic and cladistic analyses of the phylogenetic relationship within the genus Daphnia worldwide. Hydrobiologia 140: 105–124.

Benzie, J. H. A., 1988. The systematics of Australian Daphnia (Cladocera: Daphniidae). Multivariate morphometrics. Hydrobiologia 166: 163–182.

Bonhomme, I., 1986. Molecules, population and species evolution in the genus Mus (Mammalia, Rodentia). In Iwatsuki, K., P. H. Raven & W. J. Bock (eds), Modern Aspects of Species. Univ. Tokyo Press: 125–143.

Buth, D. G., 1984. The application of electrophoretic data in systematic studies. Annu. Rev. Ecol. Syst. 15: 501–522.

Claugher, D. (ed.), 1990. Scanning Electron Microscopy in Taxonomy and Functional Morphology. Syst. Assoc. Special Volume 41. Clarendon Press, Oxford: 315 pp.

Cole, C. J., 1985. Taxonomy of parthenogenetic species of hybrid origin. Syst. Zool. 34: 359–363.

DeBry, R. W. & N. A. Slade, 1985. Cladistic analysis of restriction endonuclease cleavage maps within maximum-likelihood framework. Syst. Zool. 34: 21–34.

Dumont, H. J. & H. Segers, 1996. Estimating lacustrine zooplankton species richness and complementarity. Hydrobiologia 341: 125–132.

Edwards, M. & D. R. Morse, 1995. The potential for computer-aided identification in biodiversity research. Trends Ecol. Evol. 10: 153–158 and 416–417.

Everitt, B. S. & G. Dunn, 1992. Applied Multivariate Data Analysis. Oxford Univ. Press, New York: 304 pp.

Felsenstein, J. (ed.), 1983. Numerical Taxonomy. Springer, Berlin: 644.

Frey, D. G., 1987. The taxonomy and biogeography of the Cladocera. Hydrobiologia 145: 5–17.

Hartman, S. E., 1988. Evaluation of some alternative procedures used in numerical systematics. Syst. Zool. 37: 1–18.

Hebert, P. D. N. & R. A. Melo, 1994. Allozymic variation and species diversity in North American Bosminidae. Can J. Fish. aquat. Sci. 51: 873–880.

Hughes, A. L., 1992. Avian species described on the basis of DNA only. Trends Ecol. Evol. 7: 2–3.

Hutchinson, G. E., 1957. Concluding remarks. Cold Spring Harbour symp. Quant. Biol. 22: 415–427.

Hutchinson, G. E., 1968. When are species necessary? In Lewontin, R. C. (ed.), Population Biology and Evolution. Syracuse Univ. Press: 177–186.

Jackson, J. E., 1991. A User's Guide to Principal Components. Wiley, New York: 569.

King, C. E., 1977. Genetics of reproduction, variation, and adaptation in rotifers. Arch. Hydrobiol. Beih. 8: 187–201.

Koste, W. & R. J. Shiel, 1989. Classical taxonomy and modern methodology. Hydrobiologia 186/187: 279–284.

Minelli, A., 1993. Biological Systematics. Chapman and Hall, London: 387 pp.

Neff, N. A., 1986. A rational basis for a priori character weighting. Syst. Zool. 35: 110–123.

Nogrady, T. & R. L. Wallace, 1995. Numerical taxonomic studies of the genus *Notholca*. Hydrobiologia 313/314: 99–104.

Quieroz, K. De & Good, D. A, 1997. Phenetic clustering in biology: a critique. Q. Rev. Biol. 72: 3–30.

Rohlf, F. J., 1996. NTSYS-pc for Windows. Numerical taxonomy system. Version 2.0. Exeter Software, 47 Route 25A, Suite 2, Setauket, New York 11733–2870, U.S.A.

Rohlf, F. J. & L. F. Marcus, 1993. A revolution in morphometrics. Trends Ecol. Evol. 8: 129–132.

Romesburg, H. C. 1984. Cluster analysis for researchers. Lifetime Learning Publ., Belmont, CA: 334 pp.

Scott-Ram, N. R., 1990. Transformed cladistics, taxonomy and evolution. Cambridge Univ. Press: 238 pp.

Sneath, P. H. A. & R. R. Sokal, 1973. Numerical Taxonomy. The Principles and Practice of Numerical Classification. Freeman & Co., San Francisco: 573 pp.

Sneath, P. H. A., 1995. Thirty years of numerical taxonomy. Syst. Biol. 44: 281–298.

Snell, T. W., 1989. Systematics, reproductive isolation and species boundaries in monogonont rotifers. Hydrobiologia 186/187: 299–310.

Sokal, R. R., 1986. Phenetic taxonomy: theory and methods. Ann. Rev. Ecol. Syst. 17: 423–442.

Starkweather, P. L., 1993. Hierarchical gene trees and molecular phylogeny of the Rotifera: use of the polymerase chain reaction (PCR) to dissect ecological and evolutionary patterns. Hydrobiologia 255/256: 551–555.

Swofford, D. L. & G. J. Olsen (eds), 1990. Molecular Systematics. Sinauer Assoc., Sunderland, MA.

Walker, J. M., 1986. The taxonomy of parthenogenetic species of hybrid origin: cloned hybrid populations of Cnemidophorus (Sauria: Teiidae). Syst. Zool. 35: 427–440.

Walsh, E. J. & P. L. Starkweather, 1993. Analysis of rotifer ribosomal gene structure using the polymerase chain reaction (PCR). Hydrobiologia 255/256: 219–224.

Hydrobiologia **387/388:** 101–107, 1998.
E. Wurdak, R. Wallace & H. Segers (eds), Rotifera VIII: A Comparative Approach.
© *1998 Kluwer Academic Publishers.*

Phylogenetic relationships of phylum Rotifera with emphasis on the families of Bdelloidea

Giulio Melone[1], Claudia Ricci[1], Hendrik Segers[2] & Robert L. Wallace[3]
[1]*Dipartimento di Biologia, Universita di Milano, Via Celoria 26, 20133 Milano, Italy*
[2]*Laboratory of Animal Ecology, Zoogeography and Nature Conservation, Dept. M.S.E., University of Ghent, K.L. Ledeganckstraat 35, B-9000 Gent, Belgium*
[3]*Department of Biology, Ripon College, Ripon, WI 54971-0248, U.S.A.*

Key words: Acanthocephala, aschelminthes, cladistics, evolution, Gnathostomulida, phylogeny, pseudocoelomates, Rotifera

Abstract

We investigated phylogenetic relationships of phylum Rotifera using cladistic analysis to uncover all most-parsimonious trees from a data set comprising 60 morphological characters of nine taxa: one Acanthocephala, six Rotifera, and two outgroups (Turbellaria, Gnathostomulida). Analysis of our matrix yielded a single most-parsimonious tree. From our analysis we conclude the following: (1) Class Digononta is paraphyletic; (2) it is still premature to reject rotiferan monophyly; (3) the classification hierarchy that best conforms to this morphologically based, cladistic analysis is similar to several traditional schemes. In spite of these results, it is significant that this analysis yielded a tree that is incongruent with those trees developed from molecular data or by using the principles of evolutionary taxonomy.

Introduction

Rotiferan classification schemes and their implied phylogenies have never been congruent, e.g., compare Koste (1978) to Pennak (1989). Curiously, these inconsistencies never seemed to bother students of the phylum much. However, more than a decade has passed since Lorenzen (1985) argued that acanthocephalans and bdelloid rotifers are sister groups, thus invalidating traditional classification schema for Rotifera. Since that time others have added to the discussion strengthening the argument against using established rotifer classifications. These new challenges come from both morphological (Ahlrichs, 1997) and molecular analyses (Garey et al., 1996; Garey et al., 1998). However, except for a few passing comments suggesting that such conclusions were premature (Nogrady et al., 1993; Wallace & Colburn, 1989; Wallace & Snell, 1991), no rotifer taxonomists have seriously considered these proposals. Here we take up these questions by undertaking a morphologically based, cladistical analysis of phylum Rotifera, including two taxa as outgroups. Our purposes were to (1) estab-

lish a useful set of light microscopical, ultrastructural, and biochemical characters of these taxa, (2) develop parsimonious cladograms (hypotheses) from that data set, and (3) evaluate alternative phylogenetic hypotheses from a morphological perspective (i.e., Ahlrichs, 1997; Garey et al., 1996; Koste, 1978; Kutikova & Markevich, pers. comm.; Lorenzen, 1985; Pennak, 1989). Such efforts are necessary before a resolution of morphological and molecular phylogenies is possible.

Methods

The 60 unweighted characters examined in this study were extracted from the literature dealing with the morphology, embryology, and biochemistry of Acanthocephala, Gnathostomulida, Rotifera, and Turbellaria (Table 1; Appendix). In constructing this data set, we were particularly interested in compiling information that could be corroborated from multiple sources. We also were interested in data that showed reasonable variation of character state, but for the

Table 1. Data matrix of 60 morphological characters used in this analysis of 9 taxa: one Acanthocephala, six Rotifera and two outgroups

Taxa	Characters											
	1–5	6–10	11–15	16–20	21–25	26–30	31–35	36–40	41–45	46–50	51–55	56–60
Ancestral	00000	00000	00000	000NN	NNNNN	NN000	0000?	000?0	00?00	0000?	?0000	00 000
Turbellaria	00010	00000	00000	0P000	NNNNN	NN000	0000N	000N0	00PPP	P0002	PP10P	0 0001
Gnathostomulida	??000	00001	10001	N1100	NNNNN	NN001	P000N	000N0	00000	00002	00100	01101
Acanthocephala	011NN	NNNN1	NP001	NNNNN	NNNNN	NNNNN	NP111	010N1	00010	10001	NN00?	11111
Seisonidea	11010	NN000	10110	11101	000NN	N1001	10211	10101	00010	10111	00101	10011
Philodinavidae	1101P	P0000	11100	21?01	1N100	00100	1121?	10111	201N1	1NNNN	00111	10011
Philodinidae	11011	10100	11100	21111	1N100	00100	11211	10111	201N1	1NNNN	01111	10011
Habrotrochidae	11011	10100	11100	21?11	1N100	00110	11211	10111	201N1	1NNNN	01111	10011
Adinetidae	11010	NN000	11100	21?11	1N100	00100	1121?	10111	201N1	1NNNN	00111	10011
Monogononta	1101P	PP0P0	11110	21?P1	011P1	PP000	1121?	P01P1	111P1	11000	PP101	10011
Transformations	rdirr	rrdir	rrddi	irdrd	irdrr	rrrdr	drudr	rrdrr	arrdd	diiru	rridd	rrddr

Character descriptions correspond to descriptions provided in the Appendix. Character types: r, reversible; i, irreversible; d, dollo; o, ordered; u, unordered; a, defined stepmatrix $R_i \times C_i = 011,101,ii0$. Character states: P, polymorphic within the taxon (0 or 1); N, not applicable (?). Character states were polarized as plesiomorphic (0) or apomorphic (derived) whenever we thought it reasonable to do so.

sake of completeness we retained characters within the database even when they proved to be uninformative. In designing a strategy for our investigation we followed the general protocol used by Wallace et al. (1996) in their study of pseudocoelomate morphology. All characters were defined as strictly as possible. Characters encoded as not applicable were analyzed as if they were unknown in our analysis (Swofford, 1991). While we retained characters that eventually proved to be uninformative to recognize autapomorphies and to identify characteristics warranting further study, we also ran searches using only those characters that were applicable to all groups.

The phylogenetic program PAUP© (v3.1.1; Swofford, 1991) was used to do exhaustive searches for all most-parsimonious trees on our data matrix or portions thereof. The hypothetical ancestral states (AS) and transformation types (TT) reported in Table 1 (see also Appendix) were used for all standard searches. Searches in which the ancestral states were considered to be unrooted (ASU) and the transformation types unordered (TTU) also were run. Two bootstrap simulations of 1000 replications each were done as tests of robustness: one simulation used AS and TT while the other used ASU and TTU. Tree statistics reported here are (1) tree length (TL), (2) consistency index (CI) (excluding uninformative characters) and (3) retention index (RI). The ancestral reconstructions reported for our analysis include the two-character optimization algorithms available in PAUP for rooted trees (ACCTRAN, DELTRAN). Tree statistics for cladograms from our data set are reported where multistate

characters were treated as uncertainties. Decay indices also were calculated as described by Donoghue et al. (1992) and Garey et al. (1996). A decay index is the number of steps that must be added before a clade in the minimum length trees decay into a polytomy. We also employed the permutation tail probability (PTP) test (Faith & Cranston, 1991) to examine cladistic structure of the data set in two ways: (1) AS and TT; (2) ASU and TTU. To do this we used the shuffle option of MacClade© (v3.0; Maddison & Maddison, 1992) to randomize character states. To compare our results with previously published trees, we used the constraints methodology (exhaustive searches) to generate tree statistics for the six alternative tree topologies reported by Koste (1978), Lorenzen (1985), Pennak (1989), Garey et al. (1996), Ahlrichs (1997) and Kutikova & Markevich, pers. comm.) as optimized on our data matrix. When authors made no reference to a particular taxon we permitted PAUP to determine its final placement.

Phylogenetic analysis

Unmodified data matrix

Our analysis yielded a single most-parsimonious tree (TL=67, CI=0.824, RI=0.908) (Figure 1). This tree is a member of a skewed distribution of tree lengths: mean TL=128.2, SD=10.9, $g_1=-0.927$ (Hillis, 1991). The two 50% majority-rule, consensus trees yielded from the bootstrap simulations had the identical topology

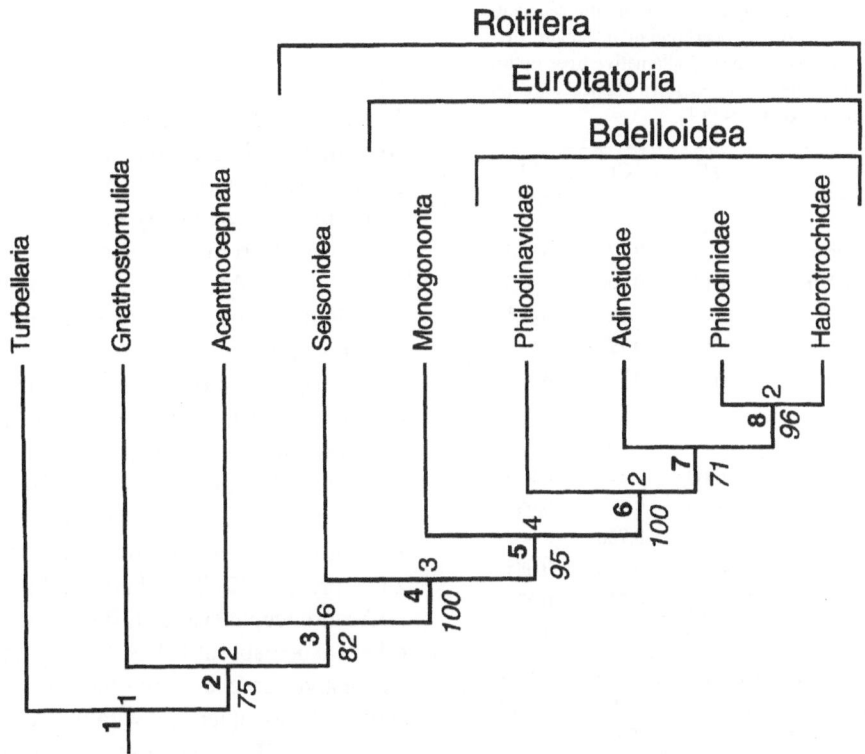

Figure 1. Cladogram of the single, most-parsimonious tree generated from our data matrix (Table 1). Because some characters were uninform-ative, not all 60 characters contibuted to this tree. Numbers in bold print (**1–8**) refer to internal nodes. Numbers in italics below each node are the percentages of 1000 bootstrap replicates supporting that portion of the tree. Numbers to the right of each node are the decay index for that node. [NB: An exhaustive search yielded the identical cladogram, with the expected changes in tree statistics, when analyzed using our data matrix as AS=unrooted and TT=unordered with or without Lorenzen's coding of Lemnisci and Rostrum, except that Node **2** was subsumed within Node **1**.] Changes in character state (#) for the rooted tree that are supported by both character optimization algorithms (ACCTRAN, DELTRAN) are reported below. These changes are annotated by node number or terminal taxon (in bold print). (**1**) Statocysts present (#60). (**2**) Pharyngeal ganglion (unpaired) present (#11); Masticatory apparatus: substructure tubular rods [NB: Acanthocephala = N] (#18); Photoreceptive eyespots absent [NB: present as Node **4**] (#58). (**3**) Chitin present (#2); ISL present (#34); Reproduction dioecious (#40); Copulation vaginal [NB: Bdelloidea=N] (#44); Permanent gonopore present (#46); Body cavity 'Pseudocoelomate' present (#56); Monociliated pits absent (#59). (**4**) ROTIFERA: Collagen absent (#1); Dorsal sensory organ (antenna) present (#13); Paired lateral antennae present [NB: absent at Node **6**] (#14); Retrocerebral apparatus: paired sacs (#16); Masticatory apparatus construction: trophi present (#20); Complete gut present (#31); Rostrum present (#36); Foot present [except obvious secondary losses] (#38); Egg diapause present (#55). (**5**) EUROTATORIA: Posterior pedal gland present (#12); Retrocerebral apparatus unpaired sacs (#16); Trophial feature: manubria present (#23); Urinary collection bladder present (#32); Gonad structure compact (#43); Vitellarium present (#45). (**6**) BDELLOIDA: Paired lateral antennae absent (#14); Trophial feature: fulcrum absent (#21); Stomach wall syncytial (#28); Adult diapause present (#54). (**7**) Masticatory apparatus position: interior (#19). (**8**) Oral ciliation differentiated (#5); Coronal pedicels present (#8); Food processing filter (#52). **Gnathostomulida**: Cephalic glands absent (#15). **Acanthocephala**: Ciliated epidermal cells absent (#3); Cephalic glands absent (#15); Adhesive glands absent [NB: see also discussion in text] (#37). **Seisonidea**: Trophial feature: few unci teeth (#27); Penis absent (#48); Male genital ducts U-shaped (#49). **Monogononta**: Manubrial chambers inflated (#25); Gonads unpaired (#42); Males haploid (#47). **Habrotrochidae**: Stomach lumen absent (solid) (#29).

of Figure 1, except that the unrooted, unordered boot-strap produced a tree which reduced nodes **1–2** into a single node. The two PTP tests did not yield any trees with TL≤67. Thus the null hypothesis, that the data defining the three most-parsimonious trees show no cladistic structure, is rejected at the 0.01 level.

Near-minimum length trees (TL≤71) also were examined to determine whether they contained in-formative trees (Table 2). Most of these clado-grams possessed rearrangements of the placement of Gnathostomulida, minor rearrangements within Bdel-loidea, or repositioning of the Monogononta, within the rotiferan clade. In no case were any other taxa placed within Rotifera. Beyond TL=72, the number of trees yielded became quite large: TL= 72 (19); TL=73 (19); TL=74(37); TL=75 (47);TL=76 (54); TL=77 (85).

We used constraint analysis (CA) to examine all five alternative phylogenies fixed on our data matrix. Necessarily, all phylogenies differing in tree topology

Table 2. Tree statistics for the single, most-parsimonious clado-gram yielded by our analysis, near-minimum length trees (TL$\leq$71), and constraint analysis of five alternative trees optim-ized on our data matrix (Table 1) using our hypothetical ancestral states (AS) and transformation types (TT).

	TL	CI	RI
No constraints			
Most-parsimonious tree	67(1)	0.824	0.908
Near-minimum length trees	68(2)	0.808	0.898
	69(2)	0.792	0.888
	70(9)	0.778	0.878
	71(5)	0.764	0.867
Constraints			
(1) Koste (1978)	67(1)	0.824	0.908
(2) Lorenzen (1985)	90(1)	0.575	0.684
(3) Pennak (1989)	72(1)	0.750	0.857
(4) Garey et al. (1996)	90(1)	0.575	0.684
(5) Ahlrichs (1997)	74(1)	0.724	0.651
(6) Kutikova & Markevich[a]	71(1)	0.764	0.867

Alternative hypotheses were derived from known trees (CAs #2, 4–6) or the logical phylogeny implicit in a classification scheme (CAs #1, 3) Exhaustive searches found minimum-length trees that matched the topology of each proposed phylogeny. TL, tree length (no. of trees found); CI, consistency index (excluding un-informative characters); RI, retention index. The indices CI and RI do not include the step matrix used for character #41
[a] Personal communication.

will require more steps than the most-parsimonious tree and will possess poorer tree statistics. Never-theless, CA permits a systematic comparison of the alternative hypotheses to highlight their differences. Only Koste's (1978) phylogeny (CA #1) was con-gruent with Figure 1 and therefore had identical tree statistics. The other alternative trees required from four to 23 more steps than Figure 1 and had poorer tree statistics (Table 2). These differences resulted from the placement of seisonids and bdelloids as sister groups (CA #3) or placement of acanthocephalans within the rotifer clade (CA #2, 4 and 5) or placement of the seisonids and monogononts as sister groups (CA #6).

Several characters warrant closer examination either because our analysis showed them to be highly variable or because of uncertainties in their condition. These include the following: multiciliated epidermal cells (#4); bilobed dorsal cerebral ganglion (#10); stomach ciliation (#30); tegument structure (#33); life cycle (#41); sperm flagellum position (#50); larval stage (#53); adnate nerve (#57). Finally, all characters establishing Node **4** (Figure 1) deserve special study as they help define placement of the acanthocephalans.

Modifications to the data matrix

Modification 1

Two characters require special attention because they are critical to Lorenzen's original hypothesis and be-cause they provide morphological support for the molecular tree of Garey et al. (1996). These char-acters are rostrum (character #36) and lemnisci (not encoded in Table 1). In our opinion the putative homologies between these structures in rotifers and acanthocephalans have not been established beyond a very superficial resemblance when viewed by light microscopy (Ricci, this volume). Nevertheless, to determine whether these two characteristics were sig-nificant in this analysis, we adjusted our data matrix to reflect Lorenzen's (1985) two assertions. (1) The Rostrum character (#36) was recoded as Acanthoceph-ala = present. (2) The character Lemnisci was added as Acanthocephala and all Bdelloida = present (A), all other taxa = absent [d]. After this adjustment we ran exhaustive searches using these character codes both separately, together, and with both eliminated from the data matrix. Two sets of searches were run: (1) as AS and TT and (2) as ASU and TTU. These searches yiel-ded expected changes in tree statistics, but all had the same tree topology as in Figure 1. While a detailed ul-trastructural analysis is warranted for these characters, until that is done we conclude that Lorenzen's asser-tion is premature and researchers seeking to validate molecular trees should not rely on these characters for support.

Modification 2

An exhaustive search using our data set with the modi-fication that the cement glands in the male mating apparatus of Acanthocephala would be treated as be-ing homologous to the adhesive glands (character #37) of the other taxa (J. Garey, pers. comm.) yielded the same tree as in Figure 1, but with one fewer step. Addi-tional exhaustive searches with this adjustment plus all the changes discussed in Modification 1 did not alter tree topology either.

Modification 3

An exhaustive search using our data set with the ad-dition of a character inadvertently omitted from our original data matrix, 'dense bodies in sperms' (Ahl-richs, 1997), yielded the same tree as in Figure 1 with the expected changes in tree statistics. Additional exhaustive searches combining Modifications 1–3 did not alter tree topology.

Modification 4

We also ran two other sets of exhaustive searches in which characters coded as non-applicable (N) in (1) the Acanthocephala or (2) in any taxa were omitted for all taxa using either AS and TT or ASU and TTU. These searches yielded trees with the expected changes in tree statistics, but in no case were the outcomes substantially different from Figure 1. As expected all trees reduced Nodes **6–8** into a single polytomy, but Acanthocephala was never placed as a sister group to Bdelloidea. The only other change involved placement of Gnathostomulida as a sister group to Turbellaria (our AS & TT) or as a sister group to the Acanthocephala (ASU and TTU).

Discussion

The most striking aspect of our cladogram is that it implies monophyly for Rotifera, with Acanthocephala as a sister group. Thus, Figure 1 supports a traditional classification hierarchy of rotifers as described by Koste (1978); Bdelloidea (Bd) and Monogononta (M) form sister groups, that together comprise a clade (Eurotatoria) that is a sister group to the Seisonidea (S). Together the Rotifera, S(Bd,M), are a sister group to the Acanthocephala (A), and the Rotifer–Acanthocephala are a sister group to the Gnathostomulida (G) (Rieger & Tyler (1995). Further, this morphologically based phylogeny supports none of the other alternative hypotheses: ROTATORIA, including Acanthocephala (Lorenzen, 1985) as (A,(Philodinidae)); DIGONONTA (Nogrady et al., 1993; Pennak, 1989) as (M,(S,Bd)); LEMNISCEA (Garey et al., 1996) as (A,(Philodinidae)); SYNDERMATA (Ahlrichs, 1997) as (G,(Bd,M),(S,A)); ROTIFERA (Kutikova & Markevich, pers. comm.) (presented at this symposium) as (Philodinidae,(S,M)).

Based on our analysis we aver three conclusions. (1) Class Digononta is paraphyletic and we urge, once again, that use of this taxon be abandoned. [NB: Digononta also is paraphyletic using the tree topology of both Lorenzen (1985) and Ahlrichs (1997).] (2) It may be premature to favor any of the alternative hypotheses that place Acanthocephala within the Rotifera. While Acanthocephala and Rotifera are clearly related, our morphologically based characters do not support the hypotheses that acanthocephalans and bdelloid rotifers (Garey et al., 1996; Lorenzen, 1985) or acanthocephalans and seisonids (Ahlrichs, 1997) are sister groups.

Nevertheless, acanthocephalans are highly specialized endoparasites with far-reaching adaptations to this mode of life, whereas most rotifers are, in the main, free-living. Because any phylogenetic analysis based on morphological features alone will be strongly influenced by their particular adaptations and, therefore, will over estimate differences between these two taxa, some caution is warranted here. (3) The classification scheme (see following) that best conforms to the morphologically based, cladistic analysis presented here is very similar to several earlier ones (e.g., Edmondson, 1959; Epp & Lewis, 1979; Koste, 1978; Melone & Ricci, 1995; Wallace & Colburn, 1989; Wallace & Snell, 1991).

Phylum Rotifera
    Superclass Seisona
        Class Seisonidea
    Superclass Eurotatoria
        Class Bdelloidea
            Order Adinetida
                Family Adinetida
            Order Philodinida
                Families Habrotrochidae, Philodinidae
            Order Philodinavida
                Family Philodinavidae
        Class Monogononta
            Orders Collothecacea, Flosculariacea, Ploimida

Conclusions

To our knowledge this is the first time a detailed, morphologically based, cladistic analysis has been attempted at this level within the Rotifera. Besides demonstrating rotiferan monophyly as (S(Bd,M)), we have shown that inclusion of rostrum and lemnisci, as defined by Lorenzen (1985), into a morphologically based data set does not provide support for any of the alternative phylogenies discussed here (see also Ricci, this volume). While the molecularly-based analyses of Garey et al. (1996), Garey et al. (1998), and some recent unpublished work (D. Mark Welch, pers. comm.) argue strongly for placement of Acanthocephala within the Bdelloidea, these studies are based on a few genes from a limited suite of taxa. Thus, we feel that it is appropriate to abstain from a more detailed comparison between morphologically and molecularly-based phylogenies. Clearly what is needed is additional molecular analyses that include more genes and more rotifer taxa and morphological study of characters critical in defining placement of the Acanthocephala.

Finally, future work needs to pay attention to the phylogeny of class Monogononta. A preliminary analysis of an expanded version of our data matrix yielded a tree in which the three Orders comprising Class Monogononta were paraphyletic:

(Seisonidea,(Collothecacea,(Flosculariacea,
(Ploimida,(Bdelloidea)))))

This may be due to the number of polymorphic characters present in Order Ploimida. We expect to address this issue in future studies.

In doing this study we attempted an in-depth survey of the recent literature, but there may be more morphological characters available for analysis; we anticipate adding them to our study as they become known to us. We also wish to enlarge our study by adding New Group 1 to our data matrix (Ahlrichs, 1997; Kristensen, 1995). As part of our work we also expect to investigate two additional phylogenetic hypotheses: Markevich (1993) (Neodermata, Rotifera) and Nielsen (1995) (Chaetognatha,(Rotifera, Acanthocephala)). We hope that this contribution to the discussion of rotifer phylogeny will inspire others to wed morphological and molecular data in a meaningful way. Without that union, knowledge of rotifer phylogeny will remain rigorously compartmentalized by discipline (morphological versus molecular) and a classification scheme acceptable to both will be unobtainable.

Acknowledgments

We thank our respective institutions for providing electronic mail services, research facilities, equipment, and research support which permitted this work to be undertaken. We also thank Jim Garey and Elizabeth Walsh for their helpful comments regarding this ms. We remain responsible for any errors that may exist within the data matrix or the analysis. Additional documentation of the data matrix may be obtained from any of the authors.

Appendix: character descriptions (p, present, a, absent)

1. *Collagen.* [p=0, a=1]. **2.** *Chitin.* [a=0, p=1]. **3.** *Ciliated Epidermal Cells.* [p=0, a=1]. **4.** *Multiciliated Epidermal Cells.* [monociliated=0, multiciliated=1]. **5.** *Oral Ciliation.* Oral ciliation uniform (u) or differentiated (d) into clusters or bands [u=0, d=1]. **6.** *Coronal Ciliation: Trochus.* Coronal ciliation consisting of 2 ciliary bands: a preoral band (trochus) and a postoral band (cingulum); mouth opens into the ciliated field, with an apical bare zone (apical field) [ciliated field=0, 2 ciliary bands=1]. **7.** *Coronal Ciliation: Pseudotrochus.* [a=0, p=1]. **8.** *Coronal Pedicels.* Corona elaborated into paired discs on pedicels [a=0, p=1]. **9.** *Coronal Infundibulum.* [a=0, p=1]. **10.** *Bilobed Dorsal Cerebral Ganglion.* [p=0, other=1]. **11.** *Pharyngeal Ganglion (unpaired).* [a=0 , p=1]. **12.** *Posterior Pedal Ganglion.* [a=0, p=1]. **13.** *Dorsal Sensory Organ (Antenna).* Sensory structure on the head bearing sensory cilia that may be elevated and finger-like or a pit with cilia (flosculum); paired or unpaired [a=0, p=1]. **14.** *Paired Lateral Antennae.* Sensory structures (paired) often similar to flosculi or sometimes finger-like unpaired [a=0, p=1]. **15.** *Cephalic Glands.* Glands with cell bodies situated behind the brain [p=0, p=1]. **16.** *Retrocerebral Apparatus.* Glandular structure composed of medial symmetrical retrocerebral sac + pair of lateral subcerebral glands [a=0, paired sacs=1, unpaired sacs=2]. **17.** *Pharynx Masticatory Apparatus.* Extracellular, sclerotized, articulating elements, secreted by ectodermal cells; connections to muscles (jaws) [a=0, p=1]. **18.** *Masticatory Apparatus: Substructure.* Jaws of parallel-layered cuticular rods [a=0, p=1]. **19.** *Masticatory Apparatus Position.* Near mouth (anterior) or posterior (interior) within body [anterior=0, interior=1]. **20.** *Masticatory Apparatus Construction.* [simple jaws=0, complex trophi=1]. **21.** *Trophial Feature: Fulcrum.* [p=0, a=1]. **22.** *Fulcrum Spacial Relationship.* Fulcrum free (f) or connected (c) to rami [f=0, c=1]. **23.** *Trophial Feature: Manubria.* [a=0, p=1]. **24.** *Manubrium Vestigial.* [normal=0, vestigial=1]. **25.** *Manubrial Chambers.* [ribbon-shaped=0, inflated cavities=1]. **26.** *Manubrial Shaft.* Manubrium elongated into a shaft permitting insertion of muscles [a=0, p=1]. **27.** *Trophial Feature: Number Unci Teeth.* [many=0, few=1]. **28.** *Stomach (Gut) Wall.* [cellular=0, syncytial=1]. **29.** *Stomach (Gut) Lumen.* Stomach with lumen or absent, with a solid syncytium [lumen=0, a=1]. **30.** *Stomach (Gut) Ciliation.* [ciliated=0, a=1]. **31.** *Complete Gut.* (disregarding autapomorphic losses) [a=0, p=1]. **32.** *Urinary Collection Bladder.* [a=0, p=1]. **33.** *Tegument Structure.* [cellular=0 , syncytial=1, both=2]. **34.** *Intracytoplasmatic Skeletal Lamina (ISL).* [a=0, p=1]. **35.** *ISL Organization.* [external=0, internal=1]. **36.**

Rostrum. Extremity located apically on head with cilia & covered by cuticular lamella [a=0, p=1]. **37.** *Adhesive Glands*. Cement glands, excluding losses in pelagic rotifers [p=0, a=1]. **38.** *Foot*. Distinct foot, disregarding obvious losses [a=0, p=1]. **39.** *Foot: toes*. [a=0, p=1]. **40.** *Reproduction*. [monoecious=0, dioecious=1]. **41.** *Life Cycle*. [biparental=0, cyclically parthenogenetic=1, obligatory parthenogenetic=2]. **42.** *Gonad Number*. [paired=0, single=1]. **43.** *Gonad Structure*. [diffuse, follicular, non-compact gonad structure=0, compact gonad structure=1]. **44.** *Copulation*. [intradermic=0, vaginal=1]. **45.** *Vitellarium*. [a=0, p=1]. **46.** *Permanent Gonopore*. Eggs or embryos released through permanent pore [a=0, p=1]. **47.** *Male Ploidy*. [diploid=0, haploid=1]. **48.** *Penis*. [p=0, a=1]. **49.** *Male Genital Ducts*. Straight (s) or U-shaped (U) [s=0, U=1]. **50.** *Sperm Flagellum Position*. [anterior=0, parallel=1, posterior=2]. **51.** *Feeding*. [microphagous=0, macrophagous=1]. **52.** *Food Processing*. [browsers/scrapers=0, suspension=1]. **53.** *Larval Stage*. True larval stage [p=0, a=1]. **54.** *Adult Diapause*. Excluding 1 case in Turbellaria [a=0, p=1]. **55.** *Egg Diapause*. [a=0, p=1]. **56.** *Body Cavity*. [acoelomate=0, perivisceral "pseudocoelomate"=1]. **57.** *Adnate Nerve Cords*. [a=0, p=1]. **58.** *Photoreceptive Eyespots*. [p=0, a=1]. **59.** *Monociliated Pits*. Possessing a collar of microvilli [p=0, a=1]. **60.** *Statocysts*. Primarily absent in Platyhelminthes [a=0, p=1].

References

Ahlrichs, W. H., 1997. Epidermal ultrastructure of *Seison nebaliae* and *Seison annulatus*, and a comparison of epidermal structures within Gnathifera. Zoomorphology 117: 41–48.

Donoghue, M. J., R. G. Olmstead, J. F. Smith & J. D. Palmer, 1992. Phylogenetic relationships of Dipsacales on RbcL sequences. Ann. MO Bot. Gard. 79: 333–345.

Edmondson, W. T., 1959. Rotifera. In Edmondson, W. T. (ed.), Fresh-Water Biology, 2nd edn. John Wiley, N.Y.: 420–494.

Epp, R. W. & W. M. Lewis, Jr., 1979. Sexual dimorphism in *Brachionus plicatilis* (Rotifera): evolutionary and adaptive significance. Evolution 33: 919–928.

Faith, D. P. & P. S. Cranston, 1991. Could a cladogram this short have arisen by chance alone? On permutation tests for cladistic structure. Cladistics 7: 1–28.

Garey J. R., T. J. Near, M. R. Nonnemacher & S. A. Nadler, 1996. Molecular evidence for Acanthocephala as a sub-taxon of Rotifera. J. Mol. Evol. 43: 287–292.

Garey J. R., A. Schmidt-Rhaesa, T. J. Near & S. A. Nadler, 1998. The evolutionary relationships of rotifers and acanthocephalans. Hydrobiologia 387/388 (Dev. Hydrobiol. 134): 83–91.

Hillis, D. M., 1991. Discriminating between phylogenetic signal and random noise in DNA sequences. In Miyamoto, M. M. & J. Cracraft (eds), Phylogenetic Analysis of DNA Sequences. Oxford University Press, N.Y.: 278–294.

Kristensen, R. M., 1995. Are Aschelminthes pseudocoelomates or acoelomates? In Lanzavecchia, G., R. Valvassori & M. D. Candia Carnevali (eds), Body Cavities Function and Phylogeny. Selected Symposia and Monographs U.Z.I. 8 Mucchi, Modena: 41–43.

Koste, W., 1978. Rotatoria. Die Rädertiere Mitteleuropas. 2 vols, Gebrüder Borntraeger, Berlin, Stuttgart, Germany.

Lorenzen, S., 1985. Phylogenetic aspects of pseudocoelomate evolution. In Conway Morris, S., J. D. George & H. M. Platt (eds), The Origins and Relationships of Lower Invertebrates. System. Assoc. Volume 28. Clarendon Press, Oxford, U.K.: 210–223.

Maddison W. P. & D. R. Maddison, 1992. MacClade: Analysis of Phylogeny and Character Evolution. Version 3.0. Sinauer Associates, Sunderland.

Markevich, G. I., 1993. Phylogenetic relationships of Rotifera to other veriform taxa. Hydrobiologia 255/256: 521–526.

Melone, G. & C. Ricci, 1995. Rotatory apparatus in bdelloids. Hydrobiologia 313/314: 91–98.

Nielsen, C., 1995. Animal Evolution: Interrelationships of the Living Phyla. Oxford University Press, Oxford.

Nogrady, T., R. L. Wallace & T. W. Snell, 1993. Rotifera. In Dumont, H. J. (ed.), Guides to the Identification of the Microinvertebrates of the Continental Waters of the World, vol. 4. SPB Academic Publishers, The Hague, The Netherlands.

Pennak, R. W., 1989. Fresh-water Invertebrates of the United States, 3rd edn. John Wiley, New York, N.Y., U.S.A.

Ricci, C., 1998. Are lemnisci and proboscis present in the Bdelloidea? Hydrobiologia 387/388 (Dev. Hydrobiol. 134): 93–96.

Rieger, R. M. & S. Tyler, 1995. Sister-group relationship of Gnathostomulida and Rotifera–Acanthocephala. Invertebr. Biol. 114: 186–189.

Swofford, D. L., 1991. PAUP: Phylogenetic analysis using parsimony, Version 3.0s Computer program distributed by the Illinois Natural History Survey, Champaign, IL.

Wallace, R. L. & R. A. Colburn, 1989. Phylogenetic relationships within phylum Rotifera: orders and genus *Notholca*. Hydrobiologia 186/187: 311–318.

Wallace, R. L., C. Ricci & G. Melone, 1996. A cladistic analysis of pseudocoelomate (aschelminth) morphology. Invertebr. Biol. 115: 104–112.

Wallace, R. L. & T. W. Snell, 1991. Rotifera. In Thorp J. & A. Covich (eds), Ecology and Classifications of North American Freshwater Invertebrates, ch. 8. Academic Press, New York, N.Y.: 187–248.

Hydrobiologia **387/388:** 109–115, 1998.
E. Wurdak, R. Wallace & H. Segers (eds), Rotifera VIII: A Comparative Approach.
© *1998 Kluwer Academic Publishers.*

Cross-mating tests re-discovered: a tool to assess species boundaries in rotifers

Roberto Rico-Martínez
Universidad Autónoma de Aguascalientes, Centro Básico, Departamento de Química,
Avenida Universidad 940, Aguascalientes, Ags. C.P. 20100, México

Key words: behavior, mating, evolution, taxonomy

Abstract

The recent isolation of a mate recognition pheromone in the marine rotifer *Brachionus plicatilis* Müller has shed new light on the mate recognition system of rotifers. One result is improved understanding of the importance of mating behavior as a highly efficient process used by rotifers to choose conspecifics. There are many differences in the main characteristics of mating behavior in members of five different families of rotifers. The present work describes the use of these characteristics to assess species boundaries, especially where boundaries between two or more species are unclear. The method proposed here can assess quantitatively the response of males of one species to females of a questionable taxon by measuring the percentage of matings initiated and the number of completed copulations. The data generated can then be used together with molecular, morphological, and other data to determine the species boundaries. This approach can help distinguish between morphological differences resulting from evolutionary divergence of species and morphological differences induced by environmental or ecological factors.

Introduction

The basic question of how animals choose their mates is today a major challenge for zoologists (Gibson & Langen, 1996). In that context, the mating behavior of rotifers is also poorly known in spite of being studied for over 30 years. Gilbert (1963) observed mating behavior in *Brachionus angularis* and *B. calyciflorus*, demonstrating that contact chemoreception is used by male rotifers to identify conspecific females. In 1983, Snell & Hawkinson described the mating behavior of *B. plicatilis* and described its different steps. Mating behavior in *B. plicatilis* is divided into five phases which correspond to; encounter, circling, coronal localization, sperm transfer and dissociation (Nogrady et al., 1993). Snell et al. (1995) have demonstrated that the mating behavior of male *B. plicatilis* is based on the recognition of a mate recognition pheromone (MRP), which is in highest concentration on the corona and at the base of the foot of females, and that this glycoprotein is necessary and sufficient to elicit male mating behavior. The mating behavior

of rotifers, therefore, represents a highly efficient mechanism to recognize conspecifics.

We have made observations of mating behavior in six different families of monogononts (Asplanchnidae, Brachionidae, Epiphanidae, Euchlanidae, Lecanidae, and Trichocercidae), eight genera and about 20 species. The two species best studied are *B. plicatilis* (Fu et al., 1993; Gómez & Serra, 1995; Gómez & Snell, 1996; Hagiwara et al., 1996; Rico-Martínez & Snell, 1995a, b; Snell & Hawkinson, 1983) and *B. rotundiformis* (Gómez & Serra, 1995; Gómez & Snell, 1996; Hagiwara et al., 1996; Rico-Martínez & Snell, 1995a, b). In these two species, male preferences for females of the same strain have been detected (Rico-Martínez & Snell, 1995a, b). A total of eight genera have been studied, including *Anuraeopsis* (Rico-Martínez & Snell, 1997), *Asplachna* (Aloia & Moretti, 1973; Birky, 1967; Gilbert et al., 1979), *Brachionus* (Gilbert, 1963; Gómez & Serra, 1995; Gómez & Snell, 1996; Hagiwara et al., 1996; Rico-Martínez, 1995; Rico-Martínez & Snell, 1995a, b, 1997; Ruttner-Kolisko, 1979; Snell,

1989; Snell & Hawkinson, 1983, *Epiphanes* (Gilbert & Williamson, 1983), *Keratella* (Rico-Martínez & Snell, 1997), *Euchlanis* (Rico-Martínez & Snell, 1997), *Lecane* (Rico-Martínez & Snell, 1997), *Plationus* (Rico-Martínez & Snell, 1997), and *Trichocerca* (Rico-Martínez & Snell, 1997). Unfortunately, we only have data on three genera (*Asplanchna*, *Brachionus* and *Lecane*) about interspecific crosses. The results of these studies showed that *Brachionus* males seem to be the most discriminative towards females of different species within the same genus (Rico-Martínez, 1995). In only one family, Brachionidae, have cross-mating tests been made between members of different genera. *Brachionus calyciflorus* males attempted mating with *Plationus patulus* females (a closely related genus recently described by Segers et al., 1993), but did not attempt to mate with females of *Keratella americana*. The percentage of *B. calyciflorus–P. patulus* mating attempts was low (0.5%), but was reproducible. In about 5 000 male–female encounters, a total of about 25 mating attempts was observed (Rico-Martínez & Snell, 1997). This result is interesting since *P. patulus* was considered a member of the genus *Brachionus* by several authors (Koste, 1978; Rao & Sarma, 1985). In all interspecific previous crosses performed with rotifers (for a review, see Rico-Martínez, 1995; Snell, 1989), no member of a genus has ever attempted to mate with a member of a different genus, including genera within the same family. The *Brachionus–Plationus* mating suggests that these taxa are very closely related; hence, the validity of the genus *Plationus* should be re-evaluated. I believe that cross-mating experiments have a great potential to clarify these evolutionary relationships, but thus far it has been used sporadically in rotifers. More cross-mating tests, along with data on post-mating reproductive barriers among the genera *Brachionus*, *Plationus* and *Platyas* are needed to assess the true phylogenetic relationships among these taxa.

In this paper, I describe how mating assays have been used in several works (Rico-Martínez & Snell, 1995a, 1997; Snell et al., 1995) and then discuss how they have been used by other authors.

Cross-mating assay

The assay I propose consists of placing four neonate females and two neonate males (both less than 6 h and both virgins) into 50 μl of synthetic freshwater (U.S. EPA, 1985), HMNK buffer (50 mM Hepes, 36 mM NaCl, 1 mM HCl and 0.1 mM MgSO$_4$, pH 7.4), or Instant Ocean artificial sea water at 15 g l^{-1} at room temperature (about 25°C). The first medium is for freshwater species and the latter for marine or brackish water species. After placing the animals in the medium, a period of 5–10 min is given to allow acclimation, and then the number of encounters, male mating attempts and completed copulations in 5 min are recorded in each of five replicates. In all experiments a dissecting scope with magnifications of 12–40× should be used. The optimal magnification varies with the species studied. An encounter is recorded each time a male contacts a female with his corona. A mating attempt is recorded in brachionids, if a male circles around a female maintaining contact with his corona. Slight variations in this pattern have been observed in non-brachionid species (Rico-Martínez & Snell, 1997). A copulation is recorded if a male attaches his penis to the female. The frequency of homogamic versus heterogamic matings is then analyzed by paired two-tailed *t*-tests using the mean percentages of the five replicate sets of two males of encounters resulting in mating attempts and mating attempts resulting in completed copulations.

Comments on the technique

There are many differences in the mating behavior of the different species investigated so far. The sites of initiating mating and penile attachment in copulation vary among all species. However, there are some similarities among members of the same genus (Rico-Martínez & Snell, 1997) and even among members of the same family (Rico-Martínez & Snell, 1997; Rico-Martínez et al., 1996). Male/female length ratios also vary from the small brachionid males with ratios of 0.5–0.74 as in *Asplanchna girodi* where males are nearly the size of females (Rico-Martínez & Snell, 1997). Swimming speeds of males and females are also quite variable. Brachionid males are fastest with speeds ranging from 0.59 to 1.04 mm/s, in spite of being the smallest of the eight species investigated. In contrast, *Lecane quadridentata* males move at a speed of 0.18 mm/s (Rico-Martínez & Snell, 1997). These differences require some variability in the observational technique, especially when comparing members of different genera. For example, two 'passive' *Lecane quadridentata* males encounter four conspecific females 8.3 times in 5 min in 50 μl, while two fast *Brachionus plicatilis* males can encounter four conspecific females 49.5 times in 5 min in the same volume (assuming no completed copulation occurs). Another source of variation is the number of copulations and their duration. The number of encounters are drastically reduced in some assays where *Brachionus* or *Plationus* males are used in spite of their being the fastest males. The reason is that time is used up in copulations. *Lecane quadridentata* males copulate with their own females in 30% of the mating attempts (or 21% of all encounters). On the other hand, *Keratella americana* males copulate with their own females in only 2% of mating attempts (or 0.30% of all encounters). Duration of copulation is also quite variable. Some *Lecane quadridentata* males spend 40 min attached to a female (Rico-Martínez, unpublished data), although the mean duration of copulation for this species is 10.1 ± 9.7 min. Males of all the other species investigated spend less than 2 min in copulation with their own females (Rico-Martínez & Snell, 1997).

The use of HMNK buffer enables researchers to study interactions among some freshwater and brackish water species. Adjusting buffer salinity to about 1 g NaCl/l allows *Brachionus calyciflorus* to be crossed with *B. plicatilis* females and vice versa. Mating attempts were observed between males and females of both species. However, other freshwater species like *Keratella americana* and *Plationus patulus* could not stand the low salinity necessary for survival of *B. plicatilis* (Rico-Martínez & Snell, 1997).

Limitations of the technique

Males are known for very few species (Pennak, 1989). This is perhaps the biggest limitation to widespread use of the cross-mating test technique to determine doubtful species boundaries in rotifers. Also, questionable species are usually rare, endemic, reported only once or a few times, and have not been cultured. Furthermore, mating behavior also has been insufficiently studied in rotifers. I described the mating behavior of eight species, adding to the mating data known for about 14 species. A big challenge to the wider application of the mating test is knowing the factors that induce males. The production of males can vary from strain to strain within a species. Of seven strains of *Brachionus calyciflorus*, that I have cultured in the lab, only one, the Kalhagen Pond strain (Rico-Martínez & Dodson, 1992) seemed unable to produce males. In the case of *Plationus patulus*, I have cultured three strains of which one failed to produce males. For

many other species we have data for only one strain. More understanding of the factors that trigger sexuality in rotifer populations both in the field and in the lab is needed to expand use of this technique to more species. Another limitation is maintaining different species in culture, and the fact that some strains may lose their ability to produce males after a short time in laboratory culture, as did *T. pusilla* (Rico-Martínez & Snell, 1997). A strain of *Lecane quadridentata* collected at Lake Chapala, Mexico, and a *Brachionus bidentatus* strain collected at Poza Ezequiel García, Calvillo, Aguascalientes, Mexico, were cultured for several weeks and produced males. However, after 2 months both strains no longer produced males under our laboratory conditions.

Despite these difficulties, cross-mating tests are applicable to many rotifer species and are powerful tools for exploring phylogenetic relationships especially when combined with morphological, biogeographical, genetic and ecological data.

Comparison with other cross-mating techniques

Nine slightly different techniques have been used to study cross-mating in rotifers, measuring the percentage of mating attempts, also called circlings or mating initiations by some authors (Gómez & Serra, 1995), copulations, and rate of fertilization. The techniques differ in several points outlined in Table 1. Several of the differences are due to the original objective for which the techniques were developed. For instance, Snell & Nacionales (1990) used freeze-killed females because they are easy to manipulate and the objective of their study was to investigate the chemical nature of the MRP. Several authors (Snell & Childress, 1987; Snell & Hoff, 1987) were more interested in obtaining data on fertilization rates and their relationship with age or other factors in one strain, rather than comparing different strains or species. The remainder of the works listed in Table 1 corresponds to assessments of pre-zygotic barriers among several rotifer species and/or strains. The work of Fu et al. (1993) is the only work that presents definitive data regarding reproductive isolation. They showed that there was no formation of resting eggs when any of seven *B. rotundiformis* strains were crossed with any of seven strains of *B. plicatilis*. The design of their mating behavior experiment is also quite different from the rest of the works in that list. Instead of a careful monitoring of the mating phases and their scoring,

Table 1. Techniques used to study cross-mating in rotifers

Reference	General conditions of the tests	Number of animals used	Measurements taken	Statistical method used
Snell & Hawkinson, 1983	50 μl Instant Ocean at 15 g l^{-1} observations during 5 min at room temperature, six replicates	One male and 25 females both neonates of *Brachionus plicatilis* and *B. rotundiformis*	Percentage of mating attempts	χ^2 contingency test and ANOVA (arc–sin transformed data) to compare intra- and inter-strain differences
Snell & Hoff, 1987	50 μl F-medium at room temperature, observation during several minutes until copulation took place	Three to five young males and one neonate male, *B. plicatilis*	Male–female encounters, time until copulation, duration of copulation	Fertilization rate of females drawn in a graph
Snell & Childress, 1987	50 μl Instant Ocean at 20 g l^{-1} and 23°C, after 24 h males were removed and after 48 h fertilized females were counted	One virgin female and two virgin males (<4 h old) of *B. plicatilis*	Percentage of fertilized females after 24 h.	Correlation between capacity for male fertilization and age
Snell & Nacionales, 1990	50 μl Instant Ocean at 15 g l^{-1} at 25°C	Two young non-ovigerous freeze-killed females and three young neonate males of *B. plicatilis*	Percentage copulation in 50 male–female encounters	χ^2 contingency test
Fu et al., 1993	5 ml seawater at 20 g l^{-1} fed 1.5 × 10^7 cells at *Nannochloropsis oculata* daily at 23°C for 6 days	Seventy-five males and about 150 females both randomly picked from cultures of *B. plicatilis* and *B. rotundiformis*	Fertilization rate and production of resting eggs, mixis rate	One-way ANOVA to compare the fertilization rate between intra- and inter-strain differences
Gómez & Serra, 1995	50 μl Instant Ocean at 10 g l^{-1}	Twenty-five randomly chosen females and six replicate males of *B. plicatilis*, *B. quadridentatus* and *B. rotundiformis*	Percentage of mating initiations (similar to percentage of mating attempts)	Adjusted *G*-test used to compare intra- and inter-strain differences
Rico-Martínez & Snell, 1995a; Snell et al., 1995; Rico-Martínez & Snell, 1997	50 μl Instant Ocean at 15 g l^{-1} or 50 μl HMNK buffer ph 7.4 or 50 μl of EPA medium ph 7.5	Four females and two males both <6 h old of 10 different species both freshwater and marine	Percentage of mating attempts, copulations recorded	Paired *t*-test on the percentage of mating attempts between intra- and inter-strain differences
Rico-Martínez & Snell, 1995b	Similar to Rico-Martínez & Snell, 1995a	Similar to Rico-Martínez & Snell, 1995a	Similar to Rico-Martínez & Snell, 1995a	Similar to Snell & Hawkinson, 1983
Gómez & Snell (1996)	Polypropilene wells (7×3×1.5 mm), 15 g l^{-1} of seawater.	Three 4 h old freeze-killed females in each side of the well and five young males	Number of males in each side of the well every 20 s during 10 min and number of copulations	Goodness of fit *G*-tests
Hagiwara et al., 1996	50 μl of seawater at 17 g l^{-1} at 25°C, three replicates	One female and three males both 1 h old	Similar to Snell & Hawkinson, 1983	Similar to Snell & Hawkinson, 1983

Fu et al. analyzed fertilization rate, mixis rate and formation of resting eggs after a 5–6-day period. The concept of reproductive isolation has been questioned by several prominent authors (e.g. Patterson, 1985; Templeton, 1989; Vbra, 1995), and the application of this approach to the species concept in rotifers has been argued by some authors (Gómez & Serra, 1995; Serra et al., 1997; Snell, 1989). The supposition of all of these species concepts is that pre-mating barriers to reproduction are of great importance and need to be assessed to understand the speciation process. On the other hand, Patterson's recognition concept does not provide a clearcut boundary between two species as Mayr's concept does when applied to rotifers. It would probably not be sufficient to find significant differences in the percentages of mating attempts (and therefore mating preferences) to determine that a particular strain belongs to a diffferent species. Gómez & Serra (1995) and Rico-Martínez & Snell (1995) found separate cases where, together with mating preferences, there were some other characteristics that suggested that a particular strain in each case did not quite fit into the expected characteristcs of *B. rotundiformis*. The importance of the mating behavior tests in these two cases is highlighted by the fact that the mating tests were able to recognize these differences. However, it is unadvisable to rely entirely on a mating behavior test as the decisive factor to separate two species.

The tests by Gómez & Serra (1995), Hagiwara et al. (1996), Snell & Hawkinson (1983), and others (see Table 1), use the mating preference of male rotifers for females as a tool to find differences among strains that cannot be detected by morphological characteristics. The unifying characteristic, and perhaps the most important, for all these is the use of the percentage of mating attempts as the parameter to be evaluated. However, this characteristic is not easy to determine in non-brachionid species. In *Asplanchna brightwelli*, *Euchlanis dilatata*, *Lecane quadridentata* and *Trichocerca pusilla*, the only non-brachionid species where a mating test using males and females of the same species has been performed (Rico-Martínez & Snell, 1997), male rotifers use their coronas in a similar way as that of the male of *B. plicatilis*. The encounter and a behavior similar to circling have been observed in these species, although there are differences in sites of mating initation (Rico-Martínez & Snell, 1997).

The age of the animals used in the mating tests is also an important issue. All the latter group of tests except Gómez & Serra (1995) used neonates or virgin males and females. Several reasons support this choice: (1) neonate males (virgin males) have a full load of sperm which is a limitation in *B. plicatilis* (Snell & Childress, 1987); (2) only very young females of B. plicatilis can be fertilized (Snell & Childress, 1987).

An important question regarding the mating behavior tests is the relationship between conditions in the laboratory compared with those found in nature. Clearly, some of the conditions that relate to production of mictic females can be manipulated in the laboratory. Population density has been outlined as an important factor in *Brachionus* (Nogrady et al., 1993). In the tests summarized in Table 1 conditions for the tests differ in several aspects like salinity, pH, temperature, etc. However, in all cases mating attempts were recorded. It seems that the range of conditions in which mating behavior of a particular species occurs is quite broad as long as the conditions for the test match those required for the culture of that particular species. Salinity is an important barrier to prevent cross-mating between brackish or marine species and those in freshwater (Rico-Martínez & Snell, 1997). However, in a defined buffer (containing about 1 g l^{-1} of salts), *B. calyciflorus* males attempted mating with *B. plicatilis* females (Rico-Martínez & Snell, 1997). There are reports of species of *B. calyciflorus, B. rotundiformis* and *Lecane quadridentata* of mating between strains recently collected from the field and some strains cultured in the lab for several years showing that similar mating behavior occurs in both kind of strains (Rico-Martínez unpublished data; Rico-Martínez et al., 1996).

The tests used to analyze the data collected (typically percentage of mating attempts or total number of attempts) seems adequate in most cases. However, in the case of Rico-Martínez & Snell (1995b) the use of a χ^2 test is probably no advisable (Gilbert, 1995, pers. commun.), considering that the mating percentages correspond to two males mating with four females. As a result of that suggestion, a better approach is to use a paired two-tailed t-test on the percentages of the five replicates (Rico-Martínez & Snell, 1995a). It is important to mention that application of the two-tailed t-tests did not change the original conclusions drawn from Rico-Martínez & Snell (1995a).

Another question that remains unanswered is about mating preferences among strains of freshwater species. The only two species for which we have extensive data are the marine/brackish water species strains of *B. plicatilis* and *B. rotundiformis*. These sibling species are probably composed of more than two species, which Gómez & Serra (1995) hypothesized to explain the mating preferences observed in their work. Similar interstrain differences in mating preferences are likely to be found among strains of freshwater rotifer species. There should be several species grouped into what we now consider only one species. The recognition of *B. plicatilis* and *B. rotundiformis* should be informative as an example of how some rotifer species may in fact represent a complex of species (Stemberger, 1979). In that context, the mating behavior tests reviewed here represent useful tools to reveal cryptic sibling species.

Acknowledgments

The author wants to thank T. W. Snell and two anonymous reviewers for critical review of the manuscript. The author also wants to thank the Mexican's Sistema Nacional de Investigadores (S.N.I.) for providing a scholarship.

References

Birky, C. W. Jr., 1967. Studies on the physiology and genetics of the rotifer *Asplanchna*. III. Results of outcrossing, selfing and selection. J. exp. Zool. 165: 104–116.

Carmona, M. J., M. Serra & M. R. Miracle, 1993. Relationships between mixis in *Brachionus plicatilis* and preconditioning of culture medium by crowding. Hydrobiologia 255/256: 145–152.

Fu, Y., A. Hagiwara & K. Hirayama, 1993. Crossing between seven strains of the rotifer *Brachionus plicatilis*. Nippon Suisan Gakkaishi 59: 2009–2016.

Gibson, R. M. & T. A. Langen, 1996. How do animals choose their mates? TIEE 11 (11): 468–470.

Gilbert, J. J., 1963. Contact chemoreception, mating behavior, and sexual isolation in the rotifer genus *Brachionus*. J. exp. Biol. 40: 625–641.

Gilbert, J. J., 1983. Rotifera. In K. G. & R. G. Adiyodi (eds), Reproductive Biology of Invertebrates, Volume VI, Part A. Oxford & IBH Publishing Co., Oxford, MA: 231–263.

Gilbert, J. J. & C. E. Williamson, 1983. Sexual dimorphism in zooplankton (Copepoda, Cladocera, and Rotifera). Ann. Rev. Ecol. Syst. 14: 1–33.

Gilbert, J. J., W. C. Birky & E. S. Wurdak, 1979. Taxonomic relationships of *Asplanchna brightwelli*, *A. intermedia*, and *A. sieboldi*. Arch. Hydrobiol. 87: 224–242.

Gómez, A. & M. Serra, 1995. Behavioral reproductive isolation among sympatric strains of *Brachionus plicatilis* Müller 1786: insights into the status of this taxonomic species. Hydrobiologia 313/314: 111–119.

Gómez, A. & T. W. Snell, 1996. Sibling species and cryptic speciation in the *Brachionus plicatilis* species complex (Rotifera). J. Evol. Biol. 9: 953–964.

Hagiwara, A., T. Kotani, T. W. Snell, M. Assava-Aree & K. Hirayama, 1996. Morphology, reproduction, genetics, and mating behavior of small tropical marine *Brachionus* strains (Rotifera). J. exp. mar. Biol. Ecol. 194: 25–37.

Koste, W., 1978. Rotatoria. Die Rädertiere Mitteleuropas, 2 vols, Gebrüder Borntraeger: 672 pp.

Nichols, H. W., 1973. Growth media – freshwater. In J. R. Stein (ed.), Handbook of Phycological Methods, Cambridge University Press, London: 7–24.

Nogrady, T., R. L. Wallace & T. W. Snell, 1993. Guides to the identification of the microinvertebrates of the continental waters of the world, Volume 4: Rotifera. SPB Academic Publishing: 142 pp.

Paterson, H. E. H., 1985. The recognition concepts of species. In E. S. Vbra (ed.), Species and Speciation. Transvaal Museum Monograph No. 4. Pretoria: 21–29.

Pennak, R. W., 1989. Fresh water invertebrates of the United States, 3rd edn. John Wiley & Sons, New York: 628 pp.

Rao, T. R. & S. S. S. Sarma. 1985. Mictic and amictic modes of reproduction in the rotifer *Brachionus patulus* Müeller. Curr. Sci. 54: 499–501.

Rico-Martínez, R, 1995. Contribution to the knowledge of speciation in Rotifera. Ph. D. Thesis. Georgia Institute of Technology. Atlanta, GA, U.S.A.: 140 pp.

Rico-Martínez, R. & S. I. Dodson, 1992. Culture of the rotifer *Brachionus calyciflorus* Pallas. Aquaculture 105: 191–199.

Rico-Martínez, R. & T. W. Snell, 1995a. Mating behavior and mate recognition pheromone blocking of male receptors in *Brachionus plicatilis* Müller (Rotifera) strains. Hydrobiologia 313/314: 105–110.

Rico-Martínez, R. & T. W. Snell, 1995b. Male discrimination of female *Brachionus plicatilis* Müller and *Brachionus rotundiformis* Tschugunoff (Rotifera) strains. J. exp. mar. Biol. Ecol. 190: 39–49.

Rico-Martínez, R., B. Dingmann & T. W. Snell, 1996. Surface glycoproteins potentially involved in mate recognition in nine freshwater rotifer species. Arch. Hydrobiol. 138: 1–10.

Ruttner-Kolisko, A., 1969. Kreuzungsexperimente zwischen *Brachionus urceolaris* und *Brachionus quadridentatus*, ein beitrag zur fortpflanzungsbiologie der heterogonen Rotatoria. Arch. Hydrobiol. 65: 397–412.

Segers, H., G. Murugan & H. J. Dumont, 1993. On the taxonomy of the Brachionidae: description of *Plationus* n. gen. (Rotifera, Monogononta). Hydrobiologia 268: 1–8.

Serra, M., A. Galiana & A. Gómez, 1997. Speciation in monogonont rotifers. Hydrobiologia 358 (Dev. Hydrobiol. 124): 63–70.

Snell, T. W., 1989. Systematics, reproductive isolation and species boundaries in monogonont rotifers. Hydrobiologia 147: 329–334.

Snell, T. W. & M. Childress, 1987. Aging and loss of fertility in male and female *Brachionus plicatilis* (Rotifera). Int. J. Invertebrate Reprod. Dev. 12: 103–110.

Snell, T. W. & C. A. Hawkinson, 1983. Behavioral reproductive isolation among populations of the rotifer *Brachionus plicatilis*. Evolution 37: 1294–1305.

Snell, T. W. & F. H. Hoff, 1987. Fertilization and male fertility in the rotifer *Brachionus plicatilis*. Hydrobiologia 147: 329–334.

Snell, T. W. & M. A. Nacionales, 1990. Sex pheromone communication in *Brachionus plicatilis* (Rotifera). Comp. Biochem. Physiol. 97A: 211–216.

Snell, T. W., R. Rico-Martínez, L. S. Kelly & T. E. Battle, 1995. Identification of a sex pheromone from a rotifer. Mar. Biol. 123: 347–353.

Stemberger, R. S., 1979. A guide to rotifers of the Laurentian Great Lakes. In C. I. Weber, (ed.), EPA-600/4-79-021. U.S. Environmental Protection Agency, Washington, DC: 185 pp.

Templeton, A. R., 1989. The meaning of species and speciation: a genetic perspective. In D. Otte & Endler, J. A. (eds), Speciation and its consequences. Sinauer Associates, Sunderland, MA 3–27.

U.S. Environmental Protection Agency, 1985. Methods for measuring the acute toxicity of effluents to freshwater and marine organisms. In W. H. Peltier & Weber, C. I. (eds), EPA-600/4-85-013. U.S. Environmental Protection Agency, Washington, DC.

Vbra, E. S., 1995. Species as habitat-specific, complex systems. In D. M. Lambert & Spencer, H. G. (eds), Speciation and the Recognition Concept. The Johns Hopkins University Press. Baltimore, MD 3–44.

Hydrobiologia **387/388**: 117–121, 1998.
E. Wurdak, R. Wallace & H. Segers (eds), Rotifera VIII: A Comparative Approach.
© *1998 Kluwer Academic Publishers.*

Preparation of rotifer trophi for light and scanning electron microscopy

Willem H. De Smet
Department of Biology, University of Antwerp, R.U.C.A.-campus, Groenenborgerlaan 171,
B-2020 Antwerpen, Belgium. E-mail: wides@ruca.ua.ac.be

Key words: Rotifera, trophi, preparation, light microscopy, scanning electron microscopy

Abstract

The methods to prepare rotifer trophi for light and scanning electron microscopy are reviewed, and the rapid method used by the author is described. Rotifers are dissolved in a minimal amount of sodium hypochlorite solution on a coverslip, and serially rinsed in distilled water. The entire procedure is done under a microscope using micropipettes.

Introduction

Trophi are one of the main descriptors used in classical taxonomy of rotifers and so constitute an important taxonomic criterion to define taxonomic units from the level of order to species. In fact, reliable identification of most illoricate and several loricate species is possible by examination of their trophi alone. However, the trophi of many species are so small that they approach the limit of resolution of light microscopy, thus providing a major handicap for the detailed study of their structure. Introduction of scanning electron microscopy (SEM) has helped significantly to overcome this problem. Three examples will suffice to make this point. Sanoamuang (1993) recently demonstrated that the morphological criteria previously used to identify *Filinia* to the level of species may be inadequate. Segers (1995) showed that apparently morphologically uniform genera may actually contain several species. The potential for trophal ultrastructure to be of use as a tool in reconstructing rotifer phylogeny was demonstrated by Markevich (1985,1989) and Markevich & Kutikova (1989).

From these examples, I argue that whenever possible a thorough examination of trophi, including light and SEM preparations, should be part of every taxonomic study. Several methods for the preparation of rotifer trophi for light microscopy and SEM have been published. Here I present a rapid and easy method to prepare trophi for light or SEM examination.

Survey of previous methods

Two methods are routinely used to study rotifer trophi. (1) They may be studied *in vivo* by examining whole animals either using compressing devices or by applying clearing agents. (2) They may be studied *in vitro* after extraction by first dissolving the soft tissues and lorica using a caustic chemical (sodium hydroxide, potassium hydroxide, sodium hypochlorite, etc.).

Living or dead specimens can be compressed with special instruments (e.g., the Taylor microcompressor) (Taylor, 1966), or by the Vaseline-spot method (Galliford, 1960). In the latter method the specimens are placed in a small drop of water in the middle of a microscope slide. A coverglass, provided with a small spot of vaseline on each corner, is placed over the drop of water. The specimen is compressed by gradually pressing down the coverglass. Although the type of trophi and its major characters can ordinarily be recognized *in vivo* with these methods, they are not suited for detailed analysis.

Whole animals can also be mounted in polyvinyl-lactophenol (Russell, 1961). In this technique, the trophi become visible after the surrounding tissues have cleared. Although the trophi types are usually easily recognized by this method, I recommend that it not be used for detailed analysis, as the trophi musculature and the adjacent cells often alter the shape of the trophi and distort their real dimensions.

In the technique developed by Myers (1937) the rotifer specimen is placed in a drop of glycerin just inside the concavity of a depression slide. Then a drop of caustic alkali or sodium hypochlorite (NaOCl) is added just outside the concavity and the depression is covered by a coverslip. By working the coverslip over the concavity, the rotifer is pushed in the angle formed by the edge of the concavity and the underside of the coverslip. When the caustic solution mingles with the glycerin drop it begins the dissolution process and after a few minutes the caustic chemical dissolves the soft tissues to leave only the hard trophi. Alternatively, the specimen can be placed in a small drop of water on a microscope slide, which is then covered by a coverslip. Then a small drop of hypochlorite solution is placed to the edge of the coverslip, and that solution is drawn underneath the coverslip by carefully touching tissue paper to the opposite side (Stemberger, 1979).

Koehler & Hayes (1969a, b) were the first to study rotifer trophi by using SEM technology. In this technique, living specimens of *Philodina acuticornis odiosa* and *Asplanchna sieboldi* are treated with full strength bleach (Clorox®) to digest all the soft tissues. Using finely drawn pipettes, the remaining trophi are washed several times with distilled water. The washing step, of course, requires that the process be carefully monitored using a high power dissecting microscope so that the trophi are not lost. Once adequately washed, the trophi are placed on a carbon-coated, 150-mesh electron microscope grid.

Salt et al. (1978) used a bulk method to study trophi of six *Asplanchna* species. In this technique, twenty specimens are placed in approximately 0.4 ml of a variety of dilutions (14–100%) of commercial sodium hypochlorite in plastic depression slides. After about 2 hr the trophi are transferred to different baths of ca. 0.4 ml distilled water. The trophi were left in the last bath for approximately a week. After the last bath the trophi are passed through seven more washes, each one hour long. The clean trophi, together with a minimum amount of water, are gathered up in a micropipette, transferred to a coverslip and placed in a dessicator until all the water has evaporated. Transfers of rotifers and trophi are done under a dissecting microscope using a micropipette.

Markevich & Koreneva (1981) prepare trophi in microfunnels on membrane filters. In this technique, rotifers are pipetted in a small drop of water onto a Nucleopore® filter (diameter 5–6 mm, pores 0.46 μm) and treated with NaOCl for 15–20 min. Low vacuum is then applied to remove the hypochlorite solution and to rinse the trophi with distilled water. After the last step the filter is dried.

Kleinow et al. (1990) dissolve rotifer tissues and the lorica with sodium dodecyl sulfate (SDS) and dithiothreiol (DTT). For a typical trophi preparation of *Brachionus plicatilis* 4–6 g (wet weight) of rotifers was needed. After homogenization, the homogenate is centrifuged and resuspended several times in SDS, treated with DTT, washed with distilled water and finally dialyzed against distilled water during 5–7 days. The dialyzed trophi material is then pipetted onto a specimen mount. This method is cumbersome and time consuming; requires large quantities of material, and, in general, is not suitable for routine analysis of trophi. Moreover, the method is not more gentle than the hypochlorite method as claimed by the authors.

Sanoamuang & McKenzie (1993) use 22 mm square glass coverslips on which a dam of plasticine, approximately 2 mm high and 10 mm in diameter, is fashioned. The enclosing dam is filled with fresh 3% NaOCl and 20–30 rotifers are pipetted into the ring and left for 1–2 hours. The degree of disintegration of the body is assessed periodically using a light microscope at 100× magnification. Trophi cleared of tissue are washed in distilled water that is changed 5–6 times at half-hourly intervals. About one third of the solution is removed at a time, using a piece of rolled filter paper as a wick. The clean trophi on the slide, in pure distilled water, are placed in a vacuum desiccator and left overnight to dry.

Segers (1993) and Segers et al. (1993) prepare trophi on a circular coverslip by dissolving tissues in 5% NaOCl and subsequent repeated washing with distilled water. After removal of most of the remaining water the trophi preparations are left to dry overnight. The entire procedure is performed under a dissection microscope, using ultrafine pipettes.

In the following section I outline the methods used in my laboratory to prepare rotifer trophi for light and SEM study.

Materials

– Micropipettes: made by drawing out non-heparinized microhematocrit tubes to a fine point in a Bunsen flame. Micropipettes with curved point are preferable for removal of washing-water. The pipettes are mounted on disposable polyethylene dropper pipettes with a tapered nozzle.

— Dissecting needles made of insect pins type 'Minuten Nadeln'.

— Glass microscope slides (25 mm × 76 mm).

— Coverslips 9 mm × 9 mm obtained by cutting 18 mm × 18 mm cover glasses into 4 pieces with a diamond pencil, or circular coverslips (diameter 10 mm).

— Glass depression slides (25 mm × 76 mm), with a depression measuring 18–20 mm × 1.75 mm.

— Compound microscope with short object 4× lens, leaving a working distance of 10–15 mm between lens and slide, or a dissection or inverted microscope.

— Glycerin.

— Deionized or distilled water, preferably filtered through a 0.4 μm membrane filter.

— Commercial sodium hypochlorite (8–10%) solution (household bleach, Clorox, eau de Javel, etc.).

— Freshly killed rotifers, or alcohol or formalin preserved specimens transferred to glycerin on a microscope slide.

Procedure

1. Apply a small droplet of glycerin in the middle of a microscope slide, and place a coverslip on top of it. The glycerin keeps the coverslip in place during the manipulations, and ensures a clear background, enhancing the visibility of the extracted trophi.

2. Place a very small droplet of hypochlorite solution in the middle of the coverslip. The volume of the droplet should be large enough to avoid formation of crystals before the digestion process is finished (ca. 0.5 μl or a droplet of ± 1–1.5 mm diameter will do for most rotifers).

3. Immediately cover the microscope slide-coverslip preparation with an upside down depression slide. This will act as a lid.

4. Pick up a rotifer specimen from the glycerin preparation with a dissecting needle. Quickly remove the lid and transfer the rotifer into the hypochlorite droplet on the coverslip. Replace lid.

5. This preparation is placed under a microscope, and the digestion process is followed at a magnification of 40× or 100×. After a few minutes the soft tissues will be completely dissolved, leaving the hard trophi. In some loricate species (e.g. *Lecane, Lepadella, Brachionus*) the lorica is only partly destroyed, but the trophi are usually expelled during step (6).

6. Remove the depression slide (lid), and gently add an excess of water with a micropipette.

7. Remove (by capillarity) most of the diluted hypochlorite solution with a micropipette with curved point, while watching the trophi through the microscope. Take care that the remaining small droplet, containing the trophi, does not dry up completely at this stage. Immediately add an excess of water again (otherwise the trophi will be entrapped in the forming crystals and break up after addition of water). *Alternative*: instead of removing the solution after the first addition of excess water, the trophi can be picked up with a micropipette in a small quantity of solution, and transferred to a previously prepared microscope slide with coverslip containing a large drop of water.

8. Repeat the rinsing process about 10 times, in order to remove all hypochlorite solution.

9. Add some water again and note upper and lower sides of the trophi. Manipulate them into desired position by cautiously stirring with a dissecting needle (do not touch the trophi!). Alternatively, one may do this by gently aspirating and expirating small quantities of water with a micropipette.

10. Remove all water until completely dry (especially suitable when the trophi are at the edge of the water-droplet), or remove most of the water, leaving trophi in a small quantity of water and allow to air dry. Strong turbulencies occur during evaporation, which may turn the trophi into an unwanted position. This can be overcome by counteracting the tensions: a dissecting needle is dipped into the droplet, and cautiously moved to or away from the trophi. This has the function of redirecting the forces produced by evaporation.

11. The position of the air-dried trophi is marked with a marking pen.

12. Remove the coverslip with the prepared trophi from the microscope slide. For light microscopy: invert the coverslip and place the surface containing the trophi on a droplet of mountant medium (e.g. glycerin, glycerin-gelatin, etc) on a microscope slide. For electron microscopy: mount the coverslip with double sided adhesive tape on an aluminum stub and immediately sputter-coat with gold as usual (or store in a desiccator).

With a little bit of skill and experience it is possible to make a trophi preparation in 15–20 min. The results are shown in Figures 1–10.

120

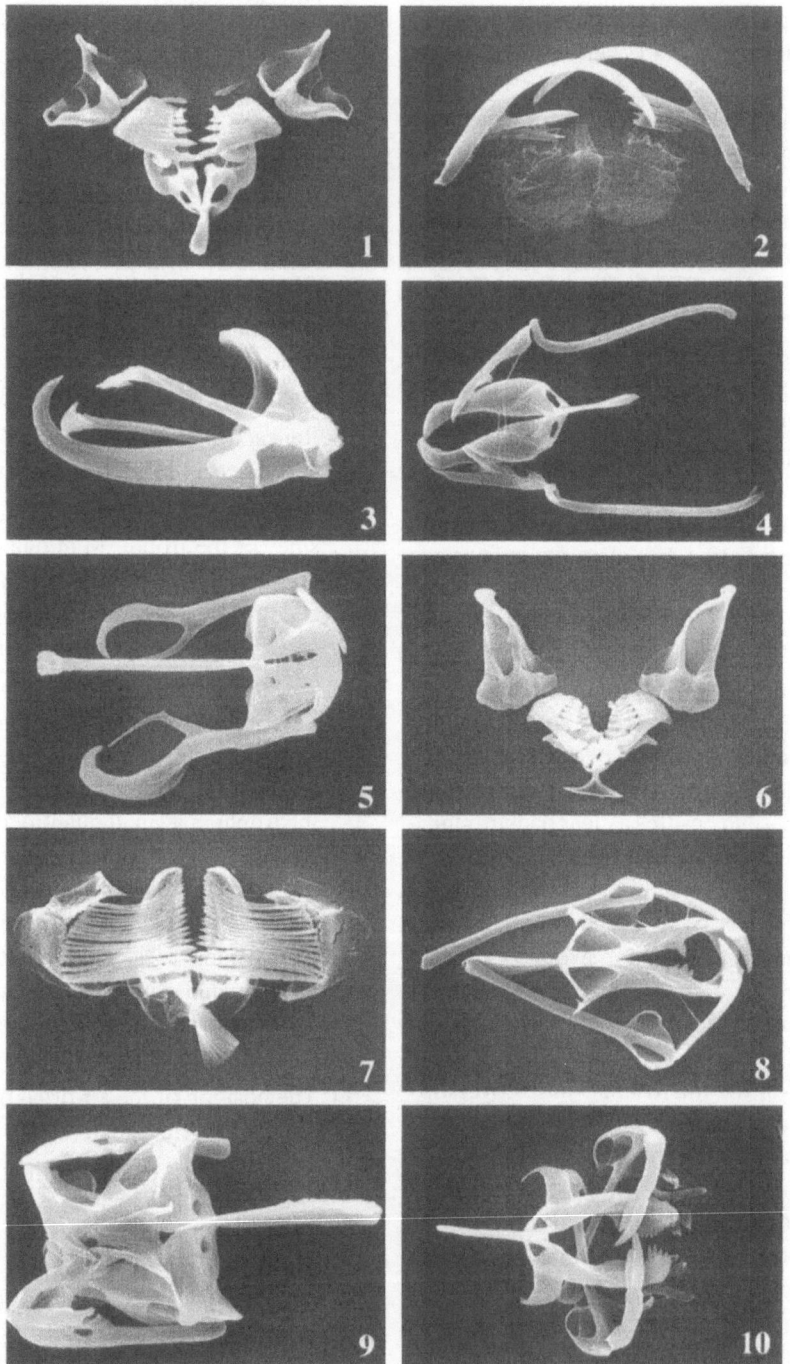

Figures 1–10. SEM photographs illustrating trophi of rotifers. 1, *Epiphanes senta*; 2, *Cupelopagis vorax*; 3, *Ascomorpha dumonti*; 4, *Aspelta psitta*; 5, *Cephalodella catellina*; 6, *Lepadella ovalis*; 7, *Testudinella elliptica*; 8, *Dicranophorus haueri*; 9, *Eosphora najas*; 10, *Itura myersi*.

Acknowledgements

I wish to express my gratitude to Drs E. Wurdak and Robert Wallace for improving the English text.

References

Galliford, A. L., 1960. How to begin the study of rotifers. Part (2). Methods of collection and study. Country Side, N.S. 19: 188–194.

Kleinow, W., J. Klusemann & H. Wratil, 1990. A gentle method for the preparation of hard parts (trophi) of the mastax of rotifers and scanning electron microscopy of the trophi of *Brachionus plicatilis* (Rotifera). Zoomorphology 109: 329–336.

Koehler, J. K. & T. L. Hayes, 1969a. The rotifer jaw: a scanning and transmission electron microscope study. I. The trophi of *Philodina acuticornis odiosa*. J. Ultrastruct. Res. 27: 402–418.

Koehler, J. K. & T. L. Hayes, 1969b. The rotifer jaw: a scanning and transmission electron microscope study. II. The trophi of *Asplanchna sieboldi*. J. Ultrastr. Res. 27: 419–434.

Markevich, G. I., 1985. The main trends of idioadaptive evolution of rotifers. Jaws. In Proceedings of the Second All-Union Symposium on Rotifers. Leningrad: 17–37 (in Russian).

Markevich, G. I., 1989. Morphology and principal organization of the sclerite system of the mastax in rotifers. In Biologiya sistematika i funktsdionalnaya morfologiya presnovodnykg zhivotnykh. Trudy Inst. biol. vnutr. vod 56(59): 27–82.

Markevich, G. I. & E. A. Koreneva, 1981. On the technique of preparing of the mastax of rotifers for electron microscopy. Zool. Zhurn. 60: 1562–1564 (in Russian).

Markevich, G. I. & L. A. Kutikova, 1989. Mastax morphology under SEM and its usefulness in reconstructing rotifer phylogeny & systematics. Hydrobiologia 186/187: 285–289.

Myers, F. J., 1937. A method of mounting rotifer jaws for study. Trans. am. Micros. Soc. 56: 256–257.

Russell, C. R., 1961. A simple method of permanently mounting Rotifera trophi. J. Quekett Microsc. Club, Ser. 4,5: 377–378.

Salt, G. W., F. G. Sabbadini & M. L. Commins, 1978. Trophi morphology relative to food habits in six species of rotifers (Asplanchnidae). Trans. am. Micros. Soc. 97: 469–485.

Sanoamuang, L., 1993. Comparative studies on scanning electron microscopy of trophi in the genus *Filinia* Bory De St. Vincent (Rotifera). Hydrobiologia 264: 115–128.

Sanoamuang, L. & J. C. McKenzie, 1993. A simplified method for preparing rotifer trophi for scanning electron microscopy. Hydrobiologia 250: 91–95.

Segers, H., 1993. Rotifera of some lakes in the floodplain of the River Niger (Imo State, Nigeria). I. New species and other taxonomic considerations. Hydrobiologia 250: 39–61.

Segers, H. H., 1995. A reappraisal of the Scaridiidae (Rotifera, Monogononta). Zool. Scr. 24: 91–100.

Segers, H., G. Murugan & H. J. Dumont, 1993. On the taxonomy of the Brachionidae: description of *Plationus* n.gen. (Rotifera, Monogononta). Hydrobiologia 268: 1–8.

Stemberger, R. S., 1979. A guide to rotifers of the Laurentian Great Lakes. U.S. Environmental Protection Agency, Ohio. 186 pp.

Taylor, H. L., 1996. The Taylor microcompressor MK III variable volumetric counting version. Microscope 44: 137–140.

Hydrobiologia **387/388**: 123–129, 1998.
E. Wurdak, R. Wallace & H. Segers (eds), Rotifera VIII: A Comparative Approach.
© *1998 Kluwer Academic Publishers.*

Stereopictures of internal structures and trophi of rotifers

W. Kleinow
Zoologisches Institut der Universität zu Köln, Weyertal 119, D-50923 Köln, Germany
E-mail: Walter.Kleinow@Uni-Koeln.De

Key words: trophi, mastax, stereo pictures, *Brachionus, Asplanchna, Filinia*

Abstract

SEM pictures were taken before and after tilting the specimen mount with the object at angles of 10–15°. Observing pairs of the resulting photographs through a suitable stereoscope (or by simpler techniques) gives a three-dimensional view of the object which assists in the interpretation of morphological structures. Examples are presented that show details of internal structures in the region between wheel organ and mastax of *Brachionus plicatilis* and compare the structure of trophi in different rotifers.

Introduction

Methods have been developed for obtaining SEM pictures from internal structures of *Brachionus plicatilis* (Kleinow & Wratil, 1995; Kleinow et al., 1991) and for isolating trophi clean and unscathed (Kleinow et al., 1990). By combination of these methods it has been possible to ascertain how the trophi are orientated within the soft tissues of the mastax (Kleinow & Wratil, 1996). Interpretation of the three-dimensional structure was much facilitated by the possibility of observing pairs of photographs by means of a reflecting stereoscope.

A clear comprehension of three-dimensional structure is necessary to understand the functional morphology of moving parts, like the trophi of the rotifer mastax. Likewise, the three-dimensional morphology of the region between the wheel organ and the mastax of rotifers needs to be clarified in order to understand the mechanism of food intake and its control. The morphology of this region seems to be very complicated. It contains several different tissues and structures, some of unknown function, and it is therefore necessary to clarify their three-dimensional alignment. To obtain such knowledge it is very helpful to get a three-dimensional view of the object. For large objects this can be achieved using a low-power stereo-microscope, while for smaller objects, corresponding 3D pictures may be obtained from SEM photographs. The application of this technique to rotifer trophi was described

some time ago (Kleinow et al., 1990), but up to now no stereo images of rotifer trophi have been published.

Here we provide some examples of how such 3D pictures can aid in the understanding of structures in the region between corona and mastax, and how they can be used to compare different types of trophi. A modification of our method for isolating the trophi will be described allowing preparations from relatively small numbers of rotifers.

Materials and methods

Brachionus plicatilis was reared and collected from the culture medium as described elsewhere (Kühle & Kleinow, 1985, 1989). Methods for sectioning rotifers and examination by SEM were performed as described previously (Kleinow & Wratil, 1996; Kleinow et al., 1991).

Living *Brachionus rubens* and *Brachionus calyciflorus* were kindly provided by Prof. Dr K.O. Rothhaupt (Konstanz) and by Dr N. Walz (Berlin). In order to remove algal food (in the case of *B. calyciflorus*) the samples were first centrifuged for 20 min at $800 \times g$, 25°C and then, after chilling the sample to 4°C, the rotifers were collected by centrifuging for 20 min at $800 \times g$ at 4°C. Small numbers of various rotifer species were obtained from a pond in Erftstadt. These animals were either prepared without fixation or fixed in 70% ethanol.

124

Cleaning and preparation of trophi for SEM

Disintegration buffer was prepared about an hour before use, by adding 20 mg/ml DTT (=0.13 M) and 0.5 mg/ml proteinase K to 5 ml buffer stock solution (2% SDS=0.07 M; 0.03 M NH_4HCO_3, pH 8). Then, 5 ml disintegration buffer were added to 1 ml of medium containing the rotifers. If necessary, the rotifers were concentrated beforehand, by chilling them at 4°C and centrifuging for 20 min at $800 \times g$. In this case the upper part of the supernatant was gently removed until a volume of about 1 ml remained. The disintegration mixture had to stand for 60–90 min, during which time it was shaken for 10 s by a Vortex mixer every 5–10 min. From about 40 min on, the preparation was observed through the microscope (about every 20 min) to see whether the trophi were already sufficiently clean. The trophi were then collected by centrifuging for 20 min at $3000 \times g$ in Corex centrifuge tubes. The trophi were washed by resuspending them in 25% ethanol and centrifuging for 20 min at $3000 \times g$. In order to ensure removal of all traces of SDS on the SEM preparations the washing was repeated until foaming was no longer observed (at least three to four times). The fluid containing the trophi was then made up to a ca 60% ethanol solution by adding 96% ethanol. After centrifuging, most of the supernatant was removed, leaving ca 0.5 ml. The sample was then distributed over (three to four) specimen mounts previously covered with 'Tempfix-glue' (Neubauer Chemikalien, Münster) and prepared for SEM as described previously (Kleinow et al., 1990).

Alternatively, for small (20–30) numbers of rotifers, the disintegration was performed in wells of a microtitre plate with $50\mu l$ sample volumes to which $100–200\mu l$ disintegration buffer was added. In this case, instead of centrifuging, the trophi were allowed (for about 60 min) to sediment to the bottom of the well by gravity alone (in order to collect them and to wash them). After each sedimentation of the trophi, the supernatant was gently removed down to a residual volume of about 50 μl. This was achieved by gentle suction of the fluid from the uppermost layer by means of a narrow polyethylene tube under close observation through a magnifying glass. Polyethylene tubes with sufficiently small ends were prepared by gently heating the middle portion of pieces of tubing measuring 2 mm in diameter and 50–60 mm in length over a small flame and drawing out their ends until the desired thinning had been achieved.

Preparing and observing of stereo images

A series of pictures was prepared by taking successive photographs with the SEM, under identical magnification, and by tilting the SEM specimen mount 5° for every new picture. A three-dimensional view of the structures is achieved if two photographs, taken at angles differing by 10–15°, are arranged side by side in the direction of tilting and observed through a reflecting stereoscope (VCH Verlagsgesellschaft, Weinheim). If the pictures are not too large or too far apart, they may even be viewed by 'naked-eye stereopsis' (McKeon & Gaffield, 1990). This procedure may be facilitated by holding two convex lenses (about +4.5 diopters), one in each hand, and looking somewhat obliquely through their inner halves onto the pictures. After some adjustment of the two lenses, overlap of the pictures and a stereo effect will be achieved.

Results and discussion

Structures in the anterior part of Brachionus

We have started an examination of the region between the wheel organ and the mastax. In this region one can detect several structures and tissues intermingled in a complicated way. In the following, we present some preliminary results on the morphology of this region.

The picture pairs Figures 1 and 2 show the internal side of the buccal tube over its whole length. Figure 1 shows the ventral part of the buccal tube, i.e. as seen from a dorsal position. In the upper part of Figure

---

Figures 1–4. **Figure 1.** Frontal section of the region of buccal tube and mastax of *Brachionus plicatilis*. Dorsal view. m, mastax; v, mastax vestibulum; w, wheel organ. Tilting angles of the SEM specimen mount: left half-picture, 12.5°; right half picture, 0°; calibration bar=13 μm. **Figure 2.** Frontal section of the region of buccal tube and mastax of *Brachionus plicatilis*. Ventral view. c, cilia-free cushion in the dorsal part of the mouth; m, mastax; v, mastax vestibulum. Tilting angles of the SEM specimen mount: left half picture, 12.5°; right half picture, 0°; calibration bar=12.5 μm ('k', label indicating direction of tilting in SEM). **Figure 3.** Neural structures behind the wheel organ of *Brachionus plicatilis* (arrows). lo, lorica; w, wheel organ. Tilting angles of the SEM specimen mount: left half picture, 55°; right half picture 65°; calibration bar=8.5 μm. **Figure 4.** An assumed neural structure circumventing the buccal tube of *Brachionus plicatilis* in the region between wheel organ and mastax (arrows). b, brain; g, bulb of the dorsal mastax gland; m, musculus adductor mallei; n, cell body of nervus pharyngeus; u, uncus; w, wheel organ. Tilting angles of the SEM specimen mount: left half picture, 45°, right half picture, 30°; calibration bar=8.7 μm.

125

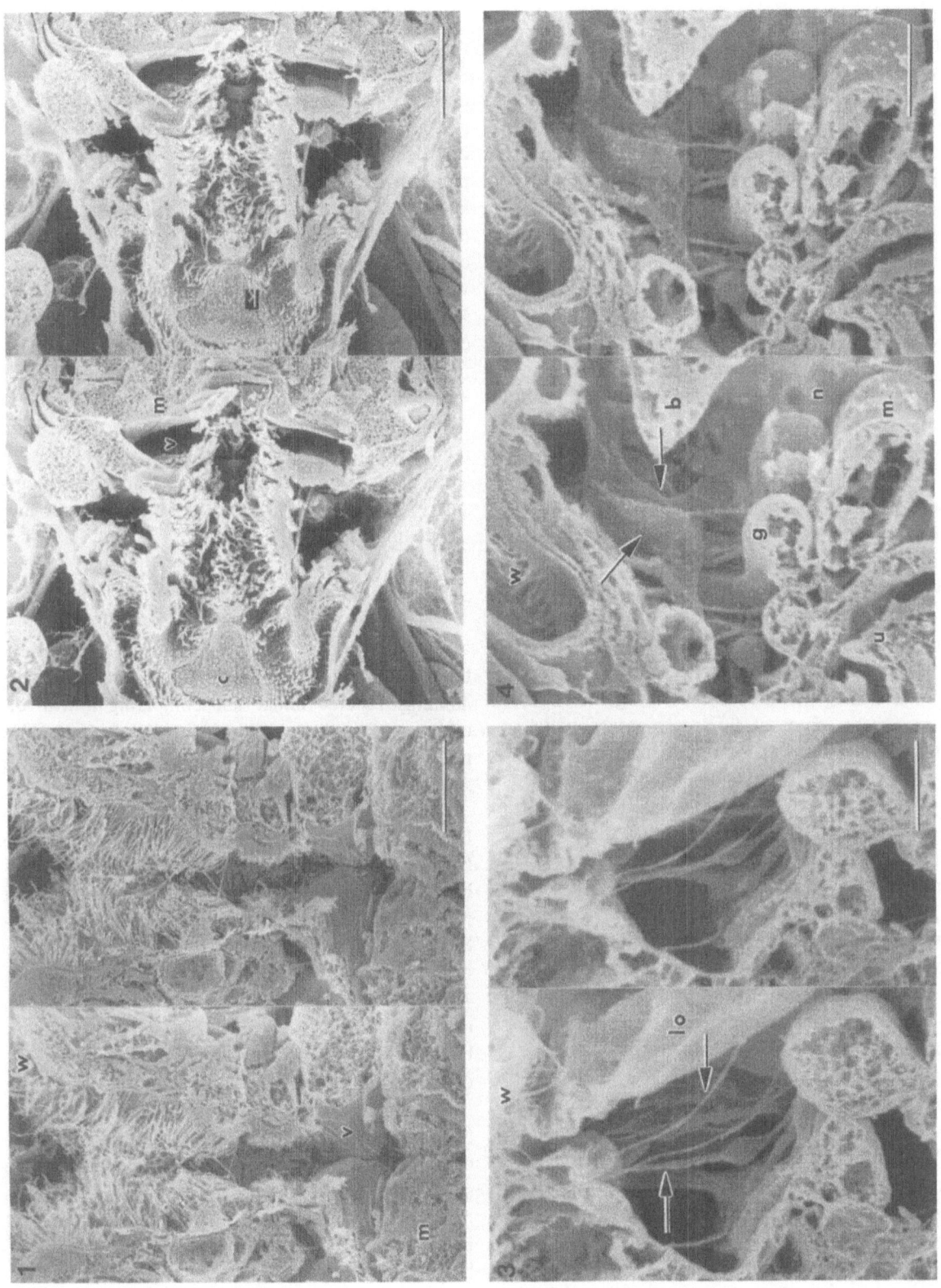

Figures 1–4. See p. 124.

1, the wide entrance of the buccal tube, the mouth, is seen, which then narrows to the next part (the pharynx). The whole buccal tube is evenly covered by cilia except the region just before its entrance to the mastax: the 'mastax vestibulum' (Kleinow & Wratil, 1996). In the lower part of the picture the beginning of the mastax (m) can be recognized. In this preparation the mastax is cut frontally, just above the tips of the uncus teeth. On the right side, one of the 'mastax bubbles' is cut open, whose proximal walls contribute to the lining of the mastax vestibulum (v). Figure 2 shows the same region, buccal tube and the beginning of the mastax as seen ventrally. The mastax (m) is seen on the right. It is cut at the plane of the uncus or subuncus teeth, which are seen from below. The wheel organ (seen on the left of this picture), in the dorsal part of the mouth, yields a cushion or a plate, from which, in this preparation, the cilia have been removed (c). Again the whole buccal tube is seen, evenly covered by cilia. The cilia free entrance of the mastax vestibulum (v) is seen as a deep cleft in this view. Remarkably, in such SEM pictures, no equivalent to the structure described as 'buccal velum' could be identified, which is observed on TEM pictures of sections (Clement et al., 1980; Kleinow, unpubl.).

Figure 3 shows a picture pair of neural structures (arrows) extending behind the wheel organ. These structures might represent (afferent) nerves which send information from sense organs of the corona to the CNS. Judging by the diameters (0.2–0.4 μm) of these processes, these neural structures might be single axons. It is also noteworthy that these processes vary in thickness along their length. In thin sections viewed by transmission electron microscopy, corresponding structures were detected, which corroborate the view that these processes are single axons. The somewhat thicker parts seem to contain additional organelles, such as mitochondria, besides the neurofibrils. Moreover, some bulbous parts can be observed (diameter 2 μm) which are probably nerve cell bodies.

Figure 4 shows what may be a more complicated neural structure of this region. It provides a view of the buccal tube from the 'outside', i.e. from the pseudocoel. The rotifer in this preparation was cut along a parasagittal plane, i.e. a plane on the left, parallel to the median plane of its body. In the upper part of Figure 4 one recognizes the cut wheel organ (w), in the lower part the mastax. The mastax is cut somewhere through the left uncus (u). In front of the mastax the musculus adductor mallei (m) is seen in cross-section along with the the cell body of the (left) nervus

pharyngeus (n) which extends by its process into the 'dorsal cleft'. Before the mastax, bulb-like structures of the dorsal mastax glands are to be seen. One of the glandular structures (g) has a thick arm, protruding in an anterior direction. In this region between the dorsal mastax gland and wheel organ, a flat, band-like or bracelet-like structure (arrows) was detected on SEM pictures. This appeared to more or less circumvent the buccal tube. The thick arm of the dorsal mastax gland extends below this structure, and nerve-like processes can be seen stretching out from it, forwards and backwards. The function of this apparently undescribed structure is unclear. Perhaps it is a neural structure controlling the functioning of the pharynx region.

In order to clarify this issue, we plan (1) to do 3D reconstructions from TEM pictures of serial sections and (2) to study more transverse sections of this region which may be obtained if it becomes feasible to cut several *Brachionus* specimens embedded alongside each other in the same block and in the same orientation.

Stereo pictures of isolated trophi

In the following section we present stereo-images of three different types of rotifer trophi. By closely examining a stereo image and comparing it with the corresponding single pictures, it is easy to see how the stereo view aids in the interpretation of the trophi structure.

For the malleate type, trophi of two closely related rotifers, *B. rubens* and *B. calyciflorus* (Figures 5–8) were prepared for SEM analysis in order to show how small variations within a single trophi type are revealed in stereo pictures. The views in Figures 5–8 correspond to orientations in relation to the rotifer or mastax (as in *B. plicatilis*, see Kleinow & Wratil, 1996) from frontal (Figures 5, 7) or from behind (Figures 6, 8). Several differences, which can already be detected by comparing the single pictures, are recognized more readily by looking at the 3D images, since these are more clear and less ambiguous than the corresponding single pictures. Features which concern the three-dimensional shape of the parts or their spatial relationships are hardly detectable without stereoscopic help: thus, by comparing the stereo images, it can be deduced that the angle between the rami-hinge (arrows) and the anterior edge of the rami is more acute in *B. calyciflorus* (Figures 5, 6), than in *B. rubens* (Figures 7, 8). This contributes to the different orientations of the two types of trophi on the

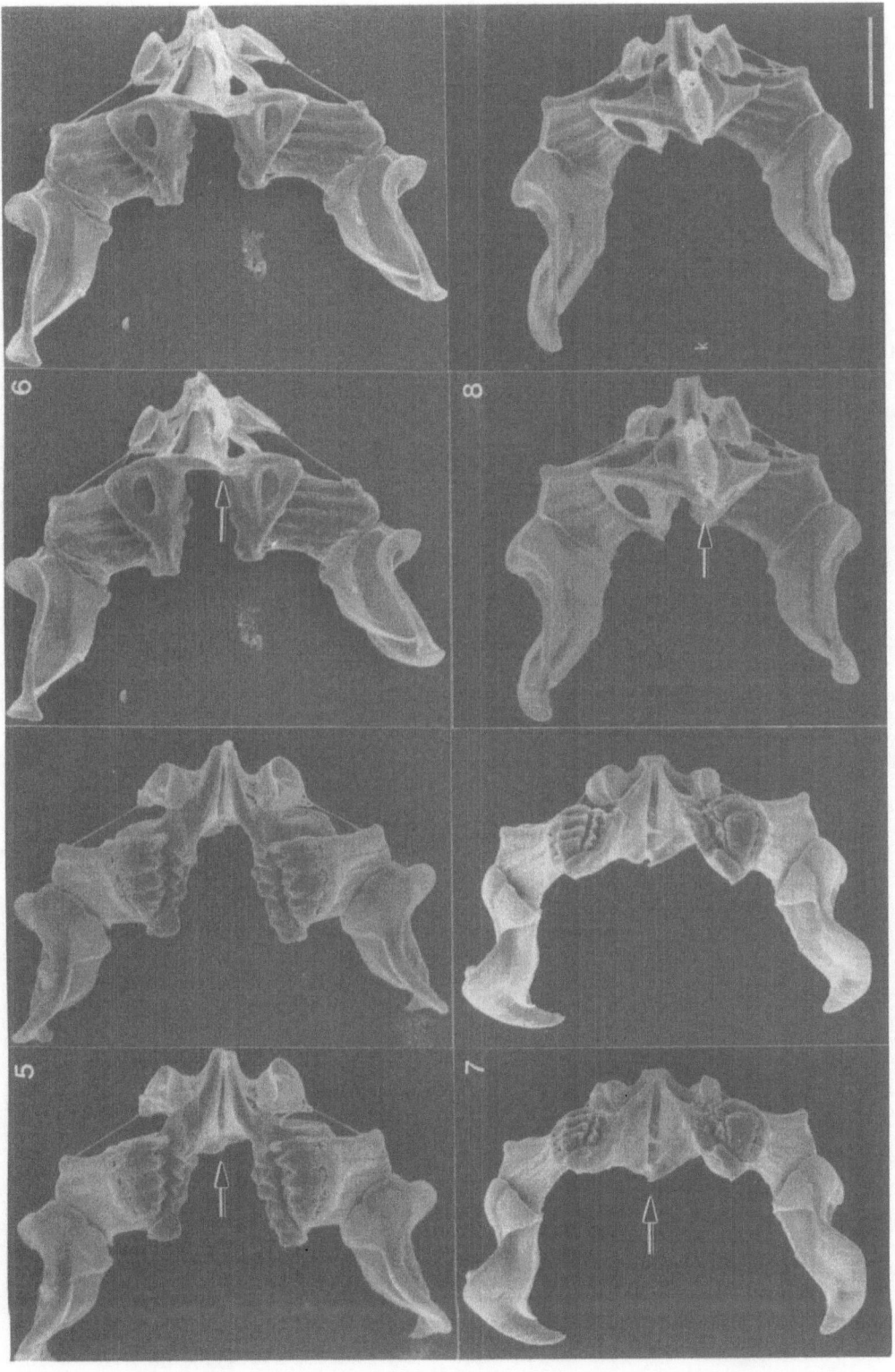

Figures 5–8. Comparison of trophi of *Brachionus calyciflorus* (Figures 5, 6) and of *Br. rubens* (Figures 7, 8). The photographs on the left side (Figures 5, 7) show anterior views, those on the right (Figures 6, 8) posterior views. The arrows point to the rami-hinges. Tilting angles of the SEM specimen mount: Figure 5, left half picture, 0°; right half picture, 10°; Figure 6, left half picture, 0°; right half picture, 10°; Figure 7, left half picture, 10°; Figure 8, left half picture, 0°; right half picture, −5°; right half picture, 12.5°; calibration bar (for Figures 5–8)=20 μm.

128

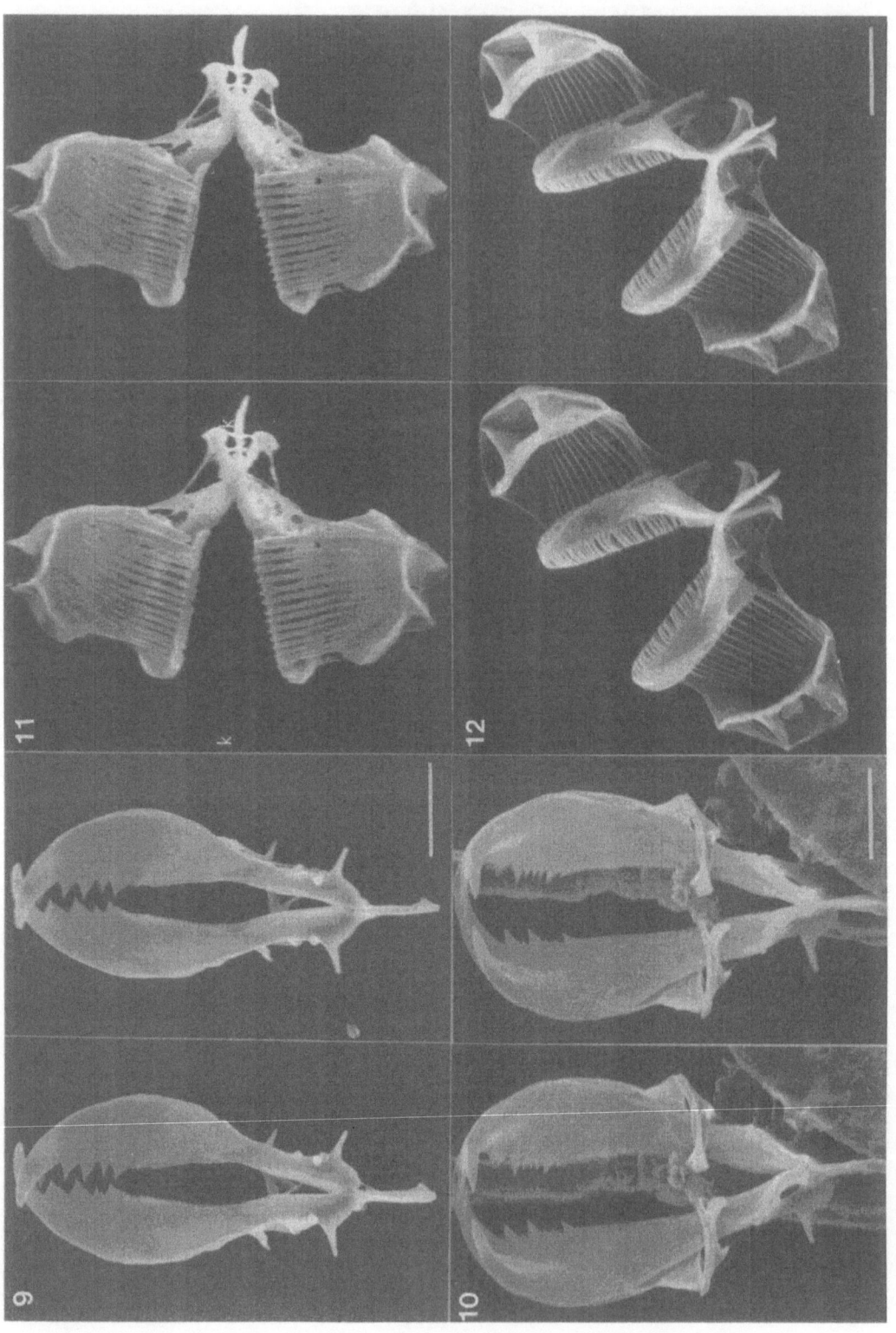

Figures 9–12. **Figures 9, 10.** Trophi of *Asplanchna priodonta* in the two orientations occurring on the SEM stage. Figure 9: tilting angles of the SEM specimen mount: left half picture, 0°, right half picture, 10°; calibration bar=15 μm; Figure 10: tilting angles of the SEM specimen mount: left half picture, 0°; right half picture, 10°; calibration bar=13 μm. **Figures 11, 12.** Trophi of *Filinia* in the two orientations occurring on the SEM stage. Tilting angles of the SEM stage. Figure 11: left half picture, 12.5°; right half picture, 0°; Figure 12: left half picture, 0°; right half picture, 12.5°; 'k', label indicating direction of tilting in SEM; calibration bar for (Figures 11–12)=9 μm.

SEM stage which is especially obvious in the 'posterior view' (Figures 6, 8). These orientations of the trophi are related to the shapes of their individual parts and how they are connected by membraneous hinges and to the centers of gravity of the parts as well as of the whole structure. Stereo pictures will be indispensible if relations between structure and function are to be clarified.

The picture pairs in Figures 9–12 provide stereo images of trophi from different rotifers caught in small numbers from ponds.

Figures 9 and 10 show the incudate type of trophi of *Asplanchna priodonta* in both possible orientations on the SEM stage. Trophi of several species of *Asplanchna* were among the first to be examined by SEM in detail (Koehler & Hayes, 1969b; Salt et al., 1978). It is quite clear that additional features are revealed in three-dimensional images. In the single pictures the rami appear be more or less flat, but the stereo images show that they are indeed twisted. Thus, in Figure 9, in the upper third, the outer edges of the rami are turned toward and the inner edges (especially both first rami teeth) away from the paper plane. In the view in Figure 10 the rami have a spoon-like appearance and two pairs of the foremost rami teeth are seen to stretch out perpendicular to the plane of the rami. This twisting will stabilize and stiffen the rami structure, whereas the spreading out of the foremost and largest of the rami teeth seems to be significant for the task of grasping the prey.

Figures 11 and 12 show examples of the malleoramate type of trophi from the genus *Filinia*. In this case the stereo-view of Figure 11 shows clearly the curvature of the unci which differs as one progresses from the larger to the smaller unci teeth. Likewise in Figure 12, the spatial relationships between unci, rami and manubriae becomes much clearer if the picture-pair is observed in stereo view.

These examples show that 3D pictures reveal additional features by which trophi from various rotifers may be characterized. However, the pictures of trophi through the light microscope give a somewhat different impression from the SEM pictures, especially if the parts of the trophi have been separated out. In some cases it might even be difficult to recognize trophi in SEM which are known from light microscopic views. It may therefore be worth compiling a catalogue of trophi in 3D SEM pictures in order to correlate SEM pictures with light microscopic pictures of the same trophi.

Acknowledgements

We wish to thank Mr Helmut Wratil for excellent technical assistance and Mrs Frances Wharton for correcting the English in the manuscript. This work was supported by the 'Verein der Freunde und Förderer der Universität zu Köln'.

References

Clement P., J. Amsellem, A.-M. Cornillac, A. Luciani & C. Ricci, 1980. An ultrastructural approach to feeding behavior in *Philodina roseola* and *Brachionus calyciflorus* (Rotifers). I. The buccal velum. Hydrobiologia 73: 127–131.

Kleinow, W., J. Klusemann & H. Wratil, 1990. A gentle method for the preparation of hard parts (trophi) of the mastax of rotifers and scanning electron microscopy of the trophi of *Brachionus plicatilis* (Rotifera). Zoomorphology 109: 329–336.

Kleinow, W., H. Wratil, K. Kühle & B. Esch, 1991. Electron microscope studies of the digestive tract of *Brachionus plicatilis*. Zoomorphology 111: 67–80.

Kleinow, W. & H. Wratil, 1995. SEM of internal structures of *Brachionus plicatilis* (Rotifera). Hydrobiologia 313/314: 129–132.

Kleinow, W. & H. Wratil, 1996. On the structure and function of the mastax of *Brachionus plicatilis* (Rotifera), a scanning electron analysis. Zoomorphology 116: 169–177.

Koehler, J. K. & T. L. Hayes, 1969. The rotifer jaw: A scanning and transmission electron microscope study. II. The trophi of *Asplanchna sieboldi*. J. ultrastruct. Res. 27: 419–434.

Kühle, K. & W. Kleinow, 1985. Measurements of hydrolytic enzymes in homogenates from *Brachionus plicatilis* (Rotifera). Comp. Biochem. Physiol. (B) 81 B: 437–442.

Kühle, K. & W. Kleinow, 1989. Localization of hydrolytic enzyme activities within cellular fractions from *Brachionus plicatilis* (Rotatoria). Comp. Biochem. Physiol. (B) 93B: 565–574.

McKeon, T. A. & W. Gaffield, 1990. Viewing stereopictures in three dimensions with naked eyes. Trends biochem. Sci. 15: 412–413.

Salt, G. W., G. F. Sabbadini & M. L. Commins, 1978. Trophi morphology relative to food habits in six species of rotifers (Asplanchnidae). Trans. am. Microsc. Soc. 97: 469–485.

Hydrobiologia **387/388**: 131–134, 1998.
E. Wurdak, R. Wallace & H. Segers (eds), Rotifera VIII: A Comparative Approach.
© *1998 Kluwer Academic Publishers.*

The rotifer corona by SEM

Giulio Melone

Department of Biology, University of Milan, Via Celoria 26, 20133 Milano, Italy

Key words: SEM, Rotifera, methodology, anesthesia, deciliation

Abstract

The scanning electron microscope (SEM) is a powerful tool to observe any surface at the ultrastructural level. During the last 15 years, I developed techniques to process rotifer specimens for SEM observation, in order to obtain images of preserved specimens that simulate their natural appearance. A characteristic feature in Rotifera is the rotatory apparatus (corona) and SEM is appropriate for studying its organization. The organization of the corona is better understood if the rotatory apparatus can be examined after the cilia have been removed. A method to prepare the rotifers for observation by SEM is presented.

Introduction

The corona is a distinctive feature for the whole phylum Rotifera. It is known that the morphology of the corona is strictly related to the life style of the rotifers, and different models of corona are described in rotifer literature (i.e. Hudson & Gosse, 1889; de Beauchamp, 1907, 1965; Remane, 1929–33).

Generally, the corona is described after observations carried out by light microscope, mainly on living rotifers. Because of the difficulty in performing good and precise observations on living (moving!), minute animals under a light microscope, and because of the weak depth of field of these instruments, it is hard to obtain precise descriptions of coronas.

In the last 30 years the availability and use of the scanning electron microscopes (SEM) has improved the possibility to investigate the fine morphology of a large number of biological samples. SEM allows us to obtain images with a large depth of field, a three-dimensional appearance and a 20 to 50 times higher resolution than that of light microscopes.

Since SEM works under vacuum, it is evident that the biological samples need to be prepared in a suitable way to be observed in this way: animals cannot be examined alive, but have to be killed by fixation in such a way that the morphology of the living animal is conserved as much as possible. For this, different methods have been suggested for treating different an-

imal groups, especially soft-bodied invertebrates. The protocols vary, but all include three steps:

(1) anesthesia and relaxation of the samples using drugs, in order for these to retain their natural appearance;
(2) fixation of specimens avoiding reaction or contraction;
(3) dehydration of the specimens to be mounted on SEM stubs, and observation under vacuum.

On the basis of the results obtained during the past 15 years working on different invertebrate taxa, a method to prepare and observe rotifer corona by SEM was developed.

Methods and comments

The three main steps in the protocol for preparation of rotifers for SEM preparation are as follows.

(1) Anesthesia and relaxation of the animals

A list of drugs used to anaesthetize rotifers was provided by Nogrady et al. (1993). I experienced that bupivacaine is the best chemical presently available anaesthetizing rotifers, although it seems to work better with Seisonidea and Monogononta than Bdelloidea, which react in variable ways when exposed

132

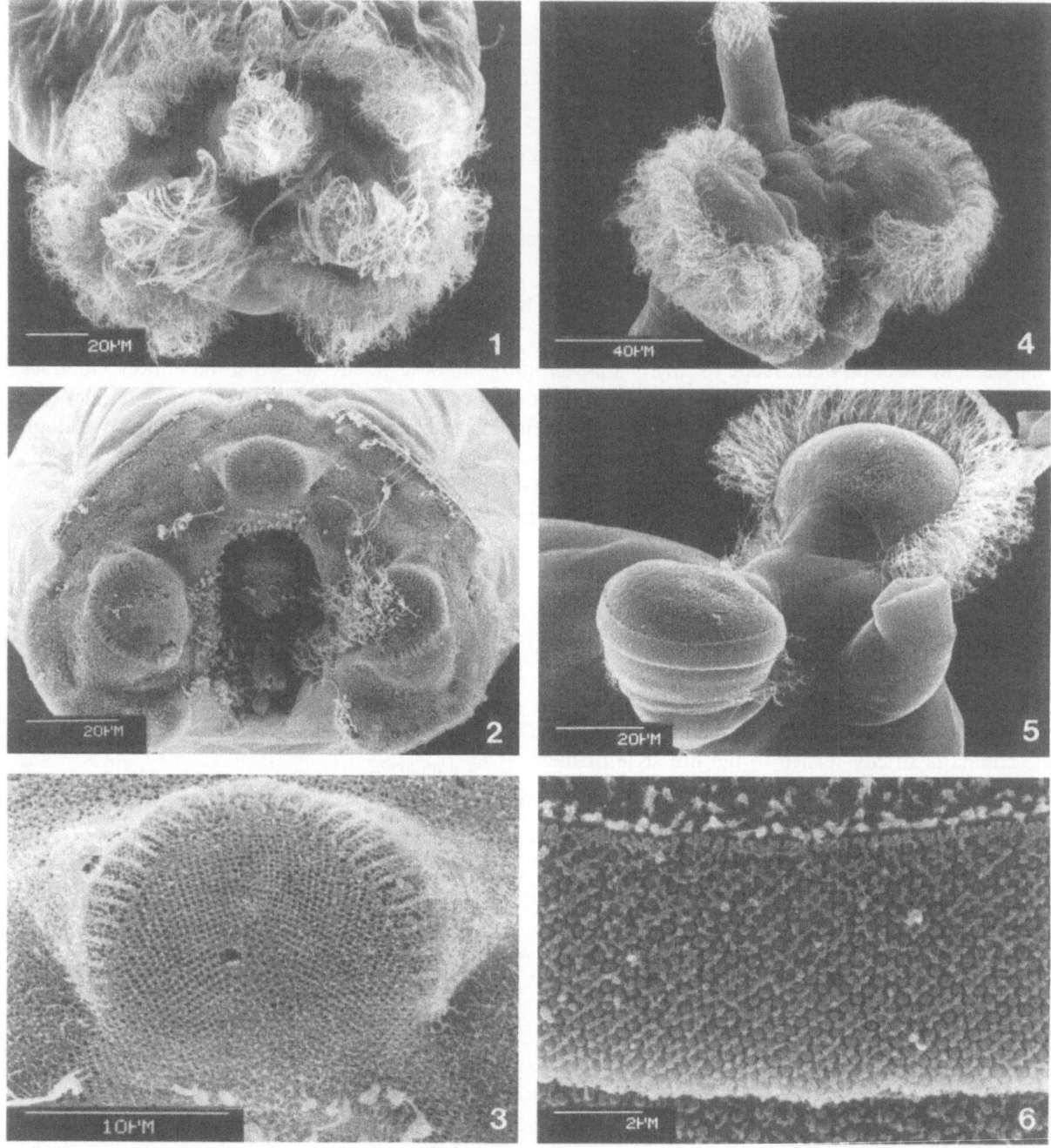

Figures 1–6. SEM images of rotifer coronas. **Figure 1:** *Brachionus plicatilis*. **Figure 2:** *B. plicatilis*, deciliated. **Figure 3:** *B. plicatilis*, detail of deciliated pseudotrochus (each hole is the mark of a cilium). **Figure 4:** *Rotaria macrura*. **Figure 5:** *R. macrura*, deciliated. **Figure 6:** *R. macrura*, detail of deciliated trochus (the round bodies, obliquely arranged, are remains of the cilia).

to the chemical. In some cases, bdelloids that are not relaxed by bupivacaine react positively to propranolol. In any case, I slowly added (one drop every three or four minutes) some drops of the drug solution (0.5% in water) to a small volume of medium (<2 ml) contain-ing the rotifers, while constantly monitoring the effect of the drug under a stereomicroscope. I recommend careful cleaning of the rotifers before anesthesia, in order to remove microscopic debris or organisms that adhere to the rotifer surface. For this, the material is

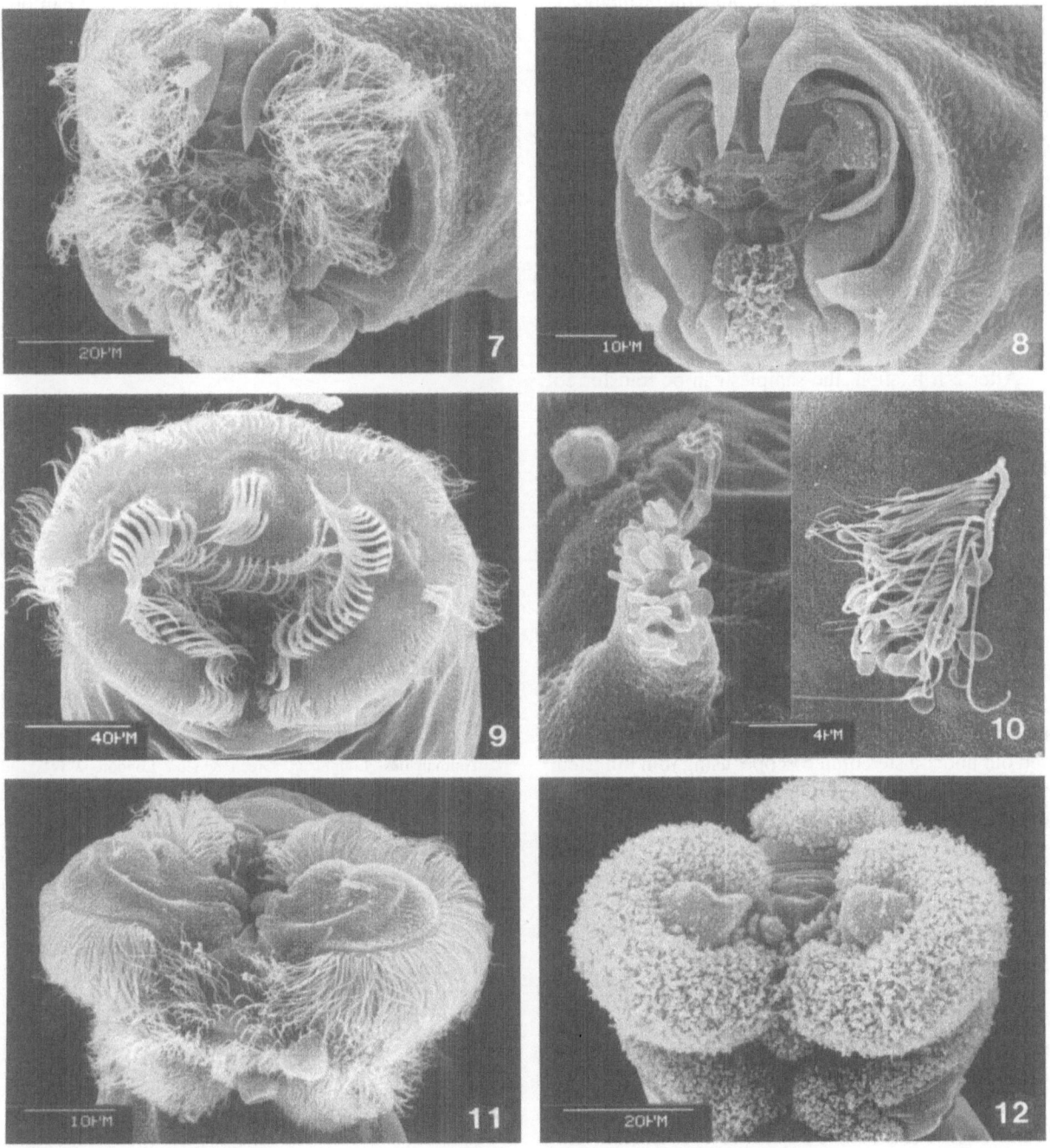

Figures 7–12. SEM images of rotifer coronas. **Figure 7:** *Mytilina mucronata.* **Figure 8:** *M. mucronata*, deciliated. **Figure 9:** *Epiphanes senta.* **Figure 10:** *Synchaeta* sp., details of corona (left) and dorsal antenna (right) with artefacts on cilia tips. **Figure 11:** *Macrotrachela quadricornifera.* **Figure 12:** *M. quadricornifera*, artefacts on the corona.

washed two or three times by replacing the medium after stirring the liquid by means of a pipette before each change.

(2) Fixation of the specimens

Fixation is another delicate point. In my experience, a good fixative solution useful for rotifers samples is PAF (Picric Acid–Formaldehyde) at 200–250 mosM for freshwater species and at 1200 mosM for marine

134

species (see the Appendix for the preparation method). The PAF solution works well on specimens after anesthetization. I recommend testing the reaction of the animals under anesthesia before adding the fixative. Usually I fixate the animals when they are still alive (they can swim because the cilia are beating), but insensitive to mechanic stimuli. Some species are difficult to anaesthetize and are able to react by contracting their body, even after exposure to high drug concentrations. In such cases I quickly added 1 ml of a 2% OsO_4 solution to the rotifer sample, followed by the PAF fixative. OsO_4 has a very fast fixative effect and can 'freeze' the rotifers when it reaches them.

After 2 h fixation, the samples can be transferred to small tubes with a bottom net, as described by Gray (1973) and Amsellem & Clément (1980). The samples are subsequently washed and dehydrated.

(3) Dehydration of the specimens

Dehydration is carried out by transferring the specimens through a graded ethanol series, i.e., 30%, 50%, 70%, 80%, 90%, 95%, and three changes of absolute alcohol. The samples are then transferred to a Critical Point Dryer (CPD) and processed.

The dried rotifers are attached onto double sided adhesive tape on a stub. This delicate operation is carried out under a stereomicroscope, using thin tungsten needles. The stub with the samples is coated with gold and observed by SEM.

Following the method outlined above, it is possible to obtain good samples and good images of rotifers (Figures 1, 4, 7, 9, 11). Nevertheless, the corona organization is not always clear, mainly because of the density and length of the cilia. This is the case, for instance, in the rotifers shown in Figures 1, 4, 7 and 11. Some years ago, I decided to cut off the cilia from the rotatory apparatus with methods used to detach cilia from ciliated protozoa or bivalve gills. I tried different methods and found that of Rosenbaum & Carlson (1969) to be the most successful. The deciliation should be carried out on anaesthetized and relaxed, but not yet fixed rotifers. Figures 2, 3, 5, 6 and 8 show the results of deciliation, which illustrates how this method can help to understand the fine details of the corona organization. In fact, the pattern of the distribution of cilia on the rotatory apparatus can easily be detected on deciliated rotifers.

Finally, I wish to caution that artifacts can be produced during the preparation of rotifers for SEM observation. In particular, when OsO_4 is added quickly, alterations of the cilia morphology can occur in some species (Figure 10). Similarly, heating can produce such artifacts, e.g., when the hot-water fixation technique is used, some overheated specimens have shortened and curled cilia (Figure 12).

A summary of the steps in the above outlined procedure is as follows.

(1) Washing of the living rotifers, in order to avoid that these would be covered by debris; (2) relaxation of the specimens; (2a) deciliation of part of the material; (3) fixation; (4) removal of fixative by rinsing; (5) dehydration and drying; (6) positioning of rotifers on SEM stub; (7) coating with gold and observation by SEM.

References

Amsellem, J. & P. Clément, 1980. A simplified method for the preparation of rotifers for transmission and scanning electron microscopy. Hydrobiologia 73: 119–122.
de Beauchamp, P. M., 1907. Morphologie et variations de l'appareil rotateur dans la série des Rotifères. Arch. Zool. exp. gén. IVe Série, Tome VI: 1–29.
de Beauchamp, P. M., 1965. Classe des Rotifères. In P. P. Grassé (ed.), Traité de Zoologie. Masson et C. éd., Paris. Tome IV: 1225–1379.
Gray, P., 1973. Wholemounts, zoological materials. In Encyclopedia of microscopy and microtechnique, 597–602. Van Nostrand Reinhold Company, New York.
Hudson, C. T. & P. H. Gosse, 1889. The Rotifera; or Wheel-Animalcules. Longmans, Green & Co., London. 2 vols and Supplement.
Melone, G. & C. Ricci, 1995. Rotatory apparatus in Bdelloids. Hydrobiologia 313/314: 91–98.
Nogrady, T., R. L. Wallace & T. W. Snell, 1993. Rotifera. Vol. I: Biology, Ecology and Systematics. Guides to the identification of the microinvertebrates of the continental waters of the world, 4. SPB Academic Publishers, The Hague, 142 pp.
Remane, A., 1929–33. Rotatoria. In: Bronn's Klassen und ordnungen des Tier-Reichs, 4, 2, 1: 1–576.
Rosenbaum, J. L. & K. Carlson, 1969. Cilia regeneration in *Tetrahymena* and its inhibition by colchicine. J. Cell Biol. 40: 415–425.

Appendix

Fixative solution PAF (Picric Acid – Formaldehyde) (Melone & Ricci, 1995):

Filtered solution of saturated picric acid, 75 ml; Paraformaldehyde, 10 g.

Dissolve paraformaldehyde by heating to 60 °C and stirring for 2–3 hours. Keep at 60 °C under continuous stirring and clarify it by adding a few drops of 1 N NaOH solution. Cool the solution to room temperature, without stirring, in a closed bottle. Add 0.1 M Cacodylate buffer pH 7.3–7.4 to obtain a final volume of 500 ml. The final solution will be 750–800 mosM and can be stored in the refrigerator (4 °C) for several months.

For freshwater rotifers, the PAF solution is diluted to about 200–250 mosM; for marine rotifers the PAF solution will be adjusted to about 1200 mosM by adding sucrose.

Hydrobiologia **387/388:** 135–140, 1998.
E. Wurdak, R. Wallace & H. Segers (eds), Rotifera VIII: A Comparative Approach.
© 1998 Kluwer Academic Publishers.

Catecholaminergic neurons in the brain of rotifers

Elena A. Kotikova

Zoological Institute of the Russian Academy of Sciences, 199034, St. Petersburg, Russia

Key words: nervous system, morphology, cathecholamines, GAIF method, Rotifera

Abstract

In 10 rotifer species from the subclasses Archeorotatoria (order Bdelloidea) and Eurotatoria (superorders Gnesiotrocha and Pseudotrocha) three patterns of catecholaminergic neurons are detected, namely: x-shaped, arch-shaped and ring-shaped. These brain complexes are developed independently and in a parallel fashion in different rotifer groups. The number of the brain catecholaminergic neurons varies from 6 to 11, constituting about 3–7% of the total number of the brain cells. The brain neuron pattern demonstrates a distinct bilateral symmetry.

Introduction

A general phylogenetic system of rotifers based on morphological characters has not yet been elaborated (Kutikova, 1985), although this group should be regarded as a separate phylum (Malakhov, 1994). The system of phylogenetic relationships between the orders of rotifers based on mastax structure (Markevich, 1990), is considered here to be the most realistic. The 'classical' nomenclature is given in parentheses.

Electron microscopic investigation of the rotifer nervous system revealed the presence of dense-cored synaptic vesicles, which are supposed to contain catecholamines (CA) (Villeneuve & Clement, 1971; Clement, 1977; Wurdak et al., 1983). Later, histochemical studies have localized CA in whole-mounts of rotifers (Nogrady & Alai, 1983); Keshmirian & Nogrady, 1987, 1988).

The mapping of CA distribution in the nervous systems of rotifers (Kotikova, 1994, 1995, 1997) should be regarded as a next step in these studies.

The mapping of biologically active substances which are characteristic of the simple nervous systems of invertebrates, opens new opportunities to study invertebrate nervous system structure and origin in more detail.

Materials and methods

10 species of Rotifera were studied:

subclass Archeorotatoria (order Bdelloidea)
    Order Bdelloidea
      Philodina sp.
subclass Eurotatoria (superorders Gneisiotrocha and Pseudotrocha P. Beauchamp, 1965)
    order Transversiramida
      Euchlanis dilatata Ehrenberg
      Manfredium eudactylotum (Gosse)
      Brachionus quadridentatus Hermann
      Platyias quadricornis (Ehrenberg)
      Lecane (Monostyla) arcuata (Bryce)
    order Saltiramida
      Asplanchna herricki Guerne
      Asplanchna priodonta Gosse
    order Saeptiramida
      Notommata sp.
    order Antrorsiramida
      Dicranophorus forcipatus (Muller)

CAs were revealed by the glyoxylic acid induced fluorescence (GAIF) method slightly modified by Kabotyanskyj (1985) with an increased (up to 5%) concentration of GA in the incubation medium prepared with 0.1 M Na-phosphate buffer pH 7.2.

The controls were carried out in two ways: exclusion of the GA from the incubation medium and addition of water to the dried preparations. Both reactions resulted in disappearance of the fluorescence. The green fluorescence of CA-containing elements was observed in a fluorescent microscope LUMAM-R3; the photomicrographs were taken on RF film.

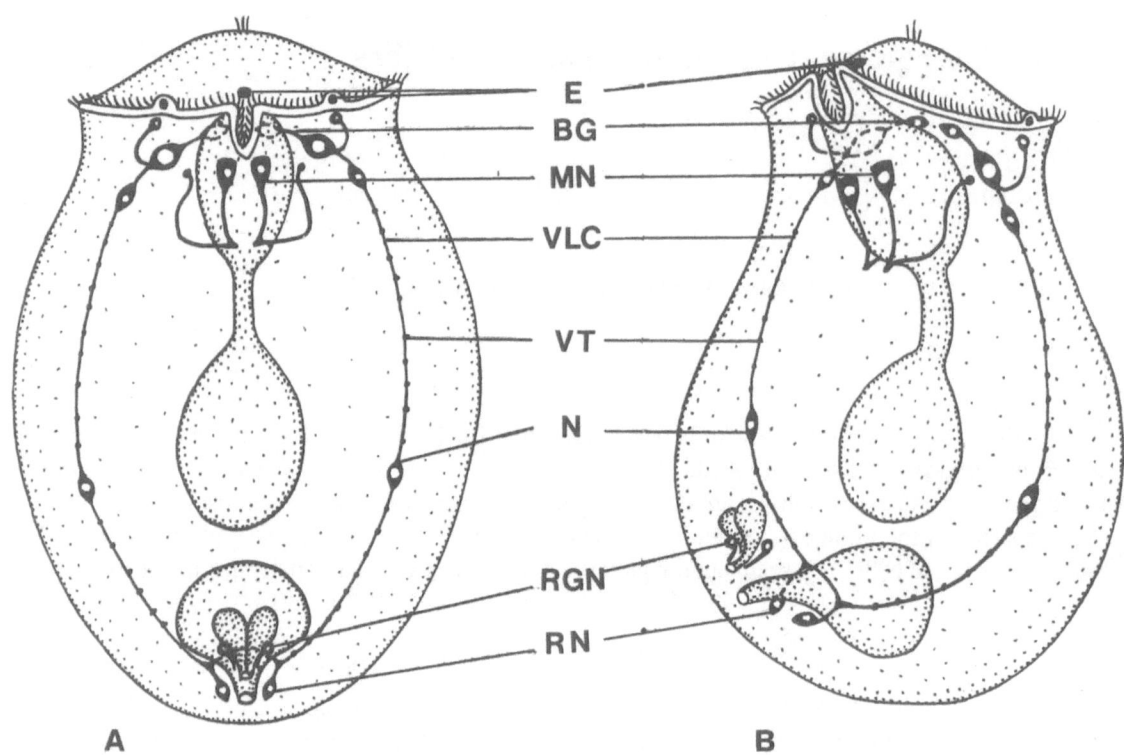

Figure 1. Schematic drawing of the catecholaminergic nervous system revealed by the GAIF method in the rotifer *Asplanchna herricki*. A: View from the ventral side. B: View from the lateral side. BG – brain ganglion; E – eye; MN – mastax neuron; N – neuron; RGN – rudimentary pedal gland neuron; RN – urinary bladder neuron; VLC – ventro-lateral nerve cord; VT – varicose thickening.

Results

Asplanchna herricki Guerne (Figures 1A–B; 2D)
The brain ganglion looks like a sharply curved arch on the dorsal side between the eyes and the pharynx. A pair of central neurons is situated on both sides of the cerebral eye spot. A second pair of larger neurons lies under the lateral eye spots. The processes of the third pair of brain neurons bend to the ventral side and give rise to the longitudinal ventro-lateral nerve cords. These cords run along the sides of the body and present varicose thickenings along their length. A pair of bipolar neurons is located at the level of the distal part of the stomach. The ventro-lateral cords join each other near the urinary bladder. A pair of unipolar neurons which send their axons into the longitudinal cord is situated at the sides of the urogenital duct. A pair of small unipolar neurons with axons directed to the lateral parts of the brain arch is located just under the lateral eye spots.

A pair of unipolar neurons lies near the mastax. Their processes first run posteriorly along the mastax at the ventral side of the body and then contour the mastax and end behind it by globular sensory terminals.

Above the urogenital opening near a pair of rudimentary pedal glands lies a pair of small unipolar neurons with their processes oriented along the ducts. All the neurons are oval-shaped or elongate; their size varies from 2 to 6 μm. The similar general pattern of the CA-ergic nervous system is characteristic for *A. priodonta* (Figure 3D) and for other species described below. However, some noticeable variations in the brain neuron pattern are observed (Figure 2).

Philodina sp. (Figures 2A, 3A)
The X-shaped brain complex is situated above the pharyngeal tube behind the cerebral eye spots and includes seven longitudinally oriented neurons. The neurons lie at three levels. At the upper level a pair of widely separated large neurons is observed. The middle level includes a pair of bipolar perikarya with apposed processes. At the lower level three oval-shaped neurons are located. Peripheral processes of the first pair of brain neurons are directed anteriorly and end at the base of corona by spherical sensory

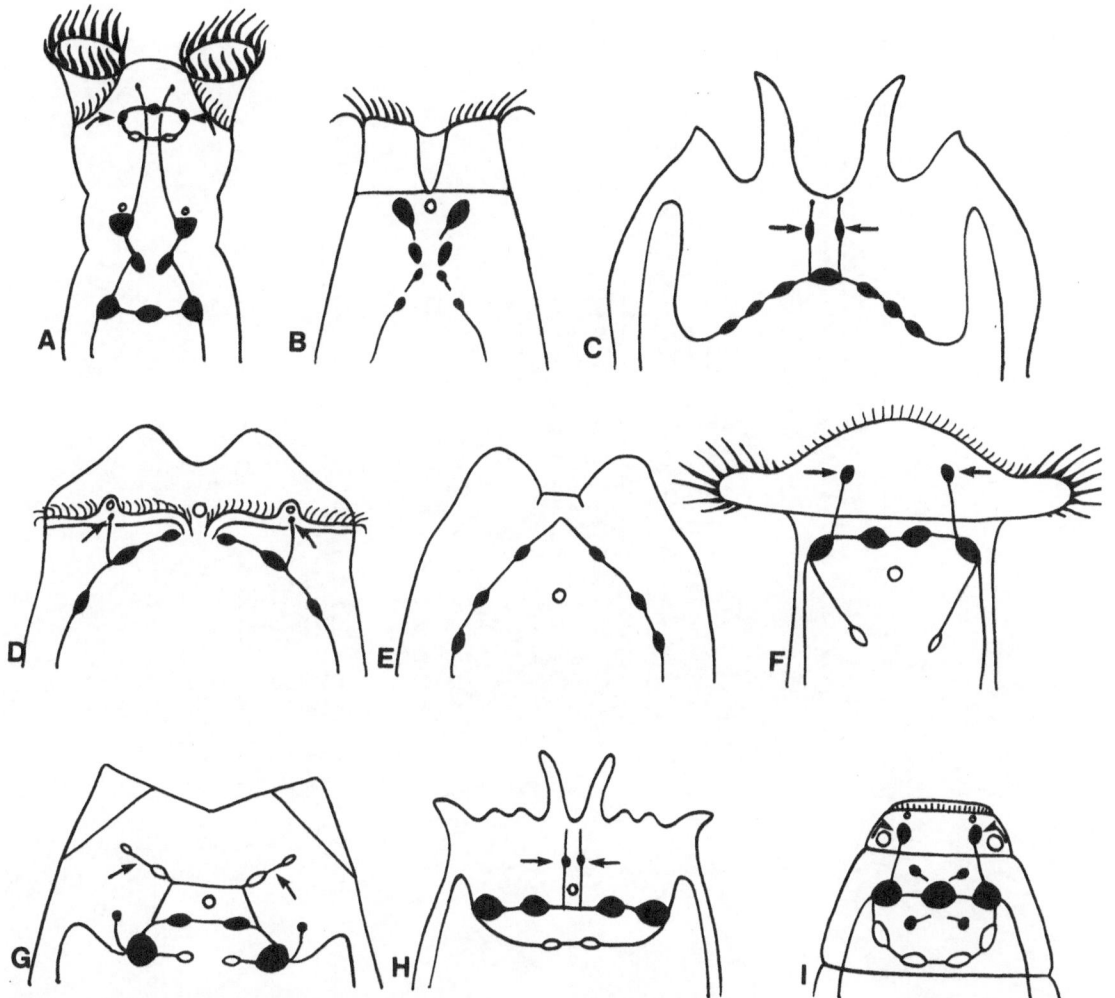

Figure 2. Scheme of the pattern of CA-ergic brain neurons and neurons of the corona in the rotifers: A – *Philodina* sp., B – *Manfredium eudactylotum*, C – *Platyias quadricornis*, D – *Asplanchna herricki*, B – *Euchlanis dilatata*, F – *Notommata* sp., G – *Lecane arcuata*, H – *Brachionus quadridentatus*, I – *Dicranophorus forcipatus*. Dorsal neurons are shown in black; ventral neurons are shown in outline. Neurons of the corona, of the anterior dorsal tentacle and the proboscis are indicated by small arrows.

terminals. Five neurons connected by their processes forming a ring-shaped complex lie around the dorsal proboscis. Three small neurons of this complex lie dorsally and two larger ones are ventrally located. The size of the neurons is 2–7 μm.

Manfredium eudactylotum (Gosse) (Figures 2B, 3B)
The brain complex of CA-ergic neurons looks like a slightly curved X-shaped structure, situated just behind the eye spot. It is formed by eight neurons. The neurons are longitudinally extended on four levels. The neurons of the two upper levels are larger (6 μm), those of lower levels do not exceed 3 μm.

Platyjas quadricornis (Ehrenberg) (Figures 2C, 3C)
The arch-shaped brain ganglion is gently curved. It includes seven bipolar neurons (5–7 μm) forming a transverse row (Kotikova, 1995).

Euchlanis dilatata Ehrenberg (Figures 2E, 3E)
The brain arch is characterized by a conical curve and consists of six bipolar neurons (4–6 μm).

Notommata sp. (Figures 2F, 3F)
CA-ergic brain neurons are arranged in a semicircular pattern. Four neurons form a nearly straight dorsal sector, two neurons (6–8 μm) are shifted to the ventral side and lie at the posterior end of the ganglion at the level of the eye spot. Axons of the two unipolar neur-

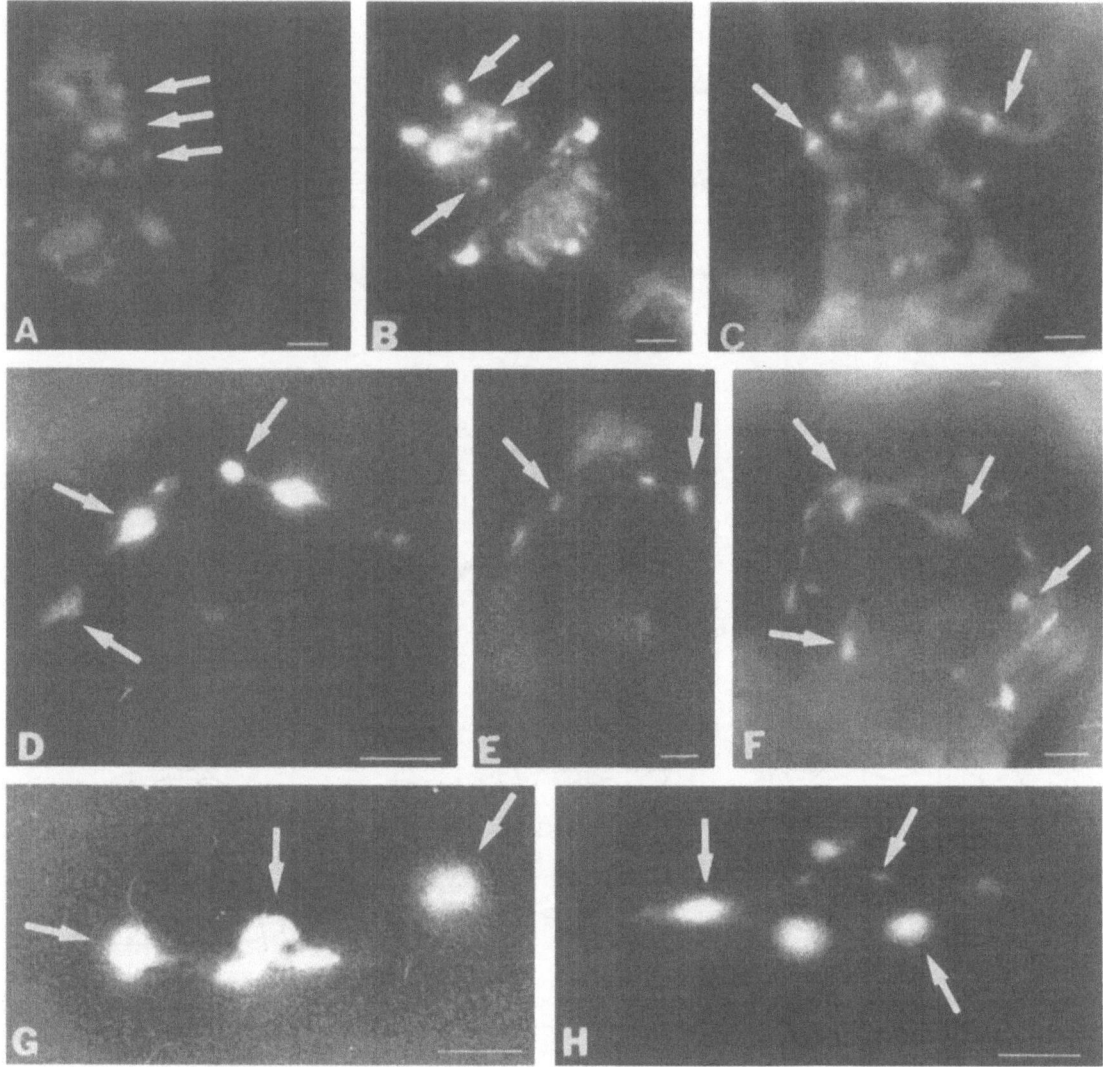

Figure 3. CA-ergic neurons in the brain of rotifers: A – *Philodina* sp., B – *Manfredium eudactylotum*, C – *Platyias quadricornis*, D – *Asplanchna priodonta*, E – *Euchlanis dilatata*, F – *Notommata* sp., G – *Lecane arcuata*, H – *Brachionus quadridentatus*. GAIF method. Scale bar = 10 μm. Brain neurons are indicated by small arrows.

ons (4 μm) lying at the level of the corona approach the lateral parts of the dorsal arch.

Lecane (Monostyla) arcuata (Bryce) (Figures 2G, 3G) *Brachionus quadridentatus* Hermann (Figures 2H, 3H)

These species are characterized by a ring-shaped brain ganglion. *L. arcuata* has a more compact ganglion which is interrupted ventrally, while *B. quadridentatus* has a ganglion extended in transverse direction. It includes six neurons (4–8 μm), four of which are located on the dorsal side and two others lie on the ventral side. In *L. arcuata* two nerves of the corona with two

pairs of neurons (4–7 μm) originate from the dorsal part of the brain. A pair of small unipolar neurons sends short axons to the dorsal part of the brain. In *B. quadridentatus*, a pair of small bipolar neurons runs from the medial part of the dorsal brain semicircle to the anterior dorsal tentacle.

Dicranophorus forcipatus (Muller) (Figure 2I)

Eleven CA-ergic neurons form a ring-shaped brain ganglion in the posterior part of the head region. A transverse dorsal part of the brain includes three main neurons and four additional ones while the curved ventral part has four neurons. Four unipolar neurons lie at the level of the frontal eye spots, a pair of dorsal neurons among them sends axons into the brain, while the ventral neurons have no processes. The size of neurons varies from 2 to 7 μm.

Discussion

All the rotifers studied, being systematically distant from each other, nevertheless show the same pattern of the CA-ergic nervous system. Main differences are observed in the location of the brain neurons, their number varying from 6 to 8 (without considering the additional perikarya, which have no influence on the ganglion shape). Three geometric patterns of CA-ergic brain neurons can be distinguished: X-shaped, arch-shaped and ring-shaped. Brain neurons can be longitudinally extended at three or four levels forming X-shaped figures of different degrees of curvature (*Philodina* sp., *M. eudactylotum*). Among the species with the arch-shaped brain, a gentle curve is recorded in *P. quadricornis*, but *A. herricki*, *A. priodonta* and *E. dilatata* are characterized by a steeper curve. Transition from the arch-shaped to the ring-shaped brain is realized through two stages. At the first stage an insignificant degree of curvature is shown in *Notommata* sp. and a further approachment of the lateral brain parts is observed in *L. arcuata*. The second stage leads to the appearance of the closed ring-shaped brain complex of *B. quadridentatus* and *D. forcipatus*. Such structural changes seem to take place independently in different groups and result in the appearance of a similar geometric pattern of CA-ergic brain neurons in representatives of different subclasses (*Philodina* sp. and *M. eudactylotum*). On the other hand, similar patterns can be observed in the representatives of the same order Transversiramida (*E. dilatata* and *P. quadricornis*) as well as among the members of the three orders Saeptiramida, Transversiramida and Antrorsiramida (*Notommata* sp., *B. quadridentatus, D. forcipatus*) which are not closely related to each other.

Among the Archeorotatoria (Bdelloidea) only one pattern of brain CA-ergic neurons is described (Kotikova, 1994, 1995), while among the Eurotatoria three patterns of it have been revealed (Kotikova, 1995, 1997).

Numerous similarities in forms of movement, feeding, structure of the corona, general morphology were described for rotifers (Kutikova, 1983). The excretory system demonstrates distinct examples of such parallelisms (Remane, 1929–32).

The pattern of CA-ergic elements of *Asplanchna* and *Brachionus* falls into the classic scheme of their nervous system (Nachtwey, 1925; Stossberg, 1932). The CA-ergic fibres are not detected in the dorsal longitudinal cords of *Asplanchna priodonta* (Nachtwey, 1925) indicating that these cords have only a motor function. Similar cords can be expected to be found in other species of rotifers.

Brain neurons of rotifers constitute 20% of the total number of cells (Martini, 1912). The brain of *Asplanchna brightwelli* numbers 200 cells, demonstrating a perfect bilateral symmetry (Ware & Lopresti, 1975). The distinct bilateral symmetry is shown also in the arrangement of CA-ergic neurons constituting in different species only 3–7% of the total number of cells in the brain.

The pattern of CA-ergic nervous system of rotifers, once it evolved, was maintained during the whole period of evolution of the group. A high degree of concentration of CA-ergic elements can be regarded as an indicator of the high level of development of the simple nervous system of rotifers.

Acknowledgements

This work was supported by the Russian Basic Research Foundation (grant N96-04-48384). I want to express my gratitude to L. Kutikova and G. Markevich for their help in the identification of rotifers and also to A. Kirejtshuk and to O. Raikova for their help in translation of the text. My sincere thanks to the Organizing Committee of the Symposium for the invitation and the financial support.

References

Clément, P., 1977. Ultrastructural research on rotifera. Arch. Hydrobiol. Beih. 8: 270–297.

Kabotyanskyj, E. A., 1985. Glyoxylate method of demonstrating the cellular monoamines on whole mounts of nervous system in invertebrates. In Sakharov, D. (ed.), Simple Nervous Systems. Kazan State University, Kazan: 81–83 (in Russian).

Keshmirian, J. & T. Nogrady, 1987. Histofluorescent labelling of catecholaminergic structures in rotifer (Aschelminthes) in whole animals. Histochem. 87: 351–357.

Keshmirian, J. & T. Nogrady, 1988. Histofluorescent labelling of catecholaminergic structures in rotifers (Aschelminthes). II, Males of *Brachionus plicatilis* and structures from sectioned females. Histochem. 89: 189–192.

Kotikova, E. A., 1994. Distribution of catecholamines in the nervous system of Bdelloida (Rotifera). In Sakharov, D. (ed.), Simple Nervous System ISIN. Pushcino: 21–22.

Kotikova, E. A., 1995. Localization and neuroanatomy of catecholaminergic neurons in some rotifer species. Hydrobiologia 313/314: 123–127.

Kotikova, E. A., 1997. Localization of catecholamines in the nervous system of the order Transversiramida. Dokl. Russ. Acad. of Science. 353: 841–843 (in Russian).

Kutikova, L. A., 1983. Parallelism in the evolution of rotifers. Hydrobiologia 104: 3–7.

Kutikova, L. A., 1985. Peculiarities of diagnostic taxa in rotifers. In Rotifers. Proc. of the Second All-Union Rotifer Symposium. Zool. Inst. Acad. Sci. USSR, Leningrad: 17–37 (in Russian).

Malakhov, V. V., 1994. Nematodes. Structure, Development, Classification and Phylogeny. Smithsonian Institution Press, Washington & London, 286 pp.

Markevich, G. I., 1990. A historic reconstruction of phylogenesis of rotifers as a basis for their macrosystem. In Rotifers. Proc. of the Third All-Union Rotifer Symposium. Zool. Inst. Acad. Sci. USSR, Leningrad: 140–156 (in Russian).

Martini, E., 1912. Studien über die Konstanz histologischer Elemente. III. *Hydatina senta*. Zeitsch. wiss. Zool. 102: 425–645.

Nachtwey, R., 1925. Untersuchungen über die Keimbbahn Organogenese und Anatomie von *Asplanchna priodonta* Gosse. Zeitschr. wiss. Zool. 126: 239–492.

Nogrady, T. & M. Alai, 1983. Cholinergic neurotransmission in rotifers. Hydrobiologia 104: 149–153.

Remane, A., 1929–33. Rotatorien. Dr. H. G. Bronn's Klassen und Ordnungen d. Tierreichs, Leipzig. Bd. IV, Abt. II, Lief. 1–4: 576 pp.

Stossberg, K., 1932. Zur Morphologie der Rädertiergattung *Euchlanis, Brachionus* und *Rhinoglena*. Zeitschr. wiss. Zool. 142: 313–424.

Villeneuve, J. & P. Clément, 1971. Le neuropile du cerveau des Rotifères: observations ultrastructurales preliminaires. J. Microsc. Fr. 11: 108.

Ware, R. W. & V. Lopresti, 1975. Three-dimensional reconstruction from serial sections. Int. Rev. Cytol. 40: 325–440.

Wurdak, E. S., P. Clément & J. Amsellem, 1983. Sensory receptors involved in the feeding behavior of the rotifer *Asplanchna brightwelli*. Hydrobiologia 104 (Dev. Hydrobiol. 14): 203–212.

Hydrobiologia **387/388**: 141–152, 1998.
E. Wurdak, R. Wallace & H. Segers (eds), Rotifera VIII: A Comparative Approach.
© *1998 Kluwer Academic Publishers.*

141

Review paper

Rotifer responses to increased acidity: long-term patterns during the experimental manipulation of Little Rock Lake

Thomas M. Frost[1], Pamela K. Montz[1], Maria J. Gonzalez[1,2], Beth L. Sanderson[1] & Shelley E. Arnott[1]
[1]*Trout Lake Station, Center for Limnology, University of Wisconsin-Madison, WI 53706, U.S.A.*
[2]*Department of Biological Sciences, Wright State University, Dayton, OH 45435, U.S.A.*

Key words: Rotifera, community, acidification, whole-lake experiments, Little Rock Lake, Wisconsin

Abstract

Little Rock Lake, Wisconsin, U.S.A. has been the site of a whole-ecosystem experiment since 1983. It was divided into a treatment basin that was acidified in three, two-year stages and a reference basin. The rotifer community in the treatment basin exhibited a variety of responses to the manipulation. Many species decreased in abundance under reduced pH conditions but other rotifers increased at the same time such that there were ultimately increases with acidification in total rotifer biomass, and quite conspicuously, in the proportion that rotifers comprised of total zooplankton biomass. Ten rotifer species decreased at some stage during the acidification (e.g., *Kellicottia longispina*, *Asplanchna priodonta* and *Keratella cochlearis*) while four species increased dramatically (e.g., *Synchaeta* sp. and *Keratella taurocephala*). Similarity indices and total rotifer biomass differences measured between the two basins exhibited very different temporal patterns of response to acidification. Similarity decreased regularly beginning with the earliest stages of acid additions while biomass was nearly the same between the basins until the late stages of the experiment. Comparisons with other nearby lakes indicate, however, that acid conditions are not the only factors generating among-lake differences in rotifer community characteristics. Changes observed with acidification in Little Rock Lake were such that its total rotifer biomass grew more similar to that in a nearby acidic-bog lake and different from that in a near-neutral-pH lake. At the same time, abundance patterns for individual rotifer species in Little Rock Lake were not particularly similar to those in the other lakes. It appears that, although they are important, acid conditions alone can not account for all observed rotifer community differences among lakes. Higher proportions of rotifer biomass and high populations of *K. taurocephala* do seem to be common features of many low pH habitats.

Introduction

Zooplankton communities vary markedly across lakes of differing acidities (e.g., Baker & Christensen, 1991; Locke, 1992; Yan et al., 1996). The rotifer components of zooplankton communities also exhibit systematic differences along pH gradients when they have been evaluated (e.g., MacIsaac et al., 1987; Brett, 1989; Siegfried et al., 1989). Despite these clear patterns, however, a number of issues remain regarding rotifers and pH including the actual nature of rotifer community responses to increasing acidity, the extent to which acidity versus other factors controls observed zooplankton community differences among lakes, and

the mechanisms that underlie among-lake differences. Understanding the effects of changing pH conditions is important because of the widespread occurrence of anthropogenic acid deposition and its effects on lakes and other ecosystems worldwide (Galloway et al., 1984; Schindler, 1988; Charles, 1991).

To examine these and other pH-related issues, we have been conducting an acidification experiment in Little Rock Lake (LRL), Wisconsin, USA since 1983 (Watras & Frost, 1989; Brezonik et al., 1993). Following a baseline period in 1984, LRL's treatment basin was acidified, in three, two-year phases from its original pH of 6.1 to a final target level of 4.7 while the lake's reference basin remained at natural pH levels. Beginning in 1991, we have been evaluating the lake's recovery from acidification (Sampson et al., 1995; Frost et al., in press). Here, we provide a detailed report of rotifer responses during the acidification phase of the experiment. We present time-series data for the entire LRL rotifer community, detailed information for a few common rotifer species, and several measures representing collective features of the entire rotifer community during the course of the experiment.

In addition we compare patterns observed in LRL with those found in two nearby lakes that have been investigated as part of the North Temperate Lakes, Long-Term Ecological Research (NTL-LTER) program (Magnuson et al., 1984). We use these comparisons to test the hypothesis that acidification of LRL's treatment basin shifted its rotifer community from one similar to that occurring in a near-neutral pH lake to one found in a acid-bog system. Accepting this hypothesis would support the notion that pH levels are the primary factor controlling the structure and dynamics of rotifer communities in north temperate lakes.

Methods

The methods used to generate the data presented in this report have been detailed in previous papers. We only summarize these techniques here and refer to the papers that describe them in greater detail.

The Little Rock Lake project was initiated in 1983. The lake's two similarly sized basins, each approximately 20 ha in surface area, were separated with a vinyl-plastic curtain in 1984 that minimized any water flow or exchange of organisms between them (Watras & Frost, 1989). Following the baseline period in 1984, acid additions were initiated in the north, treatment basin. Beginning in 1985, sulfuric acid, the dominant acid in human-influenced deposition in much of North America was mixed into the treatment basin by boat as frequently as necessary to maintain target levels during ice-free periods, usually between once every five days and once every three weeks. We maintained target-pH levels of 5.6, 5.2 and 4.7 for two years each with the last acid additions just prior to freezing in 1990 (Brezonik et al., 1993). Each of the acidification periods was initiated immediately after ice out of the first year and continued until ice out at the end of the second year. Thus the completion of the pH 4.7 period occurred in April 1991. LRL's south basin was maintained as a reference throughout the course

of the experiment to provide an indication of what would have occurred in the treatment basin without the addition of acid.

The NTL-LTER project was among the first of what is now a worldwide network of more than 20 long-term study programs funded by the US National Science Foundation to evaluate fundamental characteristics of communities and ecosystems (Magnuson et al., 1984). We compare LRL with two of the NTL-LTER primary study lakes that have been monitored since 1991 at approximately the same frequency as LRL, Crystal Lake (CL), a clear-water habitat (average Secchi depth = 7.3m) with dilute chemistry and near neutral pH (6.0), and Crystal Bog (CB), a darkly stained sphagnum-mat dominated habitat (average Secchi Depth = 1.6 m) with high dissolved organic carbon levels and with an acidic pH (5.0). Extensive information on CL, CB and five other nearby lakes is available from the NTL-LTER database through http://limnosun.limnology.wisc.edu.

Zooplankton in all lakes were sampled, using Schindler–Patalas traps equipped with 53-μm mesh buckets, at two-week intervals during ice-free periods and every five to six weeks when the lakes were ice covered. In LRL, the trap was approximately 1 m in length and samples were collected at 3 depths in each basin (Frost & Montz, 1988). Samples were processed individually but data were subsequently pooled to estimate the mean density of animals throughout the water column. For the NTL-LTER lakes, the trap was approximately 2.5 m in length and samples were usually pooled prior to counting. For these lakes, too, our data represent an estimate of the overall average density of animals within the water column (Frost & Montz, 1988). For comparisons of species among lakes, we consider only those taxa which were present on more than 5 sampling dates. Rotifer data are reported here for CL from 1984 through 1991 and for CB from 1983 through 1987.

Rotifers were identified, to species in most cases, using a dissecting microscope at 50× or 100× magnification following Ruttner-Kolisko (1974) or Stemberger (1979). Biomasses of rotifers were estimated following Ruttner-Kolisko (1974) or Downing & Rigler (1984). We report data on the abundance of individual rotifer taxa in the treatment and reference basins of LRL and emphasize comparisons between the two basins as an indication of responses to acidification. We calculated a measure of community similarity for rotifers to compare the treatment and reference basins of LRL using the technique described in Frost et al.

(1995). It is based on the total of the minimum proportion of total rotifer biomass for each species in either basin of LRL. We also report rotifer species richness as the total number of taxa observed within a year and rates of species turnover indicating the appearance or disappearance of a species in a basin in any one year. Species turnovers are calculated for each year by determining the number of rotifer taxa that were recorded in a basin at any time within a year but which were then absent during the subsequent year along with the number of taxa that exhibited the opposite pattern, appearing only during the second year of a pair. The total number of appearing and disappearing taxa was then divided by the total number of taxa present at any time in the basin during the first year plus those present during the second year. This number was then multiplied by 100 to obtain the percentage change. Both species richness and turnover measures are presented in detail in Arnott et al. (in preparation).

Assessing differences due to a treatment in a whole ecosystem experiment is a contentious area in terms of statistics. We have developed two methods specifically for testing such differences (Carpenter et al., 1989; Rasmussen et al., 1993) but even these have received some criticism (e.g., Stewart-Oaten et al., 1992). For this report we emphasize simple, primarily graphical, presentations of our data and have avoided any strictly statistical assessments of the significance of treatment effects.

Results

Responses to Acidification in LRL

A total of 26 rotifer taxa were observed on at least three sampling dates in either the treatment or reference basins of LRL during the period 1984–1991 (Table 1). Ten of these taxa appeared to decline at some stage of the acidification. Four appeared to increase, and 12 exhibited no systematic change (Table 1). Patterns of decline differed among rotifer species including those by *Kellicottia longispina*, which decreased in 1987 during the early stages of the pH 5.2 manipulation stage (Figure 1A); *Asplanchna priodonta*, which declined during the pH 5.6 stage, recovered during the pH 5.2 stage, and was essentially extirpated during the pH 4.7 stage (Figure 1B); and *Keratella cochlearis* which declined just past midway through the second year of the pH 5.6 period (Figure 2B, Gonzalez & Frost, 1994). Patterns of

increase, which frequently involved a substantially higher abundance in the treatment basin compared to the reference basin, are illustrated by *Synchaeta* sp. (Figure 3A) and, most dramatically, by *Keratella taurocephala* which shifted from a fairly minor component of overall rotifer biomass to a major portion of the rotifer community by pH 4.7 (Figure 2A, Gonzalez & Frost, 1994). The pattern for *Polyarthra vulgaris* typified those 12 species for which no discernible trends with acidification were detectable (Figure 3B).

In contrast with the patterns for individual species, there were only subtle responses to acidification in terms of the number of species observed in either basin during any of the acidification periods. Overall, the annual species richness observed in the treatment or reference basin during each period ranged between 15 and 21 taxa and both basins were generally quite similar throughout the experiment. The number of taxa observed only declined in the treatment basin during the pH 4.7 period when 17 taxa and 15 taxa occurred in year 1 and 2, respectively, compared with 21 and 20 taxa in the reference basin during the same periods. Consistent with a response to acidification, however, the overall turnover rate for species was somewhat higher in the treatment basin compared to the reference basin (Figure 4).

Overall, the net effects of changes for the entire rotifer community were such that no differences in total rotifer biomass were at all obvious between the two basins during the pH 5.6 and 5.2 phases (Figure 5A). By pH 4.7, however, the overall change was fairly dramatic and a net increase in rotifer biomass was pronounced. This pattern is even more striking, however, when the proportion of rotifers in total zooplankton biomass is considered. The increase of rotifers coupled with the decrease by other zooplankters, particularly the copepods (Frost et al., 1995), shifted the proportion of rotifer biomass from less than 20% prior to the pH 4.7 period to values that frequently exceeded 50% (Figure 6). A low proportion of rotifer biomass persisted in the LRL reference basin throughout the experiment (Figure 6). Overall, the zooplankton community decreased in total biomass by the pH 4.7 period (Frost et al., 1995) but it was increasingly dominated by rotifers as the LRL treatment basin became more acidic.

The rotifer community exhibited the same overall trends as reported previously for the total LRL zooplankton assemblage (Frost et al., 1995). Responses by individual rotifers to acidification were much more dramatic than those by collective features

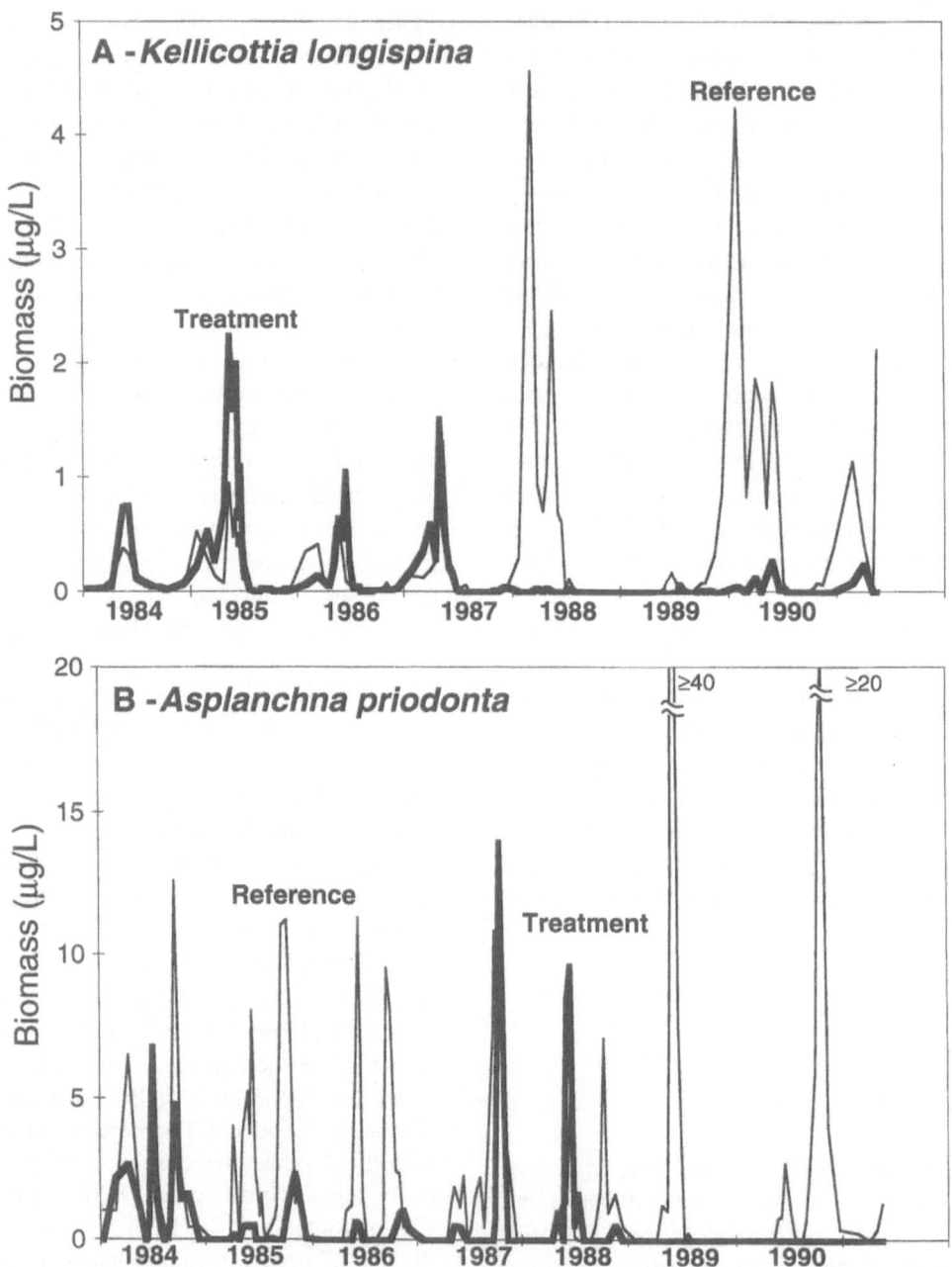

Figure 1. Average biomass (μg/L) of *Kellicottia longispina* (A) and *Asplanchna priodonta* (B) in the water columns of the treatment (thick line) and reference (thin line) basins of Little Rock Lake, Wisconsin, U.S.A.

of the total rotifer community. Rotifer community similarity between the treatment and reference basins provides perhaps the most straightforward evidence of these differences between species-based and more collectively based variables. This similarity exhibited a strong and systematic decrease through all stages of

the acidification (Figure 5B) much more pronounced than the trend in biomass difference (Figure 5A).

Comparisons among lakes

In terms of presence/absence, the rotifer assemblages in the LRL, Crystal Bog (CB), and Crystal Lake (CL)

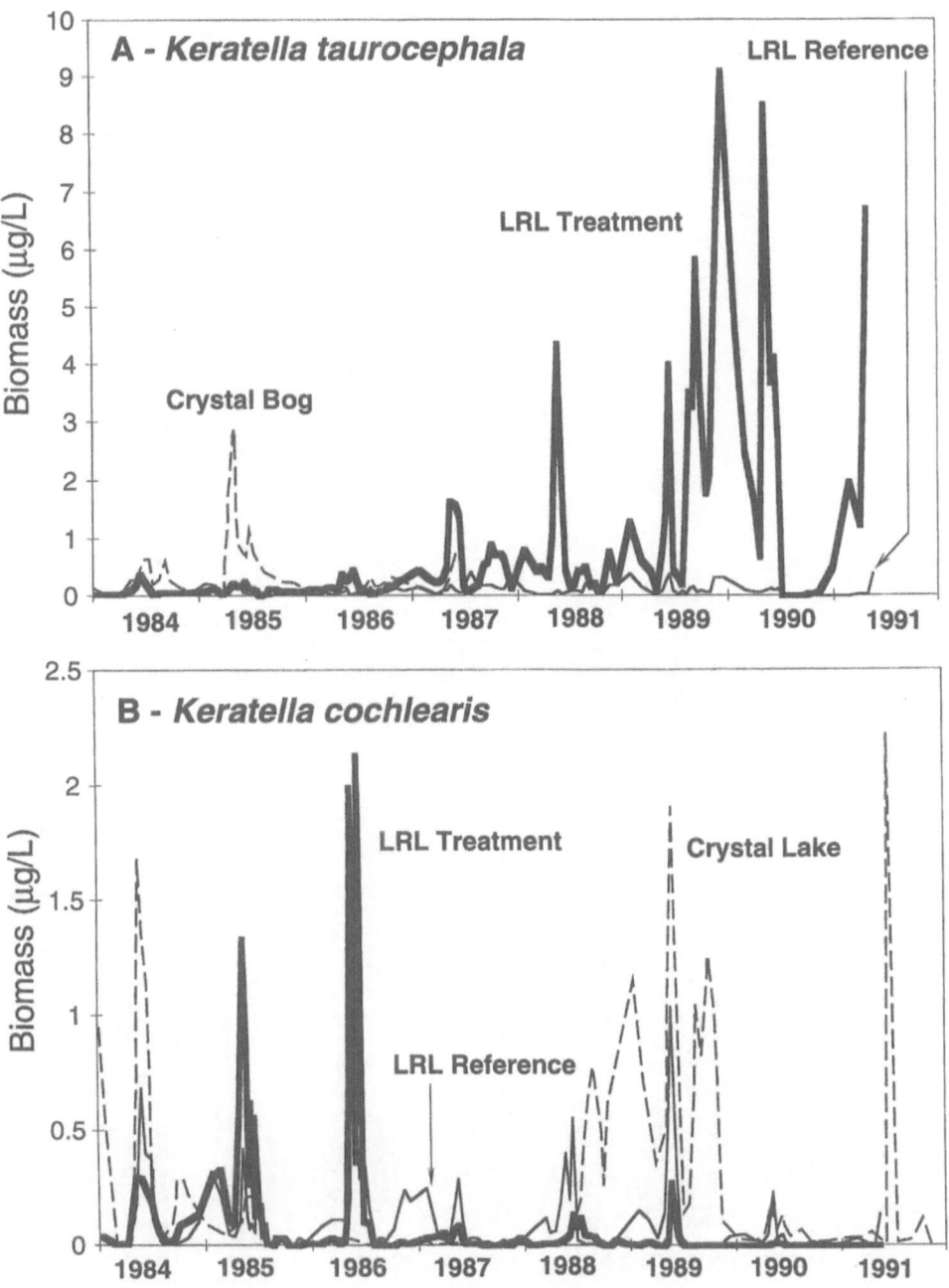

Figure 2. Average biomass (μg/L) of *Keratella taurocephala* (A) and *Keratella cochlearis* (B) in the water columns of the treatment (thick line) and reference (thin line) basins of Little Rock Lake and in Crystal Bog (A) or Crystal Lake (B) (dotted lines) in Wisconsin, U.S.A.

are largely similar (Table 1). We recorded 28 taxa in CL and 25 in CB compared with the 26 in LRL. Only two taxa, *Collotheca mutabilis* and *Notomata* sp. were present in LRL but absent in both CB and CL. Three genera were absent from LRL but present in one of the other lakes; *Anuraeopsis* sp. in CL and *Brachionus quadridentatus* and *Cephalodella* sp. in CB. Other-

wise, among-lake differences involved other taxa of genera that were common to at least two lakes, primarily *Keratella* which was more specious in CB and *Polyarthra* which had more species in CL. Overall, 15 taxa occurred in all three lakes (Table 1).

Comparisons of abundances of individual taxa did not reveal the same common conditions among

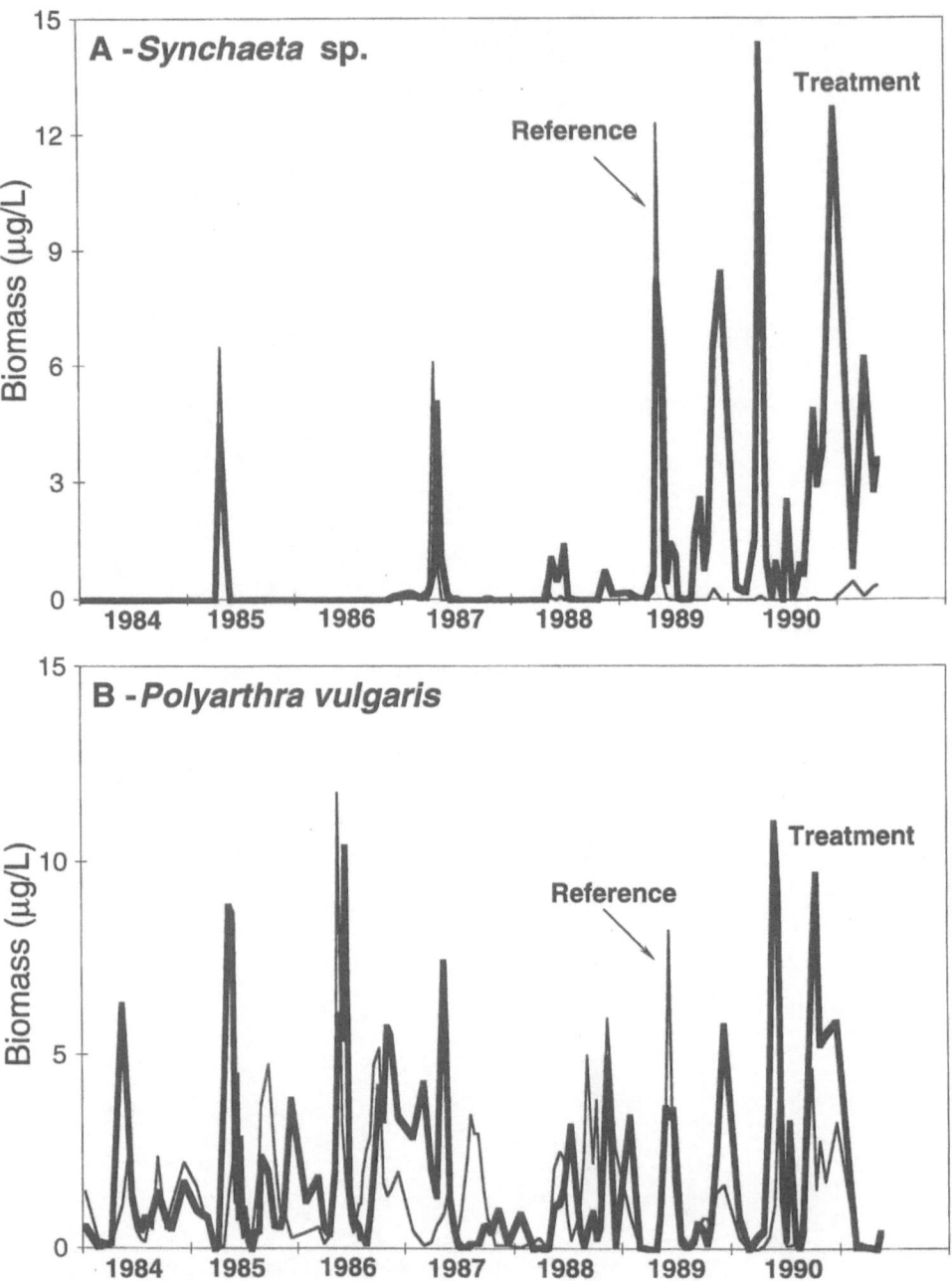

Figure 3. Average biomass (μg/L) of *Synchaeta* sp. (A) and *Polyarthra vulgaris* (B) in the water columns of the treatment (thick line) and reference (thin line) basins of Little Rock Lake, Wisconsin, U.S.A.

the lakes. For *Keratella taurocephala*, which had increased so substantially with acidification in the LRL treatment basin, we recorded abundances in CB that were intermediate between those in the treatment basin and those in the reference basin (Fig-

ure 2A). Its maximum abundance in CB was substantially lower than that occurring in the LRL treatment basin even during 1989 when its pH level of 5.2 was higher than the 5.0 average value for CB. For *K. cochlearis*, abundances in the LRL

Table 1. Annual average biomass values for rotifer species (μg/L throughout the water columns) in the treatment and reference basins during each of the periods during the experimental acidification Little Rock Lake, Wisconsin, U.S.A. These are followed by the slopes of standard linear regressions of annual average abundance vs. each year of the experiment for each species, the coefficient of determination, R^2, for this relationship, and our assessment of what the overall trend in the relationship was during the experiment. Also reported are whether each species was present (P), absent (A), or if another species in the same genus was present (G) in Crystal Bog (CB) and Crystal Lake (CL)

Period / Species	Basin	Baseline	5.6 (I)	5.6 (II)	5.2 (III)	5.2 (II)	4.7 (I)	4.7 (II)	SLOPE	R^2	Change	CB	CL
Ascomorpha sp.	T	0.050	0.028	0.019	0.007	0.0012	0.000	0.000	−0.17	0.70	0	P	P
	R	0.048	0.039	0.039	0.007	0.000	0.000	0.000					
Ascomorpha ovalis	T	0.000	0.002	0.000	0.005	0.057	0.000	0.000	−0.23	0.07	0	G	P
	R	0.000	0.000	0.001	0.003	0.018	0.002	0.004					
Asplanchna priodonta	T	1.186	0.354	0.191	1.088	1.138	0.000	0.000	−0.04	0.06	−	P	A
	R	2.005	2.685	2.129	2.459	1.179	3.137	3.244					
Collotheca mutabilis	T	0.017	0.001	0.002	0.000	0.000	0.000	0.000	−0.19	0.41	−	A	A
	R	0.010	0.011	0.025	0.021	0.006	0.027	0.047					
Conochiloides sp.	T	0.056	0.067	0.057	0.016	0.200	0.035	0.001	−1.43	0.29	−	A	P
	R	0.013	0.004	0.025	0.216	0.029	0.176	0.052					
Conochilus unicornis	T	0.647	0.538	0.443	0.091	0.245	0.002	0.001	−0.25	0.82	−	P	P
	R	0.513	0.394	0.376	0.461	0.430	0.839	0.652					
Gastropus hyptopus	T	0.022	0.158	0.048	0.052	0.026	0.002	0.003	−0.41	0.75	−	G	P
	R	0.007	0.116	0.070	0.101	0.135	0.016	0.046					
Gastropus stylifer	T	0.460	0.582	0.257	1.136	0.437	1.060	2.444	0.36	0.47	+	G	P
	R	0.610	0.372	0.532	0.821	0.540	0.594	0.633					
Kellicottia bostoniensis	T	0.056	0.029	0.100	0.021	0.056	0.017	0.001	−0.39	0.50	0	P	P
	R	0.016	0.036	0.051	0.053	0.050	0.015	0.024					
Kellicottia longispina	T	0.240	0.464	0.208	0.132	0.004	0.021	0.044	−0.39	0.077	−	P	P
	R	0.014	0.270	0.134	0.465	0.261	0.571	0.473					
Keratella cochlearis	T	0.153	0.191	0.244	0.013	0.034	0.018	0.001	−0.031	0.56	−	P	P
	R	0.137	0.073	0.222	0.053	0.104	0.114	0.016					
Keratella crassa	T	0.028	0.029	0.024	0.067	0.148	0.040	0.000	0.04	0.01	−	P	A
	R	0.044	0.012	0.048	0.093	0.046	0.015	0.000					
Keratella hiemalis	T	0.436	0.282	0.368	0.182	0.080	0.162	0.106	−0.03	0.01	0	P	P
	R	0.436	0.446	0.289	0.690	0.328	0.117	0.177					
Keratella taurocephala	T	0.100	0.111	0.205	0.601	0.718	3.028	2.988	8.60	0.61	+	P	P
	R	0.132	0.049	0.088	0.165	0.120	0.133	0.045					
Lecane sp.	T	0.022	0.010	0.005	0.007	0.011	0.009	0.014	0.73	0.55	0	P	P
	R	0.023	0.007	0.007	0.002	0.007	0.004	0.002					
Monostyla sp.	T	0.000	0.001	0.002	0.001	0.002	0.002	0.004	0.83	0.77	0	A	P
	R	0.000	0.001	0.001	0.000	0.001	0.001	0.001					
Notomata sp.	T	0.000	0.001	0.000	0.000	0.000	0.000	0.000	0.00		0	A	A
	R	0.000	0.000	0.000	0.000	0.009	0.007	0.002					
Ploesoma sp.	T	0.046	0.005	0.000	0.001	0.001	0.000	0.000	−0.22	0.99	0	A	P
	R	0.040	0.006	0.000	0.002	0.000	0.000	0.000					
Polyarthra dolichoptera	T	0.042	0.040	0.003	0.009	0.001	0.001	0.000	−0.20	0.67	−	A	P
	R	0.030	0.057	0.066	0.029	0.029	0.025	0.022					
Polyarthra renata	T	0.411	0.652	0.902	0.867	0.733	1.396	2.763	0.75	0.54	+	P	P
	R	0.433	0.489	1.130	1.003	0.441	0.214	0.661					
Polyarthra vulgaris	T	1.310	1.988	3.011	0.960	1.175	1.454	3.533	0.07	0.06	0	P	P
	R	0.863	1.124	2.286	1.055	1.763	1.060	1.370					

Continued on p. 148

Table 1. Continued.

Period / Species	Basin	Baseline	5.6 (I)	5.6 (II)	5.2 (III)	5.2 (II)	4.7 (I)	4.7 (II)	SLOPE	R^2	Change	Presence in CB	CL
Symchaeta spp.	T	0.223	0.068	0.061	0.317	0.292	2.828	2.617	3.87	0.53	+	P	P
	R	0.324	0.053	0.303	0.043	0.050	0.658	0.080					
Trichocerca sp.	T	0.017	0.002	0.002	0.000	0.000	0.002	0.001			0	G	P
	R	0.014	0.000	0.000	0.000	0.000	0.000	0.000					
Trichocerca cylindrica	T	0.221	0.203	0.256	0.024	0.010	0.004	0.011	−0.13	0.84	−	G	P
	R	0.284	0.388	0.499	0.296	0.658	0.187	0.608					
Trichocerca multicrimis	T	0.001	0.000	0.000	0.000	0.004	0.043	0.041			0	G	P
	T	0.000	0.000	0.000	0.000	0.000	0.002	0.000					
Trichocerca birostris	T	0.012	0.086	0.057	0.050	0.013	0.003	0.047	−0.28	0.24	O	G	P
	R	0.005	0.040	0.014	0.021	0.024	0.007	0.020					

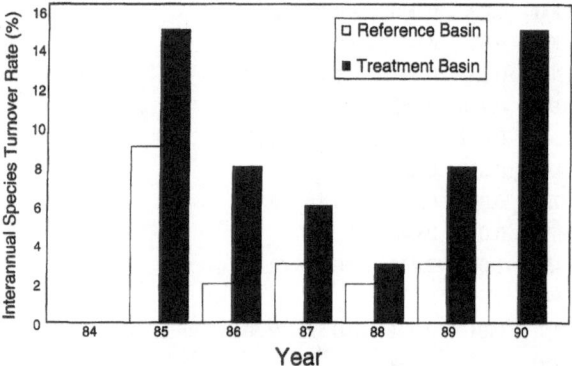

Figure 4. Inter-annual species turnover rates for the rotifer community in the treatment and reference basins of Little Rock Lake, Wisconsin, U.S.A.

treatment basin reached peaks comparable to those observed in CL during the first three years of the experiment but they were substantially lower during more acid stages of the experiment than those occurring during some CL periods (Figure 2B). *K. cochlearis* abundances in the LRL reference basin were usually lower than the peaks reached in the other two basins and were substantially lower than the high values that were recorded during some periods in CL, particularly 1988–89. For both of these *Keratella* species, which we considered to be key indicators of responses in the LRL experiment (Gonzalez & Frost, 1994), these comparisons provide little support for the hypothesis that acidification simply shifted the LRL rotifers from a situation occurring in CL to one in CB.

Considering the proportion of rotifer biomass in the zooplankton community gives a somewhat different impression, however. The proportion of rotifer biomass was fairly similar among CL, LRL reference basin, and LRL treatment basin during the initial stages of the experiment (Figures 6 and 7). The rotifer proportion of biomass was lowest in CL. Acidification did appear to shift the biomass of the total rotifer assemblage in the LRL treatment basin to be similar to that occurring in Crystal Bog (Figures 6 and 7).

Overall, we accept our hypothesis regarding the comparisons among the lakes at the community level, in terms of the proportional biomass of the rotifers in the zooplankton community, but not at the species level, because there were substantial differences in the occurrence patterns of individual rotifer taxa among the habitats we considered.

Discussion

Several rotifer taxa exhibited early and dramatic responses to the acidification of LRL. These changes were generally consistent with patterns observed in other low-pH lakes. Reduced abundance of *K. cochlearis* and high populations of *K. taurocephala*'s have been reported in several acid lakes (e.g., Brett, 1989; Siegfried et al., 1989). The inability of laboratory bioassays conducted in parallel with the LRL manipulation to predict the *K. taurocephala* response (Gonzalez & Frost, 1994) is all the more surprising given the common dominance of this species in acid situations. It certainly appears to exhibit high populations in many low pH habitats.

Measures of the collective rotifer community's responses to acidification gave a different impression than the individual-species reactions. Changes in total rotifer biomass were only evident during the most acid phase of the experiment (Figure 5A). There were no differences between the treatment and reference

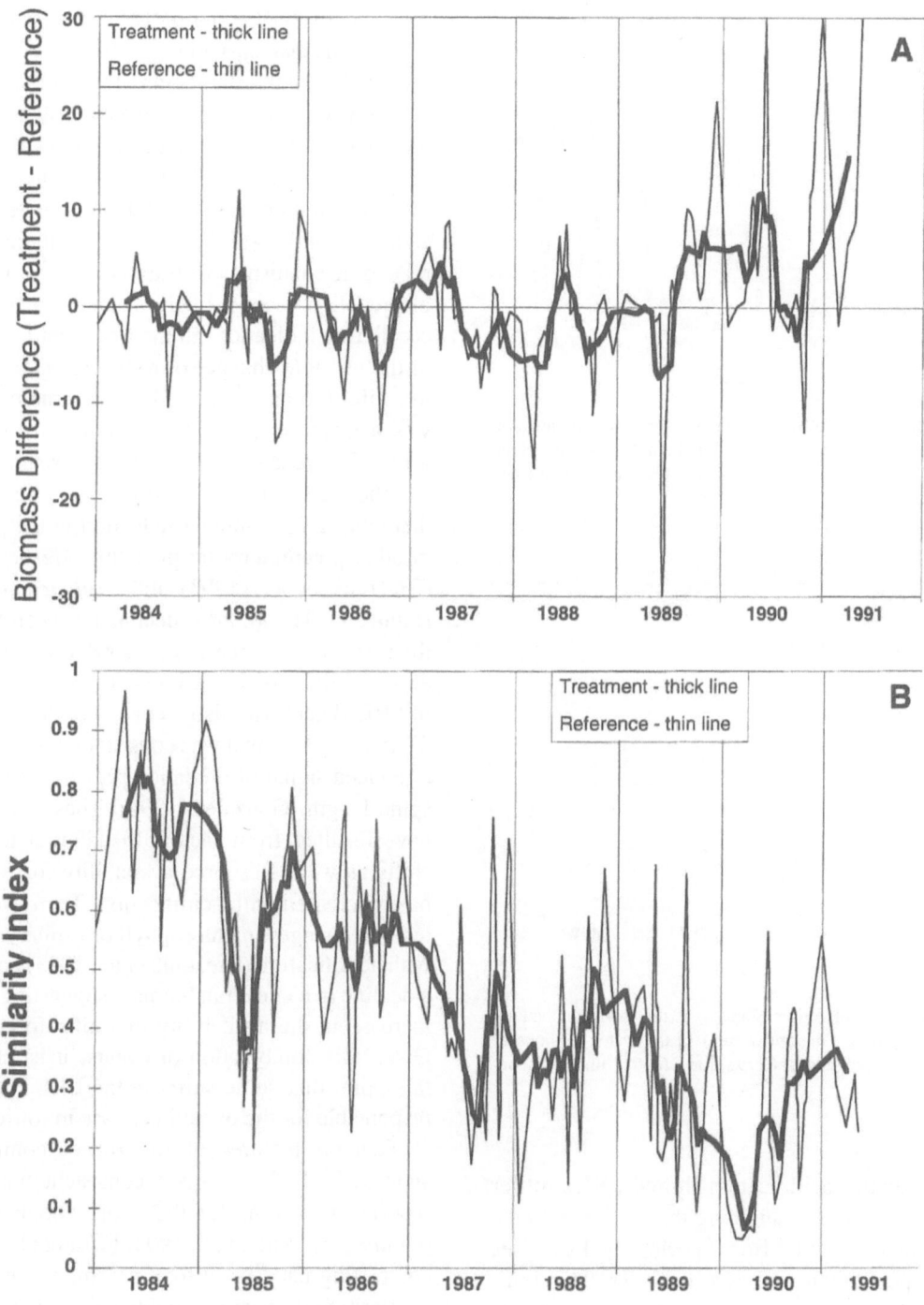

Figure 5. Five-point moving average values for (A) the difference in total rotifer biomass (mg/L) for the treatment − reference basins and (B) similarity indices calculated between the two basins of Little Rock Lake, Wisconsin, U.S.A.

basins in either the pH 5.6 or 5.2 stages but total rotifer biomass increased markedly during the pH 4.7 stage. This increase occurred despite the declines of numerous rotifer taxa and could be attributed largely to a shift in the abundance of *K. taurocephala* (Frost et al., 1995). Community-level responses supported

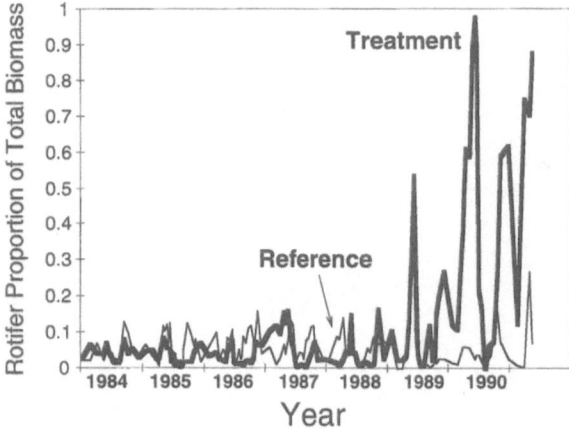

Figure 6. Proportion of rotifer biomass in the total zooplankton community (excluding *Chaoborus* spp.) in the water columns of the treatment (thick line) and reference (thin line) basins of Little Rock Lake, Wisconsin, U.S.A.

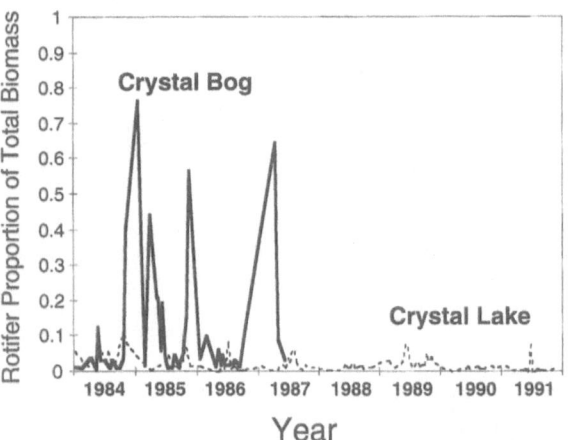

Figure 7. Proportion of rotifer biomass in the total zooplankton community (excluding *Chaoborus* spp.) in the water columns of Crystal Bog (solid line) and of Crystal Lake (dotted line), Wisconsin, U.S.A.

the hypothesis that acidification shifted LRL's rotifer assemblage from one resembling that in Crystal lake to one like that in Crystal Bog. Species-level analyses did not support this hypothesis, however. This reinforces our previous conclusions that different scales of taxonomic resolution can give very different impressions of ecosystem patterns, particularly responses to stress (Frost et al., 1995). Acid stress can shift a community to increasing rotifer dominance but pH alone can not explain the fundamental composition of rotifer communities. It is important to note that surveys of rotifers intended to detect acidification effects would have very different sensitivities depending upon

whether they were focused on individual species or on the collective properties of the rotifer community.

The mechanisms that underlie the increases we recorded in total rotifer community biomass with acidification are not clear. In general, and as reported previously for several biotic responses that were investigated during the LRL experiment, direct reactions to acidification were difficult to detect and most changes could best be explained by indirect, food-web related mechanisms (Webster et al., 1992). Decreases in predation seem a likely mechanism but there are conflicting indications in the treatment basin in terms of the possible changes of rotifer predators. Increases in rotifer biomass occurred at the same time as there was a major increase of the abundance of one major rotifer predator in the treatment basin, *Chaoborus* (Fischer & Frost, 1997). At the same time, however, there was a concomitant reduction in the populations of other potential rotifer predators, *Mesocyclops edax* (Brezonik et al., 1993) and *Asplanchna priodonta* (Figure 1). The *M. edax* decline has been linked with the *Chaoborus* increase (Fischer & Frost, 1997) indicating a shift in food-web interactions with acidification in LRL. There was also a change in the body form of *K. taurocephala* that was consistent with a response to a reduced signal of predation pressure, a reduction in spine length (Gonzalez, 1992). These changes could have resulted from factors in addition to predation shifts as well. Resource availability for rotifers may have increased with acidification. There were no systematic changes in chlorophyll or primary production with acidification (Brezonik et al., 1993) but there was a decline in cladoceran biomass suggesting a potential increase in the availability of rotifer-food resources. Given this combination of events, it is only clear at this point that some shifts in the LRL food web are responsible for the overall increase in rotifer biomass.

General features of the rotifer community responses in LRL are quite consistent with previous reports of patterns for the entire zooplankton community (Brezonik et al., 1993; Frost et al., 1995). The contrasting patterns in total community biomass and similarity indices are very much the same for the rotifers (Figure 5) and for all zooplankton. Decreases for several species but buffered responses by the total biomass of portions of the zooplankton community were reported for cladocerans and copepods as well as for rotifers (Frost et al., 1995). Only the marked increase that we report here (Figure 6) is specific to the rotifer elements of the community.

Responses to acidification in LRL are consistent with previous studies of low pH systems. Several of the patterns that we described here are similar to those reported in previous surveys of acid-stressed lakes (e.g., MacIssac et al., 1987; Brett, 1989). Systematic comparisons with other whole-lake manipulations also revealed a high degree of consistency in such large-scale experiments (Schindler et al., 1991). This suggests that it will be reasonable to predict the patterns to be expected for rotifer communities in other low-pH lakes from LRL and previous reports.

Overall, it appears that the basic importance of rotifers in lake ecosystems increases with acidity. At the same time, the nature of species-specific responses are likely to vary among habitats. Rotifer responses to low-pH conditions appear to be consistent at a community but not necessarily a species level.

Acknowledgments

This research was supported by grants from the U.S. National Science Foundation Long-Term Research in Environmental Biology and Long-Term Ecological Research Programs and by the U.S. Environmental Protection Agency. It was also supported by the Wisconsin Department of Natural Resources. We thank P. Brezonik, J. Fischer, D. Knauer, S. Knight, T. Kratz, T. Meinke, M. Sierszen, C. Watras, K. Webster, and many others for their assistance throughout this project.

References

Baker, J. P. & S. W. Christensen, 1991. Effects of acidification on biological communities in aquatic ecosystems. In D. F. Charles (ed.), Acidic Deposition and Aquatic Ecosystems: Regional Case Studies, pp 83–106. Springer-Verlag, New York, 747 pp.

Brett, M. T., 1989. The rotifer communities of acid-stressed lakes of Maine. Hydrobiologia 186/187: 181–189.

Brezonik, P. L., J. G. Eaton, T. M. Frost, P. J. Garrison, T. K. Kratz, C. E. Mach, J. H. McCormick, J. A. Perry, W. A. Rose, C. J. Sampson, B. C. L. Shelley, W. A. Swenson, & K.E. Webster, 1993. Experimental acidification of Little Rock Lake, Wisconsin: Chemical and biological changes over the pH range 6.1 to 4.7. Can. J. Fish. aquat. Sci. 50: 1101–1121.

Carpenter, S. R., T. M. Frost, D. Heisey & T. K. Kratz, 1989. Randomized intervention analysis and the interpretation of whole-ecosystem experiments. Ecology 70: 1142–1152.

Charles, D. F. (ed.), 1991. Acidic Deposition and Aquatic Ecosystems: Regional Case Studies. Springer-Verlag, New York, 747 pp.

Downing, J. A. & F. H. Rigler, 1984. A manual on methods for the assessment of secondary productivity in freshwaters. Blackwell, Oxford, England.

Fischer, J. M. & T. M. Frost, 1997. Indirect effects of lake acidification on Chaoborus population dynamics: the role of food limitation and predation. Can. J. Fish. aquat. Sci. 54: 637–646.

Frost, T. M. & P. K. Montz, 1988. Early zooplankton response to experimental acidification in Little Rock Lake, Wisconsin, USA. Verh. int. Ver. Limnol. 23: 2279–2285.

Frost, T. M., S. R. Carpenter, A. R. Ives & T. K. Kratz, 1995. Species compensation and complementarity in ecosystem function. In C. G. Jones & J. H. Lawton (eds), Linking Species and Ecosystems. Chapman and Hall, New York: 224–239.

Frost, T. M., P. K. Montz & T. K. Kratz. Zooplankton community responses during recovery from acidification: limited persistence by acid-favored species in Little Rock Lake, Wisconsin. Restoration Ecology. (In press).

Galloway, J. N., G. E. Likens, & M. E. Hawley, 1984. Acid precipitation: natural versus anthropogenic components. Science 226: 829–831.

Gonzalez, M. J., 1992. Effects of experimental acidification on zooplankton populations: A multiple-scale approach. Ph.D. Dissertation, Oceanography and Limnology Graduate Program, University of Wisconsin, Madison, WI.

Gonzalez, M. J. & T. M. Frost, 1994. Comparisons of laboratory bioassays and a whole-lake experiment: Rotifer responses to experimental acidification. Ecol. Appl. 4: 69–80.

Locke, A., 1992. Factors influencing community structure along stress gradients: zooplankton responses to acidification. Ecology 73: 903–909.

MacIssac, H. J., T. C. Hutchinson & W. Keller, 1987. Analysis of plankton rotifer assemblages from Sudbury, Ontarioa, area lakes of varying chemical composition. Can. J. Fish. aquat. Sci. 44: 1692–1701.

Magnuson, J. J., C. J. Bowser & T. K. Kratz, 1984. Long-term ecological research on north temperate lakes (LTER). Verh. int. Ver. Limnol. 22: 533–535.

Rasmussen, P. W., D. M. Heisey, E. V. Nordheim & T. M. Frost, 1993. Time-series intervention analysis: Unreplicated large-scale experiments. In: S. M. Scheiner & J. Gurevitch (eds), Design and Analysis of Ecological Experiments. Chapman and Hall, Inc., New York: 138–158.

Ruttner-Kolisko, A., 1974. Plankton rotifers, biology and taxonomy. Die Binnengewässer 26 (1) Supplement: 146 pp. E. Schweizerbart'sche Verlagsbuchhandlung, Stuttgart, Germany.

Sampson, C. L., P. L. Brezonik, T. M. Frost, K. E. Webster & T. D. Simonson, 1995. Experimental Acidification of Little Rock Lake, Wisconsin: The First Four Years of Chemical and Biological Recovery. Wat. Air Soil Pollut. 85: 1713–1719.

Schindler, D. W., 1988. Effects of acid rain on freshwater ecosystems. Science 239: 149–157.

Schindler, D. W., T. M. Frost, K. H. Mills, P. S. S. Chang, I. J. Davies, L. Findlay, D. F. Malley, J. A. Shearer, M. A. Turner, P. J. Garrison, C. J. Watras, K. E. Webster, J. M. Gunn, P. L. Brezonik & W. A. Swenson, 1991. Comparisons between experimentally- and atmospherically-acidified lakes during stress and recovery. Proc. Soc. Edinb. 97B: 193–226.

Siegfried, C. A., J. A. Bloomfield & J. W. Sutherland, 1989. Planktonic rotifer community structure in Adirondack, New York, U.S.A. lakes in relation to acidity, trophic status and related water quality characteristics. Hydrobiologia 175: 33–48.

Stemberger, R. S. 1979. A guide to the rotifers of the Laurentian Great Lakes. US EPA 600/4-79-021, 185 pp.

Stewart-Oaten, A., J. R. Bence & C. W. Osenberg, 1992. Assessing effects of unreplicated perturbations: no simple solutions. Ecology 73: 1396–1404.

152

Watras, C. J. & T. M. Frost, 1989. Little Rock Lake: Perspectives on an experimental approach to acidification. Arch. envir. Contam. Toxicol. 18: 157–165.

Webster, K. E., T. M. Frost, C. J. Watras, W. A. Swenson, M. Gonzalez & P. J. Garrison, 1992. Complex biological responses to the experimental acidification of Little Rock Lake, Wisconsin, USA. Envir. Pollut. 78: 73–78.

Yan, N. D., W. Keller, K. M. Somers, T. W. Pawson, & R. E. Girard, 1996. Recovery of crustacean zooplankton communities from acid and metal contamination: comparing manipulated and reference lakes. Can. J. Fish. aquat. Sci. 53: 1301–1327.

Hydrobiologia **387/388**: 153–160, 1998.
E. Wurdak, R. Wallace & H. Segers (eds), Rotifera VIII: A Comparative Approach.
© *1998 Kluwer Academic Publishers.*

Review paper

Rotifers in arctic North America with particular reference to their role in microplankton community structure and response to ecosystem perturbations in Alaskan Arctic LTER lakes

Parke A. Rublee
Biology Department, University of North Carolina at Greensboro, Greensboro, NC 27412, U.S.A.

Key words: Rotifera, microplankton, arctic, Alaska

Abstract

Growing interest in the development of mineral and recreational resources, along with the recognition that arctic ecosystems may be among those most affected by global change, has stimulated the study of arctic systems in recent decades. These have included studies of rotifers. Two approaches have generally been pursued: taxonomic studies to determine the number and species of individuals, and ecological studies that have attempted to determine the trophic relationships between rotifers and other microorganisms in aquatic ecosystems. Results from studies at the Arctic Long Term Ecological Research Site in Alaska, USA are reviewed and the microbial food web is described based on empirical and literature data. Arctic systems are sites of rich opportunity for further studies, especially those which can integrate taxonomic and ecological aspects.

Introduction

Microplankton, including rotifers, have received considerable attention from aquatic ecologists in recent years because of the pivotal role they play in regulating the transfer of nutrients and energy to higher trophic levels (e.g., Rieman & Christoffersen, 1993). In the classical view, microplankton grazed primary producers and in turn were prey for crustacean zooplankton which are themselves a food resource for larger insects and fish. However, over the last several decades it has become obvious that the classical view is oversimplified and that microplankton act, not only as grazers of primary producers, but also as important consumers of bacterial secondary production (derived from exudates, leaked organic carbon and decomposing organic matter). They may also be consumers of microplankton secondary production, since protozoa and rotifers frequently feed on other protozoans and rotifers.

Another focus of recent research has been the importance of bottom up and top down controls of aquatic ecosystems (Carpenter, 1988). Numerous studies (cf. Carpenter, 1988) have demonstrated that increased nutrient inputs tend to increase biomass at each trophic level, while alteration of higher trophic levels results in a top-down cascade of trophic interactions mediated by the abundance of predators. What has been addressed only marginally is how microplankton community structure changes in response to changed controls, and whether the changes modifiy the flow of the bottom-up or top-down regulation.

Given the complexity and varying roles that microplankton play, it is difficult to determine how food webs function at the microbial level and how they respond to perturbations. Arctic ecosystems offer unique opportunities for such studies, because their extreme climate leads to reduced complexity of arctic systems relative to their temperate and tropical counterparts. This paper briefly reviews studies of rotifers in North American arctic systems, especially in relation to the progress and observations made at the Arctic Long Term Ecological Research Site at Toolik Lake, Alaska.

Previous studies

Taxonomic treatments of Canadian arctic rotifers (Table 1) were recently reviewed by De Smet & Beyens (1995), and will be only briefly mentioned here. De Smet & Beyens (1995) identified 70 taxa of roti-

Table 1. Recent studies of arctic rotifers in North America

Location	Coordinates	Number of taxa	Reference
Victoria Island, NWT, Canada	69°N, 105°W	112	De Smet (1994), De Smet & Beyens (in press)
Devon Island, NWT, Canada	75°N, 86°W	70	De Smet & Beyens (1995), Minns (1977)
Little Cornwallis Island, NWT, Canada	75°N	51	De Smet & Bafort (1990)
Ellesmere Island, NWT, Canada	81°N	34	McLaren (1964), Nogrady & Smol (1989)
Cornwallis Island, NWT, Canada	75°N	1	Rigler et al. (1974)
Central NWT, Canada	62–66°N	15	Moore (1976)
Northwest Territory & Northern Yukon, Canada & Northern Alaska, USA	above 66°N	127	Chengalath & Koste (1989)
Lakes Peters, Schrader and Wolf, NE Alaska, USA	70°N, 144°W	3	Hobbie et al. (unpublished)
Northern Alaska	above 66°N	12	Holmquist (1975)
Arctic LTER site, Toolik Lake, Alaska	68°N, 149°W	20	Rublee (1992), Rublee & Bettez (1995), Rublee & Partusch-Talley (1995) Bettez et al. (in prep.)
Point Barrow and Nome, Alaska	71°N, 156°W		De Smet (personal communication)

fers, bringing to 114 the total number of taxa that had been identified in the Canadian high arctic, but suggested that many more taxa would probably be discovered as the number and intensity of studies increased. They also noted that low species richness and diversity of Canadian high arctic rotifer fauna relative to other regions was probably due to the severe physical environment, short growing season, limited spatial and habitat heterogeneity, the short length of time since the last glaciation of northern Canada, as well as the limited number of studies. De Smet & Beyens (1995) also suggested that the Alaskan arctic might have greater species richness and diversity than the Canadian arctic since large areas of Alaska had escaped glaciation. Chengalath & Koste (1989) collected rotifers from 212 sites in arctic North America and identified 165 species, of which 127 were from arctic sites in Alaska, the Yukon, and the Northwest Territories. This study is one of only two published taxonomic records of Alaskan arctic rotifers, listing 87 species from 38 northern Alaskan sites. The other study (Holmquist, 1975) listed 12 taxa from lakes of northern Alaska. An additional study of rotifers collected from Point Barrow and Nome, Alaska is in progress (De Smet, personal communication). Chengalath & Koste (1989) also cite several older studies that emphasized species descriptions of North American arctic rotifers. Nearly all studies note that most rotifer species found in the arctic are cosmopolitan and that usually only a few species are dominant at any sampling location (De Smet & Beyens, 1995)

Studies of the role of rotifers as components of aquatic ecosystems are also relatively rare in North American arctic ecosystems. Chengalath & Koste (1989) noted 48 publications that included rotifers as members of arctic aquatic communities but gave little detail. Rigler et al. (1974) reported that *Keratella cochlearis* was numerically the second most abundant zooplankter in Char Lake, but that rotifer production was only about 1% of that of the dominant crustacean zooplankter. Hobbie et al. (unpublished data) found six rotifer species in Lakes Peters, Schrader and Wolf, in northern Alaska, but also noted that they represented less than 2% of the total zooplankton biomass. Rublee & Partusch-Talley (1995) utilized artificial substrates to determine the response of the microfaunal community to added nitrogen and phosphorus in the Kuparuk River at the Arctic Long Term Ecological Research (LTER) site in northern Alaska. Eight rotifer taxa were found on the substrates, constituting 36% of the microfauna biomass. Microfauna increased in response to nitrogen amendments, but there was no significant increase due to increased phosphorus inputs, although the lack of response may have been due to predation by a high abundance of benthic insects that were the result of a previous fertilization experiment in the river. They also found that the abundance of rotifers and other microfauna on natural rock surfaces in the river was low ($<1 \times 10^3$ rotifers m^{-2}). Fertilization of river reaches stimulated growth of mosses (Bowden et al., 1994), which are ideal substrates for rotifers (De Smet & Beyens, 1995), but these were not sampled.

The most extensive studies of the summer seasonal dynamics of rotifer populations and their trophic relationships in the plankton of arctic lakes have been conducted at the arctic LTER site at Toolik Lake in northern Alaska (Bettez et al., in prep; Rublee, 1992; Rublee & Bettez, 1995). The major focus of these studies has been to assess the abundance of microplankton (including rotifers) in the water column and to determine their response to perturbations. In unperturbed lakes at the site they reported nine species of rotifers in the plankton, although four dominated: *Keratella cochlearis*, *Kellicottia longispina*, *Polyarthra vulgaris* and *Conochilus unicornis*. The abundance of rotifers was low in unfertilized lakes (100–400 individuals l^{-1}), with highest abundance found in late summer (Rublee, 1992). The remainder of this paper will address the response of rotifers to experimental manipulations and feeding studies that have been used to determine trophic interactions of the microplankton community.

Materials and methods

Study site

The Arctic LTER site (68° 38′ N, 149° 43′ W), located in the northern foothills of the Brooks Mountain

Range of Alaska, has been under study for over two decades (O'Brien et al., 1997). Climate is extreme: the region is underlain by permafrost with a mean annual temperature of −9°C. Annual precipitation is about 31 cm, with about half falling as rain from late May through September. Ice cover, up to 2 m thick, forms in late September or October and generally thaws in late June. Water temperatures may rise to 12–15°C in the epilimnion by late summer. The combination of cold climate and limited rainfall makes nutrient input a major limiting constraint in the lakes and ponds of the LTER site. As a result, the lakes are highly oligotrophic (Miller et al., 1986) with varying algal, zooplankton and fish populations (Kling et al., 1992; O'Brien et al., 1992).

O'Brien et al. (1997) have described the biotic community of Toolik lake which is typical of the lakes at the LTER site. Algal communities are dominated by small chrysophytes, dinoflagellates, and cryptophytes. Zooplankton include the herbivores *Daphnia middendorfiana* and *Diaptomus pribilofensis*, the carnivore *Cyclops scutifer*, and the larger but much less abundant predator *Heterocope septentrionalis*. Fish, if present, include lake trout, burbot, arctic grayling, and slimy sculpin. At least eight species of chironomids are found in the benthos (Kling et al., 1992). Most data presented here are from Toolik Lake, or from Lake N1, a lake that was fertilized from 1990–1994 by weekly additions of inorganic nitrogen and phosphorus during the summer at rates (3 mM N m^{-2} day^{-1} as $(NH_4)_2SO_4$ and 0.23 mM P m^{-2} day^{-1} as H_3PO_4) which are about four to ten times the normal nutrient loading during the summer. Lake N1 was highly oligotrophic prior to fertilization (Miller et al., 1986).

Rotifers were collected for both enumeration and grazing studies by first concentrating freshly collected water samples by gentle reverse flow filtration through 20-μm mesh net (Dodson & Thomas, 1964). For enumeration, the concentrated samples were preserved with cold glutaraldehyde (1% final concentration), and stored refrigerated until counting following the method of Baldock (1986) which uses rose bengal to stain organisms. In some samples, the dimensions and spine lengths of loricate rotifers were measured.

Rotifer grazing was assessed by directly counting the number of fluorescently labeled food resource analogues that were ingested by individual grazers (Rublee & Gallegos, 1989; Sherr et al., 1987). The food resource analogues included fluorescently labeled bacteria (FLB), fluorescently labeled algae

Table 2. Fluorescently labeled food analogues

Type	Species	Diameter/dimensions (μm)	Source
FLB	*Pseudomonas fluorescens*	1×2	Freshwater
FLA	*Nanochloris occulata*	2	Brackish
	Nanochloris sp.	3×7	Brackish
	LJ3B (unidentified green)	4–6	Freshwater
	Chlamydomonas rheinhardtii	10	Freshwater
	Chlorella vulgaris	6–8	Freshwater
	Phaeodactylum tricornutum	3×20	Marine
	Prorocentrum minimum	15×18	Marine
FLY	*Saccharomyces cerevisiae*	4×5	Freshwater
	Latex particles	0.49, 2.17, 2.5, 4.3, 5.7, 9.3	

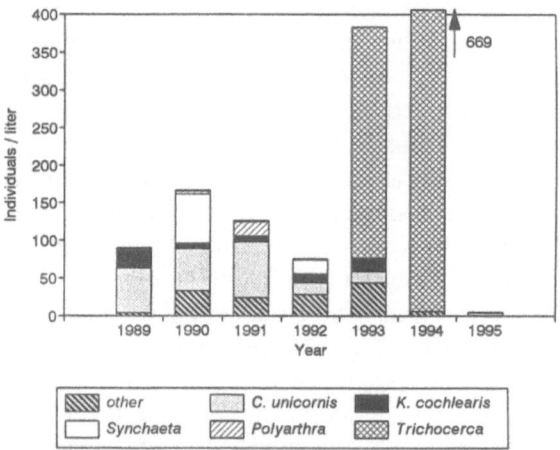

Figure 1. Abundance of rotifers during fertilization experiment of Lake N1 at the Arctic LTER site in Alaska. Lake N1 was fertilized by weekly additions of nitrogen and phosphorus from July to August, 1990–1994.

(FLA), fluorescently labeled yeast (FLY) and fluorescently labeled latex particles (Table 2). Individual microplankters were then hand picked from freshly concentrated samples using glass pasteur pipettes with finely drawn tips, and transferred to filtered lakewater (0.45-μm GFC filters), in blood dilution vials which had been pre-soaked in 10% HCl and rinsed in filtered lakewater. Food resource analogues were then added to the vials and incubated in water baths. Total water volume in the vials was 20 ml, and the number of microplankton generally ranged from 10 to 50 individuals per vial, a range of concentrations that is within the natural ranges found for microplankton in temperate lakes, but does exceed, in some cases, that found for microplankton in arctic lakes. The concentration of food analogues added was either at a level meant to simulate the natural concentration for particles of that size, or spanned a range in order to assess feeding response curves for a particular grazer. Incubation time ranged from several minutes to several hours, and was derived empirically from prior time course sampling to determine optimum balance between enough ingestion to register statistically significant counts, but not so much ingestion that counting ingested particles was difficult.

Following incubation, rotifers were collected on a 20-μm mesh net, washed with filtered lake water to remove excess labeled particles, and then collected on 5.0-μm black polycarbonate filters. Filters were mounted on slides with a 43% sucrose solution and examined by microscopy. Individual microplankters were located on the filter using transmitted light and low magnification (100×–200×) followed by enu-

meration of ingested particles which were visualized under epifluorescent illumination at higher magnifications (200×–1000×).

A two-step grazing method (Dolan & Coats, 1991) was also used on one occasion to determine if crustacean zooplankton preyed on the rotifer *Conochilus unicornis*. Briefly, two samples were incubated, both containing fluorescent latex particles and the suspected crustacean predator. One sample also contained the microplankton prey. The method relies on the ingestion of particles by the microplankton, which then appear in the predator if it grazes on the microplankton prey.

Results and discussion

Rotifer species response to fertilization of Lake N1

The response of rotifer species to fertilization in lake N1 was complex. There was a slight increase in rotifer abundance during the first 2 years of fertilization, a decline during the third year, and a dramatic increase in rotifer abundance during the fourth and fifth years (Figure 1). The rotifer abundance dropped to its lowest value after lake fertilization stopped. There was a significant change in rotifer community structure, with a shift from prefertilization dominance by *Conochilus unicornis* and *Keratella cochlearis*, to dominance by a *Synchaeta* sp. during the first year of fertilization, and overwhelming dominance by a *Trichocerca* species during the fourth and fifth years of fertilization

(Figure 1). Rotifer abundance declined dramatically during the year following fertilization, which may reflect predatory losses due to increased zooplankton densities as a result of fertilization (Bettez et al., in prep.). Species richness also changed during the fertilization experiment, due to the appearance of two species, *Conochilus natans* and *Trichocerca* sp., which had not been seen prior to fertilization, and the disappearance of several species during the last year of fertilization.

Rublee (1992) had commented that in general, there tended to be an increase in microplankton biomass with increased trophic status (as estimated by chlorophyll *a* concentrations), although there was no clear relationship between rotifer abundance to trophic status in a suite of nine Arctic LTER lakes over several summers. This lack of relationship between chlorophyll *a* concentration and rotifer abundance remains when data from the fertilization experiment in Lake N1 and from Toolik Lake in 1995 are added to the data set ($R = 0.113$, 15 d.f., NS). The lack of a clear response suggests that top-down controls on rotifer abundance may be at least as important as food resources in arctic lakes.

Rotifer grazing

Grazing experiments conducted with various food resources for a variety of rotifer species demonstrated functional response curves, as well as size selectivity. Rotifers displayed Type II/Type III functional responses when presented with FLB or FLA (Figure 2). Clearance rates were variable, ranging from 0 to 20.7 μl ind.$^{-1}$ h^{-1} (Table 3). When presented with fluorescent latex particles, both *C. unicornis* and its congener, the larger *C. natans*, exhibited clearance rates up to 8 μl ind.$^{-1}$ h^{-1}, with highest rates on 5.7- and 4.3-μm diameter particles, respectively (Figure 3). *K. cochlearis* had highest clearance rates on particles of 2.17 μm diameter. *K. quadrata* also selected for small particles, although highest clearance rates were found with 2.5-μm diameter particles. Two other rotifers, *Kellicottia longispina* and *Filinia terminalis*, which are known to be small particle feeders, showed grazing responses on latex particles similar to that of the *Keratella* spp. as expected.

The clearance rates reported here should be considered as low estimates because they were determined using food analogues rather than natural food. However, the results of these grazing studies compare

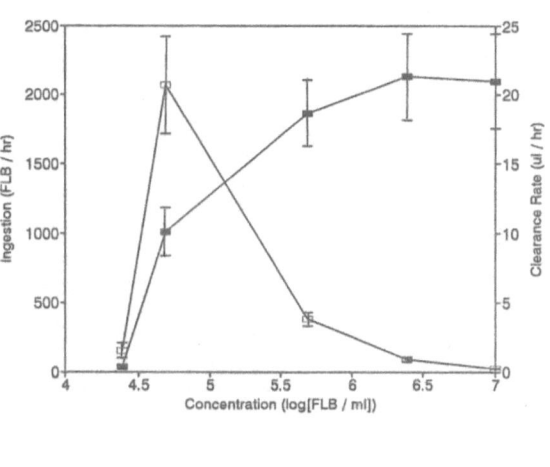

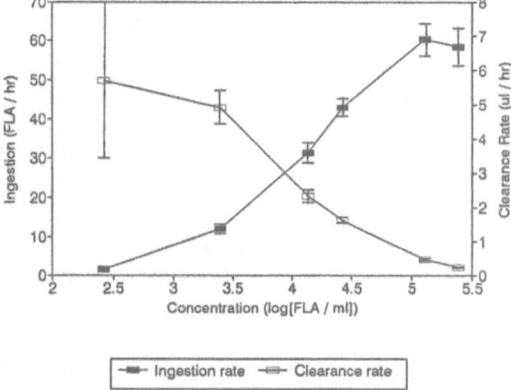

Figure 2. Examples of grazing studies on fluorescent food analogues by arctic rotifers. Upper: *Conochilus unicornis* grazing on 1 × 2-μm fluorescently labeled bacteria (FLB). Lower: *Keratella cochlearis* grazing on 5-μm fluorescently labeled alga (FLA). Error bars, ±1 SE.

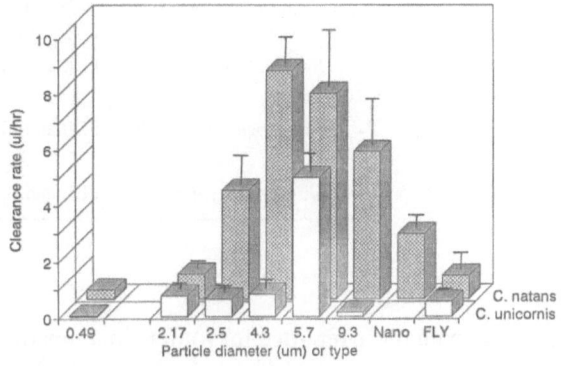

Figure 3. Example of clearance rates of fluorescent latex particles, fluorescently labeled yeast and *Nanochloris* FLA for *Conochilus unicornis* and *Conochilus natans*. Error bars, ±1 SE for *C. unicornis*; *C. natans* data is from a single experiment.

Table 3. Experimentally determined clearance rates of rotifers from the Toolik Lake LTER. site

Rotifer	Prey	Clearance rate (μl h^{-1})	Number of experiments
Keratella cochlearis	FLB	0.05–0.46	7
	FLA – *Nanochloris*	0.02–3.7	5
	FLA – LJ3	0.5–5.8	4
	FLA – *Chlamydomonas*	0.25–1.7	6
	FLA – *Chlorella*	0.00–0.56	2
	FLA – *Phaeodactylum*	0.5–15	4
	FLY – *S. cerevisiae*	0.07	1
Keratella quadrata	FLB	0.21	1
	FLA – *Nanochloris*	2.94	1
	FLA – *Chlorella*	0.01	1
Kellicottia longispina	FLB	0.07–0.7	4
	FLA – *Nanochloris*	0.01–0.17	5
	FLA – *Chlamydomonas*	0.28–1.43	5
	FLA – *Chlorella*	0.03–0.29	2
	FLA – *Phaeodactylum*	0.39–1.89	3
	FLY – *S. cerevisiae*	0.05	1
Conochilus unicornis	FLB	0.14–20.7	5
	FLA – *Nanochloris*	0.01–5.5	10
	FLA – *Chlamydomonas*	0.00–0.11	2
	FLA – *Chlorella*	0.10–2.8	3
	FLA – *Phaeodactylum*	4.9–5.5	3
	FLY – *S. cerevisiae*	0.5	1
Conochilus natans	FLA – *Nanochloris*	0.13	1
	FLA – *Chlamydomonas*	0.2	1
	FLA – Phaeodactylum	5.5	1
	FLA – *Prorocentrum*	2.6	1
	FLY – *S. cerevisiae*	0.7	1
Polyarthra vulgaris	FLB and FLA	0	10
Synchaeta sp.	FLB and FLA	0	7
Chromogaster ecaudis	FLB and FLA	0	10
Gastropus stylifer	FLB and FLA	0	8

favorably with those reported for rotifers in temperate systems which have generally used other methods to measure grazing rates (e.g., Gilbert & Bogdan, 1984).

Crustacean zooplankton *grazing on* Conochilus unicornis

Crustacean zooplankton ingestion was assayed once with *Conochilus unicornis* serving as a potential food for *Cyclops scutifer* and *Heterocope septentrionalis*.

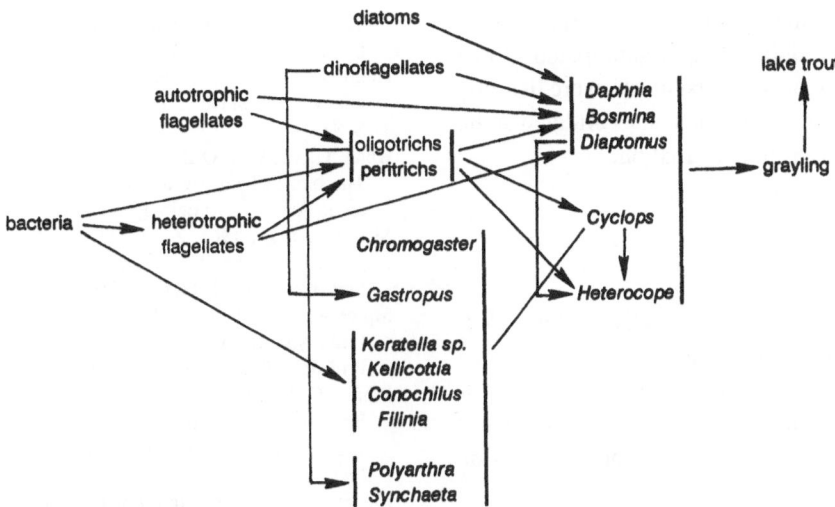

Figure 4. Generalized pelagic microbial food web for arctic LTER lakes.

Both crustaceans consumed *C. unicornis,* with the larger *Heterocope* demonstrating that it is a voracious predator when presented with this rotifer as its only food source.

Microbial food web

The grazing and ecosystem manipulation experiments at the Toolik Lake LTER provide empirical evidence for trophic interactions at a finer level of detail than previously known for these arctic planktonic systems. Insight into nutrition of other rotifer species has been derived from literature studies (e.g., Dumont, 1977; Gilbert & Bogdan, 1984; Pourriot, 1977) and limited additional experimental work at Toolik Lake. For example, *Gastropus stylifer* is known to have a unique feeding habit – it pierces the theca of dinoflagellates and sucks out the cytoplasm. In contrast, *Chromogaster ecaudis* is an autotroph. *Polyarthra vulgaris* and a *Synchaeta* sp. are known to be carnivorous. In numerous experiments with all types of fluorescent food particles, no ingestion was ever observed for any of these species. However, in a single incubation of *Chromogaster ecaudis* with ^{14}C-labeled bicarbonate, significant uptake of isotope occurred over killed controls. Combined, this information allows construction of a more detailed microbial food web for Alaskan arctic lakes, at least to our finest taxonomic level of organism identification (Figure 4).

Conclusions

Although the number of studies of rotifers and their role in arctic aquatic ecosystems is limited, they have provided numerous insights which also present many opportunities for further study. First, taxonomic studies to date suggest that there are many species yet to be identified. Particularly in the Alaskan arctic, which represents a glacial refugium, few taxonomic studies have been conducted, and benthic and littoral habitats as well as riverine habitats have yet to receive attention. Such studies may be especially valuable for comparative studies both to: (1) compare species diversity across landscapes with differing glacial histories as suggested by De Smet & Beyens (1995), and (2) to document existing fauna and the changes which are likely to occur in response to global warming and eutrophication. Integration of genetic characterization into classical taxonomic approaches might also prove extremely valuable since many arctic species are cosmopolitan and thus genetic variability can be compared across latitudinal gradients.

The extreme climate of arctic systems, manifested via limited nutrient input into lakes which keeps them highly oligotrophic also suggests that there are numerous opportunities for ecological studies. Species diversity is limited and food webs are less complex relative to temperate and tropical systems, although the information presented here and from earlier studies suggests that food webs are not simple. Such sites are ideal for studying trophic interactions and regulatory control of aquatic ecosystems, especially at the microbial level, since they represent more tractable food webs in contrast to the much greater complexity of systems at lower latitudes. Again, anticipated global changes which are expected to be most severe in arctic regions present opportunities for studies directed at the response of ecosystems to perturbation.

Acknowledgments

I thank G. Kipphut, J.E. Hobbie, G.W. Kling, W.J. O'Brien, M.C. Miller, N.D. Bettez, M. McDonald and A.E. Hershey for sharing data, insight, and discussion regarding the Arctic LTER research. This work was supported in part by National Science Foundation Grant DPP-8821679.

References

Baldock, B. M., 1986. A method for enumerating protozoa in a variety of freshwater habitats. Microb. Ecol. 12: 187–191.

Bowden, W. B., J. C. Finlay & P.E. Maloney, 1994. Long term effects of PO4 fertilization on the distribution of bryophytes in an arctic stream. Freshwat. Biol. 32: 445–454.

Carpenter, S. R. (ed.), 1988. Complex Interactions in Lake Communities. Springer, New York: 283 pp.

Chengalath, R. & W. Koste, 1989. Composition and distributional patterns in arctic rotifers. Hydrobiologia 186/187 (Dev. Hydrobiol. 52): 191–200.

De Smet, W. H., 1994. Lepadella beyensi (Rotifera Monogononta: Colurellidae), a new species from the Canadian High Arctic. Hydrobiologia 294: 61–63.

De Smet, W. H. & J. M. Bafort, 1990. Contribution to the rotifers of the Canadian High Arctic. I. Rotifers from Little Cornwallis Island, Northwest territories. Naturaliste Can. (Rev. Ecol. Syst.) 117: 253–261.

De Smet, W. H. & L. Beyens, 1995. Rotifers from the Canadian High Arctic (Devon Island, Northwest Territories). Hydrobiologia 313/314: 29–34.

De Smet, W. H. & L. Beyens, 1998. Rotifers from Victoria Island (Northwest Territories, Canadian Arctic). Hydrobiologia, in press.

Dodson, A. & W. Thomas, 1964. Concentrating plankton in a gentle fashion. Limnol. Oceanogr. 9: 455–459.

Dolan, J. R. & D. W. Coats, 1991. A study of feeding in predacious ciliates using prey ciliates labeled with fluorescent microspheres. J. Plankton Res. 13: 609–628.

Dumont, H. J., 1977. Biotic factors in the population dynamics of rotifers. Arch. Hydrobiol. 8: 98–122.

Gilbert, J. J. & K. G. Bogdan, 1984. Rotifer grazing: In situ studies on selectivity and rates. In Meyers, D. G. & R. Strickler (eds),

Trophic Interaction within Aquatic Ecosystems. AAAS Selected Symposium 85: 97–133.

Holmquist, C., 1975. Lakes of northern Alaska and Nordwestern [sic] Canada and their invertebrate fauna. Zool. Jb. Syst. Bd 102: S. 333–484.

Kling, G. W., W. J. O'Brien, M. C. Miller & A. E. Hershey, 1992. The biogeochemistry and zoogeography of lakes and rivers in arctic Alaska. Hydrobiologia 240 (Dev. Hydrobiol. 78): 1–14.

McLaren, I. A., 1964. Zooplankton of Lake Hazen and a nearby ponds, with special reference to the copepod Cyclops scutifer Sars. Can. J. Zool. 42: 613–629.

Miller, M. C., G. R. Hater, P. Spatt, P. Westlake & D. Yaekel, 1986. Primary production and its control in Toolik Lake, Alaska. Arch. Hydrobiol./Suppl. 74. 1: 97–131.

Minns, C. K., 1977. Limnology of some lakes on Truelove Lowland. In Bliss, L. E. (ed.), Truelove Lowland, Devon Island, Canada: a High Arctic Ecosystem. The University of Alberta Press, Edmonton: 569–585.

Moore, J. W., 1978. Composition and structure of zooplankton communities in eighteen arctic and subarctic lakes. Int. Rev. ges. Hydrobiol. 63: 545–565.

Nogrady, T. & J. P. Smol, 1989. Rotifera from five high arctic ponds (Cape Herschel, Ellesmere Island, N.W.T.). Hydrobiologia 173: 231–242.

O'Brien, W. J., A. E. Hershey, J. E. Hobbie, M. A. Huller, G. W. Kipphut, M. C. Miller, B. Moller & J. R. Vestal, 1992. Control mechanisms of arctic lake ecosystems: a limnocorral experiment. Hydrobiologia 240 (Dev. Hydrobiol. 78): 143–188.

O'Brien, W. J., M. Bahr, A. Hershey, J. Hobbie, G. Kipphut, G. Kling, H. Kling, M. McDonald, M. Miller, & P. A. Rublee, 1997. The limnology of Toolik Lake. In Milner, A. M. & M. W. Oswood (eds), Chapter 3, Alaskan Freshwaters. Springer Verlag: 61–103.

Pourriot, R., 1977. Food and feeding habits of rotifers Arch. Hydrobiol. 8: 243–260.

Rieman, B. & K. Christoffersen, 1993. Microbial trophodynamics in temperate lakes. Mar. Microb. Food Webs 7: 69–100.

Rigler, F. H., M. E. MacCallum & J. C. Roff, 1974. Production of zooplankton in Char Lake. J. Fish. Res. Bd Can. 31: 637–646.

Rublee, P. A., 1992. Community structure and bottom-up regulation of heterotrophic microplankton in arctic LTER Lakes. Hydrobiologia 240 (Dev. Hydrobiol. 78): 133–142.

Rublee, P. A. & N. Bettez, 1995. Change of microplankton community structure in response to fertilization of an arctic lake. Hydrobiologia 312: 183–190.

Rublee, P. A. & C. L. Gallegos, 1989. Use of fluorescently labeled algae (FLA) to estimate microzooplankton grazing. Mar. Ecol. Progr. Ser. 51: 221–227.

Rublee, P. A. & A. Partusch-Talley, 1995. Microfaunal response to fertilization of an arctic tundra stream. Freshwat. Biol. 34: 81–90.

Sherr, B. F., E. B. Sherr & R. D. Fallon, 1987. Use of monodispersed, fluorescently labeled bacteria to estimate in situ protozoan bacterivory. Appl. environ. Microbiol. 53: 958–965.

Hydrobiologia **387/388:** 161–170, 1998.
E. Wurdak, R. Wallace & H. Segers (eds), Rotifera VIII: A Comparative Approach.
© *1998 Kluwer Academic Publishers.*

Rotifer vertical distribution in a strongly stratified lake: a multivariate analysis

X. Armengol, A. Esparcia & M.R. Miracle
Departament de Microbiologia i Ecologia, Universitat de València, 46100 Burjassot, Valencia, Spain

Key words: Rotifera, vertical distribution, temperature, oxygen, diversity, PCA, karstic lake

Abstract

The main source of variation of rotifer species distributions in lake Arcas-2, a small karstic lake near Cuenca (Spain), was explored by means of principal components factor (PCA) and canonical correlation (CCA) analyses. PCA was performed using rotifer densities and CCA using rotifer densities plus physical and chemical parameters. Factor 1 of PCA separated summer species from winter–spring species and Factor 2 accounted for the variation in the vertical profile. Three summer species with different food habits (*Polyarthra dolichoptera, Hexarthra mira* and *Asplanchna girodi*) were grouped together at the positive end of Factor 1, while Factor 2 separated the two hypolimnetic species (*Filinia hofmanni* and *Anuraeopsis fissa*) from the rest. The relative position of rotifer species in the space determined by the CCA was roughly the same. The most significant environmental factors that became paired with rotifer distribution in the CCA were temperature and oxygen, and parameters related to water inflow. Segregation of filter-feeding species in the spatio–temporal subenvironments is clearly shown by the multivariate analysis.

The low diversity of rotifer species found in Lake Arcas-2 is attributed to the reduced dimensions of the lake and its morphology. This lake resembles a sinkhole with an abruptly sloping shoreline and poor development of the littoral zone. This morphology favors a strong oxygen stratification. Since midsummer the oxic–anoxic boundary is located in the upper metalimnion, the vertical structure of the oxygenated water column is simplified. This low rotifer diversity contrasts with a high ciliate diversity in the anoxic waters.

Introduction

Many studies have demonstrated seasonal and vertical variation of the rotifer planktonic community within a lake. Rotifers frequently exhibit non-random specific distributions, but the question is how are the species combinations assembled from a common species pool? Rotifer communities are dynamic systems of interacting populations and, at a given time and depth, a distinct assemblage occurs. In order to identify these patterns of change in space and time, the rotifers of a small lake, Arcas-2, possessing a high degree of stratification were studied. In this lake, stratification can be established early in the year and an anoxic hypolimnion can develop by early spring (Vicente et al., 1991). When the thermocline descends in the second half of the summer, the oxicline coincides with it and the epilimnion almost contacts the anoxic hypolimnion.

In this oxic–anoxic interface light is still available so photosynthetic organisms develop and a high diversity of ciliates is found (Esteban et al., 1993; Finlay et al., 1991). This contrasts with the low diversity of zooplankton in the oxygenated upper waters. The aim of this paper is to describe the distributions of rotifers in this lake and to explore the interrelationships of the rotifer species with the main physical and chemical parameters.

Site description and methods

Lake Arcas-2 (UTM 30SWK 732276) is a small lake located in a field of dolines near the town of Arcas (Cuenca, Spain) developed by dissolution of gypsum-rich marls. Arcas-2 consists of two flooded circular dolines connected by a short, 1-m deep channel. It is a warm monomictic hard water lake (SO_4:CO_3:Cl approx. 100:10:1), with a mean conductivity of 2.5 mS cm^{-1} and an alkalinity of 4–5 meq l^{-1}. The waters are slightly turbid with a secchi depth ranging from 2 to 3.6 m. An almost constant water level is maintained by the water table and a small surface outflow. All samples were collected at the center of the larger basin, which is deep relative to its surface area (surface area, 0.16 ha; maximum depth, 14 m). This site corresponds with a small secondary sink hole situated near the center of this basin, which deepens the lake from 12 to 14 m. The boat was fixed at the intersection of two perpendicular ropes attached to the lake shores.

Detailed vertical profiles of temperature, conductivity and oxygen were obtained in situ using WTW meters. Secchi disk depth and light penetration were also measured. Redox and pH were measured by pumping the water through ORION meters and water samples were collected using a bi-conical inlet device connected to a peristaltic pump, as described in Miracle et al. (1992). In this way we were able to sample at short depth intervals (10 cm) when the lake was stratified. Zooplankton samples were obtained by filtering 2–4 l of the pumped water through 30-μm nylon mesh. During mixing (November and February), zooplankton samples were taken by filtering through the same mesh the contents of a double Van Dorn bottle (5.6 l). In both cases, the water with living animals was filtered slowly, at low pressure. The samples were preserved in 4% formalin and counted with an inverted microscope at 100 and 200× magnification. The Shannon–Wiener diversity index for each sample was calculated.

A principal components factor analysis (PCA) without rotation was performed on logarithmically transformed values of rotifer densities and a canonical correlation analysis (CCA) also was made using these variables along with log-transformed measurements (with the exception of pH) of main physical and chemical parameters. The 4M and 6M BMDP programs were used, respectively (Dixon et al., 1983). Pearson, pair-wise, correlation coefficients were used to show relationship between a larger set of environmental factors and the species densities, as well as the principal components derived from their interactions.

Results

Rotifers were always dominant, comprising more than 60% of the total zooplankton density during most of the year, except in winter and midsummer. Copepods, especially nauplii and copepodites, made up most of the remaining zooplankton. However, rotifer diversity in Arcas-2 is relatively low. There are only six numerically important species in the plankton: *Anuraeopsis fissa*, *Filinia hofmanni*, *Keratella quadrata*, *Asplanchna girodi*, *Hexarthra mira* and *Polyarthra dolichoptera*. Along with *P. dolichoptera* some individuals belonging to other species of the *P. dolichoptera-vulgaris* group were probably counted. However, all individuals of *Polyarthra* which duly identified belong to *P. dolichoptera*. Due to the difficulty to differentiate between the species of the *Polyarthra dolichoptera-vulgaris* group in fixed specimens, we merged all individuals into the *P. dolichoptera–vulgaris* group.

Some other rotifer species appeared in the plankton samples in low numbers. They are *Cephalodella forficula*, *Colurella obtusa*, *Colurella uncinata*, *Lecane bulla*, *Lecane closterocerca*, *Lecane flexilis*, *Lecane luna*, *Lecane lunaris*, *Lophocharis salpina*, *Notholca acuminata*, *Synchaeta oblonga*, and some unidentified Bdelloids.

The physical and chemical parameters showed marked gradients in the vertical profile (Figures 1a–c). Temperature was homogeneously distributed in winter (ca 6°C in February) and varied through the year, reaching 24°C at the surface in summer, but showing a strong variation with depth. Oxygen also remained homogeneously distributed in winter (ca 9 mg l^{-1} during the mixing period), but during the early stratification period its concentration in the metalimnion became supersaturated (>140%) and then dropped drastically to zero at the hypolimnion. As summer advanced, the thermocline descended, becoming sharper, so that by the end of the stratification period the oxygen profile was a typical clinograde curve. At this time, the oxicline was located at the beginning of the thermocline. In this way, an oxic–anoxic interface was established around a depth of 8 m in 1987, and 9 m in 1988. The conductivity in the lake is relatively high (mean conductivity, ca 2.5 mS cm^{-1}). Conductivity varied slightly in the vertical profile, with a small relative increase in the anoxic hypolimnion. It showed a marked variation during the year, with lower values during rainy periods. This variation is especially due to changes in sulfate and calcium concentrations. A deep chlorophyll a maximum is characteristic of this lake (Figure 1c), and consists mainly of *Cryptomonas* and *Oscillatoria*. However, this algal biomass is largely unavailable to the rotifer population, which is usually located above it.

163

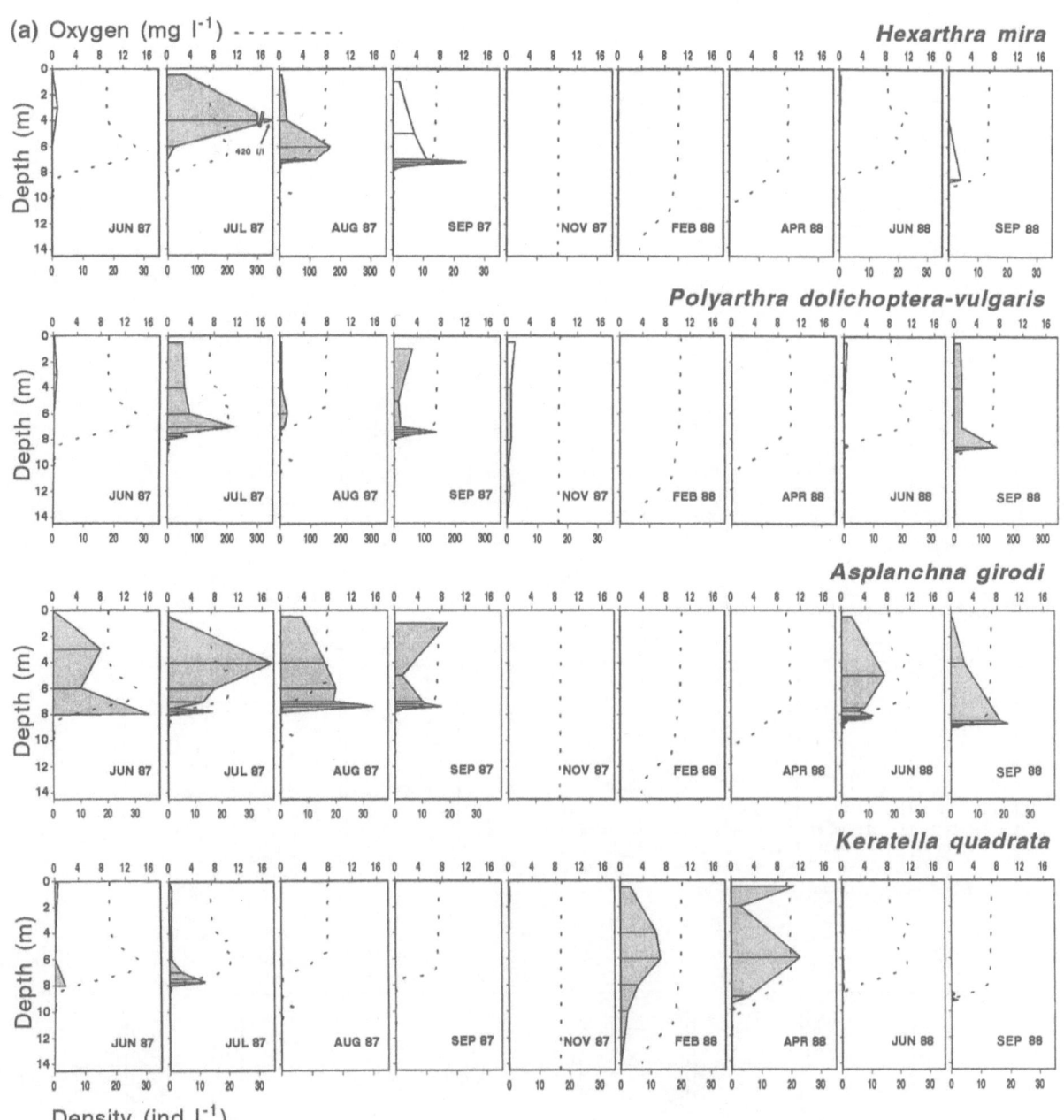

Figure 1a.

Figure 1. Vertical distribution of the main (a) epi- and metalimnetic rotifer species and (b) hypolimnetic ones at different dates from June 1987 to October 1988. For each species, non shaded areas indicate a smaller scale than the usual, absence of scale means that the species was not found in that date. Vertical profiles of oxygen are shown with dotted lines. (c) Vertical profiles of temperature, conductivity and chlorophyll *a* at the sampling dates of the years 1987 and 1988. (d) Vertical distribution at the sampling dates, from June 1987 to October 1988, of diversity (solid lines) and rotifer densities (dotted lines).

164

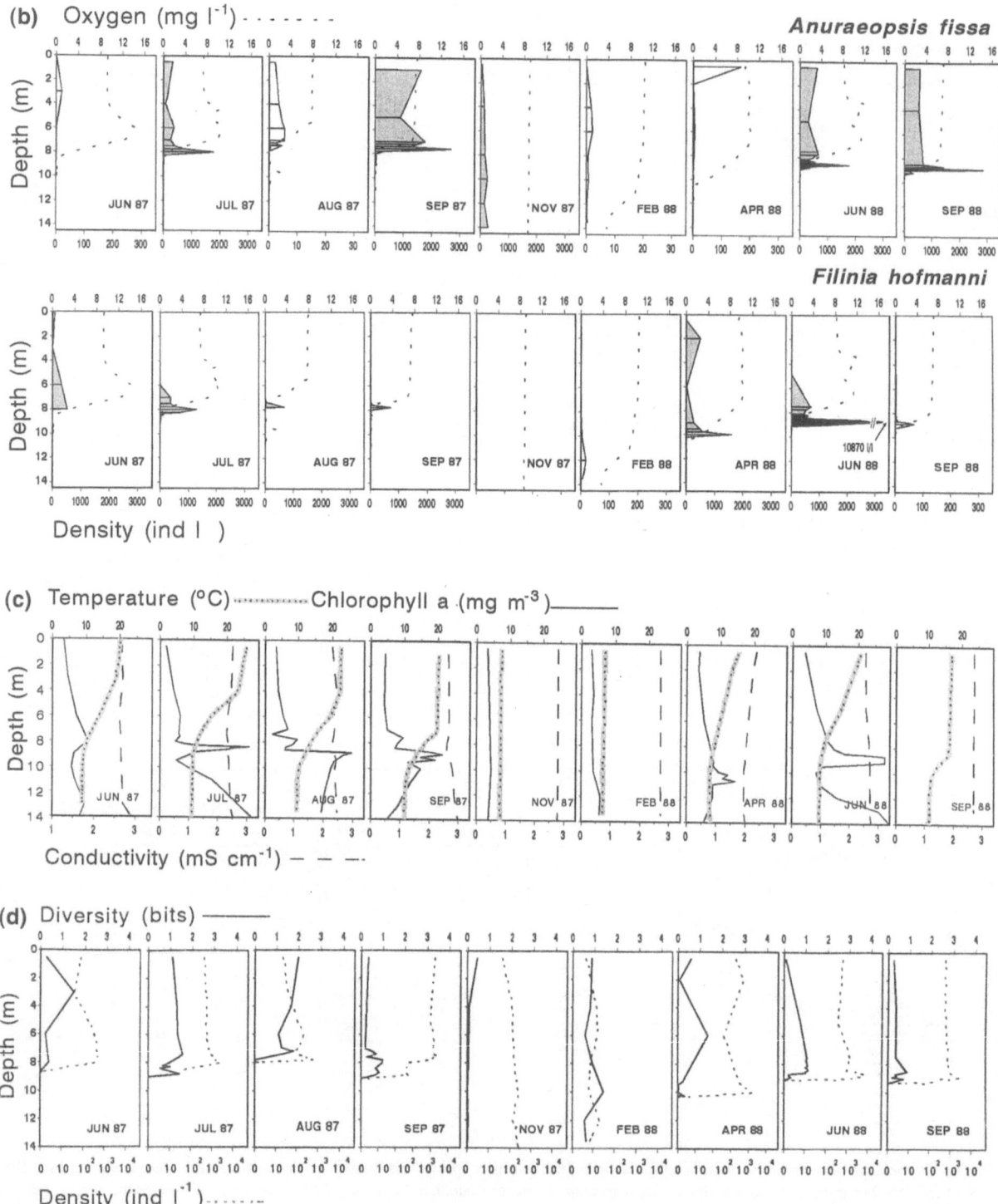

(b) Oxygen (mg l^{-1}) - - - - - -   *Anuraeopsis fissa*

Depth (m)

JUN 87   JUL 87   AUG 87   SEP 87   NOV 87   FEB 88   APR 88   JUN 88   SEP 88

Filinia hofmanni

Depth (m)

JUN 87   JUL 87   AUG 87   SEP 87   NOV 87   FEB 88   APR 88   JUN 88   SEP 88

10870 l/l

Density (ind l^{-1})

(c) Temperature (°C) ·········· Chlorophyll a (mg m^{-3}) ———

Depth (m)

JUN 87   JUL 87   AUG 87   SEP 87   NOV 87   FEB 88   APR 88   JUN 88   SEP 88

Conductivity (mS cm^{-1}) – – –

(d) Diversity (bits) ———

Depth (m)

JUN 87   JUL 87   AUG 87   SEP 87   NOV 87   FEB 88   APR 88   JUN 88   SEP 88

Density (ind l^{-1}) · · · · · ·

Figures 1b–d. For legend see p. 163.

Table 1. Correlation coefficients (r) between the main planktonic rotifer species plus the factors from PCA and some physical and chemical variables

Category	KERA	POLY	ANUR	HEXA	FILI	ASPL	Factor 1	Factor 2
Temperature	−0.22	**0.62***	**0.31**	**0.57***	−0.24	**0.50***	**0.57***	−0.05
Oxygen	**0.37**	0.15	−0.14	0.24	**−0.57***	0.16	0.09	**−0.56***
pH	**0.34**	−0.17	−0.21	−0.04	**−0.41**	0.01	−0.12	**−0.39***
Redox	0.11	0.18	0.09	0.20	−0.14	**0.28**	0.19	**−0.27**
Conductivity	−0.17	0.07	**0.42***	−0.14	−0.25	0.10	0.22	0.16
Alkalinity	−0.11	**−0.28**	−0.06	−0.20	−0.01	0.04	−0.16	0.08
Ca	0.03	**0.33**	−0.05	**0.32**	−0.02	0.13	0.26	−0.12
Cl	**−0.51***	**0.36**	**0.48***	0.24	0.01	0.14	**0.38**	**0.46***
K	0.23	**−0.47***	**−0.37**	**−0.35**	−0.23	−0.22	**−0.41**	**−0.34**
Mg	0.04	0.15	−0.13	0.15	0.18	0.14	0.09	−0.02
Na	−0.02	**0.32**	0.06	0.24	0.13	0.04	0.20	−0.05
NH$_3$	−0.10	**0.49***	0.18	**0.42***	0.12	0.20	**0.41**	0.11
NO$_2$	−0.03	**−0.30**	−0.22	**−0.29**	0.31	**−0.43***	**−0.42***	0.10
NO$_3$	**0.33**	−0.14	−0.18	−0.05	−0.18	−0.04	−0.11	**−0.35**
P soluble	−0.22	**−0.26**	−0.11	**−0.37**	0.23	**−0.52***	**−0.41**	0.23
Silic	0.16	**0.35**	0.23	**0.39**	**−0.34**	0.19	**0.41**	−0.22

Values printed in bold were significant ($P \leq 0.5$) in pair-wise correlations; values marked with an asterisk were significant ($P \leq 0.5$) in a Bonferroni sequential test.
KERA, *Keratella quadrata*; POLY, *Polyarthra dolichoptera–vulgaris*; ANUR, *Anuraeopsi fissa;* HEXA, *Hexarthra mira*; FILI, *Filinia hofmanni*; ASPL, *Asplanchna girodi*.

The vertical distribution of the six most numerous rotifer species, together with the oxygen profiles is shown in Figure 1a, b. The strong stratification of the lake during summer is reflected in an associated stratification of the zooplankton, especially rotifers. In this study, most species showed a marked tendency to develop peak abundances near the oxicline.

Filinia hofmanni and *Anuraeopsis fissa* were the most abundant species in this lake and formed important density peaks near the oxicline. The two nevertheless had segregated distributions. *F. hofmanni* is a late spring, early summer species which reached its maximum densities at the onset of the stratification period, whereas *A. fissa* was present during the whole study period, but with maximum density at the end of the stratification period. *A. fissa* also was more widely distributed over the vertical profile while *F. hofmanni* was more restricted to the hypolimnion. Both species presented peak abundances near the oxicline, but the maximum of *A. fissa* was generally above that of *F. hofmanni*.

Polyarthra dolichoptera–vulgaris also showed marked peak abundances at the oxicline in summer. Their density was relatively important during the autumn overturn. *Hexarthra mira* occurred in the upper metalimnetic waters during summer. *Keratella quad-*

rata was the least abundant species, and occurred all over the oxic layer during winter and spring; it also occurred at low densities during early summer, where its maximum density was reached above the layer with maximum density of *A. fissa*.

Asplanchna girodi was present from late spring until the autumn overturn. It dwelled mainly in epilimnetic waters, but also formed peaks in the hypolimnion during stratification periods, where it preyed on the dense *A. fissa* and *F hofmanni* population, as testified by its stomach contents.

In Figure 1d, we show the vertical distribution and seasonal variation of diversity compared with the density of rotifers. In each profile we can see an inverse relationship between both parameters. Diversity clearly increases near the bottom during stratification, although in the oxiclinal layers there was a marked drop in diversity corresponding with the density peaks found at these layers. After the autumn overturn, diversity decreased probably due to the lower temperature and the mixing with anoxic hypolimnetic waters. In winter, density decreased and diversity recovered, although the number of species present remained low.

Temperature was the parameter that showed the highest correlation coefficients with the abundance of all rotifer species (Table 1). It was negatively cor-

Table 2. Weighted averages for several physical and chemical parameters in relation to the densities of each rotifer species during the studied period

Species	Temp.	Oxygen	pH	Redox	Conductivity	Chlor *a*
A. girodi	15.8	4.7	7.57	361	2.54	6.2
K. quadrata	10.6	7.9	7.81	330	2.41	4.7
H. mira	20.5	6.6	7.55	358	2.43	4.8
P. dolichoptera–vulgaris	17.5	4.9	7.50	354	2.58	4.6
A. fissa	14.6	3.5	7.50	310	2.64	7.4
F. hofmanni	9.4	1.1	7.49	205	2.59	25.8

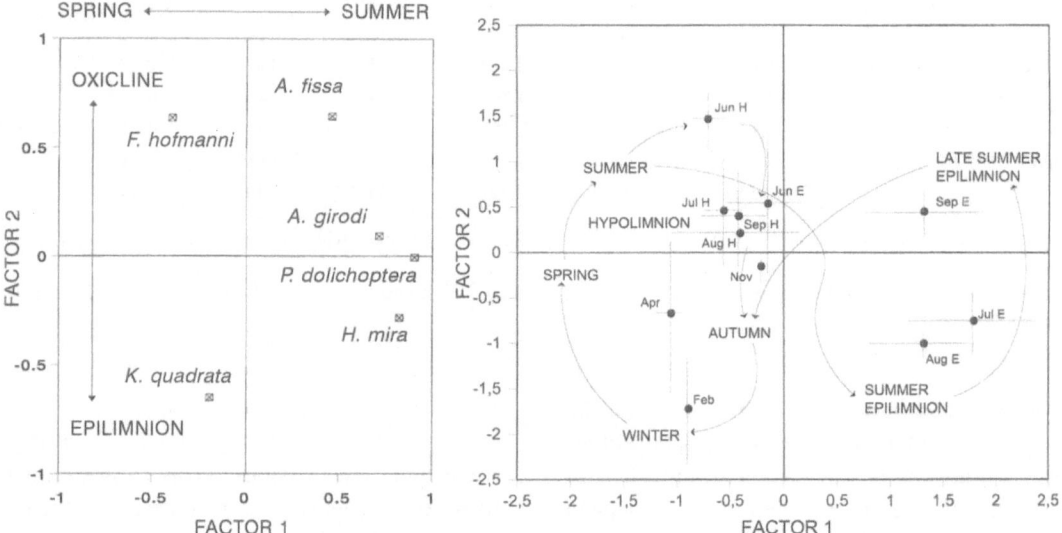

Figure 2. (a) Relative position of each rotifer species in the plane dimensioned by the first and second factors extracted by PCA, performed on rotifer densities during the studied period. Coordinates are the factor loadings or correlation coefficients. (b) Ordination of the samples in the plane dimensioned by the first and second principal factors; the points are the means with their standard deviations of samples grouped by months and by strata during stratification: epimetalimnion (E) and hypolimnion (H). June (Jun) and September (Sep) are means of samples taken in both years, 1987 and 1988.

related with the winter species (*K. quadrata*), and the species restricted to the oxicline (*F. hofmanni*), and positive correlated with summer species and with species occurring mainly in the epilimnion. Oxygen was positively correlated with abundance of epilimnetic species (*K. quadrata* and *H. mira*), and negatively correlated with abundance of *F. hofmanni*.

Temperature, oxygen, and chlorophyll *a* were the parameters with greatest differences in weighted average between the species. A typical hypolimnetic species, *F. hofmanni*, presented the lowest value for temperature (Table 2), second was *K. quadrata*, a species occurring mostly in winter and spring. On the other hand, *H. mira* (a summer species) had the highest weighted average for temperature. The oxygen clearly

separated *F. hofmanni* from the rest of species that showed a gradient of oxygen averages. Also, large differences were found with respect to chlorophyll *a*, separating *Filinia* from the rest of species, because of the correlation between chlorophyll and depth.

The PCA on 77 samples defined by species densities indicated that there were two eigenvalues greater than unity. So, only the first two factors, accounting for 40 and 22% of the total variance, respectively, were considered. Factor 1 separates summer species from winter–spring species and Factor 2 accounts for the variation of rotifer species distribution in the vertical profile (Figure 2a). Three summer species with different food habits: *P. dolichoptera, H. mira* and *A. girodi*, became grouped together in the principal factors space

Table 3. Canonical correlation of the three main cannonical variates obtained in CCA indicating also their redundancy index for species given abiotic factors

Canonical variates	Canonical correlation	Redundancy index
1	0.89	0.31
2	0.69	0.08
3	0.61	0.06

and the other three positioned at the opposite extremes of the space, indicating their differential distribution in the spatial and temporal subenvironments. Factor 2 clearly separates hypolimnetic species, *F. hofmanni* and *A. fissa*, from the more epilimnetic ones. This is confirmed by the correlations of these factors with physical and chemical parameters as shown in Table 1. Factor 1, positively correlated with temperature and negatively with soluble phosphorous, other nutrients and chlorophyll, is associated with the prevailing conditions in the summer epilimnion. Factor 2, negatively correlated with oxygen, pH and redox and positively with nitrates and some mineral components, may be associated with prevailing conditions in hypolimnetic layers. The analysis of the values of the components for individual samples shows that variability in rotifer assemblages occurs mainly in the epi-metalimnion, while changes are smaller in the layer near the oxic–anoxic interface. In order to simplify its representation, sample values have been pooled together by month and strata during stratification, and the means (± standard deviations) corresponding to these groups of samples, have been plotted in the space of the principal factors (Figure 2b). The samples in the space dimensioned by the first two factors from this analysis become clearly ordered by temperature in the first axis and by oxygen in the second axis. Temperature varies through the year but also presents a strong variation through the vertical profile in summer when thermal stratification occurs. So, the positive part of this axis contains summer epi- and metalimnetic samples, whereas oxiclinal and winter samples are in the negative part. In the positive part of the second axis are located the samples where decomposition importance is higher (i.e., hypolimnion and late summer).

The results of the CCA indicate that there are three pairs of canonical variates with high correlations (Figure 3, Table 3). The first canonical relationship has a high redundancy index or overlap, as seen from the side of rotifer distribution as added to an already available physical and chemical distribution. This index evaluates what proportion of variance in rotifer counts is explained by the first canonical variate derived from the abiotic parameters. Inspection of the loadings of the two sets of variables with the canonical variates represented in Figure 3, reveals that the first canonical variates are mainly related to temperature. Thus, *K. quadrata*, a typical winter–spring species in this work, and *F. hofmanni*, a spring–summer species restricted to the hypolimnion, were represented at the negative side, whereas the rest of the species were placed at the positive end. This first canonical variate for rotifers coincides with the first principal component of Figure 2. The second and third pairs of canonical variates although related by high coefficients, do not have high redundancy values, because the large canonical correlation coefficient is the result of a strong correlation of just a few rotifer species with just one or two variables in the abiotic factors set. The second canonical relation pairs inversely oxygen and pH with the hypolimnetic species *F. hofmanni*. The third canonical pairs conductivity with *A. fissa*, a typical species for the end of the stratification period. The second factor of PCA roughly combines these two canonical relations. The second and third canonical variates mainly indicate the species distributions in the vertical profile in spring–early summer and summer–autumn, respectively.

Discussion

The vertical heterogeneity during stratification promotes an increase in diversity in the gradient layers even when sampled in short depth intervals as in this study (Figure 1d). However, the species distributions are clearly separated spatially or seasonally. The steep gradients of the main abiotic parameters in the lower metalimnion permit a separation of the maxima of the different coexisting species. The spatio–temporal separation of rotifer species can clearly be seen in PCA results (Figure 2a) and in Table 2 where it is shown that the weighted mean of the main abiotic parameters is different for each species. The species relative position in the space, dimensioned by the first two factors from PCA, is quite similar to their position in the space dimensioned by the first and third canonical variate. The similarity is greater if we integrate the results of second canonical variate accounting for the high inverse relation of *F. hofmanni* with oxygen. This

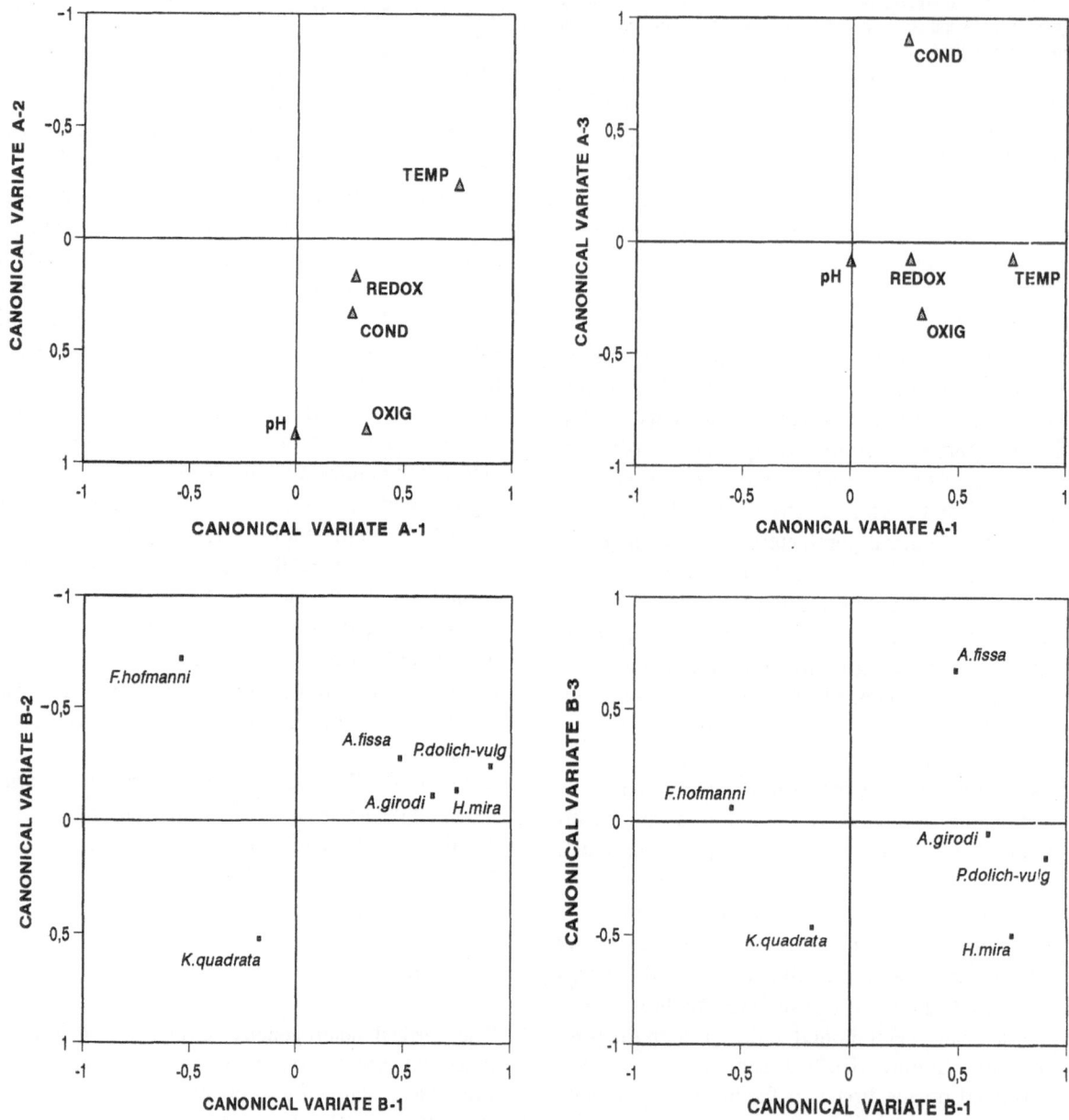

Figure 3. Relative position of each species and main environmental variables (temperature, conductivity, oxygen, redox and pH) in the plane dimensioned by the first three pairs of canonical variates. Coordinates are the respective loadings or correlation coefficients.

indicates that the underlying factors, which account for most of the variability of the rotifer distributions, are the same determinants of the seasonal and vertical variation in abiotic conditions. The different species which have a slightly different optima are segregated in the space or succeed each other in time and are positioned on opposite sides of the principal components space. The same results were obtained in other lakes (Miracle, 1974).

The first two factors obtained through PCA are correlated mainly with temperature and oxygen, both parameters are the main markers for explaining winter–summer and surface–oxicline variations. Temperature and oxygen have been found to be the most important parameters for the explanation of spatial and temporal variation in rotifer (Bogaert & Dumont, 1989; Mikschi, 1989). These parameters were seen as determinants in the segregation of congeneric species of rotifers (Armengol et al., 1993; Esparcia et al., 1991). However, they are only markers of the seasonal and vertical variation, which have other changes associated, derived from succession and from the separation between production and decomposition in the water column. May (1983) studied the relation between the seasonal occurrence of rotifer species and found that temperature affected the temporal segregation of species. However, among cold-water species there were eurytherms showing seasonal variations (e.g., *K. quadrata*). Such species normally occur in winter–spring, but may occasionally be found during summer. In Arcas-2, *K. quadrata* always shows a winter–spring occurrence, in addition to a small hypolimnetic peak in early summer. In other lakes of the same region, the species occurs in late summer, with maximum near the oxic–anoxic interface (Armengol et al., 1993; Miracle et al., 1993). May (1983) pointed out that other factors like food availability may account for the species distributions. Maybe competition with other filter-feeding taxa should also be considered in this respect. This may be the case here for *A. fissa*, which is almost always present in the plankton, but only attains great abundance at the end of summer, when the oxicline reaches its highest temperature, and *A. fissa* can dominate over *F. hofmanni*. Furthermore, at this time strong stratification develops in deep waters and favors mineral gradients. Water inflow–outflow is insignificant and evaporation important. This accounts for the relation of *A. fissa* with conductivity, whereas *F. hofmanni* and *Keratella* dwell in periods of higher water flux.

This structure of the rotifer populations appears quite constant, as it was also evident during a similar study in 1990–1991 (Miracle & Armengol, 1995). The latter study dealt mainly with the population dynamics of *A. fissa* and *F. hofmanni*, and yielded results on the spatial segregation of the two that are essentially similar to the ones presented here (see Miracle & Armengol, 1995).

Lake Arcas-2 shows a low species diversity, possibly due to: (1) its conditions of mineralization with high sulfate contents, (2) the isolation of the wetland within a dry region, (3) human impact with the establishment of agricultural land almost to its shore and the introduction of fish, and (4), most importantly, its small dimensions and particular morphology. This morphology, with steep slopes, does not permit the development of macrophyte vegetation, and promotes development of stratification starting in early spring. As the lake is not very deep, the hypolimnion easily becomes anoxic, a situation which occurs in mid-summer when the thermocline deepens and becomes more pronounced. At the same time, an oxicline develops in the upper part of the thermocline. The peculiar conditions of the metalimnion, with anoxic water almost in contact with the epilimnion, favors microphagous, *r*-strategist species like as *A. fissa* and *F. hofmanni*. This simplification of the oxygenated water column could be one of the main reasons for the low diversity in Lake Arcas-2. It is especially interesting to note the almost complete absence of congeneric species with pronounced niche specificity, a feature frequently found in many lakes (Miracle, 1977). Furthermore, a wide layer of anoxic water could make colonization through resting eggs in summer difficult, and restrict the development of the typical succession of summer species, some of which belong to congeneric clusters (*Polyarthra*, *Hexarthra*, *Synchaeta*). The low rotifer diversity contrasts with a high ciliate diversity in the anoxic waters, due to the important development of photosynthetic stratifying alga and bacteria which show small scale vertical distributions in the oxic–anoxic boundary (Esteban et al., 1993; Finlay et al., 1991; Vicente et al., 1991).

References

Armengol, J., A. Esparcia, E. Vicente & M. R. Miracle, 1993. Vertical distribution of planktonic rotifers in a karstic meromictic lake. Hydrobiologia 255/256: 381–388.

Bogaert, G. & H. J. Dumont, 1989. Community structure and co-existence of rotifers of an artificial crater lake. Hydrobiologia 186/187: 167–179.

Dixon, W. J., M. B. Brown, L. Engelman, J. W. Frane, M. A. Hill, R. I. Jennrich & J. D. Toporek, 1983. BMDP statistical software. Printing with additions. Univ. California Berkeley, 773 pp.

Esparcia, A., J. Armengol, E. Vicente & M. R. Miracle, 1991. Vertical distribution of Anuraeopsis species as related to oxygen depletion in two stratified lakes. Verh. int. Ver. Linmnol. 24: 2745–2749.

Esteban, G., B. J. Finlay & T. M. Embley, 1993. New species double the diversity of anaerobic ciliates in a Spanish lake. FEMS Microbiol. Lett. 109: 93–100.

Finlay, B. J., K. J. Clarke, E. Vicente & M. R. Miracle, 1991. Anaerobic ciliates from a sulphiderich solution lake in Spain. Eur. J. Protistol. 27: 148–159.

May, L., 1983. Rotifer occurrence in relation to water temperature in Loch Leven Scotland. Hydrobiologia 104: 311–315.

Miracle, M. R., 1974. Niche structure in freshwater zooplankton: a principal components approach. Ecology 55: 1306–1316.

Miracle, M. R., 1977. Migration, patchiness, and distribution in time and space of planktonic rotifers. Arch. Hydrobiol. Beich. 8: 19–37.

Miracle, M. R. & J. Armengol, 1995. Population dynamics of oxiclinal species in lake Arcas-2 (Spain). Hydrobiologia 313/314: 291–301.

Miracle, M. R., E. Vicente, R. L. Croome & P. A. Tyler, 1991. Microbial microcosms of the chemocline of a meromictic lake

170

in relation to changing levels of PAR. Verh. int. Ver. Limnol. 24: 1139–1144.

Miracle, M. R., E. Vicente & C. Pedrós-Alió, 1992. Biological studies of spanish meromictic and stratified karstic lakes. Limnetica 8: 59–77.

Miracle, M. R., J. Armengol & M. J. Dasi, 1993. Extreme meromixis determines strong differential planktonic vertical distributions. Verh. int. Ver. Lirrmol. 25: 705–710.

Mikschi, E., 1989. Rotifer distribution in relation to temperature and oxygen content. Hydrobiologia 186/187 (Dev. Hydrobiol. 52): 209–214.

Vicente E., M. A. Rodrigo, A. Camacho & M. R. Miracle, 1991. Phototrophic procaryotes in a Karstic sulphate lake. Verh. int. Ver. Limnol. 24: 998–1004.

Hydrobiologia **387/388**: 171–178, 1998.
E. Wurdak, R. Wallace & H. Segers (eds), Rotifera VIII: A Comparative Approach.
© *1998 Kluwer Academic Publishers.*

Influence of environmental factors on the rotifer assemblage in an artificial lake

Miloslav Devetter

Faculty of Biology, University of South Bohemia, Branišovská 31, 370 05, České Budějovice * *and Hydrobiological Institute, Czech Academy of Science, Na sádkách 7, 370 05 České Budějovice, Czech Republic*

Key words: Rotifera, redundancy analysis, reservoir, bottom-up factors, top-down factors, competition, multidimensional analysis, time series

Abstract

Seasonal changes of the plankton rotifer community in an eutrophic Czech reservoir were evaluated in relation to 46 environmental variables. To do this, data of rotifer abundance from three growing seasons (1993 – 1995) were analyzed. The seasonal dynamics of rotifers in all three years were characterized by two distinctive aspects: (1) the spring peak, with both maximum density and maximum species diversity, was dominated by *Keratella cochlearis*, *K. hiemalis*, *K. quadrata* and *Polyarthra dolichoptera*; (2) the summer-autumnal peak (or several lower peaks) of about half the intensity of the spring one, was composed mainly of *Keratella cochlearis*, *Trichocerca similis* and *Polyarthra vulgaris*. The separation between these two peaks coincided with the decline of phytoplankton and development of a clear-water phase in this reservoir. In redundancy analysis, species-abundance data for rotifers were related to all measured environmental variables. Date, abundance of *Cyclops vicinus*, total nitrogen, primary production, surface temperature, and density of heterotrophic nanoflagellates were identified as the most important variables. Partial redundancy analysis was used to assess the significance of pure and date-structured environmental factors influencing rotifers during the season. Date-structured environmental factors (such as physical and chemical variables, food, competition, and predation) significantly affected the rotifer community. This study shows that the rotifers in the reservoir are controlled by both abiotic and biotic factors.

Introduction

Study of ecosystem regulation is a perennial theme in ecology, and from such studies we know that communities are influenced by a variety of physical, chemical, and biological factors (Lehman, 1991). One of the main problems in ecology is to untangle the interactions among these factors and to measure their relative importance.

The spatial structuring of natural communities has an inherent problem: what is the relative contribution of different environmental factors whose interactions often result in acomplex spatial effect (Borcard et al., 1992)? The relative role of different ecological forces may vary among biological systems or within the same system (Hunter & Price, 1992). For rotifers their intermediate trophic position within the food web of lake also makes them responsive to bottom-up as well as to top-down ecosystem processes (Hunter &

Price, 1992; McQueen et al., 1986; Pinel-Alloul et al., 1995). Here I explore these community-level questions by describing a study that combined time, and abiotic and biotic variables into a single model. The aims of the present study are three-fold: (1) to analyze the influence and relative impact of these factors on the rotifer community; (2) to compare bottom-up and top-down controlling factors, and (3) to study competitive interactions of the rotifer community. The relative importance of these factors in determining rotifer community structure is explored and tested. Tracking a community in a time series as a primary data set for multivariate analysis is not widely used for assessing the importance of environmental variables. However, this technique has the potential to separate effects of time from other environmental variables. A similar approach was used for an assessment of phytoplankton communities in times-series by Lepš et al. (1990).

Materials and methods

Site description

The influence of environmental variables on rotifer assemblages was investigated in the Římov Reservoir, a meso-eutrophic, dimictic reservoir on the Malše River in South Bohemia, Czech Republic. The reservoir, located at an altitude of 470 in, has an area of 2.06 km^2 and a canyon-like shape (long and steep-sided). The reservoir has a volume of 34.5×10^6 m^3, a mean depth of 16.5 m and a maximum depth of 45 m. The mean retention time of the reservoir is about 100 days. The reservoir is regularly monitored by the Hydrobiological Institute, Czech Academy of Science since its filing in 1979 (Sed'a & Kubečka, 1997). The sampling site is located about 250 m from the dam at the deepest site of the reservoir profile. A detailed description of the study area is given by Brandl et al. (1989).

Sampling methods

Rotifers were sampled at 3-week intervals, during the whole season (excluding the winter period) in 3 years (1993–1995). Quantitative samples of microzooplankton were taken by vertical hauls from the bottom, using mesh plankton net (40 μm mesh size, mouth diameter 20 cm) with Apstein conus. Large zooplankton were sampled simultaneously using a net with 200 μm mesh size (Sed'a & Kubečka, 1997). Samples were preserved in 4% formaldehyde solution and processed by direct microscopic counting of several subsamples.

At the same time other characteristics of the reservoir environment were sampled or measured: dissolved oxygen, temperature, and pH were measured in the field from the surface layer by WTW oxygen electrode, termistor, and pH comparator, respectively. The Secchi transparency also was measured. Total nitrogen is shown as the sum of NO$_3$-N, NO$_2$-N, NH$_3$-N, and organic-N (N-org). Methods of sampling and processing of NO$_3$-N and NO$_2$-N are described in Procházková (1959), of NH$_3$-N in Kopáček & Procházková (1993) and of N-org in Procházková (1960). Total phosphorus was analysed as in Kopáček & Hejzlar (1993), soluble reactive phosphorus (SRP) as in Murphy & Riley (1962), SO$_4^{2-}$ as by Procházková (1961), Cl$^-$ colorimetrically with HgCl$_2$, and Ca$^+$ and K$^+$ by ICP spectrometry. Particulate organic carbon (POC) and dissolved organic carbon (DOC) were measured using a liquid TOC Foss Heraeus analyzer. Methods of sampling and processing of BOD and direct numbers of bacteria are described in Straškrabová et al. (1993), of phytoplankton biomass and densities of phytoplankton groups in Komárková (1993) and of protozoans in Šimek et al. (1996). The data of fish assemblage was not included in this analysis (Matěna, 1995). Rotifers were identified using Koste (1978).

Data analyses

Statistical analyses were based on 36 samples, 30 species, and 44 environmental variables (i.e., 19 physical and chemical; 15 food; 5 competition; 5 predator). The environmental variables represent factors (nutrients, crustacean species, predator species, etc.) potentially influencing the rotifer community. Data were analysed using the redundancy analysis (RDA) method (performed with the CANOCO 3.12 software: ter Braak, 1994; ter Braak & Prentice, 1988; ter Braak & Verdonschot, 1995). The effect of environmental variables on the whole rotifer data matrix was studied. Log transformation and standardization by samples of the rotifer species densities were applied. In all of the analyses, influence of three different years was partialled-out (as a covariable). Because of the large number of independent variables involved in the analysis, a forward selection procedure method was used to find the smallest set of environmental variables that might explain the composition of the rotifer community. In this approach, the amount of variability explained by individual environmental variables was assessed. The significance of this variability was tested by the Monte Carlo permutation test (restricted for time series) with 999 random permutations. As alternative approach, the most important environmental factors were chosen for the model. In this way, two alternative sets of environmental variables were selected: all significant variables and variables chosen by forward selection (this method is more conservative, because it removes collinearity between variables).

Partial redundancy analysis was used to compare the influence of the effect of physical–chemical environmental factors with potential food effects (bottom-up), predator effects (top-down), and competition effects. The approach is described in Borcard et al. (1992). In this method, the effects of each set of variables (as covariables) are alternatively removed and the ordination is performed with the other set as explanatory variables. This allows estimation of the percentage of variation explained by individual types of effects. The significance of influence of each data set was estimated by the Monte Carlo permuta-

---

173

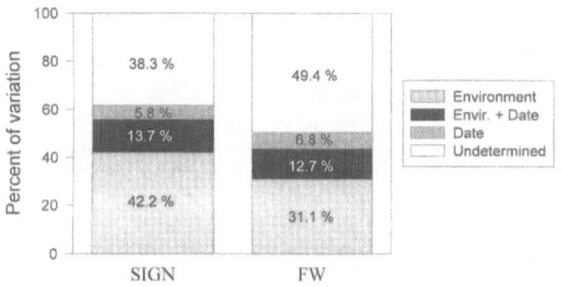

Figure 1. Variation partitioning of rotifer data matrix, using two different sets of environmental variables. SIGN, approach using all the significant environmental variables; FW, approach using the environmental variables, retained by forward selection method.

tion test, always applied with correction for temporal autocorrelation and the blocks repeated by the season.

Results

The seasonal composition of the rotifer community species was mainly characterized by consecutive dominance of *Polyarthra dolichoptera, Keratella cochlearis, Conochilus* sp. and *Trichocerca similis* followed by *Keratella cochlearis* again. The seasonal dynamics of rotifers is usually bimodal, with highest abundances in the spring (about 50 000 ind/dm^2) and a smaller peak in the autumn (about 20 000 ind/dm^2).

Influence of environmental variables

After the forward selection procedure on the spatial and environmental data matrices, two sets of individual environmental factors were selected. The six physical–chemical factors, four zooplankton factors, seven food factors (phytoplankton, protozoa, and bacteria) and date were found to have a significant impact (Table 1). The forward selection selected date, total nitrogen, temperature, pH, density of HNF, density of *Cyclops vicinus* and colour-less algae as the variables explaining most of the variance in the rotifer composition data. Figure 1 shows the relative importance of the pure environmental effect, the date-structured environmental effect, the pure date effect (a percentage of variability, explained when the whole effect of date is removed), and of the unexplained variability, for sets of environmental variables determined by the two alternative approaches. In the set of significant environmental variables, the contribution of the pure effect of the 17 environmental variables was 42.2% of variability ($P<0.01$). Total environmental contribution (the

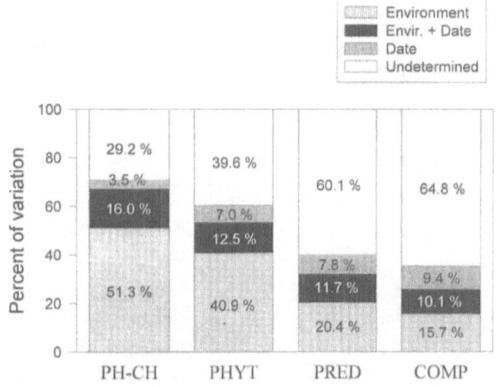

Figure 2. Variation partitioning of rotifer data matrix based on four independent data sets. PH-CH: set with the physical and chemical variables; PHYT, set with the phytoplankton (including Protozoa and Bacteria) variables; PRED, set with the invertebrate predators; COMP, set with the crustacean competitors.

influence of date not excluded) was 55.9%. The contribution of the pure date effect was 5.8% ($P<0.01$). In the set of environmental variables selected by forward selection approach, the situation is comparable: i.e., the pure contribution of six environmental variables is 31.1% of the explained variability ($P<0.01$), the total environmental contribution is 43.8% ($P<0.01$), and the pure date effect contribution is 6.8% ($P<0.01$). This indicates, that part of the effect of environmental variables on rotifers is independent of date. Nevertheless, a large part of the variability is unexplained (38.3 and 49.4%, respectively). It might depend on the other external factors (spatial heterogeneity, community dynamics, local factors etc.), not taken into account in this study.

Comparison of top-down and bottom-up effects

The independent effects of each set of environmental variables and their interactions with date were also analysed. Figure 2 illustrates the relative importance of the pure environmental effect, date-structured environmental effect, pure date effect, and undetermined variability. The total effect of physical and chemical variables explain 67.3% ($P<0.01$), of phytoplankton variables explain 53.4% ($P<0.05$), competition variables explain 25.8% ($P<0.01$), and of predation variables explain 32.1% ($P<0.01$) of variability of the whole rotifer species-composition data matrix. After excluding the effect of date, only the effect of predation was significant ($P<0.01$). The fact that the pure environmental effects of physical and chemical factors, phytoplankton, and competition were not

174

Table 1. Variance explained by each environmental variable, tested by RDA method

Type of environmental factors	Variable	Explained variance	Significance P value	FW selected
Temporal	Day of year	19.8	0.01 **	X
Physical and chemical	Total N	16.7	0.01 **	X
	Alkalinity	15.6	0.05 *	X
	pH	11.5	0.01 **	
	POC	9.4	0.02 *	
	N-org	9.4	NS	
	Temperature	9.4	0.05 *	X
	DOC	6.3	NS	
	Total P	6.3	NS	
	Dissolved oxygen	4.2	NS	
	BOD	4.2	NS	
	Soluble reactive P	3.1	NS	
	Transparency	3.1	NS	
	Conductivity	2.1	NS	
	SO_4^{2-}	2.1	NS	
	Cl^-	2.1	NS	
	Mg^{2+}	2.1	NS	
	Ca^{2+}	2.1	NS	
	Na^+	1.0	NS	
	K^+	1.0	NS	
Phytoplankton, protozoa and bacteria	Colour-less algae	9.4	0.01 **	X
	Chrysophyceae	8.3	0.01 **	
	Ciliata	6.3	0.02 *	
	Phytoplankton <20 μm	6.3	0.03 *	
	HNF	6.3	0.04 *	X
	Desmidiaceae	6.3	0.04 *	
	Total phytoplankton biomass	5.2	0.03 *	
	Volvocaceae	5.2	NS	
	Cyanophyceae	4.2	NS	
	Bacillariophyceae	3.1	NS	
	Bacteria	3.1	NS	
	Euglenophyceae	3.1	NS	
	Cryptophyceae	2.1	NS	
	Dinophyceae	2.1	NS	
	Chlorococaceae	2.1	NS	
Invertebrate predators	*Cyclops vicinus*	16.7	0.01 **	X
	Leptodora kindtii	9.4	0.01 **	
	Mesocyclops leuckarti	6.3	NS	
	Cyclops strenuus	5.2	0.04 *	
	Thermocyclops crassus	5.2	NS	
Crustacean competitors	*Diaphanosoma brachyurum*	11.5	0.01 **	
	Eudiaptomus gracilis	7.3	NS	
	Ceriodaphnia sp.	5.2	NS	
	Bosmina sp.	3.1	NS	
	Daphnia galeata	2.1	NS	

The results of forward selection are shown in the last column.
**$P \leq 0.01$; *$P \leq 0.05$; NS, not significant; X, selected by forward selection.

significant indicates that the rotifer variation is generally significantly linked with date-structured environmental factors. However, the impact of predation on the rotifer community is clearly independent of date. Influence of the pure date variable was not significant only for the physical and chemical factors (others $P<0.01$). The most important set influencing the rotifer community was the total physical and chemical factors (67.3%) and the least important was competition (25.8%). The unexplained fraction remains relatively high (from 29.2 to 64.8%) in all four cases.

Results of the RDA method for each set of environmental variables are shown in Figure 3. In the RDA ordination biplots, both species (thin arrows) and environmental variables (thick arrows) are depicted. Factors with significant influence are depicted with solid arrows, the others as dashed arrows. These species–environment biplots allow one to predict the responses of each rotifer species to the main gradients of the environment. In the ordination of physical and chemical factors (Figure 3a), the first two ordination axes explained 35.3% of species–environment variability. In the ordination diagram, the band of arrows in upper left corner represents a group of significant factors (N-org., POC, alkalinity and pH), which are probably correlated to primary production in the reservoir. The biplot shows, that the species *Synchaeta* sp. (Synch), *Notholca squamula* (N. squam), *Polyarthra dolichoptera* (P. dolich), *Brachionus angularis* (B. angul), and both *Keratella quadrata* (K.quad) and *K. hiemalis* (K.hiem), occurring in spring, are negatively correlated with high primary production by Desmidiaceae and blue-greens (alkalinity) and water temperature. Conversely *Trichocerca similis* (T.sim) and *Pompholyx triloba* (Pomph) are correlated positively with these variables. *Conochilus* sp. (Conoch) and *Polyarthra vulgaris* (P.vulg) are correlated with total nitrogen and *Lepadella ovalis* (Lep.ov). *Polyarthra major* (P.maj), *Filinia longiseta* (F.long) and *Lecane lunaris* (L.lunar) are probably correlated with temperature as a complex factor. The other taxa are depicted as shown (*Kellicotia longispina* (Kell.lon), *Colurella* sp. (Colurel), *Keratella ticinensis* (K.tic).

Figure 3b shows ordinations with variables describing the potential food resources. The first two ordination axes in this case explain 26.1% of the species–environment variability. The positions of *Keratella quadrata*, *Synchaeta* sp., *Notholca squamula* and *Polyarthra dolichoptera* in the ordination shows their preference for small algae, such as Chrysophyceae and colour-less algae. *Trichocerca similis*,

Lepadella ovalis and *Ptygura* sp. (Ptyg) are occurring in conjunction with blooms of blue-green bacteria and with the peak of *Staurastrum planktonicum* (Desmidiaceae) present in these years during the summer stratification. *Polyarthra major* and *Keratella cochlearis* (K.coch) might be correlated with heterotrophic nanoflagellate abundances.

In the ordination, based on predation factors (Figure 3d), the first two ordination axes explain 23.9% of the variability. *Keratella quadrata*, *K. hiemalis*, *Polyarthra dolichoptera*, *Notholca squamula* and *B. angularis* are associated with *Cyclops vicinus*. Conversely, *Filinia longiseta*, *Conochilus* sp. and *Polyarthra vulgaris* seem to show a negative relation with *Cyclops strenuus*. *Trichocerca similis*, *Pompholyx triloba*, *Polyarthra major* and some others are related with *Leptodora kindtii*.

For the competition factors (Figure 3c), the first two RDA axes explain 19.0% of the total species–environment variability. In this set of variables, only the *Diaphanosoma brachyurum* variable was significant. It is correlated with *Keratella cochlearis*, *Trichocerca similis*, *Pompholyx triloba*, *Ptygura* sp., *Lecane lunaris* and *Polyarthra major*.

The first ordination axis is significant for all these ordinations ($P<0.01$, under 999 time-restricted permutations).

Discussion

Influence of environmental variables

Results of this study indicate that date, nutrients, primary production, temperature, abundance of predators and competitors, and potential food resources are important factors influencing the structure of the rotifer community. Date is a variable of seasonality and is clearly correlated with many season-dependent factors each of which may influence rotifer population dynamics. Among nutrients the concentration of total nitrogen was most important. This variable indicates the ecosystem trophic state, and may partly determine the species presence and dynamics of rotifer assemblage (Bérziņš & Pejler, 1989b). Surprisingly, the effect of phosphorus as a limiting nutrient was found not to be significant (cf. Stemberger, 1995; Stemberger & Lazorchak, 1994). Alkalinity and pH are often found as variables predicting the zooplankton community composition, based on data from a large number of different water bodies (Bérziņš & Pejler, 1987; Hessen et

176

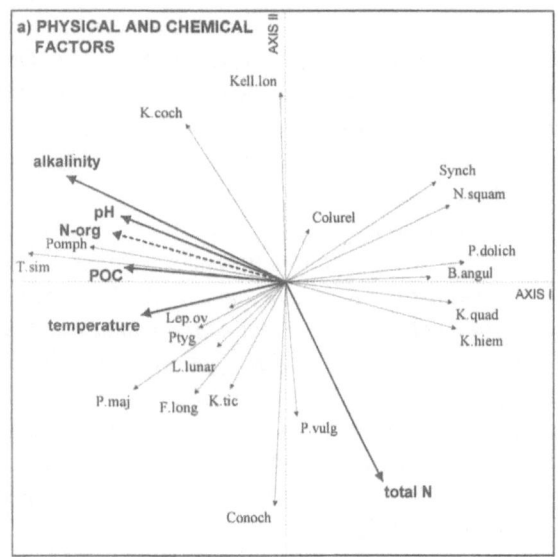

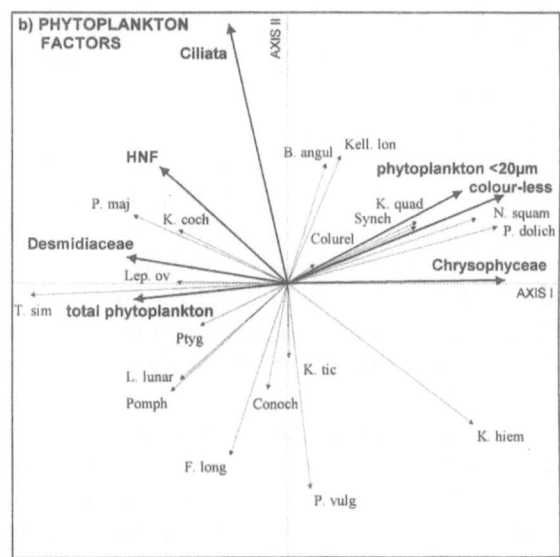

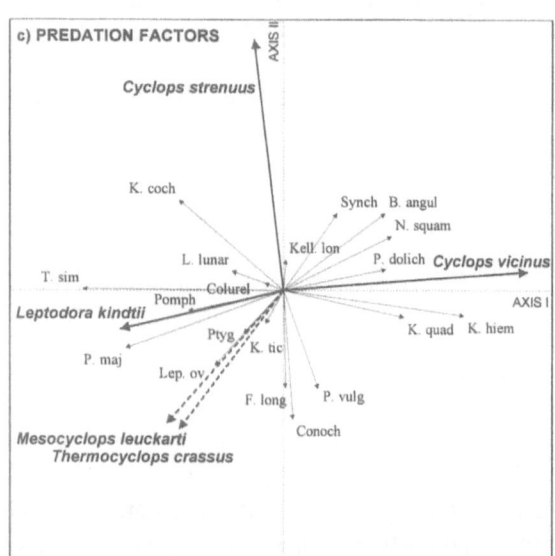

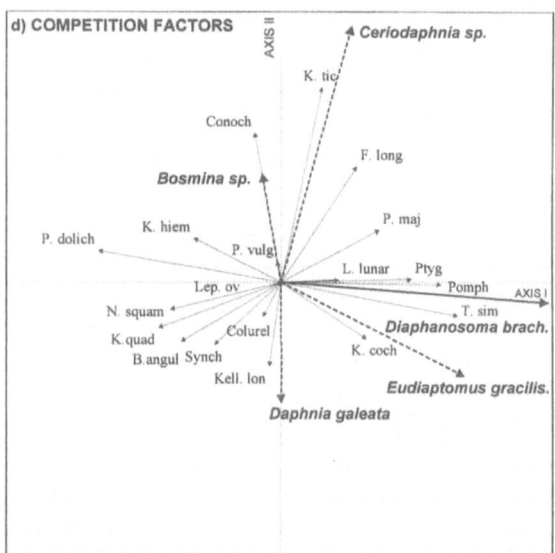

Figure 3. Redundancy analysis (RDA) biplots of each set of environmental factors and rotifer species. (a) Physical and chemical factors; (b) phytoplankton factors (including protozoa and pacteria); (c) invertebrate predation factors; (d) crustacean competition factors. The factors with significant effect are depicted as solid arrows, others as dashed arrows, species as thin arrows. The full rotifer species names are shown in text.

al., 1995; Pichlová, pers. comm.; Pinel-Alloul, 1995). Because the environmental data from this study describe the situation from one locality, they fell within a much narrower range and show other trends then those mentioned above (e.g., Pinel-Alloul et al., 1995; Stemberger, 1995). Nevertheless, alkalinity and pH are variables directly influenced by primary production. Particulate organic carbon and organic nitrogen, as characteristics of organic seston, are related to primary production as well. The correlations between these

four variables are described in Reynolds (1984) Effect of water temperature on rotifer community is well known (Bérziņš & Pejler, 1989a; Hessen et al., 1995). Concentration of the anions (SO_4^- and Cl^-) and the cations (Mg^{2+}, Ca^+, Na^+ and K^+) explained little of the variability. This result stands in contrast to the results of Pinel-Alloul (1995). Biomass of colour-less algae and Chrysophyceae seems to be important to the rotifer community. It corresponds mainly with the food preferences of the spring peak *Keratella, Synchaeta* and *Polyarthra* species. Conversely, the second most preferred food, Cryptophyceae, seems not to be a restrictive factor (Pourriot, 1977; Walz, 1995). The autumnal rotifer peak, formed mainly by bacterivorous *Keratella cochlearis*, may be in conjunction with the collapse of dense summer *Staurastrum* (Desmidiaceae) populations and that of bluegreen bacteria. The number of Bacteria is not a good variable, because it does not reflect bacterial activity. Many authors (e.g., Sanders et al., 1994; Walz, 1995) show that *Synchaeta, Keratella, Kellicottia* and *Polyarthra* spp. can feed on HNF and ciliates, which agrees with the results of this study.

Only one competitor species was found to significantly affect the rotifer community. *Diaphanosoma brachyurum* densities and their competition activities are highest in the autumn, and they occur at the time when *Staurastrum* and blue-green bacteria populations collapse. The main food source in this case are probably bacteria. The main rotifers present at this time are *Keratella* spp. No relationship has been found between the rotifer community and *Daphnia galeata* or *Bosmina* sp., the most abundant cladoceran herbivores. This agrees with the conclusions of Wickham & Gilbert (1991), but disagrees with MacIsaac & Gilbert (1991).

The most important variable of the rotifer community data is presence of *Cyclops vicinus*. Predation pressure by *C. vicinus* and *C. strenuus* is probably the main factor reducing rotifers in this reservoir. The ability of *Cyclops* spp. to affect rotifer population dynamics is well known (Brandl & Fernando, 1981; Stemberger & Evans, 1984). *Leptodora kindtii* also has a significant effect, even if their influence is problematic. Nevertheless, Kulikov et al. (1992) show that the juveniles of *Leptodora kindtii* feed on rotifers significantly.

Explained variation

The percentage of explained variance in both, significant and forward selected sets of the total model (61.7 and 50.6%) was comparable with similar analyses of other authors (Bersier & Meyer, 1994; Borcard et al., 1992; Pinel-Alloul et al., 1995). The amount of unexplained variation in the set, 38.3 and 49.4%, was relatively high. Much of this variation may be linked with stochastic fluctuations (Pinel-Alloul et al., 1995). The comparison of four sets of environmental variables shows, that the main part of variability is explained by the physical and chemical variables. Conversely, the smaller part of the variability is explained by the set with competition factors. The fact that the variation explained by each set of environmental variables (except predation) was significant only if its effect started with date was evaluated, may be explained by the seasonally tied influence of some variables (e.g., possible competition with *Diaphanosoma brachyurum* during the late summer and autumn). The effect of predation factors was significant, even if the influence of date was removed. This indicates that predation is very important in shaping the rotifer community in this reservoir, and that competition factors are probably less important.

Acknowledgements

I thank J. Sed'a, V. Straškrabová, J. Komárková, and others from HBI for their environmental data and valuable advise and P. Šmilauer and J. Š. Lepš for helpful advise with data processing. My attendance at the VIIIth International Rotifer Symposium was supported by the Hydrobiological Institute Czech Academy of Science, Faculty of Biology University of South Bohemia, University of South Bohemia Fund and Mattoni Awards for Studies of Biodiversity and Conservation Biology.

References

Bersier, L. F. & D. R. Meyer, 1994. Bird assemblages in mosaic forests: the relative importance of vegetation structure and floristic composition along the successional gradient. Acta Oecologica 15: 561–576.

Bérziņš, B. & B. Pejler, 1987. Rotifer occurrence in relation to pH. Hydrobiologia 147: 107–116.

Bérziņš, B. & B. Pejler, 1989a. Rotifer occurrence in relation to temperature. Hydrobiologia 175: 223–231.

Bérziņš, B. & B. Pejler, 1989b. Rotifer occurrence in relation to trophic degree. Hydrobiologia 182: 171–180.

Borcard, D., P. Legendre & P. Drapeau, 1992. Partialling out the spatial component of ecological variation. Ecology 72: 1045–1055.

Brandl, Z. & C. H. Fernando, 1981. The impact by cyclopoid copepods on zooplankton. Verh. Int. Ver. Limnol. 21: 1573–1577.

Brandl, Z., B. Desortová, J. Hrbáček., J. Komárková, V. Vyhnálek, J. Sed'a & M. Straškraba, 1989. Seasonal changes of zooplankton and phytoplankton and their mutual relations in some Czechoslovak reservoirs. Arch. Hydrobiol. Beih. Ergebn. Limnol. 33: 597–604.

Hessen, D. O., B. A. Faafeng & T. Andersen, 1995. Replacement of herbivore zooplankton species along gradients of ecosystem productivity and fish predation pressure. Can. J. Fish. aquat. Sci. 52: 733–742.

178

Hunter, M. D. & P. W. Price, 1992. Playing chutes and ladders: heterogeneity and the relative roles of bottom-up and top-down forces in natural communities. Ecology 73: 724–732.

Komárková, J., 1993. Cycles of phytoplankton during long-term monitoring of Římov and Slapy reservoir (Czech Republic). Verh. Int. Ver. Limnol. 25: 1187–1191.

Kopáček, J. & J. Hejzlar, 1993. Semi-micro determination of total phosphorus in fresh waters with perchloric acid digestion. Int. J. environ. anal. Chem. 53: 173–183.

Kopáček, J. & L. Procházková, 1993. Semi-micro determination of ammonia in water by the rubazoic acid method. Int. J. environ. anal. Chem. 53: 243–248.

Koste, W., 1978. Rotatoria. Die Rädertiere Mitteleuropas. Begründet von Max Voigt, Vol. I, II. Borntraeger, Berlin, 673 pp., 234 pl.

Kulikov, A. S., A. O. Shkute & L. V. Polishchuk, 1992. Feeding in juvenile *Leptodora kindtii* (Focke). Sov. J. Eco. 22: 202–205.

Lehman, J. T., 1991. Interacting growth and loss rates: the balance of top-down and bottom-up controls in plankton communities. Limnol. Oceanogr. 36: 1546–1554.

Lepš, J. S., M. Straškraba, B. Desortová & L. Procházková, 1990. Annual cycles of plankton species composition and physical chemical conditions in Slapy Reservoir detected by multivariate statistics. Arch. Hydrobiol. Beih. Ergebn. Limnol. 33: 933–945.

MacIsaac, H. J. & J. J. Gilbert, 1991. Discrimination between exploitative and interference competition between Cladocera and *Keratella cochlearis*. Ecology 73: 927–937.

Matěna, J., 1995. The role of ecotones as feeding grounds for fish fry in a Bohemian water supply reservoir. Hydrobiologia 303: 31–38.

McQueen, D. J., J. R. Post & E. L. Mills, 1986. Trophic relationships in freshwater pelagic ecosystems. Can. J. Fish. aquat. Sci. 43: 1571–1581.

Murphy, I & J. P. Riley, 1962. A modified single-solution method for the determination of phosphate in natural waters. Anal. Chim. Acta 27: 31–36.

Pinel-Alloul, B., T. Niyonsenga & P. Legendre, 1995. Spatial and environmental components of freshwater zooplankton structure. Ecoscience 2: 1–19.

Pourriot, R., 1977. Food and feeding habits of Rotifera. Arch. Hydrobiol. Beih. Ergebn. Limnol. 8: 243–260.

Procházková, L., 1959. Bestimmung der Nitrate im Wasser. Z. Anal. Chem. 167: 254–260.

Procházková, L., 1960. Einfluss der Nitrate und Nitrite auf die Bestimmung des organischen Stickstoffs und Ammonimus im Wasser. Arch. Hydrobiol. 56: 179–185.

Procházková, L., 1961. Über die Anwendung des Chloranilsäureverfahrens zur kolorimetrischen Sulfatbestimmung im Wasser. Z. Anal. Chem. 183: 103–107.

Reynolds, C. S., 1984. The Ecology of Freshwater Phytoplankton. Cambridge Studies in Ecology. Cambridge University Press, Cambridge.

Sanders, R. W., D. A. Leeper, C. H. King & K. G. Porter, 1994. Grazing by rotifers and crustacean zooplankton on nanoplanctonic protists. Hydrobiologia 288: 167–181.

Sed'a, J. & J. Kubečka, 1997. Long-term biomanipulation of Římov reservoir (Czech Republic). Hydrobiologia 345: 95–108.

Stemberger, R. S., 1995. The influence of mixing on rotifer assemblages of Michigan lakes. Hydrobiologia 297: 149–161.

Stemberger, R. S. & M. S. Evans, 1984. Rotifer seasonal succession and copepod predation in Lake Michigan. J. Great Lakes Res. 10: 417–428.

Stemberger, R. S. & J. M. Lazorchak, 1994. Zooplankton assemblage responses to disturbance gradients. Can. J. Fish aquat. Sci. 51: 2435–2447.

Straškrabová, V., J. Komárková & V. Vyhnálek, 1993. Degradation of organic substances in reservoirs. Wat. Sci. Tech. 28: 95–104.

Šimek, K., M. Macek, J. Pernthaler, V. Straškrabová & R Psenner, 1996. Can freshwater planktonic ciliates survive on a diet of picoplankton? J. Plankton Res. 18: 597–613.

ter Braak, C. J. F. 1994. Canonical community ordination. Part 1: Basic theory and linear methods. Ecoscience 1: 127–140.

ter Braak, C. J. F. & I. C. Prentice, 1988. A theory of gradient analysis. Adv. Ecol. Res. 18: 271–317.

ter Braak, C. J. F. & P. F. M. Verdonschot, 1995. Canonical correspondence analysis and related multivariate methods in aquatics ecology. Aquat. Sci. 57: 255–289.

Walz, N., 1995. Rotifer populations in plankton communities: Energetics and life history strategies. Experientia 51: 437–453.

Wickham, S. A. & J. J. Gilbert, 1991. Relative vulnerabilities of natural rotifer and ciliate communities to cladocerans: laboratory and field experiments. Freshwat. Biol. 26: 77–86.

Hydrobiologia **387/388**: 179–197, 1998.
E. Wurdak, R. Wallace & H. Segers (eds), Rotifera VIII: A Comparative Approach.
© *1998 Kluwer Academic Publishers.*

Rotifers in relation to littoral ecotone structure in Lake Rotomanuka, North Island, New Zealand

Ian C. Duggan[1,*], John D. Green[1], Keith Thompson[1] & Russell J. Shiel[2]
[1]*Department of Biological Sciences, The University of Waikato, Private Bag 3105, Hamilton, New Zealand;*
*Fax: [+64] 7 838-4324; Tel: [+64] 7 856-2889; E-mail: icd@walkato.ac.nz (*author for correspondence)*
[2]*Murray-Darling Freshwater Research Centre, P.O. Box 921, Albury N.S.W. 2640, Australia*

Key words: rotifera, artificial substrates, littoral, ecotone, macrophytes, New Zealand

Abstract

The spatial and temporal dynamics of rotifers in the littoral ecotone of Lake Rotomanuka (37° 55′ S, 175° 19′ E) were studied from February to November 1994. Rotifers were sampled with artificial substrates at two or three weekly intervals from eight sites chosen with respect to macrophyte species distribution from near shore to deeper water. 58 rotifer species were found, a high diversity in comparison to that of New Zealand limnetic communities, which usually have less than ten species. Rotifers had peak abundances in summer within emergent and submerged vegetation, when shallow regions were dry. *Lecane bulla* (Gosse) and *Testudinella parva* (Ternetz) generally had the highest numerical densities. Three major temporal groupings of species were distinguished by cluster analysis and canonical correspondence analysis (CCA): summer–autumn (e.g. *Lecane hornemanni* (Ehrenberg), *L. bulla*), winter-spring (e.g. *Mytilina mucronata* (Müller), *Trichocerca porcellus* (Gosse)), and late autumn to mid spring (e.g. *T. parva, Polyarthra vulgaris* (Carlin)). Rotifer species composition appeared to depend on seasonal change of water level and the associated shift from heterogeneous to homogenous physical and chemical conditions across the ecotone. Temporal variability in the abundance of zindividual rotifer species was far greater than their spatial variability. CCA indicated that temperature and pH were the factors most strongly associated with temporal variation in abundances of rotifer species. Macrophytes appeared to play the major role in determining spatial distribution, both because of differences in physical structure between species (affecting microhabitat diversity) and by causing variations in physical and chemical conditions (e.g. oxygen and pH) by inhibiting mixing.

Introduction

The littoral region is often the most diverse part of the lake ecosystem, and usually supports a variety of macrophytes, their associated microflora and a large number of animal species (Winterbourn & Lewis, 1975). Microfaunal species are often especially abundant and rotifers in particular have their greatest diversity in the littoral region (Pennak, 1966; Havens, 1991). Macrophytes are presumably a major influence governing the diversity and spatio-temporal structuring of the animal communities by providing a diverse and seasonally variable array of surfaces for colonization and feeding, various interstices for concealment from predators, and by causing spatial and temporal variability of water chemistry, oxygen, pH, and temperature as a res-

ult of photosynthesis, decomposition and prevention of water mixing (e.g. Dvorak, 1970; Vitt & Bayley, 1984). Such linkages have been established for some macrofaunal groups whose abundance and diversity shows considerable spatial and temporal variation within the littoral, influenced both by the composition and physical structure of the vegetation, and physical and chemical factors (e.g. Dvorak, 1970; Dvorak & Best, 1982). Littoral rotifers should be affected similarly, as differences in the density and diversity of rotifers between lakes are known to be correlated with pH, temperature and oxygen concentration (e.g. Roff & Kwiatkowski, 1977; Berzins & Pejler, 1987, 1989a,b; Mikschi, 1989) and all of these factors are influenced by macrophytes within littoral areas. Also, the varied architectures of littoral macrophytes should increase the range of microhabitats available to rotifers by enhancing, for example, the variety of feeding positions and possibilities for predator avoidance (Wallace, 1977a). However, while sessile rotifers do show selection for specific substrate types (e.g. Edmondson, 1944; Wallace, 1977b), for most (non-sessile) littoral rotifer species there is little direct information concerning the effects of macrophytes on diversity and spatial and temporal distribution.

We examined the ecology of the littoral monogonont rotifers of Lake Rotomanuka, a small lake with a well developed littoral in the Waikato region, North Island, New Zealand. The major aims were to characterize the diversity and abundance of rotifers, and to examine the influence of vegetation type and physical and chemical conditions on their spatial and temporal dynamics.

Lake Rotomanuka

Lake Rotomanuka (surface area 0.17 km^2; shoreline 1.5 km; maximum depth 8.7 m; Secchi depth 2.25–4.8 m; authors' unpublished data) (Figure 1) is a mesotrophic lake in a predominantly pastoral catchment. Marginal vegetation is a narrow band surrounding the lake, and the dominant vegetation types are common to most lakes in the Waikato region (Champion et al., 1993; De Winton & Champion, 1993).

Methods

The sampling location (Figure 1) had a gradation of littoral vegetation typical of the lake, and the eight sampling sites (Figure 2) represented the dominant macrophyte types across the ecotone, including gradations between them.

Sampling was fortnightly from 4 February to 6 June 1994, and thereafter every three weeks until 21 November 1994. Samples were obtained by wading. Sites were visited twice on each sampling occasion, with visits spaced three days apart. Samples were not taken from sites 2 and 4 between 14 March and 11 April and site 1 on 7 February, and 14 March to 9 May as they were dry or had extremely low water levels. Water level rose between 23 May and 6 June, allowing sampling at 250 mm depth at all sites thereafter. Sites 5 and 6 were not sampled on 8 August as extreme water depth made them inaccessible.

From 24 April 1994, the macrophyte species within a 1m radius and depth from each site were recorded and visual weightings (given by a value between 1 and 100) assigned to apparent contribution of submerged surface area for each macrophyte type. Macrophytes were identified using Johnson & Brooke (1989) and the Waikato Herbarium.

pH, temperature (C), and oxygen concentration (DO) and saturation (% O) were measured between 10.30 and 12.30 hours at the beginning and end of each sampling occasion, and these values were averaged to give a single number for each occasion. pH and temperature were sampled from 4 February 1994, while oxygen concentration was sampled from 21 February.

Temperature, DO and % O were measured *in situ* using a YSI Model 50 meter. % O was directly measured beginning 22 April and for earlier dates calculated from a nomogram (Wetzel & Likens, 1979). pH was measured within 5 minutes of collection with a Horiba Cardy Twin B-113 pH meter from water samples taken at depths of ca. 250 mm below the surface of the water, or in the water available if the depth was shallower. Samples taken at sites 1 to 4 from 7 February to 23 May were from depths less than 250 mm. Measurements were taken from all eight sites wherever possible.

Rotifers were sampled using artificial substrates similar to those that have been successful in sampling littoral Chydoridae (e.g. Whiteside, 1974). This method was chosen in preference to removal of macrophytes or portions of macrophytes because it provided a quantitative method for sampling periphytic animals that was non-destructive to an area which was to be sampled over a prolonged period of time. A non-destructive method using a hand-operated vacuum pump for suction of volumes of water from the surface of macrophytes (cf. Campbell et al., 1982) was also tried, but this proved inadequate due to problems with replication. Substrates were made by cutting Carlin International scouring pads into thirds (97 × 48 × 5 mm). Preliminary studies undertaken to determine the number of days required to leave substrates for colonization showed that 80% of all species found on substrates over 12 days were present by the third day of colonization. The number of substrates required to give an adequate sampling variability was determined following procedures described by Elliott (1971). Ten substrates were left at a single site for three days and the rotifers counted. Three substrates were sufficient to sample at least 89% of all rotifer species found on

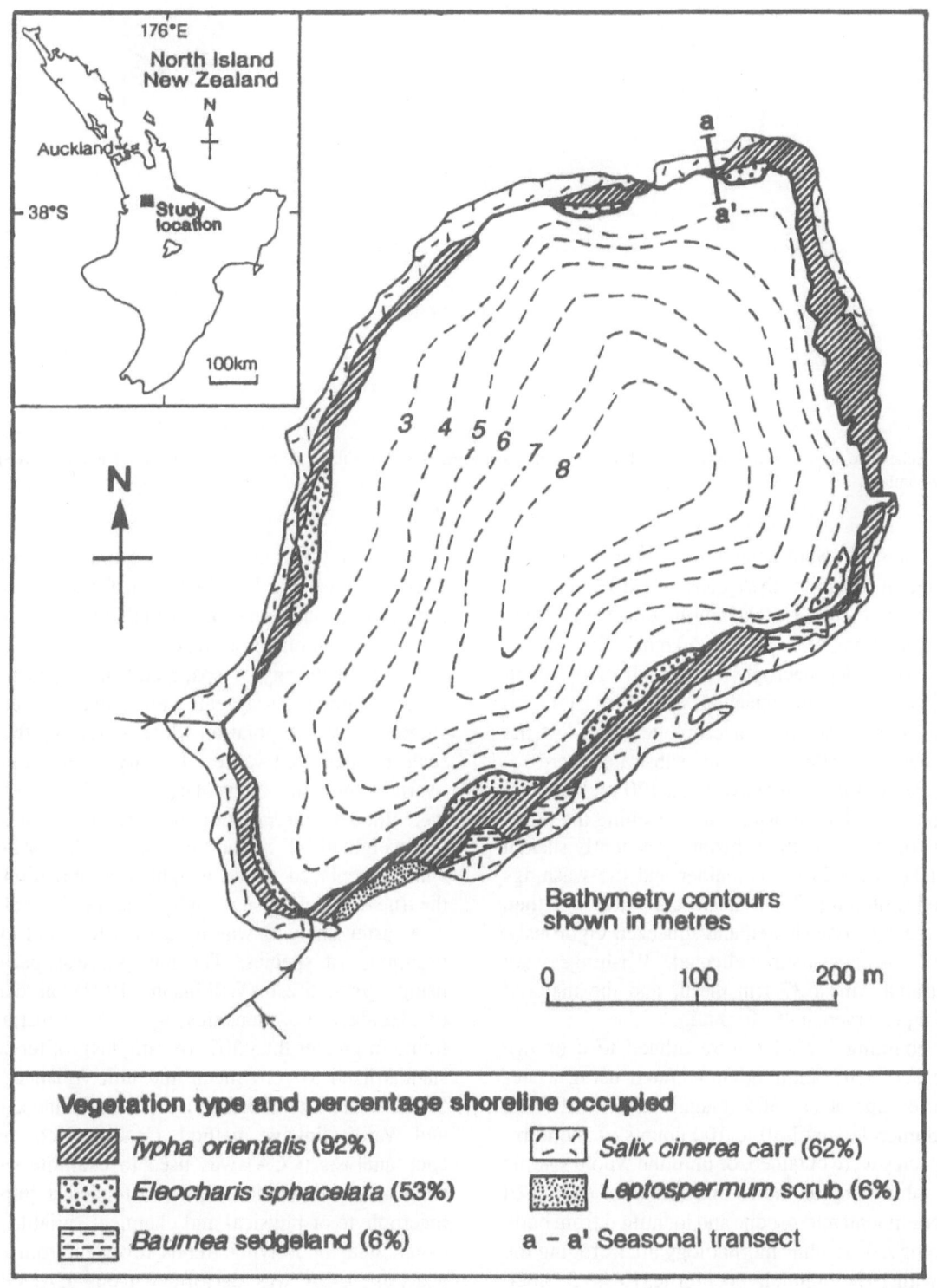

Figure 1. Map showing position of study site on the northern edge of Lake Rotomanuka. Inset shows location of Lake Rotomanuka, North Island, New Zealand (after Irwin, 1992; Champion et al. 1993).

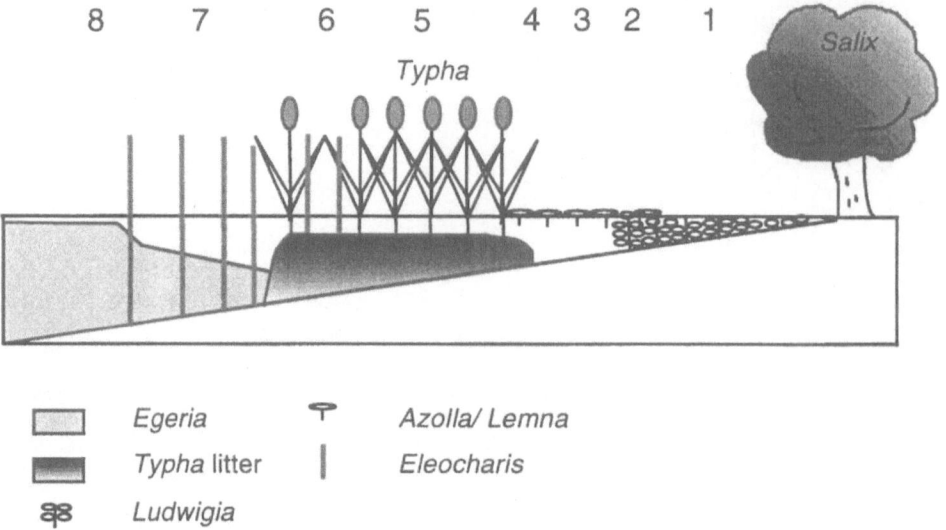

Figure 2. Schematic diagram showing generalized profile of marginal vegetation across the Lake Rotomanuka littoral. Numbers correspond to study site numbers.

all substrates and were enough to give an acceptable coefficient of variation (SD/mean) of $\leq 20\%$.

On the first visit of each sampling occasion, three artificial substrates were placed 250 mm below the water surface among macrophytes at each of the eight sites. Substrates were attached with a twist tie to a stable object (usually a macrophyte stem). At the end of each sampling occasion, substrates were retrieved by carefully manoeuvering a 120 ml screwtop container around the substrate and attaching the lid.

To extract rotifers the substrate was gently shaken in the lake water in the container and the washings collected. Filtered (37 μm mesh) tap water was then added, the substrate shaken and squeezed vigorously, and the washings again collected. Washings were concentrated with a 37 μm mesh, and the material collected preserved in 4% formalin.

For counting, samples were diluted to a known volume and 5 ml aliquots enumerated using a stereo microscope at ca. $30\times$ magnification. Aliquots were enumerated until 50 to 100 counts of the dominant species were obtained, or until the whole sample was counted. Any unknown animals were examined with a compound microscope and identified from body morphology, or trophus morphology after eroding the rotifer with 10% sodium hypochlorite (Koste & Shiel, 1991). Taxonomic literature used included Ruttner-Kolisko (1974), Koste (1978), and Shiel & Koste (1993) and references therein.

Macrophyte, chemical and animal distributions were plotted using Spyglass Transform^{TM} (Spyglass Inc., 1993), a computer program that produces intensity and/or contour plots from 3D arrays of numbers (i.e. animal density $\times$ space and time). Each density is assigned a grayscale value, and the image is smoothed by interpolation while preserving the grayscale of measured values. Density contours may be overlaid onto the interpolated plot. Missing values were filled by interpolation from true data values using the 'weighted fill' method in which each missing data value is replaced with a weighted combination of all the true data values surrounding the missing value.

Cluster analysis was used to detect and classify groupings of species. The analysis was performed using Systat 5.2.1 (Wilkinson, 1992) on densities of abundant (i.e. densities > 15 per substrate, or found in greater than 50% of samples) rotifer species standardized to zero mean and unit variance, using Standardized Euclidean Distance as a distance metric and Ward's linkage method. Canonical correspondence analysis (CCA) was used to examine species-environmental relationships and to detect important macrophyte or physical and chemical variables associated with underlying trends revealed from cluster analysis. CCA was performed using CANOCO^{TM} (Ter Braak, 1988) on non-standardized data with a log transformation ($\ln(1 + x)$).

Results

Spatial and temporal variation in macrophyte density

In early December 1993, there was a distinct zonation of five dominant macrophyte types (Figure 2). Nearest to land *Ludwigia palustris* L. dominated and extended from dry land into shallow water. Beyond this was a zone of floating vegetation (mainly *Azolla pinnata* R. Br.) replaced further lakeward by belts of emergent *Typha orientalis* C. Presl, and then *Eleocharis sphacelata* R. Br. Some submerged *Egeria densa* Planchon was associated with *Eleocharis*, and beyond the *Eleocharis* belt *Egeria* formed a dense monospecific stand which extended to a depth of ca. 4 m.

Egeria was abundant at site 8 throughout the study, and generally found in lower densities at sites 5–7 (Figure 3). *Eleocharis* was present only at sites 5–7, being least abundant in winter when densities of dead *Eleocharis* were greatest. *Typha* was found at sites 4–6, but occurred at low density or was absent during winter. Dead material from *Typha* was always abundant, particularly in site 5. *Azolla* and *Lemna minor* L. were absent or in low amounts until late winter, becoming abundant on the inner margin of *Typha*. *Ludwigia* was most abundant in shallow sites, except from winter to mid-spring when *Myriophyllum propinquum* Cunn. became the most abundant species. Small quantities of *Lycopus europeus* L., and *Salix cinerea* L. (willow) leaves and blossom were occasionally noted.

Physical and chemical conditions

Through late summer and autumn (December–June) pH was highest at deeper sites (Figure 4), but with a rise in water level from early June (early winter) became more even throughout the ecotone. There was little spatial variability in pH until late August when the gradient between shallow and deep sites began to re-establish.

Temperature (Figure 4) was slightly warmer in deeper than shallow sites from late summer until mid to late autumn. Differences between sites were negligible through winter until early spring when a gradient formed again. Maximum temperatures were recorded in February (range 23.6° to 24.6°C), and minimum temperatures in mid-June at sites 1–3 (7.6°, 7.75°, 8.0°C, respectively), and mid-July at sites 4–8 (8.6°, 8.3°, 8.55°, 8.6°, 8.65°C, respectively).

DO and % O had similar spatial and temporal patterns. Oxygen was generally higher at deeper sites, with a marked gradient over sites 5–8 from late summer until the start of winter. The highest values during the study occurred at shallow sites immediately after inundation (maximum of 14.82 mg l^{-1}, 145.35% on 23 May at site 3). DO was more even across the ecotone from winter until mid-spring when shallower sites (1–4) began to become depleted, and by November large gradients were found.

Rotifer diversity

58 species of monogonont rotifers were recorded from 27 genera (Table 1). Two varieties of *T. rattus* were also found. Fifteen species, plus *T. rattus carinata*, were never found at abundances greater than one per substrate, and only single individuals of *M. collinsi* and *Floscularia* sp. were recorded. Bdelloid rotifers were also recorded, but none could be identified to species, and are thus not included in the analysis.

The number of species (α) from any site on any sampling date ranged from 4 to 26 (Figure 5). Site 4, when inundated, generally had the highest α (12–25 spp. mean 19.4). Site 1 had the highest maximum α for any date (26 spp. on 10 October), though there were generally never more than 16 species at this site (mean 12.9). Both minimum and maximum α generally decreased with distance from site 4, both into deeper and shallower water, with sites 1 (7–26 spp., mean 12.9), and site 8 (5–18 spp., mean 12.5) having the lowest diversities of all sites.

In February α was generally high but dropped considerably by autumn, and then increased again from June (winter) until the end of the study. In sites 1 and 2 α was low immediately after inundation (< 10 spp.), while sites 3 and 4 were immediately comparable with the deeper sites. α was in general lowest in autumn and highest in spring.

Shannon-Wiener diversity (Figure 5) showed similar trends to α, though there were some differences. Most strikingly, sites 1–4 generally had higher diversity (normally > 2) than sites 6–8 (normally < 2). The lowest value was 0.75 at site 8 in early winter, while the maximum diversity was found at site 2 in mid spring (2.47). Generally, diversity was lowest in autumn and winter.

Equitability generally opposed α, indicating that when α was high few species were present in high numbers, and when α was low abundances were more evenly spread amongst species. Equitability was highest in autumn and in shallow sites when inundated.

184

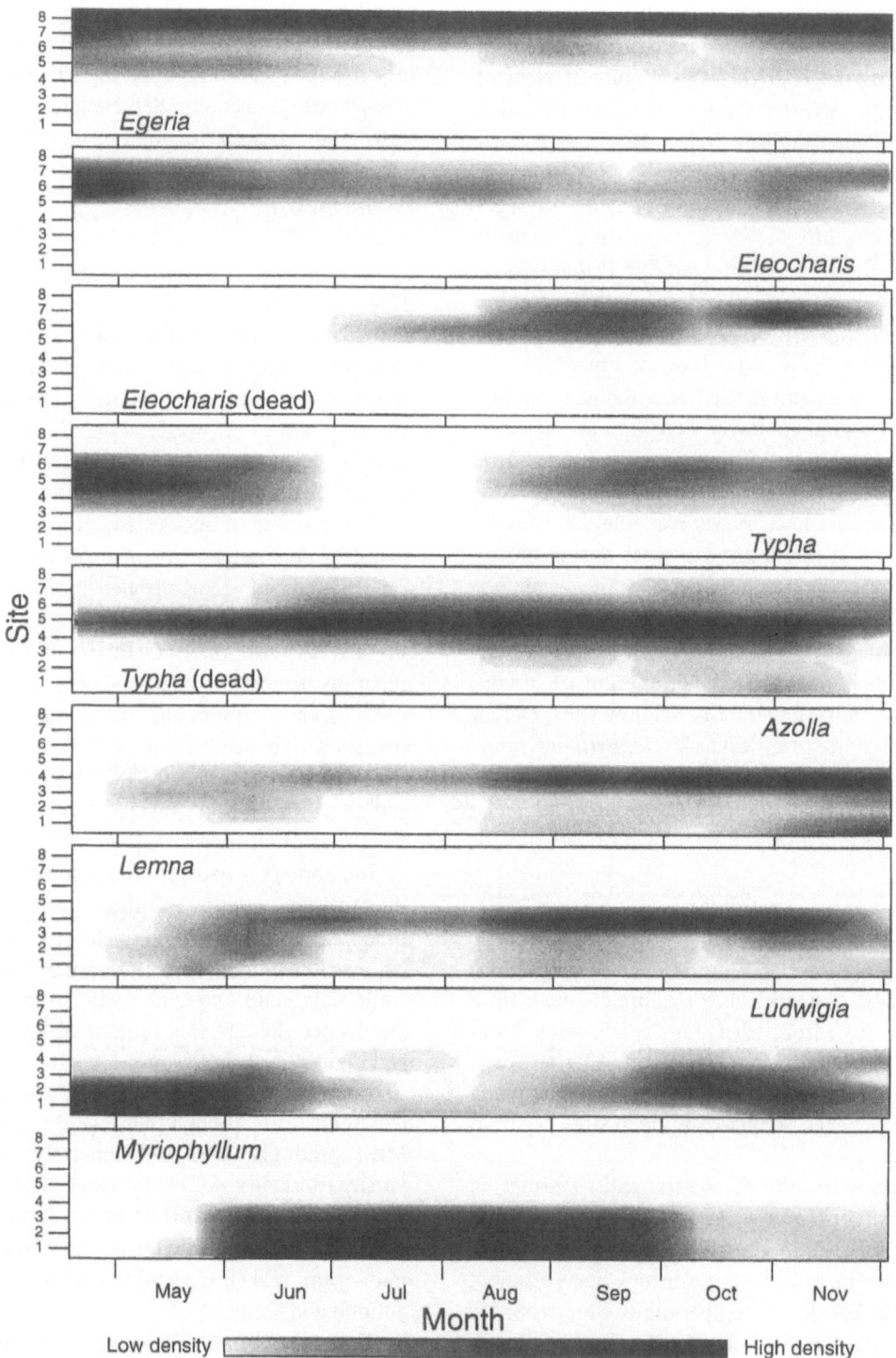

Figure 3. Spatial and temporal distribution of macrophyte species. Greyscale gives a qualitative indication of vegetation density.

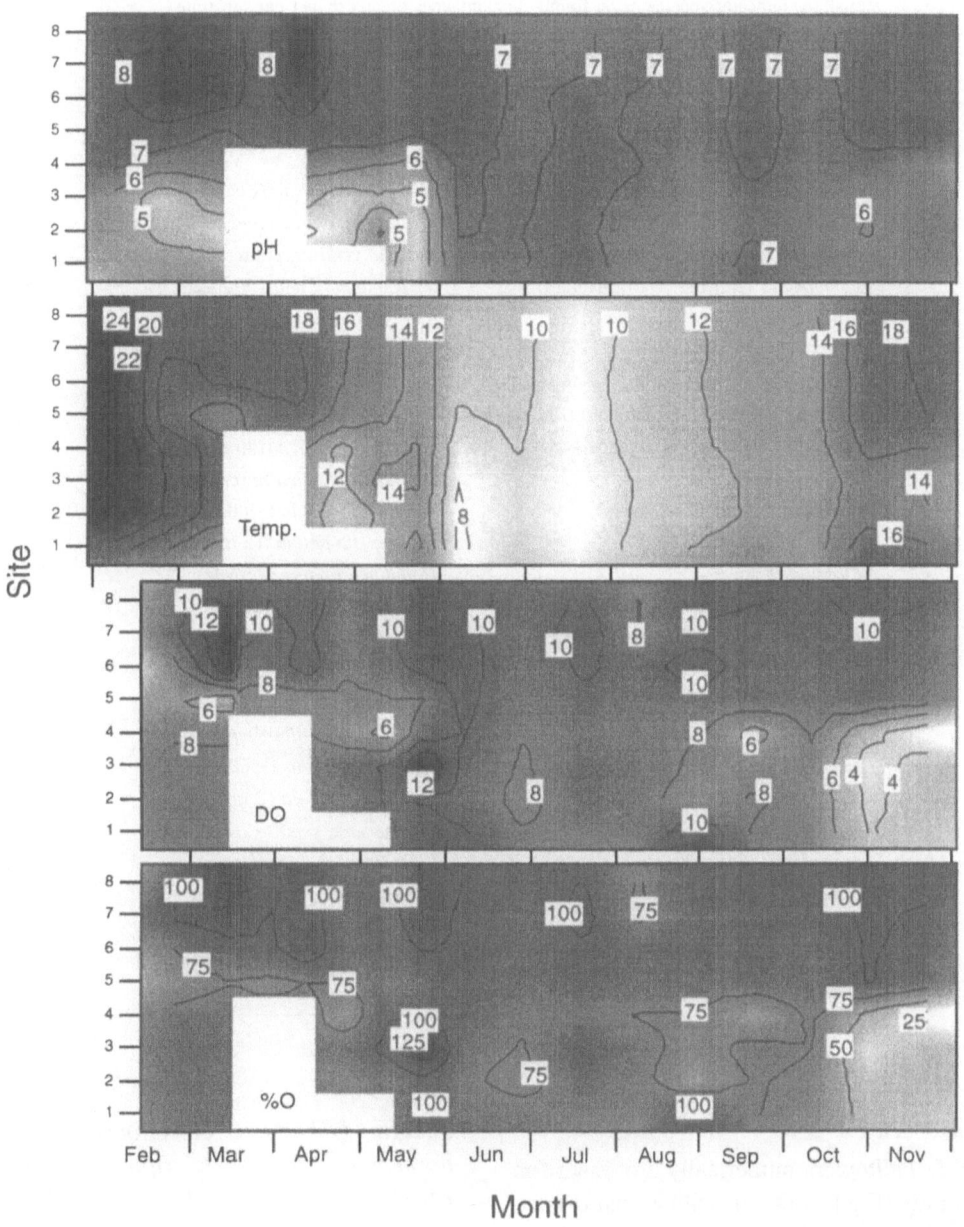

Figure 4. Spatial and temporal distribution of physical and chemical variables. White areas indicate periods and sites where samples were not taken due to lack of water.

Seasonal and spatial dynamics

(i) Total numbers

Total monogonont rotifer densities (Figure 5) were highest in late summer, on the first sampling date, when all sites had densities > 300 per substrate. Densities then declined, being lowest at sites 5–8 during April and May (mid–late autumn), but increased in deeper sites in mid-May. Small peaks occurred in early July at site 7, late August at site 6, and at sites 5, 6 and 7 on 10 October. After inundation of sites 1–4, densities generally remained low. Densities at sites 1–3 were always < 100, while site 4 had values > 100 per substrate from late August, peaking on 21 November. Highest densities occurred at high temperatures and pH values of *ca.* 7, but over a wide range of DO (cf. Figure 4).

186

Table 1. List of rotifer species collected from artificial substrates during the seasonal study. An asterisk (∗) indicates species never found at abundances greater than 1 per substrate

∗Anuraeopsis fissa (Gosse)	*∗Lepadella ovalis* (Müller)
Ascomorpha ovalis (Bergendal)	*Lepadella triptera* (Ehrenberg)
Ascomporpha saltans Bartsch	*∗Macrochaetus collinsi* (Gosse)
Cephalodella eva Gosse	*Monommata* sp.
∗Cephalodella exigua (Hudson and Gosse)	*∗Mytilina bisulcata* (Lucks)
Cephalodella gibba (Ehrenberg)	*Mytilina mucronata mucronata* (Müller)
∗Cephalodella sterea (Gosse)	*Mytilina ventralis* (Ehrenberg)
Cephalodella ventripes (Dixon-Nuttall)	*Notommata allantois* Wulfert
∗Collotheca pelagica (Rousselet)	*Notommata tripus* (Ehrenberg)
∗Collotheca sp.	*∗Platyias quadricornis* (Ehrenberg)
Colurella obtusa (Gosse)	*Polyarthra vulgaris* (Carlin)
∗Colurella uncinata f. *bicuspidata* (Ehrenberg)	*∗Ptygura* sp.
Dicranophorus lütkeni (Bergendal)	*Squatinella mutica* (Ehrenberg)
Euchlanis dilatata Hauer	*Synchaeta oblonga* Ehrenberg
∗Euchlanis c.f. *calpidia* (Myers)	*Synchaeta tremula* (Müller)
E. ?meneta Myers	*Testudinella patina* (Hermann)
∗Floscularia sp.	*Testudinella parva* (Ternetz)
Itura myersi Wulfert	*Trichocerca bidens* (Lucks)
Keratella cochlearis (Gosse)	*Trichocerca brachyura* (Gosse)
∗Keratella procurva (Thorpe)	*Trichocerca longiseta* (Schrank)
Lecane (M.) closterocerca Schmarda	*Trichocerca porcellus* (Gosse)
Lecane (M.) bulla (Gosse)	*Trichocerca pusilla* (Jennings)
Lecane (M.) decipiens (Murray)	*Trichocerca rattus* (Müller)
Lecane flexilis (Gosse)	*∗T. rattus carinata* (Ehrenberg)
Lecane hornemanni (Ehrenberg)	*∗Trichocerca similis* (Wiezejski)
Lecane ludwigii (Eckstein)	*Trichocerca tenuior* (Gosse)
Lecane luna (Müller)	*Trichocerca vernalis* (Hauer)
Lecane (M.) lunaris (Ehrenberg)	*Trichcerca ?gracilis* (Tessin)
Lepadella acuminata (Ehrenberg)	*Trichotria tetractis* f.*similis* (Ehrenberg)
Lepadella oblonga (Ehrenberg)	

(ii) Individual species

T. parva and *L. bulla* were numerically dominant for most of the study (Figures 6a–c), with *L. bulla* generally being most important through summer and autumn, and *T. parva* from late autumn. *L. closterocerca* dominated at all sites at least once during the study, but was usually associated with site 5 (*Typha*). Other species to be numerically dominant during the study were *E. dilatata*, *T. patina*, *L. lunaris*, *C. eva*, *T. bidens*, *T. rattus*, *L. acuminata*, *T. tenuior*, and *M. mucronata*. Most of these dominated at sites 1 to 4 in winter after inundation when total abundances were low, and were hence found only in low densities. 24 species (Figures 6a–c) were found in samples on more than 50% of the sampling occasions or at densities greater than 15 per substrate at some time. Species with greatest densities generally in summer and autumn were *E.*

dilatata, *L. bulla*, *L. closterocerca*, *L. decipiens*, *L. hornemanni*, *L. luna*, *L. triptera*, *T. longiseta*, *T. tetractis*, and *T. rattus*. Of these, *L. closterocerca* and *L. decipiens* were most strongly associated with site 5 (*Typha*), *E. dilatata*, *L. bulla*, *L. triptera*, and *T. rattus* generally with sites 6 and 7 (*Eleocharis*), and *L. luna*, *L. acuminata*, *T. longiseta*, and *T. tetractis* with site 8 (*Egeria*). Species most abundant in early winter (May–July) were *P. vulgaris* (site 8), *S. mutica* (site 5), and *T. parva* (sites 6 and 7). Species most abundant in late winter/spring (August–late December) were *C. obtusa*, *L. lunaris*, *L. ludwigii*, *M. mucronata*, *Monommata* sp., and *T. porcellus*. *C. obtusa*, *L. lunaris* and *Monommata* sp. appeared to have no preference for site, while *L. ludwigii*, *M. mucronata*, and *T. porcellus* appeared to be most strongly associated with site 4 (*Azolla* and *Lemna*).

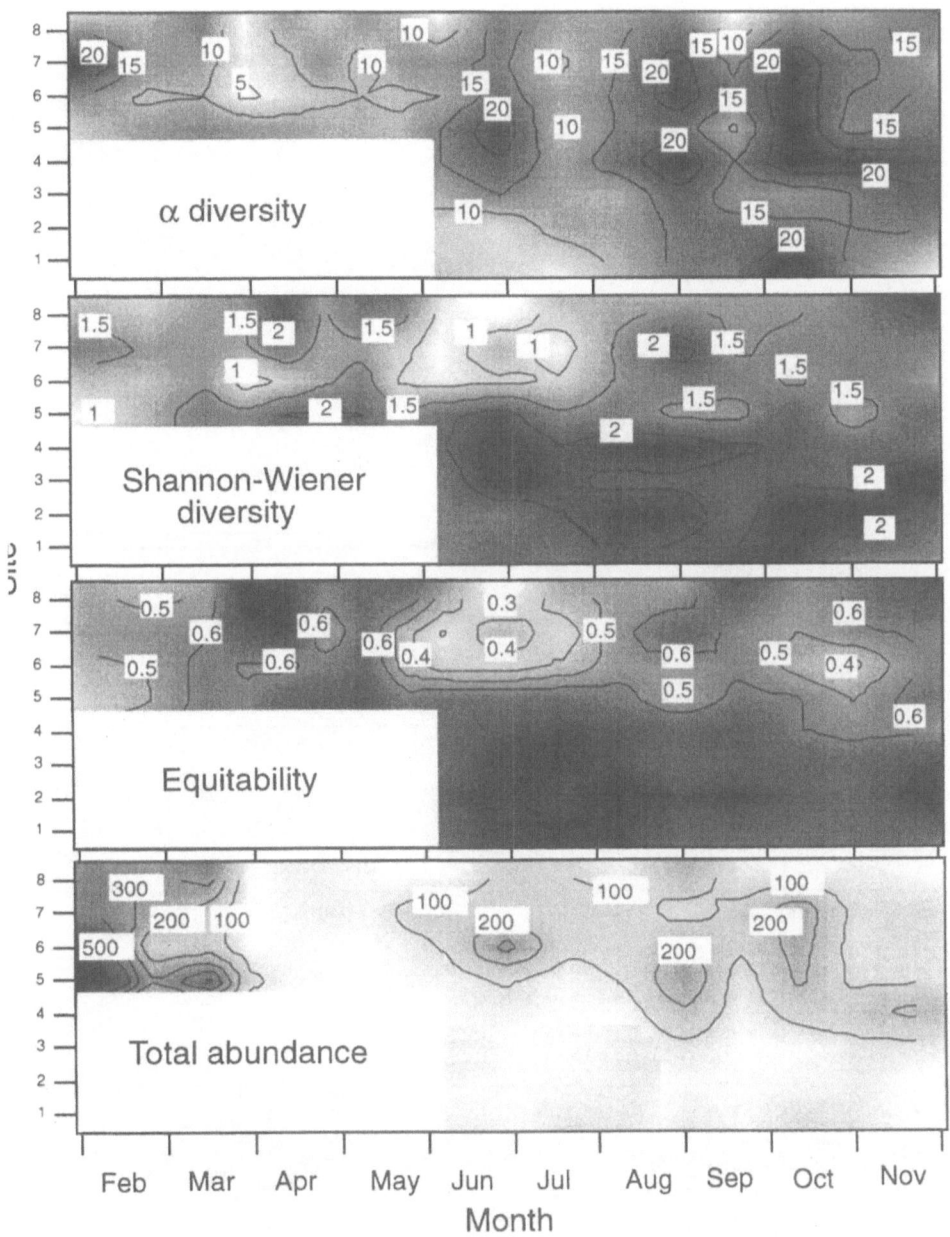

Figure 5. Spatial and temporal distribution of α diversity, Shannon-Wiener diversity, Equitability, and total abundance of rotifers (ind. substrate^{-1}). White areas indicate periods and sites where samples were not taken due to lack of water.

Multivariate analysis

(i) Cluster analysis

Cluster analysis (Figure 7) revealed 3 major groupings. Cluster 1 contained mainly samples from July to November, Cluster 2 samples from February to April, and Cluster 3 samples from May, June and October. Cluster 1 therefore corresponds to a winter

and spring community, Cluster 2 to a summer and autumn community, and Cluster 3 to a grouping of autumn–winter and mid-spring samples. Major divisions thus appear to be determined by season rather than site. Cluster 1 can be separated into two distinct sub-clusters. Cluster 1a is made up mainly of sites 4–8, and 1b only sites 1–3, indicating that there was a spatial separation of rotifer communities in winter

188

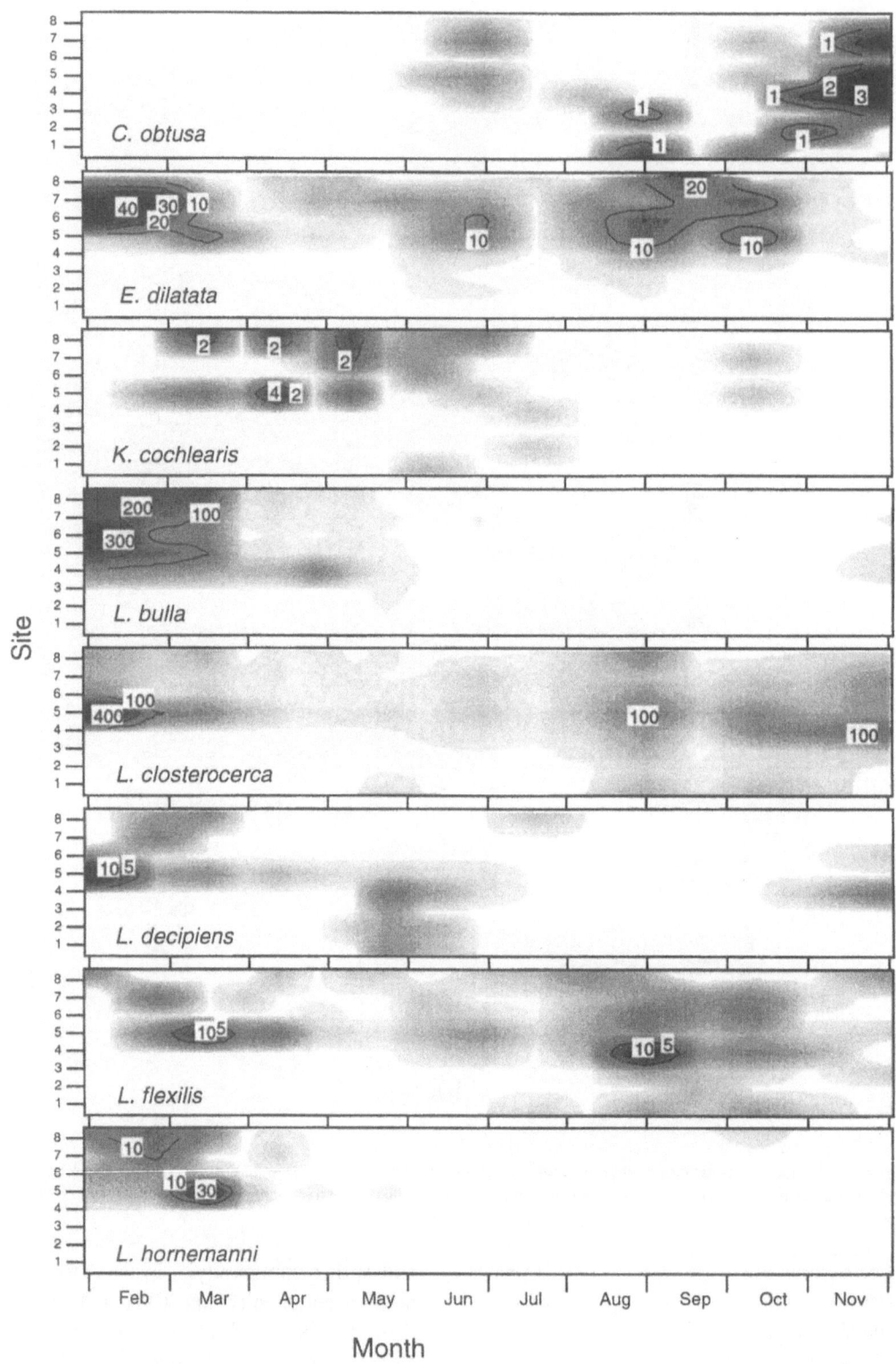

Figure 6a. Spatial and temporal distribution of abundances (ind. substrate $^{-1}$) of common and abundant species.

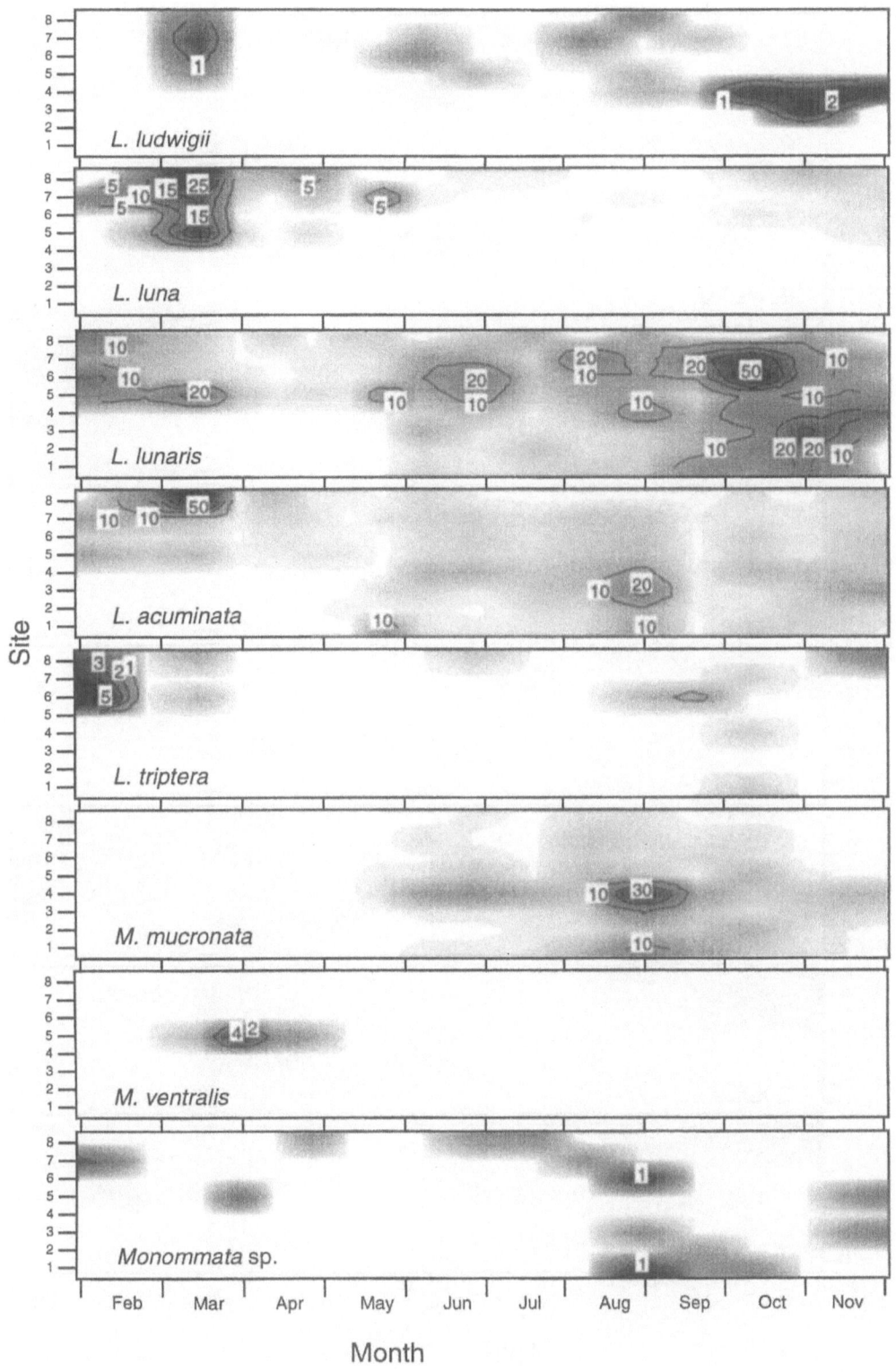

Figure 6b. Spatial and temporal distribution of abundances (ind. substrate $^{-1}$) of common and abundant species.

190

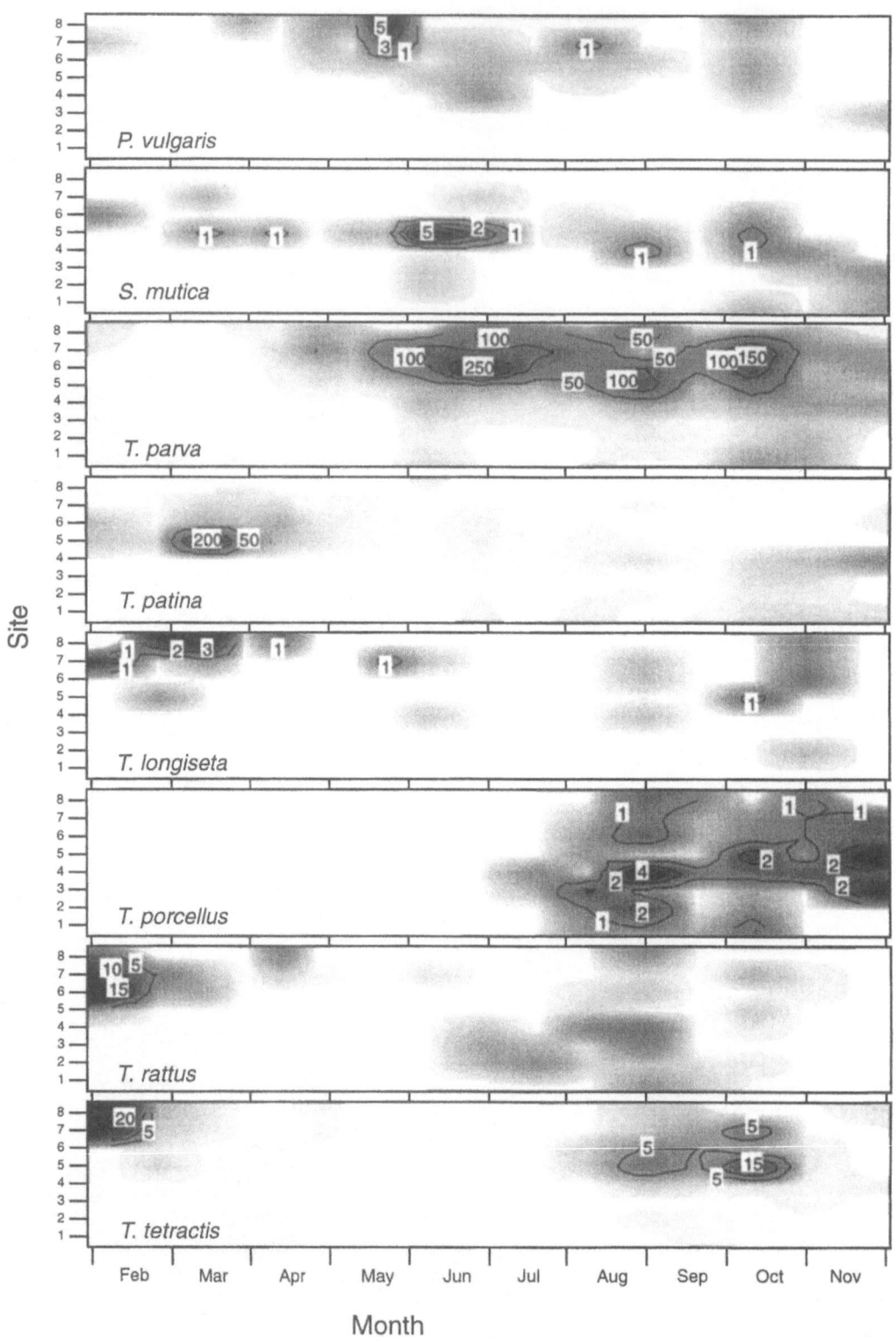

Figure 6c. Spatial and temporal distribution of abundances (ind. substrate $^{-1}$) of common and abundant species.

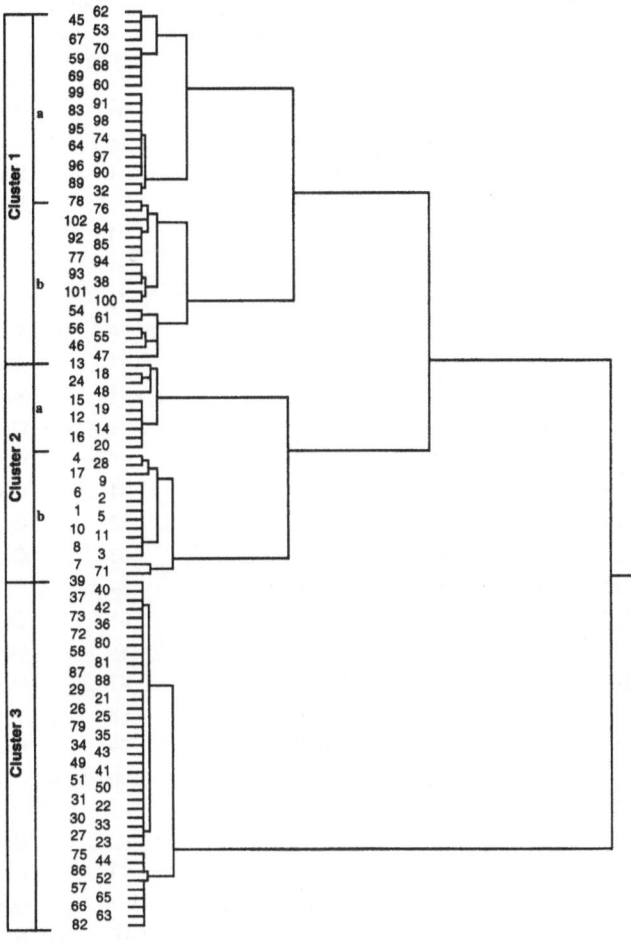

Figure 7. Cluster analysis of abundant rotifer species from seasonal samples. Numbers correspond to sample numbers. The date and site of each sample number are given in Table 2 (see p. 192).

and spring, even though physical and chemical conditions were most uniform throughout the ecotone at these times. Cluster 2 could also be divided into two groups, 2a: February–March, and 2b: March–April, showing that there was a temporal differentiation of communities through this period.

(ii) Canonical correspondence analysis
The results of CCA are presented as biplots (Figure 8) on which the sample, species and environmental loadings on the two major axes are graphed together (Ter Braak, 1987). The sample scores are plotted as points and the species and environmental loadings are represented by arrows, with the point of each arrow located at the intersection of the loading for that species on each axis. The direction of the arrows thus indicate the association between samples, species, and environmental variables and the length of the arrows the strength of the associations. Only Axis 1 (eigenvalue

0.21) and Axis 2 (eigenvalue 0.08), the two most important axes, are used in the biplots.

The spread of sample scores corresponded in general with the groupings found in cluster analysis. Cluster 1 (winter/spring) samples are mainly negatively associated with both Axis 1 and 2, Cluster 2 (summer/autumn) samples are generally positively associated with Axis 1 and negatively with Axis 2, and Cluster 3 (autumn/winter, mid-spring) samples are negatively associated with Axis 1, and mainly positively with Axis 2.

M. ventralis, L. bulla, L. luna, L. triptera, L. decipiens, K. cochlearis, and *T. longiseta* are strongly positively associated with Axis 1 (right side of ordination), and *M. mucronata, T. porcellus, C. obtusa* were strongly negatively associated with Axis 1 (left hand side of ordination). *P. vulgaris* had strong positive associations with Axis 2, and *M. ventralis* and *L.*

Table 2. Key to sample numbers (date and site) for the Cluster analysis and CCA ordination

Number	Date	Site	Number	Date	Site	Number	Date	Site
1	February 7	8	35	June 6	6	69	August 29	2
2	February 7	7	36	June 6	5	70	August 29	1
3	February 7	6	37	June 6	4	71	September 19	8
4	February 7	5	38	June 6	3	72	September 19	7
5	February 24	8	39	June 6	2	73	September 19	6
6	February 24	7	40	June 6	1	74	September 19	5
7	February 24	6	41	June 27	8	75	September 19	4
8	February 24	5	42	June 27	7	76	September 19	3
9	March 14	8	43	June 27	6	77	September 19	2
10	March 14	7	44	June 27	5	78	September 19	1
11	March 14	6	45	June 27	4	79	October 10	8
12	March 14	5	46	June 27	3	80	October 10	7
13	March 28	8	47	June 27	2	81	October 10	6
14	March 28	7	48	June 27	1	82	October 10	5
15	March 28	6	49	July 18	8	83	October 10	4
16	March 28	5	50	July 18	7	84	October 10	3
17	April 11	8	51	July 18	6	85	October 10	2
18	April 11	7	52	July 18	5	86	October 10	1
19	April 11	6	53	July 18	4	87	October 30	8
20	April 11	5	54	July 18	3	88	October 30	7
21	April 25	8	55	July 18	2	89	October 30	6
22	April 25	7	56	July 18	1	90	October 30	5
23	April 25	6	57	August 8	9	91	October 30	4
24	April 25	5	58	August 8	7	92	October 30	3
25	May 9	8	59	August 8	4	93	October 30	2
26	May 9	7	60	August 8	3	94	October 30	1
27	May 9	6	61	August 8	2	95	November 21	8
28	May 9	5	62	August 8	1	96	November 21	7
29	May 23	8	63	August 29	8	97	November 21	6
30	May 23	7	64	August 29	7	98	November 21	5
31	May 23	6	65	August 29	6	99	November 21	4
32	May 23	5	66	August 29	5	100	November 21	3
33	June 6	8	67	August 29	4	101	November 21	2
34	June 6	7	68	August 29	3	102	November 21	1

decipiens had the strongest negative associations with Axis 2.

The ordination indicated that most of the sample groupings determined from cluster analysis were associated with various groupings of species: Cluster 1 (winter/spring samples) was associated with *M. mucronata, T. porcellus, C. obtusa, L. lunaris, S. mutica, L. flexilis,* and *Monommata* sp. (Species Group A (SGA)); Cluster 2 (summer/autumn samples) was associated with *L. bulla, L. hornemanni, L. luna, L. triptera, L. decipiens, T. longiseta, T. tetractis, M. ventralis,* and *T. patina* (Species Group B (SGB)); and

Cluster 3 (autumn/winter, mid-spring samples) was associated with *T. parva, P. vulgaris* (Species Group C (SGC)). Comparing these species groups with the spatial and temporal dynamics of species (Figure 6a–c), SGA was most abundant through August to September and SGB consists of species having bimodal seasonal distributions with major peaks in February and minor peaks in August and September. SGC was most abundant in site 6 to 8 from May to October. Groupings of species thus mainly reflect seasonal distribution of species, with some spatial effect.

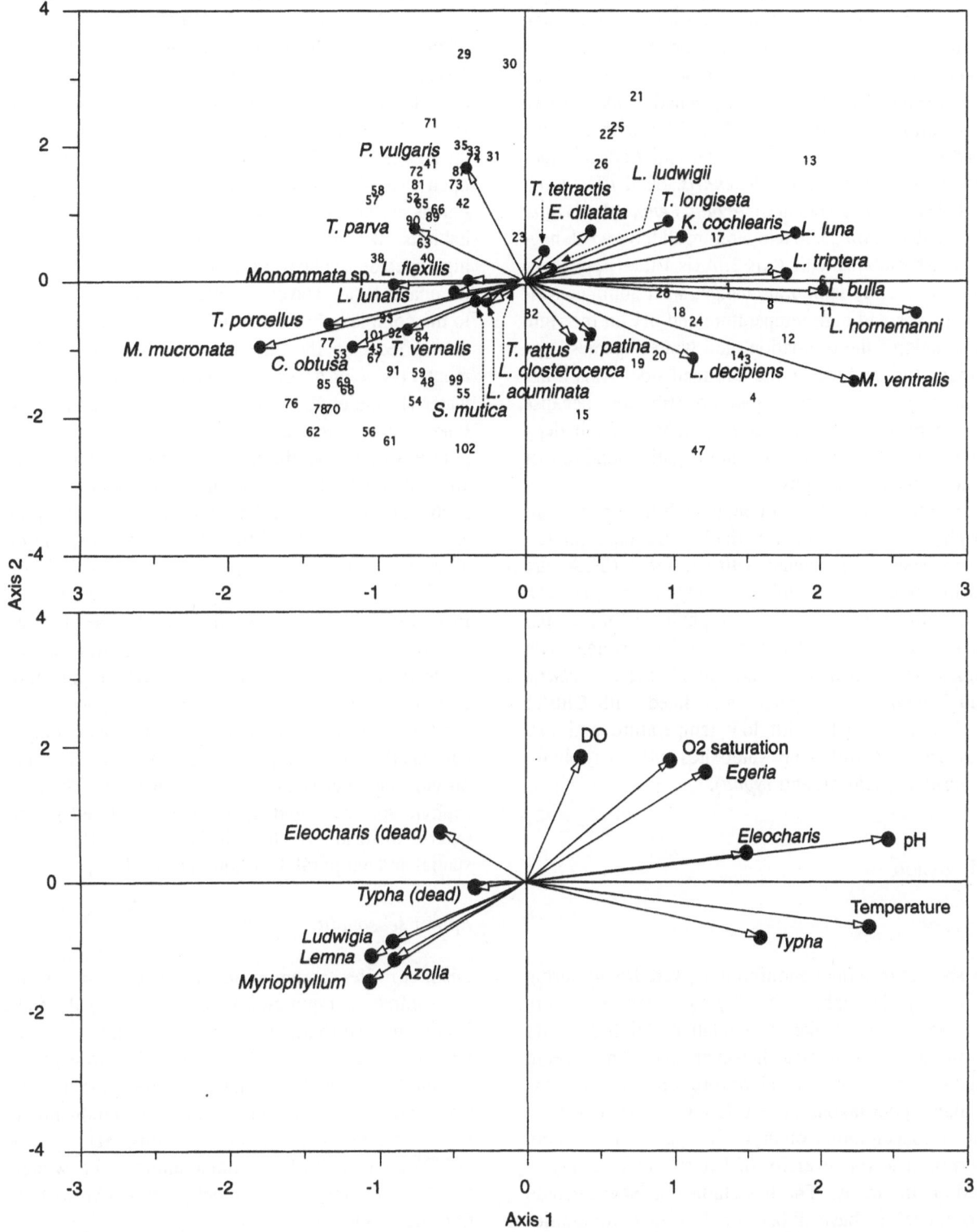

Figure 8. Ordination diagram based on canonical correspondence analysis (CCA) of rotifer species scores with respect to physical, chemical and macrophyte variables. Sample scores of rotifers species, macrophytes, physical and chemical variables were scaled to fit the sample site ordination. Numbers correspond to sample numbers. The date and site of each sample number are given in Table 2. Some sample numbers have been omitted from the ordination for clarity.

In the ordination of environmental variables, pH and temperature are strongly positively associated with Axis 1, *Typha, Eleocharis* moderately so, and *Egeria,* DO, and % O weakly associated. *Myriophyllum, Lemna, Ludwigia,* and *Azolla* had weak negative associations with this axis. Variables with positive loadings on Axis 2 are DO, %O, and *Egeria* (moderately associated), and *Eleocharis,* dead *Eleocharis* and pH (weakly associated). *Azolla, Lemna, Ludwigia, Typha,* dead *Typha* and temperature all have weak negative associations with Axis 2. Axis 1 thus appears to indicate mainly temporal change, with summer rotifer species and high temperature and pH on the right hand side of the ordination, and winter species, low temperature and pH on the left hand side. Axis 2 appears to indicate mainly spatial distribution, because of the positive loadings of macrophytes from deep sites (e.g. *Egeria*) and DO, and negative loadings of shallow water macrophytes.

Comparing rotifer groupings with the ordinations of physical and chemical variables and macrophytes, rotifer species associated with Cluster 1 (SGA) are positively correlated with low temperature, pH, DO, and with shallow water macrophytes. Rotifer species associated with Cluster 2 (SGB) correlate with high temperature and pH, and with *Egeria, Eleocharis* and *Typha.* Rotifer species associated with Cluster 3 (SGC) associate with low temperature and pH, high DO, and with deep water sites and macrophytes (*Egeria, Eleocharis,* and *Typha*).

Discussion

Diversity

58 species of monogonont rotifers were found during this study, the highest diversity so far recorded from any New Zealand lake. This relatively high diversity is probably not atypical however, and it may seem high only because of a lack of rotifer research in New Zealand, poor taxonomic resolution in earlier studies, and a concentration of these on open waters where there is a lower diversity of habitats than in vegetated environments. The few studies of New Zealand rotifer ecology have all been on limnetic communities and have usually found no more than 10 species per lake (Green, 1974, 1976; Forsyth & McCallum, 1980; Chapman et al., 1985). Rotifer species commonly found to numerically dominate limnetic communities in New Zealand, e.g. *Asplanchna* and *Conochilis* species (Chapman, 1973; Green, 1976; Forsyth & McCallum, 1980), were absent, probably due to avoidance of the shore by these forms (Pennak, 1966; Preißler, 1977; Saunders-Davies, 1989). Sanoamuang (1992) recorded 44 species in the plankton of Lake Grasmere in the South Island of New Zealand, though most of these were benthic or littoral species present in the open water due to the shallowness of the lake. Rotifer faunas are generally found to be diverse in vegetated habitats, with the littoral region found to support a high species richness compared with the open water (Pennak, 1966; Havens, 1991). This is largely due to the complex physical structure of macrophytes and their effects on the physical and chemical environment, producing a variety of habitats not found in open water (Irvine et al., 1990; Pontin & Shiel, 1995). In Lake Rotomanuka, diversity was spatially and temporally variable but did not appear to be controlled by any individual physical or chemical feature. Pennak (1966) and Pontin & Shiel (1995) found species preferences of rotifers for different macrophytes, and the diversity of crustacean zooplankton (e.g. Quade, 1969; Shiel, 1976) and macroinvertebrates (e.g. Dvorak and Best, 1982; Talbot and Ward, 1987) has been shown to differ between macrophyte species. These studies relate faunal distribution to macrophyte architecture, with macrophytes of complex structure generally being found to support a greater diversity of periphytic animals than less complex macrophytes because they provide a greater diversity of microhabitats. The macrophyte species found in this study differed greatly in growth form, which is likely to have affected the spatial and temporal distribution of rotifer species.

Rotifer seasonality

Distinct rotifer 'communities' appeared to occur during summer–autumn and winter–spring. In the Lake Rotomanuka littoral, changes in community coincide with the rise in water level, and therefore also a change from marked gradients in physical and chemical conditions in summer and autumn, to more homogenous conditions in winter and spring. Sanoamuang (1992) defined similar summer–autumn and winter–spring communities for limnetic rotifers from Lake Grasmere, South Island.

CCA indicated that temperature and pH were the most important physical and chemical variables associated with the temporal distribution of species. High densities of the summer–autumn species were associated with high pH and temperature, and vice versa for winter–spring species. Species associated with high (e.g. *L. luna, L. bulla*), or low (e.g. *M. mucronata, T. porcellus, C. obtusa*) pH and temperature in Lake Rotomanuka appear to have similar preferences in Europe and Australia (cf. Berzins & Pejler, 1987; Shiel & Koste, 1993 and references therein). CCA also indicated a relationship between rotifer distribution and macrophyte species, so temporal variability in macrophyte structure may also be important in determining seasonality of rotifers in some sites.

There have been few studies on the temporal distribution of New Zealand rotifers, and all have been on planktonic species. There appears to be no consistency in timing of the seasonality of communities, or of individual species. Peak total abundances of limnetic rotifers have been found to occur in spring (Chapman et al., 1975; Green, 1976), autumn (Forsyth & McCallum, 1980), autumn and winter (Jolly & Chapman, 1977), and in both autumn and spring (Sanoamuang, 1992). In this study, major peaks in total abundances occurred in summer, with minima in autumn, with a wide amplitude of fluctuation. A lack of obvious seasonality is characteristic of New Zealand planktonic crustacea and is thought to be due to the country's oceanic climate which leads to mild limnetic temperatures and therefore less severe climatic constraints on zooplankton reproduction and growth (Chapman & Green, 1987). The seasonal temperature change in the Lake Rotomanuka littoral was typical of New Zealand lakes, which have lowest temperatures between June and August and maxima in January and early February. The yearly temperature variation recorded ranged from 15.4 °C to 16.1 °C between sites, higher than the yearly average surface temperature ranges for limnetic regions in North Island lakes (13 ± 3.1 °C), although still within the range expected (Green et al., 1987). However, it seems likely that peak temperatures may not have been recorded as they were expected in either January or early February before sampling commenced. Wider seasonal temperature ranges within the littoral than the limnetic region may result in seasonality of littoral rotifers being constrained more by temperature than those of the open water.

Spatial pattern of rotifer abundance

Although rotifer species abundances differed spatially, spatial effects were less important in determining cluster analysis groupings than temporal effects. Cluster analysis showed no apparent separation in the summer–autumn community, though it is clear from the distributions of individual species (e.g. Figures 6a–c) that some spatial variability did occur at this time. Oxygen and pH showed marked spatial gradients through summer and autumn. This variability was probably largely a result of macrophyte, periphyton and phytoplankton photosynthesis in the deeper sites, particularly amongst the dense beds of *Egeria*. Shading from willows reducing photosynthesis, limited water movement, and decomposing plant litter are likely to have reduced oxygen in the shallower sites at this time, while pH was probably affected greatly by the varying effects of humic acids from the peaty margins and decomposition of plant litter. Heterogeneous conditions are probably also due to isolation of water between sites, caused by the macrophytes forming barriers to water movement. Variability of physical and chemical conditions is typical of vegetated littorals (e.g. Dvorak, 1970; Pieczynska, 1972), while little or no variability is found in lake margins without vegetation (e.g. Sollberger and Paulson, 1992). However, the marked spatial variability of oxygen and pH during summer and autumn did not appear to affect rotifer distribution, and it was in the winter-spring period when physical and chemical conditions were most homogenous across the ecotone that cluster analysis indicated spatial variability of species to be greatest as shown by separation of sites 1–3 from 4–8. Similarly, CCA separated shallow and deep sites as indicated by groupings of shallow and deep macrophyte associations, with *M. mucronata, T. porcellus, T. vernalis* associated with macrophytes from the near shore sites (*Myriophyllum, Ludwigia, Lemna* and *Azolla* (sites 1–4 only)), *M. ventralis* and *L. hornemanni* with the emergent *Typha* (site 5), and *L. luna, K. cochlearis* and *T. longiseta* with *Egeria* (site 8). CCA indicated that rotifer spatial variability was also associated with oxygen, although this may be an effect of greater oxygen levels generally associated with *Egeria*. Greater mixing, and hence a homogeneity of oxygen and pH within the ecotone, occurred at this time when water depths were greater and the dieback of emergent vegetation removed barriers for water movement. These results perhaps indicate that the characteristics of different macrophyte species were more responsible for spatial variability than the physical or chemical features. Such characteristics may include architecture and surface characteristics (affecting the ability of rotifer species to move over surfaces, the density of periphytic food, and refugia from predators), growth rates (affecting rate of appearance of new surfaces for colonization by rotifers), and the release of chemical compounds.

It is apparent that in Lake Rotomanuka the littoral ecotone is a complex and dynamic system with respect to physical, chemical, and biotic factors and that various attributes of this system affect the distribution of rotifer species. This is likely to be common to all lakes with well-developed littoral regions due to the complexity and variability brought about by the presence of macrophytes within the system. While most rotifer species appeared to have associations with one or more of the measured physical and chemical variables or macrophyte species, it may be that it is not these factors *per se* but factors that co-vary with them that are important. For example, food availability, which is thought to limit populations of limnetic zooplankton species in New Zealand (Chapman & Green, 1987; Chapman, 1973; Green, 1976), may have a significant effect in the littoral also, especially when shading by marginal trees and emergent species are taken into account. Similarly, predator densities are also likely to vary spatially and temporally with respect to heterogeneity and so affect rotifer densities.

Acknowledgments

We are grateful to Mr M. Sanson for allowing access to the lake. I.C.D. thanks his parents for financial support throughout, his father for aid in the construction of a boardwalk, and Adrienne Grant, Nic Brady, Polly Morrow, Bethany Hodgetts, Judith Pryor, Lissa McKinney, David Speirs, Tracie Dean, and Keith Hamill for assistance in the field and/or support. Lee Laboyrie provided valuable field assistance, Frank Bailey drafted and Mary de Winton provided useful comments on Figure1. Catherine Beard of the Waikato Herbarium aided with identification of macrophyte species.

References

Berzins, B. & B. Pejler, 1987. Rotifer occurrence in relation to pH. Hydrobiologia 147: 107–116.

Berzins, B. & B. Pejler, 1989a. Rotifer occurrence in relation to temperature. Hydrobiologia 175: 223–231.

Berzins, B. & B. Pejler, 1989b. Rotifer occurrence in relation to oxygen content. Hydrobiologia 183: 165–172.

Campbell, J. M., W.C. Clark & R. Kosinski, 1982. A technique for examining microspatial distribution of Cladocera associated with shallow water macrophytes. Hydrobiologia 97: 225–232.

Champion, P. D. M. D. de Winton & P. J. de Lange, 1993. The vegetation of the Lower Waikato lakes. Volume 2, Vegetation of thirty-eight lakes in the Lower Waikato. NIWA Ecosystems Publication 7.

Chapman, M. A., 1973. *Calamoecia lucasi* Copepoda, Calanoida and other zooplankters in two Rotorua, New Zealand, lakes. Int. Rev. ges. Hydrobiol. 58: 79–104.

Chapman, M. A. & J. D. Green, 1987. Zooplankton Ecology. In A. B. Viner (ed.), Inland Waters of New Zealand. New Zealand DSIR Bulletin 241, Wellington: 225–263.

Chapman, M. A. J. D. Green & V. H. Jolly, 1975. Zooplankton. In V. H. Jolly & J. M. A. Brown (eds), New Zealand Lakes. Auckland University Press, Auckland: 209–230.

Chapman, M. A. J. D. Green & V. H. Jolly, 1985. Relationships between zooplankton abundance and trophic state in seven New Zealand lakes. Hydrobiologia 12: 119–136.

de Winton, M. D. & P. D. Champion, 1993. The vegetation of the Lower Waikato lakes. Volume 1. Factors affecting the vegetation of lakes in the Lower Waikato. NIWA Ecosystems Publication 7.

Dvorak., J. 1970. Horizontal zonation of macrovegetation, water properties and macrofauna in a littoral stand of *Glyceria aquatica* L. Wahlb. in a pond in South Bohemia. Hydrobiologia 35(1): 17–30.

Dvorak, J. & P. H. Best, 1982. Macro-invertebrate communities associated with the macrophytes of Lake Vechten: structural and functional relationships. Hydrobiologia 95: 115–126.

Edmondson, W. T., 1944. Ecological studies of sessile Rotatoria. Part 1. Factors affecting distribution. Ecol. Monogr. 14: 31–66.

Elliott, J. M., 1971. Some methods for statistical analysis of samples of benthic invertebrates. Freshwater Biological Association Scientific Publication 25.

Forsyth, D. J. & I. D. McCallum, 1980. Zooplankton of Lake Taupo. N.Z. J. mar. Freshwat. Res. 14(1): 65–69.

Green, J. D., 1974. The limnology of a New Zealand Reservoir, with particular reference to the life histories of the copepods *Boeckella propinqua* Sars and *Mesocyclops leuckarti* Claus. Int. Rev. ges. Hydrobiol. 59(4): 441–487.

Green, J. D., 1976. Plankton of Lake Ototoa, a sand dune lake in Northern New Zealand. N.Z. J. mar. Freshwat. Res. 10(1): 43–59.

Green, J. D., A. B. Viner & D. J. Lowe, 1987. The effect of climate on lake mixing patterns and temperatures. In A. B. Viner (ed.), Inland Waters of New Zealand. New Zealand DSIR Bulletin 241, Wellington: 65–96.

Havens, K. E., 1991. Summer zooplankton dynamics in the limnetic and littoral zones of a humic acid lake. Hydrobiologia 215: 21–29.

Irvine, K., H. Balls & B. Moss, 1990. The entomostracan and rotifer communities associated with submerged plants in the Norfolk Broadland – Effects of plant biomass and species composition. Int. Rev. ges. Hydrobiol. 75: 121–141.

Irwin, J., 1982. Lake Mangahia: Lake Rotomanuka: Lake Rotomanuka South bathymetry, 1:2000. N.Z. Oceanogr. Inst. Chart, Lake Series.

Johnson, P. & P. Brooke, 1989. Wetland plants in New Zealand. DSIR Publishing, Wellington.

Jolly, V. H. & M. A. Chapman, 1977. The comparative limnology of some New Zealand lakes. 2. Plankton. N.Z. J. mar. Freshwat. Res. 11: 307–340.

Koste, W., 1978. 'Rotatoria. Die Rädertiere Mitteleuropas. Bestimmungswerk begründet von Max Voigt'. II. Tafelband. Gebrüder Borntraeger, Stuttgart.

Koste, W. & R. J. Shiel, 1991. Rotifera from Australian inland waters. VII. Notommatidae (Rotifera: Monogononta). Trans. r. Soc. S. Aust. 115: 111–159.

Mikschi, E., 1989. Rotifer distribution in relation to temperature and oxygen content. Hydrobiologia 186/187: 209–214.

Pennak, P. W., 1966. Structure of zooplankton populations in the littoral macrophyte zone of some Colorado lakes. Trans. am. microsc. Soc. 85(3): 329–349.

Pieczynska, E., 1972. Ecology of the Eulittoral zone of lakes. Ekol. Pol. 20: 637–732.

Pieczynska, E., 1990. Lentic aquatic-terrestrial ecotones: their structure, functions and importance. In R.J. Naiman & H. Decamps (eds), The Ecology and Management of Aquatic-terrestrial Ecotones. Man and Biosphere series 4: UNESCO Paris and the Parthenon Publishing Group.

Pontin, R. M. & R. J. Shiel, 1995. Periphytic rotifer communities of an Australian seasonal floodplain pool. Hydrobiologia 313/314: 63–67.

Preißler, K., 1977. Do Rotifers Show 'Avoidance of the Shore'? Oecologia 27: 253–260.

Quade, H. W., 1969. Cladoceran faunas associated with aquatic macrophytes in some lakes in Northwestern Minnesota. Ecology 50(2): 170–179.

Roff, J. C. & R. E. Kwiatkowski, 1977. Zooplankton and zoobenthos communities of selected northern Ontario lakes of different acidities. Can. J. Zool. 55: 899–911.

Ruttner-Kolisko, A., 1974. Plankton Rotifers, Biology and Taxonomy. Die Binnengewasser. 26(1) Suppl: 1–46.

Sanoamuang, L. O., 1992. The ecology of mountain lake rotifers in Canterbury, with particular reference to Lake Grasmere and the genus *Filinia* Bory de Vincent. Ph.D thesis, University of Canterbury, New Zealand.

Saunders-Davies, A. P., 1989. Horizontal distribution of the planktonic rotifers *Keratella cochlearis* Bory de St Vincent and *Polyarthra vulgaris* Carlin in a small eutrophic lake. Hydrobiologia 186/187: 153–156.

Shiel, R. J. 1976. Associations of Entomostraca with weedbed habitats in a Billabong of Goulburn River, Victoria. Aust. J. mar. Freshwat. Res. 27: 533–549.

Shiel, R. J., 1993. Rotifera from Australian inland waters. IX. Gastropodidae, Synchaetidae, Asplanchnidae (Rotifera: Monogononta). Trans. r. Soc. S. Aust. 117: 111–139.

Sollberger, P. J. & P. J. Paulson, 1992. Littoral and limnetic zooplankton communities in Lake Mead, Nevada-Arizona, USA. Hydrobiologia 237: 175–184.

Spyglass Inc., 1993. Spyglass Transform: Quick tour and reference, Version 3.0. Champaign.

Talbot, J. M. & J. C. Ward, 1987. Macroinvertebrates associated with aquatic macrophytes in Lake Alexandrina, New Zealand. N.Z. J. mar. Freshwat. Res. 21: 199–213.

Ter Braak, C. J. F., 1988. CANOCO – A FORTRAN program for Canonical Community Ordination. Ithaca, New York, USA.

Uhelova, B. & K. Pribil, 1978. Water chemistry in the fishpond littorals. In D. Dykyjova & J. Kvet (eds), Pond Littoral Ecosystems; Structure and Functioning; Methods and Results of Quantitative Ecosystem Research in the Czechoslovakian IBP Wetland Project. Ecological Studies 28, Springer-Verlag: 126–140.

Vitt, D. H. & S. Bayley, 1984. The vegetation and water chemistry of four oligotrophic basin mires in northwestern Ontario. Can. J. Bot. 62: 1485–1500.

Wallace, R. L., 1977a. Adaptive advantages of substrate selection by sessile rotifers. Arch. Hydrobiol. Beih. Ergebn. Limnol. 8: 53–55.

Wallace, R. L., 1977b. Distribution of sessile rotifers in an acid bog pond. Arch. Hydrobiol. 79: 478–505.

Wetzel, R. G. & G. E. Likens, 1979. Limnological Analyses. W. B. Saunders.

Wilkinson, L., 1992. SYSTAT: Statistics, Version 5.2. Evanston, USA.

Winterbourn, M. J. & M. H. Lewis, 1975. Littoral fauna. In V. H. Jolly & J. M. A. Brown (eds), New Zealand Lakes. Auckland University Press, Auckland: 271–280.

Hydrobiologia **387/388**: 199–206, 1998.
E. Wurdak, R. Wallace & H. Segers (eds), Rotifera VIII: A Comparative Approach.
© *1998 Kluwer Academic Publishers.*

Zooplankton succession and thermal stratification in the polymictic shallow Müggelsee (Berlin, Germany): a case for the intermediate disturbance hypothesis?

Birgit Eckert & Norbert Walz*
*Institute of Freshwater Ecology and Inland Fisheries, Department of Lowland Rivers and Shallow Lakes, Müggelseedamm 260, 12562 Berlin, Germany (*author for correspondence)*

Key words: cladocerans, Rotifera, seasonal succession, diversity, wind velocity

Abstract

From the end of May to November 1995 the succession of rotifers and cladocerans was investigated in Müggelsee with samples taken twice a week. *Keratella cochlearis* was the only rotifer which was found on every sampling day and this species also showed the highest abundances. During summer, when frequencies of strong wind events were low and water was strongly stratified, three small cladocerans were dominant (*Daphnia cucullata, Chydorus sphaericus, Eubosmina coregoni*). Food supply was the main limiting factor for *Keratella* spp. and *Synchaeta* spp. In autumn, however, when the intervals between strong winds were shorter, rotifers with shorter periods for population development prospered. Zooplankton diversity first increased and subsequently decreased after disturbances. The results do not support the intermediate disturbance hypothesis in its present formulation.

Introduction

The zooplankton populations in shallow lakes are much more exposed to changing conditions than those inhabiting deeper well stratified lakes. Due to the stochasticity of weather patterns this leads to unpredictability of species succession (Sommer et al., 1986). In shallow lakes during mictic periods r_{max}-strategists should be favoured, which of course involves most rotifer species (Walz, 1995). On the other hand, during periods of stratification biotic interactions should prevail and rotifers are often outcompeted, mostly by large *Daphnia* species with which they often show alternative fluctuations (e.g., Lampert & Rothhaupt, 1991). Competition may be triggered by mechanical interference or exploitative competition (Gilbert, 1988). Further rotifers are often the target of intense invertebrate predation (Williamson, 1983). However, when physical conditions are changing and fluctuating, rotifers should be able to evade these limiting factors. Therefore, the domain of rotifers should be fluctuating conditions.

In contrast to the well developed hypotheses of plankton regulation under equilibrium conditions

(e.g., Tilman's (1982) resource competition theory), non-equilibrium theories have not been so well developed. One idea that has been explored is the intermediate disturbance hypothesis, IDH (Connell, 1978; Reynolds, 1988; Reynolds et al., 1993). This hypothesis states that if the intervals between disturbances are too long, certain species become extirpated. Conversly, other species are not able to respond to very high frequencies of disturbance. Thus, only disturbances of intermediate frequencies permit species to coexist, and these species are sandwiched between the millstones of competitive exclusion and high frequency of disturbance. It is here that the highest diversities are to be expected there.

So far most tests of the IDH have focused on phytoplankton. For example, only a few papers (e.g., Weider, 1992) attempt to apply this hypothesis to zooplankton. This may be due to the fact that most of the papers dealing with this subject are reports of laboratory observations; most evidence comes from semicontinuous cultures (Sommer, 1995). The reason for this may be that such experiments are much more difficult to perform with zooplankton than with phytoplankton. However, this hypothesis also waits for

200

additional verification by phytoplankton field studies (Padisák et al., 1993).

The objective of this study was to test if the IDH does apply to zooplankton and to test of r_{max}-strategists (like many rotifers) are favoured by higher disturbance frequencies. Two periods with contrasting weather conditions in summer 1995, especially with different storm frequencies, gave the opportunity to test this. From the beginning of July up to the end of August an extraordinary long stratification period established. Later the weather was stormy. In other years thermally stable conditions from April to October generally lasted for only 1–2 weeks (Behrendt et al., 1993). Thus, from end of May to mid of November we analyzed zooplankton succession as a function of the biotic and abiotic steering factors.

Methods

The Mügelsee is a shallow (mean depth 4.9 m, max. depth 8 m), eutrophic, polymictic lake with a surface area of 7.2 km^2 situated in the southeast of Berlin (Germany). The River Spree flows through the lake producing a retention time of about 42 days. For details on the morphology and hydrology of the lake see Driescher et al. (1993).

Water temperature, conductivity, pH, and O_2 concentration were recorded in vertical profiles at every sampling day with a multiparameter, water-quality data transmitter (Hydrolab H20®). Wind velocities and wind directions were measured every 5 min at a station on the north shore 4 m above the lake surface and averaged over 1 day. Total chlorophyll a concentrations and the chlorophyll a fraction < 30 μm were determined according to DIN 38 412 part 16. For further details of methods see Eckert (1996).

All samples were taken twice a week from June to November 1995 using a 3.4 l Friedinger sampler. The 3 sampling points were situated in a triangle of 50 m, approximately 80 m off the northern shore, just above the 4 m counter interval. Samples were collected from 0, 1, 2, and 3 m depth and afterwards mixed to an integrated sample (40.8 l). A 20 l sample of water was screened through a 30 μm mesh and fixed with formaldehyde to a final concentration of 4%. Zooplankton individuals and eggs (attached and detached ones) were counted under a compound microscope. Rotifers were identified according to Koste (1978), cladocerans according to Flössner (1972) and copepods according to Einsle (1993). The detached rotifer

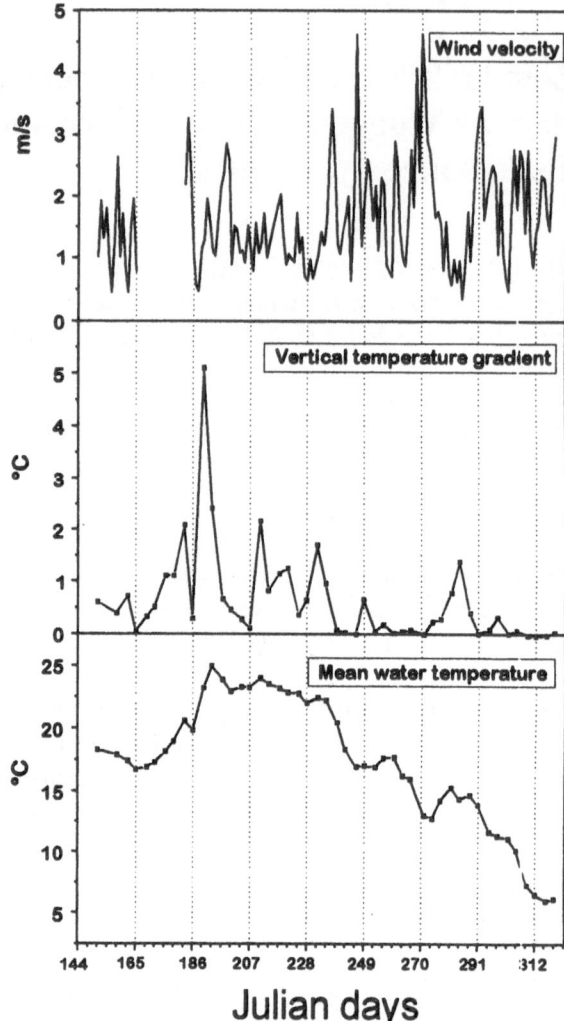

Figure 1. Daily means of wind velocities (upper panel), temperature difference between surface and 4 m depth (middle panel) and mean water temperatures (0–4 m) (lower panel) in Müggelsee 1995.

eggs in the samples were identified to species based on shape and size. The two *Synchaeta* species, *S. oblonga* and *S. kitina*, were counted as one category, because they could not be distinguished in the routine counting. The female egg ratio of the rotifers was calculated by including the detached eggs. Embryonic development times (D_e) were calculated for *K. cochlearis* according to Walz (1983), *K. quadrate* according to Herzig (1983), *S. oblonga* according to Zoufal (1989), *P. sulcata* according to Baker (1979), *D. cucculata*, *C. sphaericus*, *E. coregoni*, *D. brachyurum*, *L. kindti* according to Vijverberg (1980).

Mean periods of population developed (PDP) were calculated for each zooplankton species as the averaged intervals between maxima of egg ratios and following maxima of abundance peaks. Diversity was calculated according to Hurlbert's probability of interspecific encounter (Hurlbert, 1971):

$$PIE = N/(N + 1)(1 - \sum p_i^2),$$

where N = total number of individuals, p_i = share of species i on total number of individuals. Statistical tests were applied according to the statistical software package SPSS or according to Sachs (1992).

Results

Wind velocity and thermal stratification

Between June and November in 1955 wind velocities over the Müggelsee were characteristically different during two time periods (Figure 1). A calmer period lasted from June until the end of August, during which a wind speed of 2.5 m/s was exceeded only three times The mean length of the period between these disturbances was 23.8 days ($N = 3$) (wind-1-periods). During these intervals a temperature gradient between the surface and 4 m depth was established. This gradient disappeared only temporarily after the strong winds on July 4 and July 18. At the end of June no wind velocity measurements were available, but this time was quiet as indicated by the steady warming trend (Figure 1). In contrast the weather between September and November was stormy and the mean interval between wind speedings was significantly lower (13.6 days, $N = 6$, t-test, $P < 0.05$, wind-2-periods). Except for a period in the middle of October, when a weak stratification developed, the water was isothermal and the water temperature decreased continuously.

Chlorophyll and Secchi depth

Starting at a minimum during the clear water phase in May, chlorophyll a concentrations increased during the first calm period from June 8 and decreased to very low levels after the first strong wind event at the beginning of July (Figure 2). Further chlorophyll development was retarded by the strong wind event on July 18, but total chlorophyll, especially the larger sized fraction (> 30 μm; difference between both curves) increased during the next long calm period (until August 26). In August the biomass of cyanobacteria was highest in this fraction. On August 10 a bloom of *Aphanizomenon flos-aquae* (Sigrid Hoeg, IGB, personal communication) coincided with the shallowest

Secchi depth recorded for the sampling period (Figure 2). In contrast, in August the smaller size fraction of phytoplankton (< 30 μm) decreased to very low concentrations probably due to grazing by cladocerans. Chlorophyll a concentrations rapidly decreased after strong winds at the end of August and the beginning of September. During a 14-day calm period (after the disappearance of the cladocerans) the phytoplankton increased again (September 10) until a stormy period (end of September) when the phytoplankton decreased. In the following 21-day period of calm stratified waters (starting at September 28), chlorophyll increased to its autumnal maximum in October. However, this peak almost disappeared during the stormy phase beginning in November. Overall, there was a strong negative correlation between wind velocity and total chlorophyll concentrations (Spearman rank correlation $r_s = -0.41, N = 40, P < 0.01$).

Zooplankton

Larger daphnids (*D. galeata, D. galeata x cucculata, D. hyalina, D. lonispina*) were very rare during the sampling period. *D. galeata* and *Bosmina longirostris* dominated during the clear water phase (up to June 15). *D. galeata* increased again only in the end of October. For this reason in the summer the smaller cladocerans, *D. cucculata, Chydorus sphaericus*, and *Eubosmanina coregona*, could dominate but not before the water was stratified and smaller phytoplankton (< 30 μm) had recovered their population from the clear water phase (Figure 3). The abundances of the smaller cladocerans were significantly correlated to temperature gradients and presence of cyanobacteria (mainly *Aphanizomenon flos-aquae*, Table 1). A succession could be observed from *D. cucullata* to *Chydorus sphaericus* and *Eusbosmina coregoni. Diaphanosoma brachyurum* was rare during this time. *Leptodora kindti* was present throughout this period but at low densities (< 1.5 individuals/l) but disappeared in mid September.

The abundance of adult cyclopoid copepods (*Mesocyclops leuckarti, Thermycyclops oithonoides, Acanthocyclops robustus*) increased from the beginning of July to 65 individuals/l in the middle of August and decreased at the end of August. During this time the cyclopoid copepods co-occurred with the smaller cladocerans (Table 1).

During the sampling period, 24 rotifer species were found, but only 6 species comprised 80% of the total rotifer numbers on 43 of 48 sampling days. On

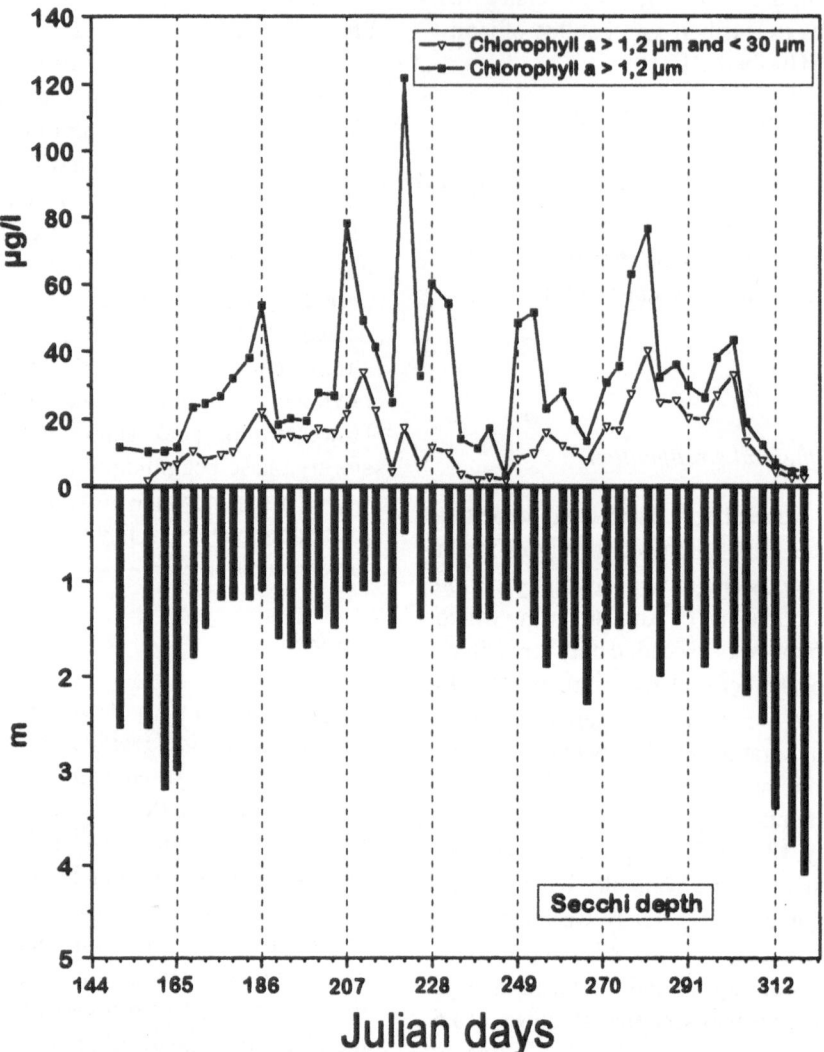

Figure 2. Total chlorophyll *a* and chlorophyll *a* of the size fraction <30 µm (upper panel) and Secchi depths (lower panel) in Müggelsee 1995.

Table 1. Spearman rank correlations between zooplankton and abiotic parameters

	Chlorophyll *a* <30 µm	Cyclopoid copepods	Cyanobacteria	Temperature gradient
Keratella cochlearis	0.56***			
Keratella quadrata	0.63***			
Synchaeta oblonga/kitina	0.45***	−0.45***		
Trichocerca pusilla	0.72**			
Trichocerca similis	−0.44*		0.73***	
Pompholyx sulcata		0.64***		0.45*
P. sulcata (in August)		−0.77*		
Daphnia cucculata 0.52*	0.8***	0.62*		
Chydorus sphaericus		0.58*	0.92***	0.60*
Eubosmina coregoni	−0.64***			

*P < 0.05; **P < 0.01; ***P < 0.001.

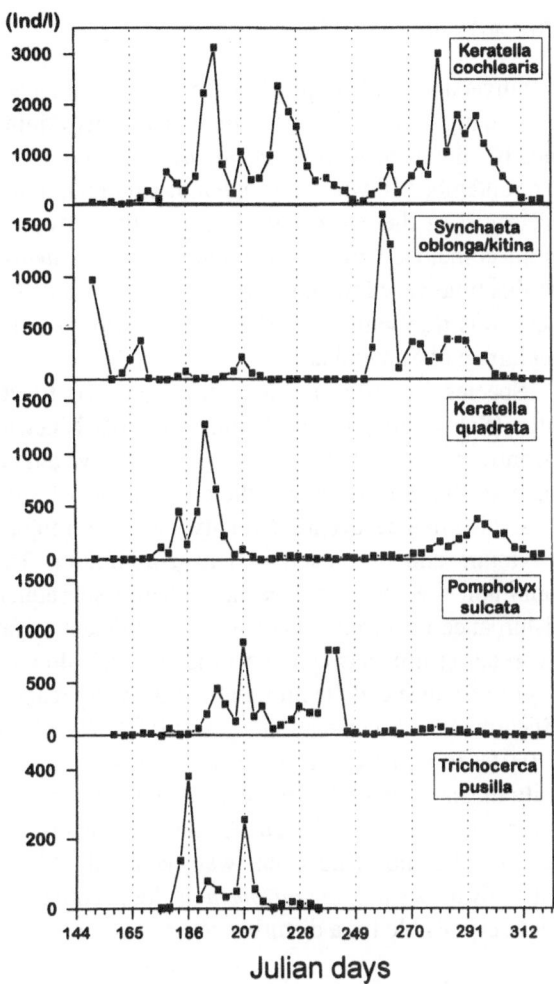

Figure 3. Abundances of the most frequent cladocerans in Müggelsee 1995.

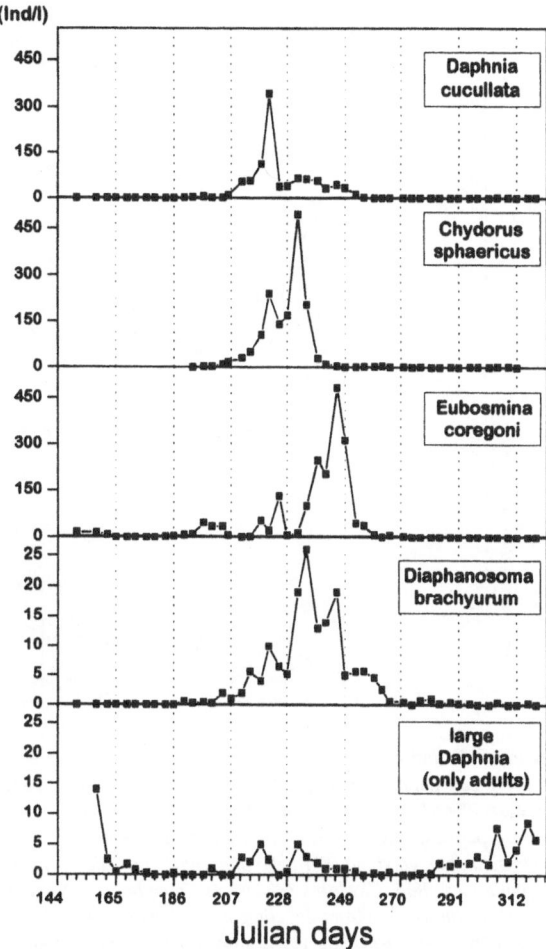

Figure 4. Abundances of the most frequent rotifers in Müggelsee 1955.

the other days the rotifer fauna plankton was more diverse. *Keratella cochlearis* was the only species which was present throughout the study and amounted > 50% of the rotifer numbers in 31 sampling days (Figure 4). This means that this species had growth phases before, within, and after the occurrence of the smaller cladocerans in July/August. *Synchaeta oblonga/kitina* and *Keratella quadrata* developed only before and after this long calm period, while *Trichocera pusilla* developed only before. In contrast, *Pompholyx sulcata* and *Trichocerca similis* were abundant at times when smaller cladocerans were also present.

Duration of developmental periods and wind velocities

PDP of a species represented the length of time necessary to build up a population (Figure 5, upper panel). Of course, PDP were not independent from environmental parameters. In contrast embryonic development times (D_e, Figure 5, lower panel) were only dependent on temperature and were characteristic for a species. As D_e is a major factor influencing PDP, both parameters were correlated to each other (Spearman rank correlation, $r_s = 0.77$, $N = 7$, $P < 0.025$). A nonparametric Kuskal-Wallis-test showed that the PDP of the zooplankton species in Figure 5 differed significantly ($P < 0.001$).

PDP fit to the frequencies of strong wind events. The species with the shortest periods, *Synchaeta oblonga/kitina*, *Keratella quadrata* and *Keratella cochlearis*, were abundant at times when winds occurred frequently (wind-2-periods, Figure 5, upper panel). On the other hand, the cladocerans, *Daphnia cucullata*, *Chydorus sphaericus*, and *Eubosmina coregoni*, and some rotifers, *Pompholyx sulcata* and *Tricocerca similis*, developed greater populations only during wind-1-periods.

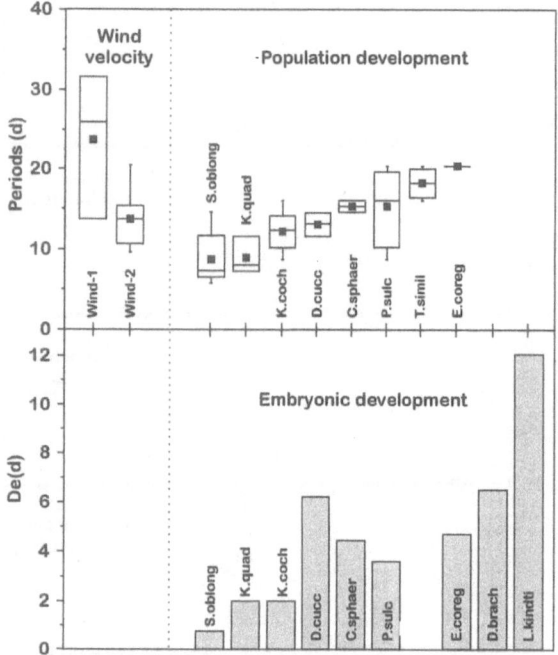

Figure 5. Duration of periods between strong wind events and of the mean duration of population developments (upper panel), duration of embryonic development, D_e at 20 °C (lower panel). Wind-1-periods: rarely storms, wind-2-periods: frequently storms. *Synchaeta oblonga/kitina, Keratella quadrata, Keratella cochlearis, Daphnia cucculata, Chydorus sphaericus, Pompholyx sulcata, Trichocerca similis, Eubosminia coregoni, Diaphanosoma brachyurum, Leptodora kindti.* Squares: arithmetic means; bottom and top of the boxes: 25th and 75th percentiles; median line in boxes: 50th percentile; end of vertical line: 5th and 95th percentiles.

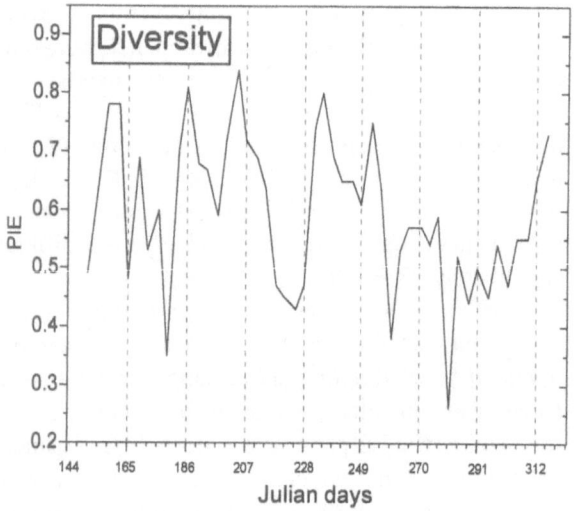

Figure 6. Zooplankton diversity based on number of individuals (copepods, cladocerans, rotifers) in Müggelsee 1955.

Diversity and wind velocity

The diversity of all zooplankton species in Müggelsee showed high fluctuations (Figure 6). Diversity minima mostly coincided with *K. cochlearis* maxima except for September 18 when a *S. oblonga/kitina* maximum occurred. At the end of June the first diversity minimum coincided with a submaximum of *K. cochlearis.* At this time very few other species were present. The same was true with an *Eusbosmina* maximum in the beginning of September.

Species diversities did not correlate to wind velocities (Spearman rank correlation, $P > 0.10$). Likewise the introduction of time lags of 3–4 and 7 days did not increase the correlation coefficients. However, after most disturbance events diversity passed through a maximum and then decreased to a minimum. This minimum seemed to be reached when a subsequent disturbance occurred. There was a significant negative relationship between the time interval after the last disturbance and the minimum of diversity reached (Spearman rank correlation, $r_s = -0.82$, $N = 7$, $P < 0.01$). In all cases strong wind events (> 2.5 m/s, daily means, Figure 2) were considered a disturbance except for August 10 when obviously a disturbance occurred but no wind event was recorded. At this date only the strong *Aphanizomenon* bloom could have been responsible for a disturbance.

Discussion

As in stratified lakes zooplankton succession in the polymictic Müggelsee in calmer periods followed the model by Sommer et al. (1986). The populations were regulated by biotic factors (bottom-up and top-down). In contrast to most other years, large cladocerans, which dominated the clear water phase, seemed to be depressed during the summer by blooms of cyanobacteria (Gliwicz & Lampert, 1990) or by increasing summer fish predation (Barthelmes et al., 1995). Compared to the larger daphnids in Müggelsee, *D. cucculata* revealed very low numbers in gut contents of planktivorous fish (Arndt et al., 1993). Invertebrate predation by *Leptodora* or cyclopoid copepods (Gliwicz & Umana, 1994) on cladocerans was not found (Table 1). Hence there was only evidence for invertebrate predation on rotifers. According to Karabin (1978) *Pompholyx* and *Synchaeta* are the preferred prey of *Mesocylops. Synchaeta oblonga/kitina* was negatively correlated to cyclopoid copepods and

P. sulcata showed a negative relationship in August (Table 1).

Equilibrium models (Wiens, 1984) assume that population densities are food limited and follow fluctuations of their resources. Such assumptions always yield outcomes that predict exclusion of specific species. The succession of the small cladocerans in August was accompanied by a reduction of edible algae due to grazing. Given grazing rates of 1.5 ml ind.$^{-1}$ d^{-1} (Gulati et al., 1991) the smaller cladocerans could have been able to clear 50% of the water per day. Although rotifers cleared only up to 30% (clearance rates up to 0.2 ml ind.$^{-1}$ d^{-1}; Telesh et al., 1955), they could have excluded by exploitative competition as their food concentration thresholds are higher than those of cladocerans (Stemberger & Gilbert, 1987; Gliwicz, 1990). In experimental conditions MacIsaac & Gilbert (1989) showed that *K. cochlearis* could coexist with smaller cladocerans because there was no mechanical interference. In Müggelsee *K. cochlearis* was clearly food limited. Egg ratios were dependent on chlorophyll concentrations ($< 30\mu$m) according to a saturation function (Eckert, 1996). With the decrease of chlorophyll at the end of August and the beginning of September the lowest egg ratios were found. All other dominant rotifers were likely to be food limited as shown by strong correlation coefficients with chlorophyll *a* ($<30 \mu$m) Table 1.

However, we still do not know the reason why smaller cladocerans dominated only in the period between end of July and beginning of September and the rotifers in the periods before and after. In June, with the recovery of the phytoplankton after the clear water phase, rotifers were able to build up their populations earlier than the cladocerans due to higher growth rates and shorter generation times (Boersma & Vijverberg, 1994; Walz, 1995). However, this cannot explain the decline in September, when chlorophyll was recovering again (Figure 2). Bertilsson et al. (1955) found all 4 dominant cladoceran species in high abundances at temperatures above 13 °C. Excluding fish predation which was not studied here, the only known reason for the change was in the weather conditions, especially the frequency of strong winds that increased in September.

Non-equilibrium models (DeAngelis & Waterhouse, 1987) explain succession and diversity by fluctuations of biotic and abiotic parameters. According to the IDH intermediate frequencies will support both, species with pioneer character (high r_{max}, short generation times) and competitive superior species (but competitive exclusion cannot take place because of subsequent disturbance). In the stormy autumn, only *K. cochlearis*, *K. quadrata*, and *S. oblonga/kitina* with short development times (Figure 6) were able to develop substantial populations. In experimental conditions (semicontinuous cultures) phytoplankton diversity was maximal at 3.5 to 7 day intervals between dilutions: i.e. at about three generation times (Gaedeke & Sommer, 1986; Sommer, 1995). For zooplankton the IDH could not be confirmed in this form but there was a correlation between the length of time after the last disturbance and the minimum of diversity reached. If the prevailing time interval between frequent storms had produced this correlation, about 14 days may be considered intermediate. This is longer than the intervals reported for phytoplankton (Padisák, 1993), but it fits for *K. cochlearis* (Walz, 1983), and that time would permit 2–3 generations. Future research is needed to investigate the influence of a broader range of disturbance frequencies on zooplankton species diversity (experimental approaches; e.g., field enclosures, laboratory experiments).

The nature of disturbance is very difficult to define in field studies (Rojo & Álvares Cobelas, 1993). Chorus & Schlag (1993) reported rather sudden stratifications in polymictic lakes as disturbance events for phytoplankton development. In Müggelsee only the inclusion of the strong *Aphanizomenon* bloom led to the above correlation. For this reason the danger is high to succumb to an argument of circular reasoning.

Acknowledgement

We thank Renate Rusche for help with the taxonomic determinations and Rüdiger Biskupek and Bernd Schütze for support with the excursions. Thomas Hintze helped with the graphs and was responsible for the wind measurements. Further, we thank two anonymous reviewers for their valuable advice.

References

Arndt, H., M. Krocker, B. Nixdorf & A. Köhler, 1993. Long-term annual and seasonal changes of meta- and protozooplankton in Lake Müggelsee (Berlin): Effects of eutrophication, food conditions and the impact of predation. Int. Rev. ges. Hydrobiol. 78: 739–402.

Baker, R. L., 1979. Birth rates of planktonic rotifers in relation to food concentrations in a shallow, eutrophic lake in Western Canada. Can. J. Zool. 57: 1206–1214.

Barthelmes, D., F. Fredrich, T. Mattheis, U. Grosch & M. Sommer, 1995. Starke Verbesserungen des Cyprinidenwachstums in anthropogen hypertrophierten, vormals natürlich eutrophen Flußseen: ein ungewöhnlicher Langzeittrend in Gewässern von Berlin (Deutschland). Limnologica 25: 251–275.

Behrendt, H., B. Nixdord & W.-G. Pagenkopf, 1993. Phenomenological description of polymixis and influence on oxygen budget and phosphorus release in Lake Müggelsee. Int. Rev. ges. Hydrobiol. 78: 411–421.

Bertilsson, J. B., B. Berzins & B. Pejler, 1995. Occurrence of limnic micro-crustacea in relation to temperature and oxygen. Hydrobiologia 299: 163–167.

Boersma, M. & J. Vijverberg, 1994. Resource depression in *Daphnia galeata, Daphnia cucculata* and their interspecific hybrid: life history consequences. J. Plankton Res. 16: 1741–1758.

Chorus, I. & G. Schlag, 1993. Importance of intermediate disturbances for the species composition in two very different Berlin lakes. Hydrobiologia 249: 67–92.

Connell, J. H., 1978. Diversity in tropical rainforests and coral reefs. Science 109: 1304–1310.

DeAngelis, D. L. & J. C. Waterhouse, 1987. Equilibrium and nonequilibrium concepts in ecological models. Ecol. Mono. 57: 1–21.

Driescher, E., H. Behrendt, G. Schellenberger & R. Stellmacher, 1993. Lake Müggelsee and its environment – natural conditions and anthropogenic impacts. Int. Rev. ges. Hydrobiol. 78: 327–343.

Eckert, B., 1996. Populationsdynamik der Rotatorien und Cladoceren im Großen Müggelsee von Juni bis November 1995. Diploma thesis, Math. Nat. Faculty, University of Bonn, 89 pp.

Einsle, U., 1993. Crustacea, Copepoda. Calanoida and Cyclopida. G. Fischer, Stuttgart/Jena, 208 pp.

Flössner, D., 1972. Krebstiere, Crustacea: Kiemen- und Blattfüßer, Branchiopoda. Fischläuse, Brachiura. In E. Dahl & F. Plus (eds), Die Tierwelt Deutschlands. 60. Teil, G. Fischer, Jena, 501 pp.

Gaedeke, A. & U. Sommer, 1986. The influence of frequency of periodic disturbances on the maintenance of phytoplankton diversity. Oecologia 71: 25–28.

Gilbert, J. J., 1988. Suppression of rotifer populations by *Daphnia*: A review of the evidence, the mechanisms, and the effects on zooplankton community structure. Limnol. Oceanogr. 33: 1286–1303.

Gliwicz, Z. M., 1990. Food thresholds and body size in cladocerans. Nature 343: 638–640.

Gliwicz, Z. M. & W. Lampert, 1990. Food thresholds in Daphnia species in the absence and presence of blue-green filaments. Ecology 71: 691–702.

Gliwicz, Z. M. & G. Umana, 1994. Cladoceran body size and vulnerability to copepod predation. Limnol. Oceanogr. 39: 419–424.

Gulati, R. D., C. Vuik, K. Siewertsen & G. Postema, G., 1991. Clearence rates of *Bosmina* species in response to changes in trophy and food concentration. Verh. int. Ver. Limnol. 24: 745–750.

Herzig, A., 1983. Comparative studies in the relationship between temperature and duraction of embryonic development of rotifers. Hydrobiologia 104: 237–246.

Hurlbert, S. H., 1971. The nonconcept of species diversity: A critique and alternative parameters. Ecology 52: 577–586.

Karabin, A., 1978. The pressure of pelagic predators of the genus *Mesocyclops* (Copepoda, Crustacea) on small zooplankton. Ekol. Polska 26: 241–257.

Koste, W., 1978. Rotatoria. Die Rädertiere Mitteleuropas. Ein Bestimmungswerk begründet von Max Voigt. Bornträger, Stuttgart. 2nd ed. Vol. 1: Text, 673 pp., Vol. 2: Tables, 234 pp.

Lampert, W. & K. O. Rothhaupt, 1991. Alternating dynamics of rotifers and *Daphnia magna* in a shallow lake. Arch. Hydrobiol. 120: 447–456.

MacIsaac, H. J. & J. J. Gilbert, 1989. Competition between rotifers and cladercerans of different body sizes. Oecologia 81: 295–301.

Padisák, J., 1993. The influence of different disturbance frequencies on the species richness, diversity and quitability of phytoplankton in shallow lakes. Hydrobiologia 249: 135–156.

Padisák, J., C. S. Reynolds & U. Sommer (eds), 1993. Intermediate Disturbance Hypothesis in Phytoplankton Ecology. Developments in Hydrobiology 81. Kluwer Academic Publishers, Dordrecht, 200 pp. Reprinted from Hydrobiologia 249.

Reynolds, C. S., 1988. The concept of ecological succession applied to seasonal periodicity of freshwater phytoplankton. Verh. int. Ver. Limnol. 23: 683–691.

Reynolds, C. S., J. Padisák & U. Sommer, 1933. Intermediate disturbance in the ecology of phytoplankton and the maintenance of species diversity: a synthesis. Hydrobiologia 249: 183–188.

Rojo, C. & M. Álvares Cobelas, 1993. Hypertrophic phytoplankton and the Intermediate Disturbance Hypothesis. Hydrobiologia 249: 43–57.

Sachs, L., 1992. Angewandte Statistik. Anwendung statistischer Methoden. Springer Verlag, Berlin, 7th edn. 846 pp.

Sommer, U., 1995. An experimental test of the intermediate disturbance hypothesis using cultures of marine phytoplankton. Limnol. Oceanogr. 40: 1271–1277.

Sommer, U., Z. M. Gliwicz, W. Lampert & A. Duncan, 1986. The PEG-model of seasonal succession of planktonic events in fresh waters. Arch. Hydrobiol. 106: 433–471.

Stemberger, R. S. & J. J. Gilbert, 1987. Rotifer threshold food concentrations and the size-efficiency hypothesis. Ecology 68: 181–187.

Telesh, I. V., A. L. Ooms-Wilms & R. D. Gulati, 1955. Use of fluorescently labelled algae to measure clearence rate of *Keratella cochlearis*. Freshwater Biol. 33: 349–355.

Tilman, D., 1982. Resource competition and community structure. Princeton University Press, Princeton N.J., 296 pp.

Vijverberg, J., 1980. Effect of temperature in laboratory studies on development and growth of Cladocera and Copepoda from Tjeukemeer, The Netherlands. Freshwater Biol. 10: 317–340.

Walz, N., 1983. Individual culture and experimental population dynamics of *Keratella cochlearis* (Rotatoria). Hydrobiologia 107: 35–43.

Walz, N., 1955. Rotifer populations in plankton communities: Energetics and life history strategies. Experientia 51: 437-453.

Weider, L. J., 1992. Disturbance, competition and maintenance of clonal diversity in *Daphnia pulex*. J. evol. Biol. 5: 505–521.

Wiens, J. A., 1984. Resource systems, populations, and communities. In P. W. Price, C. N. Slobodchikoff & W. S. Gaud (eds), A New Ecology. Novel Approaches to Interactive Systems John Wiley & Sons, New York: 397–436.

Williamson, C. E, 1983. Invertebrae predation on planktonic rotifers. Hydrobiologia 104: 385–396.

Zoufal, W., 1989. Development times of *Synchaeta oblonga* eggs from the Danube (Austria). Hydrobiologia 186/187: 163–165.

Hydrobiologia **387/388**: 207–214, 1998.
E. Wurdak, R. Wallace & H. Segers (eds), Rotifera VIII: A Comparative Approach.
© *1998 Kluwer Academic Publishers.*

Comparison of processes regulating zooplankton assemblages in new freshwater pools

Therese A. Holland & David G. Jenkins
Department of Biology, University of Illinois at Springfield, Springfield, IL 62794, U.S.A.

Key words: zooplankton, rotifers, colonization history, disturbance, nutrient enrichment

Abstract

We compared the relative importance of colonization history to other known regulators of freshwater zooplankton assemblages (i.e. disturbance and nutrient enrichment) during this six-month study of initial colonization in artificial freshwater pools. We experimentally manipulated 16 small (1.5 m diameter) pools for permanence (permanent vs. temporary) and resource availability (± nitrogen & phosphorus). Nitrogen and phosphorus were added to high resource level pools in concentrations typical of eutrophic waters, while low resource level pools did not receive added nutrients. Permanent pools were maintained with added tap water and temporary pools dried out naturally for one month.

Zooplankton colonization was limited to only 10 rotifers (species of *Brachionus*, *Cephalodella*, *Lecane*, *Lepadella*, *Rotaria*, and *Trichocerca*) and 2 crustaceans. Treatments significantly affected physical-chemical variables, colonization curves (species richness through time), and mean cumulative species number. Results indicate that local conditions (habitat permanence and resource availability) had the greatest effect on zooplankton species richness. However, low species diversity and little treatment effect on species relative abundance patterns suggest that colonization history (dispersal) was also important. Therefore, colonization history and local conditions were jointly responsible for structuring zooplankton assemblages in this study. Colonization history may have lasting effects on zooplankton composition in older, natural systems as well, but may be overlooked at some scales of measurement.

Introduction

Successful colonization of isolated bodies of water is the combined result of an organism arriving and then persisting within a particular body of water. Regional processes of dispersal that determine colonization history increase species richness by supplying communities with new species and new individuals of existing species. On the other hand, local processes (biotic interactions) and environmental conditions tend to increase mortality, thereby decreasing species richness (Ricklefs, 1987). Although regional and local processes are expected to affect community composition (Ricklefs, 1987), the relative roles of each have been examined in only a few systems (Underwood et al., 1983; Roughgarden et al., 1987; Tilman, 1997).

This study was designed to evaluate the relative roles of regional (colonization history) and local processes (disturbance and resource availability) during

initial colonization of freshwater pools. We experimentally manipulated local processes and used strong treatments (drying and high nitrogen and phosphorus levels) with the rationale that if colonization was important despite strong treatments, it may also be important given less extreme local processes. In addition, we reasoned that colonization should be most important to community composition during initial colonization: if colonization history (dispersal) was unimportant initially, its influence in established communities would likely be unimportant as well.

Materials and methods

The experiment was performed on the campus of the University of Illinois at Springfield. Sixteen plastic wading pools (1.5 m diameter and 0.3 m deep) were manipulated for permanence and resource availability,

BLOCK

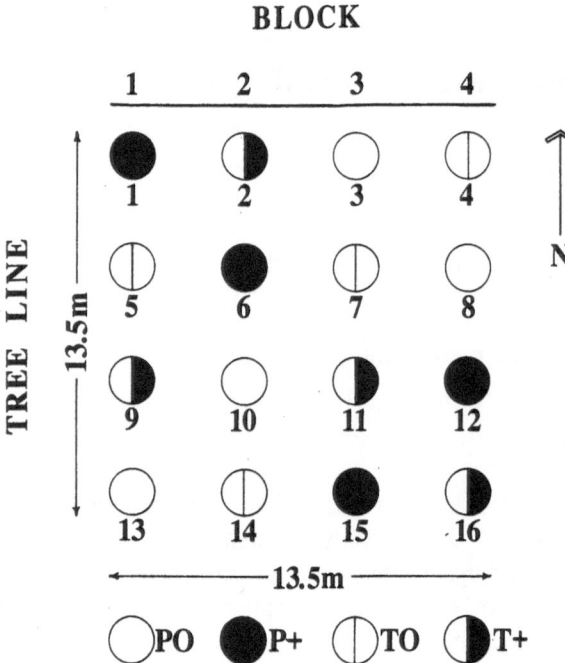

Figure 1. Diagram of experimental randomized block design. P = permanent pools, T = temporary pools, (+) = with N & P, (0) = without N & P.

The experiment ran for 25 weeks, from 25 May 1995 through 17 November 1995. Pools were sampled in random order on the same day every week and at approximately the same time each day. All equipment used for sampling was cleaned with alcohol and rinsed with distilled water between pool samples. Water temperature, dissolved oxygen (DO), and pH were measured in each pool weekly using appropriate meters. A 500 ml water sample (125 ml from each quadrant) was also collected for the following chemical analyses: chlorophyll *a*, total nitrogen (TN), total phosphorus (TP), soluble-reactive phosphate (SRP), total alkalinity, conductivity, and total hardness. All analytical procedures were performed according to APHA (1989).

In addition, a 2-l water sample (500 ml from each quadrant) was collected from each pool for zooplankton analysis. The pooled water sample was concentrated onto 35-μm mesh plankton netting, rinsed into a 20 ml scintillation vial, and preserved with 4% buffered sugar-formalin (Haney & Hall, 1973; Steedman, 1976). Approximately 10% of the preserved volume was examined microscopically in Sedgwick-Rafter chambers for species identification and relative abundance. Zooplankton relative abundance scores were based on the number of organisms observed in the entire Sedgwick-Rafter chamber at 100× magnification. Scores ranged between 1 and 5 (1 = rare, 5 = abundant). The following keys were used for zooplankton species identification: Harring & Myers (1924, 1926), Brooks (1966), Edmondson (1966), Yeatman (1966), and Stemberger (1979).

Treatment effects on zooplankton colonization curves and physical-chemical variables were assessed using a two-way, repeated measures ANOVA for a randomized block design (SAS, 1985). Two-tailed *t*-tests were used to compare mean cumulative number of zooplankton species among treatments (Borland, 1991). Canonical discriminant analysis (SAS, 1985) was used to explore zooplankton relative abundance and physical-chemical variable patterns through time. For purposes of the statistical analyses, data were entered as zeros when temporary pools were dry (weeks 15 through 21). Following the experiment, leaf litter was collected from each pool, dried, and weighed. Leaf litter biomass results were analyzed using a two-way ANOVA for a randomized block design and were used to evaluate the effect of leaf fall on treatments.

thus establishing different habitat conditions. Tap water was added as necessary to one half of the pools (permanent habitat), while the other half of the pools dried out naturally (temporary habitat). High resource availability was established by adding nitrogen (as NH_3Cl) and phosphorus (as P_2O_5) in concentrations typical of eutrophic waters. Minimum nitrogen (N) and phosphorus (P) concentrations in nutrient enriched pools were 2.4 mg l^{-1} and 0.08 mg l^{-1}, respectively (Wetzel, 1983). Furthermore, in an attempt to limit growth of cyanobacteria, N and P were manipulated to obtain TN:TP ratios of 30:1 (Smith, 1983). Low resource level pools, on the other hand, did not receive added nutrients.

In all, there were 4 treatment combinations: permanent / +NP, permanent / 0NP, temporary / +NP, and temporary / 0NP (hereafter referred to as P+, P0, T+, and T0, respectively). All treatment combinations were replicated 4 times. To account for a possible treeline effect, the pools were arranged in a randomized block design (Figure 1). The blocks were aligned with the tree line so that successive blocks were progressively farther away from the tree line. One of each treatment combination was randomly assigned to each block.

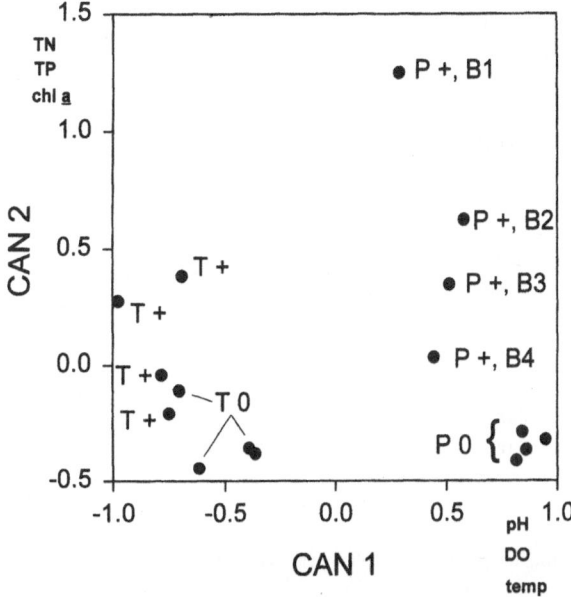

Figure 2. Canonical discriminate analysis of physical-chemical variables. CAN 1 and CAN 2 = 48% and 21%, respectively, of total data set variation. Overall canonical variate, $p = 0.0001$. Variables listed on an axis were strongly correlated with that canonical variate. Note separation of permanent (P) and temporary (T) pools on CAN 1, +NP (+) and 0NP (0) pools on CAN 2, and the block effect (B) as a gradient on CAN 2.

Results

Physical-chemical variables

As expected, treatment differences were found for most physical-chemical variables. Temporary pools dried completely by 27 September (week 18) and remained dry until 25 October. Drying had the most pronounced effect on physical-chemical variables, but effects due to nutrient enrichment and block were also evident (Figure 2). Significantly ($p = 0.0003$) more leaves fell into the pools in blocks closest to the tree line (Figure 1). This result explained the significant block effect ($p < 0.05$) on important variables such as N, P, and chlorophyll a (Figure 2). Time significantly affected all variables ($p < 0.001$).

Drying significantly affected ($p < 0.05$; Table 1) water temperature, pH, dissolved oxygen (DO), total nitrogen (TN) and total phosphorus (TP). Higher water temperature and lower DO concentrations were obvious consequences of lower water volumes in temporary pools (Table 2). Although drying affected both TN and TP, the effect was most evident following leaf fall, especially for phosphorus. Leaf fall also affected DO concentrations, presumably due to the increased respiration accompanying leaf decomposition.

As intended, nutrient enrichment resulted in significantly greater ($p < 0.01$; Table 1) concentrations of TN, TP, and chlorophyll a and a higher TN:TP ratio ($p = 0.0001$) compared to non-enriched treatments (Table 2). Also as intended, mean TN:TP ratios in nutrient-enriched pools were greater than 30:1 (mean = 32), while TN:TP ratios in non-enriched pools averaged 15. Cyanobacteria were never observed in any pool. Therefore, greater phytoplankton abundance (chlorophyll a) suggests that nutrient enrichment increased the resource quantity available for colonizing zooplankton species. The apparent absence of cyanobacteria also suggests that resource quality was adequate, although phytoplankton composition was not specifically manipulated for edibility given the uncontrolled zooplankton assemblages developing during this natural colonization study.

Zooplankton species richness and abundance

Colonization was limited to only 10 rotifer species, 1 cladoceran, and 1 cyclopoid copepod (Table 3). Colonization patterns were evaluated at three scales of measurement: (1) overall mean cumulative species richness per treatment, (2) species richness through time (colonization curves), and (3) zooplankton relative abundance through time.

A comparison of cumulative species richness among treatments (Table 3) shows a significant reduction ($p < 0.01$) in species richness in temporary treatments compared to permanent ones, although species richness was low in all treatments. Cumulative species richness was not significantly affected by nutrient enrichment.

Similarly, drying had the greatest effect on colonization curves (Figure 3), or species richness through time. There was a significantly greater number of species ($p < 0.05$) present in permanent pools compared to temporary ones. However, in contrast to results of cumulative species richness, colonization curves were also significantly affected ($p < 0.05$) by nutrient enrichment, most notably in permanent pools. Differences in colonization curves were apparent by approximately week 12, when temporary pools began to dry (Figure 3). Species richness gradually decreased in temporary pools during the drying period, while in permanent pools, species richness continued to increase (Figure 3). In general, the most successful colonizers in temporary pools arrived prior to week 10 (Table 4).

Table 1. P-value results from repeated measures ANOVA of physical-chemical variables. Time significantly affected all variables ($p < 0.001$)

Variable	Treatment			
	Block (df = 3)	Permanence (df = 1)	Nutrient Enrichment (df = 1)	Permanence X Nutrient Interaction (df = 1)
Temperature	0.03	0.0001	0.26	0.62
pH	0.11	0.0001	0.0001	0.009
Dissolved oxygen	0.03	0.02	0.82	0.12
Total nitrogen	0.001	0.04	0.0001	0.85
Total phosphorus	0.0001	0.0001	0.01	0.97
TN:TP	0.0001	0.62	0.0001	0.003
Chlorophyll *a*	0.033	0.17	0.002	0.93

Table 2. Mean level ($\pm$ SD) of physical-chemical variables per treatment: P = permanent, T = temporary, (+) = +NP, (0) = 0NP

Variable	Treatment			
	P0	P+	T0	T+
Temperature °C	20.5 (6.5)	20.8 (6.6)	23.0 (6.9)	22.4 (7.6)
pH	8.1 (0.74)	7.9 (0.9)	7.8 (0.8)	7.1 (1.2)
Dissolved oxygen mg l^{-1}	8.08 (3.1)	8.33 (3.0)	7.70 (3.0)	6.97 (3.4)
Total nitrogen mg l^{-1}	2.6 (2.1)	5.9 (3.5)	2.0 (2.1)	5.6 (4.1)
Total phosphorus mg l^{-1}	0.21 (0.26)	0.37 (0.48)	0.45 (1.0)	0.69 (1.2)
TN:TP	17.2 (14.6)	27.4 (16.0)	10.8 (11.8)	36.9 (33.2)
Chlorophyll *a* μg l^{-1}	26.7 (23.3)	83.8 (108)	29.6 (22.3)	95.2 (219.6)

Table 3. Colonization summary by pool. (X) Indicates species presence at any time during the study period. Pools are grouped by treatment: P+ = Perm / +NP, P0 = Perm / 0NP, T+ = Temp / +NP, T0 = Temp / 0NP

Species	Treatment																No. pools per species
	P+				P0				T+				T0				
Pool no.:-	1	6	12	15	3	8	10	13	2	9	11	16	4	5	7	14	
Rotaria sp. Scopoli	X	X	X	X	X	X	X	X	X	X	X	X	X	X	X	X	16
Cephalodella gracilis (Ehrenberg)	X	X	X	X	X	X	X	X	X	X	X	X	X	X	X	X	16
Lepadella patella (Müller)	X	X	X	X	X	X	X	X	X		X	X	X	X	X		14
Cephalodella catellina (Müller)	X	X		X	X	X	X	X	X	X	X	X	X		X	X	14
Brachionus bidentata Anderson	X		X	X	X	X	X	X						X			8
Colurella obtusa (Gosse)	X	X	X	X	X	X	X	X									8
Eucyclops agilis (Koch)	X	X	X		X	X	X		X								7
Lecane closterocerca Schmarda			X	X	X	X	X	X									6
Brachionus angularis Gosse			X	X	X												3
Illoricate rotifer	X						X										2
Trichocerca porcellus (Gosse)	X																1
Alona costata Sars	X																1
Cumulative species richness per pool	10	6	8	8	9	8	9	7	5	3	4	4	5	3	4	3	

Table 4. Number of pools that a single species was recorded in for the specified week

Species	Week												
	1	3	5	7	9	11	13	15	17	19	21	23	25
Lepadella patella (Müller)	1						1	3	7	7	8	12	8
Rotaria sp. Scopoli		2	6	7	10	14	15	11	14	9	8	9	8
Cephalodella gracilis (Ehrenberg)			1	8	9	8	10	9	7	3	4	4	1
Cephalodella catellina (Müller)			3	12	5	1	3						
Eucyclops agilis (Koch)				2	1		1	1	2	2	4	3	5
Trichocerca porcellus (Gosse)				1	1								
Brachionus bidentata Anderson							1	2	6	5	3	2	
Colurella obtusa (Gosse)							2	3	5	4	6	5	2
Illoricate rotifer							1			1	1	1	
Lecane closterocerca Schmarda								1	1	5	4	5	3
Alona costata Sars									1	1	1	1	1
Brachionus angularis Gosse									1	1	1		

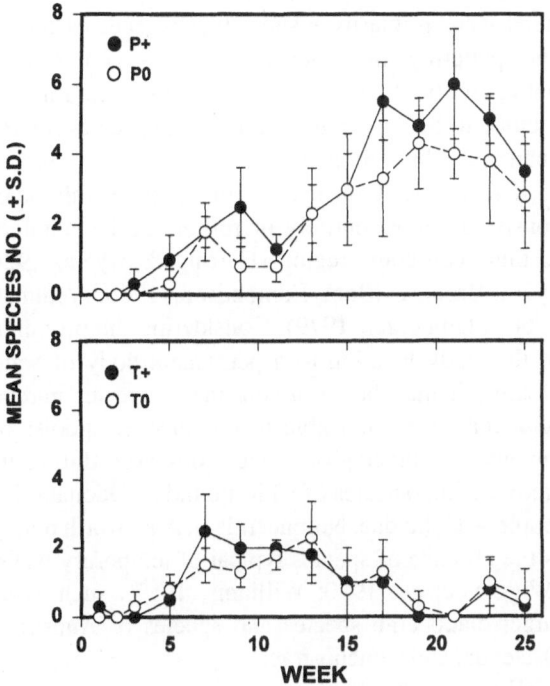

Figure 3. Colonization curves, showing number of species per treatment through time. P+ = Perm/+NP, P0 = Perm/0NP, T+ = Temp/+NP, T0 = Temp/0NP.

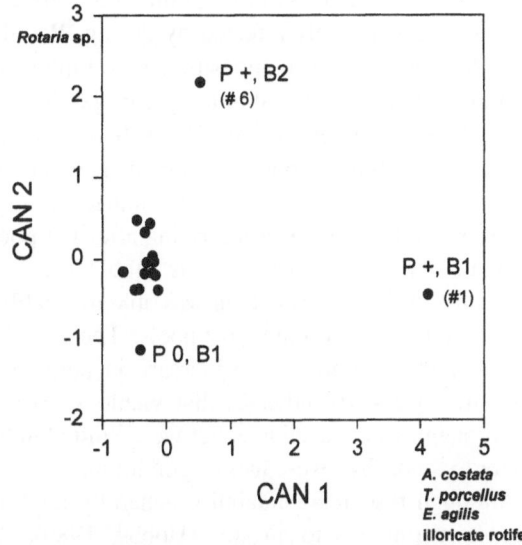

Figure 4. Canonical discriminate analysis of zooplankton relative abundance through time. CAN 1 and CAN 2 = 49% and 19%, respectively, of total data set variation. Overall canonical variate, *p* = 0.0001. Variables listed on an axis were strongly correlated with that canonical variate. Symbols are the same as in Figure 2.

Despite marked differences in physical-chemical characteristics and species richness among treatments, zooplankton relative abundance patterns (Figure 4) were generally unaffected by treatment, with two exceptions. Pool #1 had the greatest number of species throughout the study period (Table 3) and Pool #6 differed, not so much in species composition, but in colonization timing and persistence of species present.

In summary, cumulative species richness and colonization curves (species richness through time) were negatively affected by drying. Nutrient enrichment had no effect on cumulative species richness per treatment, but did positively affect colonization curves. However, colonization history (dispersal) was important to community assembly, indicated by limited species richness (only 12 species in all pools) and confounded treatment effects on zooplankton relative abundance patterns.

Discussion

The effect of disturbance on community structure is dependent on the severity and frequency of the disturbance (Sousa, 1984). Loss of habitat represents a severe disturbance. In temporary environments, species survival depends on the ability of established species to adapt to a dry environment (Wiggins et al., 1980). Species assemblages in temporary habitats are limited and dependent on life history strategies (Wiggins et al., 1980) and habitat duration (Schneider & Frost, 1996), and species richness tends to increase with increasing hydroperiod (Schneider & Frost, 1996).

As expected, both mean cumulative species richness and species richness through time (colonization curves) were negatively affected by drying. We also found that community composition was limited by the timing of species arrival in temporary pools. Half of all species observed arrived in the first 7 weeks, prior to the temporary pools drying. In general, the early-arrivals were more successful colonizers in temporary pools than the species arriving after the pools began to dry. Of the late-arrivals (Table 4), only *Brachionus bidentata* Anderson was able to establish viable populations in temporary pools. The fact that species richness continued to increase in permanent pools after week 10 indicates that viable zooplankton propagules were available, but were limited in the temporary pools by severe habitat conditions.

Although resource availability generally tends to increase recruitment to an area (Hobbs, 1989), the effect of nutrient enrichment on phytoplankton and zooplankton species richness and abundance is not easy to predict (Hall et al., 1970; O'Brien & DeNoyelles, 1974; Lynch & Shapiro, 1981). In the present study, nutrient enrichment appeared to increase the availability of high-quality resources for colonizing zooplankton species based on increased levels of chlorophyll *a* in the absence of cyanobacteria. Increased resources positively affected species richness through time, but did not affect cumulative species richness. Although two species, *Trichocerca porcellus* (Gosse) and *Alona costata* Sars, both present in a single P+ pool (Table 3), may have contributed to increased species richness in permanent enriched pools, overall species composition varied little with nutrient enrichment (Table 3).

In contrast, species persistence varied considerably among species and treatment (Table 4). Several species, notably *Eucyclops agilis* (Koch), *Colurelia obtusa* (Gosse), and *B. bidentata*, demonstrated greater persistence in nutrient enriched pools compared to non-enriched ones. Therefore, in this study, nutrient enrichment increased species richness in permanent pools due to increased persistence of some species, rather than to a greater diversity of low-persistence species.

Species relative abundance patterns through time reflect an undetermined combination of seasonal timing and colonization history. However, if observed species were uniformly present via colonization (dispersal), then all species present should have undergone seasonal successions at similar timings, but at amplitudes determined by treatments. In other words, treatment differences should have been most evident (unconfounded by colonization) if local conditions were primarily responsible for species abundance patterns. Since abundance patterns generally did not reflect treatment differences, colonization history seemed to be a confounding factor in species relative abundance through time.

The zooplankton species observed in this study consisted almost entirely of rotifer species common to this geographic region (Harring & Myers, 1924; 1926; Brooks, 1966; Edmondson, 1966; Yeatman, 1966; Stemberger, 1979). Considering the proximity of the study location to a permanent body of water (1 km), it may be surprising that a greater number of species were not able to colonize the pools: relatively few other groups were observed during the study period (species of Chironomidae, Odonata, Coleoptera, Culicidae, Notonectidae). Also worth noting, is the absence of species typical of temporary waters (Wiggins et al., 1980; Williams, 1987), such as the larger cladoceran species and species of Anostraca, Ostacoda, and Conchostraca.

However, the low species diversity in this study is consistent with results in intertidal (Underwood et al., 1983; Roughgarden et al., 1987) and terrestrial environments (Tilman, 1997) where species recruitment was limited by long-distance passive dispersal of larvae or seeds. Recruitment of zooplankton to new pools depends on similar processes. For many zooplankton species, resting and anhydrobiotic stages not only allow for species survival during unfavorable periods (Pennak, 1989; Wallace & Snell, 1991), but they also provide the means by which species are passively dispersed via wind, rain, aquatic insects or migratory waterfowl (Maguire, 1959, 1963; Proctor & Malone, 1965; Lampert & Sommer, 1997). While the belief that many zooplankton species have cosmopolitan distributions is based on the supposed ease of dispersal of resting and anhydrobiotic stages (Pennak, 1989; Pejler, 1995), in actuality, dispersal mechanisms and efficiency of dispersal for many zooplankton species remain largely unknown.

While this experiment was not designed to evaluate dispersal mechanisms, the low species diversity and type of species present in this study seem to indicate that cosmopolitan distributions are slowly developed, and dispersal is neither uniform nor easily predicted. While some zooplankton may be cosmopolitan, others are not (Frey, 1986; Wallace & Snell, 1991) and some species are more readily dispersed than others (Boileau & Hebert, 1991; Jenkins, 1995).

We have attempted to show that during initial colonization in this study, zooplankton species assemblages were regulated by a combination of local conditions and colonization history. Local conditions affected species richness and persistence, but colonization history was a constraint on species richness and a confounding variable for relative abundance patterns through time.

Although this study was limited to colonization in new pools, the importance of colonization history has been demonstrated in other established, natural environments (Underwood et al., 1983; Tilman, 1997). Additional studies comparing local processes and colonization history are needed for a better understanding of the relative importance of the processes regulating community structure in natural, freshwater environments.

Acknowledgments

This study was supported by the Dept. of Biology, University of Illinois at Springfield (UIS). We thank Joan Buckles and the Grounds Maintenance Department at UIS for their assistance with this project.

References

APHA. American Public Health Association, 1989. Standard Methods for the Examination of Water and Wastewater. APHA, 17th edn. Washington, DC, 1527 pp.

Boileau, M. G. & P. D. Hebert, 1991. Genetic consequences of passive dispersal in pond-dwelling copepods. Evolution 45: 721–733.

Borland Int., Inc., 1991. Borland Quattro Pro for Windows user's guide, version 5.0. Scotts Valley (CA), 638 pp.

Brooks, J. L., 1966. Cladocera. In W. T. Edmondson (ed.), Freshwater Biology, 2nd edn. Wiley & Sons, New York: 587–656.

Edmondson, W. T., 1966. Rotifera. In W. T. Edmondson (ed.), Freshwater Biology, 2nd edn. Wiley & Sons, New York: 420–494.

Frey, D. G., 1986. The non-cosmopolitanism of chydorid cladocera: implications for biogeography and evolution. In R. H. Gore & K. L. Heck (eds), Crustacean Biogeography. Balkema, Rotterdam: 237–256.

Hall, D. J., W. E. Cooper & E. E. Werner, 1970. An experimental approach to the production dynamics and structure of freshwater animal communities. Limnol. Oceanogr. 15: 839–928.

Haney, J. F. & D. J. Hall, 1973. Sugar-coated *Daphnia*: a preservation technique for cladocera. Limnol. Oceanogr. 8: 331–333.

Harring, H. K. & F. J. Myers, 1924. The rotifer fauna of Wisconsin II. A revision of the notommatid rotifers, exclusive of Dicranophorinae. Trans. Wis. Acad. Sci. 21: 415–549.

Harring, H. K. & F. J. Myers, 1926. The rotifer fauna of Wisconsin III. A revision of the genera Lecane and Monostyla. Trans. Wis. Acad. Sci. 22: 315–423.

Hobbs, R. J., 1989. The nature and effects of disturbance relative to invasions. In J. A. Drake, H. A. Mooney, F. di Castri, F. J. Kruger, M. Refmánek & M. Williamson (eds), Biological Invasions: a Global Perspective. J. Wiley & Sons, New York: 389–405.

Jenkins, D. G., 1995. Dispersal-limited zooplankton distribution and community composition in new ponds. Hydrobiologia 313/314: 15–20.

Lampert, W. & U. Sommer, 1997. Limnoecology: the Ecology of Lakes and Streams. Trans. J. F. Haney. Oxford Univ. Press, New York, 382 pp.

Lynch, M. & J. Shapiro, 1981. Predation, enrichment, and phytoplankton community structure. Limnol. Oceanogr. 26: 86–102.

Maguire, B., Jr., 1959. Passive overland transport of small aquatic organisms. Ecology 40: 312.

Maguire, B., Jr., 1963. The passive dispersal of small aquatic organisms and their colonization of isolated bodies of water. Ecol. Mono. 33: 161–185.

O'Brien, W. J. & F. DeNoyelles, Jr., 1974. Relationship between nutrient concentration, phytoplankton density, and zooplankton density in nutrient enriched experimental ponds. Hydrobiologia 44: 105–125.

Pejler, B., 1995. Relation to habitat in rotifers. Hydrobiologia 313/314: 267–278.

Pennak, R. W., 1989. Fresh-water Invertebrates of the United States, 3rd edn. Wiley-Interscience, New York, 628 pp.

Proctor, V. W. & C. R. Malone, 1965. Further evidence of the passive dispersal of small aquatic organisms via the intestinal tract of birds. Ecology 46: 728–729.

Ricklefs, R. E., 1987. Community diversity: relative roles of local and regional processes. Science 235: 167–171.

Roughgarden, J. S. Gaines & S. Pacala, 1987. Supply-side ecology: the role of physical transport processes. In P. Giller & J. Gee (eds), Organization of Communities: Past and Present. Blackwell Scientific, Oxford: 491–518.

SAS Institute, Inc., 1985. SAS user's guide: statistics, version 5. Cary (N.C.), 956 pp.

Schneider, D. W. & T. M. Frost, 1996. Habitat duration and community structure in temporary ponds. J. N. Am Benthol Soc 15: 64–86.

Smith, V. H., 1983. Low nitrogen to phosphorus ratios favor dominance by blue-green algae in lake phytoplankton. Science 221: 669–670.

214

Sousa, W. P., 1984. The role of disturbance in natural communities. Annual review of ecology and systematics 15: 353–391.

Steedman, H. F., 1976. General and applied data on formaldehyde fixation and preservation of marine plankton. In H. F. Steedman (ed.), Zooplankton Fixation and Preservation. Unesco Press, Paris: 103–154.

Stemberger, R. S., 1979. A Guide to Rotifers of the Laurentian Great Lakes. U.S. Environmental Protection Agency, Cincinnati, Ohio. EPA-600/4-79-021, 186 pp.

Tilman D., 1997. Community invasibility, recruitment limitation, and grassland biodiversity. Ecology 78: 81–92.

Underwood, A. J., E. J. Denley & M. J. Moran, 1983. Experimental analyses of the structure and dynamics of mid-shore rocky intertidal communities in New South Wales. Oecologia 56: 202–219.

Wallace, R. L. & T. W. Snell, 1991. Rotifer. In J. H. Thorp & A. P. Covich (eds), Ecology and Classification of North American Freshwater Invertebrates. Academic Press, Inc., San Diego: 187–248.

Wetzel, R. G., 1983. Limnology, 2nd edn. Saunders, Philadelphia. 767 pp.

Wiggins, G. B., R. J. Mackay & I. M. Smith, 1980. Evolutionary and ecological strategies of animals in annual temporary pools. Arch Hydrobiol Suppl. 58: 97–206.

Williams, D. D., 1987. The Ecology of Temporary Waters. Timber Press, Portland, 93 pp.

Yeatman, H. C., 1966. Free-living copepoda, cyclopoida. In W.T. Edmondson (ed.), Freshwater Biology, 2nd edn. Wiley & Sons, New York: 795–815.

Hydrobiologia **387/388**: 215–223, 1998.
E. Wurdak, R. Wallace & H. Segers (eds), Rotifera VIII: A Comparative Approach.
© *1998 Kluwer Academic Publishers.*

Effects of turbidity and biotic factors on the rotifer community in an Ohio reservoir

A. I. Pollard[1,*], M. J. González[1], M. J. Vanni[2] & J. L. Headworth[2]
[1]*Wright State University, Dayton, OH 45435, U.S.A. (*present address: Center for Limnology, 680 N. Park Street, Madison, WI 53706, U.S.A.)*
[2]*Miami University, Oxford, OH 45056, U.S.A.*

Key words: reservoir, Rotifera, spatial distribution, temporal variation, turbidity

Abstract

In reservoirs physical horizontal gradients may affect zooplankton distributions as well as the biotic interactions that potentially regulate zooplankton abundance and species composition. We examined patterns of rotifer abundance and population dynamics along a turbidity gradient over a 4-year period in an Ohio reservoir. To analyze the effect of turbidity on rotifer populations we compared rotifer abundance patterns, species composition, birth and death rates at two sites with high turbidity (river site) and low turbidity (dam site) conditions. Because of the potentially important biotic interaction between rotifers and cladocerans, we also compared cladoceran abundance patterns and species composition. Our results suggest no effect of turbidity on rotifers in Acton Lake. Rotifer and cladoceran abundance patterns were similar at low and high turbidity sites. Similarity indices revealed few differences in rotifer and cladoceran species composition between sites. Rotifer birth and death rates were also similar at low and high turbidity sites. In contrast to these homogeneous spatial patterns, among year comparisons indicate high temporal variability in all parameters measured. Mean rotifer densities were similar from 1993 to 1995, but in 1996 density increased 4-fold. Rotifer species assemblages were dominated by *Brachionus* spp. from 1993 to 1995, while *Keratella cochlearis* and *Polyarthra* spp. were numerically dominant in 1996. Mean cladoceran density also increased in 1996 compared to previous years. Cladoceran species composition was dominated by *Diaphanosoma birgei* from 1993 to 1995, while *Daphnia parvula* and *Bosmina longirostris* dominated the 1996 cladoceran community. Comparison of rotifer population parameters in years of contrasting *D. parvula* abundance suggests that exploitative competition may be an important mechanism regulating rotifer communities in Acton Lake. Interannual variation in *Daphnia* abundance may in turn be controlled by variation in fish biomass.

Introduction

River inflows, differing in temperature, nutrient availability, and suspended solid load, often create physical horizontal gradients in reservoirs (Hart, 1990; Thornton et al., 1990). These variations in water characteristics may affect the distribution and occurrence of organisms in a reservoir. Furthermore, interactions between biotic factors and the physical features of reservoir environments may be crucial for interpreting community dynamics (Betsill & Van Den Avyle, 1994; Hayward & Van Den Avyle, 1986; Lewis, 1978; Threlkeld, 1982, 1983; Urabe, 1989, 1990). In this paper we document rotifer abundance, composition and

population parameters along a turbidity gradient over a 4-year period in an Ohio reservoir.

Increased turbidity has been shown to have a variety of influences on biota, affecting characteristics such as ecological condition, resource availability, and species interactions (Hart, 1990). For example, in the absence of turbidity, laboratory experiments illustrate asymmetrical exploitative competition between rotifers and *Daphnia*, leading to *Daphnia* dominance of zooplankton communities (Gilbert, 1985). However, inorganic turbidity inhibited the competitive abilities of *Daphnia*, and this competitive inhibition may have lead to a decline of cladoceran abundance, causing a competitive release of rotifer populations (Kirk & Gilbert, 1990). A horizontal turbidity gradient can impact spatial heterogeneity of zooplankton communities by influencing competitive interactions of zooplankton.

In addition to this direct effect of suspended solids, increased turbidity alters predator efficiency, thereby indirectly impacting zooplankton community dynamics. Turbidity reduces the reactive distance of visually zooplanktivorous fish, thus inhibiting foraging abilities (Bruton, 1985; Miner & Stein, 1993; Vinyard & O'Brien, 1976). Vinyard & O'Brien (1976) demonstrated that high turbidity reduced the ability of bluegill to locate *Daphnia pulex*. Therefore, turbidity may affect zooplankton communities by impacting the accessibility of prey items to predators. Moreover, these species interactions in the context of a horizontal turbidity gradient may result in spatially heterogeneous zooplankton distribution.

In this paper we present an investigation of rotifer community dynamics. We describe spatial distribution along the longitudinal axis of the reservoir to examine the effects of turbidity on rotifer community dynamics. We explore temporal dynamics by examining year-to-year variations in the rotifer community. Comparison of rotifer population dynamics among years with their competitor, *Daphnia parvula*, abundance and an omnivorous predator, gizzard shad (*Dorosoma cepedianum*), biomass provided insight into the effects of biotic interactions on the rotifer community.

Study site

We performed this study in Acton Lake, a reservoir located in Hueston Woods State Park in southwestern Ohio (39° 31′ latitude, 84° 46′ longitude). Four-Mile Creek, a third order stream, was impounded in 1957 to form Acton Lake; the resultant body of water has a long narrow form following the contours of the flooded stream valley. Acton Lake has a mean width of 0.6 km and a shoreline length of 14 km. The maximum depth is 8 m; however, ~60% of the reservoir is less than 3 m deep (Kissick, 1987). Daniel (1972) describes the physical limnology of Acton Lake. Gizzard shad is the most abundant fish species in Acton Lake (Schaus et al., 1997, Vanni et al., unpublished). Gizzard shad biomass was estimated at 400 kg/ha in 1994 (Schaus et al. 1997), while the biomass in 1996 was 50 kg/ha (Vanni et al., unpublished).

Methods

Samples were taken every 2 weeks from two permanent sampling stations, one upstream shallow site (1.5 m) near river inflows (hereafter, 'river' site) and a second downstream deeper site (8 m) near the dam (hereafter, 'dam' site), from April 1993 through November 1996. Water transparency was measured with a standard secchi disk, as an indicator of turbidity levels. Turbidity estimates were corroborated from 1994 through 1996 with a measurement of inorganic non-volatile suspended solid concentrations (NVSS) in the water column. To calculate NVSS a known volume of water was filtered onto pre-weighed, pre-ashed glass fiber filters, and organic material was ashed in a muffle furnace. Filters were weighted again, to yield the mass of NVSS (Wetzel & Likens, 1991).

Zooplankton samples were obtained with a Schindler–Patalas trap (53-μm mesh) at meter depth increments at each station (river samples 0, 1, 1.5 m; dam samples, 0–8 m). Zooplankton were immediately anesthetized with effervescing CO_2 tablets, and preserved with a buffered formalin solution (5% to sample volume) (Prepas, 1978). Samples from each depth were pooled giving a single river and dam sample for each date. We then sub-sampled using two 1-ml aliquots from a Henson–Stemple pipette, and examined the aliquots in a Sedgewick–Rafter cell. Zooplankton were identified to species and enumerated. We used Stemberger (1979) to identify rotifers and Balcer et al. (1984) to identify crustaceans. The number of eggs per female was estimated for rotifers with external eggs.

Species composition of rotifer communities and cladoceran communities were compared using the following similarity index (Sale, 1984):

$$C_{ij} = 1 - 0.5 \cdot \sum_{k=l}^{s} |P_{ik} - P_{jk}|,$$

where C_{ij} represents the similarity between communities i and j, and varies from 0.0 to 1.0, with 1.0 meaning the two communities have identical compositions. P_{ik} and P_{jk} are the proportions of individuals present in communities i and j, respectively, that comprise the kth species.

We calculated population parameters of the most abundant rotifer species having external eggs in Acton Lake: *Brachionus angularis*, *B. calyciflorus*, *B. caudatus*, *Keratella cochlearis* and *Polyarthra* sp. We used egg ratio equations described by Edmondson

(1977) to calculate instantaneous population parameters:

Birth rate $\quad b = \ln(E + 1)/D$;

Growth rate $\quad r = (\ln N_t - \ln N_0)/T$;

Death rate $\quad d = b - r$;

where E is the egg ratio (number of eggs per individual) and is obtained through direct egg counts. D is the time (days) required for an egg to develop into a free-swimming individual. N_0 and N_t are the population sizes at time 0 and t, respectively. T represents the number of days between sampling dates. Abundance, egg ratios, and temperatures determined during routine sampling were used in the rate calculations. Lake temperatures were taken at 1-m increments from the lake surface to the bottom at each site with a YSI model 57 dissolved oxygen and temperature meter. These values were then averaged, giving a mean daily temperature for each sampling station. We calculated development time (D) using the equation described by Herzig (1983):

$$D = a(t - b)^{-c},$$

where t is the mean temperature (°C) in Acton Lake on each sampling date, and a, b and c are constants, which varied for each species (Herzig, 1983).

Results

Acton Lake has a distinct turbidity difference from the shallow river area to the dam. Secchi depth and non-volatile suspended solid data show that the river site has a consistently shallower secchi depth and higher amount of suspended inorganic material than the dam site (Figure 1a, b). The range of NVSS varied between 1.0 and 288.7 mg/l with a mean of 22.7±8.4 (SE) at the river and varied between 0.75 and 90.25 mg/l with a mean of 9.2±3.5 (SE) at the dam.

Abundance patterns

Total rotifer abundance was similar at river and dam sites (Figure 2a). Annual abundance patterns exhibited both spring and late summer peaks. Rotifer abundance peaked early each spring (May) and was followed by a decline in June. The second abundance peak was of a similar magnitude in 1993, 1994, and 1995, with mean rotifer density peaking around 500 ind/l each year in river and dam sites. However, in 1996, the second

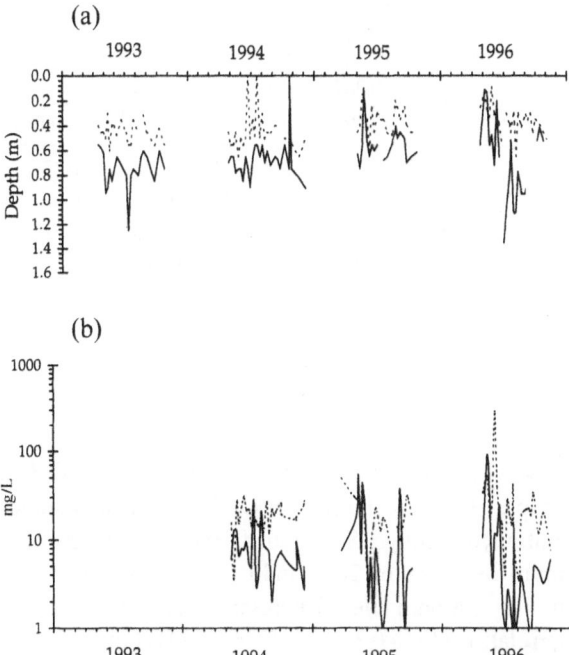

Figure 1. Secchi depth (m) (a) and LN non-volatile suspended solids (mg/l) (b) at river (dashed line) and dam (solid line) sampling stations.

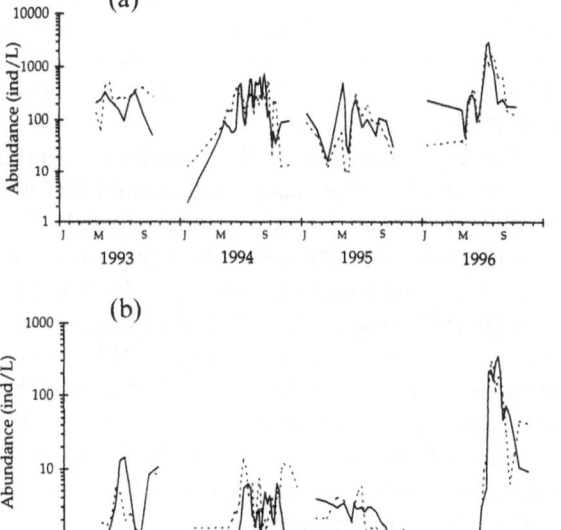

Figure 2. Total rotifer (a) and cladoceran (b) abundance (ind/l) at river (dashed line) and dam (solid line) sampling stations. Note log scale.

Table 1. Similarity index comparing rotifer and cladoceran community assemblages at river and dam sites (spatial patterns) and among years (temporal patterns)

	River vs Dam			Year-to-Year	
	Rotifers	Clado-cerans		Rotifers	Clado-cerans
1993	0.75	0.42	93 vs 94	0.67	0.35
1994	0.80	0.90	93 vs 95	0.60	0.64
1995	0.76	0.61	93 vs 96	0.50	0.67
1996	0.75	0.95	94 vs 95	0.61	0.55
			94 vs 96	0.34	0.02
			95 vs 96	0.36	0.47

peak was considerably higher in both river and dam samples than the initial peak. This high abundance continued through September 1996, reaching maximum densities of 2000 ind/l at the river and 3000 ind/l at the dam sites. Regression analysis shows no correlation ($Y= -0.32\chi +212$, $r^2=0.003$) between total rotifer abundance and NVSS in Acton Lake.

Cladoceran abundance was generally similar at river and dam sites (Figure 2b). Annually, the abundance patterns exhibited a mid-summer (June) peak in the cladoceran population. Mean cladoceran densities were similar from 1993 to 1995, peaking between 5 and 11 ind/l in each year at river and dam sites. However, cladoceran abundance increased in 1996, reaching densities of 350 ind/l in the dam and 300 ind/l in the river. Regression analysis shows no correlation ($Y=0.01\chi +3.33$, $r^2=0.008$) of cladoceran abundance and NVSS in Acton Lake.

Brachionus angularis, *B. calyciflorus*, and *B. caudatus* all had abundance peaks around 100 ind/l in all years (Figure 3a–c, respectively). *Polyarthra* sp. had similar population abundance patterns to *Brachionus* spp., with peaks around 100 ind/l from 1993 through 1995 (Figure 3d). The abundance of *Polyarthra* sp. peaked slightly higher in 1996 than in previous years. *Keratella cochlearis* generally had lower abundance than the *Brachionus* spp. and *Polyarthra*, with the exception of the 1996 abundance, when *K. cochlearis* was more abundant than any other rotifer species (Figure 3e).

Species composition

The rotifer community was numerically dominated by *Brachionus angularis*, *B. calyciflorus*, *B. caudatus*, *Keratella cochlearis*, *Polyarthra* sp. and *Synchaeta* sp.

Similarity values ranged from 0.75 to 0.80 when comparing annual rotifer species assemblages between river and dam sites (Table 1). In contrast, comparisons of interannual variation in species assemblages were more variable (Table 1). The most striking year-to-year variability was found in the species assemblages between 1994 and 1996, having a similarity index of 0.34. *Brachionus angularis*, *B. calyciflorus*, *B. caudatus* and *Synchaeta* sp. dominated species assemblages from 1993 to 1995. *Keratella cochlearis* and *Polyarthra* sp. numerically dominated the composition of the rotifer community in 1996.

The cladoceran community of Acton Lake includes *Diaphanosoma birgei*, *Daphnia parvula* and *Bosmina longirostris*. Comparisons of cladoceran community similarity at river and dam sites reveal variations in species assemblages greater than that of rotifers (Table 1). The largest variation in community assemblages between river and dam sites occurs in 1993, yielding an index value of 0.42. In this year the river cladoceran community was dominated by *D. birgei*, while the dam community was dominated by *B. longirostris*. Interannual comparisons of cladoceran species assemblages show a wide range of variability in cladoceran community composition (Table 1). The most striking community dissimilarity occurred between 1994 and 1996 (similarity, 0.02). The cladoceran community was dominated by *D. birgei* in 1994 and 1995. In contrast, *B. longirostris* and *D. parvula* dominated the 1996 cladoceran community.

Rotifer population parameters

In general, the various species of zooplankton had similar population patterns. The population dynamics of *Brachionus calyciflorus* are described as a typical example of rotifer life history patterns. The seasonal abundance dynamics for most rotifers were similar in all years, with abundance increases in May. Birth rates peaked in May, declined in June/July, and then increased again in the fall in each year, with the exception of 1995, during which the birth rates failed to decline through the year (Figure 4a). This period of continual birth is reflected in abundance, which does not decrease in late summer in 1995, as they do in all of the other years. In 1996, the birth rate was generally low. Death rate patterns were similar to birth rate patterns (Figure 4b). The death rate was highest in 1993 through 1995, while in 1996 the death rate was lower.

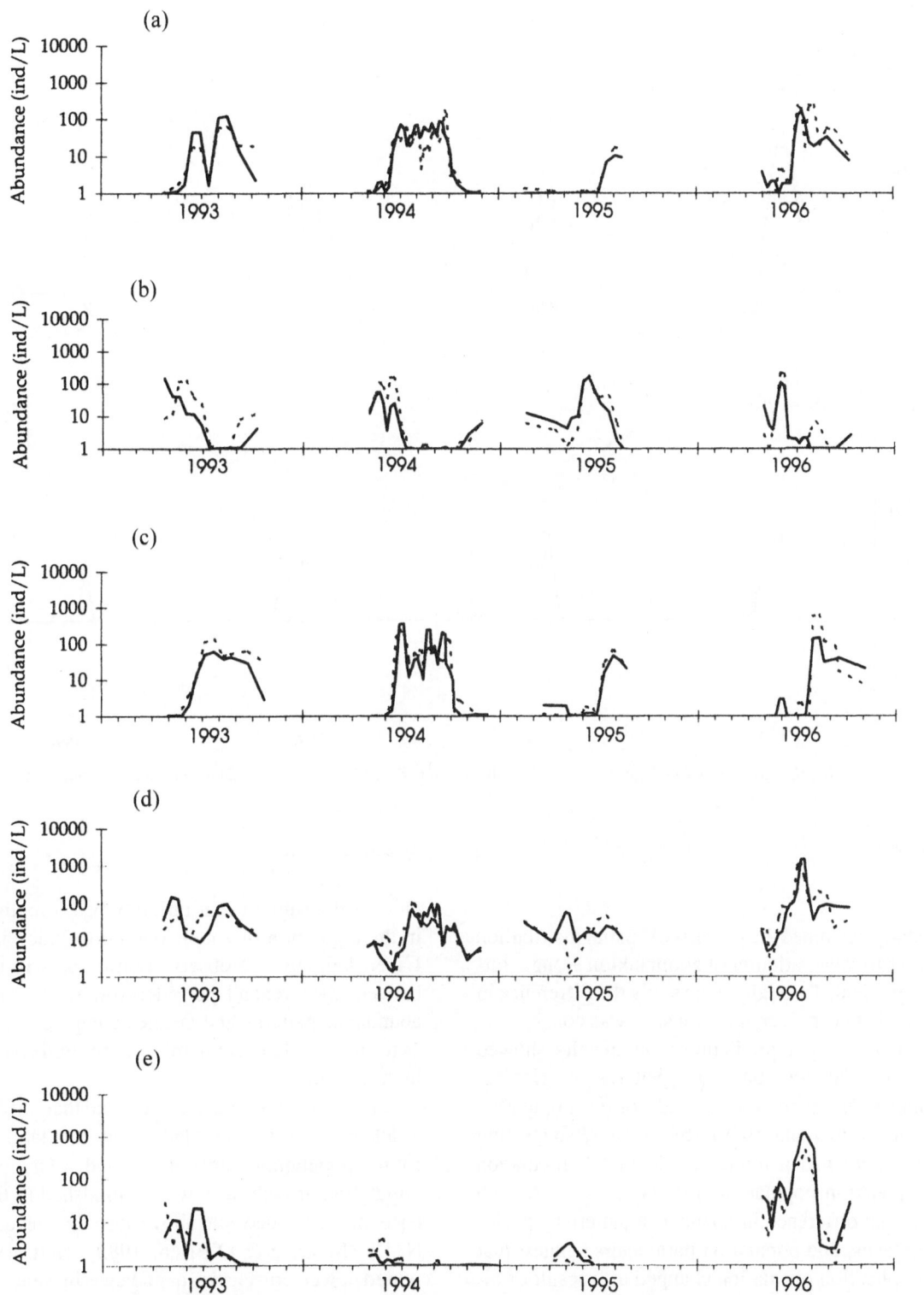

Figure 3. Abundance (ind/l) of five rotifer species: (a) *Brachionus angularis*, (b) *B. calyciflorus*, (c) *B. caudatus*, (d) *Polyarthra* sp., (e) *Keratella cochlearis* at river (dashed line) and dam (solid line) sampling stations in Acton Lake. Note log scale.

220

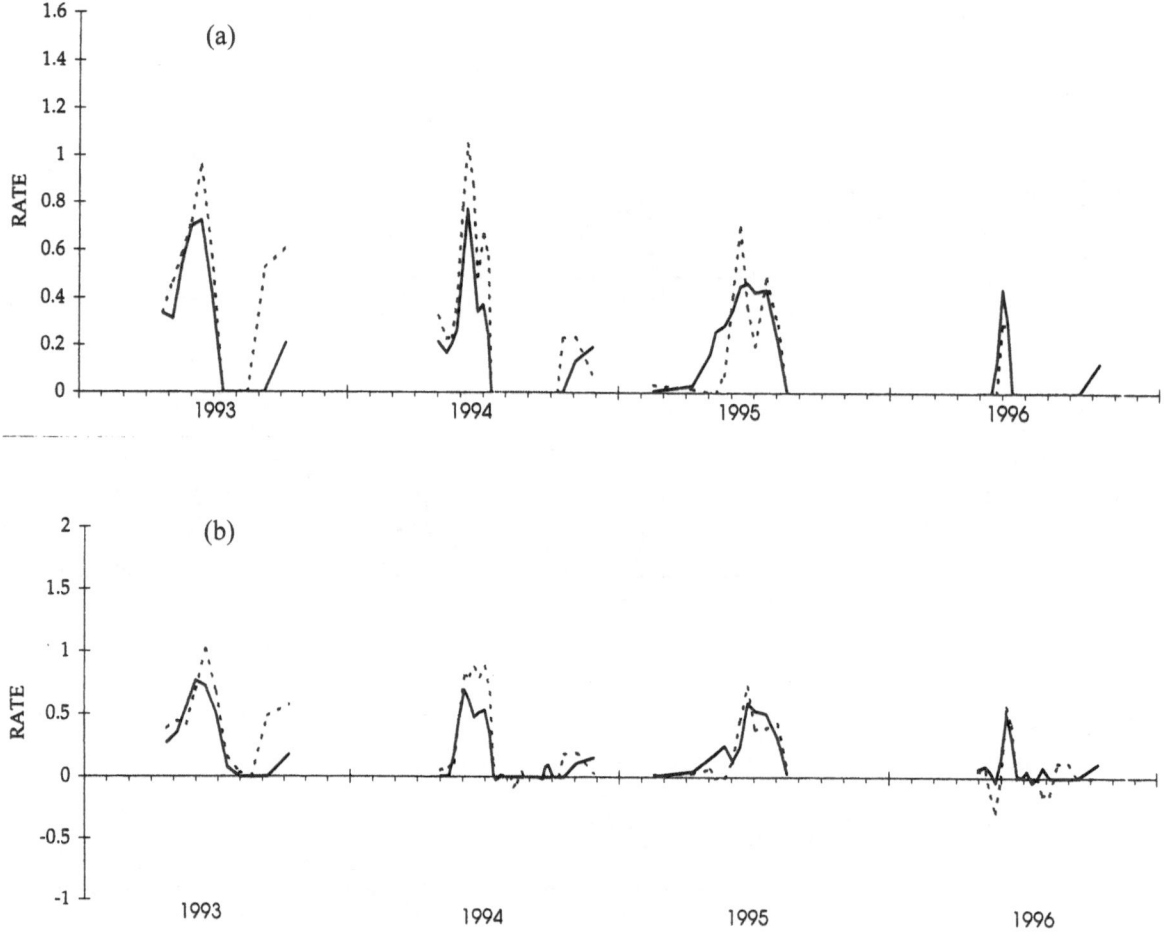

Figure 4. *Brachionus calyciflorus* birth rate (a), and death rate (b) at river (dashed line) and dam (solid line) sampling stations.

Discussion

This study examined the patterns of spatial distribution and year-to-year variation of zooplankton along a turbidity gradient. Throughout the study the difference in turbidity between river and dam sites was consistent.

Contrary to our predictions, our results showed similar zooplankton abundance patterns, species assemblages of communities, and rotifer population dynamics at low and high turbidity sites, suggesting that turbidity had a minimal role in the regulation of zooplankton populations in Acton Lake. A high interannual difference in abundance patterns, species assemblages, and population parameters suggest that the zooplankton population changed as a result of biotic influences. The spatially homogeneous character of Acton Lake's zooplankton community make these results unique among previous investigations of the impact of turbidity on zooplankton communities.

Spatial variation

Our results suggest that turbidity had a minimal role in the regulation of zooplankton population in Acton Lake. Although we observed differences in turbidity between the river and dam sites, rotifer and cladoceran abundance patterns and species composition, as well as rotifer population dynamics, were similar at low and high turbidity sites.

Previous studies have determined that turbidity has a deleterious effect on cladoceran populations while rotifer populations are not affected. *Daphnia pulex* population growth rates were diminished in the presence of suspended silt above turbidity levels of 10 NTUs (McCabe & O'Brien, 1983). Hart (1987) reported lower crustacean abundance in years of high turbidity (Lake le Roux, South Africa). In enclosure experiments, crustacean mean density decreased with the addition of clay (Cuker, 1987). Kirk & Gilbert (1990) showed that addition of coarse suspen-

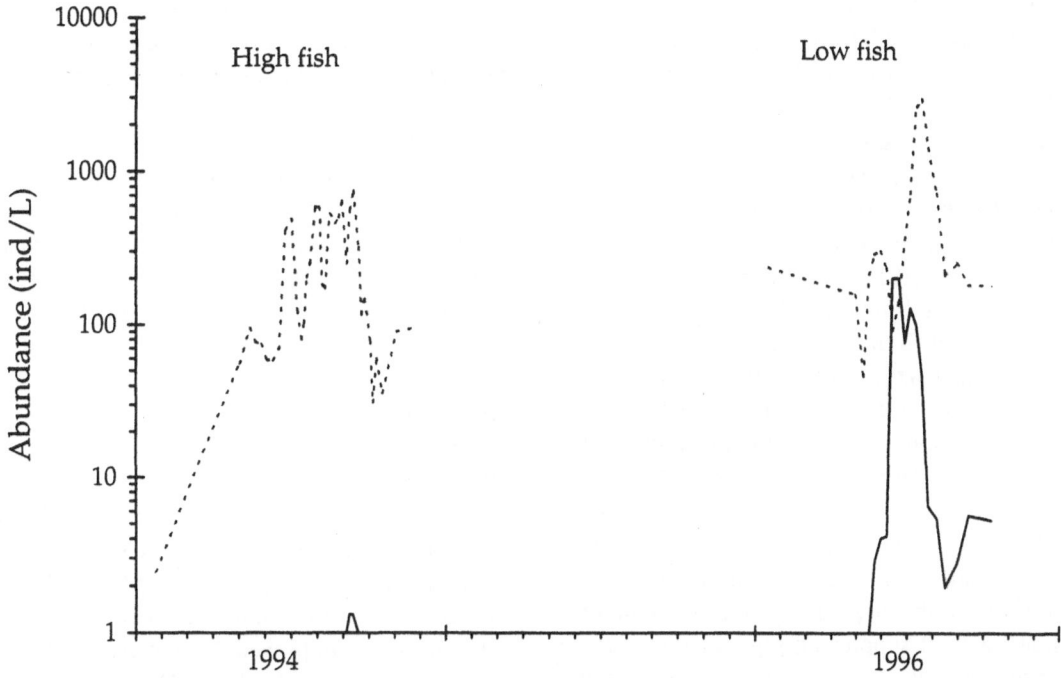

Figure 5. Rotifer (dashed line) and *Daphnia parvula* (solid line) abundance (ind/l) at the dam sampling station. Fish information (1994) taken from Schaus et al. (1997, 1996) taken from Vanni et al., unpublished. Note log scale.

ded clays to cladoceran cultures reduced population growth rates, but rotifer populations were not inhibited. Furthermore, in mixed culture of rotifers and cladocerans addition of clay reversed competitive outcome such that rotifers out-compete cladocerans due to the negative effect of high turbidity on cladoceran growth rates.

Turbidity levels in Acton Lake may not have been high enough to influence zooplankton competitive interactions. Mean non-volatile suspended solid levels in Acton Lake were 22.7±8.4 (SE) mg/l in the river and 9.2±3.5 (SE) mg/l in the dam. Laboratory studies focusing on the impact of turbidity on competition used a higher concentration (50 mg/l) of coarse-grained suspended solids (Kirk & Gilbert, 1990). These higher levels of turbidity used in experimental manipulations may be harmful to *Daphnia*, while suspended solid concentrations in Acton Lake were not high enough to have a deleterious impact. Therefore, due to similarity of zooplankton communities at high and low turbidity sites, we conclude that turbidity levels in Acton Lake did not mediate competitive interactions.

Turbidity levels potentially alter predatory interactions. Urabe (1989, 1990) found that heterogeneous zooplankton distribution in Ogochi Reservoir was due to dissimilar distribution of zooplanktivorous fish.

Suspended solids may provide a prey refuge from visual predators (Vinyard & O'Brien, 1976), thereby facilitating the coexistence of large zooplankton with predators. Although larval gizzard shad are heterogeneously distributed in Acton, stomach content analysis suggests that selective predation is homogeneous, despite the turbidity gradient in Acton Lake (Pollard, 1996). Furthermore, spatial distribution of juvenile and adult gizzard shad was similar in river and dam areas (Schaus et al., 1997). These findings, in addition to the similar patterns of zooplankton abundance, birth and death rates at river and dam sites, suggest that differences in turbidity along the long axis of Acton Lake did not provide a differential prey refuge.

Temporal variation

Rotifer and cladoceran dynamics showed high temporal variability, in contrast to their homogenous spatial distribution. Such temporal variability was mainly caused by drastic increases in rotifer and cladoceran mean densities in 1996 compared to the three previous years, and suggests that competitive interactions may mediate zooplankton community dynamics in Acton Lake. To discuss the implication of such drastic temporal changes in more detail we will compare the rotifer and cladoceran abundance and species composition in 1994 and 1996. We will focus on these two years because additional information about fish populations is available (Schaus et al., 1997, Vanni et al., unpublished).

The low similarity between 1994 and 1996 cladoceran communities was caused by differences in *D. parvula* populations. Yearly mean density of *D. parvula* was 0.5 ind/l in 1994 and 54 ind/l in 1996 over the entire study period for each year. In 1996, rotifer abundance peaked upon the *D. parvula* decline, suggesting a release from competition with *Daphnia*. The relatively small size of *D. parvula* in Acton Lake suggests that exploitative competition, rather than interference competition, was the important competitive interaction between *Daphnia* and rotifers (Burns & Gilbert, 1985). Population parameter analysis indicated that rotifer birth and death rates were generally lower in 1996 than 1994. Despite the high abundance of rotifers, low birth rates suggest that exploitative competition for food suppressed reproductive potential for rotifers in 1996. At the same time, lower death rates for rotifers indicate that predation pressure was lower in 1996 than in 1994.

Differences in zooplankton communities in 1994 and 1996 correspond to differences in the biomass of the dominant fish in Acton Lake. Gizzard shad are visual zooplanktivores as larvae (<25 mm TL), after which they switch to omnivorous pump filter feeders (Dettmers & Stein, 1992). Although omnivorous, juvenile and adult gizzard shad are important regulators of zooplankton by way of both direct effects of predation and indirect effects on phytoplankton (DeVries & Stein, 1992, Drenner et al., 1986). Therefore, changes in gizzard shad biomass can be used as an indicator of predation pressure.

Our results combined with those reported by other authors (Figure 5) suggest the following scenario for the fish–cladoceran–rotifer interaction in Acton Lake: in 1994, high fish density resulted in intense predation pressure on large zooplankton, *Daphnia* in particular, and rotifer populations dominated the 1994 zooplankton community of Acton Lake. In 1996, however, lower fish biomass resulted in reduced predation pressure on *Daphnia*. This release from predation pressure allowed the *D. parvula* population to reach high densities. During this temporal window of high *Daphnia* abundance, rotifer populations were suppressed, which was most likely the result of exploitative competition (Figure 5). Further evidence for this competitive relationship is inferred from the increase in rotifer densities upon *Daphnia* decline in 1996.

In summary our results suggest that turbidity has a minimal role on the distribution and species composition of rotifers in Acton Lake. However, competitive interactions with *D. parvula* may be an important regulator of rotifer populations in Acton Lake. Furthermore, the combination of our findings with information about temporal changes in fish biomass suggest that, similar to natural temperate lakes, fish predation may be an important mechanism regulating rotifer–cladoceran interactions in this Ohio reservoir.

Acknowledgments

We would like to thank Keetha Beelick, Danielle Crossen, Amy Lane, Ray Trimmer, Valerie Trimmer, Stephanie Wolfer, and Joe Woyche for their assistance in the collection and processing of samples. We are appreciative for the comments of two anonymous reviewers, which improved the manuscript. Funding for this study was provided by an Ohio Board of Regents Research Initiative Grant to M. J. Gonzalez and National Science Foundation Grant DEB93-18452 to M. J. Vanni.

References

Balcer, M. D., N. L. Korda, & S. I. Dodson, 1984. Zooplankton of the Great Lakes: A Guide to the Identification and Ecology of the Common Crustacean Species. The University of Wisconsin Press, Wisconsin.

Betsill, R. K. and M. J. Van Den Avyle, 1994. Spatial heterogeneity of reservoir zooplankton: a matter of timing? Hydrobiologia 277: 63–70.

Burns, C. W. & J. J. Gilbert, 1986. Direct observations of the mechanisms of interference between *Daphnia* and *Keratella cochlearis*. Limnol. Oceanogr. 31: 859–866.

Bruton, M. N., 1985. The effects of suspensoids on fish. Hydrobiologia 125: 221–241.

Cuker, B. E., 1987. Field experiment on the influences of suspended clay and P on the plankton of a small lake. Limnol. Oceanogr. 32: 840–847.

Daniel, P. M., 1972. Acton Lake: biology of its benthos and notes on its physical limnology 1959–1970. Ohio J. Sci. 72: 241–253.

Dettmers, J. M. & R. A. Stein, 1992. Food consumption by larval gizzard shad: zooplankton effects and implications for reservoir communities. Trans. am. Fish. Soc. 121: 494–507.

DeVries, D. R. & R. A. Stein, 1992. Complex interactions between fish and Zooplankton: quantifying the role of an open-water planktivore. Can. J. Fish. Aquat. Sci. 49: 1216–1227.

Drenner, R. W., S. T. Threlkeld & M. D. McCracken, 1986. Experimental analysis of the direct and indirect effects of an omnivorous filter-feeding clupeid on plankton community structure. Can. J. Fish. aquat. Sci. 43: 1935–1945.

Edmondson, W. T, 1977. Population dynamics and secondary production. Arch. Hydrobiol. Beih. Ergebn. Limnol. 8: 56–65.

Gilbert, J. J., 1985. Competition between rotifers and *Daphnia*. Ecology 66: 1943–1950.

Hart, R. C., 1987. Population dynamics and production of five crustacean zooplankton in a subtropical reservoir during years of contrasting turbidity. Freshwat. Biol. 18: 287–318.

Hart, R. C., 1990. Zooplankton distribution in relation to turbidity and related environmental gradients in a large subtropical reservoir: patterns and implications. Freshwat. Biol. 24: 241–263.

Hayward, R. S. & M. J. Van Den Avyle, 1986. The nature of zooplankton spatial heterogeneity a non-riverine impoundment. Hydrobiologia 131: 261–271.

Herzig, A., 1983. Comparative studies on the relationship between temperature and duration of embryonic development of rotifers. Hydrobiologia 104: 237–246.

Kirk, K. L. & J. J. Gilbert., 1990. Suspended clay and the population dynamics of planktonic rotifers and cladocerans. Ecology 71: 1741–1755.

Kissick, L. A., 1987. Prey selectivity and feeding periodicity of larval logperch in Acton Lake, Ohio. Envir. Biol. Fishes 20: 155–160.

Lewis W. M. Jr., 1978. Comparison of temporal and spatial variation in the zooplankton of a lake by means of variance components. Ecology 59: 666–671.

McCabe, G. D. & W. J. O'Brien, 1983. The effects of suspended silt on feeding and reproduction of *Daphnia pulex*. Am. Midl. Nat. 110: 324–337.

Miner, J. G. & R. A. Stein, 1993. Interactive influence of turbidity and light on larval bluegill (*Lepomis macrochirus*) foraging. Can. J. Fish. aquat. Sci. 50: 781–788.

Pollard, A. I., 1996. Effects of turbidity and selective predation by larval gizzard shad on a reservoir zooplankton community. Master's Thesis, Wright State University, Dayton, OH.

Prepas, E., 1978. Sugar-frosted *Daphnia*: an improved fixation technique for Cladocera. Limnol. Oceanogr. 23: 557–559.

Sale, P. F., 1984. The structure of communities of fish on coral reefs and the merit of a hypothesis-testing approach to ecology. In Strong, D. R. Jr., D. Simberloff, L. G. Abele & A. B. Thistle (eds), Ecological Communities: Conceptual Issues and the Evidence. Princeton University Press, Princeton, NJ: 478–490.

Schaus, M. H., M. J. Vanni, T. E. Wissing, M. T. Bremigan, J. A. Garvey & R. A. Stein, 1997. Nitrogen and Phosphorus excretion by a detritivorous fish (the gizzard shad, *Dorosoma cepedianum*) in a reservoir ecosystem. Limnol. Oceanogr. 42: 1386–1397.

Stemberger, R. S. 1979. A Guide to Rotifers of the Laurentian Great Lakes. EPA-600/4-79-021. U.S. Environmental Protection Agency, OH.

Thornton, K. W., B. L. Kimmel & F. E. Payne, 1990. Reservoir Limnology: Ecological Perspectives. Wiley and Sons, New York.

Threlkeld, S. T., 1982. Water renewal effects on reservoir zooplankton communities. Can. Wat. Res. J. 7: 151–167.

Threlkeld, S.T., 1983. Spatial and temporal variation in the summer zooplankton community of a riverine reservoir. Hydrobiologia 107: 249–254.

Urabe, J., 1989. Relative importance of temporal and spatial heterogeneity in the zooplankton community of an artificial reservoir. Hydrobiologia 184: 1–6.

Vrabe, J., 1990. Stable horizontal variation in the zooplankton community structure of a reservoir maintained by predation and competition. Limnol. Oceanogr. 35: 1703–1717.

Vinyard, G. L. & W. J. O'Brien, 1976. Effects of light and turbidity on the reactive distance of Bluegill (*Lepomis macrochirus*). J. Fish. Res. Board Can. 33: 2845–2849.

Wetzel, R. G. & G. E. Likens, 1991. Limnological Analyses, 2nd edition. Springer-Verlag, New York.

Hydrobiologia **387/388**: 225–230, 1998.
E. Wurdak, R. Wallace & H. Segers (eds), Rotifera VIII: A Comparative Approach.
© *1998 Kluwer Academic Publishers.*

Differences in rotifer populations of the littoral and sub-littoral pools of a large marine lagoon

A. Saunders-Davies

8 Kingfisher Close, Walton on Thames, Surrey KT12 4LF, U.K.

Key words: marine lagoon, littoral, sub-littoral, sediment loading, quantitative sampling

Abstract

The rotifer populations of littoral and sub-littoral pools of a large marine lagoon are described. Significant differences ($P < 0.05$) were found between the rotifer populations of the chlorophyte algae, the phaeophyte algae of the high littoral pools, and the algae of sub-littoral hollows. Large populations of *Proales reinhardti* (Erhenberg) were found in the phaeophyte alga (*Pilayella* sp.) during spring, but were mostly absent in the pools containing chlorophytes (*Percursaria* sp., *Blidingia* sp. and *Enteromorpha* sp.). *Encentrum marinum* (Dujardin) – the only other species to be found in appreciable numbers – and *Lindia torulosa* (Dujardin) were mainly found in the chlorophyte pools. Few rotifers were found in the sub-littoral algae, where sediment loading was found to be significantly higher ($P < 0.05$). *Proales reinhardti* populations disappeared from all pools at the beginning of May. Laboratory experiments showed that *P. reinhardti* had a high reproductive rate at 8 °C but did not survive at 18 °C, which suggests that rising temperatures in Spring are the main factor in controlling *P. reinhardti* populations.

Introduction

The great majority of quantitative work on rotifers has been done with planktonic species, for reasons which are easy to understand. Where the open water is concerned a unit volume of water provides an easy method of comparison. For periphytic rotifers arriving at a suitable unit of measurement is more difficult. I used the method previously described (Saunders-Davies, 1994) of expressing rotifer numbers per unit weight of alga, as suggested by Peters et al. (1993). The aim was to support objectively the original observation made from preliminary sampling that there appeared to be differences between the rotifer populations of (1) the algae of littoral pools containing largely chlorophyte algae, (2) those containing phaeophyte algae almost exclusively, and (3) the algae in sub-littoral sand hollows.

The site

The Fleet is a large (13 km long by 2 km wide) macrotidal (tides <2.0 m) lagoon in Dorset, Southern England (site (Ordnance Survey Map Reference SY666 756, Figure 1, see Saunders-Davies, 1994). At the end open to the English Channel where tidal range is maximum (>2 m) there is a mixture of sand and pebble flats that contain littoral pools, a little below the mean high water spring tide (MHWS, Figure 1). In one area the filamentous algae contained in the pools are almost entirely chlorophytes, consisting of *Percursaria percursa* (Rosenvinge), *Enteromorpha prolifera* (J. Argardh) and *Blidingia* sp. In another area there are pools associated mostly with *Fucus* sp. and containing the phaeophyte alga *Pilayella* sp. almost exclusively. Below the extreme low water level (ELWL, Figure 1) there are subtidal hollows in the sand containing clumps of algae. Identification of the algae was done by reference to Newton (1931) and Burrows (1991).

Field methods

A form of stratified sampling was employed similar to that described by Elliot (1971). Three randomly replicated samples of approximately 30 ml volume of the algae were taken into a 100 ml bottle. This was repeated with 3 bottles. On return to the laboratory three samples of approximately 300 mg wet algae,

226

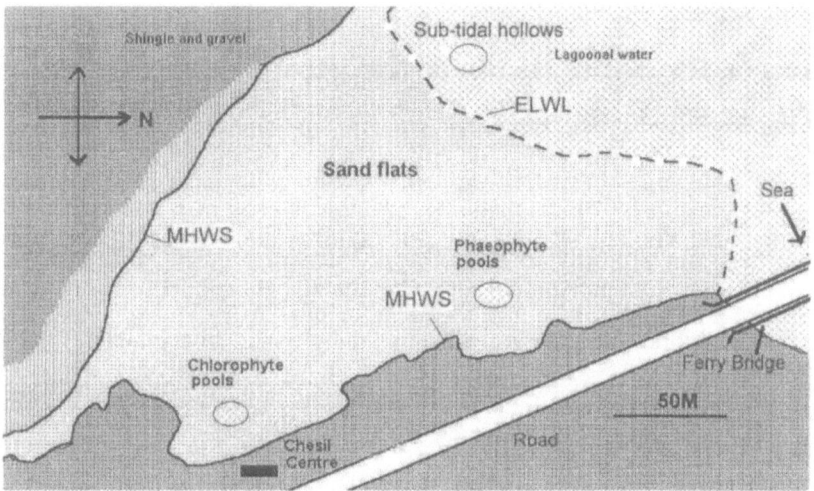

Figure 1. General view of the study site.

weighed on an Ohaus 2-place electronic balance, were taken from each bottle, making a total of 9 samples, and the rotifers counted. Other taxa recorded were nematodes, copepods and their nauplii. Ostracods, planarians and protista were not counted. The sample was then washed, dried and weighed on an Adams Electronics 4-place balance, and the numbers of rotifers and other taxa expressed as numbers per 100mg of dry algae.

Laboratory methods

Sediment loading was measured by taking approximately 10 g of wet algae, shaking vigorously with 50 ml of distilled water and filtering off the sediment through a 0.5 mm stainless steel mesh. The algae and the sediment were then dried at 40 °C and weighed. The fraction of sediment loading (S) was then expressed as

$$S = \frac{W_s}{W_{(a+s)}},\qquad(1)$$

where W_s = weight of sediment, and $W_{(a+s)}$ = weight of dry algae + sediment. Species dominance was measured by the Berger-Parker index (d) (Berger & Parker, 1970; May, 1975): $d = N_{max}/N$, where N_{max} = the number of individuals in the most abundant taxon, and N the total number of individuals in all taxa.

The population dynamics of *P. reinhardti* was examined at two temperatures. Six 6 individuals were placed in 8 ml of sea water together with 300 mg wet weight of the alga *Enteromorpha flexuosa*, which had the epiphytic diatoms used as food by the rotifers. One

Table 1. Total numbers of rotifers found in phaeophye, chlorophyte and low tide areas in Spring 1996 and Spring 1997

	Phaeophyte	Chlorophyte	Low Tide
1996			
P. reinhardti	805	162	45
P. halophila	26		18
E. marinum		125	1
E. putorius	75	18	9
L. torulosa	10		
Colurella spp.		13	
1997			
P. reinhardti	711	312	67
E. marinum		29	

set of 5 replicates was kept at a temperature of 8° ± 2°C, another set of 5 replicates was kept at 18° ± 2°C. Both were kept under a light/dark regime of 10 hours light/14 hours dark, using fluorescent tubes 45 cm above the dishes. Rotifers were counted daily under a Zeiss GZS stereo microscope fitted with a mechanical stage.

Results and discussion

The total number of rotifers found for the Spring of both 1996 and 1997 are shown in Table 1. Both species richness and total abundances were lower in 1997 than in 1996. This may have been due to the excep-

Table 2. Dominance Indices for sampling dates and locations for Spring 1996, and Spring 1997. Ph = phaeophyte pools, Ch = chlorophyte pools, LT = sub-littoral, *d* = Berger-Parker dominance index

Berger-Parker Dominance Indices

1996				1997			
Site	Date	Taxon	*d*	Site	Date	Taxon	*d*
Ph	19/01/96	*P. reinhardti*	0.58	Ph	14/01/97	Nauplii	0.37
Ch		Nauplii	0.37	Ch		Nauplii	0.30
LT		Copepods	0.35	LT		Nauplii	0.46
Ph	08/02/97	*P. reinhardti*	0.45	Ph	27/01/97	Nauplii	0.47
Ch		Copepods	0.55	Ch		*P. reinhardti*	0.38
LT		Copepods	0.45	LT		N/A	N/A
Ph	21/02/96	*P. reinhardti*	0.47	Ph	12/02/97	*P. reinhardti*	0.63
Ch		*P. reinhardti*	0.45	Ch		*P. reinhardti*	0.50
LT		Copepods	0.58	LT		N/A	N/A
Ph	15/04/96	*P. reinhardti*	0.78	Ph	25/02/97	*P. reinhardti*	0.79
Ch		Nauplii	0.31	Ch		*P. reinhardti*	0.51
LT		Nematoda	0.58	LT		Nematoda	0.85
				Ph	09/03/97	*P. reinhardti*	1.00
				Ch		*P. reinhardti*	0.51
				LT		N/A	N/A
				Ph	25/03/97	*P. reinhardti*	0.35
				Ch		*P. reinhardti*	0.45
				LT		N/A	N/A

Table 3. Abundance of *Proales reinhardti* on phaeophyte or chlorophyte algae, together with the Mann-Whitney *U* statistic and probabilities of the null hypothesis. INS = Not significant

Date	Plant	*U*	*P*
21/01/96	Phaeophyte	2	<0.05
09/02/96	Phaeophyte	11	<0.0001
21/02/96	Phaeophyte	76	<0.01
15/04/96	Phaeophyte	3	<0.01
14/01/97	Phaeophyte	6	<0.05
27/01/97	Phaeophyte	9	INS
12/02/97	Chlorophyte	8	<0.01
25/02/97	Phaeophyte	2.5	<0.001
09/03/97	Phaeophyte	56.5	<0.001
25/03/97	Phaeophyte	4	<0.001

tionally low temperatures during January 1997, at or below 0 °C for several weeks, and which combined with strong, persistent easterly winds, when the blind landward (and less saline) end of the lagoon froze over. The low precipitation during this time resulted in relatively high levels of salinity. Saunders-Davies (1994), Green (1995) and Thane-Fenchel (1968) found that species richness and abundances declined with rising salinities.

During much of the Spring *P. reinhardti* was the dominant taxon in both the phaeophyte and chlorophyte invertebrate communities (Table 2; there were no vertebrates present). This is even more remarkable since the taxa of other groups counted are taken at a higher taxonomic level than species. On most sampling dates the population of *P. reinhardti* was significantly denser in the phaeophyte pools than the pools containing mostly chlorophytes (Table 3). Figures 2 and 3 show the numbers at all three sites for

228

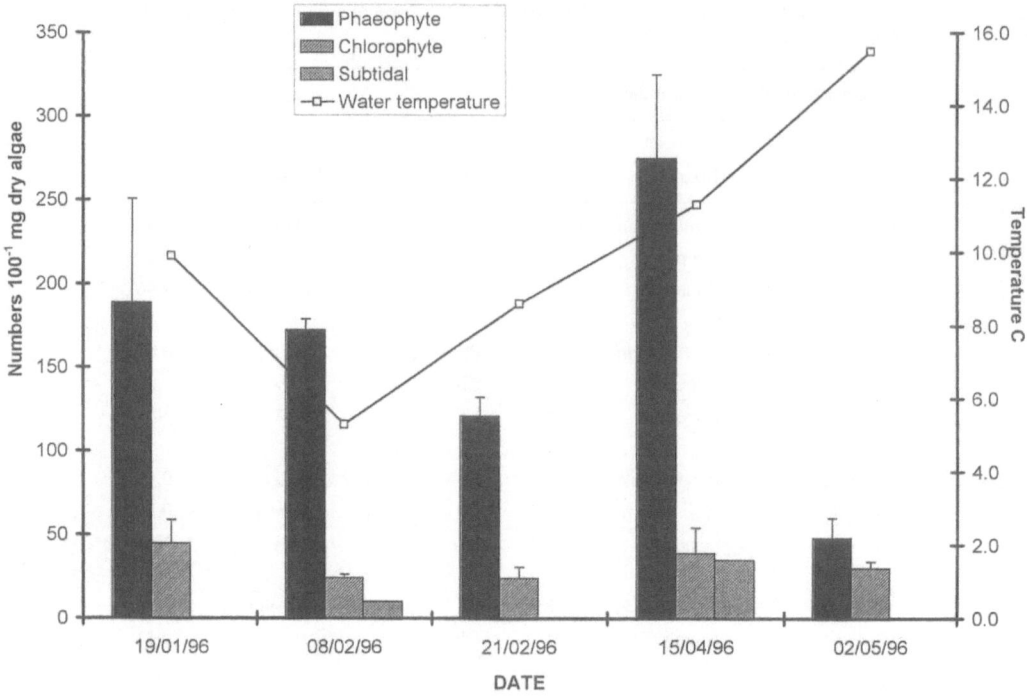

Figure 2. *Proales reinhardti* numbers in the phaeophyte, chlorophyte and subtidal areas together with midday temperatures, for Spring 1996. Error bars are for one SE.

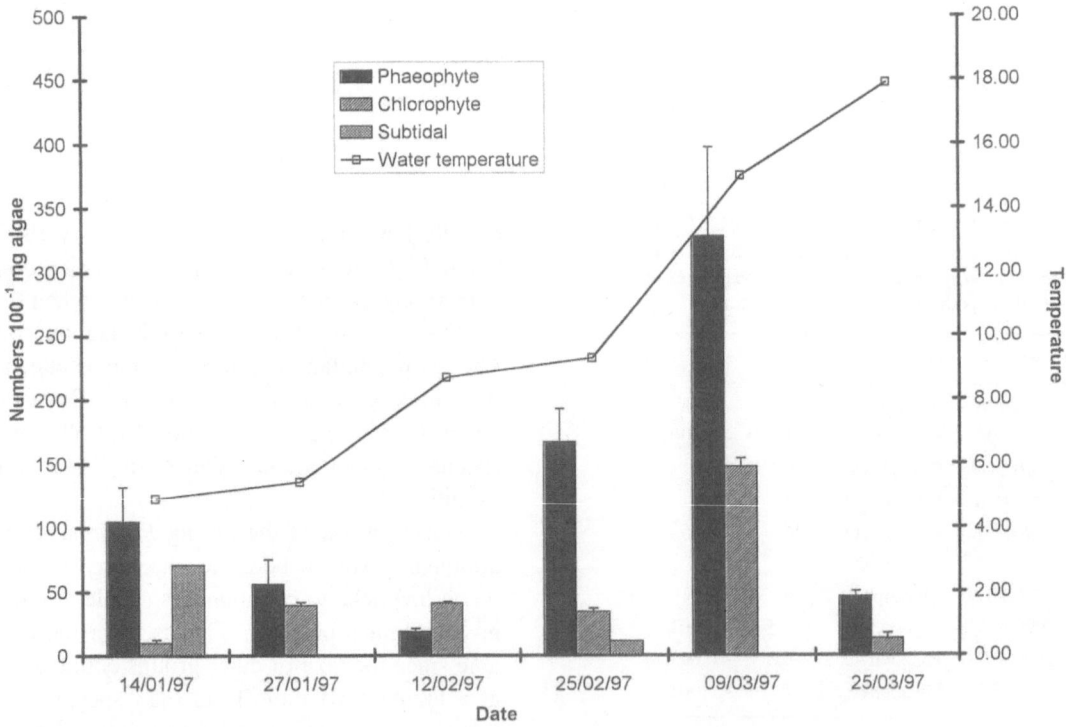

Figure 3. *Proales reinhardti* numbers in the phaeophyte, chlorophyte and subtidal areas together with midday temperatures, for Spring 1997. Error bars are for one SE.

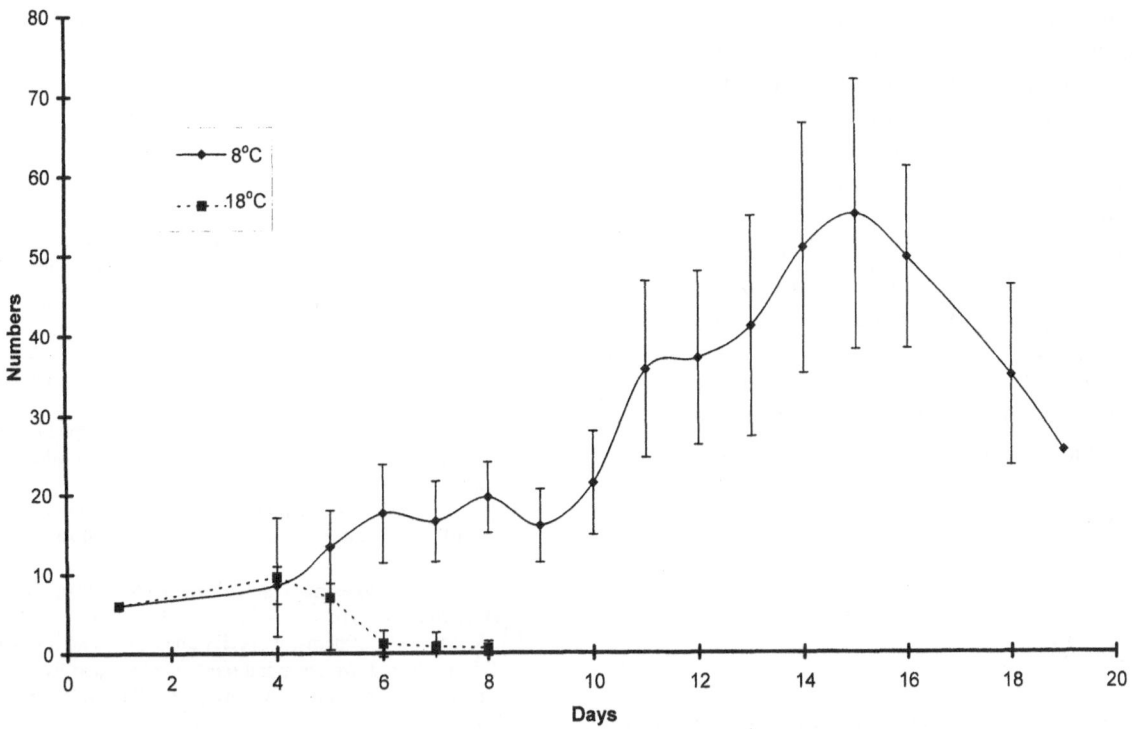

Figure 4. *Proales reinhardti* numbers in samples kept under a low-temperature regime of 8 ± 2 °C, and a high-temperature regime of 18 ± 2 °C

Table 4. Sediment loading as a decimal fraction of total mass of algae + sediment for chlorophyte area (Ch), phaeophyte area (Ph), and below Low Tide (LT) together with the *t*-statistic and probability of the null hypothesis (P). Spring 1996 combines samples from both phaeophyte and chlorophyte pools

| | Spring 1996 | | Spring 1997 | | |
	Ph & Ch	LT	Ph	Ch	LT
Mean of 5 samples	0.44	0.61	0.60	0.60	0.79
Standard Deviation	0.05	0.03	0.03	0.03	0.02
t	2.35		1.9 (For both Ph or Ch to LT)		
P	<0.01		<0.01		

1996 and 1997. Only on one occasion out of 11 were there significantly higher numbers of *P. reinhardti* in the chlorophyte pools. The reason for this is not clear, but it was noted that some of the chlorophyte algae, particularly *P. percursa*, were slippery to the touch, because of secreted mucilage, and it is possible that this inhibited the normal feeding behaviour of *P. reinhardti*. Generally this rotifer anchors itself to a point on an algal filament and browses on epiphytic diatoms either side of the anchorage point. It then moves to another point on the same or another, nearby filament and continues feeding. The species spends less than 10% of the time free-swimming.

The temperature related population dynamics of *P. reinhardti* were examined by keeping samples at different temperatures. Figure 4 shows that animals kept at 8 °C reproduced freely and reached an average of 52 individuals in each dish by day 15, while samples kept at 18 °C, after an initial slight increase, died out completely by day 6. It may be that the ultimate decline in the low temperature samples was due to exhaustion of silica resulting in an associated drop in diatom numbers: *P. reinhardti* is known to feed on pinnate diatoms (Hollowday, 1947; Koste, 1978), which is here confirmed by direct observation. In this case, mostly *Synedra* spp. were selected. In both years for which data are presented, *P. reinhardti* disappeared almost completely by late April or early May. This would seem to correlate with rising temperatures in the pools. The laboratory data appear to support the suggestion that rising temperatures in the lagoon are responsible for the decline in *P. reinhardti* numbers in late spring and early summer.

Very low numbers of rotifers where found in the sub-tidal (LT) hollows, and on occasion no rotifers at all were present. It was noted that the algae there were covered in detritus and silt. Sediment loading was measured on representative dates and was significantly higher in the sub-tidal pools than in the littoral pools (Table 4). The high levels of organic and inorganic (silt) material might fill the interstices between the algal filaments and deny the rotifer opportunities to move and feed.

Conclusions

Because of their shallow nature the pools of the littoral experience strong diurnal temperature changes. In particular temperatures rise quite quickly at the end of April and the beginning of May. This appears to correlate with the sharp decline in the numbers of *P. reinhardti* at this time. The experimental data supports the hypothesis that *P. reinhardti* is not viable at the higher temperatures encountered in late spring and summer.

The scarcity of animals in the sublittoral algae correlates with the high degree of sediment loading. The tidal nature of this habitat results in a twice daily de-

position of fine inorganic silt. This could inhibit the feeding behaviour of most rotifers by hampering their movement through the algae.

References

Berger, W. J. H. & F. L. Parker, 1970. Diversity of planktonic Foraminifera in deep sea sediments. Science 168: 1345–1347.
Burrows, E. M., 1991 Seaweeds of the British Isles, Vol. 2: Chlorophyta. HMSO.
Elliot, J. M., 1977. Some Methods for the Statistical Analysis of Benthic Invertebrates. Freshwater Biological Association Publication No. 25.
Green, J., 1995. Associations of planktonic and periphytic rotifers in a Malaysian estuary and two nearby ponds. Hydrobiologia 313/314: 47–56
Hollowday E. D., 1947. A preliminary survey of the marine and brackish water rotifers of Portsmouth. J. mar. biol Ass. 28: 239–253
Koste, W., 1978. Rotatoria. Die Rädierte Mitteleuropus. Borntraeger, Berlin, Stuttgart.
May, R. M., 1975. Patterns of species abundance and diversity. In M. L. Cody & J. M. Diamond (eds), Ecology and Evolution of Communities, Harvard University Press, Cambridge, MA: 81–120.
Newton, L., 1931. A Handbook of the British Seaweeds British Museum (Natural History).
Peters, U., W. Koste & Westheide, 1993. A quantitative method to extract mossdwelling rotifers Hydrobiologia 255/256: 339.
Saunders-Davies, A., 1995. Factors affecting the distribution of benthic and littoral rotifers in a large marine lagoon, together with the description of a new species. Hydrobiologia 313/314: 69–74.
Thane Fenchel, A., 1968. Distribution and ecology of nonplanktonic brackishwater rotifers from Scandinavia. Ophelia 5: 273–297.

Hydrobiologia **387/388**: 231–240, 1998.
E. Wurdak, R. Wallace & H. Segers (eds), Rotifera VIII: A Comparative Approach.
© *1998 Kluwer Academic Publishers.*

Review paper

Rotifers in interstitial sediments

J. M. Schmid-Araya

School of Biological Sciences, Queen Mary & Westfield College, Mile End Road, London E1 4NS, United Kingdom

Key words: Rotifera, interstitial, hyporheos, psammon, bed sediments

Abstract

Rotifers have long been known to inhabit interstitial sediments, thus confirming the high species richness of the group in a variety of habitats. This paper reviews the ecological role of rotifers within the interstitial environment (e.g. hyporheos, psammon, bed sediments) in lakes and running waters. Population densities, assemblage structure, patterns of colonization and drift are examined within riverine ecosystems.

Introduction

Rotifers can occur in an endless variety of aquatic, semiaquatic or terrestrial environments. Many exhibit distinctive morphological characteristics, through which they are well adapted to their habitat, and from which it is often possible to discern between littoral, pelagic and benthic rotifers (Klimowicz, 1972; Ruttner-Kolisko, 1974). Benthic rotifers are mostly elongate in form, allowing them to move within the narrow crevices or interstitial spaces of sediments. Most species have a ventral ciliated field which is used to scrape the biofilm (bacteria, fungi and diatoms) from the surface of organic particles. Other adaptations include a well-developed foot with adhesive glands to resist water currents (e.g., *Proales theodora* (Gosse)). Bdelloid rotifers which are a very important ecological component of soil, river and stream ecosystems, also show a peculiar ability to resist dessication and undergo parthenogenetic reproduction only (Pourriot, 1979).

Stream-dwelling rotifers are known to feed on biofilm, and in turn, are major food items for predatory chironomid larvae and stonefly nymphs (Schmid, 1994; Schmid & Schmid-Araya 1997). They also form part of the diet of other meiofaunal predators such as Nematoda, Microturbellaria, other Rotifera and large Protozoa (Schmid-Araya & Schmid, 1995b).

Interstitial spaces were first recognized as refugia for small organisms by Racovitza (1907). Such interstitial habitats are found in aquatic and terrestrial

environments, and sometimes in the transition zone between these, such as the shore region of lakes (Pennak, 1940) and the riparian zone of rivers. In freshwater biotopes, the biological terminology used to identify these habitats can be confusing, and sometimes contradictory, among different research schools (e.g., Giere, 1993).

The objectives of this review are: (1) to summarize the early work on interstitial rotifers (namely in the psammon of lakes and rivers) and combine it with the present state of rotifer research in stream and river ecosystems; (2) to collate rotifer densities, species richness and drift rates; (3) to compare species presence between freshwater and marine psammon and the streambed and hyporheic habitats.

Psammon rotifers

The term psammon was coined by Sassuchin (1926) for the community of organisms living in sandy shores and was originally considered as a transitional zone between aquatic and soil habitats. The psammon community members are groups that presently fall into the meiofauna category. These include rotifers, gastrotrichs, microturbellarians, nematodes, and Diptera larvae, among others. In 1935, Neiswestnowa-Shadina recorded 19 psammic rotifer species in the Oka River, Russia, and described the faunal differences in the sand from the midstream to the banks of the river. She was the first to relate the distribution of psammon organisms inhabiting the sandy bottom of rivers to wa-

232

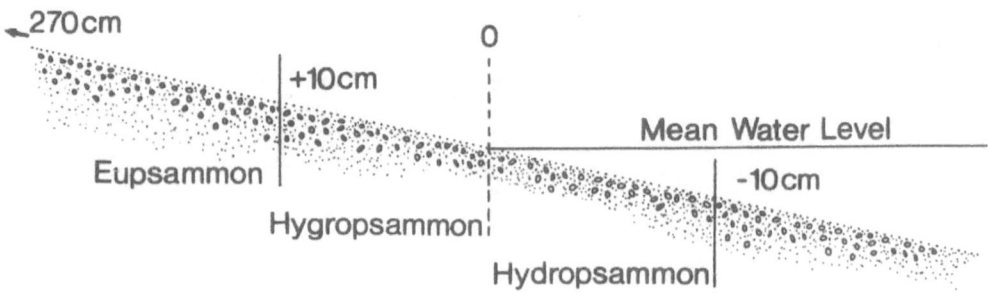

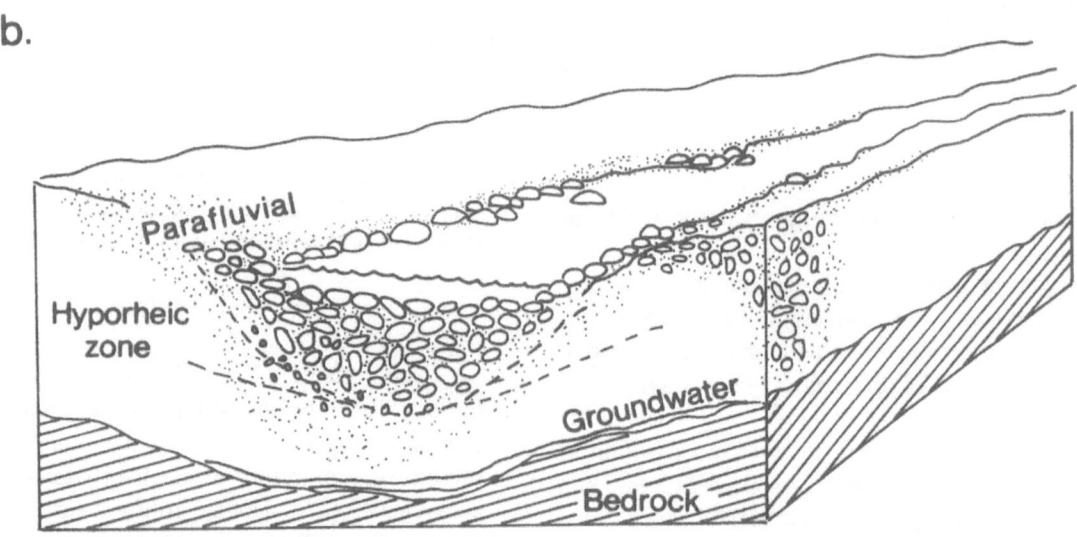

Figure 1. Diagrammatic representation of (a) freshwater and marine psammon following Wiszniewski (1937a,b) and (b) the hyporheic habitat within a stream ecosystem (adapted from Schmid, unpubl.).

ter velocity and to a series of related processes such as the accumulation of organic matter, and oxygen depletion. In Poland, Wiszniewski (1934, 1937a,b) considered the psammon as a special type of environment; the psammolittoral, which included the shallows and beaches of lakes and streams (Figure 1a). He further sub-divided this area into three categories: (a) the hydropsammon, or submerged sand along the edge of the water body, (b) the hygropsammon, sand which is above and adjacent to the water level and is completely saturated by capillary action and wave action and (c) the eupsammon, or outer boundary of the latter partially saturated, and submerged only during periods of high water (Figure 1a).

Myers (1936) and Pennak (1939c, 1940) studied rotifers from a diverse community of micrometazoans living among the sand grains of several lake beaches in the United States. Ruttner-Kolisko (1953, 1954, 1961) also carried out extensive studies in the psammon of Swedish lakes and Austrian rivers (Ruttner-Kolisko, 1961), where she found differences between grain size, water content, capillarity and evaporation among the hydropsammon, the hygropsammon and the eupsammon. Varga (1957) noted the marked contrast between the species-rich hygropsammon and the species-impoverished eupsammon community in Lake Balaton (Hungary).

The extensive studies of Pennak (1940), Neel (1948) and Ruttner-Kolisko (1953, 1954) showed that the organisms were mostly distributed in the upper 1 cm of sand. However, they also found reduced faunal densities in the top 1 cm of sand at some sites. Vertical distribution seemed to be triggered by wave action, with individuals moving into deeper (2–3 cm) layers of sand to avoid the stress from this source. Ruttner-Kolisko (1954) found rotifers as deep as 3 cm below the surface of the sand, while nematodes were found down to depths of 10 cm in Lake Erken (Sweden). The horizontal distribution of psammon fauna may also be affected by shoreline processes (Neel, 1948; Ruttner-Kolisko, 1953).

Evans (1982) reported that 60% of the psammon fauna were found in the 2–3 cm depth fraction at sites in the wave-influenced zone at the water's edge. Evans' work on sandy beaches in Lake Erie emphasized that each rotifer species was affected differently by environmental factors. Furthermore, Evans (1984) found a random spatio-temporal distribution of rotifer species and suggested that there was no evidence of competitive exclusion in an unpolluted stream.

Undoubtedly, psammolittoral organisms inhabit very unstable and unique surroundings, largely depending on the effects of factors outside the sand. The physical factors influencing the vertical and horizontal, as well as the temporal, distribution of the psammon include: exposure to wave action, quantity of organic matter available as food, sand grain morphometry, amount of water in the interstitial spaces, oxygen concentration, pH, and temperature among others (Pennak, 1940; Neel, 1948; Ruttner-Kolisko, 1953; Evans, 1982).

In marine habitats, the ecology of interstitial fauna was extensively studied in the 1950's and 1960's, however, few researchers discussed the importance of rotifers. Most of the early literature on marine rotifers dealt with taxonomy and morphology, until Thane-Fenchel (1968) carried out a quantitative study in Scandinavian waters. She showed that rotifers played a modest numerical role in marine and brackish water sands (<10 specimens cm^{-2}). She also gave a list of food items of marine rotifers, which contrasted with the common belief that rotifers are mainly detritus feeders. Later, Tzschaschel (1983), contributed further to the understanding of benthic marine rotifer population dynamics.

Despite the time elapsed since Pennak (1951) published his ecological comparison of the interstitial fauna of freshwater and marine beaches, many of his

observations may be still valid. He concluded that: (a) most organisms are confined to the uppermost 5 cm of sand in lake beaches and the uppermost 12 cm in marine beaches, (b) spatial and temporal density fluctuations are more pronounced in freshwater, (c) the range of ecological conditions is much more severe in lake than in marine beaches, and (d) rotifers are rare in marine beaches but abundant in freshwater.

Table 1 summarizes the rotifer species richness and abundance from studies in lake and marine psammon since 1934. It is clear that freshwater psammon rotifers are more diverse and more abundant than their marine counterparts. There are also slight differences in rotifer species richness among freshwater habitats, although these may be attributable to differences in sampling methods, the use of fixatives, substrate differences and the extent of sampling.

Hyporheic rotifers

In running waters, the habitat designated as the hyporheic zone (Orghidan, 1959) consists of the interstices that are formed by the mixture of coarse sand and gravel found in the rithron region of streams (Figure 1b). Schwoerbel (1961) named this habitat the hyporheic interstitial and the term hyporheos was later used to refer to its fauna (Williams & Hynes, 1974). There has been considerable debate over the definition and extension of the hyporheic zone but the most widely accepted definition is that of Orghidan (1959) and Schwoerbel (1961) who argued that it is the middle zone bordered by surface water above and true groundwater below (Figure 1b).

The current lack of knowledge of rotifer assemblages in hyporheic habitats is probably due to the difficulty of sampling this zone. The few records of hyporheic rotifers which do exist give widely differing results. Some authors argue that these organisms are poorly represented in the hyporheos (Schwoerbel, 1965; Boulton et al., 1992), while others have found them to be an abundant and species-rich group (Palmer, 1990a,b, 1992; Schmid-Araya, 1997, 1998). In my view, this contention needs to be examined by experts on hyporheic rotifers. Palmer (1990a,b) argued that many stream ecologists used mesh nets too coarse to retain meiofaunal size. Also the use of fixatives which cause contraction of many taxa such as bdelloids makes the task more difficult. In my view, mainly because they are small and difficult to identify, rotifers have not been explored thoroughly in stream ecology. However, the increasing awareness for roti-

Table 1. Rotifer species richness and density ranges from studies in freshwater (F) and marine (M) psammon and deep lake sediments.

	Author	Locality	Species richness	Density ranges (ind ml^{-1}, unless otherwise stated)
F	Wiszniewski (1934)	Lake Wigry (Poland)	82	–
F	Neiswestnowa-Shadina (1935)	River Oka (Russia)	21	–
F	Wiszniewski, (1935)	River Czarna (Poland)	27	–
F	Myers (1936)	Lakes Lanape & Union	145	–
F	Pennak (1940)	Wisconsin (20 lake beaches)	6–26	1150
F	Neel (1948)	Lake Douglas	38	0–104
F	Ruttner-Kolisko (1953, 1954)	Lapland lakes	19–21	1–300
F	Varga (1957)	Lake Balaton	45	–
F	Pejler (1962)	Lapland & other	10–21	
M	Thane-Fenchel (1968)	Baltic & Danish waters	34	<10 ind cm^{-2}
F	Klimowicz (1972)[a]	Mikolajskie & Taltowisko lakes	27–29	180–170
M	Tzschaschel (1980)	North Sea	–	0.3–0.6
F	Evans (1982)	Western Lake Eire	19	0–23
M	Tzschaschel (1983)	Island of Syl, North Sea	13	375 ind per ?
F	Evans (1984)	Raccoon Creek, USA	31	0–8.6
F	Strayer (1985b)	Mirror Lake	56	up to 57 000 m^{-2}
M	Turner (1990)	Florida beach	11	–
M	Turner (1993)	Okalossa, Island, Florida, USA	18	0–6
F	Turner (1995)	Ninnescah River, USA	24	0.2
F	Radwan & Bielawska-Grabner (pers. comm.)	Diverse lakes in South Poland	10–117	0–2217 ind dm^{-3}

[a] Deep lake sediment sampled with cylinder corer.
[b] Gyttja sediments sampled with multiple corer.

fers inhabiting the sediments of streams is confirmed by an increase in the number of publications in recent years (Palmer 1990a,b, 1992; Schmid-Araya, 1993a, 1995, 1997, 1998; Eisenmann, in press; Opravilová, in press).

Table 2 summarizes rotifer studies in rivers and streams of various orders and substrate types. Pawlowski (1956, 1972), Donner (1964, 1970, 1972a) and Koste (1976) gave listings with a high rotifer species richness. These studies were extensive and intensive and certainly included benthic as well as planktonic species. Furthermore, Pawlowski (1956, 1972) sampled 36 sites within the River Grabia catchment area. Braioni & Gottardi (1979), using a Bou Rouch pump found 67 species down to sediment depths of 30 cm in the River Adige (Table 2). Turner & Palmer (1996), reported over 80 rotifer species (Table 2) in a sandy bottom stream, while Schmid-Araya & Schmid (1995a) have listed 101 rotifer species in a gravel stream in Austria (Table 2). Undoubtedly, these accounts show that rotifers play an important role within

the benthic community by enhancing biodiversity in streams and rivers.

There are few quantitative rotifer studies in running waters (Table 2), and the variety of sampling methods used makes it difficult to compare results. Nevertheless, the highest rotifer densities of 4×10^6 ind m^{-2} have been found in Goose Creek using a standpipe corer by Palmer (1990a,b). Rotifers constituted between 35% and 85% of the meiofaunal community. Similar percentages can be calculated from Schwank (1985) in the Breitenbach, where rotifers accounted for 19%–75% of the meiofaunal assemblage. In addition, in a second order alpine gravel stream, rotifers were also among the most important groups (27%) of the meiofauna (Schmid-Araya, 1997). The latter used a modified Hess sampler for rotifers at the streambed and standpipe traps for hyporheic rotifers. Eisenmann (in press), estimated total rotifer densities using sediment chambers (Eisenmann et al., 1997) which were exposed for a specific time interval within the sediments of the River Necker (Switzer-

Table 2. Rotifer species richness and abundances in the streambed surface and/or hyporheic zone of running waters

Author	Locality	No. species	Density ranges (ind m^{-2}, unless otherwise stated)
Pawlowski (1956, 1972)	Grabia & tributaries, Poland	216	–
Pejler (1962)	Swedish Lapland streams	16–34	–
Donner (1964)	Danube (km 1900), Austria	97	–
Donner (1970)	Salzach and tributaries, Austria	154	–
Donner (1972)	Danube (kms 2240–2148), Austria	185	–
Koste (1976)	Hase, Germany	187	–
Braioni & Gottardi (1979)[2]	Adige, Italy	67	–
Zullini & Ricci (1980)	Carrega Wood stream, Italy	18	
Evans (1984)	Raccoon Creek, U.S.A.	31	0–8.6 ind ml^{-1}
Schwank (1985)	Breitenbach, Germany	48	2–8 ind ml^{-1}
Palmer (1990)[b]	Goose Creek, U.S.A.	–	4×10^6
Turner & Palmer (1996)		77	–
Schmid-Araya & Schmid (1995a); Schmid-Araya (1998)[b]	Oberer Seebach, Austria: streambed surface	69–101	up to 30 400
	hyporheic zone		11.5×10^6
Jones, Shiozawa & Schmid-Araya (unpubl.)	Colorado & Green Rivers, U.S.A.	35	–
Eisenmann (in press)[b]	River Necker, Switzerland	–	0–126 ind ml^{-1}
Opravilová (in press)[b]	Svratka River, Czech Republic	16	2–58 ind ml^{-1}
Schmid-Araya (unpubl.)	Afon Mynach, U.K.	43	0–50 000
Schmid-Araya (unpubl.)	Broadstone, U.K.	33	600–40 000

[a] 36 sites sampled.

[2] Indicates the studies including the hyporheic zone (0–40 cm).

land). Similarly, Opravilová (in press) also estimated rotifer densities using sediment chambers down to 20 cm in depth.

These findings demonstrate that rotifers may play an important role in the functioning of the benthic communities, although the relative importance of small organisms in stream and rivers remains unexplored.

Vertical distribution

Although stream studies have reported rotifers as abundant taxa (Hummon et al., 1978; Evans, 1984; Palmer, 1990a,b) few have explored their vertical distribution in the hyporheic zone. Palmer (1990a,b) reported rotifer densities at all depths and occasionally down to 60 cm in the sediments. The depth distribution of Rotifera was more variable in a riffle than at a pool site in the well-oxygenated gravel stream Oberer Seebach (Schmid-Araya, 1997). In the hyporheic zone, the depth distribution differed among the two rotifer groups, bdelloids occurred in highest densities between 0 and 30 cm sediment depth, while monogononts were most abundant at greater depths (Schmid-Araya, 1998). Mean water temperature, mean current velocity and surface discharge are important hydrophysical variables influencing the vertical distribution of rotifers in running waters.

Rotifer drift

The phenomenon of drift, or downstream transport, of benthic invertebrates in the colonization and recolonization of denuded substrates in streams has been studied intensively (see Brittain & Eikeland, 1988). Many studies have shown that drift is an important aspect of the population dynamics of many lotic invertebrates. Willer (1974) hypothesized the existence of a colonization cycle wherein newly-hatched insect larvae drift downstream because they are restricted in resources of food and space and adults fly upstream to oviposit and complete the cycle. Obviously, this hypothetical cycle cannot be applied to meiofaunal components due to their homotopic life cycle. Although benthic meiofaunal groups can be extremely abundant, few meiofaunal drift studies have been carried out, and even fewer with rotifers (Braioni, 1981; Sandlund, 1982; Palmer, 1992).

236

Table 3. The relative abundance of common meiobenthic taxa in the drift, and taxa drift rates (in brackets)

Study	Copepods (ind m^{-3})	Nematodes (ind m^{-3})	Chironomids (ind m^{-3})	Rotifers (ind m^{-3})
Pawlowski (1968)	10.1% (–)	–	–	88.0%
Koste (1976)	– (–)	– (–)	– (–)	– (40 000)
Sandlund (1982)	– (–)	– (–)	– (300–1000)	- (200–400)
Palmer (1992)	1.5% (–)	2.4% (–)	6.9% (–)	86.2% (2500)
Schmid-Araya (unpubl.)	7.5% (–)	4.1% (–)	10.1% (–)	69.2% (300)

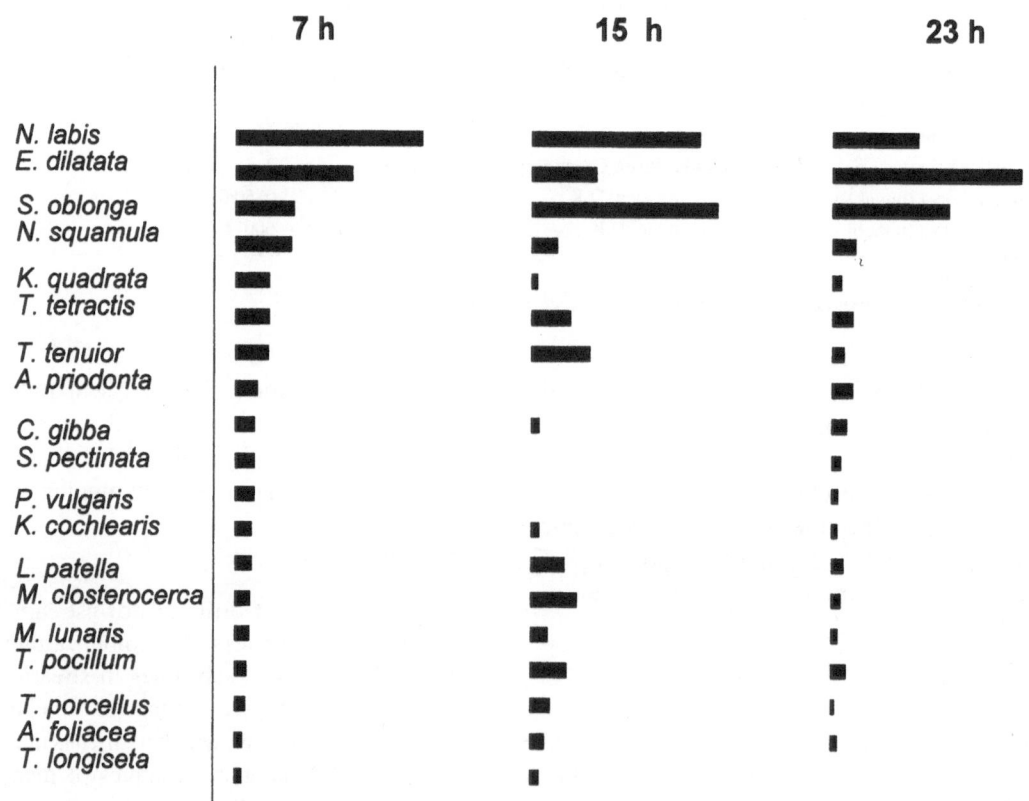

Figure 2. Frequency distribution of the most abundant rotifer species in the drift at 0700, 1500 and 2300 h in the River Grabia (Poland). Redrawn from Pawlowski (1968).

Drift is recognized as a major means of dispersal at the streambed surface. The percentage of benthic rotifers drifting in the same gravel stream can reach up to 3.1% of benthic densities (Schmid-Araya, unpublished results). At the streambed surface the temporally variable structure of rotifer assemblages may be due to drift while, in the hyporheic zone, rapid species replacement may be the result of individual species movements (Schmid-Araya, 1998).

Braioni (1981) found 26 rotifer species in the drift of the River Adige, and the benthic forms from Braioni & Gottardi (1979) were well represented in the flow.

Differences between dates in the diel drift patterns of rotifers were found by Palmer (1992), and it seems likely that drift patterns vary between species. In Goose Creek, Palmer (1992) found that rotifers exhibited higher drift rate during the night than during daytime. However, much earlier Pawlowski (1968) had shown specific diel patterns in the River Grabia (Poland). He sampled drift for 4 weeks during the summer of 1954. He found up to 65 rotifer species over a 24-h period, and distinguished the diurnal planktonic *Synchaeta oblonga* Ehrb. and *Notholca labis* Gosse from a nocturnal *Euchlanis dilatata* Ehrb. (Figure 2). The drift rates of rotifers in streams seem to be substantial (Table 3), and flow dependent (Palmer, 1992). It is clear that rotifer species in the drift vary according to the type of river. Pawlowski (1968), Braioni (1981) and Sandlund (1982) showed a mixture of plankton and benthic species. Obviously, the species composition will depend on the stream/river order.

Soil rotifers

Few quantitative accounts of soil rotifers have been published since the review of Pourriot (1979). In soil studies, rotifers are often collected together with nematode samples and their population densities are rarely reported. Rotifers can have high population densities in a variety of soil types. Estimates range from 300 000 to 2.1×10^6 ind m^{-2} (Schulte, 1954; Sohlenius, 1982; Anderson et al., 1984). This result has lead Anderson et al. (1984) to suggest that rotifers may play a significant role in nutrient cycling in the soil community. However, none of these studies established the systematic identity of these rotifers. According to Pourriot (1979), rotifers inhabiting mostly the topsoil layer, are usually bdelloids (Class Digononta) which have many adaptations to life in soils. Bdelloids are microphagous, feeding on bacteria, fungi and microscopic algae. I have observed, using video techniques, that species of the genus *Adineta* search for individual organic particles (plus attached biofilm). Once encountered, the particle is rolled and browsed upon using the ciliated ventral plate. Simultaneous movements of the mastax suggest that this genus feeds actively on bacteria attached to these particles.

In soils, many abiotic and/or biotic factors seem to influence the presence and abundance of groups such as rotifers. Schulte (1954) already summarized these as follows: humidity, temperature, the structure of soil, oxygen and carbon dioxide concentrations, light, pH, and food resources.

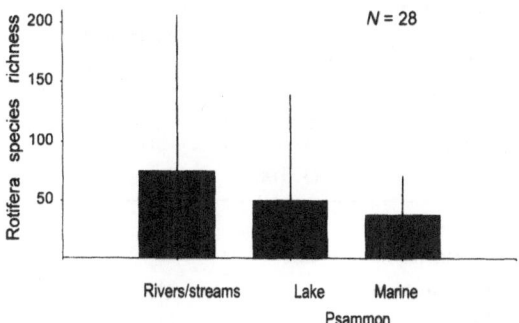

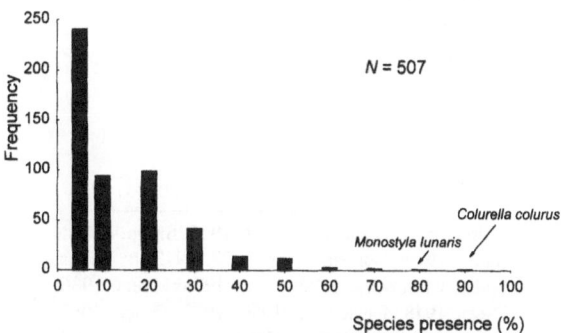

Figure 3. Results of a comparison based on 28 studies dealing with interstitial rotifers since 1934. Upper panel: mean rotifer species richness among different habitats (the line indicates the range). Lower panel: the frequency distribution of 507 species found in 28 studies between lake and marine psammon and river/streams.

Comparison among habitats

I compiled species lists from 28 published works (listed in references) on freshwater and marine psammon and lotic streambed and hyporheos. My comparison identified 507 benthic rotifer species. It is not surprising to find that 92% of the species are freshwater, since Rotifera have been widely studied in freshwater habitats, and more work has been carried out on freshwater habitats. Between lakes and lotic systems, 237 species (47%) have been reported exclusively in running waters, while 137 species are confined to lake psammon. Furthermore, the average species richness was 68 (range: 15–216) in running waters, 40 (range: 10–139) in lake psammon and 30 (range: 9–71) in marine psammon (Figure 3). It can be suggested that lotic habitats are more heterogeneous and, somewhat more disturbed, they support a higher species richness than benthic lake habitats.

The most common species among these habitats were *Colurella colurus* Ehrb. (in 90% of examined species lists), *Monostyla lunaris* (Ehrb.) (80%), *Monostyla closterocerca* Schmarda and *Cephalodella gibba* (Ehrb.) (both in 70% of the species lists, Figure 3).

Table 4. Percentage contribution of rotifer species to the most common families found in lake psammon by Wiszniewski (1937a) and those compiled in this study

Rotifera family	Wiszniewski (1937a)	This study[a]
Notommatidae	30.8	18.1
Dicranophoridae	23.7	17.4
Philodinidae	–	13.2
Lecanidae	16.1	12.2
Colurellidae	4.9	8.7
Trichocercidae	14.7	7.1
Proalidae	–	3.7
Euchlanidae	–	2.2
Testudinellidae	–	2.2
Bdelloidea	4.9	–
Brachionidae	1.2	–
Other groups	3.7	10.7

[a]References included: Althaus, 1956; Braoni & Gottardi, 1979; Donner, 1954, 1970, 1972a,b; Evans, 1982; Klimowicz, 1972; Koste, 1976; Myers, 1936; Neel, 1948; Pawlowski, 1956, 1972; Pejler, 1962; Ruttner-Kolisko, 1953, 1954; Schmid-Araya, 1993; Schwank, 1985; Thane-Fenchel, 1968; Turner, 1993, 1995; Tzschaschel, 1983; Varga, 1957; Wiszniewski, 1934, 1937a; Zullini & Ricci, 1980.

241 rotifer species occurred in 5% of the habitats examined, which indicates either (a) specialists or (b) poor dispersal. At the family level, Notommatidae was represented by 92 species (18.1%), Dicranophoridae with 88 species (17.4%), Philodinidae (Class Bdelloidea) with 67 species (13.2%) and Lecanidae with 62 species (12.2%, Table 4). These findings are consistent with an earlier examination on psammon rotifers in lakes (Wiszniewski, 1937a; Table 4). However, my compilation shows the contribution of bdelloids to rotifer benthic assemblages. Common species between lake psammon and running waters accounted only for 26.6% and, as expected, a lower percentage of common species were found between lake and marine psammon (10.3%), and lotic systems and marine psammon (14.2%).

The similarities and differences among these communities need to be explored further. A great amount of work has been carried out on the taxonomy of interstitial rotifers. However, even where quantitative data have been produced, the lack of examination of species patterns remains. Interstitial rotifers are exposed to disturbances such as wave action in the littoral psammon of lakes and marine beaches and spates in rivers and streams. These changing environments seem to be ideal for a group of organisms that reproduce quickly, occupy vacant habitats rapidly and exhibit a wide range of feeding habits.

Acknowledgements

This review is dedicated to Neiwestnowa-Shadina, Wisznieski, Pennak, Neel, Pawlowski and Ruttner-Kolisko, who produced such fascinating papers on the psammon rotifers. I thank also P. E. Schmid for his contribution. Two anonymous referees made many useful suggestions to improve the manuscript.

References

Althaus, B., 1956. Faunistisch-ökologische Studien an Rotatorien salzhaltiger Gewässer Mitteldeutschlands. Wiss. Z. Univ. Halle, Math.-Nat. 6: 117–158.

Anderson, R. V., R. E Ingham, J. A Trofymow & D. C. Coleman, 1984. Soil mesofaunal distribution in relation to habitat types in a shortgrass prairie. Pedobiologia 26: 257–261.

Boulton, A. J., H. M. Valett & S. G. Fischer, 1992. Spatial distribution and taxonomic composition of the hyporheos of several Sonoran Desert streams. Arch. Hydrobiol. 125: 37–61.

Braoni, M. G., 1981. The drift of rotifers in the River Adige: preliminary communication. Boll. Zool. 48: 305–310.

Braoni, M. G. & M. Gottardi, 1979. 1 Rotiferi dell'Adige: Confronto tra il popolamento interstiziale e quello bentico-perifitico. Boll. Museo Civ. St. Nat. Verona 6: 187–219.

Brittain, J. E. & T. J. Eikeland, 1988. Invertebrate drift – A review. Hydrobiologia 166: 77–93.

Donner, J., 1954. Zur Rotatorienfauna Südmährens. Österr. zool. Z. 5: 30–117.

Donner, J., 1964. Die Rotatorien-Synusien submerser makrophyten der Donau bei Wein und mehrerer Alpenbäche. Arch. Hydrobiol. Suppl. 27: 227–324.

Donner, J., 1970. Die Rädertierbestände submerser Moose der Salzach und anderer Wasser-Biotope des Flussgebietes. Arch. Hydrobiol. Suppl. 34: 109–254.

Donner, J., 1972a. Die Rädertierbestände submerser Moose und weiterer Merotope im Bereich der Starräume der Donau and der deutsch-österreichischen Landesgrenze. Arch. Hydrobiol. Suppl. 36: 109–254.

Donner, J., 1972b. Die Rädertierbestände submerser Moose und weiterer Merotope im Bereich der Starräume der Donau and der deutsch-österreichischen Landesgrenze. Arch. Hydrobiol. Suppl. Donauforsch. 44(5): 49–114.

Donner, J., 1972c. Rädertiere der Grenzschicht Wasser-Sediment aus dem Neusiedler See. Sitz.-Ber. Österr. Akad. Wiss., Abt. 1, Bd 180: 49–63.

Eisenmann, H., 1998. Community structure of selected micro-and meiobenthic organisms in sediment chambers from a prealpine river (Necker, Switzerland). Adv. River Bottom Ecol. (In press).

Evans, W. A., 1982. Abundances of micrometazoans in three sandy beaches in the island area of Western Lake Eire. Ohio Acad. Sci. 82: 246–251.

Evans, W. A., 1984. Seasonal abundances of the psammic rotifers of a physically controlled stream. Hydrobiologia 108: 105–114.

Giere, O., 1993. Meiobenthology The Microscopic Fauna in Aquatic Sediments. Springer-Verlag, Berlin, 328 pp.

Hummon, W. D., W. A. Evans, M. R. Hummon, F. G. Doherty, R. H. Wainberg & W. S. Stanley, 1978. Meiofaunal abundance in sandbars of acid mine polluted, reclaimed, and unpolluted streams in southeastern Ohio. In H. H. Thorp & J. W. Gibbons (eds), Energy and Environmental Stress in Aquatic Ecosystems. DOE Symposium Series, Nat. Tech. Info. Service, Springfield: 188–203.

Klimowicz, H., 1972. Rotifers of the near bottom zone of lakes Mikolajski and Taltowisko. Pol. Arch. Hydrobiol. 19: 167–178.

Koste, W., 1976. Ober die Rädertierbestände (Rotatoria) der oberen und mittleren Hase in den Jahren 1966–1969. Osnabrücker Naturw. Mitt. 4: 191–263.

Müller, K., 1974. Stream drift as a chronobiological phenomenon in running water ecosystems. Ann. Rev. Ecol. Syst. 5: 309–323.

Myers, F. E., 1936. Psammolittoral rotifers of Lenape and Union Lakes New Jersey. Am. Mus. Novitates 830: 1–22.

Neel, J. K., 1948. A limnological investigation of the psammon in Douglas lake. Trans. am. micr. Soc. 67: 1–53.

Neiwestnowa-Shadina, K., 1935. Zur Kenntnis des rheophilen Mikrobenthos. Arch. Hydrobiol. 28: 555–582.

Opravilova, V., 1998. Micro and meiobenthic organisms in the hyporheos of artificial brooks. Adv. River Bottom Ecol. (In press).

Orghidan, T., 1959. Ein neuer Lebensraum des unterirdischen Wassers: Der hyporheische Biotop. Arch. Hydrobiol. 55: 392–414.

Palmer M. A., 1990a. Temporal and spatial dynamics of meiofauna within the hyporheic zone of Goose Creek, Virginia. J. n. am. Benthol. Soc. 9: 17–25.

Palmer, M. A. 1990b. Understanding the movement dynamics of a stream-dwelling meiofauna community using marine analogs. Stygologia 5: 67–74.

Palmer, M. A., 1992. Incorporating lotic meiofauna into our understanding of faunal transport processes. Limnol. Oceanogr. 37: 329–341.

Pawlowski, L. K., 1956. Première liste des Rotifères trouvés dans la Rivière Grabia. Bull. Soc. Sci. Lett. Lódz 7(4): 1–54.

Pawlowski, L. K., 1968. Nouvelles observations sur les rotifères de la Rivière Grabia. Bull. Soc. Sci. Lodziensis, Section 111, 103: 4–52.

Pawlowski, L. K., 1972. Liste supplémentaire des rotifères trouvés dans la Rivière Grabia. Bull. Soc. Sci. Lett. Lódz 21: 1–10.

Pejler, B., 1962. On the taxonomy and ecology of benthic and periphytic Rotatoria. Zool. Bidr. Från Uppsala 33: 327–422.

Pennak, R. W., 1940. Ecology of the microscopic metazoa inhabiting the sandy beaches of some Wisconsin lakes. Ecol. Monogr. 10: 537–615.

Pennak, R. W., 1951. Comparative ecology of the interstitial fauna of fresh-water and marine beaches. Ann. Biol. 27: 217–480.

Pourriot, R., 1979. Rotifères du sol. Rev. Écol. Biol. Sol. 16: 279–312.

Racovitza, E. G., 1907. Essai sur les problemes biospéléologiques. Biospeol. 1. Arch. Zool. exp. gén. 6: 371–488.

Radwan, S. & I. Bielawska-Grabner, 1998. Preliminary investigations on the bionic structure of the psammic rotifers in the water bodies of south-eastern Poland. Pers. comm.

Ruttner-Kolisko, A., 1953. Psammonstudien I. Das Psammon des Torneträsk in Schwedisch-Lapland. Sitz.-Ber. Österr. Akad. Wiss., Abt. 1, Bd 162/3: 129–161.

Ruttner-Kolisko, A., 1954. Psammonstudien II. Das Psammon des Erken in Mittelschweden. Sitz.-Ber. Österr. Akad. Wiss., Abt. 1, Bd 163/22: 301–324.

Ruttner-Kolisko, A., 1961. Biotop und Biozönose des Sanufers einiger österreichischer Flüsse. Verh. int. Ver. Limnol. 14: 362–368.

Ruttner-Kolisko, A., 1974. Plankton rotifers – Biology and taxonomy. Binnengewässer Suppl. 1: 1–146.

Sandlund, O. T., 1982. The drift of zooplankton and microzoobenthos in the River Strandaelva, western Norway. Hydrobiologia 94: 33–48.

Sassuchin, D. N., 1926. Zur Frage der Bodenprotozoen. Russ. Arkh. Protistol. 5: 241–246.

Schmid, P. E., 1994. Is prey selectivity by predatory Chironomidae a random process? Verh. int. Ver. für theoret. u. angew. Limnol. 25: 1656–1660.

Schmid, P. E. & Schmid-Araya, J. M., 1997. Predation on meiobenthic assemblages: resource use of a tanypod guild (Chironomidae, Diptera) in a gravel stream. Freshwat. Biol. 38: 61–91.

Schmid-Araya, J. M., 1993. Benthic Rotifera inhabiting the bed sediments of a mountain gravel stream. Jahresb. Biol. St. Lunz 14: 75–101.

Schmid-Araya, J. M., 1995. New records of rare Bdelloidea and Monogononta Rotifera in gravel streams. Arch. Hydrobiol. 135: 129–143.

Schmid-Araya, J. M., 1997. Temporal and spatial dynamics of meiofaunal assemblages in the hyporheic interstitial of a gravel stream. In J. Gilbert, J. Mathieu & F. Fournier (eds), Groundwater/Surface Water Ecotones: Biological and Hydrological Interactions Options. Cambridge University Press: 29–36.

Schmid-Araya, J. M., 1998. Small-sized invertebrates in a gravel stream: community structure and variability of benthic rotifers. Freshwat. Biol. 39: 25–39.

Schmid-Araya, J. M. & P. E. Schmid, 1995a. The invertebrate species of a gravel stream. J. Ber. Biol. St. Lunz 15: 11–21.

Schmid-Araya, J. M. & P. E. Schmid, 1995b. Preliminary results on diet of stream invertebrates species: the meiofaunal assemblages. J. Ber. Biol. St. Lunz 15: 23–31.

Schwank, P., 1985. Differentiation of the coenoses of Helminthes and Annelida in exposed lotic microhabitats in mountain streams. Arch. Hydrobiol. 193: 535–543.

Schwoerbel, J., 1961. Ober die Lebensbedingungen und die Besiedlung des hyporheischen Lebensraumes. Arch. Hydrobiol. Suppl. 25: 182–214.

Schwoerbel, J., 1965. Bemerkungen Ober die interstitielle hyporheische fauna einiger bäche der südöichen Vogesen. Vie Milieu 16: 475–485.

Schulte, H., 1954. Beiträge zur Ökologie und Systematik der Bodenrotatorien. Zool. Jb. 82: 497–652.

Shiozawa, D. K., 1986. The seasonal community structure and drift of microcrustaceans in Valley Creek, Minnesota. Can. J. Zool. 64: 1655–1664.

Sohlenius, B., 1982. Short-term influence of clear-cutting on abundance of soilmicrofauna (Nematoda, Rotatoria and Tardigrada) in a Swedish pine forest soil. J. appl. Ecol. 19: 349–359.

Strayer, D., 1985. The benthic micrometazoans of Mirror Lake, New Hampshire. Arch. Hydrobiol. Suppl. 72: 287–426.

Thane-Fenchel, A., 1968. Distribution and ecology of nonplanktonic brackish-water rotifers from Scandinavian waters. Ophelia 5: 273–297.

Turner, P., 1990. Some interstitial Rotifera from a Florida USA beach. Trans. Am. Microsc. Soc. 109: 417–421.

Turner, P., 1993. Distribution of rotifers in a Floridian salwater beach, with a note on rotifer dispersal. Hydrobiologia 255/256: 435–439.

Turner, P., 1995. Notes on the hyporheic Rotifera of the Ninnescah River, Kansas, USA. Trans. Kans. Acad. Sci. 98: 92–101.

Turner, P. & M. A. Palmer, 1996. Notes on the species composition of the rotifer community inhabiting the interstitial sands of Goose Creek, Virginia with comments in habitat preferences. Queckett J. Micros. 37: 552–565.

Tzschaschel, G., 1980. Verteilung, Abundanzdynamik und Biologie mariner interstitieller Rotatoria. Mikro. Meeresb. 81: 1–56.

Tzschaschel, G., 1983. Seasonal abundance of psammon rotifers. Hydrobiologia 104: 275–278.

Varga, L., 1957. Neurere daten Ober die Mikrofauna des Balatonpsammons. Ann. Biol. Tihany 24: 271–282.

Williams, D. D. & H. B. N. Hynes, 1974. The occurrence of benthos deep in the substratum of a stream. Freshwat. Biol. 4: 233–256.

Wiszniewski, J., 1934. Recherches écologiques sur le psammon. Arch. Hydrobiol. Ryb. 8: 149–272.

Wiszniewski, J., 1935. Notes sur le psammon. 11 Riviere Czarna aux environs de Varsovie. Arch. Hydrobiol. Ryb. 3–4: 221–238.

Wiszniewski, J., 1937a. Diftérentiation écologique des rotifères dans le psammon d'eaux douces. Ann. Mus. Pol. 13: 1–13.

Wiszniewski, J., 1937b. Der feuchte Sand als Lebensmilieu. Mikrokosmos 31: 34–38.

Zullini, A. & C. Ricci, 1980: Bdelloid rotifers and nematodes in a small Italian stream. Freshwat. Biol. 10: 62–72.

Hydrobiologia **387/388**: 241–249, 1998.
E. Wurdak, R. Wallace & H. Segers (eds), Rotifera VIII: A Comparative Approach.
© *1998 Kluwer Academic Publishers.*

Effects of contrasting land use on free-swimming rotifer communities of streams in Masurian Lake District, Poland

Jolanta Ejsmont-Karabin[1,*] & Marek Kruk[2]

[1] *Institute of Ecology, Polish Academy of Sciences, Hydrobiological Station, Lesna str. 13, 11-730 Mikolajki, Poland*
(*author for correspondence)

[2] *Academy of Agriculture and Technology, Department of Water Protection and Inland Fisheries, Oczapowskiego 5, 10-957 Olsztyn, Poland*

Key words: Rotifera, streams, forest and agricultural areas, species diversity

Abstract

The influence of land-use patterns on rotifer communities of streams in Masurian Lake District, Poland, was examined. Four streams in a forest–marsh area, two in forest–meadows, and two streams in agricultural areas were sampled monthly from March 1995 to October 1996. In all 41 genera and 139 species (i.e. ca. 30% of all rotifer species reported in Poland) were collected and identified, four of these are new to Poland.

Rotifer numbers in the forest–marsh streams were 2 times higher than in forest–meadow streams and 10 times higher than in the agricultural streams. The forest streams also had higher numbers of species, both in the total number recorded in particular streams (74 in forest–marsh, 53 in forest–meadow, and 27 in agricultural ones), as well as the mean number of species recorded in one sample (17, 11, and 5 species, respectively). The values of Shannon's diversity index were markedly higher in the forest–marsh and forest–meadow (2.42 and 2.49, respectively) streams than streams from agricultural areas (1.49). The streams differed also in their dominant species. The forest–marsh stations were dominated by planktonic rotifers, whereas littoral-planktonic species dominated in streams in the forest–meadow area, and littoral species were most abundant in agricultural areas.

Introduction

Communities of free-swimming rotifers in running waters often exhibit low species numbers compared to lentic habitats (e.g., Nogrady et al., 1993; Phillips, 1995), because water currents, turbidity, and lack of food provide an unfavourable environment. Members of lotic communities usually derive from shore impoundments and littoral habitats (Nogrady et al., 1993). Although lakes can supply large amounts of zooplankton to out-flowing streams, large and soft-bodied rotifers are rapidly reduced because of strong current – so that only 5–50% remain 1–2 km downstream. Size is an important factor in the downstream fate of zooplankton – larger organisms sink more readily (Ward, 1975; Sandlund, 1982). It has been also shown (Uhlmann & Benndorf, 1980; Ejsmont-Karabin et al., 1993) that abundant and species-rich zooplankton communities can develop only in river

impoundments with retention times of 10 days or more.

Stream communities of rotifers in our study can thus be expected to be rather poor in individuals and species and composed of animals carried off by water currents from habitats such as littoral areas, bottom deposits or numerous and differentiated small lakes and ponds more or less connected with the studied streams. A rotifer community developed in this way cannot be specific for particular streams and should not show any persistence in time of composition, abundance and diversity.

Investigations of benthic invertebrate assemblages in streams near Hanmer Springs, South Island (Harding & Winterbourn, 1995) have shown that they can be strongly influenced by different land-use activities. The structure of the stream communities was highly modified in streams draining agriculturally developed catchments. Data describing the influence of

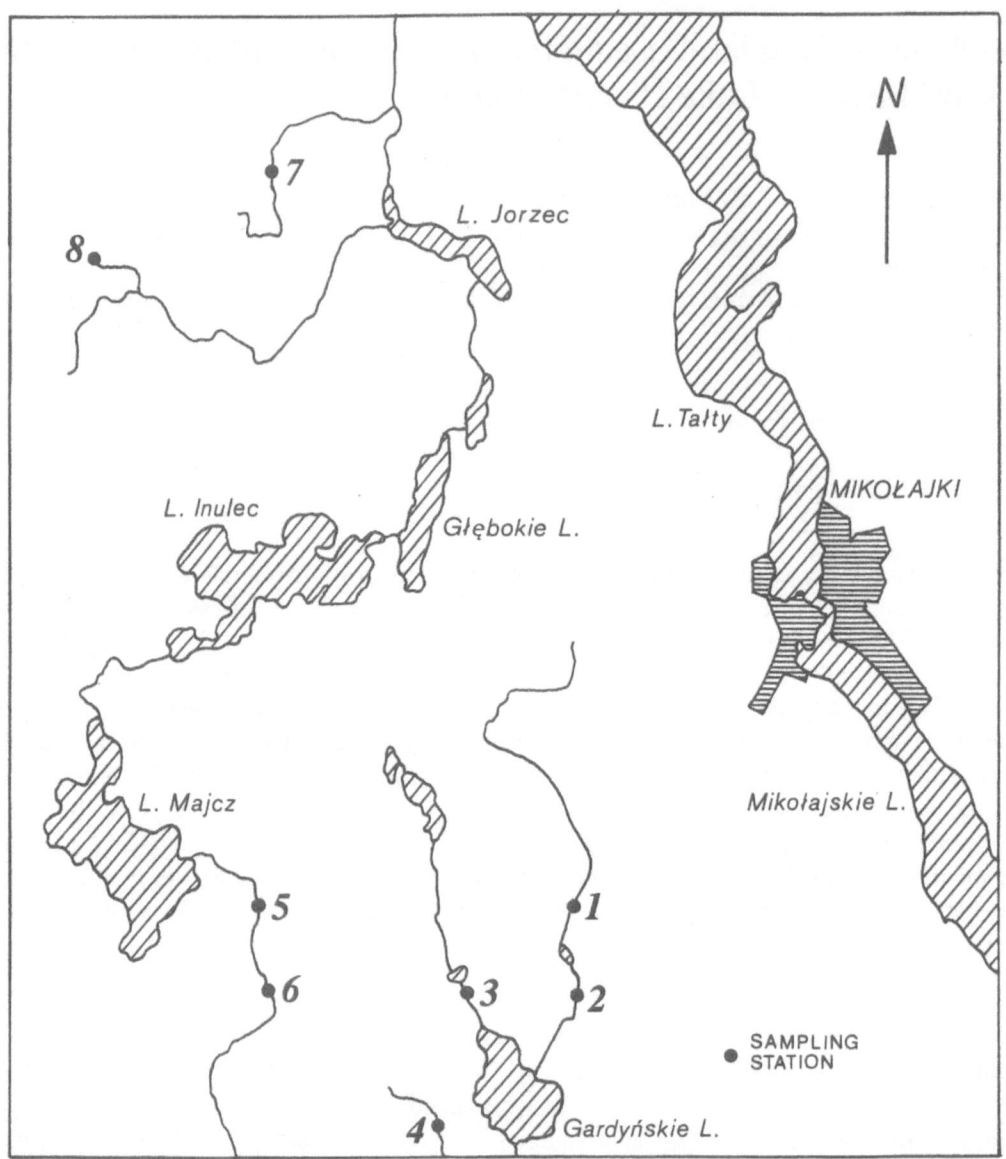

Figure 1. Location of the study streams (Masurian Lake District, northeastern Poland). See Table 1 for characteristics and code numbers.

landscape-scale watershed features on rotifer communities in streams are lacking. This study attempts to quantify the importance of land use in structuring rotifer communities.

Area, material and methods

Studies were conducted in eight streams in Masurian Lake District, Poland (Figure 1). Four streams in forest–marsh and two streams in forest–meadow area (catchment of the River Krutynia) and two streams in

agricultural area (catchment of the River Jorka) were sampled monthly from March 1995 to October 1996. The streams included in the analysis are characterized in Table 1. The streams were generally narrow and shallow. Beavers were observed to influence physical features and biota in two of the studied streams. They appeared in stream No. 2 in June 1996 and drained it completely two months later. In stream No. 3 they were present throughout the study, and the sampling station was located downstream of their dam.

Table 1. Description of the study streams

Stream no.[1]	Width cm	Depth cm	Flow rate[2] l. sek.$^{-1}$	Surroundings	Description
1	110	50	26	Meadow/forest	Dense stands of macrophytes
2	80	20	30	Forest Marsh	Devoid of macrophytes Outflow from a dystrophic lake
3	100	20	30	Forest/marsh	Modified by beavers outflow from a dystrophic lake
4	40–50	12	11	Forest/meadow	In ditch
5	280	35	115	Forest/marsh	Outflow from a meso-eutrophic lake
6	230	25	113	Forest/marsh	Devoid of macrophytes
7	40	20	11	Pasture/arable	Dense stands of shore vegetation
8	54	35	4	Arable land	Devoid of macrophytes

[1] See Figure 1.
[2] Values average for the study period.

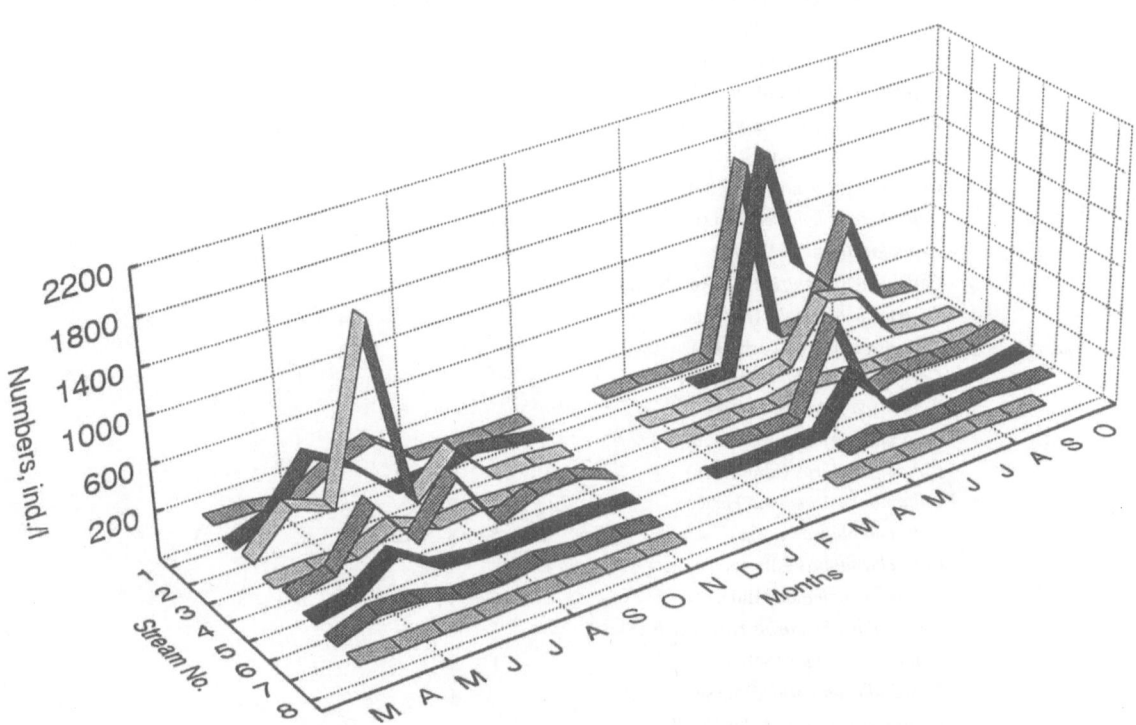

Figure 2. Seasonal changes in rotifer abundance in 8 studied streams during the years 1995 and 1996.

Rotifer material was collected with an 8-liter fixed-volume sampler. The samples were condensed by filtering them through a 30-μm mesh net, and fixed immediately with Lugol's solution, and then (in a laboratory) with formalin. In all cases, net samples were collected as well and studied alive on the day of sampling to determine species that are difficult to identify in a fixed condition.

The Shannon-Weaver diversity index (Margalef, 1957) was used to characterize the stream communit-

Table 2. List of rotifer species and the number of records of particular species in samples collected in the two-years' study. (See Figure 1 for location of streams.)

Species	Stream no. 1	2	3	4	5	6	7	8
Anuraeopsis fissa (Gosse)	2	6	5	3		1	2	1
Ascomorpha ecaudis Perty	2	3	3		10	8		
Ascomorpha saltans Bartsch	2	2	2		5	3		
Asplanchna priodonta Gosse	1	1	2		5	1	1	
Beauchampiella eudactylota (Gosse)			2					
Bipalpus hudsoni (Imhof)					1			
Brachionus angularis Gosse		3		1				1
Brachionus calyciflorus Pallas		2						
Brachionus quadridentatus Hermann		3						
Cephalodella apocolea Myers	1				1			
Cephalodella auriculata (Müller)		2				1	1	
Cephalodella catellina (Müller)	2	3	2	2	3		4	1
Cephalodella eva (Gosse)	1				1	2		
Cephalodella exigua (Gosse)	9	8	7	1	6	7	3	1
Cephalodella forficula (Ehrenberg)				5	4	4		
Cephalodella gibba (Ehrenberg)	3	3	4	1	6	2	8	
Cephalodella limosa Wulfert			1					
Cephalodella misgurnus Wulfert	1		2	2	2			
Cephalodella nana Myers		1	1					
Cephalodella pachyodon Wulfert			1					
Cephalodella sterea (Gosse)				1	1			
Cephalodella tenuior (Gosse)	1							
Cephalodella tenuiseta (Burn)		1	2	1	1	1	1	
Cephalodella tinca Wulfert			1	1		2	1	
Cephalodella ventripes (Dixon-Nuttall)	3	1	5	2	2	3	5	1
Chromogaster ovalis (Bergendal)					4	1		
Collotheca libera (Zacharias)					2			
Collotheca mutabilis (Hudson)					1	1		
Collotheca pelagica (Rousselet)					2			
Colurella adriatica Ehrenberg	7	6	9	8	7	5	8	4
Colurella colurus (Ehrenberg)	1		1	1		3	2	1
Colurella gastracantha Hauer		1	1					
Colurella obtusa (Gosse)	6	5	8	2	6	8	6	3
Colurella subtilis Althaus	1							
Colurella uncinata (Müller)	11	6	14	9	6	8	4	3
Conochilus dossuarius coenobasis (Skor.)		1						
Conochilus natans (Seligo)		1	1			2		
Conochilus unicornis Rousselet			1		6	5		
Dicranophorus forcipatus (Müller)	1				1			
Dicranophorus longidactylum Fadeev	2			3		2		
Encentrum kozminskii Wiszniewski							1	
Encentrum plicatum (Eyferth)			1	1				
Encentrum putorius Wulfert					1	1		
Erignatha clastopis (Gosse)	2		1	1	1			1
Euchlanis deflexa Gosse	2		1					
Euchlanis dilatata Ehrenberg	6	2		6	2	1	2	

Continued on p. 245

Table 2. Continued

Species	Stream no.							
	1	**2**	**3**	**4**	**5**	**6**	**7**	**8**
Euchlanis incisa Carlin					1	1		
Filinia longiseta (Ehrenberg)	2	4	3					
Filinia terminalis (Plate)	1	1	1		2	1		
Gastropus stylifer Imhof		1		11	6			
Itura aurita Ehrenberg)						1		
Kellicottia longispina (Kellicott)			4	1	6	5		
Keratella cochlearis (Gosse)	3	11	9	2	16	17	3	1
Keratella hiemalis Carlin			3					
Keratella quadrata (Müller)	1	5	2	1	6	6		
Keratella testudo (Ehrenberg)	6	3	7	5			1	
Lecane arcuata (Bryce)	5	2	4	2		4		1
Lecane bulla (Gosse)	1	1	1		3	4		
Lecane closterocerca (Schmarda)	7	7	13	5	6	6	11	2
Lecane curvicornis (Murray)			2					
Lecane elsa Hauer		1						
Lecane flexilis (Gosse)	1	2	5	1	3	7	3	
Lecane galeata (Bryce)				1		1		
Lecane gwileti Tarnogradsky	1	1		1				1
Lecane hamata (Stokes)	1	1	12	2	1	3	5	
Lecane intrasinuata (Oloffson)			1			1		
Lecane ludwigii (Eckstein)					2	1		
Lecane luna (Müller)	2		1					
Lecane lunaris (Ehrenberg)	7			1	2	4		
Lecane monostyla (Daday)						1		
Lecane opias (Harring & Myers)	1	1						
Lecane stenroosi (Meissner)						1		
Lecane subulata (Harring & Myers)		1						
Lecane tenuiseta Harring		1						
Lepadella acuminata (Ehrenberg)	3	2	10		2	6	5	1
Lepadella astacicola Hauer			1					
Lepadella ovalis (Müller)	8	5	6	2	3	4	3	2
Lepadella patella (Müller)	11	6	13	7	8	9	4	2
Lepadella triptera Ehrenberg	1				3	2	1	
Lindia truncata (Jennings)					1			
Lophocharis oxysternoon (Gosse)	1							
Lophocharis salpina (Ehrenberg)	4	6	9		3	8		
Microcodides chlaena (Gosse)		2	2					
Monommata longiseta (Müller)	3		1		4	1		
Monommata maculata (Harring & Myers)						1		
Mytilina crassipes (Lucks)			3					
Mytilina mucronata (Müller)	3	5	3	1			1	
Mytilina trigona (Gosse)			6					
Mytilina ventralis (Ehrenberg)	1	2	3		1		1	
Notholca acuminata (Ehrenberg)			1		1	2		
Notholca foliacea (Ehrenberg)					1	1		
Notholca labis Gosse					1	2		
Notholca squamula (Müller)	2		1	1	1	2	3	
Notommata copeus Ehrenberg					1	1		
Notommata glyphura Wulfert			2					
Notommata tripus Ehrenberg	2							

Continued on p. 246

Table 2. Continued

Species	Stream no.							
	1	2	3	4	5	6	7	8
Platyias quadricornis (Ehrenberg)		2	2		1	1		
Pleurotrocha petromyzon Ehrenberg	1		2	2	6	2	2	
Pleurotrocha robusta (Glasscott)	1		1			2		
Polyarthra dolichoptera Idelson	5	10	11	2	4	5		
Polyarthra euryptera Wierzejski		1						
Polyarthra major Burckhardt			1		6	2		
Polyarthra remata Skorikov	4	8	7	2	11	11	1	
Polyarthra vulgaris Carlin					11	7		
Pompholyx complanata Gosse		1				1		
Pompholyx sulcata Hudson					6	5		
Proales theodora (Gosse)				1				
Proalinopsis caudatus (Collins)		1			1			
Proalinopsis squamipes Hauer	1	1			1			
Squatinella mutica (Ehrenberg)	2		1		3	1	1	
Squatinella rostrum (Schmarda)	2		1		1		1	
Synchaeta kitina Rousselet					2			
Synchaeta lakowitziana Lucks		2	1					
Synchaeta oblonga Ehrenberg	1	3	3		1			
Synchaeta pectinata Ehrenberg	4	6	4		7	6		
Synchaeta stylata Wierzejski		1						
Synchaeta tremula (Müller)	3	4	8	1	1	1		
Taphrocampa annulosa Gosse	1		1					
Testudinella carlini Bartos					1			
Testudinella elliptica (Ehrenberg)	1							
Testudinella parva (Ternetz)		1			1			
Testudinella patina (Hermann)	1	1	9	2	5	1		
Testudinella truncata (Gosse)	2		1				1	
Trichocerca brachyura (Gosse)							1	
Trichocerca capucina (Wierzej. & Zach.)		1			2			
Trichocerca dixon-nuttalli (Jennings)					2	1		
Trichocerca intermedia (Stenroos)						1		
Trichocerca longiseta (Schrank)					3	2		
Trichocerca porcellus (Gosse)	5	2	1		3	2	1	
Trichocerca pusilla (Lauterborn)					1			
Trichocerca rattus (Müller)	1	1			1			
Trichocerca rousseleti (Voigt)		2			2			
Trichocerca similis (Wierzejski)		1	1		8	5		
Trichocerca sulcata (Jennings)	1	2	1		3	1		
Trichocerca tenuior (Gosse)	1	1			1	2		
Trichocerca tigris (Müller)					1	2		
Trichocerca weberi (Jennings)		1			1			
Trichotria pocillum (Müller)	6	1	3		2		1	
Trichotria tetractis (Ehrenberg)						1		
Total number of samples	17	14	17	17	17	17	16	16

ies. A second index used in this paper was the percentage similarity between communities (Whittaker & Fairbanks, 1958):

$$Psc = 100 - 0.5 \sum (a_i - b_i) = \sum \min(a_i, b_i),$$

Table 3. Comparison of quantitative and qualitative features of rotifer communities in streams draining lands of different use types

Structural parameters	Type of land-use		
	Forest–marsh	Forest–meadow	Agricultural
Range of cumulative species numbers per stream	67 to 82	41 to 66	17 to 36
Average species number per sample: range	5 to 31	2 to 29	0 to 16
Mean for streams	17 ± 7	11 ± 6	5 ± 4
Average numbers of individuals per litre and sample	80 to 279	31 to 181	6 to 36
Species diversity index, D range of values	0.58 to 3.99	1.37 to 3.44	0.00 to 3.11
Mean for streams	2.42 ± 0.75	2.49 ± 0.63	1.49 ± 1.00
Number of species new	3	3	2
and rare in Poland	8	5	3
Species dominating in the community numbers	*Keratella cochlearis*	*Anuraeopsis fissa*	*Lecane closterocerca*
	Polyarthra dolichoptera	*Trichocerca porcellus*	*Colurella adriatica*
	Anuraeopsis fissa	*Lecane closterocerca*	*Colurella colurus*
	Keratella testudo	*Colurella uncinata*	*Colurella obtusa*
	Polyarthra vulgaris	*Polyarthra remata*	*Colurella uncinata*
	Polyarthra remata	*Cephalodella exigua*	

where *a* and *b* = percentage of each species in streams A and B, respectively.

Results

The course of seasonal changes in rotifer numbers was generally similar in most streams (Figure 2). In autumn and winter rotifer densities were low, up to a dozen or so individuals in 1 liter. Spring and summer abundances were substantially higher, with peak values of 1731 ind. l^{-1} in stream No. 3 in June 1995 and 1627 and 1836 in streams No. 1 and 2 (respectively) in May 1996.

In general, rotifer densities were ca. 10 times higher in forest-marsh streams (No. 2, 3, 5, 6) than in streams in agricultural lands (No. 7 and 8) and twice the values in the forest-meadow streams (No. 1 and 4) (Figure 2). The range of peak values was from 438 (stream No. 6) to 1836 ind. l^{-1} (No. 2) in the forest-marsh, from 147 (stream No. 4) to 1627 ind. l^{-1} (stream No. 1) in the forest-meadow and 18 (stream No. 8) to 111 ind. l^{-1} (No. 7) in streams of agricultural area. The average values for streams during the study period were 80 to 279 ind. l^{-1} (streams No. 6 and 2,

respectively) in the forest–marsh and 31 to 181 ind. l^{-1} in the forest–meadow catchment and 6 to 31 ind. l^{-1} in the agricultural area.

In total, 41 genera and 139 species (i.e., ca. 30% of all records of rotifer species in Poland) were found throughout the years 1995 and 1996. The highest number of species was recorded in stream No. 3 (73 species), No. 5 (82) and No. 6 (74). The lowest numbers of rotifer species were 41, 36 and 17 in streams 4, 7 and 8, respectively (Table 2).

Among the species occurring in the studied streams, *Keratella cochlearis* and *Colurella uncinata* (recorded in 47% of all samples) were most common as well as two species of the genus *Lepadella* (*L. patella* (46%) and *L. ovalis* (25%)), *Cephalodella* (*C. exigua* (32%) and *C. ventripes* (17%)), and *Colurella* (*C. adriatica* (41%) and *C. obtusa* (34%)) and one *Lecane*, *L. closterocerca* (44% of samples). Except *Keratella cochlearis*, these are all common species in littoral habitats.

The rotifer assemblages are different qualitatively as well as quantitatively in the streams in the three land-use types. The most numerous rotifers encountered in forest-marsh streams included primarily pelagic species, e.g., *Keratella cochlearis* (28 ind.

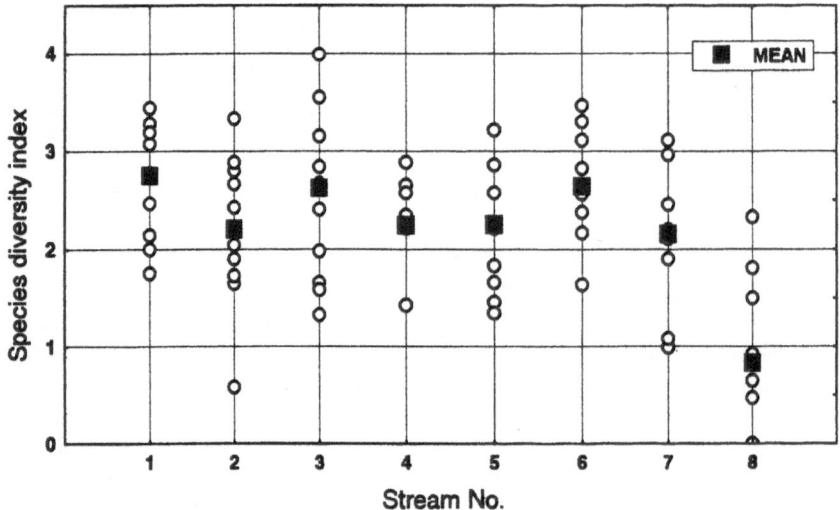

Figure 3. Values of species diversity index for the rotifer fauna of 8 studied streams.

l^{-1} sample^{-1}), *Polyarthra dolichoptera* (25 ind.) and *Anuraeopsis fissa* (20 ind.). In streams with forest-meadow catchments both pelagic (*A. fissa*: 17 ind. l^{-1}sample^{-1} *Polyarthra remata*: 4 ind. l^{-1}sample^{-1}) and littoral (*Lecane closterocerca* and *Colurella uncinata*, both 5 ind. l^{-1} sample^{-1}) species predominated whereas in two agricultural-area streams the most abundant species were the littoral *Lecane closterocerca* (2 ind. l^{-1} sample^{-1}) and *Colurella* spp, i.e. *C. adriatica, C. colurus, C. obtusa, C. uncinata* (1 ind. l^{-1} sample^{-1}).

In general, values of Shannon-Wiener's diversity index were high in both types of forest streams and low in streams in agricultural areas (Figure 3). The most diverse rotifer communities were those in streams No. 1 ($D = 1.75 - 3.37$, mean = 2.75), No. 3 ($D = 1.33 - 3.99$, mean = 2.62) and No. 6 ($D = 1.64 - 3.46$, mean = 2.64). In the case of the two agricultural streams, rotifer diversity was noticeably higher in stream No. 7 ($D = 0.99 - 3.11$, mean = 2.15), than in stream No. 8 ($D = 0.00 - 2.32$, mean = 0.88).

Most species recorded here are relatively common littoral and pelagic taxa. However, four species are new records for Poland: *Cephalodella misgurnus, C. tinca, Encentrum kozminskii,* and *Proalinopsis squamipes*. Nine species are rare in the country (*Cephalodella pachyodon, Encentrum plicatum, E. putorius, Erignatha clastopis, Lecane gwileti, L. intrasinuata, L. monostyla, L. subulata* and *Trichocerca sulcata*.).

Discussion and conclusions

The results of this study seem to contradict the standard views of qualitatively and quantitatively poor rotifer assemblages in such ecosystems. Relatively high values of species diversity as well as high densities of summer zooplankton show that there are many species that can find suitable conditions for the development of dense populations in stream habitats. However, it is difficult to ascertain what part of these communities is autochthonous. The studied forest streams flow through different ecosystems, create small floods, and meet with bogs and small pools. Each of these habitats can enrich communities carried with stream waters. It is surprising that these communities were devoid of sessile rotifers, even in streams with dense patches of macrophytes and strong current. Bogs did not appear to influence the taxonomic structure of the stream communities as species typical of this type of Masurian land, such as *Keratella serrulata* (Ehrenberg), *Lecane acus* (Harring), *L. stichaea* Harring, *Brachionus sericus* Rousselet, did not occur.

Our observations seem to provide an arguement that even in streams with strong water currents rotifer communities can develop and maintain characteristic patterns, specific to particular streams. This conclusion also is supported by the persistence over time of some species in particular streams (Table 2). Even streams draining catchments dominated by similar land use activities have different and specific rotifer assemblages.

Even more significant differences are observed between rotifer communities of streams draining different land use types, particularly for those dominated by pine forests or rowcrop agriculture. Comparisons made among the three land-use types indicate (Table 3) that simplification of landscape structure due to agricultural activity has an impact on stream rotifers, decreasing both their density and diversity. Streams from agricultural areas are characterized by lower numbers of rotifer species, either in one sample or when summed over the entire study period. The most numerous species are different in these habitats as well. In forest–marsh streams, pelagic forms prevail. In forest-meadow streams both pelagic and littoral species are abundant. In streams from agricultural lands littoral and benthic rotifers dominate.

Similar differences have been observed by Harding & Winterbourn (1995) in the occurrence of benthic invertebrates (insect larvae, molluscs, chironomids, etc.) of streams in a Canterbury (New Zealand) river system draining a catchment dominated by different land-use activities, and by Johnson et al. (1995) in the Saginaw River watershed (Michigan, USA). The taxonomic structure of the communities changed along a gradient, from streams in forested areas to streams in agriculturally developed catchments.

It is striking that, despite the generally depauperate diversity of rotifers in streams draining agricultural areas, they remain a valuable source of rare and even new species for the Polish rotifer fauna. The reason could be that such streams have a priori been treated as uninteresting from faunistic point of view, and consequently have received only little attention.

References

Ejsmont-Karabin, J., T. Weglenska & R. J. Wisniewski, 1993. The effect of water flow rate on zooplankton and its role in phosphorus cycling in small impoundments. Wat. Sci. Tech. 28: 35–43.

Harding J. S. & M. J. Winterbourn, 1995. Effect of contrasting land use on physico-chemical conditions and benthic asemblages of streams in a Canterbury (South Island, New Zealand) river system. N. Z. J. mar. Freshwat. Res. 29: 479–492.

Johnson L. B., C. Richards & G. Host, 1995. Land use and superficial geology effects on water chemistry, stream habitat and macroinvertebrate assemblages in the Saginaw River watershed, Michigan, U.S.A. In: Proceedings of the 38th Conference of the International Association of Great Lakes Research. 2200-Bonisteel Boulevard, Ann Arbor, Internat. Ass. Great Lakes Res.: 110–111.

Margalef, R., 1957. Information theory in ecology. Gen. Syst. 3: 36–71.

Nogrady, T., R. L. Wallace & T. W. Snell, 1993. Rotifera, Vol.1. Biology, Ecology and Systematics, Guides to the Identification of the Microinvertebrates of the Continental Waters of the World, Vol. 4. SPB Academic Publishing, The Hague, 241 pp.

Philips, E. C., 1995. Comparison of the zooplankton of a lake and stream in Northwest Arkansas. J. Freshwat. Ecol. 10: 337–341.

Sandlund, O. T., 1982. The drift of zooplankton and microzoobenthos in the river Ştrandaelva, western Norway. Hydrobiologia 94: 33–48.

Uhlmann, D. & J. Benndorf, 1980. The use of primary reservoirs to control eutrophication caused by nutrient inflows from non-point sources. In N. Duncan & J. Rzoska (eds), Land Use Impact on Lake and Reservoir Ecosystems, Project 5 Workshop MAB, Facultas-Verlag, Vienna: 152–188.

Ward, J. V., 1975. Downstream fate of zooplankton from a hypolimnial release mountain reservoir. Verh. Int. Ver. Limnol. 19: 1798–1804.

Whittaker, R. H. & C. W. Fairbanks, 1958. A study of plankton copepod communities in the Columbia Basin, Southeastern Washington. Ecology 39: 46–65.

Hydrobiologia **387/388**: 251–257, 1998.
E. Wurdak, R. Wallace & H. Segers (eds), Rotifera VIII: A Comparative Approach.
© *1998 Kluwer Academic Publishers.*

A study of rotifers in the River Thames, England, April–October, 1996

Linda May[1] & Jonathan A.B. Bass[2]
[1]*Institute of Freshwater Ecology, Bush Estate, Penicuik, Midlothian EH26 0QB, U.K.*
[2]*Institute of Freshwater Ecology, River Laboratory, East Stoke, Wareham, Dorset B20 6BB, U.K.*

Key words: Rotifera, River Thames, Chlorophyll *a*

Abstract

More than 30 species of rotifer were recorded in the River Thames between Inglesham and Reading from April to October, 1996. Seven of these were relatively abundant. These were *Keratella cochlearis* (Gosse), *Synchaeta oblonga* (Müller), *Polyarthra dolichoptera* Idelson, *Keratella quadrata* (Müller), *Brachionus angularis* Gosse, *Euchlanis dilatata* Ehrenberg and *Brachionus calyciflorus* Pallas.

In early spring, there was little variation in rotifer density along the river, but a marked downstream increase in abundance developed later in the year. Mean rotifer densities ranged from 24 ind. l^{-1} at the upstream site to 700 ind. l^{-1} at the most downstream site. A maximum total rotifer density of 4160 ind. l^{-1} was recorded at Reading on 29 July 1996. In general, the downstream increase in rotifer abundance seemed to parallel similar increases in chlorophyll *a* concentration in the river water. Losses due to invertebrate predation were probably low, but fish gut analyses from an earlier study had suggested that rotifers may be an important food source for larval fish.

Throughout the study, rotifer samples were collected and prepared for counting by two different methods. The results show that estimates of rotifer density may be significantly affected by sampling method.

Introduction

Zooplankton are often abundant in the main channels of major rivers. Many studies have shown that rotifers form an important component of these zooplankton communities (Admiraal et al., 1994; Blackwell et al., 1995; Ferrari et al., 1989; Gosselain et al., 1994; Kowalczewski et al., 1985; Pillard & Anderson, 1993; Sabri et al., 1993; Sandlund, 1982; Shiel, 1985; van Dijk & van Zanten, 1995). It has been suggested that the apparent dominance of rotifers in rivers may be due to their relatively short generation times compared to the larger crustacean zooplankton (van Dijk & van Zanten, 1995). Rotifer populations in the Rhine, for example, sometimes doubled at twice the rate (0.89 day^{-1}) of crustacea (0.45 day^{-1}) (De Ruyter van Stevenick et al., 1992). Such rapid rates of reproduction would give rotifers a clear advantage over crustacea in an environment where they are constantly being transported downstream. The present study was carried out to provide baseline information on rotifer populations in the River Thames, a relatively slow flowing and nutrient laden river in southern England

where rotifers are an important food source for larval fish (Mann et al., 1995, 1997).

The River Thames

The River Thames rises some 5 km south-east of Gloucester at a point about 110 m above sea level. It then travels 338 km eastwards before discharging into the North Sea. The river catchment covers almost 1×10^6 hectares in area and contains about 4000 km of river. This study is concerned with a 100 km section of the river which lies between Inglesham (50 km from its source) and Reading (150 km from its source) (Figure 1). At Reading, the river is approximately 45 m wide, 4.5 m deep and has a cross-sectional area of about 125 m^2.

Between April and October 1996, the rate of discharge of the river decreased steadily from about 10 m^3 s^{-1} to about 2 m^3 s^{-1}. Although downstream velocities were not measured during this period, the rates of discharge suggest that these probably varied between <0.2 m s^{-1} and 0.8 m s^{-1} (see Bottrell,

252

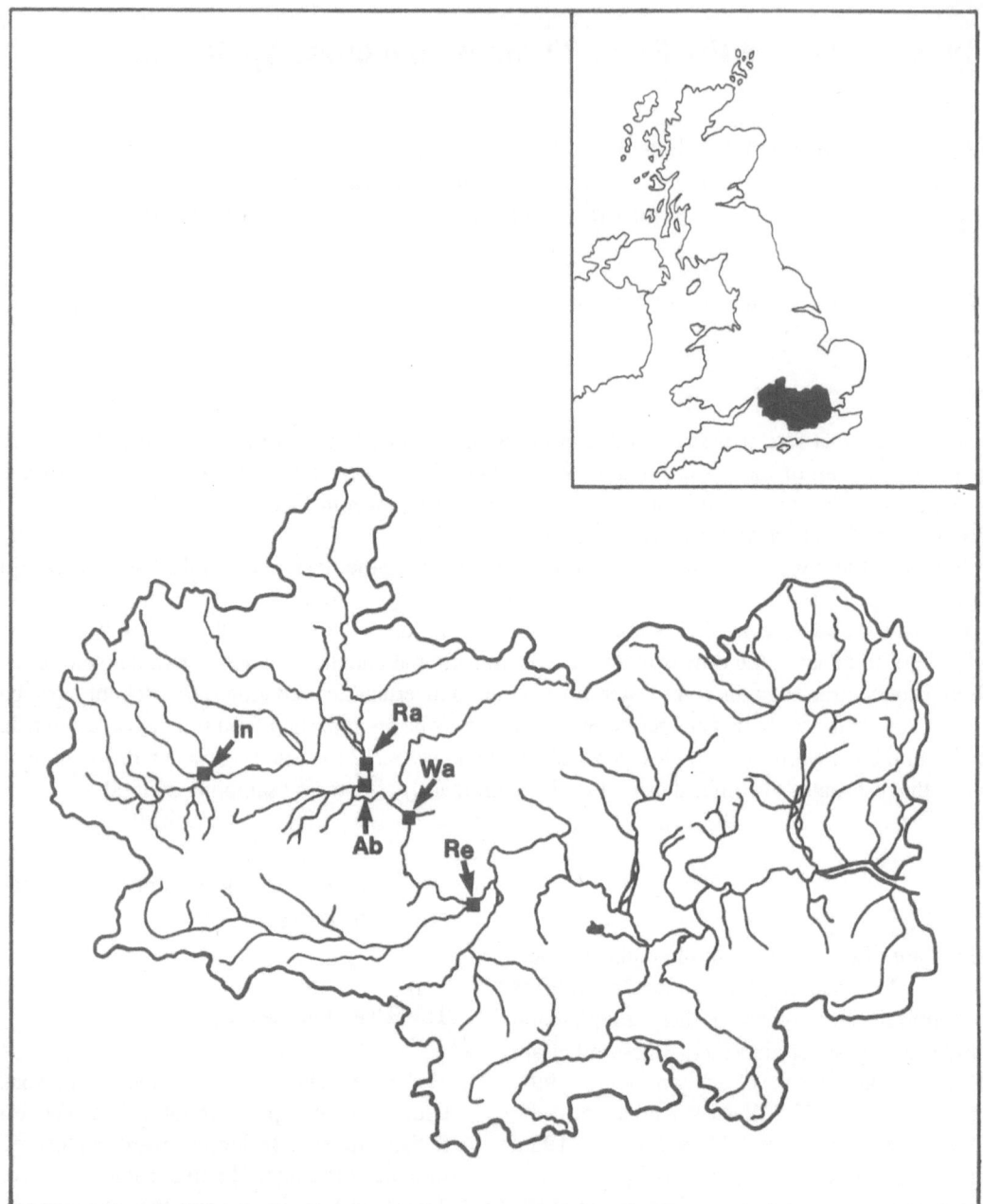

Figure 1. Catchment of the River Thames, England, showing the position of sampling sites at Inglesham (In), Radley (Ra), Abingdon (Ab), Wallingford (Wa) and Reading (Re).

1977). If so, the hydraulic travel times between the most upstream and the most downstream sites would have ranged between 1 and 6 days. However, the flow of this river is not entirely natural. As the river approaches Reading, it is regulated by a series of locks and weirs.

During the period of study, the water temperature ranged between 8°C and 22°C.

Methods

Rotifer samples were collected at 2-week intervals between April and October 1996, from five locations along the middle reaches of the River Thames (Inglesham, Radley, Abingdon, Wallingford and Reading, see Figure 1). On each sampling occasion, three types of samples were collected. These were: (1) quantitative, whole-water samples (500 ml) collected with a wide-necked, polyethylene container which was dipped into the river water; (2) quantitative samples consisting of 20 l of river water collected with a small submersible bilge pump and concentrated through a 63 μm mesh sieve (Bass & May, 1997); (3) whole-water samples (100 ml) collected by dipping a small, leakproof, polyethylene container into the river water.

Soon after collection, rotifers in the first set of samples were relaxed and killed by the addition of sufficient procaine hydrochloride ($NH_2 \cdot C_6H_4 \cdot COO \cdot CH_2 \cdot CH_2 \cdot N(C_2H_5)_2 \cdot HCl$) to give a final concentration of 0.2 g l^{-1} (May, 1985). The samples were then preserved in 4% formaldehyde, concentrated using a sedimentation technique, subsampled and enumerated at $\times 100$ magnification in the counting chamber of an inverted microscope. Rotifers in the second set of samples were preserved in 70% alcohol, subsampled and counted at $\times 100$ magnification in a Sedgewick–Rafter cell. Rotifers in the third set of samples were kept cool and examined live within 36 h of collection. Live specimens were identified according to Koste (1978).

Chlorophyll *a* concentration, determined by the methanol extraction method (Standing Committee of Analysts, 1983), was recorded on each sampling occasion and water chemistry data (PO_4–P and SiO_2 concentrations) were provided by the Environment Agency (Thames Region).

Results

Rotifer species composition and abundance

More than 30 species of rotifer were recorded in the River Thames between April and the end of October, 1996 (Table 1). Seven of these were relatively abundant, achieving mean population densities in excess of 30 ind. l^{-1} and maximum densities of up to 3700 ind. l^{-1}. These were, in order of importance: *Keratella cochlearis* (Gosse), *Synchaeta oblonga* (Müller), *Polyarthra dolichoptera* Idelson, *Keratella quadrata*

Table 1. Species list of rotifers found in the River Thames, April to October 1996

Brachionidae
Rhinoglena frontinalis Ehrenberg
Brachionus angularis Gosse
Brachionus calyciflorus Pallas
Brachionus urceolaris (Müller)
Keratella cochlearis f. *tecta* (Gosse)
Keratella cochlearis f. *typica* (Gosse)
Keratella quadrata (Müller)
Notholca acuminata Ehrenberg
Notholca squamula (Müller)
Anuraeopsis sp. Lauterborn
Euchlanis dilatata Ehrenberg
Euchlanis dilatata f. *larga* (Kutikova)
Trichotria tetractis (Ehrb.)
Colurella adriatica Ehrenberg
Lepadella sp. Bory de St Vincent

Lecanidae
Lecane ?candida Harring & Myers
Lecane lunaris (Ehrb.)

Notommatidae
Cephalodella gibba (Ehrenberg)

Trichocercidae
Trichocerca ?cylindrica (Imhof)
Trichocerca pusilla (Lauterborn)

Synchaetidae
Synchaeta oblonga (Müller)
Synchaeta ?pectinata Ehrenberg
Polyarthra dolichoptera Idelson

Testudinellidae
Testudinella patina (Hermann)
Filinia brachiata (Rousselet)
Filinia ?longiseta (Ehrenberg)

(Müller), *Brachionus angularis* Gosse, *Euchlanis dilatata* Ehrenberg and *Brachionus calyciflorus* Pallas. The remaining species were relatively scarce.

Many rotifer species occurred at all of the sites sampled, especially the more numerous species. Most of those species which were not recorded at every site were found in very low numbers, apart from *Trichocerca pusilla* which, in late July, was absent from the upstream sites (Inglesham, Radley, Abingdon) and rel-

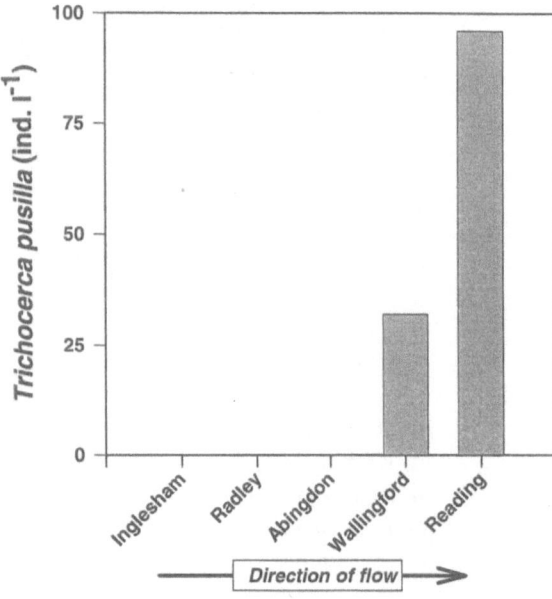

Figure 2. Downstream increase in abundance of *Trichocerca pusilla* in the River Thames on 27 July 1996.

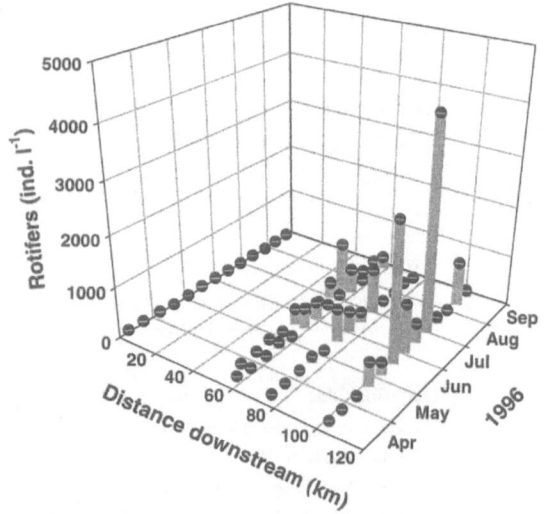

Figure 3. Downstream increase in total rotifers in the River Thames, April to October 1996.

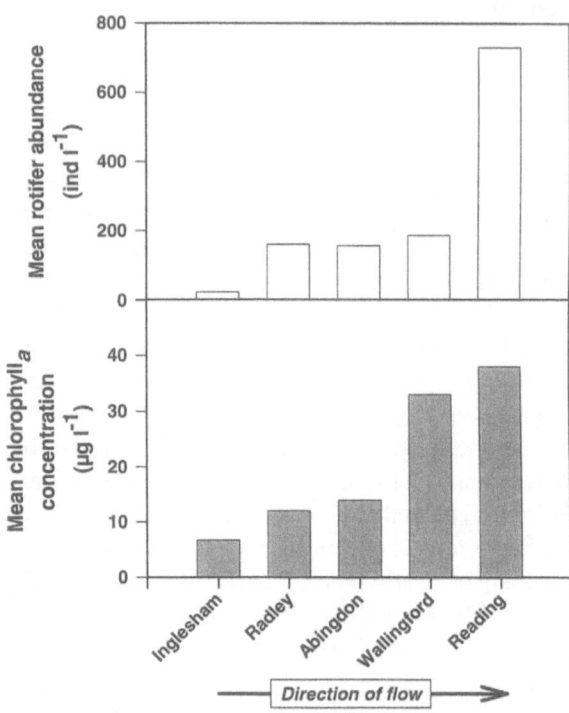

Figure 4. Mean chlorophyll*a* concentration and mean rotifer abundance at each sampling site, April to October 1996.

atively abundant at the downstream sites (Wallingford, Reading) (Figure 2). The reason for this was unclear.

In spring, when rotifer numbers were low, there was little variation in abundance among the sites sampled (Figure 3). However, later in the year, a marked downstream increase in abundance was observed. In general, this seemed to parallel a similar pattern of increase in chlorophyll *a* concentration in the river water (Figure 4). This appeared to be due to the combined effects of growth rate and increased travel times (allowing more time for growth) to the downstream sites, rather than to increased nutrient availability further downstream. This is because nutrient concentrations at the downstream were usually similar to, or slightly less than, those recorded at the upstream sites.

Although the mean rotifer density at the most upstream site (Inglesham) was only 24 ind. l^{-1}, a mean rotifer density of more than 700 ind. l^{-1} was recorded at the most downstream site (Reading). Species dominance also changed between the upstream and downstream sites (Figure 5). *S. oblonga* was dominant upstream (Inglesham, Radley, Abingdon, Wallingford), while *K. cochlearis* was dominant downstream (Reading). Population densities were very high for some species on some sampling occasions. This was especially true for *K. cochlearis* and *S. oblonga* which reached maximum population densities of 3660 ind. l^{-1} on 29 July 1996 and 1370 ind. l^{-1} on 17 June 1996, respectively. The maximum total rotifer density recorded was 4160 ind. l^{-1} on 29 July 1996. These population maxima were all recorded at the Reading sampling site.

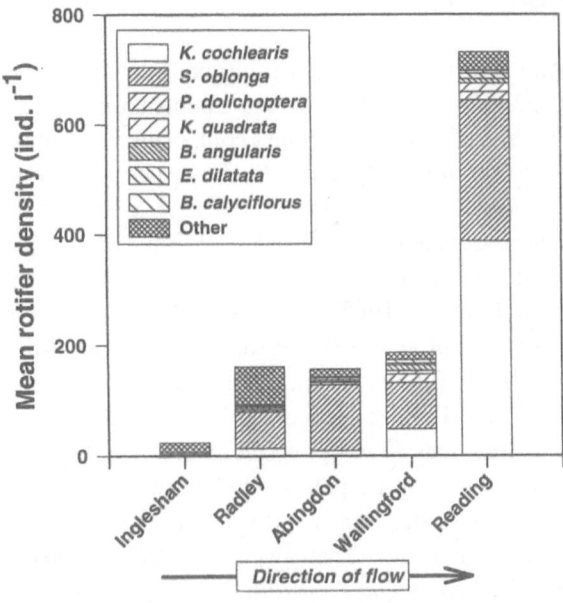

Figure 5. Mean rotifer density at each sampling site on the River Thames, showing relative abundance of the most important species.

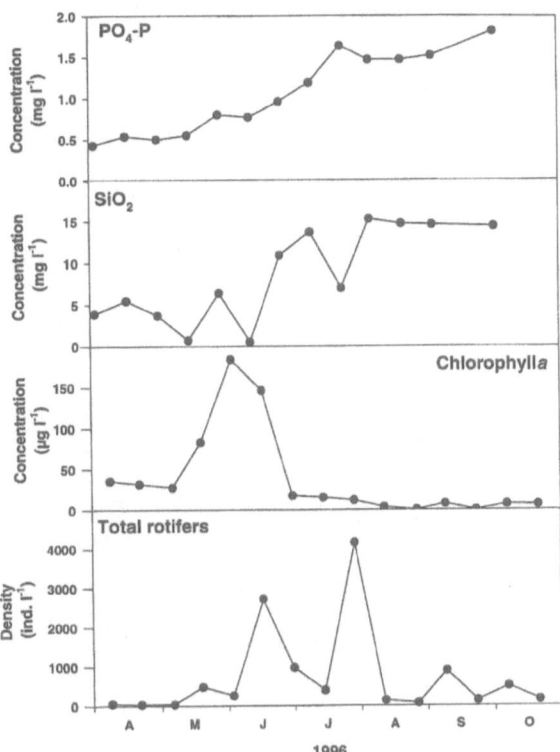

Figure 6. Temporal variation in nutrient concentration, chlorophyll*a* levels and rotifer abundance in the River Thames at Reading, April to October 1996.

Predatory planktonic invertebrate numbers in the River Thames were low (less than 0.1 ind. l^{-1}), so these are unlikely to have reduced rotifer numbers significantly. However, fish gut analyses have confirmed that rotifers are an important food source for larval fish in this river (Mann et al., 1995, 1997). The occurrence of spined forms of the rotifer *Brachionus calyciflorus* from mid-June until mid-July coincides with appearance of larval fish, tending to support this hypothesis (Stemberger & Gilbert, 1987). The preponderance of small rotifers in June and July may also have resulted from size-selective predation by juvenile fish.

Controlling factors

The data collected at the Reading site were examined to determine which factors may have been influencing their species composition and abundance. Two peaks in abundance were recorded here (Figure 6). The first (almost 3000 ind. l^{-1} in mid-June) seemed to follow a peak in chlorophyll *a* levels, with rotifer numbers declining shortly after a fall in chlorophyll *a* concentrations. This suggested a strong link between rotifer numbers and algal abundance early in the year, when the phytoplankton community was dominated by small centric diatoms. The decline in diatom abundance in early June, which seems to have precipitated a similar decline in rotifer densities, was almost certainly brought about by the low levels of dissolved

silica present in the water in late spring. Phosphorus levels remained high throughout the study (Figure 6), and are unlikely to have limited algal productivity.

The peak in rotifer abundance later in the year (4000 ind. l^{-1} in late July) is more difficult to explain (Figure 6). At this time, the rotifer community was dominated (88%) by *K. cochlearis* and algal abundance was very low. This suggests that the rotifers were feeding on an alternative food source, such as bacteria, heterotrophic nanoplankton or ciliates. Although data on these components of the plankton were collected, no obvious relationship between their abundance and that of the rotifers could be found.

Comparison of rotifer sampling techniques

Figure 7 compares the rotifer densities estimated at each sampling site using two different sampling methods. Method (1) involved collecting the sample with a pump and concentrating the rotifers by passing the water through 63 μm mesh net. Method (2) involved the collection of whole-water samples which were concentrated by a sedimentation method prior

256

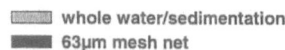

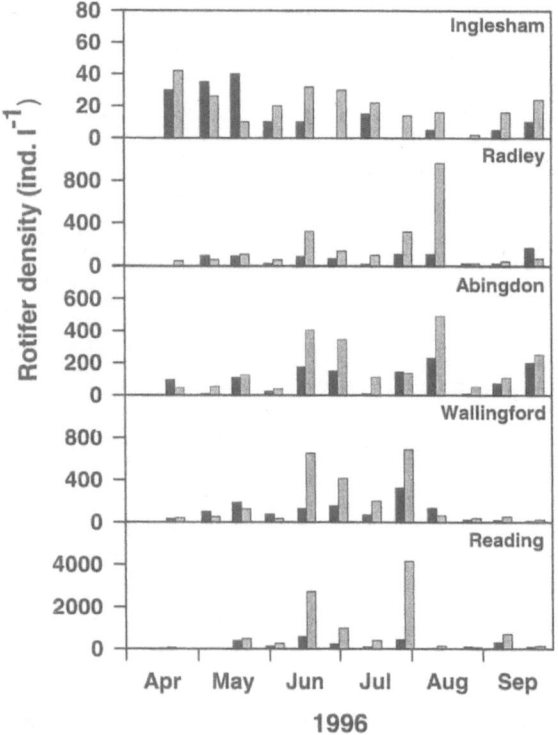

Figure 7. Comparison of rotifer density estimates obtained by two different methods (see text for details).

Discussion

Although rotifers are generally considered to be an important component of the zooplankton in large rivers, studies of their species composition and abundance are relatively rare. Such studies in lowland European rivers include the Danube (Vranovský, 1995), Guadalquivar (Guisande & Toja, 1988), Loire (Lair, 1980; Pourriot et al., 1982), Vistula (Kowalczewski et al., 1985), Po (Ferrari al., 1989), Rhine (Admiraal et al., 1994; de Ruyter van Steveninck et al., 1990; Tubbing et al., 1994; van Dijk & Zanten, 1995) and Thames (Bottrell, 1977).

More than 30 rotifer species were recorded in the Thames during the present study, but most of these were rare or occasional species. Only seven were relatively abundant. These were *Keratella cochlearis*, *Synchaeta oblonga*, *Polyarthra dolichoptera*, *Keratella quadrata*, *Brachionus angularis*, *Euchlanis dilatata* and *Brachionus calyciflorus*. All of these species are typical of those recorded in other European rivers (Admiraal et al., 1994; Ferrari et al., 1989; van Dijk & Zanten, 1995), apart from *S. oblonga* which, though abundant in the River Thames, has rarely been recorded in other rivers.

The maximum and mean rotifer densities recorded in the River Thames at Reading (4160 and 730 ind. l^{-1}, respectively) fall within the range of values recorded for other European rivers, including the Rhine (1300 and 110 ind. l^{-1}, respectively) (Admiraal et al., 1994) and the Po (6660 and 2000 ind. l^{-1}, respectively) (Ferrari et al., 1989). However, such comparisons are of limited use because many different methods have been used to collect and process the samples and this is known to affect estimates of rotifer abundance. Of particular concern are the mesh sizes of nets used to concentrate the samples, prior to counting. Collecting such small animals with coarse nets and sieves may lead to significant underestimates in population densities (Bottrell et al., 1976; Shiel et al., 1982). For example, Bottrell et al. (1976) report that up to 80% of smaller species may be lost through mesh sizes as small as 45 μm; many of the earlier studies have used even larger mesh sizes (up to 200 μm see Harris et al., 1992) to collect and concentrate their samples.

to counting Method (2) often resulted in population density estimates which were more than double those recorded using method (1), and sometimes an order of magnitude greater (e.g. 29 July 1996 at Reading) (Figure 7). The number of rotifers collected by method (1) occasionally exceeded those collected by method (2), but this only occurred when rotifer densities were low.

The maximum rotifer densities recorded in the River Thames near Reading between 1970 and 1971 was 69 ind. l^{-1} (Bottrell, 1977). That recorded for 1996 during the present study was 4050 ind. l^{-1}. The data seem to suggest that rotifers were far more abundant in 1996 than during the earlier study. However, the samples from the earlier study were collected with a 125 μm mesh plankton net, which may have led to an underestimate of population densities (Bottrell et al., 1976).

Acknowledgements

We are grateful to L. Ruse and A. Love who kindly provided water chemistry and phytoplankton data for this study, and to G. Collett who helped collect samples. This work was part funded by the Environment Agency, Thames Region.

References

Admiraal, W., L. Breebart, G. M. J. Tubbing, B. Van Zanten, E. D. de Ruyjter van Steveninck & R. Bijerk, 1994. Seasonal variation in composition and production of planktonic communities in the lower River Rhine. Freshwat. Biol. 32: 519–531.

Bass, J. A. B. & L. May, 1997. Zooplankton interactions in the River Thames. Report to the Environment Agency (Thames Region): 66 pp.

Blackwell, B. G., B. R. Murphy & V. M. Pitman, 1995. Suitability of food resources and physicochemical parameters in the Lower Trinity River, Texas, for Paddlefish. J. Freshwat. Ecol. 10: 163–173.

Bottrell, H. H., 1977. Quantitative Studies on the Cladocera, Copepoda and Rotifera of the River Thames and River Kennet at Reading. Unpubl. PhD Thesis, University of Reading.

Bottrell, H. H., A. Duncan, Z. M. Gliwicz, E. Grygierek, A. Herzig, A. Hillbricht-Ilkowska, H. Kurasawa, P. Larsson & T. Weglenska, 1976. A review of some problems in zooplankton production studies. Norw. J. Zool. 24: 419–456.

Berzins, B. & B. Pejler, 1989. Rotifer occurrence and trophic degree. Hydrobiologia 182: 171–180.

de Ruyter van Steveninck, E. D. & W. Admiraal, 1992. Plankton in the River Rhine: structural and functional changes observed during downstream transport. J. Plankton. Res. 14: 1351–1368.

de Ruyter van Steveninck, E. D., W. Admiraal & B. Van Zanten, 1990. Changes in plankton communities in regulated reaches of the Lower River Rhine. Regulated Rivers 5: 67–75.

Ferrari, I., A. Farabegoli & R. Mazzoni, 1989. Abundance and diversity of planktonic rotifers in the Po River. Hydrobiologia 186/187: 201–208.

Guisande, C. & I. Toja, 1988. The dynamics of various species of the genus Brachionus (Rotatoria) in the Guadalquivar River. Arch. Hydrobiol. 112: 579–595.

Gosselain, V., J.-P. Descy & E. Everbecq, 1994. The phytoplankton community of the River Meuse, Belgium: seasonal dynamics (year 1992) and the possible incidence of zooplankton grazing. Hydrobiologia 289: 179–191.

Harris, J. H., G. Scarlett & R. J. MacIntyre, 1992. Effects of a pulp and paper mill on the ecology of LaTrobe River, Victoria, Australia. Hydrobiologia 246: 49–67.

Koste, W., 1978. Rotatoria: Die Rädertiere Mitteleuropas. Publ. Gebrüder Borntraeger, Berlin: 673 pp.

Kowalczewski, A., M. Perlowska & J. Przyluska, 1985. Seston of the Warsaw reach of the Vistula River in 1982 and 1983. III. Phyto- and Zooplankton. Ekol. Pol. 33: 423–438.

Lair, N., 1980. The rotifer fauna of the River Loire, France, at the level of the nuclear power plants. Hydrobiologia 73: 153–160.

Mann, R. H. K, G. D. Collett, J. A. B. Bass & L. C. V. Pinder, 1995. River Thames 0 Group Fish Gut Contents Survey 1995. Report to National Rivers Authority, Thames Region: 44 pp.

Mann, R. H. K, G. D. Collett, J. A. B. Bass & L. C. V. Pinder, 1997. River Thames 0 Group Fish Gut Contents Survey 1996. Report to Environment Agency, Thames Region.

May, L., 1985. The use of procaine hydrochloride in the preparation of rotifer samples for counting. Verh. int. Verein. Limnol. 22: 2987–2990.

Pillard, D. A. & R. V. Anderson, 1993. Longitudinal variation in zooplankton populations in pool 19, Upper Mississippi River. J. Freshwat. Ecol. 8: 127–132.

Pourriot, R., D. Benest, P. Champ & C. Rougier, 1982. Influence de quelques factors du milieu sur la composition et la dynamique saisonnière du zooplancton de la Loire. Acta Oecol. Occol. Gen. 3: 353–371.

Sabri, A. W., Z. H. Ali, S. F. Shawkat, L. A. Thejar, T. I. Kassim & K.A. Rasheed, 1993. Zooplankton population in the River Tigris: effects of Samarra impoundment. Regulated Rivers: Res. Manage. 8: 237–250.

Sandlund, O. T., 1982. The drift of zooplankton and microzoobenthos in the river Strandaelva, western Norway. Hydrobiologia 94: 33–48.

Shiel, R. J., 1985. Zooplankton of the Darling River System. Verh. int. Verein. Limnol. 22: 2136–2140.

Shiel, R. J., K. F. Walker & W. D. Williams, 1982. Plankton of the lower River Murray, South Australia. Aust. J. Mar. Freshwat. Res. 33: 301–327.

Standing Committee of Analysts, 1983. The determination of chlorophyll a in aquatic environments 1980. In Methods for the Examination of Waters and Associated Materials. Her Majesty's Stationary Office, London: 26 pp.

Stemberger, & J. J. Gilbert, 1987. Defenses of planktonic rotifers against predators. In Kerfoot, W. C. & A. Sih (eds), Predation. Direct and Indirect Impacts on Aquatic Communities. Publ. University Press of New England, Hanover: 227–239.

Tubbing, G. M. J., W. Admiraal, D. Backhaus, G. Friederich, E. D. de Ruyter van Steveninck, D. Müller & I. Keller, 1994. Results of the international plankton investigation on the River Rhine. Wat. Sci. Tech. 29: 9–19.

van Dijk, G. M. & B. van Zanten, 1995. Seasonal changes in zooplankton abundance in the lower Rhine during 1987–1991. Hydrobiologia 251: 275–284.

Vranovsk, M., 1995. The effect of current velocity upon the biomass of zooplankton in the River Danube side arms. Biol. Bratislava 50: 461–464.

Hydrobiologia **387/388**: 259–265, 1998.
E. Wurdak, R. Wallace & H. Segers (eds), Rotifera VIII: A Comparative Approach.
© 1998 Kluwer Academic Publishers.

Planktonic rotifers of Samborombón River Basin (Argentina)

Beatriz E. Modenutti
Department of Ecology, Centro Regional Universitario Bariloche, Universidad Nacional del Comahue, Unidad Postal Universidad, 8400 Bariloche, Argentina

Key words: Rotifera, species composition, La Plata estuary basin, salinity, river

Abstract

The rotifer fauna of the river Samborombón and its tributaries (La Plata river basin) was analysed, and 47 species of monogonont rotifers were identified. Results indicate that differences in salinity and ion composition between waters of the main river and that of its tributaries account for differences in the species composition.

Introduction

The La Plata River system, including the upper Paraná–Paraguay, middle Paraná–Paraguay and Uruguay (Welcome, 1985) is one of the most important hydrological basins of South America. The system has floodplains composed of lakes and low gradient channels with variations caused mainly by changes in water level. As a consequence, physical and chemical parameters exhibit strong variations and thus modify the composition of aquatic communities. Several studies report changes in zooplankton community structure related to the physical and chemical characteristics of this system (Bonecker & Lansac-Toha, 1996; Bonecker et al., 1994; José de Paggi, 1978, 1983; Martinez & Frutos, 1986; Paggi & José de Paggi, 1990), but they refer mainly to the upper and middle sections of the rivers Paraná and Paraguay. Thus, rivers flowing into the Rio de la Plata estuary are not well investigated. In previous studies (Modenutti, 1991; Modenutti & Claps, 1988) the planktonic and periphytic rotifers of the inner estuary tributaries were analysed. Here I report the spatial and temporal distribution of planktonic rotifers of the Samborombón river and its tributaries located in the Pampasic region that flow into the outer Rio de La Plata estuary.

Study area and sampling sites

The Samborombón river lies in the depressed Pampa, in Buenos Aires province (Frenguelli, 1950). Its basin lies NW–SE and drains 5090 km². The main river course is approximately 140 km long. It is a plain river (Solari & Claps, 1996) and its gentle gradient (0.13 m km⁻¹) produces flooded areas, especially along the middle and lower course. Its flow is regulated by rainfall (923 mm year⁻¹), with great floods in spring, while in summer it is fed only by underground waters.

Ten sampling stations were established (Figure 1). Five of these were fixed on the river Samborombón from its headwaters in Brandsen to its mouth in La Plata River estuary (main course stations called El, E2, E3, E4 and E5). The hydrophytes *Potamogeton striatus* Ruiz et Pavón and *Chara* sp were found in headwaters, and *Salicornia ambigua* Michaus and *Spartina* sp in backwaters. The other five stations were located on tributaries: San Vicente, San Carlos, Manantiales, Dulce, and Saladillo (E6, E7, E8, E9, E10). These streams were colonized by different macrophytes, *Althernanthera* sp in San Vicente Stream, *Schoenoplectus californicus* (Meyer) Steud in Dulce, San Carlos, Manantiales and Saladillo Streams, *Scirpus americanus* and *Sagittaria* sp in San Carlos Stream and *Ludwigia* sp in Dulce Stream.

Methods

Samples were taken for each season in November 19, February 11, May 18, and August 10, 1987 corresponding to spring, summer, autumn, and winter, respectively. I took duplicate plankton samples from the middle course of the river channels, by filtering 50 l through a 30-μm mesh net. Samples were fixed with 4% formaldehyde. Water temperature, trans-

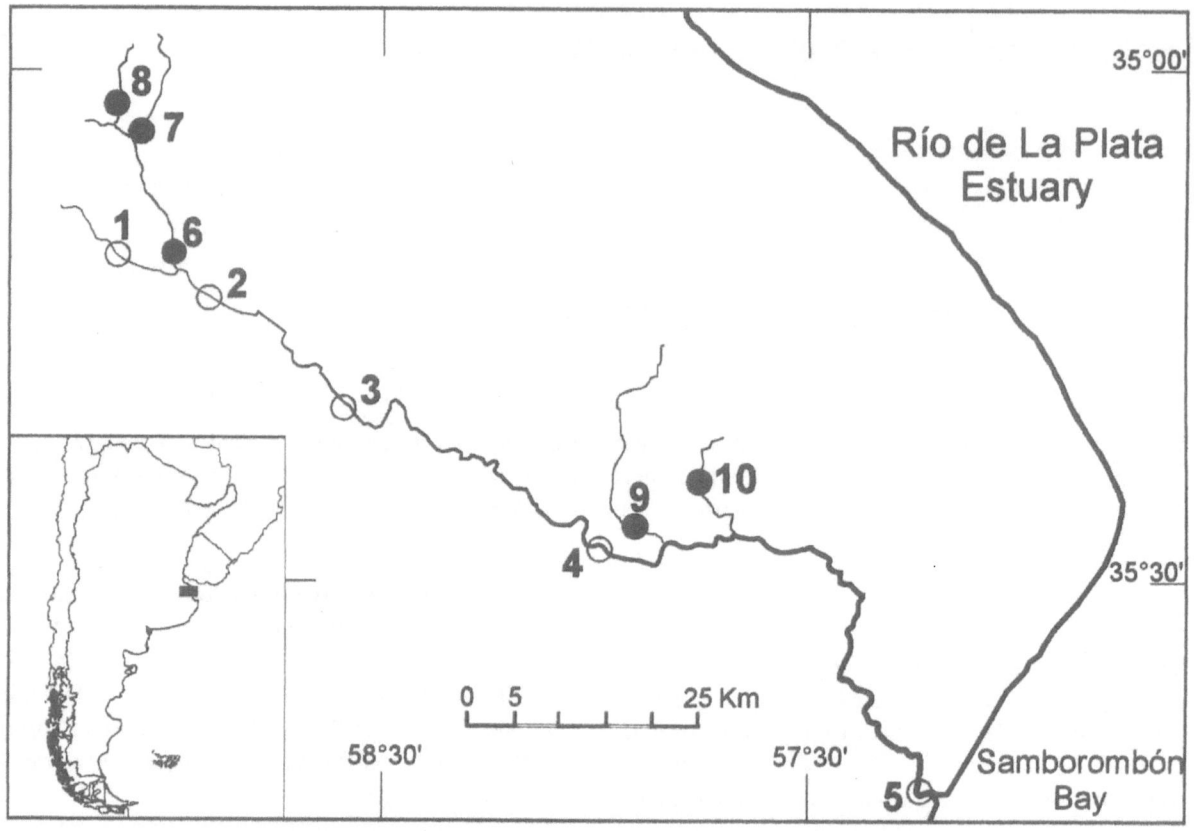

Figure 1. Map of the studied area. References: 1–5, main course sampling stations in the Samborombón River; 6–9, Tributary streams sampling stations; 6, San Vicente stream; 7, San Carlos stream; 8, Manantiales stream; 9, Dulce stream; 10, Saladillo stream.

parency, pH, conductivity, ion concentrations and dissolved oxygen were measured for each sample (Laboratorio de Química, Instituto de Limnología, La Plata, Argentina). Zooplankton was counted in a 1-ml Sedgwick–Rafter chamber under compound microscope. Biovolumes for rotifers were calculated using the formulas by Ruttner-Kolisko (1977). Presence/absence data were analysed using the percent of dissimilarity distance index and grouped by the average linkage method.

Results

Results showed that the physical and chemical parameters varied between the main river course and its tributaries (Table 1). Conductivity and pH values were much higher in the main course than in the tributaries. In both cases, the large variations in these values are related to factors affecting salt content. Conductivity values were highly variable between the four sampling

dates during the course of the year. Maximum fluctuations occurred in E5, located at the river mouth in the La Plata estuary. These variations reflect effects of rainfal, runoff, and high tides from the outer estuary.

Ion content concentration, mainly Na^+ and Cl^-, also showed a marked increase towards the river mouth (Solari & Claps, 1996). These differences could be clarified through the analysis of the weight ratios $Na/(Na + Ca)$ and $Cl/(Cl + HCO_3)$ relative to total dissolved salts (Gibbs, 1970). The relationships indicate that the main river and its tributaries are controlled by distinct processes (Figure 2). According to Gibbs' classification, the evaporation-fractional, crystallization process controls the main Samborombón course; while rock dominance controls its tributaries. Dissolved oxygen concentration and transparency did not show great differences between the main river and tributary stations. San Vicente (E6) had an oxygen concentration lower than the other sampling stations, which, in summer, even decreased to undetectable levels (Table 1). Water temperature was similar at all

Table 1. Mean values of the physical and chemical parameters of the river Samborombón and its tributaries

	Main Course					Tributaries				
	1	2	3	4	5	6	7	8	9	10
Conductivity (μS cm^{-1})										
Mean	3673	3023	3493	4292	4485	1026	1530	594	602	561
Minimum	1003	545	1230	729	655	604	750	282	500	255
Maximum	7689	5141	7689	8800	15750	1286	2400	1218	809	809
pH										
Mean	8.35	8.45	8.64	8.53	8.30	7.70	7.78	7.67	7.72	7.74
Minimum	8.18	8.14	8.20	8.20	8.18	7.47	7.50	7.35	7.68	7.65
Maximum	8.50	8.72	8.96	8.74	8.36	7.90	8.30	7.86	7.76	7.94
Oxygen (mg l^{-1})										
Mean	7.70	7.70	7.90	7.20	7.30	3.80	5.70	6.50	7.00	7.50
Minimum	5.60	6.60	6.60	4.10	4.70	0.00	3.80	5.00	4.40	3.00
Maximum	9.00	10.00	9.60	9.40	9.40	6.00	7.40	7.40	9.10	9.00
Transparency (cm) Secchi disc										
Mean	17	18	21	20	18	14	14	17	19	15
Minimum	15	15	18	15	10	10	10	10	12	12
Maximum	19	20	30	30	32	20	20	42	20	20

References as in Figure 1.

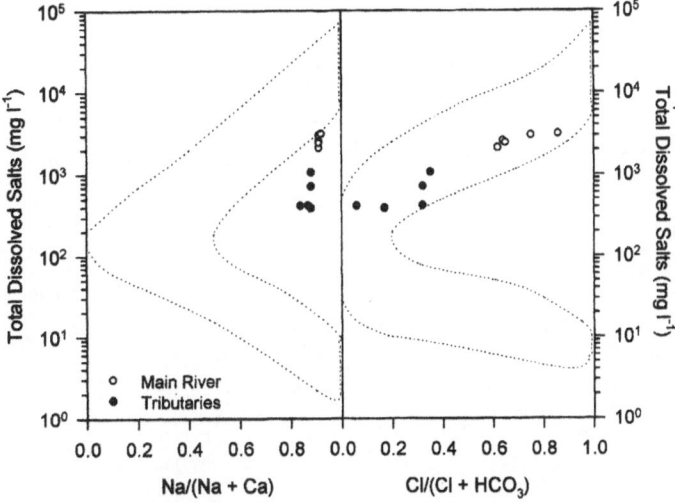

Figure 2. Relationship between total dissolved solids and Na/(Na + Ca) and Cl/(Cl + HCO$_3$) weight ratios for the river Samborombón and its tributaries.

sampling stations, averaging 13°C in autumn, 11°C in winter and 25°C in spring and summer.

In total, 47 taxa of monogonont rotifers were found (Tables 2 and 3). Bdelloids were present in the samples but could not be identified. Rotifers were the most important zooplankton group representing 57.3% of the total zooplankton species diversity. The family Brachionidae was represented by the highest number of species (16) followed by Lecanidae with nine species. The total number of rotifer species found in one sampling station varied from 18 to 29 (Figure 3a). In the main river, E4 was the site with the lowest number of species (Figure 3a). At the mouth of the river, E5, rotifers were absent in summer (Figure 3b). The summer plankton samples from this station were dominated by estuarine crustaceans such as the calanoid *Acartia tonsa* Dana and nauplii of the barnacle *Balanus*. In the tributaries, only E6 (San Vicente) did not contain rotifers in the autumn samples (Figure 3b), and only a few ciliates and two species of testate rhizopods were found in these samples.

Table 2. List of rotifer species and their presence in the river Samborombón and its tributaries sampling stations

Species	Main Course					Tributaries				
	1	2	3	4	5	6	7	8	9	10
Anuraeopsis fissa (Gosse)									Sp	
Asplanchna (A.) brightwelli (Gosse)		Su			Sp					
Brachionus angularis Gosse			SpSu	SpSu	Sp	Su	Su	Sp		SpSu
B. bidentata Anderson					Sp	Sp				Sp
B. budapestinensis Daday	Su		Su							Sp
B. calyciflorus Pallas	SuW	SpSuW	SpSu	SpSuW	W	SpSuW	Su			Sp
B. caudatus astrogenitus Ahlstrom					Sp					
B. caudatus insuetus Ahlstrom	SpSu	SpSu	SpSu	Sp		Su	Su			SpSu
B. patulus (Müller)						Su			SuW	W
B. plicatilis (Müller)	Su	Su	Su	SpSu						
B. quadridentatus Hermann	Sp	SpSuA	Sp		AW	SpW		SpSuW		
B. urceolaris nilsoni (Ahlstrom)		Su		SpW						
B. urceolaris urceolaris Müller	Sp	SpSu	SpSu		Su		Su			
Cephalodella sp 1	Sp	SpSuW	SuAW			SpSu	SpSu	SpA	SuAW	AW
Cephalodella sp 2	Sp	Sp				W			W	
Colurella uncinata (Müller)	SpSuA	SpA	SpAW	A	SpAW	Sp		SuW	SpW	W
Epiphanes senta (Müller)			W			W				
Euchlanis dilatata Ehr.						Su	Sp		A	
Filinia longiseta (Ehr.)	SuW	SpSu	SpSu	SpSu	SpAW	SuW	Su	Su	W	SpSuW
Gastropus sp		W							W	W
Hexarthra fennica (Levander)	Su	SuA	Su	SpSu		Su		Su		
Horaëlla thomassoni Koste									Sp	Sp
Keratella cochlearis (Gosse)					Sp			Su	A	
K. lenzi (Hauer)	W				AW	W	W	W	Su	W
K. tropica (Apstein)	SuW	W		A	SpAW	W		W	SpAW	W
Lecane (L.) hastata (Murray)							Sp		SpA	SuA
L. (L.) luna (Moller)		Sp								
Lecane (L.) sp		A				Su	SpSu	Su		A
Lecane (M.) bulla Gosse	SpA		Sp	A		Su		W	Sp	A
L. (M.) closterocerca Schmarda					W					W
L. (M.) cornuta (Müller)				A	A			Sp	SuW	A
L. (M.) hamata (Stokes)	AW	A			A				W	A
L. (M.) lunaris (Ehr.)										W
L. (M.) pyriformis (Daday)	W	A			A			W	W	
Lepadella ovalis (Müller)	W	AW	A			W	SpW	SpW	SpW	AW
Lophocaris salpina (Ehr.)	A				Sp					Su
Mytilina ventralis (Ehr.)									Sp	W
Notholca acuminata (Ehr.)	AW	AW	W	AW	W		W	W	W	
N. squamula (Müller)		W		W	W					
Platyas quadricornis (Ehr.)	W				Sp					
Polyarthra vulgaris Carlin	SuW	SuAW	SpSuA	A	SpAW	SuW	SuW	SpSuW	SpSuAW	SpSuW
Pompholyx sulcata Hudson	SuW				SpA			Su	Su	
Synchaeta sp	W	W	W	W	AW		AW	AW	AW	W
Testudinella patina (Hermann)	W	Sp			SpA		Sp	W	A	W
Trichocerca (T.) rattus (Müller)	Su	SpSuAW	SuA	A	AW	SuW	SuW	SpSuAW	SpSuAW	SpW
Trichotria pocillum (Müller)										W
Wolga spinifera Western			A	A				SpW	A	

References: Sampling stations as in Figure 1. Seasons: Sp, Spring; Su, Summer; A, Autumn; W, Winter.

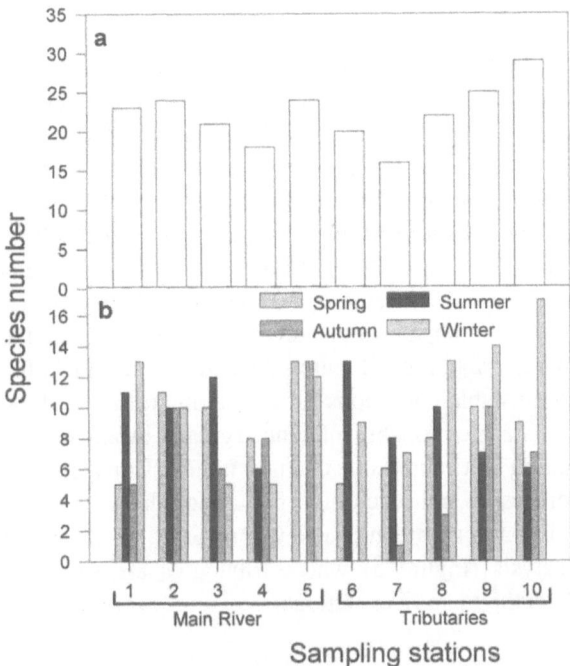

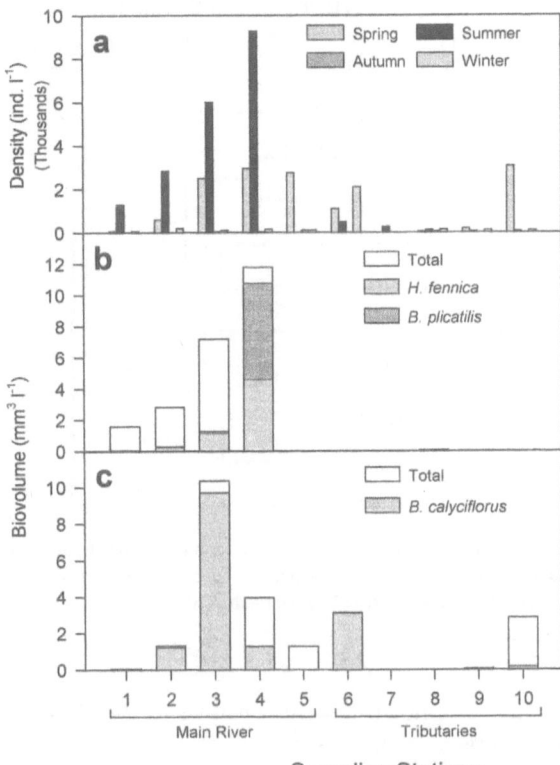

Figure 3. Number of rotifer species found in the Samborombón sub-basin. (a) Total number; (b) seasonal variation in species number. References as in Figure 1.

Figure 4. (a) Seasonal variation of the rotifer density in the Samborombón sub-basin. (b) Total rotifer and dominant species biovolume in summer. (c) Total rotifer and dominant species biovolume in spring. References as in Figure 1.

Table 3. Comparison of number of species of planktonic rotifers between inner estuary and outer estuary Rio de la Plata tributaries and pampasic lentic water bodies of Buenos Aires province

Categories	Outer estuary tributaries[1]	Inner estuary tributaries[2]	Lobos pond[3]	Chascomús system[4]
Total rotifers	47	59	38	39
Brachionidae	22	18	17	16
Lecanidae	9	14	4	6
Total zooplankton	82	123	71	74
% of rotifers	57.3	47.9	53.5	52.7

Data: 1, This study; 2, Modenutti (1991) and Modenutti & Claps (1988); 3, Boltovskoy et al. (1990); 4, Ringuelet et al. (1967).

Total rotifer density was higher in the main course than in the tributaries (Figure 4a). Remarkably, a few species are responsable for the high spring and summer densities. *B. calyciflorus* made up 90% of total rotifer biovolume at E3 during spring, while in E4 *B. plicatilis* and *Hexarthra fennica* dominated the summer zooplankton (Figure 4b, c). The latter two species were not found in E5 or in the tributary streams.

Rotifer species composition varied with season. In the Samborombón basin, rotifers can be classified according to temperature as perennial, summer

and winter species. Most of the *Brachionus* species and *Hexarthra fennica* were found in spring and summer at water temperatures ranging from 20 to 28°C. Ruttner-Kolisko (1974) also identifies these species to be thermophilic. *Notholca acuminata* and *Synchaeta* sp were recorded during autumn and winter at temperatures below 15°C. *N. squamula* was found only in winter at temperatures below 13°C. Other rotifers, e.g., *Polyarthra vulgaris*, *Keratella tropica* and *Keratella cochlearis*, showed no preferences; they are considered to be perennial species with eurythermal behaviour.

On the basis of the dendrogram, the 10 sampling stations can be divided into main river (El, E2, E3 and E4) and tributary (E6, E7, E8 and E9) groups. In addition to these, two isolated stations E5 (main river mouth) and E10 (Saladillo stream) can be recognized (Figure 5). Each group can be described in terms of its limnological characteristics (Table 1, Figure 2). This analysis, based on rotifer species composition, showed

264

Percent of Dissimilarity

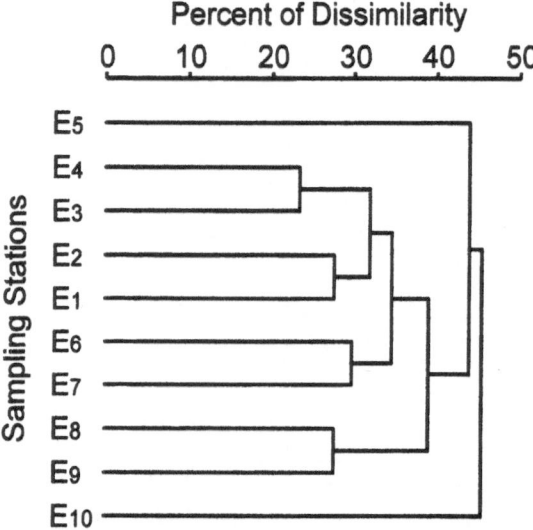

Figure 5. Dendrogram of the Samborombón sub-basin resulting from the rotifer species presence–absence distance matrix and grouped through an average linkage method. References as in Figure 1.

that the assemblages were sensitive to the different characteristics of the basin.

Discussion

Rotifer composition can be related to conductivity, since salinity is a chemical limitation of rotifer communities (Ruttner-Kolisko, 1971, 1974). The Samborombón river represents a complex salinity gradient as fluctuations in salinity can occur over short time periods because of diluting effects of rain and the influence of high tides from the outer La Plata estuary. These factors were more evident at the river mouth. The rotifers found at E5 (a variable and saline station) probably represent an ephemeral freshwater association scoured from tributaries and headwaters. This assumption is supported by the absence of species associated with inland saline waters such as *Brachionus plicatilis* and *Hexarthra fennica*, which were found at the main river stations El, E2, E3 and E4, all with high conductivity values (Table 1). These two species tolerate high salinity but are probably sensitive to the fluctuating salinity of estuaries (Green, 1993, 1995) as was observed at E5. In Argentina, *Hexarthra fennica* has been recorded in alkaline waters (José de Paggi, 1990) and in Patagonian water bodies with high contents of chloride (Kuczynski, 1987) while large populations of *Brachionus plicatilis* are restricted to saline waters (José de Paggi, 1990).

Rotifers are opportunistic organisms whose densities reflect temporal variations related to environmental conditions (Allan, 1976). Rotifer densities were higher in the middle section of the Samborombón river (E3 and E4; Figure 4a). In this section, a good development of planktonic and benthic algae also was recorded, due to greater availability of nutrients, slower stream velocity and minimal variations in water level (Solari & Claps, 1996). Thus, rotifer populations can attain high densities. On the other hand, the low rotifer densities and the high number of species registered in the tributaries (Figures 3b and 4a) would indicate unstable conditions for planktonic rotifers. These streams are probably influenced greatly by any precipitation that cause large changes in flow. Consequently, rotifers do not reach high population densities there. This particular feature also was revealed by the cluster analysis (Figure 5) where tributaries are separated from the main river stations.

Acknowledgments

I thank Lic. Victor Conzonno for providing me the chemical data of Samborombón river and Dr M.C. Claps and Dr Lía Solari for their amusing help in field work.

References

Allan, J. D., 1976. Life history patterns in zooplankton. Am. Nat. 110: 165–180.

Bonecker, C. C. & F. A. Lansac-Tôha, 1996. Community structure of rotifers in two environments of the upper River Paraná floodplain (MS)-Brazil. Hydrobiologia 325: 137–150.

Bonecker, C. C., F. A. Lansac-Tôha & A. Staub, 1994. Qualitative study of rotifers in different environments of the high Paraná river floodplain (MS)-Brazil. Unimar 16: 1–16.

Frenguelli, J., 1950. Rasgos generales de la morfología y geología de la provincia de Buenos Aires. MOPBA, La Plata, 72 pp.

Gibbs, R. J., 1970. Mechanisms controlling world water chemistry. Science 170: 1088–1090.

Green, J., 1993. Zooplankton associations in East African lakes spanning a wide salinity range. Hydrobiologia 267: 249–256.

Green, J., 1995. Associations of planktonic and periphytic rotifers in a Malaysian estuary and two nearby ponds. Hydrobiologia 313/314: 47–56.

José de Paggi, S., 1978. First observations on longitudinal succession of zooplankton in the main course of the Paraná River between Santa Fé and Buenos Aires Harbour. Stud. Neotrop Fauna Environ. 13: 143–156.

José de Paggi, S., 1983. Estudio sinóptico, del zooplancton de los principales cauces y tributarios del valle aluvial del río Paraná: Tramo Goya-Diamante (I Parte). Rev. Asoc. Cienc. Nat. Litoral 14: 163–178.

José de Paggi, S., 1990. Ecological and biogeographical remarks on the rotifer fauna of Argentina. Rev. Hydrobiol. trop. 23: 297–311.

Kuczynski, D., 1987. The rotifer fauna of Argentine Patagonia as a potential indicator. Hydrobiologia 150: 3–10.

Martínez, C. C. & Frutos, S. M., 1986. Fluctuación temporal del zooplancton en arroyos y esteros del Chaco Oriental (Argentina). Ambiente subtrop. 1: 112–133.

Modenutti, B. E., 1991. Zooplancton de ambientes lóticos de la Subcuenca Delta del río Paraná (Buenos Aires, Argentina). Iheringia (sér. Zool.) 71: 67–80.

Modenutti, B. E. & Claps, M. C., 1988. Monogononta rotifers from plankton and periphyton of pampasic lotic environments (Argentina). Limnologica (Berlin) 19: 167–175.

Paggi, J. C. & S. José de Paggi, 1990. Zooplâncton de ambientes lóticos e lênticos do rio Paraná medio. Acta Limnol. Brasil 3: 685–719.

Ruttner-Kolisko, A., 1971. Rotatorien als Indikatoren für den Chemismus von Binnensalzgewässern. Sber. Akad. Wiss., math-nat. KI., Abt 1, 179: 283–298.

Ruttner-Kolisko, A., 1974. Plankton rotifers. Biology and taxonomy. Binnengewässer 26: 146 pp.

Ruttner-Kolisko, A., 1977. Suggestions for biomass calculations of plankton rotifers. Arch. Hydrobiol. Beih. 8: 71–76

Solari, L. C. & M. C. Claps, 1996. Planktonic and benthic algae of a pampean river (Argentina): comparative analysis. Ann. Limnol. 32: 89–95.

Welcome, R. L., 1985. River Fisheries. FAO Fish. Tech. Pap. 262.

Hydrobiologia **387/388**: 267–276, 1998.
E. Wurdak, R. Wallace & H. Segers (eds), Rotifera VIII: A Comparative Approach.
© *1998 Kluwer Academic Publishers.*

Review paper

Chemical ecology of rotifers

Terry W. Snell

School of Biology, Georgia Institute of Technology, Atlanta, GA 30332-0230, U.S.A.
Tel: [+1] (404) 894-8906; fax: [+1] (404) 894-0519; e-mail: terry.snell@biology.gatech.edu

Key words: Rotifera, chemoreception, chemical ecology, sensory, predation, mating, mixis

Abstract

One of the primary channels of sensory input for zooplankton are chemical signals. Much zooplankton behavior is triggered by chemical stimuli, including feeding, predator defense, mating, and migration. Chemically regulated zooplankton behavior affects larger scale ecosystem processes like grazing, recruitment and secondary production. Knowledge of how chemicals transmit information about location, food quality, conspecifics, competitors, and predators is critical for understanding how aquatic ecosystems function. This paper reviews the behavioral evidence that planktonic rotifers respond to a variety of chemical stimuli. Although a rich variety of rotifer behaviors are regulated by chemical signals, little progress has been made to isolate and characterize these stimuli. If aquatic ecology is to become a predictive science, knowledge of the mechanisms causing the observed interactions is necessary. Chemical signals need to be isolated, purified, and characterized, and their causal role in regulating population and community processes needs to be demonstrated. Rotifers have chemosensory neurons in their corona and electron microscopy has revealed chemoreceptive pores in the anterior integument of several species. Some rotifers use these chemoreceptors to discriminate food particles based on the flavors on the cell surface. In *Asplanchna*, prey are discriminated by contact chemoreception. *Asplanchna* releases a waterborne signal that induces spine formation in several *Brachionus* species, *Keratella cochlearis*, *K. slacki*, and *Filinia longisecta*. The colonial *Sinatherina socialis* is defended against fish predation by warts containing unpalatable chemicals that have yet to be identified. Larval settlement in *Collotheca gracillipes* is determined by the chemistry of aquatic plant surfaces. Larvae prefer the undersurface of leaves where there is a low Ca^{++} microhabitat due to photosynthesis. Oviposition in *Euchlanis dilatata* is restricted to plant surfaces familiar to the maternal female. Hydrogen peroxide and certain prostaglandins stimulate resting egg hatching even in the dark. Sexual reproduction and polymorphism in *Asplanchna sieboldi* is regulated by dietary tocopherol. A chemical signal that allows assessment of conspecific population density is detected in conditioned media by several rotifer species. Water soluble extracts of *Brachionus plicatilis* increase mictic female production 1.7 times more than controls. Unknown compounds produced by certain bacteria also increase mixis 4–10 fold over controls. Mate recognition in *B. plicatilis* is determined by a 29 kD surface glycoprotein called the mate recognition pheromone (MRP). The MRP has been isolated, purified, and a polyclonal antibody against it has been prepared. The structure of the oligosaccharide and protein components of the MRP are currently being characterized. Elucidation of the chemicals regulating rotifer life cycles will make important contributions to the understanding of ecological processes in aquatic communities.

Introduction

Chemicals are the most ancient and universal of stimuli (Dusenbery, 1992). All organisms from bacteria to humans utilize chemical information about their environment and for most groups, like bacteria and arthropods (Carde & Bell, 1995), it may be the dominant sensory modality. The primary channels for the acquisition of sensory information in zooplankters are probably chemical and mechanical. Much of zooplankton behavior is likely affected by chemicals released into the water or on the surface of their food or other animals. The significance of chemical communication in zooplankton life histories was reviewed by

Larsson & Dodson (1993). They emphasized Clado-cera, so some opportunities for investigating the role of chemicals in shaping the ecological interactions of rotifers may have been overlooked. The purpose of this review is to illustrate the rich variety of research opportunities for investigating the chemical ecology of rotifers and the significance of chemical communication in regulating many life history events and the evolutionary fate of this group.

The nature of chemical signals used by zooplankton are poorly understood when compared to those used by terrestrial insects (Carde & Bell, 1995). Behavioral observations of zooplankton suggest that chemical signals synchronize important life history processes like feeding, predator defense, and mating. These contribute to larger scale ecosystem processes like recruitment, secondary production, and grazing. Consequently, understanding how chemicals transmit information about location and quality of food, con-specifics, competitors, and predators is critical for understanding how aquatic ecosystems work.

From the pioneering work by de Beauchamp (1952), Pourriot (1964) and Gilbert (1966), behavioral evidence was obtained that planktonic rotifers used chemical stimuli to regulate reproduction, induce predator defenses, and to selectively forage on phytoplankton. The behavioral studies clearly implicated chemicals, but little insight was obtained into the nature of these chemical signals. How is the information chemically encoded, how is species-specificity achieved, what variation exists among species in reception and transmission, and how could knowledge of the chemical sensory systems provide insight into the phylogenetic relationships of the Rotifera? A further concern is how anthropogenic stressors at very low concentrations might interfere with chemical communication. Detecting such effects requires clear understanding of the chemical communication channels utilized by rotifers. This paper is organized according to major ecological processes: feeding, predator defense, spatial distribution, reproduction, and mating. I outline the evidence for the role of chemicals in regulating the behavior, review knowledge about the chemicals likely to be involved, and describe experimental approaches likely to be useful in future investigations.

Why is it important to know chemical structure?

Behavioral observations demonstrating that animals respond to chemical signals is the first step in chemical ecology, and rotifer biologists have described evidence for a rich variety of ecological interactions mediated by chemicals. As with most other aquatic animals, however, investigations have rarely proceeded beyond the initial behavioral descriptions or preliminary characterization of the signal compound. The mechanism for a behavior, the chain of causes that explains its existence and function (Dusenbery, 1998), has not been described for any rotifer behavior. If ecology is to become a predictive science, the mechanisms regulating ecological interactions need to be elucidated. This means isolating, purifying, and characterizing the signal molecules and demonstrating their causal role in regulating ecological processes.

A variety of benefits will follow from knowledge of the chemical structure of signal molecules. Knowledge of how ecological information is transmitted chemically will help explain selective grazing, what limits species distributions, why aquatic communities are organized as they are, and how natural selection works in aquatic environments. Reception of and response to chemical signals probably drive much ecological specialization. Knowledge of the mechanisms of chemical communication between algae and herbivores will illustrate how they coevolve, the nature of algal defenses, and why certain grazers become dominant. Knowledge of anti-grazing molecules may identify compounds with novel properties that might have commercial value (Hay, 1996). Knowledge of how predators detect prey and how prey avoid them will help to understand an important mechanism of population regulation. This would provide insight into the adaptive significance of many zooplankton behaviors and non-genetic polymorphisms, as well as the rationale for the temporal and spatial organization of zooplankton assemblages. Knowledge of chemical signals will explain how mates are recognized and how life histories are regulated, opening the possibility for more effective biomanipulation. Understanding the channels of chemical communication will allow better management of aquatic communities, reducing their vulnerability to pollution. Availability of purified signal molecules will permit more rigorous experiments for the dissection of ecological relationships. If a mechanistic understanding of ecological processes that allows prediction is our goal, then detailed knowledge of chemical signals in aquatic environments is essential.

Table 1. Chemical signals responded to by rotifers

Behavior	Chemical source	Signal location
Avoidance toxic food	Cyanobacteria	Intracellular
Selective ingestion	Many algae	Cell surface
Predator defense	*Asplanchna*	Water soluble,
	Tropocyclops	Diffusible
	Sinantherina	Body surface (warts)
Selective cannibalism	*Asplanchna*	Body surface
Feeding response	*Asplanchna*	Homogenate
Mixis	*Asplanchna*	Water soluble,
	Brachionus	Diffusible
	Notommata	
Mate recognition	*Brachionus*	Surface glycoprotein
Larval settlement	Aquatic plants	Cell surface
Oviposition	*Euchlanis*	Conspecific eggs

Chemical sensors

The electron microscopy of Clement and colleagues demonstrated that rotifers are endowed with many sensory neurons, mostly concentrated in the coronal region (Clement et al., 1983; Clement & Wurdak, 1991). The most abundant of these are mechanoreceptors and chemoreceptors which are in contact with the outside medium and make direct connection to the brain. Chemoreceptive pores have been observed in the anterior integument beneath the cingulum of *Brachionus calyciflorus, B. sericus, B. plicatilis*, and *Notommata copeus*. Feeding behavior in *Brachionus* and *Philodina* appears to be regulated by tactile and chemoreceptors controlling muscles beneath the pseudotrochal cilia. Although rotifers are sensitive to the quantity, quality, direction and duration of light, they are not capable of image formation. Thus the sensory details of their environment are probably derived primarily from tactile and chemical signals. Rotifers respond to a wide variety of chemical signals (Table 1) and these are discussed in the following pages.

Feeding

The ability of rotifers to detect algal chemicals is not clear, although it is probably less than that of copepods and Cladocera (Rothhaupt, 1991). These latter zooplankters seem able to use mechano- and chemoreception to discriminate algal species based on their surface chemistry (Poulet & Marsot, 1978; Leiger-Visser et al., 1986), but there is no evidence

that copepods can detect algal cells from a distance (Demott & Watson, 1991). *Daphnia galeata × hyalina* hybrids are able to detect algal odors in a Y-tube olfactometer and move to the arm with the edible algal species (van Gool & Ringelberg, 1996). Similar experiments are possible with rotifers to clarify whether they are attracted by remote chemoreception of algal odors and the distance from which cells are perceived.

Rotifers live in an environment where there are many types of suspended particles differing in size, texture, and chemical composition. Algae, detritus, sand and clay particles vary substantially in their ingestibility, digestibility, and nutrient content. Chemical cues on the surface of these particles could provide information on their nutritional quality and improve rotifer foraging efficiency. Selective feeding occurs in rotifers, but the relative contribution of cell size, shape, and chemical composition is not understood (Starkweather, 1980). Rotifer species differ markedly in their feeding mechanism and dietary specialization (Gilbert & Bogdan, 1984). In experiments with flavored polystyrene spheres, Demott (1986) found that *Brachionus calyciflorus* fed preferentially on 12 μm spheres, but did not discriminate those with adsorbed algal flavors. In contrast, *Filinia terminalis* fed preferentially on 6 μm spheres flavored with algae. Feeding in at least some rotifer species is therefore under chemosensory control.

Fluid flow patterns in the corona and food manipulation by *Brachionus calyciflorus* were analyzed by Starkweather (1995). Individual algal cells were selectively removed from feeding currents by direct interception by single cilia. Algal cells were retained in the inner pseudotrochal space for up to 500 ms, ample time for sensory examination of their physical and chemical properties before ingestion. Algal contact was not required for sensing their chemical characteristics, but signals received across coronal boundary layers traveled less than 25 μm. Rotifers therefore may be able to sense algal cells from a distance, but it is a very small distance indeed.

Some rotifer species are susceptible to toxic cyanobacteria whereas other species are resistant (Gilbert, 1990; 1994). The cyanobacterium *Anabaena flos-aquae* produces a neurotoxic alkaloid, anatoxin-a, that when ingested suppressed survival and fecundity of *Brachionus calyciflorus* at 0.5 μg dry weight ml^{-1} and *Keratella cochlearis* at 4 μg ml^{-1} (Gilbert, 1994). Neither *B. calyciflorus* nor *Asplanchna girodi* were inhibited by filtrates from *A. flos-aquae* cultures, so there is no release of extracellular toxin and ingestion is required for the toxic effect. Evidence that some rotifers are able to avoid ingestion of toxic *Microcystis* has been provided by Rothhaupt (1991). *Microcystis* is differentially toxic to rotifer species (Fulton & Paerl, 1987; Starkweather & Kellar, 1987; Smith & Gilbert, 1995), but the mechanism of toxicity is unknown. The differential toxicity of cyanobacteria to rotifer species raises the question of whether some use chemoreception to detect and avoid ingestion of toxic strains. This differential toxicity could lead to substantial changes in the composition of zooplankton assemblages (Gilbert, 1996).

As in herbivores, chemoreception seems to play a role in the detection of prey by predators. The feeding response of *Asplanchna sieboldi* campanulate females was induced by homogenates of saccates and campanulates (Gilbert, 1977). This response was similar to the feeding response with live prey, so the implication was that females were detecting feeding stimulant chemicals in the homogenates. Moreover, feeding was triggered solely by chemical stimuli; tactile stimuli were not necessary. There was no detection from a distance; stimulus contact with chemoreceptors in the corona was required. Cannibalism by *A. intermedia* may be extremely selective, with attacks on certain clonemates rare (Gilbert, 1976). It seems that saccate females have different surface properties than cruciform or campanulate morphotypes and that kin and perhaps conspecifics are distinguished by chemical signals on their body surface.

Chemical communication between zooplankton and algae is bidirectional. Zooplankton sense chemicals on algal surfaces to determine edibility, but zooplankton also release chemicals into the water that signal their presence and trigger algal chemical defenses. Several cases of toxic or inhibitory compounds produced by algae in response to herbivory have been documented in marine environments (Hay, 1996; Verity & Smetacek, 1996). Few active compounds have been identified, but they are likely to be secondary metabolites sequestered in algal cells to deter ingestion. Another example of chemical signals from zooplankton triggering algal defenses is from Hessen & Donk (1993). They found that the presence of *Daphnia magna* caused the green alga *Scenedesmus subspicatus* to change from single-celled to an eight-celled growth form. Filtrate of *D. magna*-conditioned water induced algal colony formation, implying that the signal was chemical. Lampert et al. (1994) extended these observations to *Scenedesmus acutus*, showing that active feeding by *D. magna*

was required for colony induction. Homogenates of *S. acutus* and *D. magna* as well as ammonium and urea had no inducing effect. Chemical analysis of the *D. magna* soluble factor revealed that it is a nonvolatile, organic, small (< 500 Da), moderately lipophilic, heat-stable, pH resistant, and pronase E-resistant compound. It would be interesting to investigate whether rotifer grazers have similar effects on *Scenedesmus* colony formation. The work of Lampert et al. (1994) further illustrates how bioassay-directed fractionation can provide considerable information on the chemical nature of zooplankton signals with relatively simple methods.

Daphnids suppress feeding of their own and other species by excretion of inhibitory chemicals (Matveev, 1993). These allelopathic chemicals in conditioned culture water were stable after three days; however, they were destroyed by heating to 100°C. There have been no investigations to determine whether rotifers also excrete compounds with similar alleolopathic effects.

Predator defenses

Some of the first observations that waterborne chemicals signal the presence of predators in pelagic environments were with rotifers. Spine induction in response to compounds released by *Asplanchna* was described by de Beauchamp (1952) for *Brachionus calyciflorus*, Pourriot (1964) for *B. bidentatis*, *B. urceolaris* and *Filinia longiseta*, Gilbert (1966) for *B. calyciflorus*, and Stemberger & Gilbert (1984) for *Keratella slacki*. Gilbert (1967) showed that both *A. sieboldi* and *A. girodi* induced spines in *B. calyciflorus*. The inducing compound was present in *Asplanchna*-conditioned medium and in water-soluble *Asplanchna* homogenates. It was heat-stable after 5–10 minutes at 100°C, but lost activity after heating 60–80 minutes. Spine-inducing activity was lost when conditioned medium was filtered through a 0.45 μm pore size membrane filter, but not when ultracentrifuged at 105,400 g for three hours. Charcoal filtration did not adsorb the activity and the compound was not dialyzable. It was resistent to periodate and ribonuclease, but activity was lost when exposed to 0.01% pronase for two hours at 37°C. These observations suggest that the spine-inducing stimulus is a heat-stable protein.

Gilbert & Stemberger (1984) refined our understanding of the effects of compounds from *A. brightwelli* and *A. girodi* on *Keratella slacki*. They demonstrated that exposure to *Asplanchna*-conditioned medium caused *K. slacki* to develop into about 15% larger adults with about 30% longer anterior spines and 130% longer right posterior spines. Left posterior spines also are longer, but not as long as the right spines. These morphological changes reduced the susceptibility of *K. slacki* to capture and ingestion after *Asplanchna* attack by about two-fold and five-fold, respectively. Two species of predatory copepods did not induce spines in *K. slacki*, so spine induction appears to be a specific response to *Asplanchna* predation. A different result was found for spine induction in *K. cochlearis* (Stemberger & Gilbert, 1984). The predatory copepods *Tropocyclops prasinus* and *Mesocyclops edax* as well as the rotifer *A. priodonta* induced spine development in 9–55% unspined *K. cochlearis*. Therefore it seems that there are different degrees of specificity in the spine induction response of different rotifer species. A similar response to three predators suggest that *K. cochlearis* responds to general metabolic products common to copepods and *Asplanchna*.

As described above, the morphological and ecological consequences of predator-induced responses have been well characterized in rotifers. Identification of the signal molecules and their mode of action, however, has lagged that of work on daphnids. Tollrian & von Elert (1994) have chemically characterized the *Chaoborus* factor that induces cyclomorphosis in *Daphnia*. Using C-18 silica bonded sorbents, they extracted the signal compound from natural waters and concentrated it 100-fold. It was purified using reversed-phase HPLC and its characterisitics explored with several cation and anion exchange columns. These techniques and ultrafiltration demonstrated that the *Chaoborus* factor is a < 500 Da, non-olefinic, hydoxy-carboxylic acid.

Fish kairomones responsible for inducing diel vertical migration in daphnids have been characterized using similar techniques. Von Elert & Loose (1996) concentrated the active kairomone from natural waters using C-18 columns. Bioassay-directed fractionations were used with various ion exchange columns and HPLC to purify the active compound. The exact chemical has not yet been identified, but it is known to be a < 500 Da, water-soluble, temperature- and pH-stable, nonvolatile, anionic compound with hydoxyl groups that are essential for its activity. The compound is not predator-specific since it has been found in three species of fish.

The chemical defenses of marine invertebrates are well known and some of the active molecules have been characterized (Hay, 1996). Only a single report on rotifers documents the use of chemicals as an anti-predator defense (Felix et al., 1995). These authors argued that colony formation in the rotifer *Sinantherina socialis* places them in the 0.8 to 4 mm size range that is vulnerable to fish predation. They demonstrated that *S. socialis* colonies were readily attacked and captured by 14 species of fish within a five-minute test period. But most of the *S. socialis* colonies were expelled from the fish's mouth without damage. Of the more than 1000 *S. socialis* colonies offered to fish, 71–100% were rejected. In contrast, 86–100% of the *Daphnia* sp. used as a control were consumed.

These data clearly suggest that *S. socialis* is unpalatable to most small-mouthed, predatory fish. The unpalatability factor is unknown, but it is likely to be a chemical. Two pairs of 20–30 μm dark brown, elliptical bodies called warts are located just beneath the corona on the antero-ventral surface. These are not conspicuous when feeding, but when *S. socialis* is disturbed, their corona is contracted into the body, prominently exposing the warts. Freeze-dried tubifex worms are a commercial fish food. When these are rehydrated in a crude homogenate of *S. socialis* colonies, the worms become unpalatable to fish.

Spatial distribution

The spatial distribution of sessile rotifers is strongly influenced by substrate recognition and larval settlement. The larvae of *Collotheca gracilipes* settled differentially on plants with the following preferences: *Lemna* > *Elodea* > *Myriophyllum* >*Nymphaea* > plastic depressions (Wallace & Edmondson, 1986). Younger leaves were preferred to older leaves and the undersurface was preferred to the top. In an attempt to identify the settling signals, these authors found that crushing *Elodea* leaves, exposing them to dilute HCl, and heating greatly reduced their ability to induce larval settling. Crude extracts from all of the plants tested greatly enhanced larval settling on plastic depressions within 0.5–4 hours. The activity of the extracts was stable for at least 6 days at 5°C, and it could withstand heating at 100°C for 60 minutes. The settling factor was soluble in water and 80% ethanol, but it was not filterable, not diminished by dialysis, and not adsorbed by charcoal. Treatment of the extract with 40 mg l^{-1} α-amylase did not denature the signal; rather it stimulated larval settlement. Besides its enzyme activity, α-amylase has the ability to act as a chelator, binding calcium. Exposure to 500 mg l^{-1} α-amylase reduced

Ca^{2+} in solution and triggered settling. Likewise, 50 mg l^{-1} EDTA had a similar stimulatory effect on settling. When the Ca^{2+} concentration in the medium exceeded 250 mg l^{-1}, larval setting was inhibited. Wallace & Edmondson concluded tha Ca^{+2} concentration was the critical factor regulating *C. gracilipes* larval settlement. This made good ecological sense because Ca^{2+} concentration is reduced to near 0 in the unstirred microhabitat (400–800 μm) beneath the surface of photosynthesizing leaves. Calcium concentration is therefore a reliable indicator of the preferred larval microhabitat.

Larval settlement and metamorphosis in many marine invertebrates is controlled by sensory recognition of exogenous chemical signals. For example, oyster larvae settlement is triggered by a glycyl-glycyl-L-arginine tripeptide (Zimmer-Faust & Tamburri, 1994). The entire signal transduction pathway in red abalone, *Haliotis rufescens*, has been described (Morse & Morse, 1996). An algal oligopeptide that mimics the neurotransmitter γ-aminobutyric acid (GABA) acts as a morphogenetic inducer. The specific receptor of this signal has been identified and lysine as a co-stimulus is capable of increasing its sensitivity up to 100-fold.

The spatial distribution of daphnids also is strongly influenced by the reception of chemical signals. Compounds released by fish are known to stimulate diel vertical migration (De Meester, 1993; Loose et al., 1993; Larsson & Dodson, 1993). Whether rotifers likewise respond to chemicals from fish is not known and deserves further investigation.

The spatial distribution of rotifers can also be influenced by the deposition of eggs by females. Oviposition in *Euchlanis dilatata* appears to be regulated by chemical signals on the substrate (Walsh, 1989). Eggs were laid preferentially on artificial substrates containing the eggs of conspecifics. More eggs were laid on plant substrates from which *E. dilatata* was collected than on unfamiliar substrates. However, only two of the three clones tested exhibited substrate preference in oviposition.

The temporal distribution of monogonont rotifers is largely controlled by resting egg hatching. Several investigations have shown that light and temperature play a key role in hatching (Gilbert, 1974; Pourriot & Snell, 1983). Recently, chemical signals have been shown to stimulate resting egg hatching in the absence of light (Hagiwara et al., 1995). When resting eggs of *Brachionus plicatilis* were exposed to 3 μmol hydrogen peroxide in darkness at 25 °C, 12% of rest-

ing eggs hatched in three days compared to 0% in controls. Exposure to three different prostaglandins at 3×10^{-6} μmol caused 22–36% resting egg hatching as compared to 1% in controls. These authors hypothesized that active oxygen is generated by UV photolysis and oxidizes unsaturated fatty acids in the resting egg embryo. This reaction generates prostaglandins which induce hatching. These observations suggest that resting egg hatching may be chemically mediated, triggered by chemical signals in the environment.

Sexual reproduction

Chemical signals have long been implicated in the induction of sexual reproduction in rotifers. The ability of dietary tocopherol to induce polymorphism and sexual reproduction in *Asplanchna sieboldi* is well documented (Gilbert, 1980). Chemical factors associated with population density have been implicated as inducers of sexual reproduction in several rotifer species. Since the pioneering work of Gilbert (1963) with *Brachionus calyciflorus*, density related triggers for mixis have been described in *B. plicatilis* (Hino & Hirano, 1976), *Notommata copeus* (Pourriot & Clement, 1977), *Asplanchna girodi* (King & Snell, 1980), and several strains of *B. plicatilis* and *B. rotundiformis* (Snell & Boyer, 1988; Carmona et al., 1993; Hagiwara et al., 1993). Although chemical signals have been repeatedly implicated as mixis inducers, little is known about the nature of these signals. Renewal of the culture medium suppresses mixis (Hino & Hirano, 1976), presumably by diluting the chemical signal. Population densities of *B. plicatilis* as low as 147 rotifers l^{-1} are sufficient to trigger a mictic response (Snell & Boyer, 1988). Culture medium conditioned by a population growing from 1 to 80 females ml^{-1} over five days increased the mictic response of *B. plicatilis* (Carmona et al., 1993). The signal was not stable as its activity was lost after freezing. These authors suggested that the signal may be a general metabolite since medium conditioned by *Artemia salina* produced a weaker mictic response. The response to conditioned medium also differed among rotifer strains. Hagiwara et al. (1994) showed that water-soluble extracts from 1000 rotifers increased mictic rate and resting egg production 1.7 times over controls. All of this work suggests that a water-soluble factor that accumulates in the medium plays an important role in triggering mixis. The mictic bioassays are well developed for detecting its activity and the fractionation protocols are available from the work isolating similar water-soluble signals for *Daphnia* (Tollrian & Von Elert, 1994). There now appear to be no technical obstacles to isolating and identifying the mictic factor that has intrigued rotifer biologists for so long.

Other chemical signals may enhance the mictic response. Hagiwara et al. (1994) found that mixis in *B. plicatilis* could be increased with the addition of bacteria isolated from rotifer cultures. Out of the 17 bacterial strains investigated, five (3 *Pseudomonas*, 1 *Moraxella* and 1 *Micrococcus* strain) increased mixis 4–10 fold. The mechanism of mixis enhancement is not known. Bacteria could excrete water-soluble compounds into the medium that are directly taken up by rotifers, or rotifers could obtain the mixis-enhancing compounds through ingestion. The experiments necessary to test these hypotheses seem straightforward.

Evidence is accumulating that some anthropogenic chemicals can disrupt normal hormonal communication, producing harmful effects on reproduction in a wide variety of animals (Colborn et al., 1993). Environmental estrogens that disrupt endocrine regulation of reproduction are being detected in aquatic environments at concentrations thought to cause biological effects. These compounds are persistent, bioaccumulating, and bind to known estrogen receptors (Vom Saal, 1995). Endocrine disruptors have different effects on different developmental stages and the timing of exposure is critical. Effects are usually not observed until the offspring of exposed animals reach maturity. The environmental estrogen diethyl stilbesterol has marked effects on *Daphnia magna* (Baldwin et al., 1995). Exposure to 0.5 mg l^{-1} for three weeks had no effect on female survival or reproduction, although impaired testosterone metabolism was detected. First-generation offspring, however, had reduced frequency of moulting, and second-generation offspring had lower fecundity. It is conceivable that endocrine disruptors may have similar effects on rotifers, but it will take carefully-designed experiments to detect them. Mictic reproduction would be good to investigate since this phase of the life cycle seems to be more sensitive to toxicants than the amictic phase (Snell & Carmona, 1995).

The role of known vertebrate and invertebrate hormones in the regulation of rotifer metabolism has been implicated by Gallardo et al. (1997). They examined the effects of eight hormones on population growth, mictic female production, and body size of *Brachionus plicatilis*. Rotifers were inoculated at a density of 1 female ml^{-1} and cultured with a particular hormone for 48 hours. Then, every other day until day 8, they were transferred to fresh medium with algae but lacking hormones. The estradiol-17β treatment of 50 mg l^{-1} caused a 2.3-fold increase in mictic female production by day 6. Other compounds tested like growth hormone and gamma aminobutyric acid (GABA) significantly increased population growth and mictic female production over control levels. Lorica length also increased by 4% and 9%, respectively. The fact that certain hormones have activity in *B. plicatilis* suggests that endocrine disrupters mimicking endocrine signals may affect rotifer population growth, reproduction, and body size.

Mating

There is no evidence that rotifers use chemical signals to locate mates from a distance. There is a critical size range of 0.2–5 mm below which aquatic organisms are unlikely to benefit from producing a diffusible mate location pheromone (Dusenbery & Snell, 1995). Most rotifers are smaller than this critical size and in *Brachionus plicatilis*, male-female encounters are random (Snell & Garman, 1986).

Mate recognition in rotifers occurs by contact chemoreception (Gilbert, 1963; Aloia & Moretti, 1973; Snell & Hawkinson, 1983). Chemical signals are detected by males using chemoreceptors in their corona (Clement et al., 1983) upon contact with the body surface of conspecific females. The nature of this mate recognition signal has been characterized as a surface glycoprotein (Snell et al., 1988; Snell & Nacionales, 1990; Snell et al., 1993; Snell et . al., 1995) and is the first mating pheromone described for any zooplankter. A 29 kD glycoprotein was isolated from *B. plicatilis* (Russian strain) using serial lectin affinity chromatography with *Lens culinaris* followed by *Tetragonolobus purpureas* (Snell et al., 1995). A polyclonal antibody (anti-gp29) was prepared against this glycoprotein and used to localize the mate recognition pheromone (MRP) on the body surface of females. Epifluorescence microscopy revealed that the MRP was most concentrated in the coronal region of females. When females were exposed to anti-gp29, the probability that a male-female encounter would result in mating was reduced by about 80%. Newborn males were exposed to purified MRP with the hypothesis that MRP would bind to their receptors, reducing their ability to recognize females. In males exposed to 170 μg MRP ml^{-1}, the probablity that a male-female encounter resulted in mating was reduced by about 90%. It was demonstrated that the MRP

274

alone was sufficient to elicit male mating behavior by binding purified MRP to sepharose beads. In 38% of male-bead encounters, mating behavior was initiated and deglycosylation of the MRP bound beads reduced this to 2.3%.

Characterization of the MRP has continued along several lines. The ability of anti-gp29 to bind to surface glycoproteins in a variety of rotifer species has been demonstrated (Rico-Martinez et al., 1996). The monosaccharide composition of oligosaccharides hydrolized from the MRP has been reported (Snell & Rico-Martinez, 1996). The MRP oligosaccharide appears to be composed of primarily mannose, followed by glucose, N-acetylglucosamine, fucose, and N-acetylgalactosamine. Dingmann and Snell reported the amino acid composition of the MRP in this symposium along with the monosaccharide sequence of the MRP oligosaccharide. Based on these data, a structure for the MRP oligosaccharide is proposed. Work is in progress to isolate and sequence the MRP gene.

Future directions

If rotifers are to become important in the elucidation of the chemical ecology of aquatic environments, rotiferologists need to employ modern methods to isolate and characterize the signal molecules. Knowledge of the chemical signals regulating ecological interactions and life histories will be required if there is to be a clear understanding of how aquatic communities function. Bioassays currently exist for many chemical signals employed by rotifers, and the ecological significance of the responses are often well characterized. Rotiferologists need to take the next step and apply chemical fractionation methods to identify the signal compounds. If this is not done, experiments with rotifers will never gain the attention that those with daphniids command.

We can capitalize on the existence of good rotifer bioassays to guide modern fraction schemes (Figure 1). Strategies for experimental investigations are clear, beginning with conditioned medium and concentration of active fractions with C-18 columns. Next, ion exchange columns can selectively retain the active fractions, followed by HPLC to purify the compounds to a point where GC-MS might be applied for unequivocal identification. Separation methods continue to improve in resolution and sensitivity. Capillary electrophoresis permits the isolation of microquantities of signal molecules (Novotny, 1997). Centers for the application of these techniques by non-specialists are

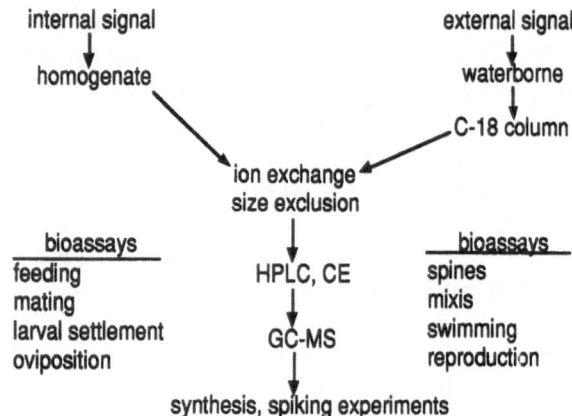

Figure 1. Bioassay-directed fractionation of rotifer chemical signals.

becoming more widely available. If we do not take advantage of such opportunities, rotifer ecology could be relegated to the sleepy backwaters of a bygone era.

A more mechanistic understanding of system behavior is required if we are to sustainably manage aquatic ecosystems. Human impacts can be better anticipated and minimized if we know how aquatic animals communicate and what kinds of chemical are likely to disrupt these communication channels. A survey is needed of how the major classes of toxicant disrupt chemical communication in rotifers. For example, do endocrine disrupters impact the sexual phase of rotifer life cycles?

It seems sensible to start with a rotifer behavior that is known to be regulated by chemical signals and for which a good bioassay exists. Examples of these include waterborne signals from *Asplanchna* that are known to induce postero-lateral spines in *Brachionus calyciflorus*, autoconditioning of the medium by *B. plicatilis* that triggers mixis, and compounds from certain bacteria that also stimulate mixis. A protocol for concentrating waterborne signals from the medium with a C-18 column, elution, and then bioassay-directed fractionation of the active compound was used successfully by von Elert & Loose (1995) for fish kairomones. This protocol with only slight modification should be applicable to rotifers. The compounds responsible for unpalatability of *Sinantherina* are particularly interesting. These could probably also be isolated and identified using the bioassay-directed fractionation approach. Chemical signals detected on contact also seem important in rotifers. Methods for the analysis of surface glycoproteins on rotifers have been described by Snell et al. (1995). Analyses of the monosaccharide sequence of the MRP, oligosaccharide, as well as analysis of the protein structure and the gene coding for the MRP are underway. The factors responsible for morphotype recognition in *Asplanchna*, and selective grazing of algal cells may respond to a similar approach.

Conclusions

Like other aquatic animals, the chemical ecology of rotifers is poorly known, but recent developments now make it possible to make significant advances. It is important to identify chemical signals in order to develop a mechanistic, predictive understanding of important ecological processes like grazing, predation, and reproduction in aquatic environments. Aquatic ecosystems can be more effectively managed if the chemicals regulating life histories are known. Behavioral observations suggest that many important ecological interactions of rotifers are chemically mediated. Several good bioassays exist that could be used to direct fractionations, permitting isolation and identification of the signal molecules. Application of modern fractionation methods is needed to characterize the chemical signals. Availability of purified molecules will improve our ability to experimentally dissect ecological relationships. Many opportunities exist for rotifers to assume a leading role in elucidating the chemical ecology of zooplankton.

Acknowledgments

This paper was improved by comments from Gary Cecchine and David Dusenbery and was supported by grants OCE 9115860 and OCE 9503521 from the National Science Foundation.

References

Aloia, R. C. & R. L. Moretti, 1973. Mating behavior and the ultrastructure of copulation in the rotifer *Asplanchna brightwelli*. Trans. Am. Microsc. Soc. 90: 371–380.

Baldwin, W. S., D. L. Milam & G. A. LeBlanc, 1995. Physiological and biochemical perturbations in *Daphnia magna* following exposure to the model environmental estrogen diethylstilbestrol. Environ. Toxicol. Chem. 14: 945–952.

Beauchamp, P. de, 1952. Un facteur de la variabilite chez les rotifere du genre *Brachionus*. Comptes Rendus Hebdomadaires des Seances de l'Academie des Sciences 234: 573–575.

Carde, R. T. & W. J. Bell, 1995. Chemical Ecology of Insects, 2. New York, Chapman & Hall, 433 pp.

Carmona, M. J., M. Serra & M. R. Miracle, 1993. Relationship between mixis in *Brachionus plicatilis* and preconditioning of culture medium by crowding. Hydrobiologia 255/256: 145–152.

Clement, P., E. Wurdak & J. Amsellem, 1983. Behavior and ultrastructure of sensory organs in rotifers. Hydrobiologia 104: 89–130.

Clement, P. & E. Wurdak, 1991. Rotifera. In: Microscopic Anatomy of Invertebrates. Wiley-Liss.: 219–297.

Colborn, T., F. S. vom Saal & A. M. Soto, 1993. Developmental effects of endocrine disrupting chemicals in wildlife and humans. Environ. Health Perspect. 101: 378–384.

De Meester, L., 1993. Genotype, fish-mediated chemicals and phototactic behavior in *Daphnia magna*. Ecology 74: 1467–1474.

DeMott, W. R. 1986. The role of taste in food selection by freshwater zooplankton. Oecologia 69: 334–340.

DeMott, W. R. & M. D. Watson, 1991. Remote detection of algae by copepods: resposnes to algal size, odors, and motility. J. Plankton Res. 13: 1203–1222.

Dusenbery, D. B., 1992. Sensory Ecology. W. H. Freeman, New York: 558 pp.

Dusenbery, D. B., 1998. A logic of behavior. Submitted.

Dusenbery, D. B. & T. W. Snell, 1995. A critical body size for use of pheromones in mate location. J. Chem. Ecol. 21: 427–438.

Felix, A., M. E. Stevens & R. L. Wallace, 1995. Unpalatability of a colonial rotifer, *Sinantherina socialis*, to small zooplanktivorous fishes. Invert. Biol. 114: 139–144.

Fulton, R. S. I. & H. W. Paerl, 1987. Toxic and inhibitory effects of the blue-green alga *Microcystis aeruginosa* on herbivorous zooplankton. J. Plankton Res. 9: 837–855.

Gallardo, W. G., A. Hagiwara, Y. Tomita, K. Soyano & T. W. Snell, 1997. Effect of some vertebrate and invertebrate hormones on the population growth, mictic female production, and body size of the marine rotifer *Brachionus plicatilis*. Hydrobiologia 358: 113–120.

Gilbert, J. J., 1963. Contact chemoreception, mating behavior and sexual isolation in the rotifer genus *Brachionus*. J. exp. Biol. 40: 625–641.

Gilbert, J. J., 1966. Rotifer ecology and embryological induction. Science 151: 1234–1237.

Gilbert, J. J., 1967. *Asplanchna* and postero-lateral spine production in *Brachionus calyciflorus*. Arch. Hydrobiol. 64: 1–62.

Gilbert, J. J., 1974. Dormancy in rotifers. Trans. Amer. Microscop. Soc. 93: 490–513.

Gilbert, J. J., 1976. Selective cannibalism in the rotifer *Asplanchna sieboldi*: Contact recognition of morphotype and clone. Proc. natn. Acad. Sci. U.S.A. 73: 3233–3237.

Gilbert, J. J., 1977. Control of feeding behaviour and selective cannibalism in the rotifer *Asplanchna*. Freshwat. Biol. 7: 337–341.

Gilbert, J. J., 1980. Female polymorphism and sexual reproduction in the rotifer *Asplanchna*: evolution of their relationship and control by dietary tocopherol. Amer. Nat. 116: 409–431.

Gilbert, J. J., 1990. Differential effects of *Anabaena affinis* on cladocerans and rotifers: mechanisms and implications. Ecology 71: 1727–1740.

Gilbert, J. J., 1994. Susceptibility of planktonic rotifers to a toxic strain of *Anabena flos-aquae*. Limnol. Oceanogr. 39: 1286–1297.

Gilbert, J. J., 1996. Effect of food availability on the response of planktonic rotifers to a toxic strain of the cyanobacterium *Anabena flos-aquae*. Limnol. Oceangr. 41: 1565–1572.

Gilbert, J. J. & R. S. Stemberger, 1984. *Asplanchna*-induced polymorphism in the rotifer *Keratella slacki*. Limnol. Oceangr. 29: 1309–1316.

Gilbert, J. J. & K. G. Bogdan, 1984. Rotifer grazing: *In situ* studies on selectivity and rates. In: D. G. Meyers and J R Strickler (eds), Trophic Interactions Within Aquatic Ecosystems. Boulder, CO, AAAS, 85: 97–133.

Hagiwara, A., K. Hamada, S. Hori & K. Hirayama, 1994. Increased sexual reproduction in *Brachionus plicatilis* (Rotifera) with the addition of bacteria and rotifer extracts. J. exp. Mar. Biol. Ecol. 181: 1–8.

Hagiwara, A., N. Hoshi, F. Kawahara, K. Tominaga & K. Hirayama, 1995. Resting eggs of the marine rotifer *Brachnious plicatilis* Muller: development and effect of radiation on hatching. Hydrobiologia 313/314: 223–229.

Hay, M. E., 1996. Marine chemical ecology: what's known and what's next? J. exp. mar. Biol. Ecol. 200: 103–134.

Hessen, D. O. & E. Van Donk, 1993. Morphological changes in *Scenedesmus* induced by substances released from *Daphnia*. Arch. Hydobiol. 127: 129–140.

Hino, A. & R. Hirano, 1976. Ecological studies on the mechanism of bisexual reproduction in the rotifer *Brachionus plicatilis*. I. General aspects of bisexual reproduction. Nippon Suisan Gakkaishi 42: 1093–1099.

Lampert, W., K. O. Rothhaupt & E. von Elert, 1994. Chemical induction of colony formation in a green alga (*Scenedesmus acutus*) by grazers (*Daphnia*). Limnol. Oceangr. 39: 1543–1550.

Larsson, P. & S. Dodson, 1993. Chemical communication in planktonic animals. Arch. Hydrobiol. 129: 129–155.

Leiger-Visser, M. F., J. G. Mitchell, A. Okubo & J. A. Fuhrman, 1986. Mechanoreception in calanoid copepods. A mechanism of prey detection. Mar. Biol. 90: 529–535:

Loose, C. J., E. von Elert & P. Dawidowicz, 1993. Chemically-induced diel vertical migration in Daphnia: a new bioassay for kairomones exuded by fish. Arch. Hydrobiol. 126: 329–337.

Matveev, V., 1993. An investigation of allelopathic effects of *Daphnia*. Freshwat. Biol. 29: 99–105.

Morse, A. N. C. & D. E. Morse, 1996. Flypapers for coral and other planktonic larvae. BioScience 46: 254–262.

Novotny, M. V., 1997. Capillary biomolecular separations. J. Chromatogr. 689: 55–70.

Poulet, S. A. & P. Marsot, 1978. Chemosensory grazing by marine calanoid copepods (Arthropoda: Crustacea). Science 200: 1403–1405.

Pourriot, R., 1964. Etude experimentale de variations mophologiques chez certaines especes de rotiferes. Bull. Soc. zool. Fr. 89: 555–561.

Pourriot, R. & P. Clement, 1977. Comparison of the control of mixis in three clones of *Notommata copeus*. Arch. Hydrobiol. Beih. 8: 174–177.

Pourriot, R. & T. W. Snell, 1983. Resting eggs in rotifers. Hydrobiologia 104: 213–224.

Rico-Martinez, R., B. Dingmann & T. W. Snell, 1996. Surface glycoproteins potentially involved in mate recognition in nine freshwater rotifer species. Arch. Hydrobiol. 138: 1–10.

Rothhaupt, K. O., 1991. The influence of toxic and filamentous blue-green algae on feeding and population growth of the rotifer *Brachionus rubens*. Int. Rev. ges. Hydrobiol. 76: 67–72.

Smith, A. D. & J. J. Gilbert, 1995. Relative susceptibilities of rotifers and cladocerans to *Microcystis aeruginosa*. Arch. Hydrobiol. 132: 309–336.

Snell, T. W. & C. A. Hawkinson, 1983. Behavioral reproductive isolation among populations of the rotifer *Brachionus plicatilis*. Evolution 37: 1294–1305.

Snell, T. W. & B. L. Garman, 1986. Encounter probabilities between male and female rotifers. J. exp. mar. Biol. Ecol. 97: 221–230.

Snell, T. W. & E. M. Boyer, 1988. Thresholds for mictic female production in the rotifer *Brachionus plicatilis*. J. exp. mar. Biol. Ecol. 124: 73–85.

Snell, T. W., M. J. Childress & B. C. Winkler, 1988. Characteristics of the mate recognition factor in the rotifer *Brachnious plicatilis*. Comp. Biochem. Physiol. 89A: 481–485.

Snell, T. W. & M. A. Nacionales, 1990. Sex pheromone communication in *Brachionus plicatilis* (Rotifera). Comp. Biochem. Physiol. 97A: 211–216.

Snell, T. W., P. D. Morris & G. A. Cecchine, 1993. Local zation of the mate recognition pheromone in *Brachionus plicatilis* (O. F. Muller) (Rotifera) by fluorescent labeling with lectins. J. exp. mar. Biol. Ecol. 165: 225–235.

Snell, T. W., R. Rico-Martinez, L. N. Kelly & T. E. Battle, 1995. Identification of a sex pheromone from a rotifer. Mar. Biol. 123: 347–353.

Snell, T. W. & M. J. Carmona, 1995. Comparative toxicant sensitivity of sexual and asexual reproduction in the rotifer *Brachionus calyciflorus*. Environ. Toxicol. Chem. 14: 415–420.

Snell, T. W. & R. Rico-Martinez, 1996. Characteristics of the mate-recognition pheromone in *Brachionus plicatilis* (Rotifera). Mar. Fresh. Behav. Physiol. 27: 143–151.

Starkweather, P. L., 1980. Aspects of feeding behavior and trophic ecology of suspension-feeding rotifers. Hydrobiologia 73: 63–72.

Starkweather, P. L. & P. E. Kellar, 1987. Combined influences of particulate and dissolved factors in the toxicity of *Microcystis aeruginosa* (NRC-SS17) to the rotifer *Brachionus calyciflorus*. Hydrobiologia 147: 375–378.

Starkweather, P. L., 1995. Near-coronal fluid flow patterns and food cell manipulation in the rotifer *Brachionus calyciflorus*. Hydrobiologia 313/314: 191–195.

Stemberger, R. S. & J. J. Gilbert, 1984. Spine development in the rotifer *Keratella cochlearis*: induction by cyclopoid copepods and *Asplanchna*. Freshwater Biol. 14: 639–647.

Tollrian, R. & E. von Elert, 1994. Enrichment and purification of *Chaoborus* kairomone from water: further steps towards its chemical characterization. Limnol. Oceanogr. 39: 788–796.

Van Gool, E. & J. Ringelberg, 1996. Daphnids respond to algae-associated odours. J. Plankton Res. 18: 197–202.

Verity, P. G. & V. Smetacek, 1996. Organism life cycles, predation, and the structure of marine pelagic ecosystems. Mar. Ecol. Prog. Ser. 130: 277–293.

Vom Saal, F. S., 1995. Environmental estrogenic chemicals: Their impact on embryonic development. Human Ecol. Risk Assess. 1: 3–15.

Von Elert, E. & C. J. Loose, 1996. Predator-induced diel vertical migration in *Daphnia*: enrichment and preliminary chemical characterization of a kairomone exuded by fish. J. Chem. Ecol. 22: 885–895.

Wallace, R. L. & W. T. Edmondson, 1986. Mechanism and adpative significance of substrate selection by a sessile rotifer. Ecology 67: 314–323.

Walsh, E. J., 1989. Oviposition behavior of the littoral rotifer *Euchlanis dilatata*. Hydrobiologia 186/187: 157–161.

Zimmer-Faust, R. K. & M. N. Tamburri, 1994. Chemical icentity and ecological implications of a waterborne larval settlement cue. Limnol. Oceanogr. 39: 1075–1087.

Hydrobiologia **387/388**: 277–281, 1998.
E. Wurdak, R. Wallace & H. Segers (eds), Rotifera VIII: A Comparative Approach.
© *1998 Kluwer Academic Publishers.*

Differential sensitivity of *Synchaeta* and *Daphnia* to nucleosides from *Anabaena affinis*

John J. Gilbert

Department of Biological Sciences, Dartmouth College, Hanover, NH 03755, U.S.A.

Key words: Anabaena, Cladocera, cyanobacteria, *Daphnia*, nucleosides, Rotifera, *Synchaeta*, toxicity

Abstract

The cyanobacterium *Anabaena affinis* contains two nucleosides responsible for its toxicity: 9-deazaadenosine 5′-α-D-glucopyranoside (compound 1) and 9-deazaadenosine (compound 2). As expected, a strain of *Daphnia pulex* inhibited by *A. affinis* also was inhibited by these nucleosides. Surprisingly, however, a strain of *D. pulex* coexisting with *A. affinis*, and not inhibited by it, was equally or more inhibited by the nucleosides. LC_{50} values for compounds 1 and 2 were, respectively, 1.33 and 0.56 μg ml^{-1} for the former *D. pulex* and 0.79 and 0.54 μg ml^{-1} for the latter. The resistant *D. pulex*, which benefits from the ingestion of *A. affinis*, may have evolved a mechanism to detoxify the nucleosides in its intestine. In contrast, *Synchaeta pectinata*, which is unaffected by *A. affinis*, was not inhibited by the nucleosides. High concentrations of compounds 1 and 2 (3.6 and 2.2 μg ml^{-1}, respectively) reduced neither survivorship nor fecundity. The resistance of this rotifer to the dissolved nucleosides may be due to its inability to absorb them across its surface membranes, to its inability to metabolize them into more toxic compounds, or to its lack of a receptor for them. An evolved resistance seems unlikely, as *S. pectinata* probably does not ingest *A. affinis*. The effect of *A. affinis* on natural zooplankton communities should be very different from that of strains of *Anabaena flos-aquae* producing the alkaloid, anatoxin-a. The *A. affinis* should be ingested by many cladocerans but not rotifers, and it contains toxins which inhibit cladocerans but not *S. pectinata* and perhaps other rotifers. The *A. flos-aquae* is ingested by rotifers as well as cladocerans, and its toxin inhibits both rotifers and cladocerans.

Introduction

The filamentous cyanobacterium *Anabaena affinis* directly affects some zooplankton taxa but not others (Gilbert, 1990). It strongly inhibited a typical strain of *Daphnia pulex* but had no effect on another, smaller-sized strain of the species with which it co-occurred in Star Lake (Norwich, Vermont). The *A. affinis* also had no effect on *Synchaeta pectinata* and four other planktonic rotifers. Experiments testing the effect of aqueous extracts of *A. affinis* on the sensitive strain of *D. pulex* and on *S. pectinata* and two of the other rotifers, and experiments comparing the survivorship of this *D. pulex* fed only *A. affinis* or no food, indicated that the cyanobacterium contained a factor toxic to the *D. pulex* but not to the rotifers. No such experiments were conducted to determine the sensitivity of the Star Lake strain of *D. pulex* to a potentially toxic factor in *A. affinis*.

The different effect of *A. affinis* on the two strains of *D. pulex* and on the rotifers could be due, at least in part, to the efficiency at which these taxa ingest the filaments of this cyanobacterium. The straight filaments are long, with a mean length of 215–427 μm, and are surrounded by a mucilaginous sheath, giving them a diameter of 19–23 μm (Gilbert, 1990). Thus, they should be most efficiently ingested by the typical, large-sized *D. pulex*, less efficiently ingested by the smaller Star Lake *D. pulex*, and not eaten at all by the rotifers. Radiotracer feeding experiments using ^{14}C-labeled *Cryptomonas* and ^{32}P-labeled *A. affinis* demonstrated just such a pattern (Kirk & Gilbert, 1992).

Since the above-mentioned studies were completed, Namikoshi et al. (1993) found that methanol extracts of *A. affinis* contained a factor toxic to murine leukemia cells, and then purified and identified two pyrrolo[3,2-*d*]pyrimidine derivatives responsible for the toxicity. Compound 1 (9-deazaadenosine 5'-α-D-glucopyranoside) and compound 2 (9-deazaadenosine) comprised 0.25 and 0.028% of the dried cell weight, respectively. The availability of these pure compounds made it possible to determine whether the differential susceptibility of various zooplankton taxa to living *A. affinis* is due to their sensitivity to these compounds. The present study compares the sensitivity of the typical *D. pulex*, the Star Lake *D. pulex* and *S. pectinata* to the two nucleosides. It shows that the typical *D. pulex*, which is inhibited by live *A. affinis*, also is inhibited by the nucleosides, while *S. pectinata*, which is not affected by live *A. affinis*, is not inhibited by the nucleosides. Most surprisingly, the study shows that the Star Lake *D. pulex*, which is not affected by the cyanobacterium, is just as sensitive to the nucleosides as the typical *D. pulex*.

Materials and methods

Clones of the typical form of *D. pulex*, with an adult body length of about 2.6 mm, were derived from an isolate from Post Pond (Lyme, NH) and from an individual in a culture from Carolina Biological Supply Company (Burlington, NC). Two clones of the small form of *D. pulex* from Star Lake, with an adult body length of about 1.5 mm, were derived from isolates collected several years apart. A clone of *S. pectinata* was derived from an isolate collected near Hanover, NH. All of these taxa were cultured in filtered (0.45 μm) lake water on *Cryptomonas* sp. (*Daphnia*) or *C. erosa* var. *reflexa* (*Synchaeta*) after Gilbert (1990).

Anabaena affinis (strain VS-1) was isolated from Star Lake, cultured in Woods Hole MBL medium, and harvested for experimentation as described elsewhere (Gilbert, 1990). A subculture sent to W.W. Carmichael was used as the source for the material from which Namikoshi et al. (1993) purified and identified the nucleoside toxins. Samples of compounds 1 and 2 were kindly sent to me by R.S. Klein and M. Namikoshi. A weighed amount of each compound was dissolved in distilled water and serially diluted for assay.

Life-table experiments testing the effect of *A. affinis* on *D. pulex* were conducted after Gilbert (1990). Details of each experiment are shown in Table 1. Life-table experiments testing the effect of the two *A. affinis* nucleosides on *D. pulex* and *S. pectinata* were conducted at 19°C (L:D, 16:8) using procedures

of Gilbert (1990) and Gilbert (1994), respectively. *D. pulex* neonates were batch-cultured individually for 3 days in 50-ml volumes containing *Cryptomonas* sp. (2×10^4 cells ml^{-1}) and a trace of cetyl alcohol to prevent animals from adhering to the surface film. Cohort size was 21–24. Experiments with the two *D. pulex* clones were conducted within a week of each other, and those comparing compounds 1 and 2 for a given clone were set up within 2 days of each other. Survivorship at day 3 was used to calculate LC$_{50}$ values (SAS Probit procedure). *S. pectinata* neonates (3–6 h of age) were batch cultured individually for 3 days in 1 5-ml volumes containing *C. erosa* (2×10^4 cells ml^{-1}). Cohort size was 15. Experiments comparing compounds 1 and 2 were set up within 1 day of each other.

Results and discussion

A. affinis severely inhibits, and is toxic to, the typical form of *D. pulex*. This was demonstrated in the present study with a clone of *D. pulex* from Post Pond (Table 1). Animals given the cyanobacterium with *Cryptomonas* had a much lower survivorship and fecundity than those fed only *Cryptomonas* (experiment 1), and those fed the cyanobacterium alone had a much lower survivorship than those given no food (experiment 3). Comparable experiments conducted previously with another clone of the typical form of *D. pulex* (Gilbert, 1990) gave similar results. In addition, Gilbert (unpublished) found that two other clones of the typical form of *D. pulex* had much lower survivorship and fecundity with the cyanobacterium and *Cryptomonas* than with *Cryptomonas* alone. One of these clones was from Storrs Pond (Hanover, NH), and the other one, used in the present study in tests with the *A. affinis* nucleosides, was from the Carolina Biological Supply Company.

In stark contrast to all four clones of the typical form of *D. pulex*, none of the several isolates of the Star Lake strain of *D. pulex* that have been cloned are inhibited by *A. affinis*. The first clone tested survived and reproduced just as well with *Cryptomonas* and *A. affinis* as with *Cryptomonas* alone. This is shown in Table 1 (experiment 2) and more extensively in Gilbert (1990). Since then, two other clones (Star Lake 2 and 3) have been shown to respond in a similar way (S.C. Fradkin, unpublished; Gilbert, unpublished). In addition, the Star Lake *D. pulex* is able to utilize *A. affinis* as food. This is demonstrated by the fact that animals given only *A. affinis* survived and reproduced

Table 1. Effect of *Anabaena affinis*, with or without *Cryptomonas* sp., on the survivorship and fecundity (offspring per female) of *Daphnia pulex*

Experiment	Clone	Age at day 0	Concentration in culture	Cohort size	Cryptomonas	Anabaena	Day x	l_x	Fecundity
			Daphnia		Food (cells ml^{-1})		Results when all in cohort dead or at end of experiment (day x)		
1	PP	0–1d	1 in 100 ml	6	10^4	0	18	1.00*	68.2
				6	10^4	5×10^3	9	0	1.3
2	SL 1	0–1d	1 in 40 ml	6	7.5×10^3	0	20	1.00	17.7
				6	7.5×10^3	5×10^3	20	0.83	17.0
3	PP	4–5d	2-3 in 50 ml	14	0	0	9	0*	0.7
				14	0	10^4	3	0	0
4	SL 2	5–6d	1 in 50 ml	14	0	0	14	0.14*	0.1
				15	0	5×10^4	14	1.00	3.5

All experiments at 20°C (L:D, 16:8). Symbols: PP, Post Pond; SL 1 and 2, Star Lake clones 1 and 2; l_x, survivorship at day x; asterisk, significant difference ($P < 0.002$) in survivorship of cohorts with and without *Anabaena* as determined by logrank and Wilcoxon tests (SAS Proc Lifetest). Experiment 2 from Gilbert (1990).

Table 2. Effect of compound 1 (9-deazaadenosine 5′-α-D-glucopyranoside) and compound 2 (9-deazaadenosine) from *Anabaena affinis* on the survival of *Daphnia pulex* (Carolina and Star Lake 2 clones) and *Synchaeta pectinata* at 19°C (L:D 16:8)

Compound	Dose ($\mu g\ ml^{-1}$)	Daphnia (Carolina)	Daphnia (Star Lake 2)	Synchaeta
		Survival at day 3		
1	0	1.0	0.89	0.87
	0.14	1.0	0.86	1.0
	0.36	1.0	0.77	1.0
	0.71	0.93	0.48	0.87
	1.42	0.48	0.10	1.0
	3.56	0	0	0.93
2	0	1.0	1.0	1.0
	0.22	1.0	1.0	1.0
	0.55	0.73	0.33	1.0
	1.10	0	0	1.0
	2.20	0	0	1.0

See text for experimental details.

Table 3. Three-day LC_{50} values (95% confidence limits) in $\mu g\ ml^{-1}$ for compounds 1 and 2 from *Anabaena affinis* on two clones of *Daphnia pulex* at 19°C (L:D, 16:8) (see Table 2)

Clone	Compound 1	Compound 2
Carolina	1.34 (1.11–1.64)	0.56
Star Lake 2	0.79 (0.62–1.02)	0.54

The Carolina clone is inhibited by the *Anabaena*; the Star Lake clone is not.

much better than those that were starved (Table 1, experiment 4). This fact, and the ability of the Star Lake *D. pulex* to ingest isotopically labeled *A. affinis* (Kirk & Gilbert, 1992), demonstrate that the resistance of the Star Lake clone to live *A. affinis* cannot be due to a failure to eat its filaments.

Surprisingly, both the typical form of *D. pulex* (Carolina clone) and the Star Lake *D. pulex* were strongly inhibited by the purified nucleosides from *A. affinis*. The effect of the concentration of compounds 1 and 2 on survivorship is shown in Table 2, and the LC_{50} values calculated for these compounds are presented in Table 3. Confidence intervals could not be calculated for the LC_{50} value for compound 2, because too few concentrations were tested. The Star Lake clone was significantly more sensitive to compound 1 than the Carolina clone, as judged by the non-overlap of the 95% confidence limits, but similarly sensitive to compound 2. Both clones were more inhibited by compound 2 than by compound 1.

The LC_{50} values reported here for *D. pulex* are similar to the 2-day LC_{50} values obtained by Namikoshi et al. (1993) with *Ceriodaphnia dubia*: 0.5 and 0.3 $\mu g\ ml^{-1}$ for compounds 1 and 2, respectively. *C. dubia* appears to be somewhat more sensitive to the compounds and, like *D. pulex*, is more inhibited by compound 2 than by compound 1. *C. dubia* was previ-

Table 4. Effect of compounds 1 and 2 from *Anabaena affinis* (see Table 2) on the fecundity of *Synchaeta pectinata* at 19°C (L:D, 16:8)

Compound	Dose (μg ml^{-1})	Fecundity	ANOVA	
			F	*P*
1	0	7.4 (1.6)	1.69	0.145
	0.14	7.7 (1.8)		
	0.36	8.5 (1.7)		
	0.71	8.0 (2.9)		
	1.42	7.8 (1.6)		
	3.56	6.4 (2.4)		
2	0	5.8 (3.4)	3.93	0.003
	0.11	8.1 (1.6)*		
	0.22	7.4 (1.5)		
	0.55	7.9 (1.6)*		
	1.10	9.3 (1.8)*		
	2.20	8.3 (2.6)*		

Fecundity is mean number of eggs per female after 3 days of life (1 SD). Asterisk indicates significant ($P < 0.05$) effect of compound compared to control (Fisher's PLSD test). Cohort size 15.

ously found to be strongly inhibited by live *A. affinis* (Gilbert, 1990).

The sensitivity of the Star Lake *D. pulex* to the *A. affinis* nucleosides indicates that its ability to tolerate and benefit from live *A. affinis* cannot be attributed to a detoxification of the nucleosides following their assimilation. Instead, a likely hypothesis is that the resistance of the *Daphnia* to the *A. affinis* is due to the detoxification of ingested toxins in the intestine. If this were not the case, intact toxins should come out with the feces and then affect the *Daphnia*. Thus, the coexistence of the Star Lake *D. pulex* with *A. affinis* may have caused this daphniid to evolve an ability to enzymatically break down the nucleosides of the cyanobacterium in its intestine, enabling it to utilize the cyanobacterium as a food source.

The rotifer *Synchaeta pectinata* was not inhibited by the nucleosides of *A. affinis*. High concentrations of compound 1 (3.6 μg ml^{-1}) and compound 2 (2.2 μg ml^{-1}) reduced neither survivorship (Table 2) nor fecundity (Table 4). The fecundity of this rotifer at the three highest concentrations of compound 2 was significantly higher than it was in the absence of the compound. However, this was due to an unexpectedly low fecundity in this control, as the fecundity in the control for compound 1 was appreciably higher. The sensitivity of other rotifers to the *A. affinis* nucleosides

was not determined due to the limited availability of the pure compounds.

The observed resistance of *S. pectinata* to the *A. affinis* nucleosides is somewhat surprising, since selective pressure for the evolution of resistance should have been minimal. *S. pectinata* is very unlikely to ingest any of the large filaments of *A. affinis*, because it is highly selective for flagellated algae (Gilbert & Bogdan, 1984; Pourriot, 1965). Thus, it should encounter the nucleosides only if they were free in the environment as a result of the lysis of *A. affinis* cells. The mechanism for the resistance of *S. pectinata* to the nucleosides is not known. While the rotifer may be able to absorb the nucleosides from its digestive tract, it may not be able to assimilate them across its surface membranes. Thus, if *S. pectinata* did ingest *A. affinis*, it would not be inhibited by it. Alternatively, *S. pectinata* may not metabolize the nucleosides into more toxic compounds. Finally, the rotifer may have no receptor for the compounds.

The results of this study suggest that blooms of living or decomposing *A. affinis* in natural communities should not directly inhibit *S. pectinata* or any other rotifers insensitive to the nucleosides. On the other hand, they may directly inhibit cladocerans. First, any cladoceran that can ingest live filaments, and also has not specifically evolved a chemical defense against the nucleosides (like the Star Lake *D. pulex*), should be severely inhibited by the cyanobacterium. This has been demonstrated by laboratory experiments with *Ceriodaphnia dubia*, *Daphnia galeata mendotae* and several clones of the typical form of *D. pulex* (Gilbert, 1990; this study). Second, all cladocerans that are sensitive to the nucleosides (including any like the Star Lake *D. pulex* able to detoxify the compounds in the intestine), whether or not they can ingest live cells, could be inhibited by decomposing *A. affinis*. While live *A. affinis* does not seem to release any toxin to the environment (Gilbert, 1990), dead cells may liberate the nucleosides during lysis.

While *S. pectinata* is unaffected by the nucleosides of *A. affinis*, it and other rotifers may be very sensitive to other toxic cyanobacteria. For example, reproduction in *Asplanchna girodi*, *Brachionus calyciflorus*, *Keratella cochlearis* and *S. pectinata* is inhibited by *Anabaena flos-aquae* and by its alkaloid neurotoxin, anatoxin-a (Gilbert, 1994). *S. pectinata* was the most sensitive of these rotifers to anatoxin-a, being severely inhibited by a concentration of 0.2 μg ml^{-1}, and was at least as sensitive to this toxin as *Daphnia pulex*.

In conclusion, the effect of toxic cyanobacteria on natural zooplankton communities can be expected to vary with the species of the cyanobacterium and with the amount and type(s) of toxin it contains and may release into the environment when alive or lysed. The effects on various zooplankton taxa depend on the ability of the animals to ingest live cells, and on the sensitivity of the animals to ingested or free toxins. *Anabaena affinis* and *A. flos-aquae* could have very different effects. *A. affinis*, with its nucleosides, is likely to be eaten only by cladocerans (Kirk & Gilbert, 1992), and should directly inhibit all cladocerans that ingest it (except those with an evolved defense) or are exposed to its nucleosides after cell lysis (including those with an evolved defense). Even when the cyanobacterium is very inefficiently ingested, it still can strongly inhibit cladocerans. For example, in mixtures of *A. affinis* and *Cryptomonas*, *Ceriodaphnia dubia* eats very little of the cyanobacterium (α for *Cryptomonas* = 0.99; Kirk & Gilbert, 1992) but is greatly suppressed by the cyanobacterium (Gilbert, 1990). *A. affinis* should not inhibit *Synchaeta pectinata* or other rotifers that are unaffected by its nucleosides. In contrast, a strain of *A. flos-aquae* containing anatoxin-a has smaller filaments that can be ingested by rotifers as well as cladocerans, and the anatoxin-a inhibits all taxa that have been exposed to it (Gilbert, 1994). Accordingly, it should inhibit a wide variety of cladocerans and rotifers able to eat it, and perhaps most if not all zooplankton taxa that become exposed to the free alkaloid after cell lysis.

Acknowledgments

I am indebted to Robert S. Klein for sending a pure sample of one of the nucleosides from *Anabaena affinis*, and to Michio Namikoshi for sending pure

samples of both nucleosides from this cyanobacterium and for suggesting interpretations for the results of my experiments with these compounds. I am most grateful to Russell Shiel and Terry Hillman for providing me with office space at the Murray-Darling Freshwater Research Centre, Albury, New South Wales, where this paper was written. I thank an anonymous referee for comments that improved the manuscript. This study was supported by U.S. Environmental Protection Agency research grant R81-9351.

References

Gilbert, J. J., 1990. Differential effects of *Anabaena affinis* on cladocerans and rotifers: mechanisms and implications. Ecology 71: 1727–1740.
Gilbert, J. J., 1994. Susceptibility of planktonic rotifers to a toxic strain of *Anabaena flos-aquae*. Limnol. Oceanogr. 39: 1286–1297.
Gilbert, J. J. & K. G. Bogdan, 1984. Rotifer grazing: in situ studies on selectivity and rates. In Meyers, D. G. & J. R. Strickler (eds), Trophic Interactions within Aquatic Ecosystems. American Association for the Advancement of Science Symposia, Vol. 85. Westview Press, Boulder, CO: 97–133.
Kirk, K. L. & J. J. Gilbert, 1992. Variation in herbivore response to chemical defenses:zooplankton foraging on toxic cyanobacteria. Ecology 73: 2208–2217.
Namikoshi, M., W. W. Carmichael, R. Sakai, E. A. Jares-Erijman, A. M. Kaup & K. L. Rinehart, 1993. 9-Deazaadenosine and its 5'-α-D-glucopyranoside isolated from the cyanobacterium *Anabaena affinis* strain VS-1. J. am. Chem. Soc. 115: 2504–2505.
Pourriot, R., 1965. Recherches sur l'écologie des rotifères. Vie Milieu Suppl. 21: 1–224.

Hydrobiologia **387/388**: 283–287, 1998.
E. Wurdak, R. Wallace & H. Segers (eds), Rotifera VIII: A Comparative Approach.
© *1998 Kluwer Academic Publishers.*

Toxicity of the Chrysophyte flagellate *Poterioochromonas malhamensis* to the rotifer *Brachionus angularis*

Joseph E. Boxhorn[1],*, Dale A. Holen[2] & Martin E. Boraas[1]
[1]*Department of Biological Sciences, University of Wisconsin-Milwaukee, Milwaukee, WI 53201, U.S.A.*
(*author for correspondence; E-mail: j.boxhorn@csd.uwm.edu)*
[2]*The Pennsylvania State University-Worthington Scranton Campus, Dunmore, PA 18512, U.S.A.*

Key words: Brachionus angularis, Poterioochromonas malhamensis, rotifers, toxicity, rotifer distribution

Abstract

The toxicity of the chrysophyte flagellate *Poterioochromonas malhamensis* to the rotifer *Brachionus angularis* was investigated. Fed rotifers exposed to the flagellate experienced a mortality rate indistinguishable from starvation. Unfed rotifers exposed to the flagellate experienced a higher mortality rate. The mortality rate appears to depend on the flagellate concentration. Higher doses of flagellates resulted in quicker rotifer death. These laboratory results are consistent with the hypothesis that the occurrence of *B. angularis* in the field may be negatively related to the presence of *P. malhamensis* and related flagellates.

Introduction

While there has been much study of the effects of toxic cyanobacteria on rotifers (Fulton & Paerl, 1987; Rothhaupt, 1991; Smith & Gilbert 1995; Gilbert, 1994, 1996a, b), there has been little study of the effects of toxic chrysophytes. Chrysophytes in the genera *Poterioochromonas* and *Ochromonas* have been known since the mid-1960s to produce toxins (Reich & Speigelstein, 1964; Speigelstein et al., 1967; Halevy & Avivi, 1968). Few studies have examined the effects of these toxins on zooplankton (Leeper & Porter, 1995).

In this paper we report the results of a laboratory study which evaluated the toxicity of the chrysophyte flagellate *Poterioochromonas malhamensis* (= *Ochromonas malhamensis*, see Lee et al., 1995) to the rotifer *Brachionus angularis*. Survivorship of *B. angularis* exposed to *P. malhamensis* was contrasted to survivorship of animals that were not exposed in both the presence and absence of food. In addition the effect of flagellate concentration on rotifer survivorship was examined.

Methods

Brachionus angularis was isolated from Lake Michigan at the Milwaukee harbor. Stock cultures were derived from a single individual. These were maintained in 75 ml of Sorokin and Krause medium (Sorokin & Krause, 1958) in 125-ml Erlenmeyer flasks under constant illumination. Every second day they were fed *Chlorella vulgaris* (UTEX 26) grown continuously under constant light in Sorokin and Krause medium.

Poterioochromonas malhamensis (UTEX L 1297) was obtained from the Culture Collection of Algae at the University of Texas at Austin (Starr, 1978). They were grown axenically in continuous culture under constant illumination in the organic *Ochromonas* Growth Medium (Starr, 1978). The growth medium was one-fourth the normal concentration. At the concentration used in these experiments this medium contains glucose, tryptone and yeast extract each at concentrations of 0.25 g l^{-1} in addition to 0.35 g l^{-1} of beef liver extract.

All experiments were conducted in 24-well tissue culture plates (Falcon 3047, Becton Dickinson & Co., Lincoln Park, NJ). Into each well 1.0 ml of spent Sorokin and Krause medium was pipetted. Spent medium is medium in which *Chlorella vulgaris* had been grown. We used spent medium to avoid changing the chemical environment that the animals were grown in. This was filtered first through a 3.0-μm membrane filter and then through a 0.4-μm filter to remove cells. Ten adult animals without eggs were transferred to each well. Before transfer animals with eggs were collected using a micropipette. The eggs were removed by passing the animals through a 25-gauge hypodermic needle (Boraas, 1983). To minimize evaporation from the wells, the plates were placed in a covered plastic container which had its bottom covered with water. The container was gently shaken on a shaker table. Periodically the number of live animals in each well was counted using a Wild M5 binocular microscope.

All counts of *Chlorella* and *Poterioochromonas* were made using an Elzone 280PC electronic particle counter (Particle Data, Inc., Elmhurst, Illinois).

The first experiment tested the effects of the flagellates and starvation upon the rotifers. There were four treatments. In two of these, about 22000 flagellates were added to each well at the beginning of the experiment. The other two received no flagellates. One treatment of each type of received 0.1 ml *Chlorella* suspension every 24 hours. This suspension contained 3.3×10^8 cells ml^{-1}. The other treatments received 0.1 ml of spent Sorokin and Krause medium. Each treatment was replicated 12 times.

A separate control was conducted with this experiment to determine whether *Ochromonas* medium is toxic to the rotifers. Rotifers were placed in either spent Sorokin and Krause medium or fresh *Ochromonas* medium. No food was added to the wells. There were 10 replicates of each treatment.

The second experiment tested the effect of flagellate concentration on rotifer survivorship. Flagellates were added to the wells in the following concentrations: 0, 20000, 40000, 80000 and 160000 cells ml^{-1}. In this experiment the animals were not fed. Each treatment was replicated 10 times.

Statistics were computed using SYSTAT, version 5.0 (SYSTAT, 1990). In the first experiment the data were log-transformed and analyzed using linear regression. The resulting lines were compared by testing for homogeneity of slopes. Differences among slopes were examined using the T'-method of unplanned comparisons (Sokal & Rohlf, 1981). This last test was computed manually. The second experiment was analyzed using repeated-measures analysis of variance. An α level of 0.05 was used for all tests.

Results

In the first experiment, the number of rotifers in the treatment without flagellates that received food in-

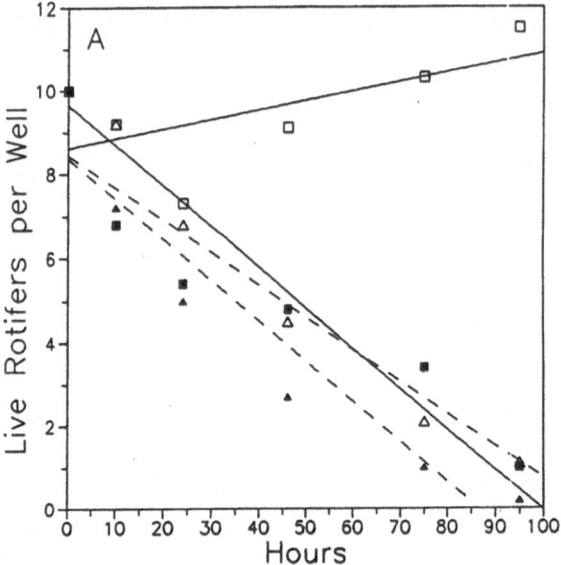

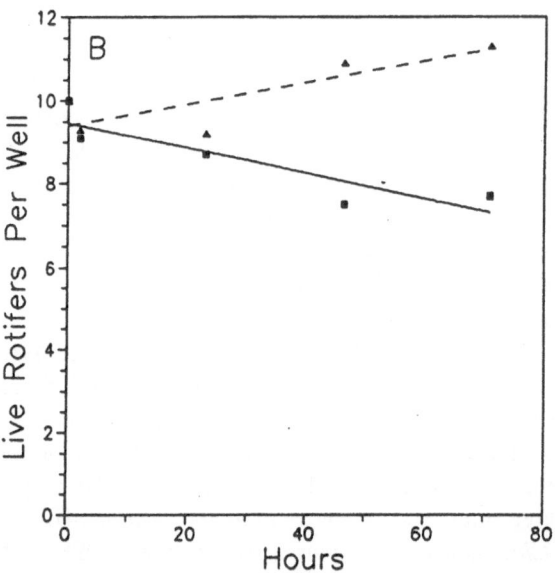

Figure 1. Brachionus angularis survivorship over time. A. Experiment testing the effects of *Poterioochromonas malhamensis* and feeding. Closed symbols and dotted lines indicate treatments with flagellates at ca. 2×10^4 cells ml^{-1}, open symbols and solid lines indicate treatments without flagellates. Squares are treatments that received food, triangles are treatments that were not fed. Each point is the mean of 12 replicates. B. Effect of culture medium on the survivorship of *Brachionus angularis*. Dashed line is fresh *Ochromonas* medium. The continuous line is spent Sorokin and Krause medium. Each point is the mean of 10 replicates.

Table 1. Repeated-measures ANOVA for effect of *Poterioochromonas malhamensis* density on *Brachionus angularis*

Source	SS	df	MS	F	P
Between subjects					
Density	473.464	4	118.366	41.387	< 0.001
Error	28.700	45	2.860		
Within subjects					
Time	1586.824	4	396.706	326.955	< 0.001
$T \times D$	37.976	16	8.624	7.107	< 0.001
Error	218.400	180	1.213		

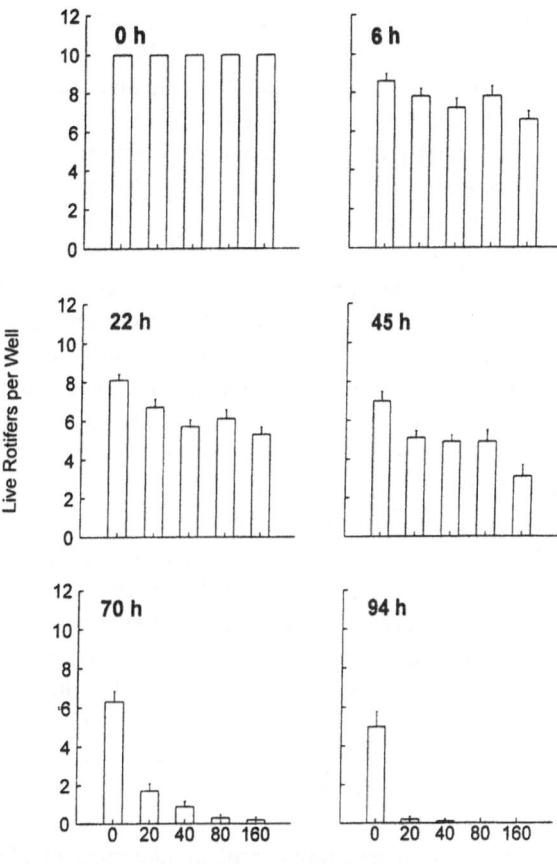

Figure 2. The effect of initial *Poterioochromonas malhamensis* density on the survivorship of starved *Brachionus angularis* over time. Each bar is the mean of 10 replicates. Error bars represent one standard error.

creased slightly. Rotifer numbers declined in each of the other treatments. Mortality of *B. angularis* occurred more rapidly in the presence of *P. malhamensis* (Figure 1A). This was seen both when the rotifers were starved and when they were fed. The most rapid decline in rotifer numbers occurred in the treatment with *P. malhamensis* that did not receive food. Tests for homogeneity of slopes showed that the slopes of these four lines were not equal ($F = 1066.02$; df = 8,277; $P < 0.0001$). Unplanned comparisons showed that, with one exception, all pairs of slopes were significantly different from each other. No difference in slopes was seen between the mortality of starved rotifers in the absence of flagellates and the mortality of fed rotifers in the presence of flagellates. Starving rotifers exposed to the flagellates experienced more rapid mortality than rotifers in either of these treatments.

Unfed rotifer survivorship was higher in fresh *Ochromonas* medium than in spent Sorokin and Krause medium (Figure 1B). Reproduction occurred in 5 of the *Ochromonas*-medium replicates.

Figure 2 shows the effect of flagellate concentration on survivorship of unfed rotifers. The rate of rotifer mortality increased with increasing initial flagellate concentration. By 70 hours a single live rotifer was found in each of three wells in the 80000 flagellate ml^{-1} treatment and in one replicate in the 160000 flagellate ml^{-1} treatment. By 94 hours there were no live rotifers in any replicates in either of these treatments. Repeated-measures ANOVA shows significant differences between the effects of different flagellate densities and significant difference by time within treatments, as well as a significant interaction between time and flagellate density (Table 1).

Discussion

B. angularis experienced increased mortality in the presence of *P. malhamensis* (Figure 1A). Fed rotifers exposed to the flagellate experienced mortality at a rate that was indistinguishable from starvation. Starving rotifers experienced a higher rate of mortality. The rate of mortality increased with increasing concentration of the flagellate (Figure 2). The growth of the rotifers in *Ochromonas* medium rules out the possibility that this mortality is due to the medium (Figure 1B). These data indicate that *P. malhamensis* is toxic to *B. angularis*.

Previous studies have shown that *P. malhamensis* produces substances that have a toxic effect on fish (Reich & Speigelstein, 1964; Halevy et al., 1971; Magazanik & Halevy, 1973), induce mortality in *Daphnia* (Leeper & Porter, 1995), lyse mammalian erythrocytes (Speigelstein et al., 1969) and inhibit bacterial growth (Hansen, 1973). The toxins have only been partially characterized. There appear to be at least two toxins (Speigelstein et al., 1969). They are soluble in water and in polar solvents and stain red with rhodamine 6G (Halevy et al., 1971).

The toxicity of these flagellates seems to be dependent upon growth conditions. *P. malhamensis* appears not to produce toxin when it is growing autotrophically. Leeper & Porter (1995) found that *Daphnia ambigua* was able to grow and reproduce while growing on *P. malhamensis* which was cultured phototrophically in inorganic medium. *P. malhamensis* which was grown heterotrophically on organic medium was toxic to *D. ambigua*. It is not known whether the flagellate produces toxin when it is growing phagotrophically. Phototrophic growth in *P. malhamensis* is considerably slower than heterotrophic growth on either organic medium or bacteria. Holen & Boraas (1995) found that growth rates for *P. malhamensis* grown phototrophically on defined inorganic medium, osmotrophically on organic medium and phagotrophically on *Escherichia coli* were 0.068, 1.3 and 2.1 d^{-1}, respectively. If the lack of toxin production in phototrophically growing cells is due to metabolic differences associated with autotrophy or to the relative lack of biomass assimilation during autotrophic growth versus heterotrophic growth, phagotrophically growing flagellates might very well produce toxins at levels similar to or exceeding those produced during osmotrophic growth.

It appears that *P. malhamensis* can excrete the toxin into the medium. Halevy et al. (1971) found that spent medium in which the flagellate was grown was toxic to the fish *Danio malabaricus*. They also found that rhodamine turns 'spent' medium red. Hansen (1973) reported that medium from 5-day-old *P. malhamensis* cultures added 1:1 to bacterial medium stopped bacterial growth. Since the flagellates had not lysed, he concluded that toxin was excreted into the medium. This excretion suggests that these toxins may affect aquatic organisms even if the cells are not ingested. Similar toxicity has been found in other Ochromonads including *Ochromonas danica*, *O. minuta*, *O. sociabilis* and an unidentified species of *Ochromonas* (Reich & Speigelstein, 1964; Halevy & Avivi, 1968).

Starving *B. angularis* experience higher mortality when exposed to toxic flagellates than fed *B. angularis*. This is consistent with the idea that low food rations may increase zooplankton sensitivity to toxins. A number of studies show that this is the case for at least some rotifer species. Rothhaupt (1991) found that *Brachionus rubens* fed only a toxic, unicellular strain of

the cyanobacterium *Microcystis aeruginosa* died more quickly than unfed controls, whereas rotifers cultured with a combination of the green alga *Monoraphidium minutum* and *Microcystis* had survival comparable to the starving controls. Rao & Sarma (1986) reported a significant interaction between the effects of food concentration and DDT concentration on the population growth rate of *Brachionus patulus*. Higher levels of food appeared to mitigate DDT toxicity at high DDT levels. Using chronic toxicity tests, Gilbert (1996) examined the effect of food availability on the toxicity of an anatoxin-a producing strain of *Anabaena flosaquae* to the rotifers *Synchaeta pectinata* and *Brachionus calyciflorus*. He found that the cyanobacterium caused a greater decrease in both the life span and fecundity of *S. pectinata* at lower food concentrations. This sort of interaction was seen only for fecundity in just one out of two clones of *B. calyciflorus* that were tested. An alternate explanation for our results is that starving *B. angularis* exposed to *P. malhamensis* ingested more of the flagellates than did fed *B. angularis*. Our data cannot distinguish between these two possibilities because we did not determine whether *B. angularis* ingested the flagellates.

The reproduction of *B. angularis* in *Ochromonas* medium was unexpected (Figure 1B). *Ochromonas* medium is an organically rich medium. It is possible that *B. angularis* may be able to take up and utilize dissolved organics at these concentrations. More likely, the rotifers in the organic medium control were consuming bacteria which were growing on the medium.

Toxins produced by *P. malhamensis* and flagellates in the genus *Ochromonas* may be a factor affecting the distribution of rotifer species. These chrysophytes are common and often abundant in lakes. Bird & Kalff (1989) reported *Ochromonas* concentrations exceeding 3000 cells ml^{-1} in the metalimnion of Lac Gilbert, Quebec. Over a 24-hour period, Abbot's Pool in Somerset, England had a mean concentration of 1139 *Ochromonas* cells ml^{-1} (Happey-Wood, 1976), with peak estimates just under 2500 cells ml^{-1}. Rosén (1981) presented results of phytoplankton analysis from surveys of 1250 lakes in Sweden. He reported concentrations as mm^3 algal volume l^{-1}. All species of the genus *Ochromonas* are pooled in his report. If a diameter of 10 μm is taken for the size of a typical ochromonad flagellate and if the shape of the cell is approximated as a sphere, very rough estimates of the number of cells ml^{-1} can be computed from Rosén's Table 4. We estimate that Rosén found one lake with about 3800–9500 cells ml^{-1}, two lakes with about 960–1900 cells ml^{-1}, fifteen lakes with about 570–960 cells ml^{-1} and eight lakes with about 380–570 cells ml^{-1}. We consider these numbers conservative as ochromonad cells have been reported as being smaller. *P. malhamensis* cells are 8–10 μm in diameter (Holen, 1994). Salonen & Jokinen (1988) reported the length of *Ochromonas* collected from Mekkojärvi in Finland as being 5–7.5 μm.

These examples suggest that densities of *Ochromonas* and *Poterioochromonas* may achieve toxic levels in lakes. Some data have been published which may suggest that this happens. Baker (1979) reported a negative correlation between the concentration of small flagellates and the birth rates of *B. angularis* in Hastings Lake in Alberta, Canada. The higher ochromonad densities are consistent with toxic densities observed in this study.

It is important to note that the densities of *P. malhamensis* used is this study are higher than the densities of ochromonads that have been reported in lakes. The lowest density we used, 20000 cells ml^{-1}, is about twice our estimate of the highest density reported. It is unknown how representative the densities reported above are, as small flagellates are often not identified to family or genus level.

Acknowledgements

The authors would like to thank Robert Wallace for confirming the identification of the rotifer species used, Tom Slawski for help with SYSTAT and John Gilbert and an anonymous reviewer for comments on the manuscript.

References

Baker, R. L., 1979. Birth rate of planktonic rotifers in relation to food concentration in a shallow, eutrophic lake in western Canada. Can. J. Zool. 57: 1206–1214.

Boraas, M. E., 1983. Population dynamics of food-limited rotifers in two-stage chemostat culture. Limnol. Oceanogr. 28: 546–563.

Bird, D. F. & J. Kalff, 1989. Phagotrophic sustenance of a metalimnetic phytoplankton peak. Limnol. Oceanogr. 34: 155–161.

Fulton, R. S., III & H. W. Paerl, 1987. Toxic and inhibitory effects of the blue-green alga *Microcystis aeruginosa* on herbivorous zooplankton. J. Plankton Res. 9: 837–855.

Gilbert, J. J., 1994. Susceptibility of planktonic rotifers to a toxic strain of *Anabaena flos-aquae*. Limnol. Oceanogr. 39: 1286–1297.

Gilbert, J. J., 1996a. Effect of food availability on the response of planktonic rotifers to a toxic strain of the cyanobacterium *Anabaena flos-aquae*. Limnol. Oceanogr. 41: 1565–1572.

Gilbert, J. J., 1996b. Effect of temperature on the response of planktonic rotifers to a toxic cyanobacterium. Ecology 74: 1174–1180.

Halevy, S. & L. Avivi, 1968. Isolation of hemolysins from *Ochromonas* spp., *Pymnesium parvum* and *Trypanosoma ranarum*. J. Protozool. 15 (Suppl.): 45.

Halevy, S., R. Saliternik & L. Avivi, 1971. Isolation of rhodamine-positive toxins from *Ochromonas* and other algae. Int. J. Biochem. 2: 185–192.

Hansen, J. A, 1973. Antibiotic activity of the chrysophyte *Ochromonas malhamensis*. Physiol. Pl. 29: 234–238.

Happey-Wood, C, M., 1976. Vertical migration patterns in phytoplankton of mixed species composition. Br. phycol. J. 11: 355–369.

Holen, D. A. & M. E. Boraas, 1995. Mixotrophy in chrysophytes. In C. D. Sandgren, J. P. Smol & J. Kristiansen (eds), Chrysophyte Algae: Ecology, Phylogeny and Development. Cambridge University Press, Cambridge: 119–140.

Lee, J. L., S. H. Hutner & E. C. Bovee, 1985. An Illustrated Guide to the Protozoa. Allen Press, Inc., Lawrence, Kansas.

Leeper, D. A. & K. G. Porter, 1995. Toxicity of the mixotrophic chrysophyte *Poterioochromonas malhamensis* to the cladoceran *Daphnia ambigua*. Arch. Hydrobiol. 134: 207–222.

Magazanik, A. & S. Halevy, 1973. Some characteristics of *Ochromonas* hemolysins. Experientia 29: 310–311.

Rao, T. R. & S. S. S. Sarma, 1986. Demographic parameters of *Brachionus patulus* Muller (Rotifera) exposed to sublethal DDT concentrations at low and high food levels. Hydrobiologia 139: 193–200.

Reich, K. & M. Spiegelstein, 1964. Fish toxins in *Ochromonas* (Chrysomonadinia). Israel J. Zool. 13: 141.

Rosén, G., 1981. Phytoplankton indicators and their relations to certain chemical and physical factors. Limnologica 13: 263–290.

Rothhaupt, K. O., 1991. The influence of toxic and filamentous blue-green algae on feeding and population growth of the rotifer *Brachionus rubens*. Int. Rev. ges. Hydrobiol. 76: 67–72.

Salonen, K. & S. Jokinen, 1988. Flagellate grazing on bacteria in a small dystrophic lake. Hydrobiologia 161: 203–209.

Smith, A. D. & J. J. Gilbert, 1995. Relative susceptibilities or rotifers and cladocerans to *Microcystis aeruginosa*. Arch. Hydrobiol. 132: 309–336.

Sokal, R. R. & F. J. Rohlf, 1981. Biometry, 2nd edn. W. H. Freeman, New York, 859 pp.

Sorokin, C. & R. W. Krause, 1958. The effects of light intensity on the growth rates of green algae. Plant Physiol. 33: 109–113.

Spiegelstein, M., K. Reich & F. Bergmann, 1967. Toxin in *Ochromonas* cultures. J. Protozool. 14 (Suppl.): 44.

Spiegelstein, M., K. Reich & F. Bergmann, 1969. The toxic principals of *Ochromonas* and related chrysomonadinia. Verh. int. Ver. Limnol. 17: 778–783.

Starr, R. C., 1978. The culture collection of algae at the University of Texas at Austin. J. Phycol. 23 (Suppl.): 1–47.

Hydrobiologia **387/388**: 289–294, 1998.
E. Wurdak, R. Wallace & H. Segers (eds), Rotifera VIII: A Comparative Approach.
© *1998 Kluwer Academic Publishers.*

Influence of abiotic and biotic factors on morphological variation of *Keratella cochlearis* (Gosse) in a small Andean lake

Maria Diéguez, Beatriz Modenutti & Claudia Queimaliños
Department of Ecology, Centro Regional Universitario Bariloche, Unidad Postal Universidad, 8400 Bariloche, Argentina

Key words: Rotifera, *Keratella cochlearis*, morphological variation, food resources

Abstract

Morphological variation of *Keratella cochlearis* was studied during a spring-summer period in a small Andean lake. Morphometry was studied in relation to temperature, food resources, cladoceran competitors and invertebrate predators. Three different morphs were recorded. We observed lack of allometric growth of the posterior spine. Absence of allometric growth could be related with the low density of a predaceous water mite and the small size of the cladocerans present in the lake. Fluctuations in lorica length and width were negatively correlated with temperature and algal biovolume. We discuss the benefits of this morphological response of *Keratella* in relation to environmental conditions.

Introduction

The morphological variation in *Keratella cochlearis* (Gosse) has been associated with different environmental conditions. In particular, presence and size of the posterior spine, and lorica length and width were related to temperature and lake trophic status (Pejler, 1962; Eloranta, 1982; Hofmann, 1983; Bielañska-Grajner, 1995). Furthermore, the development of the posterior spine was studied as a defensive mechanism against invertebrate predators (Stemberger & Gilbert, 1984, 1987; Conde-Porcuna et al., 1993).

In the Neotropics, José de Paggi (1978) observed that *K. cochlearis* had a shorter caudal spine in some tributaries of Paraná river than in floodplain lakes of the same system. This change was related to the physical conditions of the environments. There is no further information on the morphological variation of this rotifer in the region. In North Patagonian Andean lakes, *K. cochlearis* is widely distributed (Diéguez & Modenutti, 1996) and details about its populations dynamics in small Andean lakes are available (Balseiro & Modenutti, 1990; Modenutti, 1994). This rotifer is an important perennial species within the simple zooplankton assemblage of these lakes. In the present study we analyse the changes in morphological parameters of *K. cochlearis* in relation to physical and biological factors in a small Andean lake.

Study site

This study was conducted in Laguna Ezquerra (41° 3′ S, 71° 30′ W), a small and shallow North Patagonian lake (3 m Z_{max}), located at 758 m a.s.l. in the Nahuel Huapi system (Rio Negro, Argentina). In winter its surface often freezes and no stratification occurs during spring and summer. The lake is oligotrophic, with a Secchi depth between 2–3 m and the dissolved oxygen concentration values are always near 100% of saturation. Conductivity ranges from 50 to 70 μS cm^{-1}, and pH from 6.9 to 7.2.

During the spring-summer period the zooplankton constituted of seven rotifers, two cladocerans, one copepod species and a predaceous water mite (Balseiro, 1992; Balseiro et al., 1992; Modenutti, 1994). The two cladoceran species were *Bosmina longirostris* (OFM) and *Ceriodaphnia dubia* (Richard), which showed different seasonal patterns of abundance. Bosmina was present during the whole spring–summer period, with densities of 100 ind l^{-1} from October to December and an increase in late February reaching 700 ind l^{-1}. *Ceriodaphnia* showed an unimodal cycle during October–December reaching 80 ind l^{-1} in late November. The copepod *Boeckella gracilipes* Daday developed during spring from copepodite 1 to adults but these remained at very low densities during summer. The water mite *Limnesia patagonica* Lundbald was present over the spring-summer period but never exceeded densities of 9 ind m^{-3}. Water mites were the only invertebrate predators in this lake considering that copepods are represented only by an herbivorous calanoid (Balseiro, 1992).

Methods

Field observation

During a six months period (October 1988 – March 1989), we measured temperature and collected phyto- and zooplankton samples every 3–4 days. Phytoplankton was collected with a Van Dorn bottle at 1–1.5 m and 2–2.5 m deep. The samples were pooled and preserved with acid Lugol's solution. In the lab, phytoplankton samples were allowed to settle for ten days in 250 ml cylinders and the concentrate (5 ml) was counted at 400× magnification in a 8.5 μl shallow chamber (thickness 0.15 mm). Algal biovolume was estimated by approximations to the appropriate geometric figures. Food resources were estimated as the algal biovolume of the fraction smaller than 20 μm GALD (Greatest Axial Linear Dimension).

Zooplankton was sampled at the same depths in 50 l samples using a Schindler-Patalas trap, filtered with a 35 μm mesh net and preserved with 4% formalin solution. Rotifers were counted in 1 ml Sedgwick-Rafter chamber under microscope. A minimum of 30 individuals of *Keratella cochlearis* from each weekly sample were measured under microscope (Olympus BH2) at 400× using a micrometer eyepiece. Four morphometric parameters were considered: total length (hereafter TL), lorica length without considering anterior and posterior spine (hereafter LL), lorica width (hereafter LW), and posterior spine length (hereafter PS).

Correlation between morphometric parameters and environmental factors

Morphometric data were analysed using correlation tests and ANOVA. We correlated *K. cochlearis* morphometry with: (1) water temperature, (2) food resources (algal biovolume of the fraction smaller than 20 μm GALD), (3) density and biomass of main competitors (*Bosmina longirostris* and *Ceriodaphnia dubia*), and (4) density of the water mite *Limnesia patagonica*.

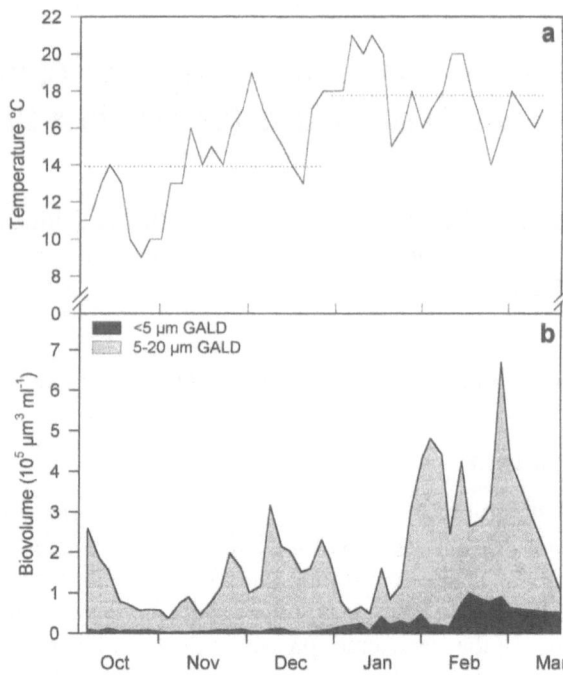

Figure 1. (a) Temperature in Laguna Ezquerra during October 1988–March 1989. Dotted lines represent mean temperature during spring and summer. (b) Biovolume of flagellate algae smaller than 20 μm GALD in Laguna Ezquerra during October 1988–March 1989.

Laboratory predation experiments

Laboratory experiments were performed in order to test if *Limnesia* was able to consume *Keratella*. Predation experiments were conducted in 50 ml Erlenmeyers flasks. Two *Keratella* densities (200 and 500 ind l^{-1}) were exposed to *L. patagonica* (20 ind l^{-1}). These densities were chosen in order to obtain reliable results, although they can not be extrapolated to field abundance of both species. The experiments were run for 4 hours, in darkness at 18°C. At the end of each exposure, we counted the number of living *Keratella*. Each experiment consisted of, at least, three replicates and three controls.

Results

Field observation

No vertical thermal stratification was observed during the study period. However there were some seasonal variations (Figure 1a). During spring temperature ranged between 11°C and 19°C (average 13.9°C), while in summer temperature ranged between 13°C and 21°C (average 17.5°C).

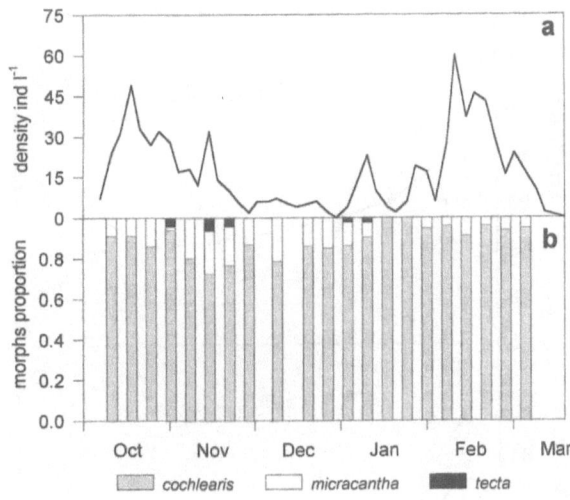

Figure 2. (a) Density of *Keratella cochlearis* in Laguna Ezquerra during October 1988–March 1989. (b) Proportion of the three morphs of *Keratella cochlearis* in Laguna Ezquerra.

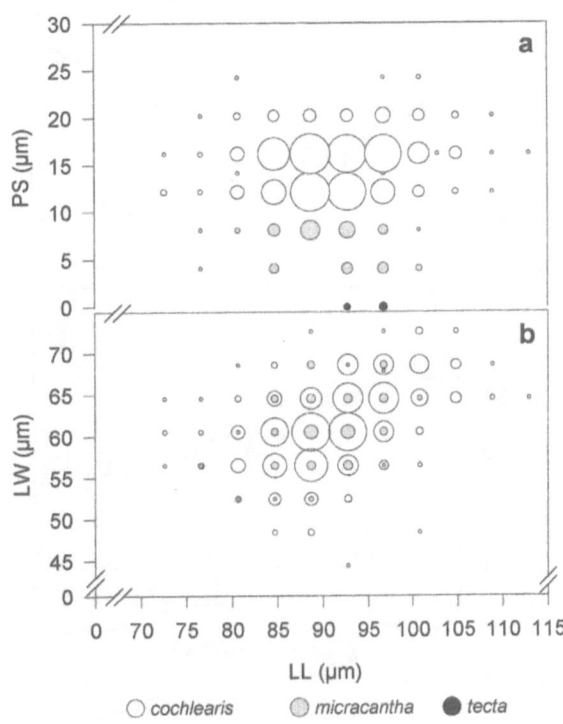

Figure 3. Relationship among lorica parameters of *Keratella cochlearis* (*cochlearis*, *micracantha* and *tecta*) in Laguna Ezquerra. (a) Lorica length vs. spine length. (b) Lorica length vs. lorica width. The area of the circles represents the frequency of the relationship.

The phytoplankton community was largely (>90%) dominated by flagellate cells under 20 μm GALD, fraction considered as the main food resource for *Keratella*. The flagellate *Rhodomonas lacustris* (Pascher & Ruttner) Javornicky predominated in the 5 to 20 μm GALD fraction, and *Chrysochromulina parva* Lackey and *Ochromonas* sp in the < 5 μm GALD fraction. Food resources, expressed as algal biovolume, showed seasonal fluctuations (Figure 1b). From October to January (spring and early summer) the biovolume fluctuated below 3×10^5 μm^3 ml^{-1}, while in mid January it decreased reaching a minimum of 0.6×10^5 μm^3 ml^{-1}. Then, during summer the biovolume steadily increased reaching 6.5×10^5 μm^3 ml^{-1} in late February (Figure 1b).

Keratella cochlearis was present during the entire study period. We observed two density peaks in its population cycle, one in spring with 50 ind l^{-1} and the other in late summer with 60 ind l^{-1} (Figure 2a). During spring the rotifer represented 99% of total rotifer density, while in summer, six other rotifer species peaked and thus *Keratella* relative importance in the community decreased (Modenutti, 1994). The minimum value of *K. cochlearis* abundance was registered in mid January (Figure 2a), after the unimodal cycle of *Ceriodaphnia dubia*.

Three morphs were distinguished: *cochlearis*, *tecta* and *micracantha*. The morph *cochlearis* was the most abundant and was present over the entire period (Figure 2b). The morph *tecta* was present in very low densities (2 ind l^{-1}) in October, November and Janu-

ary (Figure 2b). The morph *micracantha* was defined as individuals with PS length less than 10 μm and was also present over the whole studied period (Figure 2b). However, our data showed the presence of two distinctive groups of *micracantha*, one group with a PS mean length of 8 μm and another with PS mean length of 4 μm. The *micracantha* morph with the largest PS was always present whereas the morph with the shortest PS disappeared in early January.

Correlation between morphometric parameters and environmental factors

The morphs were clearly separated by the PS length (ANOVA, $P < 0.05$) (Figure 3a), nevertheless they did not differ in LL and LW (ANOVA, $P > 0.05$) (Figure 3b). No significant correlation was observed between PS and LL (Figure 3a) ($P > 0.05$), indicating an absence of allometric growth of the caudal spine. Mean values of LL and LW decreased during the study period (Figure 4a and b). On the other hand, the PS length of *cochlearis* morph decreased during spring from 16.5 μm (mean value) in October to

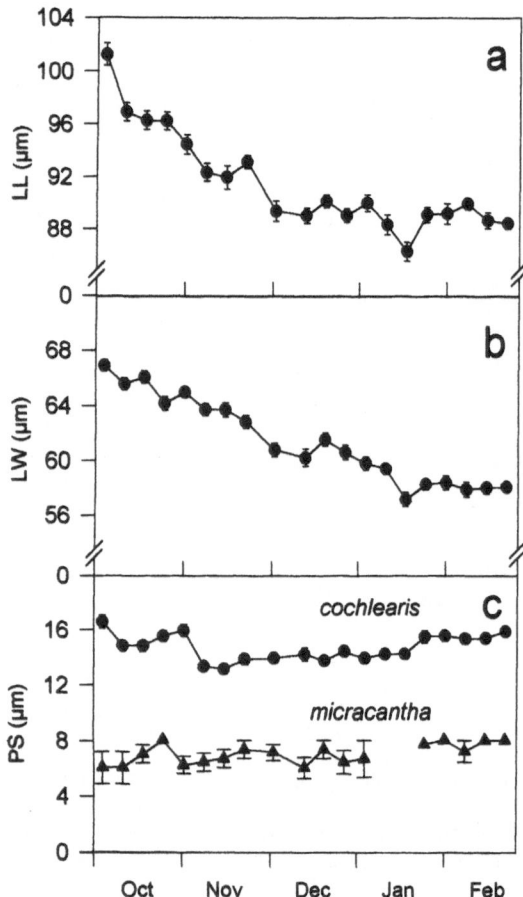

Figure 4. Temporal variation in morphometric parameters of *Keratella cochlearis* in Laguna Ezquerra during October 1988–March 1989. (a) Lorica Length. (b) Lorica Width. (c) Spine Length.

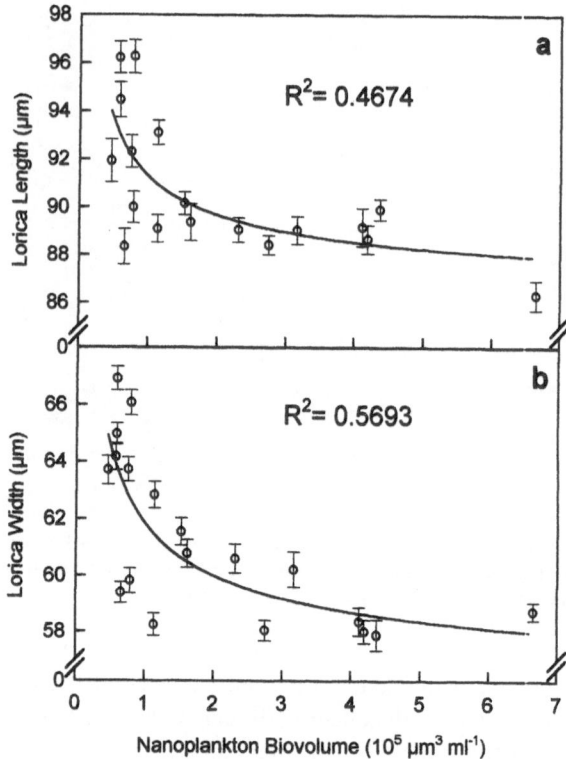

Figure 5. Relationships between lorica parameters and biovolume of flagellate algal cells. (a) Lorica Length. (b) Lorica Width.

Table 1. Number of *Keratella cochlearis* individuals dead in laboratory experiments with the predaceous water-mite *Limnesia patagonica* and controls. References: m = number of dead individuals day^{-1} (mean value); S.E. = standard error; n = replicates

	Keratella cochlearis	
	200 ind l^{-1}	500 ind l^{-1}
Predation experiments with *Limnesia patagonica*	$m = 2.08$ S.E. = ± 0.08 $n = 6$	$m = 1.75$ S.E. = ± 0.14 $n = 3$
Control without the predator	$m = 0.33$ S.E. = ± 0.14 $n = 6$	$m = 0.17$ S.E. = ± 0.08 $n = 3$

13 μm (mean value) in November (Figure 4c). Spine length increased during summer but never reached the maximum value of spring (Figure 4c). The morph *micracantha* remained with its stable mean values with a maximum of 8 μm in late October (Figure 4c).

We observed significant negative relationships between temperature and the morphometric parameters LL and LW (R^2=0.6250; R^2=0.6787 respectively; $P < 0.01$ in all cases). LL and LW also declined exponentially with the increase of nanoplankton biovolume ($P < 0.01$ in both cases) (Figure 5a and b). No relationship was found neither between PS length and temperature nor with nanoplankton biovolume ($P > 0.05$).

The main zooplankton competitors of *K. cochlearis* in Laguna Ezquerra were the cladocerans *Bosmina longirostris* and *Ceriodaphnia dubia* (Modenutti, 1994). No relationship was found between morphometric variables of *K. cochlearis* and cladoceran density or biomass $P > 0.05$).

Laboratory predation experiments

Limnesia patagonica consumed *K. cochlearis* in laboratory experiments. The water mite consumed up to 42 *Keratella* predator^{-1} day^{-1} (Table 1). Nevertheless, no significant relationship existed between morphometry of *K. cochlearis* and *Limnesia* density in the field ($P > 0.05$).

Discussion

Spined morphs of *K. cochlearis* were present during spring and summer in Laguna Ezquerra. We observed that the length of the PS was smaller and less variable than in other lakes of the world (Stemberger, 1979; Eloranta, 1982; Green, 1987; Bielañska-Grajner, 1995), being within the ranks mentioned by José de Paggi (1978) for Paraná river. Although PS decreased during spring and increased towards summer (Figure 4c), we did not observe any significant relationship between this parameter and water temperature or food resources, as many authors did (Pejler, 1962; Ruttner-Kolisko, 1974; Eloranta, 1982; Hillbricht-Ilkowska, 1983; Hofmann, 1983). In addition, PS was independent of LL (Figure 3a). This absence of allometry was previously noted in North Patagonian lakes (Diéguez & Modenutti, 1996) and this condition was also found in populations of *K. cochlearis* in Zimbabwe, Africa (Green, 1987), in the Southern Hemisphere too. The small response of the relative growth of PS in Africa could be related to the limited range of temperature (Green, 1987). However, in a temperate lake as Laguna Ezquerra, we can not assume that temperature fluctuations (Figure 1a) were narrow enough to explain the lack of allometry.

The largest individuals were observed during spring when lower temperatures were registered. In fact, we obtained a negative correlation between LL and LW, and water temperature. This agrees with the statement that low temperatures slow down growth rates resulting in larger individuals (Ruttner-Kolisko, 1974; Dumont, 1977).

Lorica size was closely related to food resources. The significant correlations obtained between LL and LW and biovolume of flagellate cells suggest that (Figure 5a and b). During spring, under low food concentration, we found the larger specimens (Figure 1b and 4a and b), while the smaller ones were observed during summer with a higher algal biovolume (Figure 1b and 4a and b). Assuming Bogdan & Gilbert's (1982) prediction, larger individuals would have higher clearance rates than smaller ones. Thus, the variation in body length implies an increase or a decrease in clearance rates as a response to the fluctuations of food resources.

Although exploitative competition and interference competition may also produce morphological changes (Gilbert, 1988), we did not find any relationship between *Keratella* morphometry and cladoceran biomass or density. This lack of correlation can be related to the small body size of *Bosmina* and *Ceriodaphnia* since the magnitude of interference competition depends on cladoceran body size (Burns & Gilbert, 1986). Nevertheless, *C. dubia* was observed to affect negatively the egg ratio of *K. cochlearis* in Laguna Ezquerra (Modenutti, 1994).

Many of the cyclomorphic adjustments, including the allometric growth of spines, can be induced by predation (Dodson, 1989). So, in Laguna Ezquerra the absence of allometry in PS may be related to the type and low densities of the invertebrate predator present in the lake. *Limnesia patagonica* can effectively consume *K. cochlearis* in laboratory experiments but we did not find any relationship between rotifer morphometry and density of the water mite. Probably, this situation was due to the low densities of natural populations of this predator (1 to 9 ind m^{-3}), since predation effect depends not only on predator feeding rate but also on predator abundance (Gliwicz & Pijanowska, 1989). This is consistent with Balseiro (1992) findings that *Limnesia* did not affect significantly *Bosmina*'s mortality in nature although this predator consumes *Bosmina* in laboratory experiments. There is an energetic cost associated with the production of spines. The sinking rate of *Keratella* increases with the presence of spines (Stemberger & Gilbert, 1984, 1987; Gilbert & Kirk, 1988), with a consequent decrease in swimming speed. So, these defensive morphs should be developed when the predator causes significant mortalities.

In summary, morphometry of *K. cochlearis* in Laguna Ezquerra clearly responded to fluctuations of temperature and food resources. Larger individuals were observed when temperature and food level were low. A larger lorica would increase the clearence rate without the energetic costs of spine development. Then, during the spring period of lower temperature and food resources the increase in body size would enhance the clearance rates allowing *Keratella* to persist in the presence of the competitors *Bosmina* and *Ceriodaphnia*.

Acknowledgements

We thank Dr Balseiro and two anonymous reviewers for their critical review of the manuscript. This study was supported by the UNC Grant B-701 and AN-PCyT PMT-PICT 0389. M. Diéguez was supported by Fundación Antorchas Grant A-1 3393/1-4.

References

Balseiro, E. G., 1992. The role of pelagic water mites in the control of cladoceran population in a temperate lake of the southern Andes. J. Plankton Res. 14: 1267–1277.

Balseiro, E. G. & B. E. Modenutti, 1990. Zooplankton dynamics of Lake Escondido (Rio Negro, Argentina), with special reference to a population of *Boeckella gracilipes* (Copepoda, Calanoida). Int. Revue ges. Hydrobiol. 75: 475–491.

Balseiro, E. G., B. E. Modenutti & C. P. Queimaliños, 1992. The coexistence of *Bosmina* and *Ceriodaphnia* in a South Andes lake: an analysis of the demographic responses. Freshwat. Biol. 28: 93–101.

Bielańska-Grajner, I., 1995. Influence of temperature on morphological variation in populations of *Keratella cochlearis* (Gosse) in Rybnik Reservoir. Hydrobiologia 313/314: 139–146.

Bogdan, K. G. & J. J. Gilbert, 1982. The effect of posterolateral spine length and body length on feeding rate in the rotifer, *Brachionus calyciflorus*. Hydrobiologia 89: 263–268.

Burns, C. W. & J. J. Gilbert 1986. Effects of daphnid size and density on interference between *Daphnia* and *Keratella cochlearis*. Limnol. Oceanogr. 31: 848–859.

Conde-Porcuna, J. M., R. Morales-Baquero & L. Cruz-Pizarro, 1993. Effectiveness of caudal spine as a defense mechanism in *Keratella cochlearis*. Hydrobiologia 225/256: 283–287.

Diéguez, M. C. & B. E. Modenutti, 1996. *Keratella* distribution in North Patagonian lakes (Argentina). Hydrobiologia 321: 1–6.

Dodson, S., 1989. Predator induce reaction norm. BioScience 39: 447–452.

Dumont, H. J., 1983. Biotic factors in the population dynamics of rotifers. Arch. Hydrobiol. 8: 98–122.

Eloranta, P., 1982. Notes on the morphological variation of the rotifer species *Keratella cochlearis* (Gosse) s.l. in one eutrophic pond. J. Plankton Res. 4: 299–312.

Gilbert, J. J., 1988. Suppression of rotifer population by *Daphnia*: a review of the evidence, the mechanisms and the effects on zooplankton community structure. Limnol. Oceanogr. 33: 1286–1303.

Gilbert J. J. & K. L. Kirk, 1988. Scape response of the rotifer *Keratella*: description, stimulation, fluid dynamics, and ecological significance. Limnol. Oceanogr. 33: 1440–1450.

Gliwicz Z. M. & J. Pijanowska, 1989. The role of predation in zooplankton succession. In U. Sommer (ed.), Plankton Ecology: Succession in Plankton Communities. Springer Verlag, London: 253–296.

Green, J., 1987. *Keratella cochlearis* (Gosse) in Africa. Hydrobiologia 147: 3–8.

Hillbricht-Ilkowska, A., 1983. Morphological variation of *Keratella cochlearis* (Gosse) in Lake Biwa, Japan. Hydrobiologia 104: 297–305.

Hofmann, W., 1983. On temporal variation in the rotifer *Keratella cochlearis* (Gosse): the question of 'Lauterborn-cycles'. Hydrobiologia 101: 247–254.

Jose de Paggi, S., 1978. Nota sobre las variaciones locales en *Keratella cochlearis* (Gosse). Comparaciones entre poblaciones de ambientes lóticos y leniticos de la llanura aluvial del rio Paraná. Neotropica 24: 27–32.

Modenutti, B. E., 1994. Spring-summer succession of planktonic rotifers in a south Andes lake. Int. Revue ges. Hydrobiol. 79: 373–383.

Pejler, B., 1962. On the variation of the rotifer *Keratella cochlearis* (Gosse). Zool. Bidr. Upps. 35: 1–17.

Ruttner-Kolisko, A., 1974. Plankton Rotifers: Biology and Taxonomy. Binnengewässer 26: 1–146.

Stemberger, R. S., 1979. A guide to rotifers of the Laurentian Great Lakes. Final Report. U. S. Environmental Protection Agency. Cincinnati, Ohio, 185 pp.

Stemberger, R. S. & J. J. Gilbert, 1984. Spine development in the rotifer *Keratella cochlearis*: induction by cyclopoids copepods and *Asplanchna*. Freshwat. Biol. 14: 639–648.

Stemberger, R. S. & J. J. Gilbert, 1987. Defenses of planktonic rotifers against predators. In W. C. Kerfoot & A. Sih (eds), Predation: Direct and Indirect Impacts on Aquatic Communities. The University Press of New England, Hannover (N.H.), Lond.: 227–239.

Hydrobiologia **387/388**: 295–300, 1998.
E. Wurdak, R. Wallace & H. Segers (eds), Rotifera VIII: A Comparative Approach.
© *1998 Kluwer Academic Publishers.*

Diel variation in the egg ratio of *Hexarthra bulgarica* in the high mountain lake La Caldera (Spain)

L. Cruz-Pizarro, J.M. Conde-Porcuna* & P. Carrillo
*Instituto del Agua, Universidad de Granada, E-18071 Granada, Spain (*author for correspondence)*
Departamento de Biología Animal y Ecología, Facultad de Ciencias, Universidad de Granada, E-18071 Granada, Spain

Key words: oviposition, egg-ratio, diel variation, *Hexarthra bulgarica*

Abstract

Hexarthra bulgarica is the dominant rotifer species in the lake La Caldera (south of Spain) both in number of individuals and in biomass. This species shows a nocturnal vertical migration and a simultaneous diel horizontal movement. In the present study, a detailed 24-h sampling program was carried out at four depths in six stations located along the two main transects of lake La Caldera. The results show the existence of diel cycles in oviposition of amictic and mictic (male and resting) eggs of this rotifer. The egg ratio for amictic and mictic eggs increases during the night. The oviposition of amictic eggs was also higher in the pelagic zone of the lake. Oviposition of amictic and mictic eggs was not related with density of *H. bulgarica* nor with density of competitors (*Mixodiaptomus laciniatus*). Our results suggest that UV radiation or an endogenous circadian rhythm of activity could be responsible of the diel cycle in oviposition of *H. bulgarica*.

Introduction

Experimental data on diel periodicity in rotifer egg ratio were first provided for *Keratella quadrata* by Edmondson (1965) who identified periodicity in oviposition as a possible source of error in egg ratio-based estimates of birth rate, and pointed out the necessity for further research on this matter.

Because of the time-consuming and tedious work, diel studies are difficult to perform. Moreover, in recent years, scant literature exists on diel egg ratio variation. Several studies have shown this phenomenon in species other than *K. quadrata* both in nature and under laboratory conditions (Champ & Pourriot, 1977; Lewis, 1979; Magnien & Gilbert, 1983; Ringelberg & Steenvoorden, 1986; Saunders, 1980; Van den Bosh & Ringelberg, 1985). According to these studies, a nocturnal maximum in oviposition seems to prevail. However, differences between species and seasons can be found. Moreover, the role played by factors other than light on this pattern, and the nature of the stimulus and its adaptive significance, is not clear. For these reasons, it is not easy to consider a single general daily

pattern. More information, particularly from field observations, seems to be required, and Edmondson's plea remains valid. In this sense, Saunders (1980) considered that the potential for the occurrence of this diel variation in egg ratio is sufficiently great that it should be tested on a case-to-case basis.

Hexarthra bulgarica is the dominant rotifer species in lake La Caldera, both in number of individuals and in biomass. For this species Cruz-Pizarro (1978, 1981) described a definite (and consistent over time) nocturnal vertical migration, and suggested a simultaneous diel horizontal movement of the population. This horizontal movement was later confirmed by Carrillo et al. (1989).

The aims of our study were two-fold. First we demonstrate the existence of a rhythmicity in oviposition of amictic and mictic eggs by *H. bulgarica* and try to investigate the suggested relationship between this process and the migratory movements of the animals (Gophen, 1978). Second, we analysed if oviposition shows a horizontal spatial pattern.

Site description

Lake La Caldera is a small (surface area, 2.4 ha) oligotrophic high mountain lake (Maximum depth, 12 m; Secchi disk visibility over 10 m; chlorophyll values range from 0.2 to 2.0 μg/l) of glacial origin, located in the Sierra Nevada (Southern Spain) at an elevation of 3050 m. The lake remains frozen for about 7–8 months each year.

The plankton community is rather simple. The phytoplankton is largely dominated by flagellates (Chrysophyceae and Dinophyceae) and by a Cyanophycean species (*Cyanarcus* sp.). Other species (mainly green algae and diatoms) are much less abundant (Carrillo et al., 1995). *Mixodiaptomus laciniatus* and *Hexarthra bulgarica* dominate in zooplankton, and a population of *Daphnia pulicaria* develops at the end of summer (Cruz-Pizarro, 1981; Carrillo, 1989).

Particularly noteworthy is the absence of a littoral rooted vegetation and of visible inlet and outlets. There are no fish.

Material and methods

A diel sampling program was carried out in August 1986 (8–9 August), during the population density peak of *H. bulgarica*. Throughout a 24-h period, samples were taken at 2–4-hourly intervals (see Figures 1 and 3), using a double Van Dorn sampler (capacity, 2 × 8 l), from six different stations located along the two main transects of the lake. Three stations were representative of the pelagic zone of the lake while the other three were representative of the littoral zone (Carrillo et al., 1989). We selected four depths at each station: surface; one-third maximum depth; two-thirds maximum depth and bottom, getting a total of 192 different spatial and temporal samples. Each sampling series was completed within half an hour.

Samples were immediately filtered through a 45-μm mesh net and preserved in 4% formaldehyde. Counting of zooplankton was done using an inverted microscope at 250× magnification. Amictic and mictic (male and resting) eggs of *H. bulgarica* were counted separately.

The complex migratory behaviour described for *H. bulgarica* in lake La Caldera (Carrillo et al., 1989), which involves well-defined horizontal and vertical movements, shows that in most sampling depths and points the number of individuals was higher than one

Table 1. Nested ANOVAs testing for the differences of amictic and resting egg ratios, *Hexarthra* density, and *Mixodiaptomus* density, between light and dark hours

Categories	df	mS	F
Log-amictic egg ratio			
Between day and night	1	1.04	17.92***
Between hours (within day and night)	6	0.09	1.60[NS]
Error	34	0.06	
Log-resting egg ratio			
Between day and night	1	1.31	11.09**
Between hours (within day and night)	6	0.23	1.9[NS]
Error	34	0.12	
Log-*Hexarthra* density			
Between day and night	1	0.84	1.61 [NS]
Between hours (within day and night)	6	0.91	1.75 [NS]
Error	34	0.52	
***Mixodiaptomus* density**			
Between day and night	1	70.14	7.81**
Between hours (within day and night)	6	69.57	7.75***
Error	34	8.98	

Dark (night) hours: 21:00 – 24:00 – 03:00 – 05:00 h
Light (day) hours: 15:00 – 18:00 – 08:00 – 12:00 h.
*** $P < 0.001$; ** $P < 0.01$; NS, not significant.

organism per litre. Consequently, reliable estimates of egg ratios could be reached.

Results

The diel variation in the number of amictic and resting eggs per *Hexarthra* female during this study is shown in Figure 1. According to this, oviposition increases during night hours. In this sense, a nested ANOVA (Bennington & Thayne, 1994; Sokal & Rohlf, 1995) was used to compare the amictic and resting egg ratios of light hours (summer time: from 15:00 to 18:00 h and from 08:00 to 12:00 hours) and dark hours, after sunset (summer time: from 21:00 to 05:00 hours). The fixed effect in this analysis was the light and dark condition, while the nested effect was the hour of the day. Homogeneity of variances and normality were checked by Bartlett's test and Kolmogorov–Smirnov test, respectively (Sokal & Rohlf, 1995). Some variables were logarithmically transformed to comply with assumptions for ANOVA. The results show that oviposition during nighttime is significantly higher than during daytime, while there are no differences between hours within day and night (Figure 2, Table 1). To test if oviposition was density related, we compared *Hexarthra* densities during

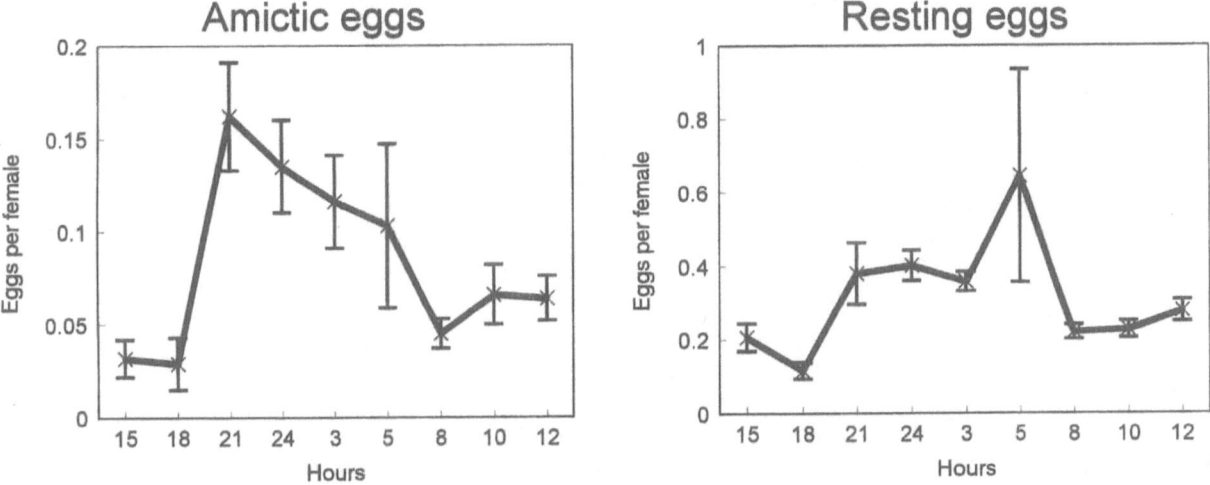

Figure 1. The diel pattern in egg ratio for amictic and resting eggs of *Hexarthra bulgarica*. Mean values were calculated from all stations and depths. Error bars, standard error between depths.

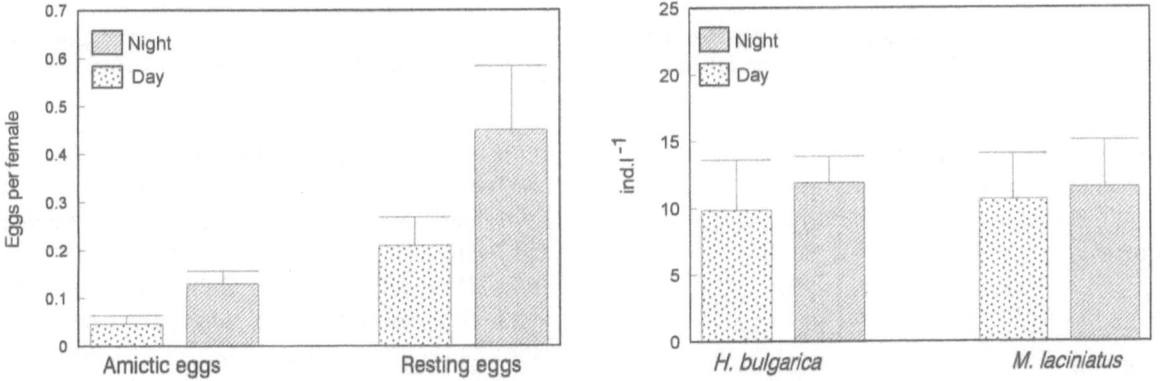

Figure 2. Mean values (± SD) during night and day in egg ratio for amictic and resting eggs of *Hexarthra bulgarica* and for densities of *Hexarthra bulgarica* and *Mixodiaptomus laciniatus*. Night hours: from 21:00 to 05:00 h. Day hours: from 15:00 to 18:00 and from 08:00 to 12:00 h.

night and day hours and found no difference (Figure 2, Table 1). In relation to potential competitors, *Mixodiaptomus laciniatus* was the only species considered in this study because it is the only important competitor for *Hexarthra bulgarica* in lake La Caldera. Densities of *M. laciniatus* were higher during the night than during the day (Figure 2, Table 1), so, although competition would have been stronger during night hours, oviposition of *Hexarthra* was higher during these hours. As *Daphnia* densities were very low during this study (<0.5 individuals l^{-1}), we did not consider this species in our study. It was not possible to look for differences in egg ratio in relation to depth because most of the eggs were unattached and, consequently, they may sink to deeper water. Indeed, we observed the highest number of eggs at the bottom of the lake (Friedman test, $P<0.05$).

To test if oviposition exhibited a horizontal pattern, we compared the number of amictic eggs per female of the pelagic zone (mean values from pelagic stations) with that of the littoral zone (mean values from littoral stations). Figure 3 shows that the egg ratio was higher in the pelagic zone than in the littoral one. A *t*-test for paired observations reveals that this difference is significant ($t=2.582$; $P<0.05$; df= 8). This pattern was not related with the density of *H. bulgarica* observed in both zones. The density of *H. bulgarica* was not different between zones ($t= 0.40$; $P= 0.70$; df=8). *M. laciniatus* was also not responsible for the horizontal pattern observed in the egg ratio because no difference in *Mixodiaptomus* density was found between the

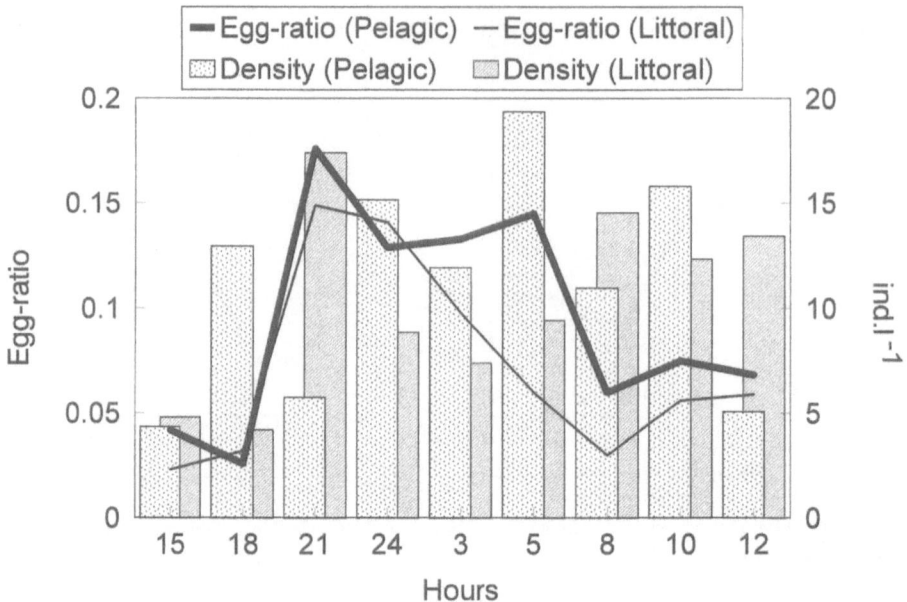

Figure 3. The diel pattern in egg ratio for amictic eggs and in density of *Hexarthra bulgarica* in the pelagic and littoral zones. Mean values of egg ratio in the pelagic and littoral zones were compared, hour by hour, using a *t*-test for paired observations: *t*= 2.582 (*P*<0.05).

pelagic and littoral zone (*t*= 0.37; *P*= 0.72; df=8). On the contrary, the number of resting eggs per female does not show a horizontal pattern (*t* = 0.31; *P*=0.77; df= 8).

Discussion

Previous studies in lakes of the Sierra Nevada have shown that *Hexarthra bulgarica* undergoes an early period of mixis, just before the population maximum (Cruz-Pizarro, 1981; Reche, 1991). In lake La Cadera, Morales-Baquero et al. (1995) observed that, although males have never been observed in the samples, there was evidence that mixis and sexual reproduction occur at a very early stage in *H. bulgarica,* because male eggs appear early in the development of the population and, consequently, numbers of resting eggs increase afterwards.

The data presented here demonstrate that oviposition fluctuation is very pronounced in *H. bulgarica.* Production of amictic and mictic eggs is higher during night hours. This production is independent of *Hexarthra* density and of the presence of potential competitors (*M. laciniatus*). The differences observed for *Mixodiaptomus* (copepodites) densities between day and night, and between hours, may be because copepodite density increased from nine to 14 individuals

per litre in 3 h (from 21:00 to 24:00 h), probably due to developing of older nauplii. Temperatures in lake La Caldera during August 1986 were homogeneous and relatively low (<13°C) (Morales–Baquero et al., 1995). However, Magnien & Gilbert (1983) suggested that high temperatures may induce diel cycles in egg ratios. Several studies have shown that for different rotifer species, no diel variation in egg ratio can be found at low temperatures (Keen & Miller, 1977; Ringelberg & Steenvoorden, 1986) but can occur at higher temperatures (>23°C) (Magnien & Gilbert, 1983; Saunders, 1980). Our results show that diel variation in egg ratio actually may accur at low temperatures as well.

Gophen (1978) linked vertical migration of copepods with egg production. He considered that the consequence of migration is that adult organisms are exposed to relatively high temperatures during most of the day and early night, with accelerated egg production as a result. This phenomenon is possible during periods when lakes are thermally stratified, providing the opportunity for populations to vary their thermal environment by migration. However, during the ice-free period in lake La Caldera, the vertical thermal structure is more or less homogeneous and a true thermocline is not established, not even in the middle of the summer (Cruz-Pizarro & Carrillo, 1996). Consequently, during our study there was no thermal heterogeneity that could have explained higher egg ratios during night. Moreover, daily variation in temperature is small in lake La Caldera (2–3°C in surface water and 0–1°C at 3 m depth; Cruz-Pizarro, unpublished results).

Other explanations have been proposed to explain egg hatching synchrony. For instance, Champ & Pourriot (1977) considered that photoperiod may control egg development time and result in hatching rhythmicity. *H. bulgarica* develops during a very short period in lake La Caldera so differences in photoperiod during this growth phase will be negligible. On the other hand, we observed that our results were not affected by *Hexarthra* densities and competitors. A horizontal spatial pattern is also shown for egg ratios in our study.

Carrillo et al. (1989) have shown that *H. bulgarica* presents a general daily trend of movement that couples vertical migration to horizontal one, and it has been suggested that light may be responsible for these movements. Intense light radiation may inhibit photosynthesis and ultraviolet light may produce harmful effects on zooplankton species (Lampert, 1989). These effects are most evident in clear air and at high altitudes (Home & Goldman, 1994). In this sense, during our study in the high mountain lake La Caldera, light intensity in the pelagic zone for day hours was higher than 1000 μE m^{-2} s^{-1} up to 2–3 m depth. This is an inhibitory light intensity for algal growth and, in the most extreme cases, entire surface algal blooms can be destroyed by too intense sunlight (Home & Goldman, 1994). This fact could explain why *H. bulgarica* (a rotifer without pigmentation) migrates to littoral waters during the day: by searching refuge between stones, the animals may try to avoid short-waved irradiation as done by *Daphnia* (Cruz-Pizarro et al., personal observation). According to this hypothesis, it should be expected that egg production will be higher during night hours at any zone of the lake, as we observed. The higher egg ratios in the pelagic zone could be because *H. bulgarica* is a pelagic species that, although it does migrate to the littoral zone, reaches its highest fitness in pelagic waters.

According to our results, we suggest that *H. bulgarica* possesses a circadian rhythm of activity, causing it to increase egg production during the night, which could be endogenous and/or triggered by UV radiation. Notwithstanding our results and the results of previous studies, further information on diel egg ratio variation of rotifers is still needed to evaluate if the induction of fluctuation in egg ratios is species specific and/or controlled by factors such as temperature and light. More investigations into the endogenous circadian rhythmicity and its possible relevance to diel vertical migration are required in order to resolve these questions.

Acknowledgements

We thank two anonymous referees for their constructive comments on a previous version of this manuscript. Financial support was obtained by CICYT Projects AMB94-0459, AMB94-021 and AMB97-0996 and by a Contract to J.M. Conde-Porcuna (University of Granada, Spain).

References

Bennington, C. C. & W. V. Thayne, 1994. Use and misuse of mixed model analysis of variance in ecological studies. Ecology 75: 717–722.

Carrillo, P., L. Cruz-Pizarro & R. Morales-Baquero, 1989. Empirical evidence for a complex diurnal movement in *Hexarthra bulgarica* from an oligotrophic high mountain lake (La Caldera, Spain). Hydrobiologia 186/187: 103–108.

Carrillo, P., I. Reche, P. Sánchez-Castillo & L. Cruz-Pizarro, 1995. Direct and indirect effects of grazing on the phytoplankton seasonal succession in an oligotrophic lake. J. Plankton Res. 17: 1363–1379.

Champ, P. & R. Pourriot, 1977. Particularités biologiques écologiques du rotifer *Sinantherina socialis* (Linné). Hydrobiologia 55: 55–64.

Cruz-Pizarro, L., 1978. Comparative vertical zonation and diurnal migration among Crustacea and Rotifera in the small high mountain lake La Caldera (Granada, Spain). Verh. int. Ver. Limnol. 20: 1026–1032.

Cruz-Pizarro, L., 1981. Estudio de la comunidad zooplanctónica de un lago de alta monta na (La Caldera, Sierra Nevada, Granada). Ph.D. Thesis. Univ. Granada: 186 pp.

Cruz-Pizarro, L. & P. Carrillo, 1996. A high mountain oligotrophic lake (La Caldera, Sierra Nevada, Spain). In Cruz-SanJulian J. J. & J. Benavente (eds), Wetlands: a Multiapproach Perspective. Water Research Institute, University of Granada: 111–130.

Edmondson, W. T., 1965. Reproductive rate of planktonic rotifers as related to food and temperature in nature. Ecol. Monogr. 35: 61–111.

Gophen, M., 1978. Errors in the estimation of recruitment of early stages of *Mesocyclops leuckarti* (Claus) caused by the diurnal periodicity of egg-production. Hydrobiologia 57: 59–64.

Home, A. J. & C. R. Goldman, 1994. Limnology. McGraw-Hill, New York, USA, 576 pp.

Lampert, W., 1989. The adaptive significance of diel vertical migration of zooplankton. Funct. Ecol. 3: 21–27.

Lewis, W. M., 1979. Zooplankton Community Analysis. Springer-Verlag, Berlin, 159 pp.

Magnien, R. E. & J. J. Gilbert, 1983. Diel cycles of reproduction and vertical migration in the rotifer *Keratella crassa* and their influence on the estimation of population dynamics. Limnol. Oceanogr. 28: 957–969.

Morales-Baquero, R., P. Carrillo & L. Cruz-Pizarro, 1995. Effects of fluctuating temperatures on the population dynamics of *Hexarthra bulgarica* (Wiszniewski) from high mountain lakes in Sierra Nevada (Spain). Hydrobiologia 313/314: 359–363.

Ringelberg, I. & I. Steenvoorden, 1986. Diel variation in the egg ratio of rotifers throughout the season (preliminary report). Hydrobiol. Bull. 19: 153–158.

Reche, I., 1991. Análsis de la sucesión fitoplanctónica en una laguna de alta montaña. Las Yeguas (Sierra Nevada). Tesis de Licenciatura. Universidad de Granada, 215 pp.

Saunders, J. F., 1980. Diel patterns of reproduction in rotifer populations from a tropical lake. Freshwat. Biol. 10: 35–39.

Sokal, R. R. & F. J. Row, 1995. Biometry, 3rd edn. W.H. Freeman and Company, New York, 887 pp.

Van den Bosch, F. & I. Ringelberg, 1985. Seasonal succession and population dynamics of *Keratella cochlearis* (Ehrb.) and *Kellicotia longispina* (Kellicot) in Lake Maarsseveen I (Netherlands). Arch. Hydrobiol. 103: 273–290.

Hydrobiologia **387/388**: 301–310, 1998.
E. Wurdak, R. Wallace & H. Segers (eds), Rotifera VIII: A Comparative Approach.
© *1998 Kluwer Academic Publishers.*

Strategic variation of egg size in *Keratella cochlearis*

J. Green
17 King Edwards Grove, Teddington, Middlesex TW11 9LY, U.K.

Key words: Keratella cochlearis, egg size variation, Rotifera

Abstract

The volume of a single amictic egg of *Keratella cochlearis* can vary between 32 000 and 132 000 μm^3. Much of this variation is related to the size of the female laying the egg. When compared to populations in Europe, those in the Southern Hemisphere (Southern Africa and New Zealand) show a smaller increase in egg volume per unit increase in lorica length. Both lorica length and egg volume show a strong negative correlation with temperature. At high temperatures the females are smaller and they lay smaller eggs, but there are differences between populations in different lakes. In oligotrophic and high altitude lakes there is less variation than in lowland eutrophic lakes.

Introduction

Eggs of aquatic animals show considerable variations in size within a single species. Hutchinson (1951) found that planktonic copepods tended to produce few large eggs when food was scarce, and numerous smaller eggs when more food was available. The adaptive nature of this response has been verified experimentaly (Jamieson & Burns, 1988).

In the cladocerans, *Simocephalus vetulus* and *Scapholeberis mucronata*, the sizes of the parthenogenetic eggs vary seasonally, and this variation is governed largely by an inverse relationship with the temperature of the water. The size of a cladoceran egg is partially determined by differing advantages at different temperatures. At low temperatures large eggs ensure that the neonate has a maximum chance of reaching maturity, reaching this state after fewer instars (Green, 1956). At higher temperatures, when the rate of development is accelerated, the larger number of smaller eggs ensures a greater rate of population growth (Green, 1966). The egg size of the two cladocerans also varied latitudinally, with northern populations producing larger eggs than more southerly populations.

From the results obtained with other planktonic animals one might expect rotifers to show similar adaptive variation in egg size. However, recent work has shown various responses. Walz & Rothbucher (1991) found that larger females of *Brachionus angularis* pro-

duced larger eggs, and that as the food concentration increased there was an increase in egg volume. This was most striking at relatively low food concentrations. When the food was increased from 0.5 mgC l^{-1} to 8.5 mgC l^{-1}, the egg volume increased from 145 000 to 172 009 μm^3, with only a small increase in the size of the females. Similarly Kirk (1997) found that the egg size of *Brachionus calyciflorus* increased with increasing food concentration, and that the eggs from clutches of more than one were actually larger than single eggs. Experiments with *Synchaeta pectinata* showed that the egg size remained constant at all food levels. This might indicate that *S. pectinata* has evolved to produce a minimum viable egg intended to enhance population growth at all food levels. It is perhaps noteworthy that *S. pectinata* is one of the few warm water species in its genus.

In a survey of 43 rotifer species (Walz et al., 1995), it was found that the larger species produced larger eggs, but the increase was not proportional to body size. The relative egg volume (i.e. as a percentage of body volume) decreased with increasing body size. Two egg volumes were given for *Keratella cochlearis*: one from a population in India (E. V. 40 856 μm^3) and one from a laboratory culture in Germany (E. V. 70 618 μm^3). In this culture the relative egg volume was 84%, so that the neonates hatching from such eggs had to increase size by only a small amount to reach adult size. This is an enormous investment for any animal to make in a single offspring.

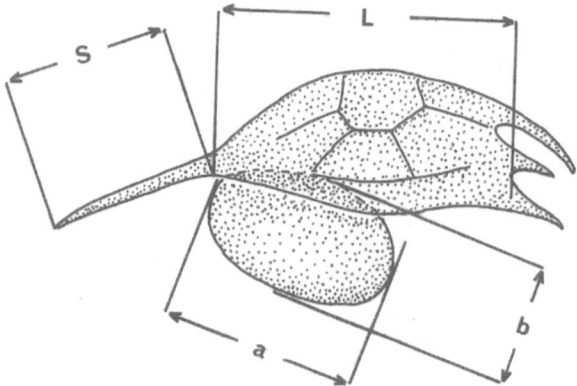

Figure 1. Diagram to show how measurements were made.

The present study examines a wide range of populations to establish just how much variation occurs, and to see if comparative field data can provide evidence of adaptive variation in egg volume.

Methods

Samples were collected with 55 μm meshed nets, either from the shore of the lake, or more usually from an inflatable boat. Where possible vertical hauls of 10 m were made, and four samples were taken from each lake, and later pooled. Secchi disc measurements were made with a 25 cm disc. Chlorophyll was estimated by taking a water sample from a depth of 0.5 m and filtering through a Whatman glass fibre filter. The filter was ground up with acetone, then centrifuged to clear, and the volume was made up to 5 ml. Absorption at 665 nm was measured in a spectrophotometer. Absorption at 750 nm was also measured to estimate scattering and residual absorption. These readings were generally low, and were deducted from the reading at 665 nm before converting the latter to an estimate of chlorophyll.

Samples of rotifers were preserved in 4% formaldehyde and examined at a magnification of ×400. Measurements were made between the limits shown in Figure 1. The spine length will not be considered in this paper, apart from considering the tecta form in relation to spined females. Lateral measurements were chosen because the eggs are often tilted with respect to the main axis of the lorica. If the ventral side of the animal is viewed, the egg can appear foreshortened and erroneous measurements can result. The egg volume

was calculated from the formula

$$EV = \frac{1}{6}ab^2,$$

where a is the length of the egg and b is the breadth in lateral view. This is not a completely accurate calculation because the egg is not circular in cross-section, but is slightly wider in ventral view than in lateral view. There is also a slight tapering of the egg towards the posterior end. A more accurate assessment of egg volume would involve several more measurements. For comparative purposes, the volumes calculated here are reasonably accurate and the errors will be similar in all samples.

Mean egg volumes shown in the figures are based on numbers varying with the availability of egg-bearing females. Most are based on 20, a few on as few as 5, and some up to 30.

Results

Variation in a single sample

Figure 2 shows the relationship between lorica length and egg volume in a sample of 20 spined females and 20 tecta females from the Mile End Canal in October. The egg volume shows considerable variation, from 35 to 95×10^3 μm^3, with the tecta showing less variation, from 58 to 78×10^3 μm^3. There is a clear relation of egg volume to lorica length, but at the same time there is clear variation in egg volume at a given lorica length. The spined females fall into two groups, with tecta lying between them.

Figure 3 shows a sample of 37 spined females from Regents Park Lake in April. No tecta were present at this time. The females fell into two groups: large, with large eggs, and medium sized with medium sized eggs. The egg volumes range from 50 to 132×10^3 μm^3. The mean for the whole sample was 69×10^3 μm^3. Again, there is a clear relationship between lorica length and egg column, but considerable variation at a given lorica length.

These two samples have been selected from many others that might have been used, to give an idea of the range of variation in a natural population, and to emphasise that in the later figures, where mean egg volumes are used, that each point must be visualised as occupying the centre of a fairly extensive cluster of points.

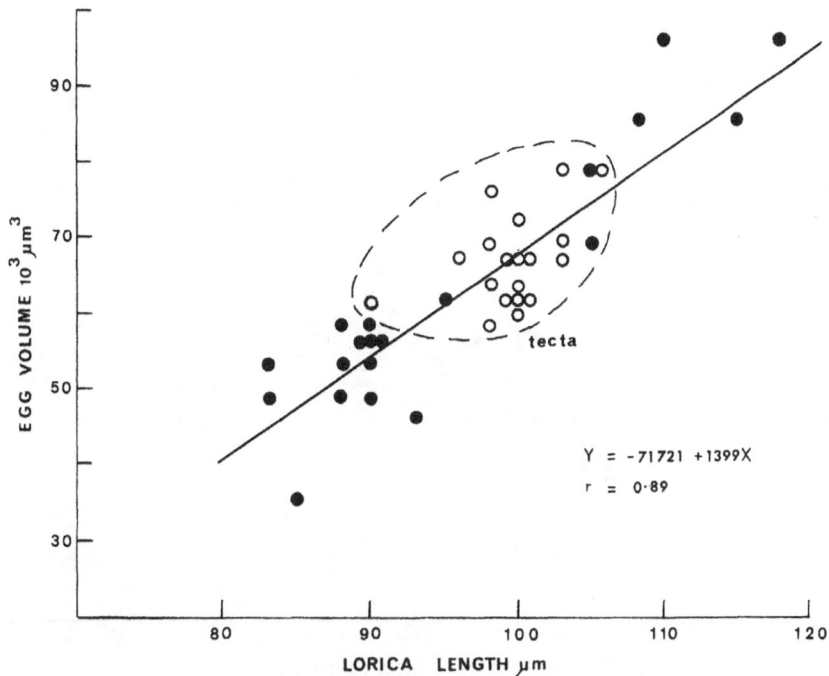

Figure 2. Keratella cochlearis: lorica length and egg volume in a single sample from Mile End Canal, 31 October 1991. Each point represents an individual. The mean values for the tecta and spined forms are not significantly different, but the distributions are very different.

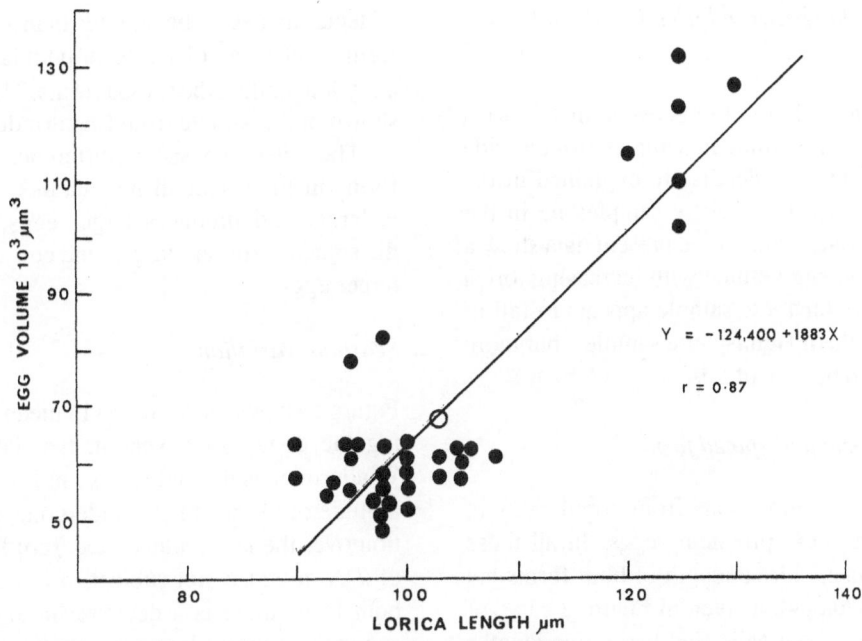

Figure 3. Keratella cochlearis: lorica length and egg volume in a sample from Regents park Lake, 7 April 1987. The overall mean is shown by the open circle.

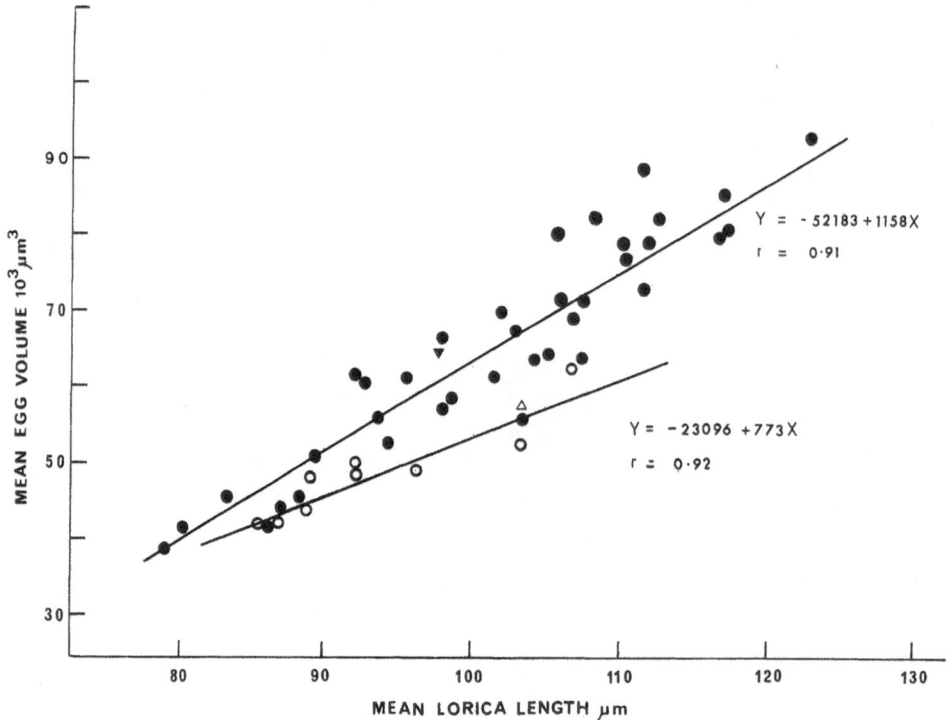

Figure 4. Keratella cochlearis: relationship between mean lorica length and mean egg volume. The solid dots are based on British and European populations. Open circles are based on Southern Hemisphere populations (Southern Africa and New Zealand). The inverted solid triangle represents the population in crater lake Kerid, Iceland, and the open triangle represents the population in Lake Akan, Hokkaido, Japan.

General relationship between lorica length and egg volume

Figure 4 shows the relationship between mean lorica length and mean egg volume in samples from a wide geographical range. The details are explained in the legend. The southern hemisphere samples lie in the lower part of the range, and in the present data show a smaller increase in egg volume with increasing lorica length. The single Japanese sample appears to fall in line with the Southern Hemisphere samples, but more data are needed to be sure of this.

Comparison of tecta and spined forms

Table 1 shows six comparisons from populations in which both forms were producing eggs. In all these populations the mean lorica length of both forms lay below 100 μm, although as seen in Figure 2 some of the spined females from Mile End had lorica lengths over 110 μm.

The lorica lengths in the table show that tecta is not merely the end of an allometric reduction in spine length. If that were so one would expect the lorica

of tecta always to be smaller than that of the spined forms, but some of the tecta population are significantly longer than the spined forms. This is most clearly shown in the sample from Llandrindod Wells.

There is no consistent difference between the two forms in these natural populations. Sometimes tecta is larger and produces larger eggs, and sometimes the spined forms are larger, and consequently produce larger eggs.

Seasonal variation

Figure 5 shows the variation in mean egg volume during the course of a year in two lakes in Cumbria. Windermere is the largest lake in England, it lies at an altitude of 39 m and has undergone some eutrophication over the last century. Red Tarn lies at an altitude of 718 m and is not subject to organic pollution. In both lakes there is a decrease in egg volume in the summer, followed by an increase in the autumn. The decrease was somewhat greater in Windermere, and the increase was not so fast in the autumn. A partial explanation of this difference may be related to the different temperature regimes of the two lakes. The

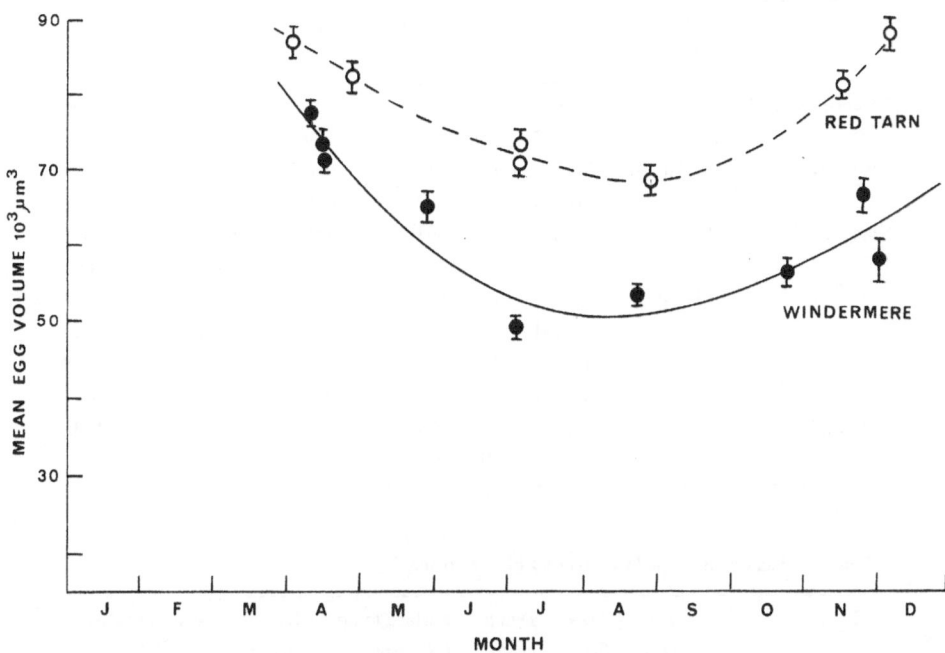

Figure 5. Keratella cochlearis: seasonal variation in egg volume in Windermere and Red Tarn. The error bars give one standard error above and one below the mean.

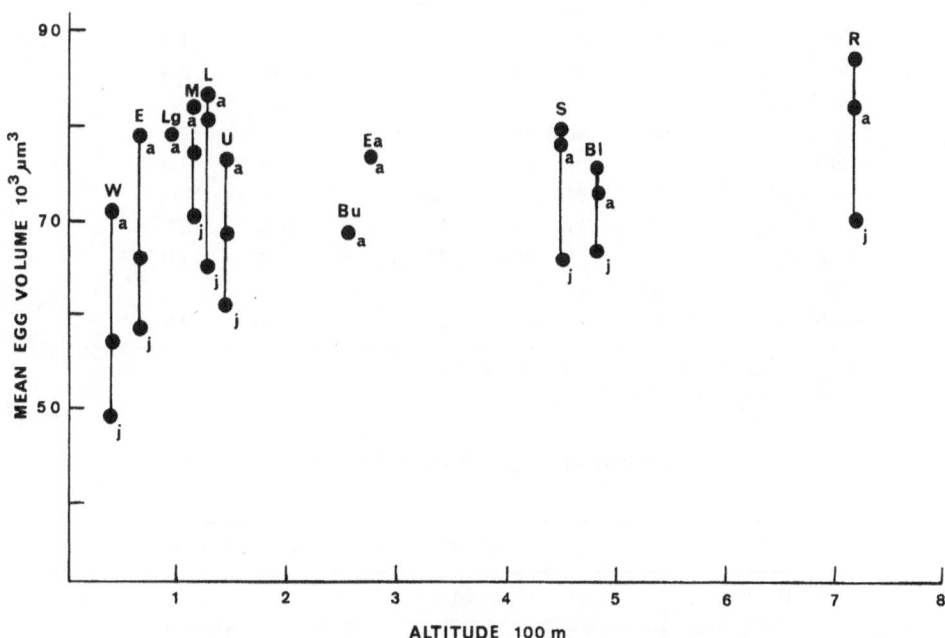

Figure 6. Keratella cochlearis: altitude and seasonal variation in egg volume in Cumbrian lakes in 1983. (a) April; (j) July; unmarked, December. Bl, Blea Water; Bu, Burnmoor Tarn; E, Esthwaite; Ea, Easedale Tarn; L, Loweswater; Lg, Loughrig Tarn; M, Mockerkin Tarn; R, Red Tarn; S, Small Water; U, Ullswater; W, Windermere.

Table 1. *Keratella cochlearis*: comparison of tecta and spined forms

Locality	Date	Form	n	Lorica L. (μm)	+ S.E.	Egg volume (μm^3)	+ S.E.
Newport Pagnell	16 Sep 81	Tecta	27	87.9	0.73	47 004	1396
		Spined	17	81.7	0.66	48 338	1236
Llandrindod	19 Aug 91	Tecta	12	91.5	0.71	49 730	798
		Spined	20	78.9	0.30	39 694	582
Llangorse	19 Aug 91	Tecta	9	89.0	1.01	46 666	3099
		Spined	20	86.8	1.25	44 326	1691
	19 Sep 89	Tecta	13	88.2	0.92	45 441	901
		Spined	5	95.2	0.80	61 132	3843
	20 Sep 91	Tecta	12	87.5	1.59	49 733	2938
		Spined	11	91.6	2.58	61 959	4497
Mile End	31 Oct 91	Tecta	20	99.6	0.69	67 145	1232
		Spined	20	95.2	2.41	62 089	3807

Table 2. Data on lakes used to construct Figures 6 and 7[a]

Lake	Alt. (m)	Area (ha)	Depth (z_m)	Surface temp. (°C) Apr.	Secchi, Apr. 83 (cm)	Chlorophyll, Apr. 83 (μg l^{-1})
Windermere	39	1477	64	5.6	540	2.5
Wastwater	61	291	76	6.0	1200	0.6
Esthwaite	65	100	16	8.8	360	8.3
Bassenthwaite	69	528	19	8.0	N.R.	12.2
Loughrig Tarn	94	9	10	N.R.	N.R.	11.1
Crummock Water	98	252	44	5.8	800	1.3
Buttermere	101	94	29	6.0	600	2.5
Mockerkin Tarn	115	4	3	8.5	170	6.7
Loweswater	121	64	16	7.0	290	17.0
Ullswater	145	894	63	6.2	400	7.0
Burnmoor Tarn[b]	254	24	13	N.R.	N.R.	N.R.
Easedale Tarn	279	11	21	6.2	1010	0.7
Small Water	452	5	16	3.8	670	1.0
Blea Water	483	17	68	2.8	910	0.7
Red Tarn	718	10	25	2.8	700	3.5

[a] Morphometric data from Carrick & Sutcliffe (1982).
[b] Used only in Figure 6.

Table 3. Indicators of trophic status in Red Tarn, Windermere, and Llangorse Lake

	Red Tarn	Windermere	Llangorse
Depth z_m	25	64	5
Conductivity[a] (μS cm 20 °C)	25–30	55–63	200–300
Secchi Disc[a] cm	700–1100	360–600	70–110
Chlorophyll (μg l^{-1})	1.5–3.5[a]	2.0–11.0[a]	up to 400[b]
'Tecta' form of *K. cochlearis*	Absent	Very rare	Abundant

[a] Based on nine sampling dates (1981–83).
[b] Data from Jones & Benson-Evans (1974).

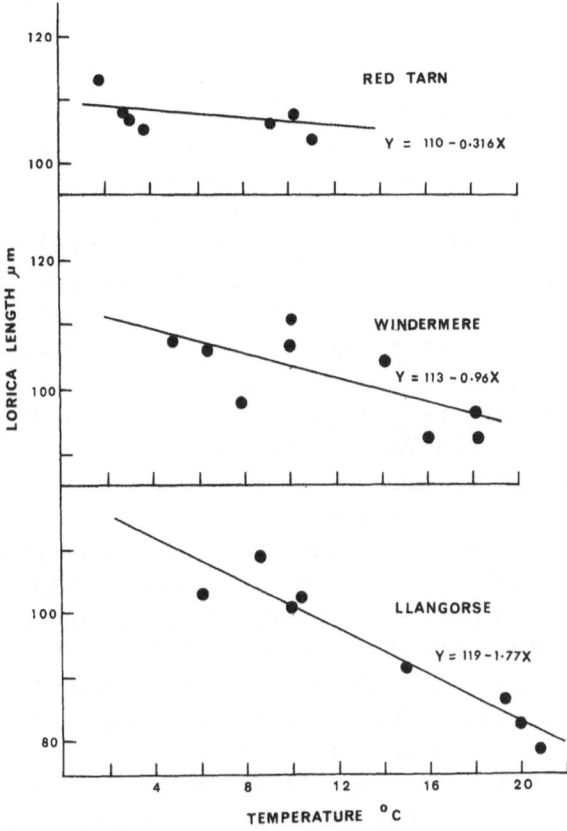

Figure 7. Keratella cochlearis: temperature and lorica length in three localities of different trophic status. Red Tarn is oligotrophic, Llangorse is eutrophic and Windermere is mesotrophic.

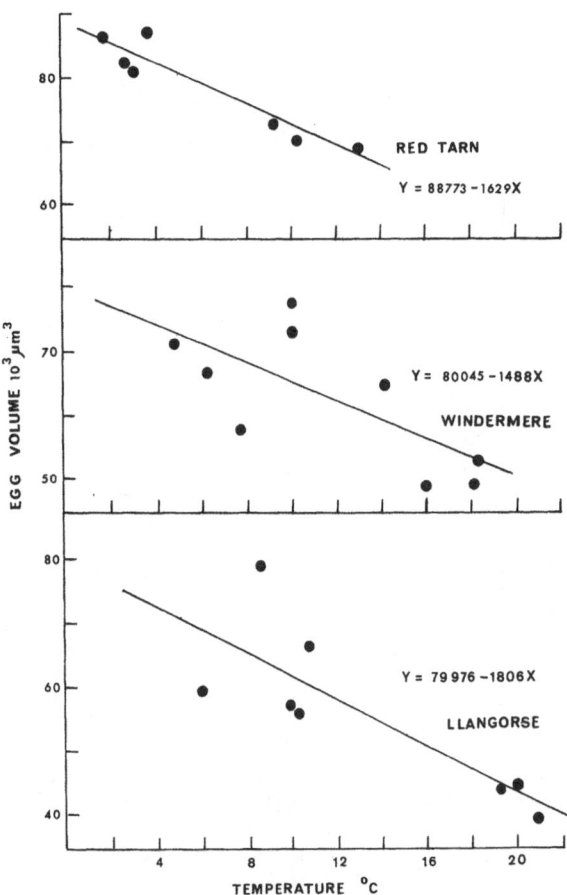

Figure 8. Keratella cochlearis: temperature and mean egg volume in three localities of differing trophic status.

altitude of Red Tarn results in a surface water temperature generally between 4 and 6 °C cooler than that of Windermere, although there are variations from year to year. When sampled in April 1981 the surface temperature of Windermere was 10 °C, and Red Tarn was 5.2 °C, but in April 1982 Windermere was at 6.4 °C and Red Tarn at 1.8 °C. The more rapid cooling of the smaller higher lake could account for the larger egg volume in Red Tarn in December 1983. At that time the surface temperature of Windermere was 7.8 °C while Red Tarn was at 3.6 °C.

Similar seasonal variation has been observed in other localities. For instance in the eutrophic gravel pits at Newport Pagnell in 1981 the mean egg volume of *K. cochlearis* fell from 61×10^3 μm^3 in May to 48×10^3 in September, and was back up to 64×10^3 μm^3 in December. The increase between September and December was accompanied by a fall in temperature from 18 to 5 °C, and a fall in chlorophyll from 46 to 18 $\mu g\, l^{-1}$.

Variation between lakes

Altitude

Figure 6 shows the variation in egg volume in 11 Cumbrian lakes in relation to altitude. Some features of the lakes are given in Table 2. In the figure the data for each lake are linked by a line when seasonal data are available. A consistent feature is that the egg volumes are lowest in July, and the mean egg volumes show a general trend to increase with increasing altitude. A group of lakes (Loweswater, Mockerkin Tarn, Loughrig Tarn and Esthwaite) show relatively high values in April, these are explained in the next section.

In the lower lying lakes the egg volumes were greatest in April, but in the three tarns above 400 m the egg volume was greatest in December. This may be attributable to rapid cooling of these relatively small lakes when compared to their lower lying, larger neighbours.

308

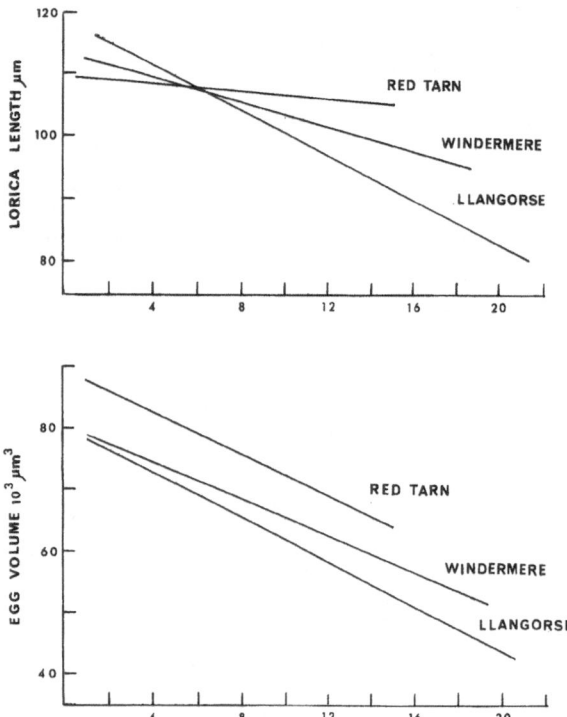

Figure 9. Keratella cochlearis: summary of the relationships between temperature, lorica length and mean egg volume in three localities of differing trophic status. Red Tarn is oligotrophic, Llangorse is eutrophic and Windermere is mesotrophic.

Temperature and trophic status

The data on seasonal variation and altitude indicate that temperature plays an important role in the determination of egg volume. This is strikingly confirmed in Figures 7–9. The three lakes are quite distinct in their trophic status. This is shown in Table 3. When the relationships between temperature and lorica length in the three lakes are compared it is seen that the negative regression in Red Tarn is much smaller than in the other two lakes (all the slopes differ significantly at *p* < 0.001). The regressions of egg volume on temperature show no significant differences in the slopes, but the intercept of Red Tarn is significantly higher than that of Llangorse (*p* < 0.001). Overall in these three lakes there is a general tendency towards smaller size at high temperatures, combined with a smaller egg volume. The egg volume is decreased further by an increase in trophic status.

It should be noted that at low temperatures the lorica length in Llangorse Lake may actually be longer than in Red Tarn, but at all temperatures *K. cochlearis*

in Llangorse lake produces smaller eggs than in Red Tarn.

Chlorophyll in Cumbrian lakes

In view of the seasonal variation, and the relationship with temperature described in the previous sections, comparisons between lakes are best made at one time of year. Figure 10 shows results from 14 Cumbrian lakes in April 1983. The lake with the highest chlorophyll had the largest eggs. The greatest variation was at the lower end of the chlorophyll range. Wastwater is a lot-altitude, unproductive lake, and it produced the smallest eggs. Higher altitude lakes, such as Blea Water, Small Water and Red Tarn, with only slightly higher chlorophyll, produced much larger eggs; this may be attributed to their lower temperatures.

The relationship with phytoplankton chlorophyll helps to explain the relatively large egg volumes in April in Loweswater, Mockerkin Tarn, Loughrigh Tarn and Estahaite shown in Figure 6. This result from Cumbria at first sight appears to conflict with the comparison of Red Tarn, Windermere and Llangorse in the previous section. But the range of chlorophyll concentrations in this sample of Cumbrian lakes is relatively small, and it is feasible for maximum egg size to occur at an intermediate trophic level, as Galindo et al. (1993) have shown for several brachionids.

Discussion

The data on variation within a single sample show that at any one time of year there can be wide variation in the sizes of eggs of *K. cochlearis*. This variability could be an adaptive feature, so that all possibilities are covered if there is a sudden change in the environment. But the situation is not entirely chaotic. There is a clear relationship between the mean size of the females and the mean size of the eggs they produce. Ricci (1995) found a similar positive correlation in the bdelloid *Macrotrachela quadricornifera*.

In *K. cochlearis* both the lorica length and the egg volume show a strong inverse relationship with temperature. At high temperatures females are smaller and product smaller eggs. There is a regular seasonal variation in egg volume, with a decrease in summer. The magnitude of this variation differs between lakes.

Food is also important. The springtime data from the Cumbrian lakes show that the egg volumes in a low altitude, unproductive lake (Wastwater) were lower than in lakes of similar productivities at higher

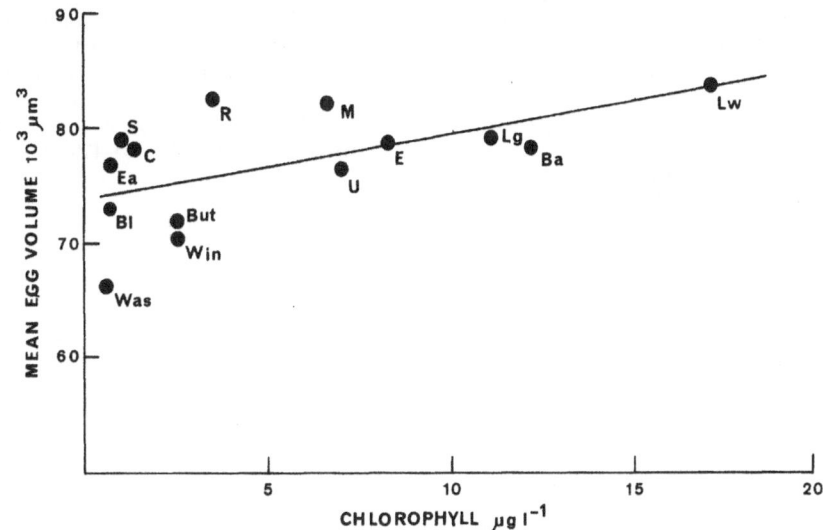

Figure 10. Keratella cochlearis: phytoplankton chlorophyll and mean egg volume in Cumbrian lakes, April 1983. The regression is significant ($p < 0.01 > 0.005$). Ba, Bassenthwaite; B1, Blea Water; But, Buttermere; C, Crummock Water; E, Esthwaite; Ea, Easedale Tarn; Lg, Loughrig Tarn; Lw, Loweswater; M, Mockerkin Tarn; R, Red Tarn; S, Small Water; U, Ullswater; Was, Wastwater; Win, Windermere.

altitudes, but also lower than in more productive lakes at similar altitudes. The comparison between oligotrophic Red Tarn, mesotrophic Windermere and eutrophic Llangorse Lake shows that at high trophic levels smaller eggs are produced at all temperatures.

In a study of natural populations, Galindo et al. (1993) found that food concentration explained most variation in egg size in *Anuraeopsis fissa* and *Brachionus calyciflorus*, while temperature explained most variation in *B. angularis* and *Keratella quadrata*. An important aspect of their findings was that maximum egg volume was found at an intermediate food level in all four species. The data on chlorophyll and egg volume at one time of year in the Cumbrian lakes, combined with the comparison of Red Tarn and Llangorse Lake, indicate that something similar happens in *K. cochlearis*.

One advantage of producing a large egg is that the neonate does not need to grow much to reach adult size. This is clearly an advantage if food is scarce, but it may also give some protection against small predators. In this study those lakes that had large females of *K. cochlearis* invariably had substantial populations of copepods. In Priest Pot, a eutrophic pool in Cumbria, where copepods are scarce, *K. cochlearis* remains relatively small throughout the year. However, this is merely anecdotal evidence, and the presence of predatory copepods needs to be put on a quantitative basis. The situation in Priest Pot is complicated by the fact that small roach (*Rutilus rutilus* (L.)) prey on *K. coch-*

learis (Hewitt & George, 1987) and this may select towards small body size.

Acknowledgments

The work on the Cumbrian lakes was done at Brathay Field Centre, where the late Mike Mortimer made every facility available. I was accompanied in the field by my stalwart friend Jon Davies; without him access to the higher tarns would not have been possible — he carried the boat! I am also grateful to two anonymous referees, whose comments helped to clarify my thoughts.

References

Carrick, T. R. & D. W. Sutcliffe, 1982. Concentrations of major ions in Lakes and Tarns of the English Lake District (1953–1978). Freshwat. Biol. Assoc. Occ. Publ. 16: 1–170.

Galindo, M. D., C. Guisande & J. Toja, 1993. Reproductive investment of several rotifer species. Hydrobiologia 255/256: 317–324.

Green, J., 1956. Growth, size and reproduction in *Daphnia* (Crustacea: Cladocera). Proc. zool. Soc. Lond. 126: 173–204.

Green, J., 1966. Seasonal variation in egg production by *Cladocera*. J. Anim. Ecol. 35: 77–104.

Hewitt, D. P. & D. G. George, 1987. The population dynamics of *Keratella cochlearis* in a hypereutrophic tarn and the possible impact of predation by young roach. Hydrobiologia 147: 221–227.

Hutchinson, G. E., 1951. Copepodology for the ornithologist. Ecology 32: 571–577.

310

Jamieson, C. & C. Burns, 1988. The effects of temperature and food on copepodite development, growth and reproduction in three species of *Boeckella* (Copepoda: Calanoida). Hydrobiologia 164: 235–257.

Jones, R. & K. Benson-Evans, 1974. Nutrient and phytoplankton studies of Llangorse Lake. Field Studies 4: 61–75.

Kirk, K. L., 1997. Egg size, offspring quality and food level in planktonic rotifers. Freshwat. Biol. 37: 515–521.

Ricci, C. 1995. Growth pattern of four strains of a Bdelloid rotifer species: egg size and numbers. Hydrobiologia 313/314: 157–163.

Walz, N. & F. Rothbucher, 1991. Effect of food concentration on body size, egg size, and population dynamics of *Brachionus angularis* (Rotatoria). Verh. int. Ver. Limnol. 24: 2750–2753.

Walz, N., S. S. S. Sarma & U. Benker, 1995. Egg size in relation to body size in rotifers: an indication of reproductive strategy? Hydrobiologia 313/314: 165–170.

Hydrobiologia **387/388**: 311–316, 1998.
E. Wurdak, R. Wallace & H. Segers (eds), Rotifera VIII: A Comparative Approach.
© *1998 Kluwer Academic Publishers.*

Relative investment in offspring by sessile Rotifera

Robert L. Wallace, Jacob J. Cipro & Ryan W. Grubbs
Department of Biology, Ripon College, Ripon, WI 54971-0248, U.S.A.

Key words: amictic (subitaneous) eggs, Atrochidae, body volume, Collothecidae, egg volume, Flosculariidae, larvae, lecithotrophic, life history strategy, planktotrophic, sessile invertebrate

Abstract

Correlations of egg and body volumes of 65 species of oviparous Rotifera from the three families of sessile rotifers (Atrochidae, Collothecidae, Flosculariidae) were determined from size measurements documented in the literature and from unpublished data (RLW). While egg volume (EV) of amictic (subitaneous) eggs increased as a function of body volume (BV) in these families, relative egg volume (REV) decreased with increasing BV indicating that relative investment per offspring is less in larger-bodied species. Regression coefficients for REV as a function of BV for these families were significantly different from each other and that of the strictly planktonic species studied by Walz et al. (1995). Thus, our statistical analysis indicates that relative investment per offspring was greatest in planktonic species, intermediate in Flosculariidae, and lowest in Collothecidae. These results suggest that the sessile families do not follow the standard pattern of EV predicted for planktotrophic and lecithotrophic larvae as is found in many marine benthic invertebrates.

Introduction

Except for a few free swimming species, sessile rotifers (Atrochidae; Collothecidae; Flosculariidae) produce mobile young (inappropriately, but conveniently referred to as larvae) that swim for a certain period of time before attaching to a substratum (Kutikova, 1995; Wallace, 1980). Based on morphology of the larval corona and that of the trophi, Wallace (1980) argued that Flosculariidae should produce larvae that are capable of feeding and Collothecidae should produce larvae that do not feed. The purpose of this study was to explore this hypothesis.

Undoubtedly the best way to investigate this question would be to examine the larval feeding biology of many species from each family. Unfortunately, unlike their planktonic counterparts, sessile rotifers are difficult to collect and culture (Wallace, 1980). Consequently, we advanced our investigation by studying larval development in sessile rotifers by comparing it to larval development in marine benthic invertebrates. Marine larvae employ several developmental strategies, each with its own advantages and disadvantages (Havernhand, 1995; Levin & Bridges, 1995; Pechenik, 1979; Strathmann, 1985; Todd & Doyle,

1981; Vance, 1973). Here we term the compilation of these different strategies the larval development hypothesis (LDH) (Table 1). Thus, to examine Wallace's (1980) hypothesis we explored the applicability of the LDH to the three taxa of sessile rotifers. In doing so, we chose to examine egg volume (EV) and body volume (BV) as a rapid way of testing the relative strength of the LDH. Our null hypothesis was that the relationship of EV to BV of the sessile taxa would be the same as in planktonic rotifers (Pauli, 1989; Pourriot & Rougier, 1991; Walz, 1997; Walz & Rothbucher, 1991; Walz et al., 1995). The alternative hypothesis was that there would be a difference in this relationship. Our specific working hypothesis was that if the Collothecidae produced non-feeding (i.e., lecithotrophic larvae), they would bear eggs with higher egg:body volume ratios across all BV.

Methods

To develop our data we extracted information on EV and BV from the literature and from unpublished data (RLW). A total of 65 oviparous species, including 53 sessile and 12 planktonic forms, were examined in our

Table 1. Larval and egg development features as predicted by the larval development hypothesis (information compiled from several sources)

Parameter	Larval developmental strategy			
	Lecithotrophic	Direct	Mixed	Planktotrophic
Relative egg size	Large	Large	Intermediate	Small
Clutch size	Few	Few	Few	Many
Egg development time	Long	Long	Intermediate	Short
Larval pelagic time	Short	None	Intermediate	Long

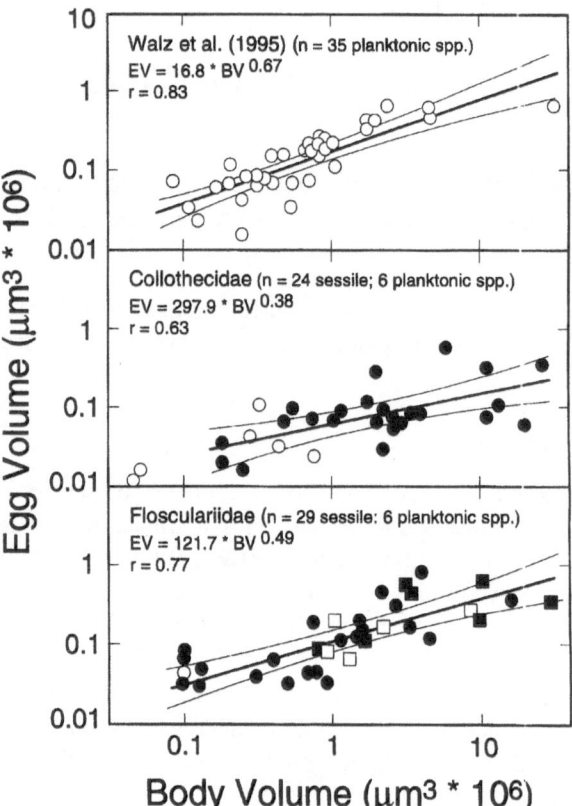

Figure 1. Relationship between egg volume and body volume in rotifers. Upper panel, planktonic species (*n*=35) from Walz et al. (1995); middle panel, families Collothecidae and Atrochidae (*n*=30); lower panel, family Flosculariidae (*n*=35). Closed figures, sessile species; open figures, planktonic (free moving) species. Circles, solitary; squares, occasionally or commonly colonial. Regression lines (±95% confidence intervals) and regression statistics were calculated for only the sessile species in the lower two panels. Regression lines are restricted to the range of data points. Overlapping data points were repositioned slightly to reveal the locations otherwise hidden.

study: Collothecidae (including Atrochidae), *n*=30; Flosculariidae, *n*=35 (Table 2). For comparative purposes we used data compiled by Walz et al. (1995: Table 1) for 35 planktonic species. (NB: Excluded from that data set were all members of Collothecidae, Conochilidae, and Flosculariidae.) Estimates of EV for amictic (subitaneous) eggs and adult BV were calculated from linear dimensions using equations for similar geometric forms (Ruttner-Kolisko, 1977; Walz et al., 1995). When a range in size was available, we used the mean value. Following the analytical methods of Walz et al. (1995) we evaluated the power function $Y = a \times X^b$ from the linear regression of the $\log_{(10)}$-transformed data: thus, $\log Y = \log a + b \times \log X$. Regression lines (±95% confidence intervals) and regression statistics were calculated only for the sessile species.

Before presenting our analysis, we must consider that using these kinds of data has several disadvantages. (1) Although we have attempted an in-depth survey, we do not consider our effort to have been exhaustive. (2) We anticipate that comparisons of our data to specific extant populations will reveal noticeable differences, as was the case when we compared the data set of Walz et al. (1995) to our own unpublished data. Therefore, while we assumed that measurements reported in the literature represented values that were close to the species' means, we have no way of confirming this. (3) Use of geometric forms to calculate biovolumes must be viewed as rough approximations of real body volume. (4) EV may vary considerably (2–4× in some monogononts) depending on a variety of environmental factors (Green, this volume; Walz & Rothbucher, 1991). (5) Data on biovolumes may not reflect true investment if yolk quality changes as a function of maternal nutritional status.

Results and analysis

Planktonic species from Walz et al. (1995)

As expected, except for minor differences in regression statistics, the results reported by Walz et al. (1995) were repeated in our analysis of a slightly smaller set of their data (Figure 1). The slope of the regression line for EV plotted on BV was significantly different from *b*=1 (*P*<0.001), but it was not significantly different from *b* = 0.66 (*P*> 0.05, two-tailed *t*-test). Thus, we felt confident in using this smaller data set of planktonic species for comparative purposes.

Table 2. Estimated egg volume (EV), adult body volume (BV), and relative egg volume% (REV) of 65 oviparous rotifers of families Atrochidae and Collothecidae (Collothecacea), and Flosculariidae (Flosculariacea)

Collothecaceae	EV	BV	REV	Flosculariidae	EV	BV	REV
Acyclus inquietus[1]	34.3	2631.3	1.3	*Beauchampia crucigera*	3.2	49.9	6.4
Collotheca ambigua	8.4	351.9	2.4	*Floscularia conifera*[3]	51.8	322.6	16.1
C. annulata	5.9	2022.3	0.3	*F. decora*	46.1	218.1	21.1
C. balatonica[2]	1.6	5.3	29.7	*F. janus*	11.4	452.3	2.5
C. bicornuta	1.9	18.6	10.3	*F. melicerta*	35.2	1596.4	2.2
C. bilfingeri	6.7	104.0	6.4	*F. noodti*	4.3	71.9	6.0
C. campanulata	10.5	1320.0	0.8	*F. ringens*[3]	52.0	322.6	16.1
C. cornetta	9.1	226.6	4.0	*Lacinularia elliptica*[2,3]	8.4	89.8	9.3
C. cyclops	6.1	296.6	2.1	*La. causeyae*[2,3]	6.5	130.9	4.9
C. edentata	8.8	117.2	7.5	*La. elongata*[3]	8.4	84.8	10.0
C. ferox	27.6	203.5	13.6	*La. flosculosa*[3]	18.8	984.2	1.9
C. gracilipes	8.4	379.8	2.2	*Limnias ceratophylli*[3]	11.1	160.8	6.9
C. heptabrachiata	6.9	76.2	9.0	*L. melicerta*	14.8	160.8	9.2
C. hoodii	56.8	817.6	9.7	*Octotrocha speciosa*	78.5	405.6	19.4
C. libera[2]	1.1	4.6	23.2	*Ptygura beauchampi*	7.5	9.9	75.6
C. longicaudata	6.4	204.3	3.1	*P. brachiata*	4.7	13.2	35.2
C. mutabilis[2]	10.3	33.5	30.8	*P. crystallina*	4.4	78.0	5.5
C. ornata	3.0	232.2	1.3	*P. furcillata*	6.4	10.1	63.4
C. pelagica[2]	3.0	44.2	6.8	*P. libera*[2]	4.4	10.0	43.4
C. quadrinodosa	9.5	55.6	17.0	*P. linguata*	12.7	151.3	8.4
C. riverai	1.6	25.2	6.2	*P. longicornis*	3.7	30.9	12.0
C. rodewaldi[2]	4.1	28.0	14.6	*P. melicerta*	3.2	94.3	3.4
C. tenuilobata	5.3	262.0	2.0	*P. mucicola*	3.0	12.6	23.4
C. trifidlobata	3.4	18.5	18.2	*P. pectinifera*	16.7	327.8	5.1
C. trilobata	7.4	1141.4	0.6	*P. pedunculata*	18.8	155.2	12.1
C. undulata[2]	2.4	77.6	3.0	*P. pilula*	10.9	115.6	9.4
C. volutata	6.4	48.4	13.1	*P. tacita*	6.0	41.6	14.3
C. weberi	7.3	260.0	2.8	*P. tridorsicornis*	3.2	9.8	32.2
Stephanocerus fimbriatus	32.8	1113.2	3.0	*P. vetata*	18.1	75.0	24.2
St. millsii	11.6	177.0	6.5	*Sinantherina aripipes*[3]	30.8	272.2	11.3
				S. procera[3]	33.9	2862.0	1.2
				S. semibulata[2,3]	26.4	853.9	3.1
Ovovivparous Atrochidae[4]				*S. socialis*[3]	59.6	1017.9	5.8
Atrochus tentaculatus[1]	4039.2	37900.0	10.7	*S. spinosa*[2,3]	16.4	223.7	7.3
Cupelopagis vorax[1]	318.0	7841.4	4.0	*S. triglandularis*[2,3]	19.4	103.5	18.7

Linear measurements of amictic (subitaneous) egg and body sizes were taken from the literature and unpublished data (RLW). Estimates of egg and body volume were calculated using equations for similar geometric forms (Ruttner-Kolisko, 1977; Walz et al. 1995). Volume estimates are reported as 10^4 μm^3. REV, relative egg volume calculated as % (egg volume/body volume).
[1] Family Atrochidae; others in the upper part of this column are in the Family Collothecidae.
[2] Planktonic species.
[3] Occasionally or commonly colonial.
[4] The two ovoviviparous Atrochidae are included for comparative purposes.

Collothecidae and Flosculariidae

As in the planktonic species, we found a large difference in EV and BV between extremes in the sessile species. In Collothecidae the ratio of largest BV (*Acyclus inquetius*) to smallest (*Collotheca lib-*

era) was 566. In Flosculariidae this ratio was 293 (largest, *Sinantherina procera*; smallest, *Ptygura tridorsicornis*). These ratios span the value reported by Walz et al. (1995) for the planktonic species (370). In Collothecidae the ratio of largest EV (*Collotheca hoodii*) to smallest (*Collotheca libera*) was 52. In Flosculariidae this ratio was 27 (largest, *Octotrocha spe-ciosa*; smallest, *Ptygura mucicola*). These ratios also encompass the value reported for planktonic species (43).

The relationship of EV to BV is reported by the allometric equations shown in Figure 1. In general these results are similar to those reported by Walz et al. (1995). The correlation coefficients (r) for both sessile families were significantly different from $r=0$ ($P<0.001$, t–test). The slopes of these regressions also were significantly different from $b=1$ ($P<0.001$, t–test) for both families. However, this was not the case for a test of a slope of $b = 0.66$. In that case, the slope of the regression was significantly different for Collothecidae ($P<0.05$, t–test), but not for Flosculariidae ($P>0.05$, t–test). Because there was no distinct pattern of the residuals (differences between predicted log Y and the measured log Y values) for both Collothecidae and Flosculariidae data sets, linearity of the log-transformed regression was accepted (see Walz et al., 1995).

An analysis of covariance among slopes rejected the null hypothesis that the slopes of the three groups are equal ($P<0.001$, $F_{0.05,(1),2,95} \approx 3.09$). In fact, a three-way, multiple comparison among slopes indicates that each is different from the other two ($P<0.001$, Tukey test). This implies that relative investment per offspring by the two sessile families and the planktonic taxa (Walz et al., 1995) is significantly different. Based on these results we rejected the null hypothesis that the relationship of EV to BV of the sessile families is the same as in the planktonic taxa. However, we point out that variation in BV explains no more that about 40% (Collothecidae) and 60% Flosculariidae) of the total variation in EV. Thus, other factors not explored in this study account for a considerable fraction of EV variability.

As was reported by Walz et al. (1995) for their data set, relative egg volume (REV=EV/BV, as %) also decreased with increasing BV for both sessile families (Figure 2). In Flosculariidae the decrease was from 75.6% for *Ptygura beauchampi* to about 1.5% for *Lacinularia flosculosa* and *Sinantherina procera*. However, in Collothecidae this decline was not as steep: from 30.8% for *Collotheca mutabilis* to >1% for three species of *Collotheca* (*C. annulata*, *C. campanulata*, *C. trilobata*). The difference in relative investment per offspring for all three data sets is illustrated in Figure 3 as REV plotted as a function of BV. This figure shows that Collothecidae produce eggs that are smaller than both Flosculariidae and planktonic species

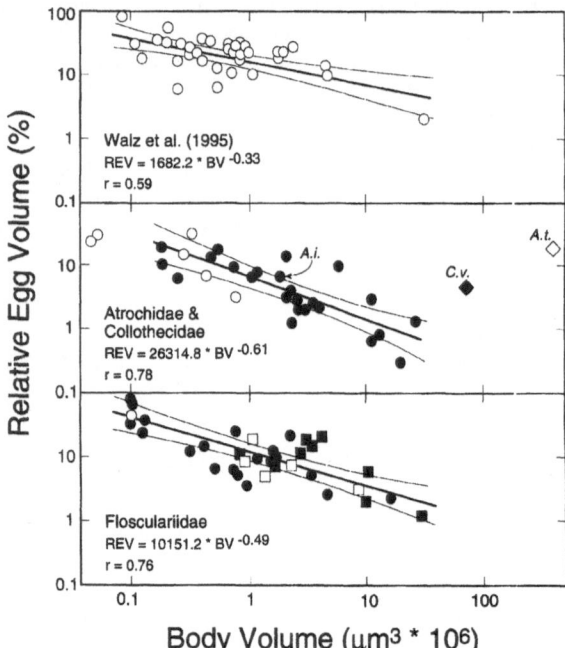

Figure 2. Relationship between relative egg volume (%) and body volume for in rotifers. Legend as in Figure 1, except that the two ovoviviparous Atrochidae (diamonds) are plotted here for comparative purposes: ovoviviparous, *Atrochus tentaculatus* (*A.t.*) and *Cupelopagis vorax* (*C.v.*). The oviparous Atrochidae, *Acyclus inquietus* (*A.i.*), is also indicated.

across all BV. Based on these results we also rejected our alternative hypothesis that Collothecidae produce eggs with higher egg:body volume ratios (i.e., the lecithotrophic strategy). In fact, we found the opposite: relative investment as measured by REV was greatest

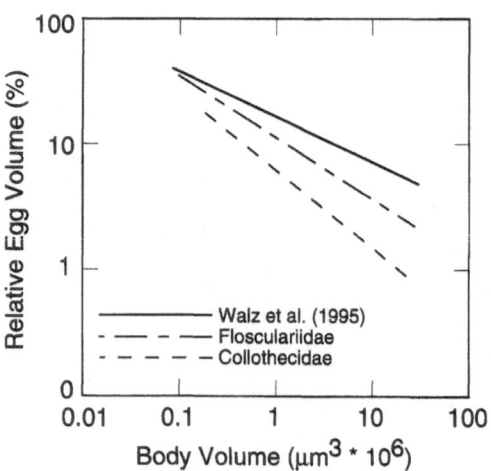

Figure 3. Comparison of regression lines of relative egg volume (%) as a function of body volume for families Collothecidae, Flosculariidae, and 35 planktonic species reported by Walz et al. (1995). Regression lines are restricted to the range of data points.

in planktonic species, intermediate in Flosculariidae, and lowest in Collothecidae.

Miscellaneous observations

Free swimming species of Collothecidae ($n=6$) were significantly smaller in BV ($P<0.01$, Wilcoxon test) and had higher egg:body volume ratios than the sessile species of this family (open versus closed symbols; Figures 1 and 2). However, BV of planktonic Flosculariidae ($n=6$) were evenly distributed ($P>0.05$ Wilcoxon test; open versus closed symbols). Further, we note that, colonial Flosculariidae ($n=14$) possessed larger BV and have smaller egg:body volume ratios than solitary species of this family ($P<0.01$, Wilcoxon test; squares versus circles; Figures 1 and 2). The two ovoviviparous species of Atrochidae are shown in Figure 2. These species possess REV that are 1–2 orders of magnitude larger than what might be expected as members of Order Collothecacea.

Discussion

Sessile families

While there is considerable variation in EV:BV ratios, our results are very similar to those reported by Walz et al. (1995), namely that EV in the sessile families increased as a function of BV, while REV decreased with increasing BV. These results indicate that relative investment per offspring is less in the larger-bodied species. However, REV differs significantly among the three groups (Figure 3). Thus, we reject both our null hypothesis and our working alternative hypothesis, and are left with the conclusion that the sessile rotifers do not follow the standard pattern of egg size predicted for planktotrophic and lecithotrophic larvae (Table 1).

Whether the eggs of sessile rotifers are planktotrophic or lecithotrophic, the adaptive significance of egg size in sessile rotifers may be to produce the largest number of small eggs possible. This fecundity may lead to increased dispersal, or it may be necessary to overcome heavy larval mortality in the plankton. The significance of Collothecidae producing the smallest eggs across all BV may represent the extreme condition, perhaps overcoming heavier losses in the plankton or increasing dispersal. Further, coupling small egg size with a brief development period should increase little-r even more, as Havenhand (1993) has showed for certain marine invertebrates.

Sessile versus planktonic and solitary versus colonial

Coronal ciliation and locomotory ability is greatly limited in the planktonic Collothecidae. Therefore, it is not surprising that the planktonic Collothecidae had small BV and concomitantly smaller REV than the sessile species of that family. The same skewed distribution is not seen in the planktonic Flosculariidae. Here coronal ciliation is strong and feeding and locomotion are probably positively related. The larger BV and concomitantly smaller REV in the colonial Flosculariidae may be related to the putative increased feeding efficiency attributed to colonial forms (Wallace, 1980).

Ovoviviparous Atrochidae

REV in the two ovoviviparous species of Family Atrochidae (*Atrochus tentaculatus* and *Cupelopagis vorax*) is much greater than the third member of their family and all members of the Collothecidae (≈ 10–$100\times$) from what might be expected (Figure 2). These two species may be employing a larval developmental strategy that approches the direct or mixed categories (Table 1). Based on the REV of *A. tentaculatus* and *C. vorax* we predict that they will have a relatively long embryonic development time while spending a short time in the plankton. One significance of this strategy should be a marked reduction of larval mortality; however, it also should greatly limit dispersal.

Conclusions

Unfortunately, there is no way of confirming lecithotrophy in Collothecidae and planktotrophy in Flosculariidae without examining larval feeding biology of numerous species and exploring other criteria predicted by the LDH (Table 1). However, based on our statistical analysis we conclude that sessile rotifers do not follow the standard pattern of larval development as formulated for marine invertebrates. Surprisingly, the pattern of EV is reversed in the two sessile families from that predicted by the LDH, namely that EV is larger in Flosculariidae and smaller in Collothecidae across all BV, and both families are smaller than that of the planktonic comparison group.

There are several unresolved issues involving the hypotheses explored here. (1) Is energy per unit EV uniform over the range of EV and across taxonomic lines (e.g., Jaeckle, 1995)? (2) Do birefringent structures in the guts of embryos and newborn young provide energy for locomotion or post-settlement development (Wallace, 1993)? (3) Does development time change as a function of EV? (4) Does planktotrophy improve habitat selection or larval dispersal? (5) How do ambient food and temperature variations affect EV:BV ratios? Studies to address these questions will take considerable effort, but they are necessary if the question of the applicability of the LDH to sessile rotifers is to be rigorously tested.

Acknowledgments

This research was supported by funds for Faculty Development from Ripon College to R.L.W. We thank D. Light, N. Walz, and an anonymous reviewer for their help in improving the manuscript.

References

Green, J., 1998. Strategic variation of egg size in *Keratella cochlearis*. Hydrobiologia 387/388 (Dev. Hydrobiol. 134): 301–310.

Havenhand, J. N., 1993. Egg to juvenile period, generation time and the evolution of larval type in marine invertebrates. Mar. Ecol. Prog. Ser. 97: 247–260.

Havenhand, J. N., 1995. Evolutionary ecology of larval types. In McEdwards, L. (ed.), Ecology of Marine Invertebrate Larvae. CRC Press, Boca Raton: 79–122.

Jaeckle, W. B., 1995. Variation in the size, energy content, and biochemical composition of invertebrate eggs: correlates to the mode of larval development. In McEdwards, L. (ed.), Ecology of Marine Invertebrate Larvae. CRC Press, Boca Raton: 49–77.

Kutikova, L. A., 1995. Larval metamorphosis in sessile rotifers. Hydrobiologia 313/314: 133–138.

Levin, L. A. & T. S. Bridges, 1995. Pattern and diversity in reproduction and development. In McEdwards, L. (ed.), Ecology of Marine Invertebrate Larvae. CRC Press, Boca Raton: 1–48.

Pauli, H. -R., 1989. A new method to estimate individual dry weights of rotifers. Hydrobiologia 186/187: 355–361.

Pechenik, J. A., 1979. Role of encapsulation in invertebrate life histories. Am. Nat. 114: 859–870.

Pourriot, R. & C. Rougier, 1991. Importance volumétrique des oeufs chez les Rotifères planctoniques. Ann. Limnol. 27: 15–24.

Ruttner-Kolisko, A., 1977. Suggestions for biomass calculation of planktonic rotifers. Arch. Hydrobiol. Beih. 8: 71–76.

Strathmann, R. R., 1985. Feeding and nonfeeding larval development and life-history evolution in marine invertebrates. Ann. Rev. Ecol. Syst. 16: 339–361.

Todd, C. D. & R. W. Doyle, 1981. Reproductive strategies of marine benthic invertebrates: a settlement–timing hypothesis. Mar. Ecol. Prog. Ser. 4: 75–83.

Vance, R., 1973. Reproductive strategies in marine benthic invertebrates. Am. Nat. 107(955): 339–352.

Wallace, R. L., 1980. Ecology of sessile rotifers. Hydrobiologia 73: 181–193.

Wallace, R. L., 1993. Presence of anisotropic (birefringent) crystalline structures in embryonic and juvenile monogonont rotifers. Hydrobiologia. 255/256: 71–76.

Walz, N., 1997. Rotifer life history strategies and evolution in freshwater plankton communities. In: Streit, B., T. Städler & C. M. Lively (eds), Evolutionary Ecology of Freshwater Animals. Birkhäuser Verlag, Basel: 119–149.

Walz, N. & F. Rothbucher, 1991. Effect of food concentration on body size, egg size, and population dynamics of *Brachionus angularis* (Rotatoria). Verh. int. Ver. Limnol. 24: 2750–2753.

Walz, N., S. S. S. Sarma & U. Benker, 1995. Egg size in relation to body size in rotifers: an indication of reproductive strategy? Hydrobiologia 313/314: 165–170.

Hydrobiologia **387/388**: 317–320, 1998.
E. Wurdak, R. Wallace & H. Segers (eds), Rotifera VIII: A Comparative Approach.
© *1998 Kluwer Academic Publishers.* .

The paradox of bdelloid egg size

Simona Orsenigo, Claudia Ricci & Manuela Caprioli
Dipartimento di Biologia, via Celoria 26, 20133 Milano, Italy

Key words: Rotifera, bdelloids, egg volume, RES, trade-off, fecundity, recovery

Abstract

Egg volumes or relative egg size (RES) of seven bdelloid species were plotted against life-history traits, and recovery rates from 7-day desiccation periods to find evidence for the costs and benefits of producing a few big, or many small eggs. Increased RES of bdelloids is correlated with decreased fecundity and longevity, increased age at maturity, increased egg developmental time and increased recovery from desiccation for both embryos and adults. The production of large eggs represents a cost for the bdelloid rotifer, which, at first sight, does not receive compensating advantages. This paradox, however, is only superficial, as it is suggested that an increase of recovery and, in particular, increased viability of late embryos compensates for the loss of fitness related to the production of large eggs.

Introduction

Trade-offs between the number and size of eggs have received considerable attention in many taxa, from Crustacea (e.g., Poulin, 1995) to Vertebrata (e.g., Elgar, 1990). For all organisms it was found that the resulting egg size is a compromise between producing many small, or few big eggs. The choice depends upon habitat pressures, physiological constraints and reproductive modalities.

Rotifera are eutelic and lack larval stages, two characteristics which limit the animals' options for egg size. Literature records on body and egg volumes of species of Rotifera Monogononta, mostly from field samples, report single eggs to constitute up to 84% of the adult volume (Pourriot & Rougier, 1991; Walz et al., 1995). No trade-off between egg size and number was assessed either across species or across environmental conditions because this approach necessitates laboratory studies to determine life-table traits.

Life-history studies of several bdelloid rotifers were carried out under controlled conditions, and it was found that net reproduction rates (R_0) differ considerably among species (Ricci, 1983, 1991). These differences are related to habitat pressures acting on different species which canalized resources to maximize either fecundity or survivorship. Since the cost of egg production seems to differ among species (Ricci &

Fascio, 1995), it can be hypothesized that the amount of resources a bdelloid invests per egg determines the trade-off between egg size and number (= mean fecundity), as in other organisms. But, while the production of large eggs is more costly, compensating advantages like increased survivorship are expected. Probably, rotifers that produce more eggs will allocate less resources per egg, which will be smaller and hatch into a smaller juvenile, while animals having a few eggs only, will be able to produce relatively larger ones, resulting in hatchlings with a store of energy improving their chances of survival. A trade-off between fecundity and survival was reported in a monogonont rotifer (Snell & King, 1977), however egg size was not considered. Most bdelloids live in unpredictable habitats which desiccate from time to time. They can survive desiccation by entering anhydrobiosis which, however, can cause dramatic mortality rates. In such a case, the extra investment required to produce large eggs could be remunerative if it enhances the probability of recovery from desiccation.

In this study we analyze the influence of body volume and life-history traits on the egg size of several species of Rotifera Bdelloidea, focusing on the costs and benefits of large eggs. Trade-offs were established by plotting either egg volumes or relative egg size (RES, Lynch, 1980) against life-history traits. In addition, we attempt to find the compensating advantage of large eggs by relating their volumes to survival of eggs, juveniles and adults after 7-day desiccation.

Material and methods

At 22° C cohorts of seven bdelloid species were studied following the life-table experimental protocol (Birch, 1948) to determine life-history traits and age-specific volumes of animals and eggs (Ricci & Fascio, 1995). The species were *Habrotrocha tranquilla, Philodina acuticornis, P. roseola* and *P. rapida*. The results of the experiments will be described in detail in a separate paper. Other species' life-table data and volumes (*P. vorax*, and several strains of *Macrotrachela quadricornifera* and *M. q. vanoyei*) recorded under the same experimental conditions are already available (Ricci, 1995; Ricci & Fascio, 1995). Results on additional strains of *P. acuticornis, M. quadricornifera* (called C, G, H) and *M. q. vanoyei* (formerly called *M. q.* Va and S), collected from different sites, are here considered separately. Volumes of eggs and animals of each species or strain were obtained by averaging of about 200 measurements.

Embryogenesis time of four species, *H. tranquilla, M. quadricornifera, M. q. vanoyei* and *P. vorax*, was longer than 25–30 h. Their eggs were desiccated at different developmental stages, kept dry for 7 days, rehydrated and tested for their viability. Egg volumes and RES (Lynch, 1980) were related to the life-history traits and recovery rates through Pearson's correlation tests.

Results

The size of the eggs differed between species, and was influenced by the volume of the animals at full size ($r=0.644$, $P=0.02$, $n=12$) and at maturity ($r=0.920$, $P<0.0001$, $n=12$). Egg sizes in different species were relativized by comparing egg volume to the volume of the animal at maturity (RES). RES values ranged between 8.4 (*P. rapida*) and 21.8 (*P. vorax*). We found a significant negative correlation between RES and fecundity (mean reproductive rate) ($r=-0.837$, $P=0.001$, $n=12$) (Figure 1a). As for longevity, a negative but insignificant correlation was found when all species were grouped together ($r=-0.504$, $P=0.09$, $n=12$) or for *Philodina* ($r=-0.916$, $P=0.083$, $n=4$), this negative correlation was significant for *Macrotrachela* ($r=-0.947$, $P=0.004$, $n=5$) (Figure 1b).

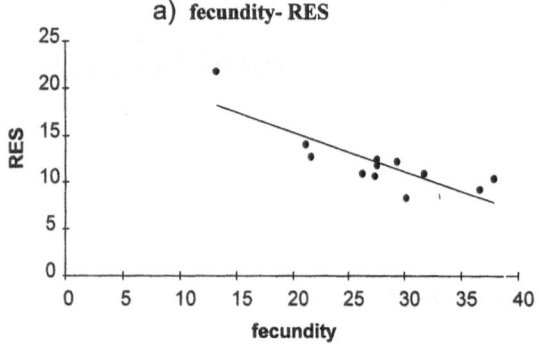

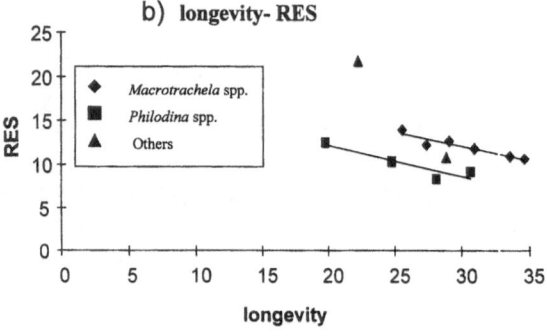

Figure 1. Trade-offs between Relative Egg Size and life-history traits of 12 species, or strains. (a) Fecundity – RES, $r=-0.837$; $P=0.001$. (b) Longevity – RES correlation coefficient for all species: $r=-0.504$; $P=0.094$. The same for *Philodina* spp.: $r=-0.917$; $P=0.083$. The same for *Macrotrachela* spp.: $r=-0.948$; $P=0.004$.

The egg volumes, expressed as RES, were not significantly correlated to age at maturity ($r=0.507$, $P=0.09$, $n=12$), or embryogenesis time ($r=0.272$, $P=0.22$, $n=12$). Two or 3-day-old eggs of *H. tranquilla, M. quadricornifera, M. q. vanoyei* and *P. vorax* had a higher probability of recovering from desiccation than younger eggs (Figure 2), indicating that the longer duration of embryogenesis increased recovery from anhydrobiosis. No statistically significant correlation was found between RES and recovery rate at all tested ages, after 7-day anhydrobiosis ($r=0.202$, $P=0.27$, $n=32$).

Discussion

The results of all correlation tests between RES and life-history traits of bdelloids reveal an apparent paradox: the rotifers that produce bigger eggs decrease their fecundity, life expectancy, increase generation time, and do not obtain compensating advantages. Possibly, the advantages do not concern enhancement of life-history traits, but improved fitness of bdelloids

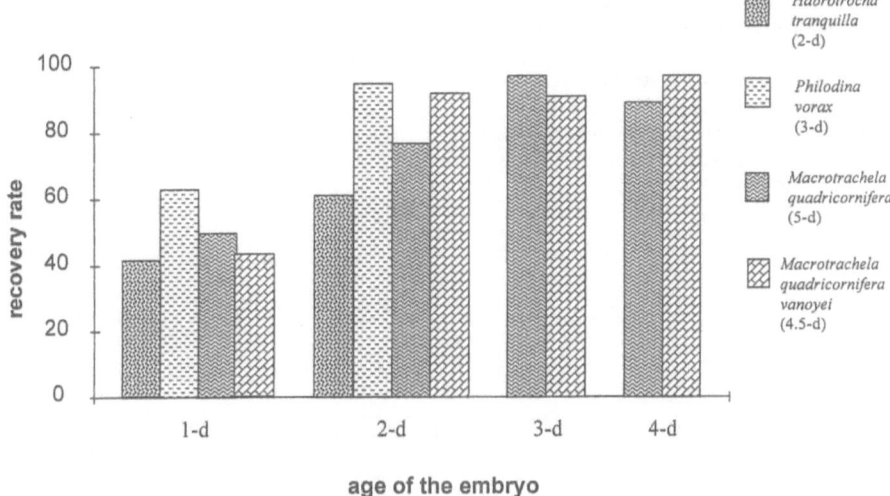

Figure 2. Histograms of the recovery rates of embryos at different ages after 7-day desiccation. The number of the embryos dried for 7 days is given on each histogram. The duration of embryogenesis for each species is given under the species name in parenthesis.

could also result from a higher capability of recovering from anhydrobiosis, which seems relevant to their survival.

Bdelloid late embryos were found more resistant to desiccation (Örstan, 1995), the viability rate of late embryos was twice that of newly laid eggs, and this can be generalized to more species. Commonly, the adults of bdelloid rotifers represent the stage with the highest probability of recovering from anhydrobiosis (Ricci, this volume), and the late embryos of a species with a long developmental time could represent an additional stage of maximal resistance. The presence of two resistant stages during the lifetime can increase the chances of a rotifer population to survive desiccation. Following this line of reasoning, after a severe drought the structure of a bdelloid population will consist of at least two age classes, one mature and almost ready to lay eggs, and the other close to hatching and starting the new generation, increasing the heterochrony of the resulting population. Furthermore, eggs possibly can be better dispersed than adults because of their relatively smaller size. Thus, another aspect that could be important and deserves attention concerns the possible interactions between egg size, dormancy and dispersal, analogous to what is known from plant seeds (Venable & Brown, 1988). According to these hypotheses, egg size could represent a correlate result of a bet-hedging strategy.

At the moment, we do not have a clear indication that the duration of embryo development compensates

for the cost of large eggs; bdelloid development time increased with RES, but the correlation coefficient was very low and not significant. The same trend can be seen on the recovery rates of rotifers at different ages. Hence, we can only speculate that some increase of recovery and the presence of an additional resistant phase could compensate for the loss of fitness caused by the production of large eggs. Such hypothesis has yet to be tested rigorously with both experimental and theoretical approaches.

References

Elgar, M. A., 1990. Evolutionary compromise between a few large and many small eggs: comparative evidence in teleost fishes. Oikos 59: 283–287

Lynch, M., 1980. The evolution of cladoceran life histories. Quart. Rev. Biol. 55: 23–42.

Örstan, A., 1995. Desiccation survival of the eggs of the rotifer *Adineta vaga* (Davis, 1873). Hydrobiologia 313/314: 373–375.

Poulin, R., 1995. Clutch size and egg size in free-living and parasitic copepods: a comparative analysis. Ecology 49: 325–336.

Pourriot, R. & C. Rougier, 1991. Importance volumétriques des oeufs chez les Rotifères planctoniques. Ann. Limnol. 27: 15–24.

Ricci, C., 1983. Life histories of some species of Rotifera Bdelloidea. Hydrobiologia 104 (Dev. Hydrobiol. 14): 175–180.

Ricci, C., 1991. Comparison of five strains of a parthenogenetic species, *Macrotrachela quadricornifera* (Rotifera, Bdelloidea). I. Life-history traits. Hydrobiologia 211: 147–155.

Ricci, C. & U. Fascio, 1995. Life-history consequences of resource allocation in two bdelloid rotifer species. Hydrobiologia 299: 231–239.

320

Ricci, C., L. Vaghi & M. L. Manzini, 1987. Desiccation of roti-fers (*Macrotrachela quadricornifera*): survival and reproduction. Ecology 68: 1488–1494

Snell, T. W. & C. E. King, 1977. Lifespan and fecundity patterns in rotifers: the cost of reproduction. Evolution 31: 882–890.

Venable, D. L., & J. S. Brown, 1988. The selective interactions of dispersal, dormancy, and seed size as adaptions for reducing risk in variable environments. Am. Nat. 131: 360–384.

Walz, N., S. S. S. Sarma & U. Benker, 1995. Egg size in relation to body size in rotifers: an indication of reproductive strategy? Hydrobiologia 313/314: 165–170.

Hydrobiologia **387/388**: 321–326, 1998.
E. Wurdak, R. Wallace & H. Segers (eds), Rotifera VIII: A Comparative Approach.
© *1998 Kluwer Academic Publishers.*

Anhydrobiotic capabilities of bdelloid rotifers

Claudia Ricci
Dipartimento di Biologia, via Celoria 26, 20133 Milano, Italy

Key words: bdelloid rotifers, anhydrobiosis, desiccation, recovery

Abstract

To test if anhydrobiotic capability is apomorphic to class Bdelloidea, I focused on the recovery from desiccation of 15 bdelloid species. The species belonged to 6 genera, represented the four bdelloid families, and were collected from water and moss environments. Eggs or embryos, prereproductive and reproductive specimens of most species were desiccated and kept dry for 7 days. The highest recovery rates were obtained rehydrating adult bdelloids of moss species, while three aquatic species did not survive anhydrobiosis. Species from aquatic and moss habitats differed in their capacity to enter anhydrobiosis and to recover successfully. This difference may be related to the different desiccation frequencies of the two habitats, although aquatic species were able to survive desiccation. It seems likely that anhydrobiotic capacity is a feature common to all bdelloids, and that was subsequently lost by some species.

Introduction

Sediments and periphyton of lotic and lentic water bodies, and also soils, mosses and lichens are common habitats of Bdelloidea species. Although the species that can be found in the different environments often differ, the wide ecological adaptability of the class has been attributed to two biological characteristics: obligatory parthenogenesis as a means to escape a critical population density and the ability to resist desiccation by entering a dormant state, called anhydrobiosis (Ricci, 1987, 1992). Anhydrobiosis, a particular form of dormancy, is caused by the loss of water for evaporation; it requires the animals to undergo several metabolic adjustments, like the synthesis of protective chemicals (Crowe, 1971; Keilin, 1959). As most bdelloids are inhabitants of unstable environments that vary rapidly between frozen, wet and dry conditions, survival through anhydrobiosis could be a feature common to the class (Gilbert, 1974). But, while animals living in terrestrial environments are permanently exposed to the selection exerted by the habitat desiccation, and are expected to be equipped with all the metabolic machinery to enter anhydrobiosis, those from large water bodies should not face desiccation frequently, and should not 'practice' anhydrobiosis commonly. If aquatic species retain a capability to

enter anhydrobiosis, it cannot be due to habitat selection but to some 'innate' capacity, possibly inherited from their ancestors.

Species belonging to several genera are reported to withstand desiccation and to recover successfully from anhydrobiosis (Table 1), but their recovery capacities were never investigated in detail (e.g., Dobers, 1915; Donner, 1956; Gilbert, 1974). Bdelloid rotifers can enter anhydrobiosis at any age during their life time, and their survival was found to differ with age (Ricci et al., 1987).

Here I report the recovery rates of 15 species after 7-d anhydrobiosis. To test if anhydrobiotic capacity is apomorphic to the whole class or specific of some taxa, the species studied belonged to different genera and represented all bdelloid families. The animals were collected from different habitats in order to test if the environmental selective pressure affects the anhydrobiotic capacities and if resistant bdelloids were limited to any particular environment.

Materials and methods

All species were cultured at 22 °C in embryo dishes with deionized water and a suspension of powdered trout pellets as food. Each embryo dish contained a

Table 1. List of the species studied arranged according to their habitat

Family	Moss and soil	Permanent water	Temporary water
Habrotrochidae	*Otostephanos macrantennus* de Koning, 1947		
Philodinidae	*Philodina rapida* Milne, 1916	*Philodina acuticornis* Milne, 1916	
	P. vorax (Janson, 1893)	*P. roseola* Ehrenberg, 1832	
	Rotaria sordida (Western, 1893)	*Rotaria neptunia* (Ehr., 1832)	
	Macrotrachela quadricornifera vanoyei Schepens, 1954	*R. neptunoida* Harring, 1913*	
		R. rotatoria (Pallas, 1766)	
Adinetidae	*Adineta vaga* (David, 1873)		
Philodinavidae		*Henoceros falcatus* (Milne, 1916)	*Abrochtha intermedia* (de Beauchamp, 1909)
		Philodinavus paradoxus (Murray, 1905)	

Newly-laid eggs, 2-d-old juveniles and 8-d-old adults were desiccated for each listed species, except the three Philodinavidae and *Rotaria* spp.

*Including embryos in gravid adults.

clonal group of 10–20 rotifers of the same age. The rotifers were desiccated at two ages, corresponding to prereproductive (2-d-old) and reproductive (8-d-old) stage. Newly-laid eggs were collected and dried, and parallel control eggs were kept hydrated and checked for viability. Viviparous species, as *Rotaria*, were desiccated as embryos (still inside the mother), newborns and adults, but only a few specimens ($\cong$ 30) for each age class were tested. The *Rotaria* adults were reproductive but their ages were not controlled. Species of family Philodinavidae were desiccated occasionally and the ages of the rotifers were unknown: only qualitative data are reported here.

Rotifers and eggs were desiccated by adding small pieces of filter paper to the embryo dish and placing the animals between the paper pieces, removing most water medium with a pipette and allowing the remaining water to evaporate. Slow dehydration is very important for animals' recovery (Jacobs, 1909), and the embryo dishes were kept in a closed container at 22 °C for about 24 h to retard evaporation. After that time, all samples were completely dry and the embryo dishes were stored in a Petri dish and kept at the same temperature until re-hydration. Of course, such procedure did not control the amount of water left in the samples, which can affect the desiccation process remarkably. When possible, 3–6 replicate dishes were desiccated for each age experiment following the same procedure and the mean value of recovery rates (%, ±S.E.) is indicated. The number of specimens of each experiment ranged between 40–100 according to the consistency of the results between the replicates.

The duration of desiccation in most experiments was 7 days. Rotifers were counted as alive if they were active 24 h after re-hydration. Eggs were checked for viability when the developmental time of control eggs was over. After re-hydration, the developmental time of all eggs was not affected by the anhydrobiosis and no delayed hatching occurred. Eggs were counted as viable if the newborns were alive 24 h after hatching (Örstan, 1995). Viability of rehydrated eggs is reported to differ with development stage (Örstan, 1995; Ricci, 1996), but here I consider the recovery of recently laid eggs only.

The species used in this study were collected from different habitats, and are listed in Table 1.

Results

The rotifers that survived 7-d desiccation resumed activity about 1 h after rehydration. Under the given conditions, three species did not recover at any stage of their lives: *Otostephanos macrantennus*, *Rotaria neptunia* and *R. rotatoria*. The other species showed different survival capacities.

Eggs of all tested species had rather low probability to survive desiccation: viability was 40–60%. Viability of the control eggs was 100% in all species except *P. rapida*. About 25–30% of the eggs of *P. rapida* under hydrated conditions failed to hatch, and the viability of the desiccated eggs was adjusted to that of controls. Only the eggs of *Otostephanos macrantennus* collapsed during desiccation, while 100% of its hydrated control eggs hatched regularly. Some gravid *Rotaria* specimens were desiccated and the recovery of both embryo and mother were considered. From a sample of 15 gravid *R. neptunoida*, 8 embryos survived desiccation, and 5 of them hatched from the dead mother. Three of the desiccated *R. neptunoida* embryos were released before drying, so that the newborns were desiccated; two newborns recovered from 7-d desiccation. These data are not significant, of course, but only indicative of the species capabilities.

Young, pre-reproductive aquatic species showed poor recoveries (20–50%). Moss species performed better; 50–100% juveniles recovered. The recovery capacity of the juveniles of three species, *P. vorax*, *P. rapida* and *H. tranquilla*, was very similar to that of the respective adults. In these species, the recovery of all stages was very high, suggesting that they are excellent anhydrobionts. Reproductive animals had the highest recoveries, and again the highest survival (> 90%) was from moss species, while aquatic species had relatively poorer recoveries (60–80%) (Figure 1). On a different occasion, out of the population of 150 *Otostephanos macrantennus* kept dry for about one month, only one adult recovered, suggesting that the species has some capacity to undergo anhydrobiosis, but that possibly the experimental conditions were not suitable for its survival. Further experiments were not performed on the species.

Almost all *Abrochta intermedia* and most *Henoceros falcatus* recovered from 7-d dryness, but no recovery percentage was calculated. A culture of about 15 specimens of *Philodinavus paradoxus* was desiccated for 7 days, and only a juvenile and an adult recovered. No experiment was run to test age-specific resistance, but eggs of all three species were able to survive desiccation.

On the whole, adult rotifers had better anhydrobiosis capabilities than juveniles and eggs (Figure 1). An exception seemed to be *Rotaria neptunoida*, which ap-

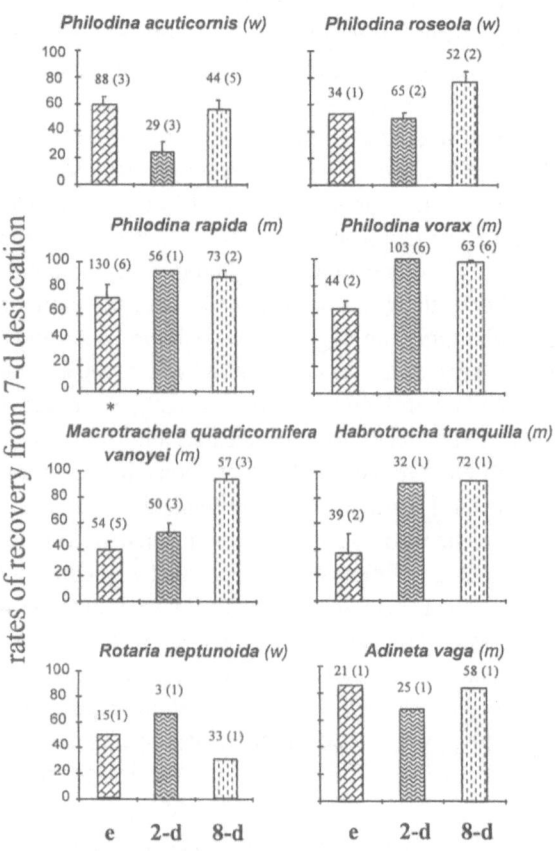

Figure 1. Recovery rates of species collected from water bodies (W) and terrestrial mosses (M). e: newly laid eggs or embryos. 2-d: 2-d-old juveniles. 8-d: 8-d-old adults. The number on the histogram refers to sample size, and in parentheses the replicate number. The histogram shows the average value of the recovery rates among the replicates and the bar indicates the standard error. * the viability percentage is adjusted to control.

parently was more resistant if desiccated as an embryo or newborn than as an adult.

Discussion

The study intended to test if anhydrobiotic capability is apomorphic to class Bdelloidea, and the recovery from desiccation of 15 bdelloid species was considered. The species belonged to 6 genera, represented the four bdelloid families, and were collected from both water and moss environments. The bdelloid rotifer's age at the time of drying affected its recovery from desiccation. This pattern occurred in all tested species, confirming other reports (Dobers, 1915; Ricci et al., 1987). Recently laid eggs had poor viabilities after 7-d desiccation, but older embryos survived

desiccation better (Örstan, 1995; Ricci, 1996 and unpublished). The viability enhancement could be either due to more efficient synthesis of protective chemicals (e.g. trehalose) and/or to finer control of the rate of water loss by the older embryos. The importance of water loss control rate is suggested by the desiccation resistance of *Rotaria neptunoida* embryos, which recovered better than adults. The embryos were provided extra protection from desiccation by the mother's body, be it dead or alive. This is probably similar to what is reported on the resistance of embryos of viviparous nematodes (Evans & Perry, 1976).

Pre-reproductive 2-d-old rotifers generally had poor resistance to desiccation, and the reproductive rotifers were more resistant to anhydrobiosis, and the species from moss had higher probabilities to recover from 7-d-desiccation. In a previous study, *Macrotrachela quadricornifera* were kept dry for 4 days, and the recovery rates at all ages were remarkably lower than those presented here for *M.q. vanoyei* or any other species from moss (Ricci et al., 1987). The main differences between this study and Ricci et al. (1987) concerned the desiccation procedure and the use of batch culture vs isolated animals to be dried; possibly both facts contributed to make desiccation faster in the old study, thus affecting the recovery capacity of the rotifers.

Three of the 15 bdelloid species tested (*R.. rotatoria, R. neptunia* and *O. macrantennus*) did not recover at any life stage under the given conditions; although *O. macrantennus* exhibited slight desiccation resistance. The failure of these species to recover from desiccation cannot be extended to the respective genera; other species of *Rotaria* did enter anhydrobiosis and recovered successfully, and *Otostephanos* species are reported to inhabit mosses and soils (Table 2), where they could not live if incapable of recovery after desiccation. *R. rotatoria, R. neptunia* and *O. macrantennus* live in water bodies that do not dry out frequently. However, other bdelloid species that typically dwell in permanent water habitats (e.g. *P. acuticornis* or *R. neptunoida*) survived desiccation quite well. The capacity to enter anhydrobiosis and to recover successfully differs between species that inhabit desiccation-prone habitats and aquatic habitats, but the latter group also seems able to survive desiccation. Different desiccation probabilities of habitats could account for the different recoveries observed, but this hypothesis requires additional tests.

Bdelloidea consists of three orders and four families (Donner, 1965; Melone & Ricci, 1995), and the

Table 2. List of the genera of class Bdelloidea. For each genus the main habitat is reported (after Donner, 1965), and its capability to recover from anhydrobiosis

Genus	Occurrence	Desiccation (+,−)	References in addition to Donner, 1965
Habrothrochidae			
Habrotrocha	Mainly moss and soil	+	Schramm, Becker, 1987; this study
Otostephanos	Moss, *Sphagnum, Lemna*	(+)	Murray, 1911a
Scepanotrocha	Moss, soil	(+)	Donner, 1975
Philodinidae			
Anomopus	Water	?	
Ceratotrocha	Moss, soil	(+)	Donner, 1975
Didymodactylos	Moss	(+)	Donner, 1975
Dissotrocha	Moss, *Sphagnum*	?	
Embata	Water	?	
Macrotrachela	Moss, water	+	Dobers, 1915; Ricci et al., 1987
Mniobia	Moss, soil	+	Dobers, 1915
Pleuretra	Moss	(+)	Murray, 1911b
Philodina	Moss, water	+	Jacobs, 1909; this study
Rotaria	Mainly water, soil	+	This study
Zelinkiella	Water	?	
Adinetidae			
Adineta	Moss, water	+	Dobers, 1915; Örstan, 1995
Bradyscela	Moss	(+)	Donner, 1975
Philodinavidae			
Abrochtha	Water	+	This study
Henoceros	Water	+	This study
Philodinavus	Water	+	This study

+: Resistance is documented. (+): Resistance is expected from a consideration of environmental dryness.
?: Unknown.

ability to survive desiccation seems common to species from all families. Does this capacity represent an apomorphy or a homoplasy, in other words is it ancestral or was it gained by several bdelloid taxa independently?

Anhydrobiotic ability implies that the animal must be capable of complex morphological and physiological adjustments in order to reduce surface/volume ratio, to lose a great deal of water, and to synthesize chemicals to protect biological structures (Crowe, 1971; Womersley, 1981). Therefore, the events that increase the probability of recovery after anhydrobiosis are rather elaborate and probably are controlled by a set of coadapted genes. It seems improbable that several species, phylogenetically related to each other, have acquired such complex capacity independently of each other. It seems more likely that single species lost anhydrobiotic capacity because it was unnecessary, than that several species 'invented' such an elaborate process independently. If anhydrobiotic capability is common to most bdelloid rotifers, as indicated by this study, it can be better assumed to be apomorphic than homoplasic to the class.

Acknowledgements

This study is part of project supported by ASI (Agenzia Spaziale Italiana).

References

Crowe, J. H., 1971. Anhydrobiosis: an unsolved problem. Am. Nat. 105: 563–573.

Dobers, E., 1915. Über die Biologie der Bdelloidea. Int. Revue ges. Hydrobiol. Hydrol. Suppl. zu Bd. 7: 1–128.

Donner, J., 1956. Über die Mikrofauna, besonders die Rotatorienfauna des Bodens. VI Congrés Int. Science du Sol. III 20: 121–124.

Donner, J., 1965. Ordnung Bdelloidea. Akademie Verlag. 297 pp.

Donner, J., 1975. Randbiotope von Fliessgewässern als orte der Anpassung von Wasserorganismen an Bodenbedingungen, gezeigt an Rotatorien der Donau und Nebenflüsse. Verh. Ges. Ökkol.: 231–234.

Evans, A. A. F. & R. N. Perry, 1976. Survival strategies employed by nematode anhydrobiotes in relation to their natural environment. In: N. A. Croll (ed.), The organization of Nematodes. Academic Press, New York: 383–424.

Gilbert, J. J., 1974. Dormancy in rotifers. Trans. Am. Microsc. Soc. 93: 490–513.

Jacobs, M. H., 1909. The effects of desiccation on the rotifer *Philodina roseola*. J. Exp. Zool. 6: 207–263.

Keilin, D., 1959. The problem of anabiosis or latent life: history and current concept. Proc. Roy. Soc. London 150: 149–191.

Melone, G. & C. Ricci, 1995. Rotatory apparatus in Bdelloids. Hydrobiologia 313/314: 91–98.

Murray, J., 1911a. Some African rotifers: Bdelloidea of tropical Africa. J. Royal Mic. Soc.: 1–18.

Murray, J., 1911b. Bdelloid Rotifera of South Africa. Ann. Transvaal Mus. 3: 1–19.

Örstan, A., 1995. Desiccation survival of the eggs of the rotifer *Adineta vaga* (Davis, 1873). Hydrobiologia 313/314: 373–375.

Ricci, C., 1987. Ecology of bdelloids: how to be successful. Hydrobiologia 147: 117–127.

Ricci, C., 1992. Rotifers: parthenogenesis and heterogony. In R. Dallai (ed.), Sex Origin and Evolution. Selected Symposia and Monographs UZI 6: 329–341.

Ricci, C., 1996. Desiccation as a switch for some microinvertebrates. Proc. VI Eur. Symp. Life Sciences Research in Space. ESA SP-390: 273–275.

Ricci, C., L. Vaghi & M. L. Manzini, 1987. Desiccation of rotifers (*Macrotrachela quadricornifera*): survival and reproduction. Ecology 68: 1488–1494.

Schramm, U. & W. Becker, 1987. Anhydrobiosis of the Bdelloid Rotifer *Habrotrocha rosa* (Aschelminthes). Z. mikrosk.-anat. Forsch. 101: 1–17.

Womersley, C. Z., 1981. Biochemical and physiological aspects of anhydrobiosis. Com. Bioch. Physiol. 70B: 669–678.

Hydrobiologia **387/388**: 327–331, 1998.
E. Wurdak, R. Wallace & H. Segers (eds), Rotifera VIII: A Comparative Approach.
© *1998 Kluwer Academic Publishers.*

Factors affecting long-term survival of dry bdelloid rotifers: a preliminary study

Aydin Örstan
13348 Cloverdale Place, Germantown, MD 20874, U.S.A.
E-mail: bdelloid1@aol.com

Key words: Rotifera, Bdelloidea, anhydrobiosis, extraction method, humidity, oxygen

Abstract

Naturally dried lichens and mushrooms were collected, stored at various relative humidities and temperatures either under air or argon, and extracted in a 0.2 M sucrose solution to determine the long-term survival of resident bdelloid rotifers. Survivorship of rotifers in samples kept at 21 °C for 8 months declined at both <1% and 76% humidities, but remained the same as the starting levels at 23% and 43% humidities. Lowering the temperature to 4 °C improved survival at both <1% and 76% humidities; at −20 °C and <1% humidity, survivorship of rotifers did not decline for up to 18 months. Storage at 21 °C under argon gas improved survival of bdelloids at <1% humidity, but not at 76% humidity. These results suggest that several processes, including oxidation reactions, may be partly responsible for death of anhydrobiotic bdelloids. To facilitate taxonomic work it is recommended that naturally dried samples containing bdelloids be stored over a desiccant at temperatures below 0 °C until they are to be rehydrated.

Introduction

Bdelloid rotifers (Rotifera; Bdelloidea), Tardigrada and Nematoda are the most abundant microscopic invertebrates present in semiterrestrial, freshwater microhabitats: regions subject to frequent drying (Fantham & Porter, 1945; Stubbs, 1989; Wright, 1991). While tolerance of tardigrades and nematodes to desiccation has been well studied (e.g., Crowe & Madin, 1974; Wright et al., 1992), relatively few studies have been done on bdelloids (e.g., Ricci et al., 1987). Previously I have examined survivorship of the eggs of *Adineta vaga* (Davis) at <1% humidity (Örstan, 1995). Here, I report a preliminary study that examines effects of temperature, humidity, and anoxic conditions on survival of naturally dried bdelloids. To do this naturally dried lichens and mushrooms were collected a few days after a rain and stored at various humidities and temperatures either in air or argon atmospheres. After various lengths of storage, portions of these materials were rehydrated, the bdelloids extracted and the survivors enumerated.

Materials and methods

Sample collection and storage

Three sets of samples were collected from a wooded park in Montgomery County, Maryland, USA: Set #1 – Foliose lichens from rocks and tree trunks collected three days after a rain; Set #2 – Polypore mushrooms from a tree trunk collected two days after a rain; Set #3 – Foliose lichens from a tree trunk.

All experiments were begun within 1–3 days after collection of the samples. During this period samples were kept at room temperature (17–22 °C) and humidity (ca. 60%). In each set, the initial sample was cut into small pieces, mixed, and divided into portions. A sample portion and a drying solution (see *Establishing constant humidity*, below) were sealed in separate vials inside either a glass jar (90 ml) with a tight plastic lid or a glass jar (150–500 ml) with a rubber gasket. The latter jars were used for storage under argon. To replace air, a jar was flushed with argon (99.998% pure, $O_2 < 0.0004\%$) from a hose attached to a pressurized argon cylinder for about 2 min before it was sealed. This process was repeated after every time such a jar was opened to remove portions for extraction. Because argon is heavier than oxygen, I assumed that it would be more effective in replacing oxygen than would a lighter gas, such as nitrogen. I did not attempt to remove the oxygen that may have been dissolved in the salt solutions used to establish constant humidity (see below).

Jars were kept at room temperature (21 ± 2 °C), in a water bath (21 ± 1 °C), in a refrigerator (4 ± 1 °C), or in a freezer (−20 ± 2 °C). Before a sample was placed in the freezer it was dried further over anhydrous calcium sulfate (see below) in a refrigerator for 3–5 days. This drying process was repeated after every time such a jar was opened to remove portions for extraction. To prevent condensation inside the jars and on the samples, jars stored in a refrigerator or freezer were warmed up to room temperature before opening.

Establishing constant humidity

Constant humidity, independent of temperature at 4–21 °C (Rockland, 1960; Rockland & Nishi, 1980), was maintained with saturated solutions of sodium chloride (76% humidity), potassium carbonate (43% humidity) and potassium acetate (23% humidity). Indicating Drierite[R] (anhydrous calcium sulfate, impregnated with cobalt chloride, W.A. Hammond Drierite Co., Xenia, Ohio, U.S.A.) was used to obtain the lowest humidity. I assumed that the relative humidity of air dried with Drierite was temperature independent in the range −20–21 °C and using the water contents of saturated air (Hodgman, 1949) and air dried with calcium sulfate (Skoog & West, 1976), calculated its value to be <1%.

Sample extraction and data collection

Rotifers were extracted from the dried samples with a 0.2 M solution of sucrose in distilled water, a modification of the water extraction method of Peters et al. (1993). Since the bdelloids remain contracted in sucrose, this method allows one to collect as many rotifers as needed, remove any debris, count the rotifers and, then, dilute sucrose with water to allow the animals to resume activity. Also, prehydration at a high osmolarity (or a high humidity) before rehydration in water may improve the recovery of bdelloids, as with dried nematodes (Crowe & Madin, 1975), bacteria (Kosanke et al., 1992) and seeds and pollen (Crowe et al., 1992).

To extract the rotifers, a 0.15–0.4 g portion from a dry sample was placed in a small plastic vial to which 2 ml of 0.2 M sucrose was added. The vial was sealed with Parafilm[R], vigorously shaken for about 25 s, and the extract was poured into a Petri dish. The sample was extracted two more times with 2 ml of 0.2 M sucrose each time. Under a stereomicroscope, the contracted bdelloids (but not their eggs) were removed with a fine-tipped, glass pipet and transferred into a watch glass containing a small amount of 0.2 M sucrose. After a number of rotifers were collected (ca. 25 or more), debris was removed from the sample, and the rotifers were counted. Rotifers remained in 0.2 M sucrose for no longer than about 40 min before it was diluted with distilled water. After a 24-hour recovery period (17–22 °C), the number of live rotifers was determined. When I was unsure if an animal was alive or not, I examined it under a compound microscope; in that case, animals with active flame cells or cloacae were considered alive.

Unless otherwise indicated, each datum in Table 1 and Figures 1 and 2 is the average of two extractions plus or minus the standard deviation. In Sets #1 and #3, at least 40 rotifers were isolated at each extraction, while in Set #2 at least 26 rotifers were isolated at each extraction. The mean percentages of bdelloids alive at the start of the study (i.e., day 0) were obtained from duplicate extractions of each sample set.

Results

List of bdelloids recovered

Set #1: *Adineta barbata* Janson; *Macrotrachela ehrenbergii* (Janson); *Macrotrachela multispinosa* Thompson; *Macrotrachela nana* (Bryce); *Mniobia russeola* (Zelinka); *Mniobia scabrosa* Murray (first U.S.A. record); *Mniobia tetraodon* (Ehrenberg); *Philodina plena* (Bryce). Set #2: *Adineta vaga* (Davis); *Habrotrocha* sp.; *Macrotrachela quadricornifera* Milne; *M. scabrosa*; *Philodina eurystephana* Schulte (first U.S.A. record). Set #3: *Macrotrachela* sp.; *M. scabrosa*; *Philodina* sp.

Survivorship

In Set #1, portions of a lichen sample were stored in air at specific humidities and temperatures indicated in Table 1. After 8 months at 21 °C, the percentages of live rotifers at 23% and 43% humidities were roughly the same as on day 0. In comparison, at 21 °C, 0%

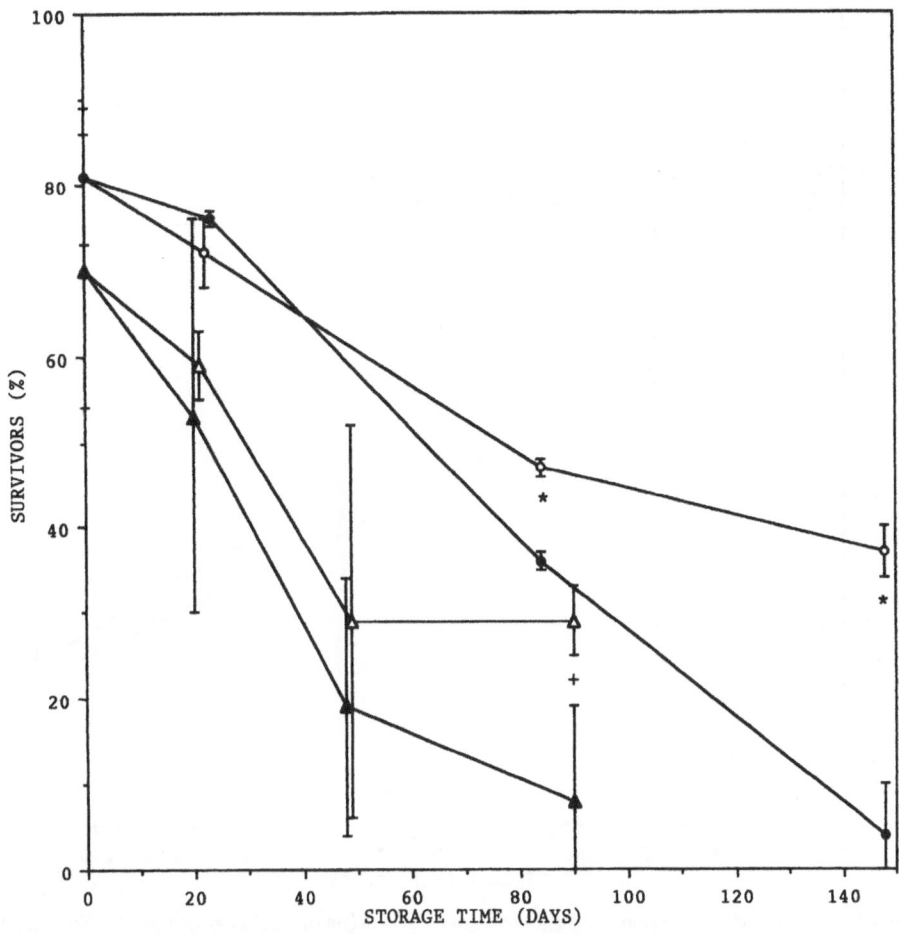

Figure 1. Survival of bdelloids in mushroom (set 2, triangles) and lichen (set 3, circles) stored in air (closed symbols) or argon (open symbols) at 21 °C, <1% humidity. (Overlapping error bars for day 20 for argon in set 2, and at day 22 for air in set 3 were shifted to the right by one day to reveal their position.) Statistical significance obtained from pairwise comparison with Student's *t*-test are indicated as follows: $p < 0.02$ = *; $0.1 < p < 0.2$ = +.

of rotifers survived at <1% humidity and only 4 ± 5% survived at 76% humidity for 8 months. However, lowering the temperature to 4 °C increased survival at both <1% and 76% humidity. Further, the percentage of survivors did not decline from day 0 over an 18-month period at −20 °C and <1% humidity (Table 1).

To duplicate some of the conditions in Table 1, portions of the lichen sample in Set #3 were stored in air at 21 °C and 43% humidity and at −20 °C and <1% humidity. After about 8 months, 74 ± 5% of rotifers at 43% humidity and 72 ± 1% of rotifers at <1% humidity were alive (Figure 2). In agreement with Table 1, these values were not significantly different (Student's *t*-test, $0.2 < p < 0.4$) than the percentage of live rotifers in Set #3 at day 0 (81 ± 8%).

The effect of replacing air with argon on survival was evaluated in Sets #2 and #3 at 21 °C. At <1% humidity, in both sets survival in argon was significantly better than that in air after about 3 months (Figure 1). However, at 76% humidity argon had no significant effect on the survival of rotifers (Figure 2).

In agreement with earlier observations (Jacobs, 1909; Hickernell, 1917), I have noticed that bdelloids in recently dried samples or in dry samples kept under conditions that do not decrease survivorship quickly become active after rehydration. For example, some bdelloids extracted from the lichen in Set #3 that had been kept at 21 °C, 43% humidity for 8 months started feeding as soon as sucrose was diluted with distilled water. Whereas, bdelloids from samples kept dry under less favorable conditions took several hours to fully revive after rehydration. It is easiest to extract the

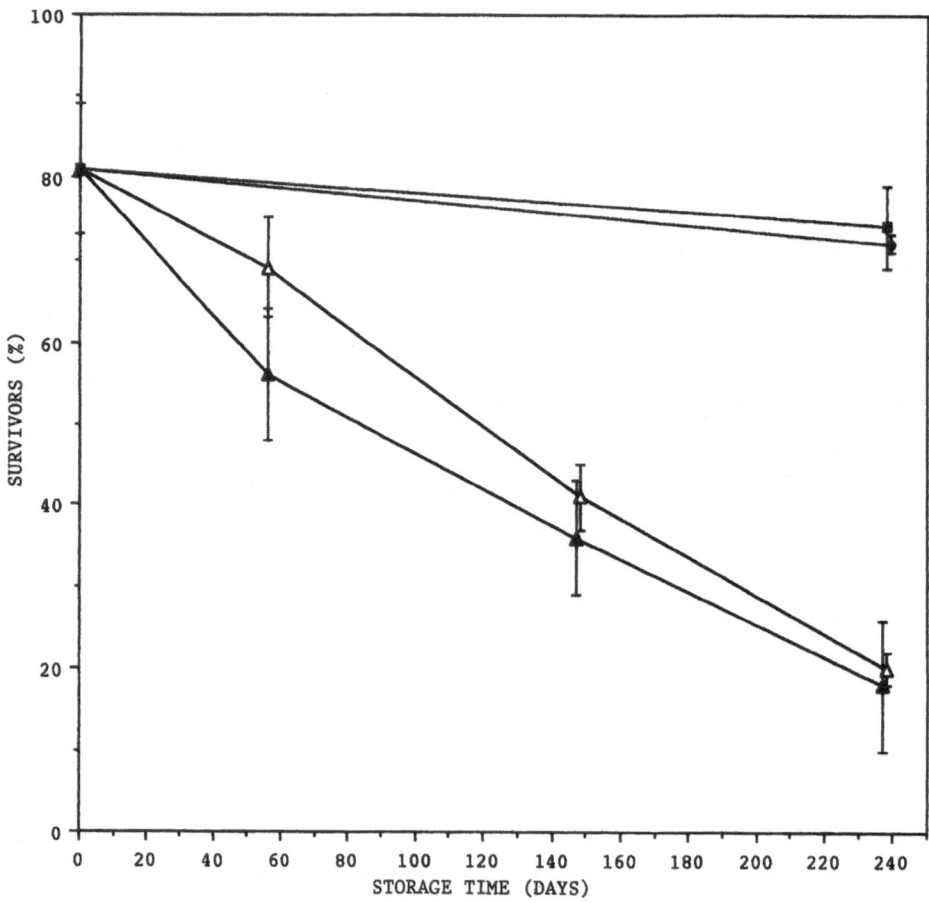

Figure 2. Survival of bdelloids in lichen (set 3) stored in air (closed symbols) or argon (open symbols) at 21 °C, 76% humidity (triangles) and at 21 °C, 43% humidity (square) and at −20 °C, <1% humidity (circle). (Overlapping error bars for argon at days 147 and 237 were shifted to the right by one day to reveal their position.)

Table 1. Percentage survivorship of bdelloids at different relative humidities and temperatures from Set #1 (sample of lichen)

Conditions [a]		Month number [b]			
Humidity (%)	°C	5	8	12	18
<1	21	5 ± 4	0	–	–
23		72 ± 11	67 ± 19	–	–
43		68 ± 8	68 ± 5	–	–
76		28 ± 1	4 ± 5	–	–
<1	4	53 ± 4	46 ± 18	51 ± 15	28 ± 15
76		77 ± 2	70 ± 16	–	–
<1	−20	73 ± 1	72 ± 4	–	75 ± 2

[a] Conditions of humidity and temperature under which the samples were held.
[b] At the start of the procedure (day 0), a mean of 69 ± 8% of bdelloids were scored as being alive in Set #1. Extractions under a given month (column) were actually done over a 1–3-day period. Number of months are approximations calculated by dividing the number of days by 30. (All values are averages of two extractions, except for the values at 23% and 43% humidity at 5 months, which are averages of 3 extractions.)

former samples with 0.2 M sucrose, while the latter may be extracted with water. To check the effect of the extraction medium on post-extraction survival of bdelloids, I extracted two portions of a lichen sample kept for 86 days under adverse conditions (i.e., 21°C, <1% humidity) with 0.2 M sucrose and two additional portions with water. In this case, there was no significant difference between the percentages of survivors 24 h after the extraction with sucrose (46 ± 1%) and water (39 ± 12%) as analyzed using the *t*-test.

Discussion

What processes are responsible for the gradual decline in the number of live rotifers in dry samples? I suggest that drying itself is not lethal to the rotifers for two reasons. (1) If removal of water killed the rotifers (perhaps by damaging their cells and organs), most would be expected to die immediately upon or soon after drying, assuming that their internal water content quickly equilibrated with the ambient humidity. However, at 76% and <1% humidities, the death rate was relatively slow at 21 °C and it could be decreased further by lowering the temperature (Table 1) or by replacing air with argon at <1% humidity (Figure 1). (2) At the intermediate humidities of 23% and 43%, the percentages of live rotifers did not decline during the 8-month period (Table 1, Figure 2).

Thus, I argue that certain humidity-controlled chemical reactions, which may use oxygen, seem to be involved in the death of anhydrobiotic bdelloids. Oxidation damage to cellular components, including lipids, has been implicated in the reduced viability of dried prokaryotes (Zentner, 1966; Potts, 1994) and dried invertebrates (Crowe & Madin, 1974). Moreover, the rate of lipid oxidation in dry foods is slowest at around the water activity of 0.2 (20% humidity) (Labuza, 1975), near the range (i.e., 23–44% humidity) where the reactions that kill the dried bdelloids also seem to be the slowest at ordinary temperatures. Therefore, I suggest that lipid oxidation may be involved in the effects of humidity on the survival of dried bdelloids. Nevertheless, bdelloids continue to die even in the presence of argon (Figure 1). If argon completely replaced the air in the jars and no air subsequently leaked back into them, the results indicate that oxidation is not the only process responsible for the death of anhydrobiotic bdelloids. So at present we do not know the mechanism(s) responsible for mortality of anhydrobiotic bdelloids.

Bdelloids are difficult to fix and preserve for study using the conventional techniques that are otherwise adequate for monogonont rotifers. This difficulty often results in the lack of useful specimens for taxonomic work. Thus, I recommend that naturally dried samples be subjected to additional care. These extra steps would include drying over a desiccant for several days in a refrigerator and long-term storage at <0 °C in the absence of oxygen.

Acknowledgements

I thank Dr James S. Clegg for helpful comments, D. F. Dodgen for argon and Nicole Örstan for proofreading the manuscript. I also thank two anonymous reviewers and the editors of the rotifer symposium for suggesting editorial changes in the manuscript.

References

Crowe, J. H. & K. A. Madin, 1974. Anhydrobiosis in tardigrades and nematodes. Trans. am. Mic. Soc. 93: 513–524.

Crowe, J. H. & K. A. Madin, 1975. Anhydrobiosis in nematodes: evaporative water loss and survival. J. exp. Zool. 193: 323–334.

Crowe, J. H., F. A. Hoekstra & L. M. Crowe, 1992. Anhydrobiosis. Ann. Rev. Physiol. 54: 579–599.

Fantham, H. B. & A. Porter, 1945. The microfauna, especially the protozoa found in some Canadian mosses. Proc. zool. Soc. Lond. 115: 97–174.

Hickernell, L. M., 1917. A Study of desiccation in the rotifer, Philodina roseola, with special reference to cytological changes accompanying desiccation. Biol. Bull. 32: 343–407.

Hodgman, C. D. (ed.), 1949. Handbook of Chemistry and Physics, 31st edn. Chemical Rubber Publishing Co., Cleveland, 1938.

Jacobs, M. H., 1909. The effects of desiccation on the rotifer Philodina roseola. J. exp. Zool. 6: 207–263.

Kosanke, J. W., R. M. Osburn, G. I. Shuppe & R. S. Smith, 1992. Slow rehydration improves the recovery of dried bacterial populations. Can. J. Microbiol. 38: 520–525.

Labuza, T. P., 1975. Oxidative changes in foods at low and intermediate moisture levels. In R. B. Duckworth (ed.), Water Relations of Foods. Academic Press, New York: 455–473.

Örstan, A., 1995. Desiccation survival of the eggs of the rotifer Adineta vaga (Davis, 1873). Hydrobiologia 313/314: 373–375.

Peters, U., W. Koste & W. Westheide, 1993. A quantitative method to extract moss-dwelling rotifers. Hydrobiologia 255/256: 339–341.

Potts, M., 1994. Desiccation tolerance of prokaryotes. Microb. Rev. 58: 755–805.

Ricci, C., L. Vaghi & M. L. Manzini, 1987. Dessication of rotifers (Macrotrachela quadricornifera): survival and reproduction. Ecology 68: 1488–1494.

Rockland, L. B., 1960. Saturated salt solutions for static control of relative humidity between 5 and 40 °C. Anal. Chem. 32: 1375–1376.

Rockland, L. B. & S. K. Nishi, 1980. Influence of water activity on food product quality and stability. Food Technol. 34(4): 42–51.

Skoog, D. A. & D. M. West, 1976. Fundamentals of Analytical Chemistry. Holt, Rinehart & Winston, New York, 804 pp.

Stubbs, C. S., 1989. Patterns of distribution and abundance of corticolous lichens and their invertebrate associates on Quercus rubra in Maine. Bryologist 92: 453–460.

Wright, J. C., 1991. The significance of four xeric parameters in the ecology of terrestrial tardigrada. J. Zool., Lond. 224: 59–77.

Wright, J. C., P. Westh & H. Ramløv, 1992. Cryptobiosis in tardigrada. Biol. Rev. 67: 1–29.

Zentner, R. J., 1966. Physical and chemical stresses of aerosolization. Bact. Rev. 30: 551–557.

Hydrobiologia **387/388**: 333–340, 1998.
E. Wurdak, R. Wallace & H. Segers (eds), Rotifera VIII: A Comparative Approach.
© *1998 Kluwer Academic Publishers.*

Some notes on *Brachionus rotundiformis* (Tschugunoff) in Lake Palaeostomi

Juta Haberman[1] & Minoru Sudzuki[2]
[1]*Institute of Zoology and Botany, Võrtsjärv Limnological Station, EE-2454 Rannu, Estonia*
[2]*Biological Laboratory, Nihon Daigaku University, Omiya-shi, Saitama-ken, Japan 330*

Key words: Rotifera, *Brachionus rotundiformis*, morphology, amphoteric females, population density

Abstract

In 1933 a connection was formed between the Black Sea and Lake Palaeostomi (Georgia) after which the latter became a brackish-water lake with water salinity up to 13‰. Water salinity is variable, depending first of all on the direction and strength of the wind. *Brachionus rotundiformis* (Tschugunoff) (with 20% or more of total zooplankton abundance) was the dominant zooplankton species in the samples collected in 1977 (July, Aug) and 1996 (July) from this lake. The morphology and ratio of physiological types of female and the population density of dominant species were studied. The lorica length of females was 105–250 μm, its widest part 95–250 μm. The average abundance fluctuated between 5 and 2,600 × 10^3 ind m^{-3}. A positive correlation was revealed between the occurrence of *B. rotundiformis* and water salinity ($r = 0.7$; $P < 0.0001$). *B. rotundiformis* formed 84% of the abundance of rotifers and 49% of the numbers of total zooplankton. Several types of egg bearing female were detected. The proportions of females with egg types (M + F), (F + M) and (D/F) were 0.13, 0 and 0.01% in 1977 and 0.14, 0.01 and 0.01% in 1996, respectively.

Introduction

Brachionus rotundiformis was originally established as a new variety of *B. mülleri* (=*plicatilis*, O. F. Müller, 1786), based on specimens from the Caspian Sea (Tschugunoff, 1921). Later, Kutikova (1970), Rodewald (1937), Wiszniewski (1954) and Sudzuki (1982, 1985, 1995, 1996) also named it *B. plicatilis rotundiformis* or *B. rotundiformis*. In contrast, Ahlstrom (1940) and Koste (1978) included it within the species *B. plicatilis*. This idea was, however, strongly opposed, mainly by some larviculture specialists (Fukusho & Okauchi, 1983; Sudzuki, 1982, 1985, 1987, 1995). According to Fukusho (1983) two distinct groups of strain (form) of *B. plicatilis* were recognized in mass-cultured populations. One population is 130 to 340 μm in length (mean of 238.9), possesses a saccate, soft lorica, and appears commonly at temperatures below 20 °C. The other population is 100 to 210 μm in length (mean of 163.3), possesses a stiffer lorica, ovoid body, and tends to occur at temperatures above 20 °C. For convenience, the small group was called 'S-type', the large 'L-type' of *B. plicat-*

ilis (Oogami, 1976). Recently, several studies (Fu et al., 1993; Hagiwara et al., 1995b; Hirayama & Hagiwara, 1995; Munuswamy et al., 1996; Segers, 1995) were conducted, which led to the conclusion that these groups represent different species. In a situation where the large variation has confused the taxonomy of *B. plicatilis* (s.l) (Carmona et al., 1989; Carmona et al., 1995; Gómez & Serra, 1995; Gómez et al., 1995; Temprano et al., 1994), data on the morphology and ecology of natural populations of the species may prove valuable, particularly considering that genotypic variation within populations is reported.

In this paper, the following topics are addressed: (1) morphology; (2) physiological types of the female and (3) population density of *B. rotundiformis*.

Description of the lake

Lake Palaeostomi (Figure 1) is located 2 km from the shore of the Black Sea in the vicinity of the town of Poti (Georgia). It is a brackish-water lake of relict character, originating from a bay of the Black Sea.

334

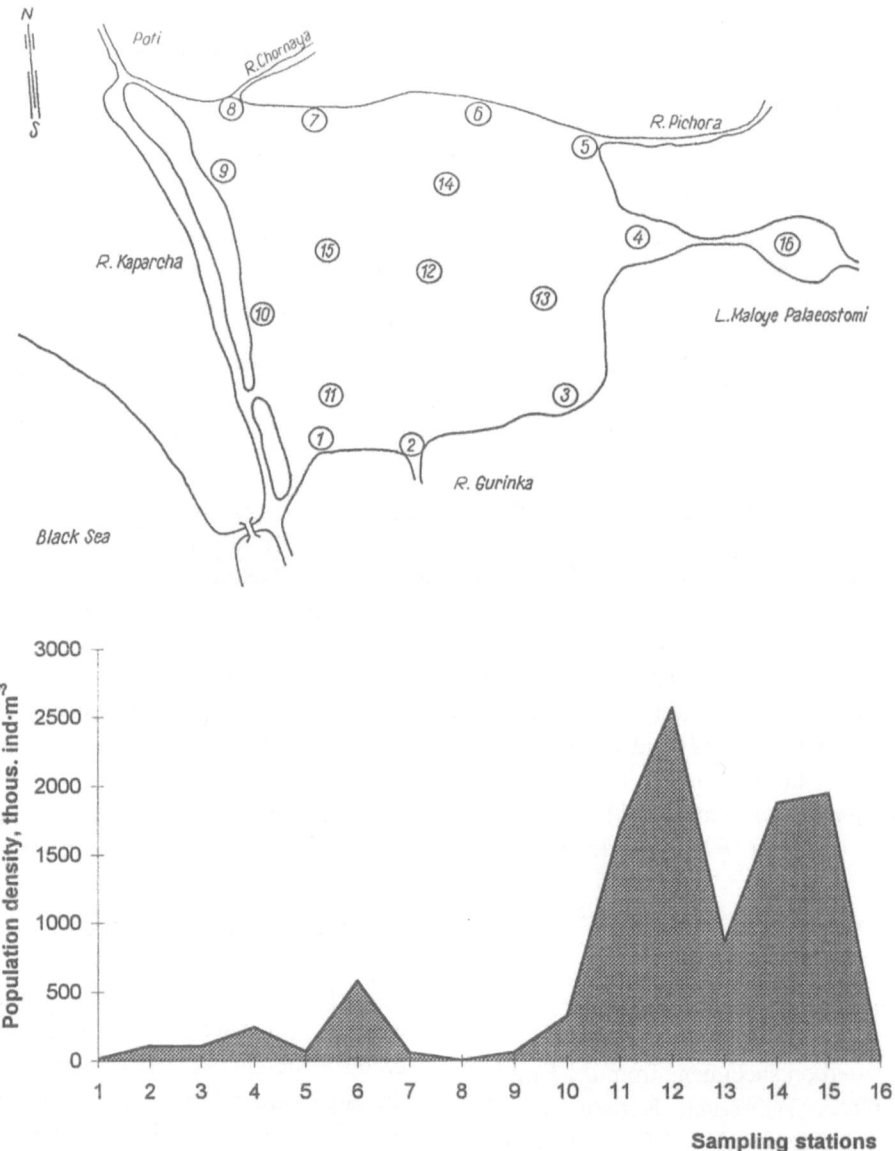

Figure 1. **Above:** Location of sampling stations in L. Palaeostomi; **Below:** Population density of *B. rotundiformis* in L. Palaeostomi, in July–August 1977.

The area of the lake is 17.3 km², average depth is 2.2 m and maximum depth is 4.0 m. Pichora, Chornaya and Gurinka rivers are the inlets and the River Kaparcha is the outlet. Lake Palaeostomi is located in a humid subtropical region where precipitation is abundant. The lake is never frozen. The temperature regime is characterized by the following average water temperatures: (1) 8–10 °C in winter; (2) 16–19 °C in May; (3) 23–29 °C in July–August; (4) 18–22 °C in Sept–Oct. Until 1933 the lake was connected with the sea only via the River Kaparcha. In 1933 the rising

of the lake water level formed a 140–160 m wide, 3–3.5 m deep and 2 km long channel connecting the lake directly with the sea. Before the formation of the direct connection the water salinity in the lake was relatively low (0.02–0.99‰) and had a clearly seasonal character (Lyatti, 1940). Later, salinity increased and remained high: up to 12‰ in the surface layer and up to 13‰ at the bottom. At present, salinity depends mainly on the direction and strength of the wind. Strong westerly winds cause large amounts of sea water to penetrate into the lake, raising its salinity, which has a relatively

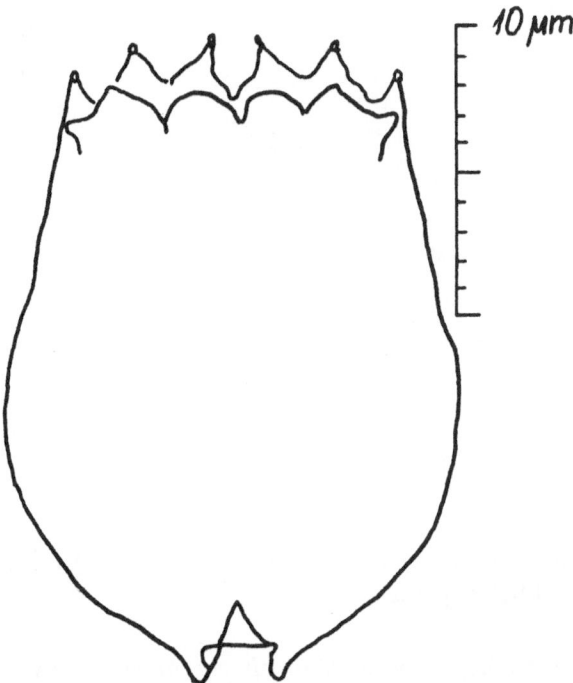

Figure 2. Female (?) of *B. rotundiformis* from L. Palaeostomi.

unstable character (Haberman, 1982). The brackish-water L. Palaeostomi is connected with the broadened part of the river called L. Maloye (Small) Palaeostomi. On the basis of phytoplankton biomass values (Randveer et al., 1982), L. Palaeostomi can be considered as eutrophic or even hypertrophic.

Materials and methods

The materials for the present paper were collected in July and August 1977 and in July 1996 in different parts of Lake Palaeostomi (Figure 1). In 1977 samples were collected from 15 sampling stations with a quantitative Juday net of 85 μm mesh size. Water temperature, pH, water salinity and phytoplankton biomass were measured (Table 1). For comparison, 4 quantitative samples were collected from nearly fresh-water L. Maloye Palaeostomi (Figure 1, sampling station 16). In 1996, 30 litres of water were filtered through a net of 45 μm mesh size at three sampling stations (1, 2, 11). In addition to quantitative samples, a total of 31 samples was gathered (21 samples in 1977 and 10 in 1996) for qualitative analyses. The samples were preserved in 4% formaldehyde and analyzed by conventional quantitative methods. To determine the egg-bearing types present about 23,000 specimens

(12,495 individuals in 1977 and 10,593 in 1996) were examined.

Results and discussion

Types of egg-bearing female in B. rotundiformis from L. Palaeostomi

The female of monogonont rotifers has been classified into 6 or 8 types such as SF, (?F), FF, (PF), CF, AF, MF and DF with respect to egg type and life history (Sudzuki, 1962a, 1962b, 1964, 1983). In the population of *B. rotundiformis* from Lake Palaeostomi, the following three rarest cases have been observed: (1) females which first produce an amictic egg, then unfertilized mictic (male) eggs (Figure 3A); (2) females which first produce a male egg, then an amictic egg (Figure 3B) and (3) females which carry both fertilized and amictic eggs (Figure 3C). All these belong to both critical (Sudzuki, 1955,1962a,1962b) and amphoteric (Bogoslowsky, 1958, 1960) females. A problem arises, however, if we accept the idea that mictic-female production is induced by extrinsic factors. In this case, we suggest the following classification: (1) (?) females which carry no eggs; (2) (F + F) females which carry amictic eggs only; (3) (F + M) females which produce first amictic, then unfertilized mictic eggs; (4) (M + M) females which carry male eggs only; (5) (M + F) females which produce first male, then amictic eggs; (6) (M + D) females which produce first male, then fertilized mictic eggs; (7) (+ D) females with a fertilized mictic egg; (8) (D + M) females which first produce a fertilized egg, then male eggs; (9) (D/F) females with both fertilized mictic and amictic eggs; (10) (F/M/D) females with three kinds of egg. All these types of female except for 6, 8 and 10 were observed in Lake Palaeostomi (Table 2). Table 2 reveals that (1) most females do not carry eggs for a long time, (2) (F + M), (M + F), and especially (D/F) females are worth studying from the viewpoint of chromosomal research, and (3) (M + F) females are more common than (F + M) females.

The existence of (F + M + D) females has also been reported in *Asplanchna* (Mrázek, 1897; Sudzuki, unpublished data) and in *Conochiloides coenobasis* (Bogoslowsky, 1960); (M + D) females, in *A. priodonta*, in *B. (Schizocerca) diversicornis*(Sudzuki, 1955, 1962a, 1962b) and in *Stephanoceros* sp. (Bogoslowsky, 1957; cited in Sudzuki, 1962a); (M/F) females, in *B. angularis*, *B. budapestinensis* and

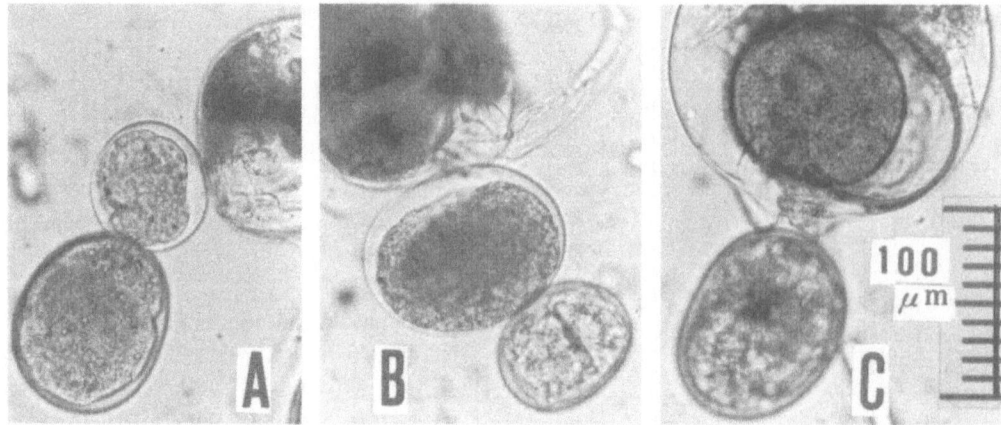

Figure 3. Three types of egg-bearing female of *B. rotundiformis*: A. (F + M) female; B. (M + F) female; C. (D/F) female.

Table 1. Characteristics of sampling stations in L. Palaeostomi

Sampling station	Depth (m)	Water temperature (°C)	pH	Salinity (‰)	Biomass of phytoplankton (gm^{-3})
1	0.4	25.8	8.4	6.6	3.45
2	0.8	25.6	7.7	4.3	5.97
4	2.0	26.4	8.4	2.8	25.00
5	3.7	26.9	9.1	2.7	23.18
6	0.9	26.7	9.7	–	134.99
8	0.6	27.2	8.7	5.3	9.42
10	0.8	25.6	9.0	–	19.16
11	2.4	25.2	8.9	6.6	57.58
12	2.6	25.4	9.1	6.5	39.53
13	2.5	26.8	8.9	–	23.6
14	2.5	26.4	9.1	–	60.38
15	2.6	26.0	9.05	–	119.35

B. forficula (Sudzuki, 1962a), in *Sinantherina socialis* (Bogoslowsky, 1958, Champ & Pourriot, 1977 in vitro) and in *A. girodi* (Snell & King, 1977 in vitro); (D/F) females, in *A. priodonta* and *B. angularis* (Sudzuki, 1962a). The ratios of amphoteric females are 0.48% and 0.53 in vitro (Snell & King, 1977). It should be mentioned here that in one and the same marsh in Urawa, near Tokyo, different types of females appeared in 1954–1955 as follows: (M + D) females on 21.04.1954 in *A. priodonta*, on 11.11.1954 in *B. (S) diversicornis*; (M/F) females on 1.04.1955 in *B. angularis*, on 13.10, 19.10 and 22.10.1954 in *B. budapestinensis*, on 7.07 and 1.08.1954 in *B. forficula*; (D/F) females on 1.04.1955 in both *A. priodonta* and *B. angularis* (Sudzuki, 1955; 1962a). The propor-

tions of females in *B. rotundiformis* may be variable at different periods.

Morphological notes on Brachionus rotundiformis from L. Palaeostomi

1. Female (?) (Figure 2): lorica length (without spines) 105–250 μm, maximum width 95–250 μm, maximum height 100–170 μm. Occipital spines: lateral 15–23 μm, intermedian 13–21 μm, median sinus 22–32 μm. Pectoral projection 10–15 μm. Females (F + F): lorica length 100–210 μm, maximum width 130–160 μm. Amictic egg 90–120 × 80–100 μm. Females (M + M): lorica length 180–200 μm, maximum width 150–160 μm, maximum height 110 μm. Hudson's thread 15 μm. Male egg 72–82 × 55–62 μm. Females (+ D): lorica length 220–242 μm, maximum width 168–172 μm, maximum height 140 μm. Fertilized mictic egg 130–148 × 90–100 μm. Critical or amphoteric females: lorica length 190–210 μm, maximum width 145–155 μm. Male egg 71–79 × 60–62 μm, amictic egg 99–101 × 81–83 μm. 2. Male (contracted): length 78–100 μm, width 62–70 μm, height 55–60 μm. Penis 25–30 μm thick at the base. 3. An example of aberrant forms: body illoricate, sacciform like that of an asplanchnid rotifer; flexible, not separated into a dorsal and a ventral plate; 65–245 μm long, 110–163 μm wide and 100 μm high; foot 75–80 μm long; toes 18–21 μm long.

Tschugunoff (1921) measured the length of his species as 126–129 μm; maximum width 112 μm. According to Kutikova (1970), Rodewald (1937) and Rudescu (1960) lorica length is 150 (105)–157 μm. The species occurring in Lake Palaeostomi form a population belonging to *rotundiformis*. However, it is clearly

Table 2. Proportions (%) of types of egg-bearing females in *B. rotundiformis* from Lake Palaeostomi

Types	+?	F + F	F + M	M + M	M + F	M + D	+ D	D + M	D/F	F/M/D
1977	89.8	7.2	0.00	2.2	0.13	0	0.6	0	0.01	0
1996	94.0	4.4	0.01	0.8	0.14	0	0.7	0	0.01	0

different from the species described by Tschugunoff, Rudescu and Kutikova in the pattern of the pectoral margin (cf. *estoniana* Sudzuki, 1985, 1987).

Population density of B. rotundiformis

In July and August 1977, *B. rotundiformis* was abundant in L. Palaeostomi. The number of *B. rotundiformis* fluctuated between 8 and 2,334 × 10^3 ind m^{-3} (average for the lake 745 × 10^3 ind m^{-3}) in July, and from 5 to 2,800 × 10^3 ind m^{-3} (average 942 × 10^3 ind m^{-3}) in August. The average of summer (July–August) fluctuated between 5 and 2,600 × 10^3 ind m^{-3}, the summer mean being 702 × 10^3 ind m^{-3} (Figure 1). *B. rotundiformis* formed 71–100% of rotifer numbers (average 84%), and 49% (5–79%) of total zooplankton numbers, representing thus the dominant species. A species is considered dominant when it forms at least 20% of the numbers or biomass of total zooplankton in a sample (Haberman, 1977). The dominance of *B. rotundiformis* in the zooplankton indicates that it is obviously adapted to the variable salinity regime of the lake. *B. rotundiformis* was more numerous in the pelagic part of the lake, where the salinity was higher (Table 1), than in the littoral, in freshened river-mouths and in the nearly freshwater L. Maloye Palaeostomi. The average population density of *B. rotundiformis* was 1,790 × 10^3 ind m^{-3} in the pelagial, 192 × 10^3 ind m^{-3} in the littoral, 106 × 10^3 ind m^{-3} in river mouths and 12 × 10^3 ind m^{-3} in L. Maloye Palaeostomi. Correlation analysis revealed a positive relationship between water salinity and the abundance of *B. rotundiformis* ($r = 0.7$; $P < 0.0001$). In 1996 the population densities of *B. rotundiformis* were studied at three sampling stations (1, 2, 11) only. In sampling station 1 it was 50 × 10^3 ind m^{-3}, in station 2 (mouth of R. Gurinka) 95 × 10^3 ind m^{-3}, and in station 11 (pelagial) 1,850 × 10^3 ind m^{-3}. Low zooplankton density (both in 1977 and 1996) in the part of the lake closest to the sea (sampling station 1) may be taken as evidence of the inhibitory effect of fluctuating salinity. Phytoplankton biomass was also the lowest in this part of the lake (Table 1). It is possible that the mod-

est population density of *B. rotundiformis* was caused, at least partially, by the abundant predatory jellyfish *Moerisia maeotica* (Ostroumov) which can consume large amounts of zooplankton (Schwarz, 1978). The influence of other interspecific interactions on the zooplankton community (competition, commensalism, ammensalism etc.) is also possible (Hagiwara et al., 1995a). *B. rotundiformis* is numerous in L. Palaeostomi. However, in comparison with the population densities of this species reported elsewhere, 15,100 × 10^3 ind m^{-3} (Carmona et al., 1995), and even 16,000 × 10^3 ind m^{-3} (Gómez et al., 1995), its abundance in the lake under study can be considered rather modest.

In the summer of 1977, water salinity fluctuated from 2.7 to 6.6‰ in different parts of the lake. *B. rotundiformis* (S-type of *B. plicatilis*) is considered a euryhaline species (Hagiwara et al., 1989). This conclusion is supported by data on the pond of Poza Sur in Spain where the species was found at salinities 5–64‰ (Carmona et al., 1995) and by data on L. Yaskhan (in Kara-Kum desert, Turkmenia) where *B. rotundiformis* was abundant at salinities of up to 3.9‰ (Haberman, 1993; Virro, 1993). Hagiwara et al. (1989) have pointed out that *B. rotundiformis* is more euryhaline than *B. plicatilis*.

Varying from 25.2 to 26.9 °C, the water temperature of L. Palaeostomi (Table 1) can be regarded as suitable for the growth of thermophilic *B. rotundiformis*. Epp & Lewis (1980) have found that a temperature range of 20° to 28 °C is optimal for the development of this species. The summer plankton of L. Palaeostomi is represented by *B. rotundiformis*, *Brachionus angularis* Gosse, *Brachionus falcatus* Zach. and *Brachionus rubens* Ehrenb. but not by *B. plicatilis* Müller. The two rotifer species, *B. rotundiformis* and *B. plicatilis* differ in their growth responses to water temperature; *B. rotundiformis* grows best at higher temperatures (Hagiwara, 1994; Hagiwara & Hirayama, 1993; Hagiwara & Lee, 1991; Hagiwara et al., 1988; Hagiwara et al., 1995b; Hirayama & Rumengan, 1993). Taking into account that *B. plicatilis* is less thermophilic than *B. rotundiformis*, its occurrence at lower temperatures (in autumn and in winter) would be expected. In the shallow pond of Poza Sur (Spain) *B. rotundiformis* was present from the beginning of the study in July 1992 until November 1992 after which it was replaced by *B. plicatilis*. The latter was present in Poza Sur until May 1993; *B. rotundiformis* reappeared in May 1993, replacing *B. plicatilis* (Carmona et al., 1995). *B. rotundiformis* and *B. plicatilis* behaved analogously in culture tanks. L-type (*B. plicatilis*) dominated a culture from April to June and from November to January when the water temperature was less than 20 °C. When the temperature exceeded 20 °C (from July to October), S-type rotifers (*B. rotundiformis*) dominated (Kokura et al., 1982, cited in Hagiwara & Lee, 1990).

The population density of *B. rotundiformis* is largely dependent on food (Yúfera & Navarro, 1995; Yúfera & Pascual, 1984). In L. Palaeostomi the species was obviously well supplied with food as the amount of phytoplankton was quite large (Laugaste & Haberman, 1983). The amount of algae (*Ankistrodesmus angustus*) Bern, *A. minutissimus* Korschik; *Cyclotella atomus* Hust., *Monoraphidium contortum* (Thur.) Kom.-Legn., *M. minutum* (Näg.) Kom.-Legn., *Nitzschia acicularis* W. Sm., *Scenedesmus* spp.) which is suitable in size for *B. rotundiformis* to consume is also relatively large (Haberman & Laugaste, 1982). Both *B. rotundiformis* (S-type) and *B. plicatilis* (L-type) can ingest *Euglena gracilis* whose cell diameter is 30–50 μm (Hamada et al., 1993).

Bérzins & Pejler (1987) have concluded that for rotifer species found in oligotrophic waters pH optima were at or below neutral (pH = 7), whereas for the species common in eutrophic waters pH optima were at or above neutral. In eutrophic L. Paleostomi, the pH fluctuated between 8.4 and 9.7 in the summer of 1977. Obviously, such a pH level is suitable for the normal growth of the *B. rotundiformis* population; it has been stated that a pH in the range of 7.7–10 (Yúfera & Navarro, 1995) stimulates the development of this species.

Considering the above data on the ecology of *B. rotundiformis*, it can be concluded that all major environmental factors (water temperature, food, water salinity, pH) in L. Palaeostomi are favourable for the normal development of this species. However, water salinity seems to be too low for the mass occurrence of *B. rotundiformis*. The inhibitory effect of the unstable salinity regime cannot be excluded, either. This is evidenced by the scarcity of plankton in the station closest to the sea. Green (1995) is of the opinion that *B. plicatilis* (s.l.) can tolerate high salinities but may not be able to cope with fluctuating salinities. However, available data on L. Palaeostomi do not allow us to draw any reliable conclusions in this respect. It has been suggested (Amspoker & McIntire, 1978) that a salinity of 5‰ is a critical limit for several organisms at which an abrupt change takes place in the dominant species. Evidently, such a change has already occurred in L. Palaeostomi, and a brackish-water zooplankton community, adapted to an unstable salinity regime, has (in general) formed. Penetration of the low salinity (12‰; Tiidor, 1982) inshore water of the Black Sea into L. Palaeostomi does not appear to bring about hazardous salinity fluctuations. This viewpoint is supported by the fact that in 1996 both the composition and amount of zooplankton were very close to those in 1977. It means that during 20 years no significant changes have taken place either in the zooplankton or in the *B. rotundiformis* population.

Acknowledgements

We wish to express thanks to Dr A. Hagiwara and to an anonymous reviewer for revising the manuscript. Thanks are also due to Mrs Ester Jaigma for the linguistic revision.

References

Ahlstrom, E. H., 1940. A revision of the Rotatorian genera *Brachionus* and *Platyias* with descriptions of one species and two new varieties. Bull. am. Mus. nat. Hist. 77: 148–184.

Amspoker, M. C. & D. C. McIntire, 1978. Distribution of intertidal diatoms associated with sediments in Yaquina Estuary, Oregon. J. Phycol. 14: 387–395.

Bérzins, B. & B. Pejler, 1987. Rotifer occurrence in relation to pH. Hydrobiologia 147: 107–116.

Bogoslowsky, A. S., 1958. New data on the reproduction of heterogonous rotifers. Observations on the reproduction of *Sinantherina socialis*. Zool. Zhur. 37: 1616–1623 in [Russian with English summary].

Bogoslowsky, A. S., 1960. Observations on the reproduction of *Conochiloides coenobasis* Skorikov and the statement of a physiological category of females new to heterogonous Rotifera. Zool. Zhur. 39: 670–677 [in Russian with English summary].

Carmona, M. J., A. Gómez & M. Serra, 1995. Mictic patterns of the rotifer *Brachionus plicatilis* Müller in small ponds. Hydrobiologia 313/314: 365–371.

Carmona, M. J., M. Serra & M. R. Miracle, 1989. Total protein analysis in rotifer populations. Biochem. Systematics and Ecology 17: 408–415.

Champ, P. & R. Pourriot, 1977. Reproduction cycle in *Sinantherina socialis*. Arch. Hydrobiol. Beih. Ergebn. Limnol. 8: 184–186.

Epp, R. W. & W. M. Lewis, 1980. Metabolic uniformity over the environmental temperature range in *Brachionus plicatilis* (Rotifera). Hydrobiologia 73: 145–147.

Fu, Y., A. Hagiwara & K. Hirayama, 1993. Crossing between seven strains of the rotifer *Brachionus plicatilis*. Nippon Suisan Gakkaishi 59: 2009–2016.

Fukusho, K., 1983. Morpohology and variation. In *Brachionus plicatilis*, its biology and mass culture. Koseikaku, Tokyo: 35–51 [in Japanese].

Fukusho, K. & M. Okauchi, 1983. Sympatry in natural distribution of two strains of a rotifer, *B. plicatilis*. Bull. Nat. Res. Inst. Aquaculture 4: 135–138.

Gómez, A. & M. Serra, 1995. Behavioral reproductive isolation among sympatric strains of *Brachionus plicatilis* Müller 1786: insights into the states of this taxonomic species. Hydrobiologia 313/314: 111–119.

Gómez, A., M. Temprano & M. Serra, 1995. Ecological genetics of a cyclical parthenogen in temporary habitats. J. Evol. Biol. 8: 601–622.

Green, J., 1995. Associations of planktonic and periphytic rotifers in a Malaysian estuary and two nearby ponds. Hydrobiologia 313/314: 47–56.

Haberman, J., 1977. Shkaly dominirovaniya, chastoty vstrechaemosti i velichiny zooplankterov (Scales of domination, occurrence frequency and weights of zooplankters). In XIX nauchnaya konferentsiya po izucheniyu vodoemov Pribaltiki i Belorussii. Minsk: 153–154 [in Russian].

Haberman, J., 1982. Kharakteristika ozera Palaeostomi (Characterization of L. Palaeostomi). In E. Rull (ed.), Produktsionno-biologicheskie svojstva i usloviya zhizni ryb v ozere Palaeostomi. Tallinn: 9–17 [in Russian with English summary].

Haberman, J., 1993. Zooplankton of Lake Yaskhan. Limnologica 23: 215–225.

Haberman, J. & R. Laugaste, 1982. On the summer phyto- and zooplankton of Lake Palaeostomi. I. Species composition of phyto- and zooplankton. Brief characterization of the lake. Proc. Acad. Sci. Estonian SSR. Biol. 31: 226–232.

Hagiwara, A., 1994. Practical use of rotifer cysts. The Israel Journal of Aquaculture-Bamidgen 46: 13–21.

Hagiwara, A. & K. Hirayama, 1993. Preservation of rotifers and its application in the finfish hatchery. TML Conference Proceedings 3: 61–71.

Hagiwara, A. & C.-S. Lee, 1991. Resting egg formation of the L- and S-type rotifer *Brachionus plicatilis* under different water temperature. Nippon Suisan Gakkaishi 57: 1645–1650.

Hagiwara, A., A. Hino & R. Hirano, 1988. Comparison of resting egg formation among five Japanese stocks of the rotifer *Brachionus plicatilis*. Nippon Suisan Gakkaishi 54: 577–580.

Hagiwara, A., M. M. Jung, T. Sato & K. Hirayama, 1995a. Interspecific relations between marine rotifer *Brachionus rotundiformis* and zooplankton species contaminating in the rotifer mass culture tank. Fisheries Sci. 61: 623–627.

Hagiwara, A., T. Kotani, T. W. Snell, M. Assava-Aree & K. Hirayama, 1995b. Morphology, reproduction, genetics, and mating behavior of small, tropical marine *Brachionus* strains (Rotifera). J. exp. mar. Biol. Ecol. 194: 25–37.

Hagiwara, A., C. Lee, G. Miyamoto & A. Hino, 1989. Resting egg formation and hatching of the S-type rotifer *Brachionus plicatilis* at varying salinities. Mar. Biol. 103: 327–332.

Hamada, K., A. Hagiwara & K. Hirayama, 1993. Use of preserved diet for rotifer *Brachionus plicatilis* resting egg formation. Nippon Suisan Gakkaishi 59: 85–91.

Hirayama, K. & A. Hagiwara, 1995. Recent advances in biological aspects of mass culture of rotifers (*Brahionus plicatilis*) in Japan. ICES mar. Sci. Symp. 201: 153–158.

Hirayama, K. & I. M. F. Rumengan, 1993. The fecundity patterns of S and L-type rotifers of *Brachionus plicatilis*. Hydrobiologia 255/256: 153–157.

Koste, W., 1978. Rotatoria. Die Rädertiere Mitteleuropas. Gebr. Borntraeger, Berlin, Stuttgart: 673 pp. 234 plates.

Kutikova, L. A., 1970. Kolovratki fauny SSSR. Nauka, Leningrad: 744 pp. [in Russian].

Laugaste, R. & J. Haberman, 1983. On the summer phyto- and zooplankton of Lake Palaeostomi. IV. Quantitative data on phytoplankton and relations with zooplankton. Proc. Acad. Sci. Estonian SSR. Biol. 32: 216–221.

Lyatti, S. Ya., 1940. Gidrologo-gidrokhimicheskij ocherk ozera Palaeostomi. (Hydrological-hydrochemical survey of L. Palaeostomi). Trudy nauchnoj rybokhozyajstvennoj i biologicheskoj stantsii Gruzii 3: 379–417 [in Russian].

Mrázek, A., 1897. Zur Embryonalen Entwicklung der Gattung *Asplanchna*. Vest. Cesk. Spol. Nauk. math.-prirod. 58: 1–11.

Munuswamy, N., A. Hagiwara, G. Murugan, K. Hirayama & H. J. Dumont, 1996. Structural differences between the resting eggs of *Brachionus plicatilis* and *Brachionus rotundiformis* (Rotifera, Brachionidae): an electron microscopic study. Hydrobiologia 318: 219–223.

Oogami, H., 1976. On the morphology of *Brachionus plicatilis*. Newsletter from Izu Branch, Shizuoka Prefectural Fisheries Research Center 18: 2–5 [in Japanese].

Randveer, A., R. Laugaste & H. Tammert, 1982. Letnii fitoplankton ozera Palaeostomi (The summer phytoplankton of Lake Palaeostomi). In E. Rull (ed.), Produktsionno-biologicheskie svojstva i usloviya zhizni ryb v ozere Palaeostomi. Tallinn: 22–35 [in Russian with English summary].

Rodewald, L., 1937. Rädertierfauna Rumäniens. II. Neue und bemerkenswerte Rädertiere aus Rumänien. Zool. Anz. 118: 235–248.

Rudescu, L., 1960. Fauna Republicii Populare Romîne. Trochelminthes (volume II, Fascicula II). Rotatoria. Editura Academici Republicii Populare Romîne, Bucuresti, 1192 pp.

Schwarz, S., 1978. Biomasseuntersuchungen am Zooplankton der ostmecklen burgischen Küstengewässer – ein Beitrag zum Eutrophiestatus. Teil 4: Zur Biomasse und Topographie der Boddenketten von Rügen. Acta hydrochim. hydrobiol. 6: 299–306.

Segers, H., 1995. Nomenclatural consequences of some recent studies on *Brachionus plicatilis* (Rotifera, Brachionidae). Hydrobiologia 313/314: 121–122.

Snell, T. W. & E. C. King, 1977. Amphoteric reproduction in *Asplanchna girodi*. Arch. Hydrobiol. Beih. Ergebn. Limnol. 8: 182–183.

Sudzuki, M., 1955. On the general structure and the seasonal occurrence of the males in some rotifers. III. Zool. Mag. 64: 189–193 [in Japanese with English summary].

Sudzuki, M., 1962a. Rotatoria. In T. Uchida (ed.), Systematic Zoology. 4, Nakayama-Shoten, Tokyo: 9–82 [in Japanese].

Sudzuki, M., 1962b. Discovery of new types of the female in the Rotifera and its significance. Zool. Mag. 71: 396 [in Japanese].

Sudzuki, M., 1964. New systematical approach to the Japanese planktonic Rotatoria. Hydrobiologia. 23: 1–124.

Sudzuki, M., 1982. Taxonomy of mixohaline Rotifera, *B. plicatilis* and its allied species. Zool. Mag. 91: 657 [in Japanese].

Sudzuki, M., 1983. Rotatoria. In Dan et al. (eds), Developmental Biology of Invertebrates. I. Baifu-kan, Tokyo: 197–209 [in Japanese].

Sudzuki, M., 1985. Some problems of taxonomic treatment in the so-called *Brachionus plicatilis*. Iden (Heredity), 39: 52–57 [in Japanese].

Sudzuki, M., 1987. Intraspecific variability of *Brachionus plicatilis*. Hydrobiologia 147: 45–47.

Sudzuki, M., 1995. Taxonomy of *Brachionus plicatilis* and its related groups. I. Discussion and considerations on the papers before 1925. Õbun Ronsõ 40: 1–21.

Sudzuki, M., 1996. Taxonomy of *Brachionus plicatilis* and its related groups. II. Discussion and considerations on the papers during 1926–1952. Õbun Ronsõ 42: 1–22.

Temprano, M., I. Moreno, M. J. Carmona & M. Serra, 1994. Size and age at maturity of two strains of the rotifer *Brachionus plicatilis* in relation to food level. Verh. int. Ver. Limnol. 25: 2327–2331.

Tiidor, R., 1982. Kratkaya gidrokhimicheskaya kharakteristika ozera Palaeostomi. [A brief hydrochemical characterization of Lake Palaeostomi]. In E. Rull (ed.), Produktsionno-biologicheskie svojstva i usloviya zhizni ryb v ozere Palaeostomi. Tallinn: 18–21 [in Russian with English summary].

Tschugunoff, N. L. 1921. Über das Plankton des nördlichen Teils des Kaspisees. Raboty Volzhskoj Biol. Stantsii 6: 120–121 [in Russian].

Virro, T., 1993. Rotifers from Lake Yaskhan, Turkmenistan. Limnologica 23: 233–236.

Wiszniewski, J., 1954. Matériaux relatifs á la nomenclature et á la bibliographie des Rotifères. Pol. Arch. Hydrobiol. 2: 7–260.

Yúfera, M. & N. Navarro, 1995. Population growth dynamics of the rotifer *Brachionus plicatilis* cultured in non-limiting food condition. Hydrobiologia 313/314: 399–405.

Yúfera, M. & E. Pascual, 1984. Influencia de la dieta sobre la puesta del rotífero *Brachionus plicatilis* en cultivo (Influence of diet on egg deposition of rotifer *Brachionus plicatilis* in culture). Inv. Pesq. 48: 549–556 [in Spanish with English summary].

Hydrobiologia **387/388**: 341–348, 1998.
E. Wurdak, R. Wallace & H. Segers (eds), Rotifera VIII: A Comparative Approach.
© *1998 Kluwer Academic Publishers.*

Effect of temperature and food concentration in two species of littoral rotifers

Ignacio Alejandro Pérez-Legaspi & Roberto Rico-Martínez*
*Universidad Autónoma de Aguascalientes, Centro Básico, Departamento de Química, Avenida Universidad 940, C.P. 20100, Aguascalientes, Ags., México (*author for correspondence)*

Key words: Lecanidae, life-table, mortality, reproduction rate, survivorship

Abstract

We performed life-table experiments with two species of the littoral rotifers *Lecane luna* (O.F. Müller, 1776) and *Lecane quadridentata* (Ehrenberg, 1832). Three different temperatures (20, 25 and 30°C) and food concentrations of *Nannochloris oculata* (1×10^7, 5×10^6, and 1×10^6 cells ml^{-1}) were investigated. We found important differences between both species in all the treatments regarding offspring sizes, hatching percentages, life span and reproductive rates. Our data on hatching percentages of asexual eggs suggested that the optimal temperature for both species is in the 20–25°C range. On the other hand, reproductive data placed the optimal temperature near 25°C. This information can be used to develop aquatic toxicology tests with littoral species.

Introduction

Temperature and food are among the most important factors influencing growth and development of rotifers, which are mostly 'r' strategists and have a high ability to rapidly occupy new niches (Nogrady et al., 1993). There are several works in the literature that have studied the effect of environmental factors on rotifer growth and development (for a review see Nogrady et al., 1993). Galkovskaya (1987) studied the ingestion rate of *Chlorella* sp. cells by *Brachionus calyciflorus* Pallas at different food concentrations and temperatures. She found three-fold increases of the maximum ingestion rate in this species when temperature was increased from 20°C to 35°C. It is well-known that the small size and impermeability of their integument make rotifers quite susceptible to physical and chemical changes. (Nogrady et al., 1993). Rotifers produce cysts to escape adverse environmental conditions. The density of cysts and their hatching percentages in natural conditions varied among the different species investigated by May (1987). Gilbert (1995) did not observe differences in the mean number of cysts produced by *Synchaeta pectinata* at temperatures of 12 and 19°C. He did find that when populations of *Synchaeta pectinata* were deprived of food in the

lab for long periods of time, they were able to produce cysts more frequently.

The above examples clearly show that severe environmental conditions can have an important effect on rotifer populations. Therefore, it is important to develop life-table data to determine the influence of some of these factors on rotifer populations. Temperature is one of the most important factors controlling natality and growth rate in several species of rotifers (Galkovskaya, 1987; Sarma, 1985; Rao & Sarma, 1988; Sarma & Rao, 1990). Life-table experiments provide us with information about dynamic changes in populations based on observations of reproduction and mortality of independent individuals (Krebs, 1985). One of the best life-table studies of rotifers was carried out by Galkovskaya (1987). She worked with three species of rotifers; *B. calyciflorus*, *B. urceus* and *Epiphanes brachionus*, and found that the mean generation time was more than double the sum of the duration of embryogenesis plus the juvenile phase (De + Dj). In her experiments, the mean generation time was reduced when the temperature increased from 30 to 40°C, and the net reproductive rate (Ro) at 39°C increased two-fold when food concentration increased. However, the effect of increasing food concentration was negligible at 37°C, and at 40°C the net reproductive rate (Ro) was reduced significantly.

Several life-table experiments have been performed so far (for a review see Nogrady et al., 1993); however, these works are mainly on planktonic species. Little information is given on littoral species. In the large genus *Lecane* this lack of information is an obstacle to understanding the biogeographic distribution of its numerous species. Hummon & Bevelhymer (1980) developed life-table experiments with *Lecane tenuiseta* using baker's yeast as food. Their results were quite interesting in spite of the low quality food they used. Therefore, it would be quite interesting to compare their data with that obtained from life-tables where a more adequate food (such as algae) is used.

Toxicity assessment is an activity that becomes more important every day. Although several rotifer species have been used to develop standard toxicity tests, most of the tests have used species of the family Brachionidae (Snell & Janssen, 1995). This may be due to their ease of culture in the lab and the advantage of ease in cysts hatching over maintenance of stock cultures (Snell & Moffat, 1992). However, these planktonic species are poor models to study sediment toxicity which can be a potential threat to rivers and water reservoirs (ReVélle & ReVélle, 1992).

Therefore, the goal of our work was to determine the influence of temperature and food concentration on the two littoral species *Lecane quadridentata* (Ehrenberg, 1832) and *Lecane luna* (O. F. Müller, 1776) using life-table experiments with a more suitable food, the alga *Nannochloris oculata*.

Materials and methods

The littoral species used in this work were collected in the field and grown in the lab for at least six months prior to experiments. *Lecane quadridentata* (Ehrenberg, 1832) was collected at Lake Chapala (Mexico's biggest natural lake) and *Lecane luna* (O. F. Müller, 1776) was collected at Los Arquitos Dam (for location of this dam see Silva-Briano, 1992). These two species were cultured in EPA medium (U.S. EPA, 1985) and fed the green algae *Nannochloris oculata* (LB2194, University of Texas Collection) grown in Bold's Basal Medium (Nichols, 1973). We studied the effect of three different temperatures (20, 25 and 30°C), and three different food concentrations of *N. oculata* (1×10^6, 5×10^6 and 1×10^7 cells ml^{-1}) in 18 combinations. The life-table started with hatching of

100 asexual eggs of each species. We observed the eggs every two hours and assigned a mean value of one hour to every individual hatched within the two hour period. Almost all animals were acclimated to the correspondent temperature for at least 48 hours prior to each experiment. There was an exception to this acclimation step. Neither species was able to initially acclimate to 30°C and produce enough asexual eggs to allow us to produce a life table. We have data for only one treatment where acclimation was possible at 30°C. These data correspond to *L. quadridentata* fed with 5×10^6 cells ml^{-1} of *N. oculata*. Therefore, in this treatment we made a comparison between acclimated and non-acclimated individuals, to analyse the effect of acclimation. Hatching percentages were recorded for up to 120 hours. Neonates were then transferred to individual wells in a 24-well polystyrene plate (Corning) with the corresponding food concentration, and then incubated at the corresponding temperature in the dark. A minimum of 15 individuals were used for each treatment. Individuals were observed every twelve hours and their neonates were counted and removed from the well. Instead of changing the original individuals to new wells with fresh food (a procedure which damaged these littoral species), half of the medium was replaced every twenty-four hours by fresh medium. The algae were counted by means of a haemocytometer. The dry weight of 10^6 cells ml^{-1} of *N. oculata* (2.13 ± 1.90 µg for 1×10^6 cells ml^{-1}; $n=10$) was determined by Standard Methods (1995) to make our results comparable with other works.

The following parameters were analyzed: the twelve hours time interval (x), mean duration of lifespan (D), mean generation time (Tc), potential reproductive rate (re), net reproductive rate (Ro), and life expectancy (e_x). All of these parameters were determined according to Krebs (1985). Reproductive value (V_x) was calculated according to Krebs (1994) and Begon et al. (1996).

To obtain the mean size of individuals of both species at different ages, they were incubated at 20 ± 2°C with 10^6 cells ml^{-1} of *N. oculata*. We measured: total length, maximum width, length of the foot, and length of the pseudoclaw, according to Stemberger (1979).

Morphometric data were analyzed by *t*-tests. Simple regression analysis between survivorship (lx) and life expectancy (e_x) for each treatment were also performed.

Table 1. Life-table data of *Lecane quadridentata* and *L. luna* under different temperatures and food concentrations of individually cultured cohorts

Temperature (°C)	Food concentration (10^6 cells ml^{-1})	Mean duration of lifespan (D) (h)	Mean generation time (T) (h)	Potential reproductive rate (r_e) (h^{-1})	Net reproductive rate (Ro) (h^{-1})
L. luna					
20	1	153.7	110.0	0.0547	67.1
	5	169.0	125.6	0.0502	71.4
	10	176.1	127.7	0.0382	70.9
25	1	175.5	94.7	0.0753	145.3
	5	167.2	114.9	0.0630	127.5
	10	91.6	87.0	0.0431	27.6
30	1	88.5	64.3	0.0834	69.2
	5	45.7	73.7	0.0126	2.4
	10	32.6	0	0	0
L. quadridentata					
20	1	163.3	115.9	0.0492	104.3
	5	138.5	123.8	0.0376	56.0
	10	196.5	143.5	0.0437	176.2
25	1	178.5	104.9	0.0771	232.4
	5	135.1	86.2	0.0677	118.6
	10	158.2	92.6	0.0664	141.8
30	1	85.4	60.5	0.0787	59.9
	5	62.8	45.2	0.0205	2.5
	10	51.3	0	0	0

Results and discussion

According to Galkovskaya (1987), it is of utmost importance to know the range of temperature tolerated by rotifer populations. Temperature has been shown to influence the duration of lifespan in several rotifer species (Halbach, 1970; Walz, 1983; Galkovskaya, 1987). This effect was particularly important for *Keratella cochlearis* (Walz, 1983), where individuals grown at 10–15°C lived for about 47–55 days, while those raised at 20–25°C lived for 15–20 days. We found a similar response in our life-tables for both species. At 20°C and the highest food concentration tested (1×10^7 cells ml^{-1}), longevity was the highest for both species. Also, at 20°C increases in food concentration produced increases in longevity in both species (Table 1). In contrast, at 25 and 30°C when food

concentration increased, longevity decreased in both species, except for *L. quadridentata* at 25°C. This result can be explained by the increases in respiratory costs due to increases in food concentration (Doohan, 1973; Rico-Martínez & Dodson, 1992). In our work, as well as in others (Walz, 1983; Galkovskaya, 1987) the effect of temperature on the mean generation time (Tc) parallels the effect on longevity. The highest values of the potential reproductive rate (re) were found at the smallest food concentration and at the highest temperature in both species, as was found by Galkovskaja although for *Brachionus calyciflorus* (1983), a similar value was also found at the lowest temperature and the highest concentration of food in her work. In our work, there is a positive relationship between the values of the potential reproductive rate (re) and the net reproductive rate (Ro) at 20 and

Table 2. Morphological characterization of individuals of different ages of *Lecane luna* and *L. quadridentata* at 25±2 °C. Measurements are given in micrometers with ± one SD

Age (hours)	Maximum length	Length of the foot	Length of the pseudoclaw	Maximum width
Lecane luna				
0	176.3±19.6	57.0±5.5	9.6±1.2	110.2±10.3
24	166.9±9.5	55.0±5.6	10.6±1.1	111.0±5.9
48	166.2±14.8	55.7±8.2	10.1±1.0	105.9±6.4
Ovigerous females*	201.4±9.1	67.7±5.5	11.1±1.3	134.5±9.1
Lecane quadridentata				
0	227.5±19.3	87.6±10.3	17.4±2.2	113.1±10.6
24	246.0±8.7	89.6±11.0	17.1±2.2	121.0±7.3
48	246.0±9.7	92.7±8.4	18.6±2.7	120.7±6.2
Ovigerous females*	288.0±15.0	108.6±7.3	23.1±2.3	145.2±4.0

*Randomly picked up from our cultures.

Table 3. Hatching percentages of two species of littoral rotifers. Hatching percentages were determined at different temperatures and times after eggs were transferred to individuals wells

Species	Temperature (°C)	Time after transfer			n*
		24	48	72	
L. luna	20	28.6±22.3	74.8±22.9	79.1±25.1	10
	25	51.7±38.0	62.4±40.2	64.1±45.8	10
	30	0	0	0	5
L. quadridentata	20	53.9±15.2	90.6±12.5	93.1±9.9	7
	25	37.7±25.5	46.1±26.3	48.4±26.4	13
	30	0.1±0.3	2.5±5.3	2.0±5.3	9

*Eggs were randomly collected from the cultures. n = Number of wells. Each well contained 10 eggs.

25°C, but there is a big discrepancy between the same parameters at 30°C. This discrepancy is explained by the difference in mean generation time, which drops drastically at 30°C. The highest Ro values for both species were thus recorded at 25°C. From the analysis of our life-table data it is clear to us that the optimal temperature for both *L. luna* and *L. quadridentata* is probably found around 25°C. Galkovskaya (1983) also found similar relationships between r_e and Ro for *B. calyciflorus*, including high values of r_e and low values of Ro at the higher temperatures.

In our life-table experiments, *L. luna* and *L. quadridentata* started reproduction after 36 hours. In one or two exceptional cases, neonates were produced at 24 hours, and at 30°C reproduction was delayed in some treatments or never appeared (Figure 1). We found that the reproductive value (V_x) peaked in the lowest food concentration treatments in *L. luna* at 25 and 30°C,

Table 4. Correlation (adjusted R^2) between survivorship (lx) and mean life expectancy (ex) for all 18 treatments

Species	Food concentration (10^6 cells ml^{-1})	Temperature		
		20°C	25°C	30°C
L. luna	1	0.7473	0.7065	0.6239
	5	0.8449	0.8768	−0.0797
	10	0.6492	0.54422	0.1428
L. quadridentata	1	0.8183	0.7454	0.6881
	5	0.7753	0.8484	0.8036
	10	0.8470	0.8463	0.8860

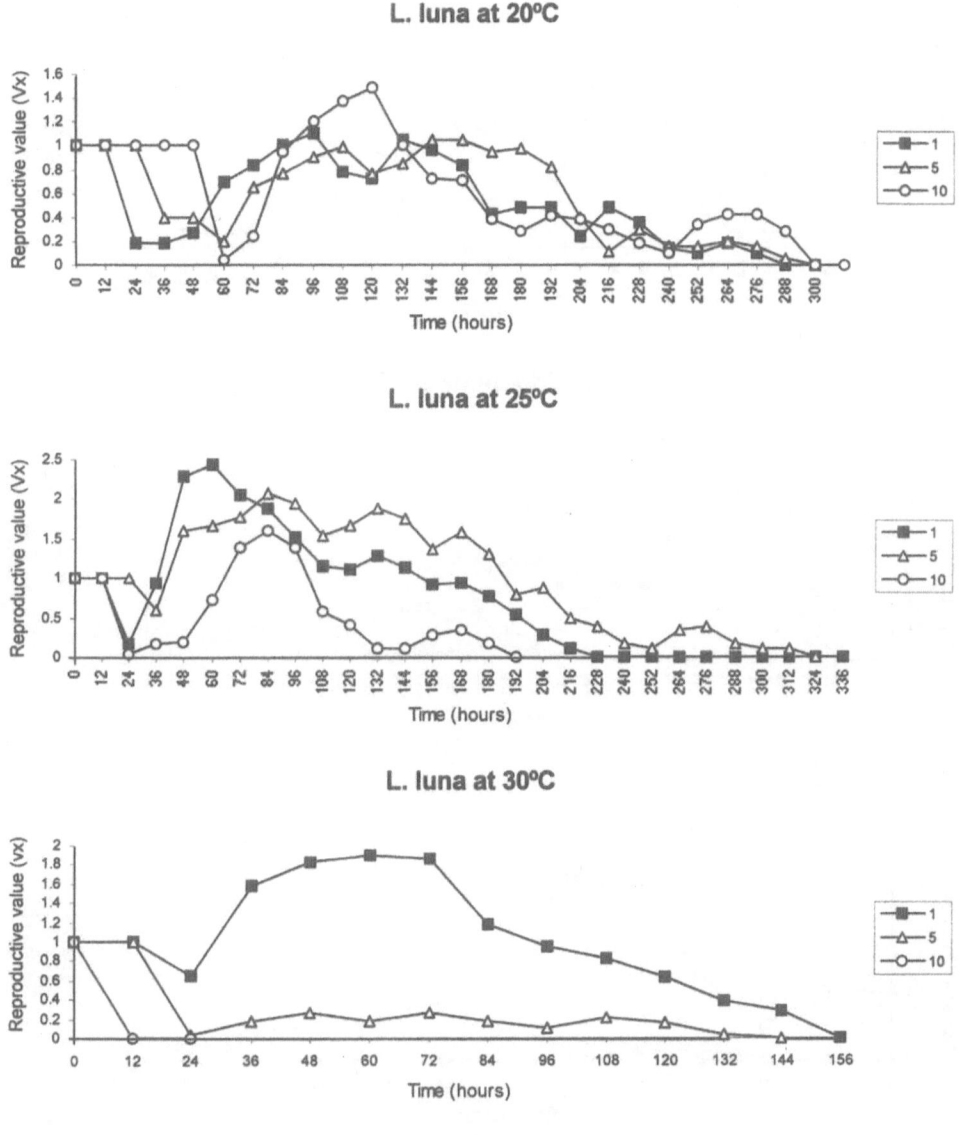

Figure 1a.

Figure 1. Influence of food concentration and temperature on the reproductive value (V_x) of (a) *L. luna* and (b) *L. quadridentata*. The numbers: 1, 5, and 10 correspond to the food concentration in 10^6 cells ml^{-1} of *N. oculata*.

and at 30°C in *L. quadridentata* (Figure 1). Also, the peak occurred between 48 and 108 hours for all treatments (including 20°C). Our highest reproductive values are slightly higher than those of Hummon & Bevelhymer (1980) and Galkovskaya (1987).

While we were developing the method to do a life table of both *Lecane* species, we encountered several problems. One was the injury or death of progenitors after transference to a new well with fresh medium. Therefore, we decided to remove neonates, allowing the progenitor to stay in the same well thus avoiding

the stress produced by the transfer. Another problem was the low hatching percentage of asexual eggs at 30°C (Table 3). For that reason we were unable to perform the life-table experiments at 30°C under the same pre-conditions as those at 20 and 25°C. To assess the effect of acclimation to the previous generation (Galkovskaya, 1983) we performed an additional life-table experiment using animals obtained from hatching asexual eggs at 30°C (produced by individuals previously acclimated at 30°C for at least 48 hours). We compared survivorship (lx) and fecundity (mx) for

346

L. quadridentata at 20°C

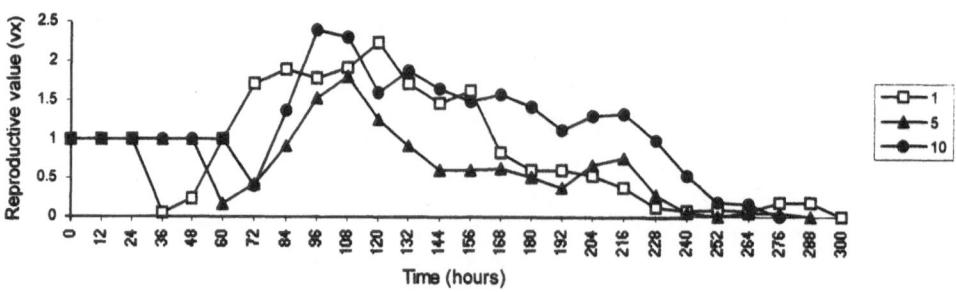

L. quadridentata at 25°C

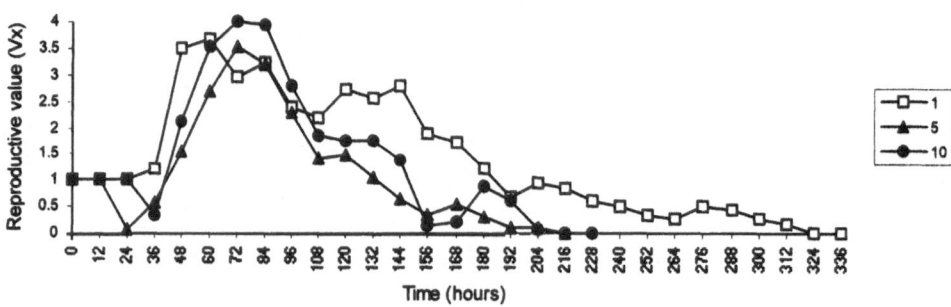

L. quadridentata at 30°C

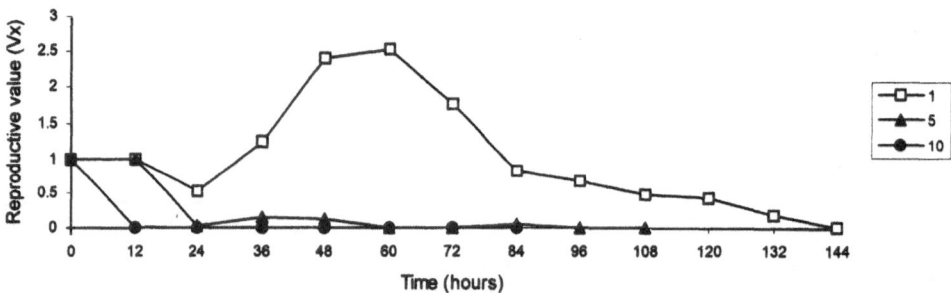

Figure 1b. For legend see p. 345.

acclimated vs. non-acclimated animals (Figure 2). Acclimated animals live longer and are more fecund. A similar result was reported by Galkovskaya (1983), who reported considerable mortality of animals at higher temperatures during acclimation. Similarly, in our experiments, after keeping the cultures of both species for up to two weeks at 30°C, all animals died. In contrast, cultures at 20 and 25°C have been kept for years in our lab. This observation confirms the inadequacy of 30°C for culturing these species. Perhaps the ability to tolerate high temperatures for up to two weeks confers on these two species an ecological advantage in occupying temporally a particular niche.

Galkovskaya (1983) discussed the probability of using 'thermal selection' for producing highly efficient rotifer cultures, even when parthenogenesis is the only mechanism of reproduction.

Hummon & Bevelhymer (1980; Figure 1) found a strong relationship between survivorship (lx) and life expectancy (e_x) for *Lecane tenuiseta*. We investigated that same relationship for *L. luna* and *L. quadridentata* in our work, and in general we found high correlations between these two parameters (Table 4). The only two exceptions were in the two highest food concentration treatments for *L. luna* at 30°C. In these two cases correlations are lower than 0.50.

L. quadridentata at 30°C

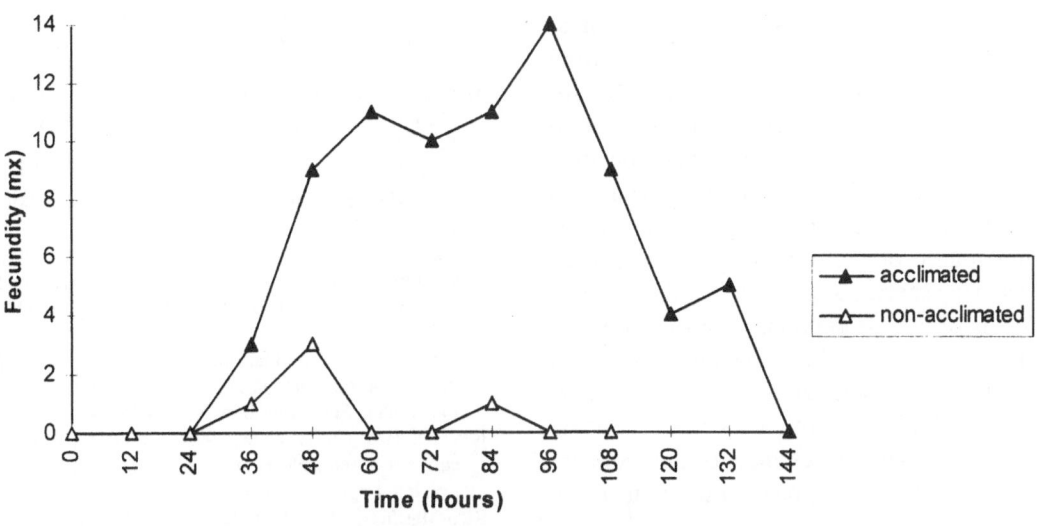

L. quadridentata at 30°C

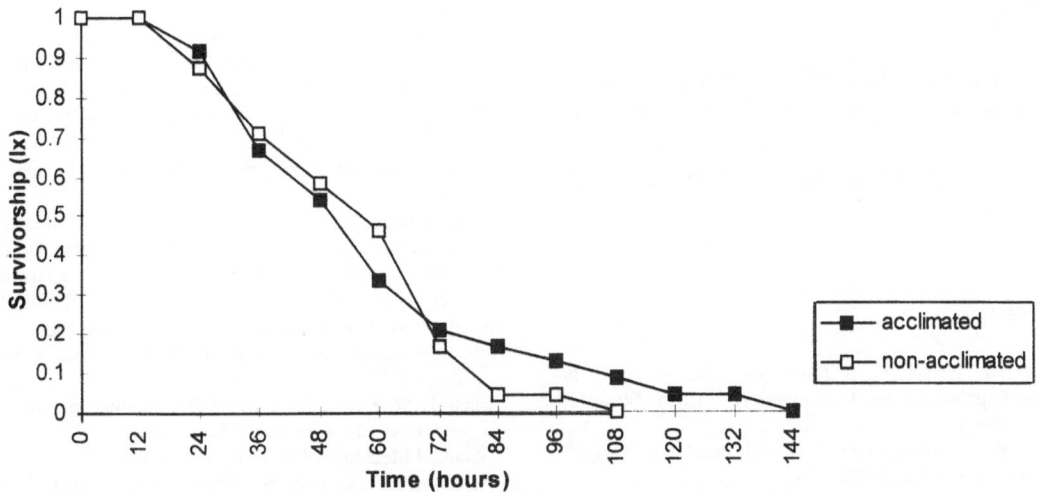

Figure 2. Comparison of acclimated vs. non-acclimated *L. quadridentata* individuals. The data correspond to the 5×10^6 cells ml^{-1} treatment at 30°C. (a) Survivorship (lx) comparison. (b) Fecundity (mx) comparison.

We performed two individual tailed t-tests to analyze the morphometric data in Table 2. As a result of that analysis, it is clear that *L. quadridentata* grows little from 24 to 48 hours. A comparison of maximum length (ML) and maximum width (MW) for *L. quadridentata* showed no significant differences between these two age groups ($p=0.99$ in both cases). However, significant differences were found between neonates and either 24- or 48-hour-old animals and between ovigerous females (adults) and all other groups ($p<0.001$). In the case of foot length and pseudoclaw length, only in adults are the differences

significant ($p<0.001$). For all other groups there was no significant growth of these structures from neonates to 48-hour individuals. In the case of *L. luna* there was no significant growth from neonates to 48-hour individuals. However, as expected, adults are bigger than any other group for all measurements taken ($p<0.001$). Comparison between both species showed that *L. quadridentata* is bigger than *L. luna* for almost all measurements taken ($p<0.001$), except for the MW of neonates ($p=0.39$).

The analysis of hatching percentages (Table 3) suggests that the optimum temperature for hatching of asexual eggs in both *Lecane* species is between 20°C and 25°C.

Our data show big differences in the life-table parameters of both *L. luna* and *L. quadridentata* from those obtained by Hummon & Bevelhymer (1980) for *L. tenuiseta*. Important differences can be inferred in the duration of lifespan and mean generation time among the three species. The fact that such differences are found among members of the same genus points out the importance of performing life-table experiments in as many species as possible within the same genus. This is of utmost importance in the case of a genus as large as *Lecane*, which contains 163 species (Segers, 1995). These data are necessary to develop toxicity assays with littoral rotifer species. These assays would allow a better assessment of the sediment toxicity, which represents a potential threat to many ecosystems.

Acknowledgements

The authors thank Dr Manuel Serra for important comments on the work, and two anonymous reviewers who made important improvements to the manuscript. This work was supported by the grant CONACyT-SIHGO RN-20/96 to R.R.

References

Begon, M., J. L. Harper, C. P. Townsend, 1996. Ecology: Individuals, Populations, and Communities, 3rd edn. Blackwell Scientific: 1068 pp.

Doohan, M., 1973. An energy budget for adult *Brachionus plicatilis* Müller. Oecologia 13: 351–362.

Galkovskaya, G. A., 1983. On temperature acclimation in an experimental population of *Brachionus calyciflorus*. Hydrobiologia 104: 225–227.

Galkovskaja, G. A., 1987. Planktonic rotifers and temperature. Hydrobiologia 147: 307–317.

Gilbert, J. J. & D. K. Schreiber, 1995. Induction of diapausing amictic eggs in *Synchaeta pectinata*. Hydrobiologia 313/314: 345–350.

Halbach, U., 1970. Die Ursachen der temporalvariation *von Brachionus calyciflorus* Pallas (Rotatoria). Oecologia 4: 262–318.

Hummon, W. D. & D. P. Bevelhymer, 1980. Life table demography of the rotifer *Lecane tenuiseta* under culture conditions and various age distributions. Hydrobiologia 70: 25–28.

Krebs, C. J., 1985. Ecología: Estudio de la distribución y la abundancia, 2nd edn. Ed. Harla. México D.F., Mexico: 753 pp.

Krebs, C. J., 1994. Ecology, the Experimental Analysis of Distribution and Abundance. 4th edn. Harper-Collins Publ: 801 pp.

May, L., 1987. Effect of incubation temperature on the hatching of rotifer resting eggs collected from sediments. Hydrobiologia 147: 335–338.

Nichols, H. W., 1973. Growth media – freshwater. In J. R. Stein (ed.), Handbook of Phycological Methods. Cambridge University Press: 7–24.

Nogrady, T., R. L. Wallace & T. W. Snell, 1993. Guides to the identification of the microinvertebrates of the continental waters of the world, Volume 4: Rotifera. SPB Academic Publishing: 142 pp.

Rao, T. R. & S. S. S. Sarma, 1988. Effect of food and temperature on the cost of reproduction in *Brachionus patulus* (Rotifera). Proc. Indian natn. Sci. Acad. B54, No. 6: 435–438.

ReVélle, P. & Ch. ReVélle, 1992. The Global Environment: Securing a Sustainable Future. Jones and Bartlett Publishers, Inc. London, England: 480 pp.

Rico-Martínez, R. & S. I. Dodson, 1992. Culture of the rotifer *Brachionus calyciflorus* Pallas. Aquaculture 105: 191–199.

Sarma, S. S. S. 1985. Effect of food density on the growth of the rotifer *Brachionus patulus* Müller. Bull. bot. Soc. Sagar 32: 54–59.

Sarma, S. S. S. & T. R. Rao, 1990. Population dynamics of *Brachionus patulus* Müller (Rotifera) in relation to food and temperature. Proc. Indian Acad. Sci. (Anim. Sci.) 99(4): 335–343.

Segers, H., 1995. Guides to the identification of the microinvertebrates of the continental waters of the world: Volume 2. Rotifera: The Lecanidae (Monogononta). SPB Academic Publishing: 226 pp.

Silva-Briano, M., 1992. Preliminary study of the zooplankton of Mexico. End of course report of the International Training Course 'Lake Management: A Tool in Lake Management'. State University of Ghent, Ghent, Belgium: 12 pp (23 illustrations).

Snell, T. W. & B. D. Moffat, 1992. A 2-d life cycle test with the rotifer *Brachionus calyciflorus*. Envir. Toxicol. and Chem. 11: 1249–1257.

Snell T. W. & C. R. Janssen, 1995. Rotifers in ecotoxicology: a review. Hydrobiologia 313/314: 231–247.

Standard Methods, 1995. 19th edn: 258 pp.

Stemberger, R. S., 1979. A guide to rotifers of the Laurentian Great Lakes. U.S. EPA publication: EPA/600/4-79/021, Washington. DC: 186 pp.

United States Environmental Protection Agency, 1985. Methods for measuring the acute toxicity of effluents to freshwater and marine U.S. EPA publication: EPA-600/4-85-013, Washington, DC.

Walz, N., 1983. Individual culture and experimental population dynamics of *Keratella cochlearis* (Rotatoria). Hydrobiologia 107: 35–45.

Hydrobiologia **387/388**: 349–353, 1998.
E. Wurdak, R. Wallace & H. Segers (eds), Rotifera VIII: A Comparative Approach.
© *1998 Kluwer Academic Publishers.*

Population growth in planktonic rotifers. Does temperature shift the competitive advantage for different species?

Claus-Peter Stelzer
Max-Planck-Institut für Limnologie, P.O. Box 165, D-24306 Plön, Germany

Key words: competition, threshold food level, Rotifera, temperature, *Brachionus*, *Synchaeta*

Abstract

The numerical response of populations to different food concentrations in an important parameter to be determined for a mechanistic approach to interspecific competition. Theory predicts that the species with the lowest food level (TFL) should always be the superior competitor if only one food source is offered. However, TFLs are not species specific constants but may change along environmental gradients such as food size or temperature.

The hypothesis that temperature differentially affects the TFLs of three planktonic rotifers (*Asplanchna priodonta*, *Brachionus calyciflorus* and *Synchaeta pectinata*) was tested in laboratory experiments. Numerical responses were assessed for all three rotifers at 12, 16, 20, 24 and 28°C with *Cryptomonas erosa* as food alga. Growth rates of all three rotifers at high food concentrations (1 mg C l^{-1}) increased as temperature increased until the limits of thermal tolerance were reached. This increase was very pronounced for *Brachionus*, but less for *Synchaeta* which already had relatively high growth rates at 12°C. Along the temperature gradient, the TFLs of *Synchaeta* increased from 0.074 to 0.66 mg C l^{-1}, whereas those of *Asplanchna* and *Brachionus* stayed relatively constant at 0.3 and 0.2 mg C l^{-1}, respectively. Hence, the zero net growth isocline (ZNGI) of *Synchaeta* crossed those of *Brachionus* and *Asplanchna* at 16 and 20.5°C, respectively. The results suggest that *Synchaeta* is better adapted to low temperatures than the other two rotifers and should be the superior competitor below 16°C.

Introduction

Resource limitation is regarded as one of the most important factors for structuring plankton communities (Gliwicz, 1985). According to the 'threshold hypothesis' (Lampert, 1977; Lampert & Schober, 1980) the superior species in a resource limited environment is the one with the lowest food requirement needed to maintain growth (threshold food level, TFL). Lampert & Schober (1980) distinguished between a threshold for the individual and a threshold for a population.

Threshold food levels in rotifers have usually been determined on the population level and are defined as the food concentration at which population growth is zero. Stemberger & Gilbert (1985) showed that TFLs in eight planktonic rotifer species varied by a factor of 17 and that small rotifers tend to have lower TFLs than large rotifers. However, TFLs are not species-specific constants and may change along various environmental gradients. For example Rothhaupt (1990) found considerable differences in the TFLs of two

Brachionus spp. along a gradient of food size, with the lowest TFLs for food algae in the most readily ingested size range for the respective species. Achenbach & Lampert (1997) determined TFLs of four cladoceran species along a temperature gradient from 16–28°C and found an increase in the TFLs above 20°C for all species. This did not change the competitive abilities among the different species because the species with the lower TFLs at 16 and 20°C still had the lower TFLs at 24 and 28°C.

The aim of this paper is to investigate how the population growth and the TFLs of three different planktonic rotifers (*Asplanchna priodonta*, *Brachionus calyciflorus*, *Synchaeta pectinata*) change along a temperature gradient of 12–28°C. According to their maximum abundance in the field, the three species may have different thermal preferences. *Synchaeta* and *Asplanchna* are most abundant at 12 and 15°, respectively, and *Brachionus* at 20°C (Berzins & Pejler, 1989). The hypothesis tested was that the 'cold wa-

350

ter' rotifers have lower TFLs than the 'warm water' rotifers at low temperatures and vice versa.

Materials and methods

The rotifers *Asplanchna priodonta* and *Synchaeta pectinata* were isolated from Schöhsee (northern Germany) and cultured as clones. *Brachionus calyciflorus* was obtained from K.O. Rothhaupt and is identical to the *B. calyciflorus* used in his work (e.g., Rothhaupt, 1990).

All rotifers were raised on *Cryptomonas erosa* var. *reflexa*, which was obtained from J.J. Gilbert, New Hampshire. In all experiments the rotifers were cultured in ADaM medium (Klüttgen et al., 1994a, b), which was supplemented with 2.2 mg l^{-1} Na$_2$EDTA and Woods Hole MBL (Guillard, 1975) medium (9:1) to improve the conditions for the *Cryptomonas* food (Kirk & Gilbert, 1990). *C. erosa* was cultured in MBL medium in semicontinuous culture (dilution rate = 0.25 day^{-1}) and was continuously illuminated (100 μmol quanta m^{-2} s^{-1}). The carbon content of *Cryptomonas erosa* cells was determined in a C/N analyser (NA 1500, Carlo Erba). Samples were filtered on precombusted (5 h at 500°C) glass-fibre filters (Whatman GF/F) and dried overnight. No corrections for associated bacteria were made. Food concentrations were determined with an electronic particle counter (CASY).

The experimental treatments were combinations of five different temperatures (12, 16, 20, 24, and 28°C) and various food concentrations ranging from 0.075 to 3 mg C l^{-1}. At least three replicates were used for each combination of food and temperature.

Populations of rotifers were acclimated to the experimental temperatures for at least 2 weeks in 2-l or 3-l Erlenmeyer flasks at high food concentrations (3 mg C l^{-1}). During the acclimation phase, the rotifers were cultured in fed batch cultures and medium was replaced every 4–5 days by filtration through 30-μm gauze, except for the 24 and 28°C treatments, where the replacement interval was 2 days. At these temperatures there was rapid population growth and enhanced grazing particularly by *Brachionus*, so that the algae were depleted much faster than at the lower temperatures. In the experiments, the rotifers were cultured in 100-ml glass test tubes sealed with a double layer of Parafilm (American Can Co.). The test tubes were placed on plankton wheels, rotating every 15 min for 2 min at 1 rpm, and set in an incubator with low

illumination (5 μmol quanta m^{-2} s^{-1}). Every day, the food algae and medium in each test tube were exchanged by filtration with a Plexiglas tube covered with a 30-μm gauze. The tube was outfitted with a device which permitted a low flow rate and left 10 ml of medium in the bottom of the tube, so that the rotifers would not become stuck to the gauze. The tube was refilled with 90 ml of fresh ADaM to ensure that most of the old algae were removed. The rotifers were then poured into a new test tube which was filled with medium and *Cryptomonas* in the desired concentrations.

After 3–4 days of acclimation to the different food concentrations, population growth was measured for another 3–4 days. Initial population sizes during acclimation were around 50 to 100 animals per test tube (100 ml), depending on expected growth at the various food concentrations. Visual inspections were made to ensure that the populations did not exceed 150–200 rotifers per test tube to prevent overgrazing during the acclimation period. At the start of the growth measurements, 50–100 rotifers were selected randomly from the acclimated populations and transferred into new test tubes. When high clearance rates were expected (low food concentrations combined with high temperatures) the initial concentration of rotifers was always 0.5 ind. ml^{-1}. On the following days, samples of ~40% volume were taken from these experimental populations, fixed with Lugol's solution, transferred to 50-ml sedimentation chambers, and enumerated under an inverted microscope. In cases of very slow population growth at low temperatures and low food concentrations samples were taken at 2-day intervals. The remaining 60% of rotifer culture medium was exchanged in the way described above. At the end of the growth measurements, all animals of one test tube were concentrated to 50 ml, fixed and enumerated. Threshold food levels were defined as the intersections of the regression lines fitted through near-zero growth rates with the x-axis (food concentration). Intersections of the 95% confidence limits with the x-axis were calculated as 'fiducial limits' according to Draper & Smith (1980).

Previous tests with all three rotifer species at 20°C showed that no animals got lost or damaged during the filtration procedure and that the population growth rates obtained by this method were comparable to those obtained by other methods (e.g., Rothhaupt, 1990). During the experiments it became obvious that the growth rates of *Synchaeta* at the low temperatures (12 and 16°C) were underestimated by the above de-

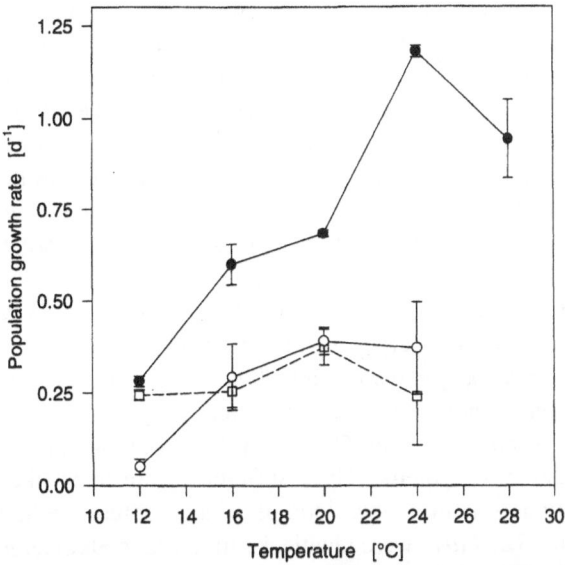

Figure 1. Growth rates of *Asplanchna priodonta* (circles), *Brachionus calyciflorus* (dots) and *Synchaeta pectinata* (squares) at a food concentration of 1 mg C l⁻¹ and at different temperatures. Means and S.D.s ($n=3$).

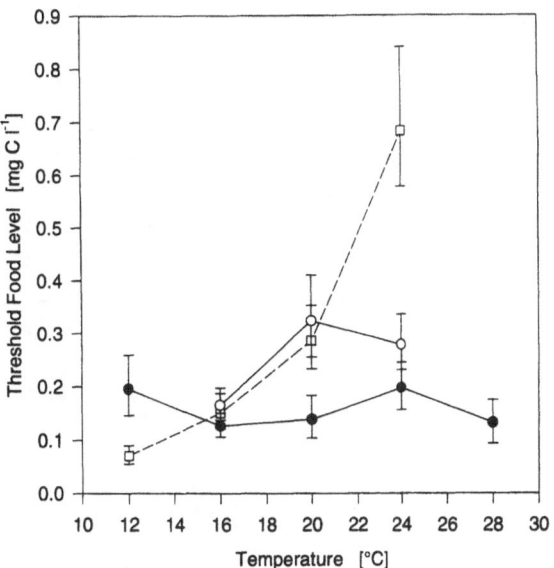

Figure 2. Threshold food levels of *Asplanchna priodonta* (circles), *Brachionus calyciflorus* (dots) and *Synchaeta pectinata* (squares) at different temperatures. TFLs are the intersections of the regression lines fitted through near-zero growth rates with the *x*-axis (food concentration) and error bars are the 95% fiducial limits.

scribed method. This was most likely because some of the eggs, which were laid freely into the culture medium, were found stuck to the walls of the culture vessel. Hence some of these eggs may have been lost during the daily medium exchange, and this effect must have been more pronounced at the low temperatures because egg development times are longer. Thus it was necessary to culture *Synchaeta* in six-well polystyrene culture plates (containing 8 ml medium), with daily transfers of all rotifers and eggs via glass pipettes (Stemberger & Gilbert 1985). To prevent overgrazing, not more than 30 rotifers were cultured per well. If necessary, the numbers of rotifers were reduced and care was taken that the egg ratio was not changed by this procedure.

Results

The carbon content of *Cryptomonas erosa* was 219 ± 8.7 pg/cell ($n=12$) and the average cell volume was 885 μm³.

Brachionus was able to grow at all of the experimental temperatures, whereas *Synchaeta* and *Asplanchna* could not be cultured at 28°C for long time periods. Even at food concentrations above 3 mg C l⁻¹ the populations died after some days. No numerical response curves could be assessed for *Synchaeta* and *Asplanchna* at 28°C.

In all three rotifer species, population growth rates at 1 mg C l⁻¹ generally increased with temperature, apart from slight decreases in *Brachionus* above 24°C and *Synchaeta* above 20°C (Figure 1). For *Brachionus* a 4.2-fold increase in the population growth rate was found from 12 to 24°C. In *Asplanchna* the increase in population growth was highest between 12 and 16°C (5.9-fold), lower between 16 and 20°C (1.3-fold), and zero between 20 and 24°C. *Synchaeta* had a relatively high growth rate at 12°C and its increase with temperature (1.5-fold from 12 to 20°C) was not as pronounced as in *Brachionus* or *Asplanchna*.

The TFL of *Asplanchna* was 0.3 mg C l⁻¹ at 20 and 24°C, but lower (0.16 mg C l⁻¹) at 16°C (Figure 2). At 12°C population growth rates in *Asplanchna* were not significantly different from zero (*t*-test, $P>0.05$) for the tested food concentrations (0.3–3mg C l⁻¹), hence no TFL could be calculated. The significantly positive value for the growth rate at 1 mg C l⁻¹ (Figure 1) is probably an artefact of the small number of replicates. For *Brachionus* the TFLs stayed relatively constant between 0.15 and 0.2 mg C l⁻¹ along the investigated temperature gradient. The small differences between the TFLs were not significant as can be seen from the overlapping 95% fiducial limits. The TFLs of *Synchaeta* increased dramatically with temperature, from 0.074 mg C l⁻¹ at 12°C to 0.66 mg C l⁻¹ at

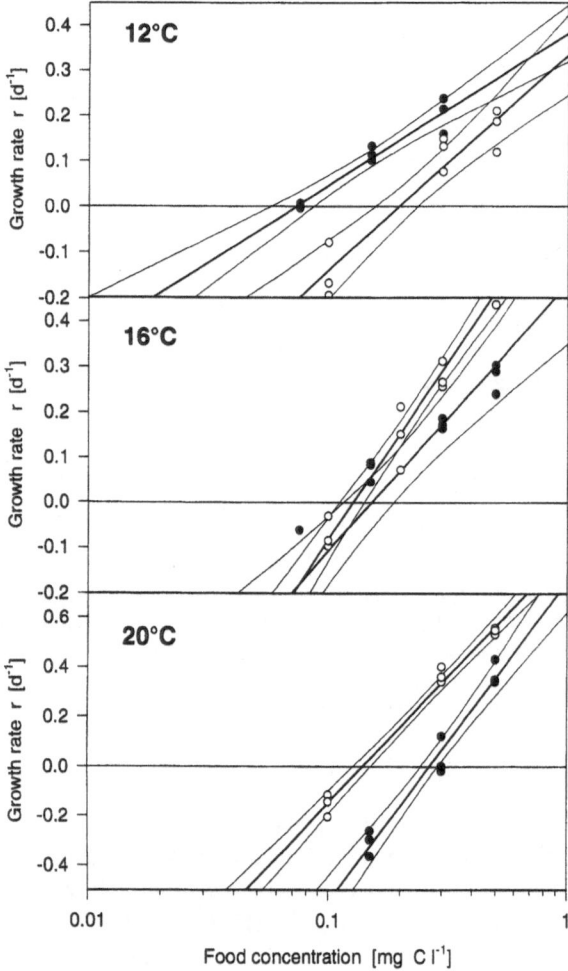

Figure 3. Population growth rates near the threshold food levels of *Brachionus calyciflorus* (circles) and *Synchaeta pectinata* (dots) at 12, 16 and 20°C. Regression lines (thick lines) and 95% confidence intervals (thin lines).

24°C. Along its increase from 12 to 24°C the zero net growth isocline (ZNGI) of *Synchaeta* intersected the ZNGIs of the other two rotifers. The intersections were at 16°C for *Synchaeta–Brachionus* and at 20.5°C for *Synchaeta–Asplanchna*. Below these intersections *Synchaeta* had the lower TFLs for *Cryptomonas* and above them the other rotifer had the lower TFLs. This situation is shown in Figure 3 for *Synchaeta* and *Brachionus*.

Discussion

Growth rates measured in this study are comparable to those found in previous studies (e.g., Rothhaupt, 1990; Seale et al., 1993; see Stemberger & Gilbert, 1985). Since most of the previous studies have been conducted at temperatures around 20°C, comparisons are restricted to these temperatures. Population growth of *Brachionus calyciflorus* is most comparable to the work of Rothhaupt (1990), since the rotifers originate from the same clone. Both, TFL and maximum growth rates (determined at 1 mg C l^{-1} in this study) are consistent with his results for *Chlamydomonas sphaeroides* (an alga of comparable size to *Cryptomonas erosa*). Assuming a 50% carbon content of dry weight, comparisons can be made to the data in Stemberger & Gilbert (1985). The TFLs of *Synchaeta pectinata* and *Asplanchna priodonta* are consistent with their data, whereas the growth rates of *Synchaeta* at 1 mg C l^{-1} and 20°C were about 50% lower than theirs. These differences are most likely clone specific, since food algae and culture methods for *Synchaeta* were identical with those in Stemberger & Gilbert's work.

Growth rates at the relatively high food concentration of 1 mg C l^{-1} generally increased with temperature (Figure 1). As rotifers are ectothermic organisms their metabolism is directly exposed to the temperature of their environment, hence the speed of biochemical reactions is enhanced at higher temperatures. The decrease in the growth rates at the highest experimental temperatures may have occurred because the upper limit of thermal tolerance was reached. Additionally, in *Synchaeta* it may also be an effect of the high TFL at 24°C (Figure 2). The relatively high growth rates of *Synchaeta* at 12°C and their small increase with temperature support the hypothesis that this rotifer is more adapted to cold temperatures than *Asplanchna* and *Brachionus*.

The most surprising result of this study was the strong increase of *Synchaeta* TFLs with temperature, leading to a ZNGI that intersected the ZNGIs of the other two rotifer species. According to theory of exploitative competition such intersections are points where superiority can switch between two competitors (Tilman, 1982). Tilman et al. (1981) demonstrated this for two diatoms, whose ZNGIs (at a dilution rate of 0.11 day^{-1}) intersected at one point along a temperature gradient. Below 20°C *Asterionella formosa* had lower requirements for silicate to maintain population growth than *Synedra ulna*, whereas above 20°C the situation was reversed. In competition experiments Tilman et al. (1981) showed that *Asterionella* displaced *Synedra* below 20°C and that *Synedra* displaced *Asterionella* above 20°C. Similar changes in competitive superiority may also occur if pairs of the rotifers used in this study were to compete at different temperatures for their food *Cryptomonas*. The order of superiority should be: *Synchaeta* > *Brachionus* > *Asplanchna* below 16°C, *Brachionus* > *Synchaeta* > *Asplanchna* between 16 and 20°C and *Brachionus* > *Asplanchna*>*Synchaeta* at 24°C. It may be difficult to test these predictions for *Asplanchna priodonta*, since this species is an omnivore and can also prey on juveniles of *Synchaeta* or *Brachionus* (Pourriot, 1977).

The competitive relationships between *Brachionus* and *Synchaeta* determined in laboratory can apply in field, but under special circumstances since both species can use other food sources than *Cryptomonas*. *Brachionus calyciflorus* can use a very broad array of food particles such as bacteria (Starkweather et al., 1979), many different kinds of planktonic algae and even small ciliates (Gilbert & Jack, 1993). *Synchaeta pectinata* seems to be restricted to large particles like *Cryptomonas erosa* or some ciliates (Gilbert & Jack, 1993). Hence, overlap in the diets of the two rotifer species is relatively small and competitive interactions in the field can only be expected for particular food conditions like dominance of cryptomonads.

In conclusion, the results of this study suggest that temperature can shift the competitive advantage for the rotifers *Synchaeta pectinata* and *Brachionus calyciflorus*. However, only well-controlled competition experiments can test the robustness of these findings. Thus this problem needs further study.

Acknowledgements

I thank M. Brewer and N. Weider for constructive comments on the manuscript and linguistic help. This work was supported by DFG Grant LA 309/12–1.

References

Achenbach, L. & W. Lampert, 1997. Effects of elevated temperatures on threshold food concentrations and possible competitive abilities of differentially sized cladoceran species. Oikos 79: 469–476.

Berzins, B. & B. Pejler, 1989. Rotifer occurrence in relation to temperature. Hydrobiologia 175: 223–231.

Draper, N. R. & H. Smith, 1980. Applied Regression Analysis. Wiley, New York.

Gilbert, J. J. & J. D. Jack, 1993. Rotifers as predators on small ciliates. Hydrobiologia 255/256: 247–253.

Gliwicz, Z. M., 1985. Predation and food limitation: an ultimate reason for extinction of planktonic cladoceran species. Arch. Hydrobiol. Suppl. 21: 419–430.

Guillard, R. R. L., 1975. Culture of phytoplankton for feeding marine invertebrates. In Smith W. L. & M. H. Chanley (eds), Culture of Marine Invertebrate Animals. Plenum, New York: 29–60.

Kirk, K. L. & J. J. Gilbert,1990. Suspended clay and the population dynamics of planktonic rotifers and cladocerans. Ecology 71: 1741–1755.

Klüttgen, B., U. Dülmer, M. Engels & H. T. Ratte, 1994a. ADaM, an artificial fresh water for the culture of zooplankton. Wat. Res. 28: 743–746.

Klüttgen, B., U. Dülmer, M. Engels & H. T. Ratte, 1994b. Corrigendum. Wat. Res. 28: 1.

Lampert, W., 1977. Studies on the carbon balance of *Daphnia pulex* de Geer as related to environmental conditions. IV. Determination of the 'threshold' concentration as a factor controlling the abundance of zooplankton species. Arch. Hydrobiol. Suppl. 48: 361–368.

Lampert, W. & U. Schober, 1980. The importance of 'threshold' food concentrations. In Kerfoot, W. C. (ed.), Evolution and Ecology of Zooplankton Communities. University Press of New England, Hanover, New Hampshire: 264–267.

Pourriot, R., 1977. Food and feeding habits of rotifera. Arch. Hydrobiol. Suppl. 8: 243–260.

Rothhaupt, K. O., 1990. Population growth rates of two closely related rotifer species: effects of food quality, particle size, and nutritional quality. Freshwat. Biol. 23: 561–570.

Seale, D. B., M. E. Boraas & J. B. Horton, 1993. Using of semi-continuous culture methods for examining competitive outcome between two freshwater rotifers (Genus *Brachionus*) growing on a single algal resource. In Walz, N. (ed.) Plankton Regulation Dynamics. Springer, Berlin: 161–177.

Starkweather, P. L., J. J. Gilbert & T. M. Frost, 1979. Bacterial feeding in the rotifer *Brachionus calyciflorus*: clearance and ingestion rates, behavior and population dynamics. Oecologia 44: 26–30.

Stemberger, R. S. & J. J. Gilbert, 1985. Body size, food concentration, and population growth in planktonic rotifers. Ecology 66: 1151–1159.

Tilman, D., M. Mattson & S. Langer, 1981. Competition and nutrient kinetics along a temperature gradient: An experimental test of a mechanistic approach to niche theory. Limnol. Oceanogr. 26: 1020–1033.

Tilman, D., 1982. Resource competition and community structure. Princeton University Press, Princeton, N.J.

Hydrobiologia **387/388**: 355–360, 1998.
E. Wurdak, R. Wallace & H. Segers (eds), Rotifera VIII: A Comparative Approach.
© *1998 Kluwer Academic Publishers.*

Carbon content of some freshwater rotifers

Irena V. Telesh[1], Minna Rahkola[2] & Markku Viljanen[2]
[1]*Zoological Institute of the Russian Academy of Sciences, 199034 St. Petersburg, Russia*
[2]*University of Joensuu, Karelian Institute, P.O. Box 111, FIN-80101 Joensuu, Finland*

Key words: Rotifers, carbon content

Abstract

Carbon content of rotifers from 14 species (*Keratella cochlearis, K. c. tecta, K. c. hispida, K. ticinensis, K. quadrata, Polyarthra remata, P. vulgaris, P. major, P. euryptera, Synchaeta sp., S. stylata, S. pectinata, Trichocerca capucina, Asplanchna priodonta*) was determined with the high temperature combustion method of Salonen (1979). Rotifers for the carbon analysis were collected from different fresh water bodies in Russia (Lake Ladoga) and Finland (lakes Pohjalampi, Varaslampi, and two small ponds in Lammi). Average individual carbon mass of rotifers varied between 0.0064 and 0.058 μg in *Keratella* spp., 0.012 and 0.051 μg in *Polyarthra* spp., 0.020 and 0.133 μg in *Synchaeta* spp., 0.162 and 0.555 μg in *A. priodonta*. The carbon level in the studied rotifer species differed 100-fold ranging from 0.31% WW in *A. priodonta* to 31.5% WW in *K. c. tecta*. Body length/carbon mass and body volume/carbon mass regressions were established for the studied rotifers.

Introduction

Rotifers form the essencial structural and functional component of pelagic communities in different freshwater habitats. In rivers and estuaries, rotifers can be responsible for 95% of total zooplankton biomass (Telesh, 1995). Even in the pelagic areas of large lakes, such as the great North-American lakes Huron, Erie, Ontario and Superior, rotifers contribute 14–20% to the total averaged for June–September zooplankton biomass; in Lake St. Clair in August 1984, 30% of zooplankton biomass were formed by rotifers (Sprules & Munawar, 1991). In some northern areas of Lake Ladoga (Russia) rotifers form 85% of zooplankton biomass during certain periods (Telesh, 1998).

Adequate comparison of the relative importance of different structural components of any community or ecosystem is possible only on the basis of common units of the chosen parameter, for example biomass. At present biomass of aquatic organisms is generally expressed as dry weight (Botrell et al., 1976; Dumont et al., 1975) or carbon mass (Duncan et al., 1985; Latja & Salonen, 1978; McCauley, 1984; Rahkola et al., 1998; Salonen et al., 1976; Vasama & Kankaala, 1990), and still often is expressed as wet weight (Balushkina & Winberg, 1979). Until recently, car-

bon analysis of rotifers has been limited due to the small size of these animals, which prevented wide application of the method.

This study presents the results of carbon content measurements in 14 common species of planktonic rotifers from a number of fresh water bodies of different types in Russia and Finland. The aim of the study was to express numerically general relationships between carbon content and length/volume/wet weight of rotifers, for a more reliable determination of rotifer biomass and adequate evaluation of their role in the trophic webs within pelagic communities.

Materials and methods

Carbon content was measured in rotifers from 14 species collected in the freshwater habitats of different types: Lake Ladoga in Russia (the largest European lake, surface area 17 891 km^2), and four small lakes and ponds in Finland (lake Pohjalampi, area 0.61 km^2; lake Varaslampi, area 0.03 km^2; reference pond in Lammi, area 160 m^2; and small natural pond in Lammi, area 25 m^2). The following rotifer species were analysed: *Keratella cochlearis* (Gosse), *K. c. tecta* (Gosse), *K. c. hispida* (Lauterborn), *K. ticinensis*

(Callerio), *K. quadrata* (Müller), *Polyarthra euryptera* Wierzejski, *P. major* Burkhardt, *P. remata* Skorikov, *P. vulgaris* Carlin, *Synchaeta* sp., *S. stylata* Wierzejski, *S. pectinata* Ehrenberg, *Trichocerca (s. str.) capucina* (Wierzejski & Zacharias), and *Asplanchna priodonta* Gosse.

Samples for the carbon analysis were collected in August 1995 (Lake Ladoga), May through August 1996 (lake Pohjalampi), June 1996 (lake Varaslampi) and May 1996 (ponds in Lammi). Samples were taken with the plankton net (mesh size 50 μm), preserved in 4% formaldehyde and frozen at $-18°C$. This method provides superior presevation for the purpose of carbon analysis for most zooplankton species (Salonen & Sarvala, 1980).

Before carbon content measurement, zooplankton samples were thawed, 300–500 individuals of each rotifer species were picked from the sample and briefly rinsed five times in small volumes of distilled water in Petri dishes. The samples were kept at about 5°C until processing on the same day. Prior to carbon analysis, length and width of 20 individuals of each rotifer species were measured and average values were calculated. Body volumes of rotifers were calculated according to the standard formulae (Ruttner-Kolisko, 1977) on the basis of mean linear dimentions and converted to wet weights assuming that 10^6 μm^3 equals 1 μg wet weight (Botrell et al., 1976).

Carbon mass of rotifers was determined with the high temperature combustion method of Salonen (1979, 1981). Rotifers were transferred into the combustion tube (+950°C) of the Universal Carbon Analyser with a Pasteur pipette in drops of a standard volume of distilled water, either one by one (*Asplanchna, Trichocerca*), or in groups of three to five (*Polyarthra, Synchaeta, K. quadrata*) or 10 animals (*Keratella*). Carbon content of the standard volume of the same water was determined in each measurement and substracted from the carbon weight of the animals (Latja & Salonen, 1979). In total, the number of rotifers analysed for the carbon content varied between 14–47 (*Asplanchna*) and 320 (*Synchaeta*). Carbon level was calculated as percent of wet weight for each rotifer species. The length/carbon mass and body volume/carbon mass relationships for rotifers were discribed by a power function: $Y = a X^b$ separately for the genera *Keratella* (five species), *Polyarthra* (four species) and *Synchaeta* (three species). For *S. stylata* it was possible to perform a series of eight sets of the length–width measurements, therefore the regressions for the three species from the genus *Synchaeta* are based on 10 pairs of mean carbon/length and carbon/volume values. General regressions for 13 rotifer species are based on the species means of carbon content and body size. The statistical tests were performed using the SPSS program.

Results

The body length of the studied rotifer species varied 10-fold (0.07–0.675 mm), and body volumes were within the range of 0.018–84.5 $\times 10^6$ μm^3 (Table 1). The organic carbon content exposed a 200-fold difference for the 14 measured rotifer species ranging between 0.0036 and 0.685 μg (Table 1). Difference between the measured carbon mass and means for each species did not exceed 30%, averaging at 20%. The measured variation in carbon mass was lowest in *K. cochlearis*, *P. remata* and *P. vulgaris* from lake Pohjalampi, due to low variation in the individual body sizes of these rotifers (Table 1). The highest variation in the carbon content was registered in *S. pectinata* from the small pond in Lammi, and was most probably caused by the naturally high variation in the size of these rotifers. *A. priodonta* was sampled almost simultaneously in two lakes, Pohjalampi and Varaslampi, and the similar-sized animals from two lakes differed considerably in carbon content (Table 1).

Carbon content was positively correlated with rotifer length and body volume. (Tables 2 and 3. Figure 1). The coefficients of determination (r^2) were high (0.75–0.96) in all cases except for the relationship between the body length and carbon content of *Synchaeta* spp. (Table 2).

Carbon level in the studied rotifer species varied 100-fold ranging between 0.31% WW in *A. priodonta* and 31.5% WW in *K. c. tecta*. Standard deviations of the results obtained for each rotifer species varied between 12 and 34% of the mean. However, the carbon level values were rather stable when congeneric species were compared: 19.8–24.0% WW in *Keratella* of the *cochlearis* type, 6.8–10.3% WW in *Synchaeta*, 2.6–3.7% WW in *Polyarthra* (Table 1). Exceptionally, the carbon level of *A. priodonta* of the similar size (51.7 and 56.2 μg) from two lakes exposed a three-fold difference (0.31 and 0.99% WW, respectively).

Carbon level was inversely related with wet weight in rotifers smaller than 2 μg WW (Figure 2A), but

Table 1. Body length (L, mm), body volume ($V \times 10^6$ μm^3), carbon content (C, μg) and mean carbon level (% of wet weight and standard deviation, SD) of rotifers

Species of rotifers	Sampling sight	Date of sampling	Body length (L, mm), mean and range	Body volume ($V \times 10^6$ μm^3), mean and range	Carbon (C, μg) mean and range	Carbon level %WW	SD	Number of measurements	Number of rotifers analysed
Keratella cochlearis tecta	Lake Pohjalampi	22.07.1996	0.079 (0.070–0.090)	0.027 (0.018–0.042)	0.0064 (0.0036–0.0085)	23.8	7.8	4	130
Keratella cochlearis	Lake Pohjalampi	14.05.1996	0.115 (0.110–0.120)	0.096 (0.091–0.100)	0.022 (0.019–0.027)	23.3	3.1	8	245
Keratella cochlearis hispida	Lake Pohjalampi	05.08.1996	0.098 (0.095–0.100)	0.064 (0.039–0.095)	0.015 (0.013–0.018)	24.0	3.1	6	139
Keratella ticinensis	Natural pond in Lammi	11.05.1996	0.120 (0.115–0.125)	0.085 (0.063–0.104)	0.017 (0.0098–0.022)	19.8	4.8	6	265
Keratella quadrata	Lake Varaslampi	07.06.1996	0.150 (0.140–0.180)	0.755 (0.604–1.283)	0.058 (0.045–0.075)	7.7	1.1	11	225
Polyarthra remata	Lake Pohjalampi	22.07.1996	0.101 (0.095–0.110)	0.304 (0.240–0.373)	0.012 (0.011–0.012)	3.7	0.1	2	60
Polyarthra vulgaris	Lake Pohjalampi	22.07.1996	0.128 (0.120–0.135)	0.559 (0.484–0.689)	0.018 (0.015–0.021)	3.2	0.4	4	68
Polyarthra euryptera	Lake Pohjalampi	22.07.1996	0.185 (0.160–0.200)	1.780 (1.147–2.240)	0.064 (0.052–0.080)	3.6	0.6	9	40
Polyarthra major	Lake Ladoga	09.08.1995	0.191 (0.180–0.215)	1.980 (1.630–2.783)	0.051 (0.041–0.064)	2.6	0.4	5	75
Synchaeta sp.	Lake Pohjalampi	05.08.1996	0.118 (0.100–0.155)	0.254 (0.188–0.326)	0.020 (0.012–0.027)	7.7	2.4	3	56
Synchaeta stylata	Reference pond in Lammi	11.05.1996	0.138 (0.105–0.180)	0.404 (0.140–1.239)	0.042 (0.030–0.057)	10.3	1.5	8	295
Synchaeta pectinata	Natural pond in Lammi	11.05.1996	0.213 (0.180–0.275)	1.956 (1.053–2.860)	0.133 (0.093–0.189)	6.8	1.6	9	320
Trichocerca capucina	Lake Pohjalampi	22.07.1996	0.268 (0.240–0.290)	1.251 (1.011–1.662)	0.057 (0.043–0.076)	4.6	0.7	8	54
Asplanchna priodonta	Lake Pohjalampi	11.06.1996	0.550 (0.500–0.600)	51.7 (43.9–57.9)	0.162 (0.047–0.224)	0.39	0.08	30	30
			0.637 (0.625–0.650)	70.3 (63.2–84.5)	0.328 (0.207–0.492)			47	47
Asplanchna priodonta	Lake Varaslampi	07.06.1996	0.450 (0.400–0.550)	15.3 (10.5–21.1)	0.063 (0.018–0.109)	0.71	0.24	14	14
			0.590 (0.575–0.600)	42.8 (35.0–50.0)	0.312 (0.148–0.448)			34	34
			0.675	56.2	0.555 (0.450–0.685)			17	17

Table 2. Carbon/length relationship for rotifers

Rotifers	L (mm) range	$a \pm$S.E.	$b \pm$S.E.	n	r^2
Keratella	0.070–0.180	0.351±0.066	0.290±0.047	5	0.93
Polyarthra	0.095–0.215	0.547±0.102	0.373±0.052	4	0.96
Synchaeta	0.110–0.275	0.200±0.041	0.119±0.063	10	0.31
Rotifers (13 species)	0.070–0.275	0.509±0.101	0.367±0.054	13	0.81

Range of length measurements (L, mm), intercept (a), the slope (b), number of observations (n) and coefficient of determination (r^2) are presented for the power function: $C = a\, L^b$.

Table 3. Carbon/body volume relationship for rotifers

Rotifers	$V \times 10^6$, μm^3 range	$a \pm$S.E.	$b \pm$S.E.	n	r^2
Keratella	0.018–1.283	42.76±33.47	1.512±0.193	5	0.95
Polyarthra	0.240–2.783	45.12±22.87	1.112±0.140	4	0.97
Synchaeta	0.188–2.860	2.466±0.947	0.570±0.116	10	0.75
Rotifers (13 species)	0.018–2.860	62.57±58.29	1.455±0.256	13	0.75

Range of body volume measurements ($V \times 10^6$ μm^3), intercept (a), the slope (b), number of observations (n) and coefficient of determination (r^2) are presented for the power function: $C = a\, V^b$.

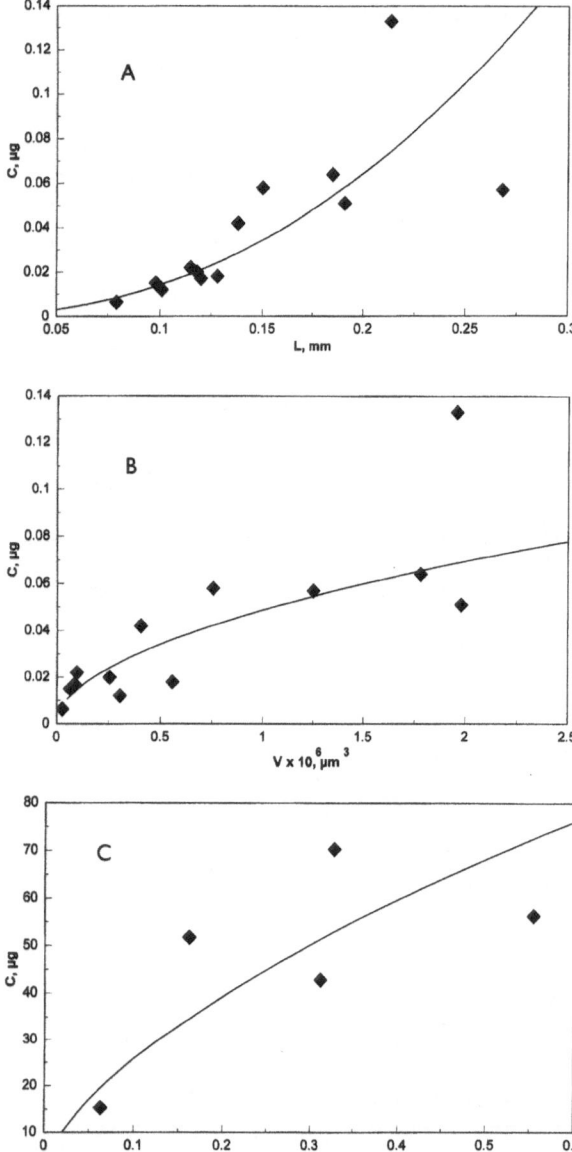

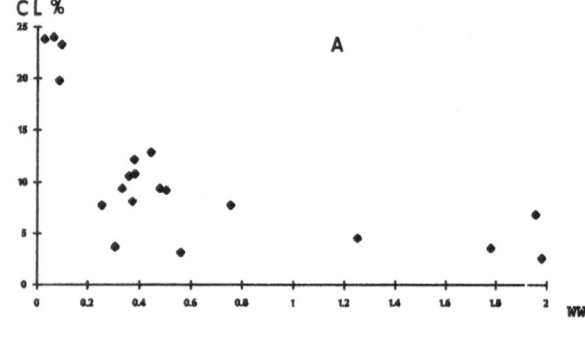

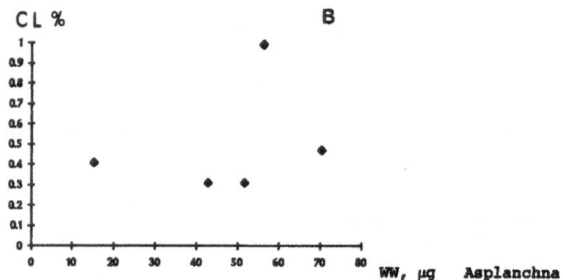

Figure 2. Relationship between wet weight (WW, μg) and carbon level (CL,%WW) for rotifers from the genera *Keratella*, *Polyarthra*, *Synchaeta* and *Trichocerca* (A), and for *A. priodonta* (B).

Figure 1. Relationships (A) between body length (L, mm) and carbon mass (C, μg), and (B) between body volume (V x 10^6, μm^3) and carbon mass (C, μg) for the studied rotifer species, excluding *A. priodonta*; (C) body volume (V x 10^6, μm^3) versus carbon mass (C, μg) for *A. priodonta*.

showed no correlation for *Asplanchna* of 10–80 μg WW (Figure 2B).

Discussion

Variation in weight estimates of aquatic organisms can depend on changing environmental conditions, types of water bodies, species-specific seasonal changes in

body composition of the animals, as well as on methods of sampling and weight determination (Bottrell et al., 1976; Duncan et al., 1985; Dumont et al., 1975; Guisande et al., 1991; Hessen & Lyche, 1991; Latja & Salonen, 1978; Rahkola et al., 1998; Sterner & Hessen, 1994). In rotifers, seasonal variation in body mass was demonstrated for *Kellicottia longispina* from Lake Pääjarvi when the rotifers carbon mass in November (0.030 μg C) was two times as high as in June (0.015 μg C, Latja & Salonen, 1978). However, population of *K. cochlearis* from the same lake demonstrated no seasonal variation of carbon content (Latja & Salonen, 1978).

In our study, most species of rotifers exhibited low intraspecific variation in carbon mass (Table 1), except for *S. pectinata* from the small pond in Lammi, Southern Finland, where a two-fold variation in carbon content probably reflected a natural size variation within the population. The low coefficient of determination for the carbon/length regression for *Synchaeta* (Table 2) is most probably due to the fact that rotifers were contracted in various degrees.

The measured individual carbon weights of *K. cochlearis*, *P. vulgaris* and *A. priodonta* are in accordance with those reported in literature, closest to values determined using the same method (Table 4). The most

Table 4. Comparison of the observed carbon weights of rotifers (C, μg) to those calculated from volumes (V), wet weights (WW) and dry weights (DW) presented in literature

Rotifer species	C, μg	Original unit	Source
Keratella cochlearis	0.004	V	Naulapaa, 1966
	0.005	V	Aasa, 1970
	0.055	DW	Dumont et al., 1975
	0.019	DW	Shindler & Noven, 1971
	0.024	C	Latja & Salonen, 1978
	0.019–0.027	C	This study
Polyarthra vulgaris	0.019	V	Naulapaa, 1966
	0.100	V	Aasa, 1970
	0.014–0.046	WW	Bottrell et al., 1976
	0.033	C	Latja & Salonen, 1978
	0.015–0.021	C	This study
Asplanchna priodonta	12.5	V	Naulapaa, 1966
	2.15	V	Aasa, 1970
	0.245–0.260	DW	Dumont et al., 1975
	0.300	DW	Narita & Mori, 1975
	0.11	DW	Makarewich & Likens, 1979
	0.22	DW	Bottrell et al., 1976
	0.115–0.504	C	Latja & Salonen, 1978
	0.15–0.66	C	Salonen & Latja, 1988
	0.018–0.685	C	This study

variable data were obtained for *A. priodonta* collected on almost the same date from two lakes in Eastern Finland (Table 1). Extreme values of carbon content of *A. priodonta* from Lake Pohjalampi varied 10-fold, in Lake Varaslampi this variation was over 40-fold. As a result, our data present a broader range of individual carbon mass values for *A. priodonta* than known from the literature. However, the maximal values of individual carbon mass of *A. priodonta* measured in our study (0.685 μg C) corresponded to the published data (Salonen & Latja, 1988).

The carbon level of rotifers estimated in our study was lowest in *A. priodonta*: it was always below 1% WW, with a mean of 0.55% WW. The same exceptionally low carbon levels (0.8% WW) for the contracted *A. priodonta* were reported by Latja & Salonen (1978). Even lower carbon level (0.2% WW) was reported by Salonen & Latja (1988) for the living animals of *A. priodonta*. These authors demonstrated an inverse relationship between carbon level and wet weight of two species, *A. priodonta* and *A. herricki,* ranging in size between 6 and 910 μg WW. In our study, wet

weight of *A. priodonta* varied between 10 and 85 μg, and no trends of size dependance of carbon level were observed within these limits. Carbon level of *Asplanchna* of similar body volumes differed significantly (three-fold) thus demonstrating large differences between the populations of *Asplanchna* and/or the specific environmental conditions in the lakes.

Carbon level of other studied rotifer species tended to be inversely correlated with wet weight (Figure 2A) reaching maximal value of 31.5% WW in *K. c. tecta.* Variation in carbon level within the genera *Keratella, Polyarthra* and *Synchaeta* was low (Table 1). However, carbon level of *Keratella* spp. and *Synchaeta* spp. was significantly higher, and that of *Polyarthra* spp. was lower than the 5% value, or outside the limits of 5-10% proposed by Latja & Salonen (1978) to convert rotifers wet weight to carbon mass. Our data on carbon content and carbon level of 14 common planktonic species of rotifers as well as the proposed general regressions linking the measured body length and volume reflecting wet weight of rotifers with their carbon mass could serve for more unbiased estimation of rotifer biomass and carbon flow through pelagic communities.

Acknowledgements

The authors thank Dr K. Salonen for fruitful discussions on methodology of carbon analysis at the early stage of the work. Two reviewers are acknowledged for valuable comments on the paper. This study was supported by the grant from the Academy of Finland to I.V.T., by the Russian Foundation for Basic Research grant No. 96-15-97875, and by the Research Support Scheme of the OSI/HESP, grant No. 1350/1997.

References

Aasa, R., 1970. Plankton i Lilla Ullevifjarden. Nat. Swedish Environm. Protection Board, Limnol. Surv. Uppsala, Rep. 33: 1–62.

Balushkina, E. V. & G. G. Winberg, 1979. Svyaz mezhdu massoi i dlinoi tela u planktonnykh zhivotnykh. /The relationship between mass and body length of the planktonic animals. In Winberg, G. G. (ed.), Obshchie osnovy izucheniya vodnykh ekosistem. Nauka, Leningrad: 169–172 (in Russian).

Bottrell, H. H., A. Duncan, Z. M. Gliwicz, E. Grygierek, A. Herzig, A. Hillbricht-Ilkowska, H. Kurasawa, P. Larsson & T. Weglenska, 1976. A review of some problems in zooplankton production studies. Norw. J. Zool. 24: 419–456.

Dumont, H. J., I. Van de Velde & S. Dumont, 1975. The dry weight estimate of biomass and a selection of Cladocera, Copepoda and Rotifera from the plankton, periphyton and benthos of continental waters. Oecologia (Berl) 19: 75–97.

Duncan, A., W. Lampert & O. Rocha, 1985. Carbon weight on length regressions of Daphnia spp. grown at threshold food concentrations. Verh. int. Verein. Limnol. 22: 3109–3115.

Guisande, C., J. Toja & N. Mazuelos, 1991. The effect of food on protein content in rotifer and cladoceran species: a field correlational study. Freshwat. Biol. 26: 433–438.

Hessen, D. O. & A. Lyche, 1991. Inter- and intraspecific variations in zooplankton element composition. Arch. Hydrobiol. 121: 343–353.

Latja, R. & K. Salonen, 1978. Carbon analysis for the determination of individual biomasses of planktonic animals. Verh. int. Ver. Limnol. 20: 2556–2560.

Makarewich, J. C. & G. E. Likens, 1979. Structure and function of the zooplankton community in Mirror lake, New Hampshire. Ecol. Monogr. 49: 109–127.

McCauley, E., 1984. The estimation of abundance and the biomass of zooplankton in samples. In Downing, J. A. & F. H. Rigler, (eds), A Manual on Methods for the Assessment of Secondary Productivity in Fresh Waters. Blackwell Scientific Publications, Boston: 228–265.

Narita, T. & S. Mori, 1975. Secondary production of zooplankton. In Mori, S. & G. Yamamoto (eds), Productivity of Communities in Japanese Inland Waters. JIBP Synthesis 10: 22–25.

Naulapaa, A., 1966. Eraiden Suomessa esiintyvien planktereiden tilavuuksien keskiarvoja. Mean volumes for some plankters found in Finland. Vesiensuojelutoimiston tiedonantoja 21: 1–26.

Rahkola, M., J. Karjalainen & V. A. Avinsky, 1998. Individual weight estimates of zooplankton based on length–weight regressions in Lake Ladoga and Saimaa lake system. Nordic Journ. of Freshwater Research, in press.

Ruttner-Kolisko, A., 1977. Suggestions for biomass calculation of planktonic rotifers. Arch. Hydrobiol. Beih. Ergebn. Limnol. 8: 71–76.

Salonen, K., 1979. A versatile method for the rapid and accurate determination of carbon by high temperature combustion. Limnol. Oceanogr. 24: 177–187.

Salonen, K., 1981. Rapid and precise determination of total inorganic carbon and some gases in aqueous solutions. Wat. Res. 15: 403–406.

Salonen, K. & J. Sarvala, 1980. The effect of different preservation methods on the carbon content of Megacyclops gigas. Hydrobiologia 72: 281–285.

Salonen, K. & R. Latja, 1988. Variation in the carbon content of two Asplanchna species. Hydrobiologia 162: 79–87.

Salonen, K., J. Sarvala, I. Hakala & M.-L. Viljanen, 1976. The relation of energy and organic carbon in aquatic invertebrates. Limnol. Oceanogr. 21: 724–730.

Sprules, W. G. & M. Munawar, 1991. Plankton community structure in Lake St. Clair, 1984. Hydrobiologia 219: 229–237.

Sterner, R. W. & D. O. Hessen, 1994. Algal nutrient limitation and the nutrition of aquatic herbivores. Annu. Rev. Ecol. Syst. 25: 1–29.

Telesh, I. V., 1995. Rotifer assemblages in the Neva Bay, Russia: principles of formation, present state and perspectives. Hydrobiologia 313/314: 57–62.

Telesh, I. V., 1998. Species diversity and distribution of rotifers in Lake Ladoga, Russia. J. Boreal Environ. Res., in press.

Vasama, A. & P. Kankaala, 1990. Carbon-length regressions of planktonic crustaceans in Lake Ala-Kitka (NE-Finland). Aqua Fenn. 20: 95–102.

Hydrobiologia **387/388**: 361–372, 1998.
E. Wurdak, R. Wallace & H. Seger (eds), Rotifera VIII: A Comparative Approach.
© *1998 Kluwer Academic Publishers.*

Review paper

Seasonal variation as a determinant of population structure in rotifers reproducing by cyclical parthenogenesis

Charles E. King[1] & Manuel Serra[2]
[1]*Department of Zoology, Oregon State University, Corvallis, OR 97331, U.S.A.*
[2]*Departament de Microbiologia i Ecologia, Universitat de Valencia, E-46100 Burjassot, Valencia, Spain*

Key words: cyclical parthenogenesis, sexual reproduction, life history evolution, competition, seasonal succession, speciation, clonal competition, mictic ratio, rotifers

Abstract

Monogonont rotifers live in habitats that display extensively variation in both biotic and abiotic components. Much of this variation is seasonal and therefore predictable for a given pond or lake. In 1972, King proposed one physiological and two genetic models presenting alternative modes of adaptation to this temporal variation. Our purpose in the present paper is to review and evaluate how our knowledge of the seasonal structure of rotifer populations has changed in the past 25 years. Seasonal changes in clone frequencies have been reported from three studies of natural populations using electrophoretic analysis of isozymes. In one of these studies there was evidence for substantial temporal *overlap* of multilocus genotypes suggesting that these clones were broad-niched generalists. By contrast, both the genetic and ecological analyses in the other two studies support a *non-overlap* model in which clonal groups are composed of narrow-niched specialists that undergo seasonal succession. In both of these studies the clonal groups appear to have achieved the status of sibling species, a phenomenon that we conclude is probably common in monogonont rotifers. Strong competition promotes reproductive isolation between successive groups of seasonal specialists. The existence of this competition has been inferred from natural populations and demonstrated by studies in the laboratory. Also required, and also supported by field observations, is a temporal separation of periods of mictic (sexual) reproduction. A final requirement of the nonoverlap model is seasonal variation in the timing of resting egg hatching. That is, clones established from hatching of resting eggs must enter a physiologically appropriate habitat if they are to increase in number and achieve a competitive advantage. Unfortunately, we still have little information on this topic. Finally, we present the results of a study analyzing the effects of variation in the mictic ratio (i.e., the relative frequency of mictic females) on the adaptive structure of rotifer populations. Mixis may shift the balance between costs and benefits of specialization thereby producing seasonally specialized populations that overlap in space but not time. Life history patterns may therefore provide fundamental insights on the adaptation of rotifers to the extensive temporal variation in their environments.

Introduction

Environmental variation across both space and time is universal in nature. Most research in population biology is an attempt to tease apart the dynamic consequences of this variation so that we can understand the forces molding the evolutionary history and potential of the species we study. These consequences are seen in both the ecological and genetic structures of their populations.

Rotifers dwelling in a lake or pond throughout a major portion of the year will be exposed to a variety of changes in their physical, chemical and biotic environments. Some of these changes may be relatively minor and the individuals in the population can adjust by acclimatization, modifications of their position in the water column, or changes in behavior. Other environmental challenges are more severe and elicit natural selection. These challenges lead to the changes in a population's genetic and ecological structures that are the focus of this paper.

When confronted with environmental variation populations respond by altering the numbers and relative frequencies of their component phenotypes through both space and time. Analyses of the two types of variation are quite different however. Spatial variation acts only on the current distribution of the population and leads to fitness increases or decreases as the environmental characteristics of the occupied area change. By contrast there are historical and predictive components to adaptation occurring in response to temporal variation because population survival requires not only adjustments to the present environment, but also appropriate adjustments to temporal states that are likely to occur in the future. One of the more predictable temporal changes is seasonal variation.

Life history evolution in response to seasonal change

Monogonont rotifers reproduce by cyclical, ameiotic parthenogenesis. Typically, amictic females produce diploid eggs by mitosis and these eggs develop into the next generation of amictic females. Environmental stimuli, which are known for only a few species, may induce the diploid eggs to develop into mictic females that differ from their amictic mothers by their mode of reproduction. Mictic females produce genetically variable haploid eggs by meiosis. These haploid eggs, if unfertilized, develop into haploid males. If haploid eggs are fertilized, they form thick-shelled resting eggs that typically have a period of diapause before they hatch into a new generation of amictic females.

In seasonal environments having environmental states that lie outside of the physiological tolerance of adults, there can be no survival from one year to the next without mixis. During these periods it is the resting egg that is the repository of the population's gene pool. In 1972, King studied the population dynamics of *Euchlanis dilatata* living in a small pond in central Illinois. Individuals were collected periodically from mid-Spring until mid-Summer and brought back to the laboratory where they were acclimated under standard conditions for 10 generations. Life table experiments were then performed under the same conditions. Clonal lines that appeared to be morphologically identical but which were established from females collected at different times were found to have significantly different rates of increase, r, net reproduction, R_0, and equilibrium population size, K. King interpreted these results to indicate the occurrence of a seasonal succession of genotypes.

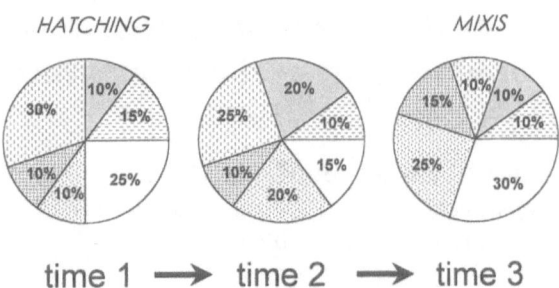

Figure 1. Overlap Model of population structure showing the relative frequencies of six clones at each of three different times (representing different environmental patches). Note that all six clones are present in each time period. Hatching is assumed to occur at time 1 and sexual reproduction at time 3. Intervening reproduction is by parthenogenesis.

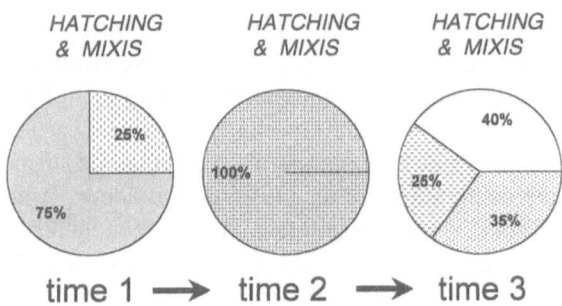

Figure 2. Nonoverlap Model of population structure showing the relative frequencies of clones at each of three different times. Note that multiple clones may or may not occur in a single time period, but there is a complete change of clonal composition between time periods. Hatching and sexual reproduction are assumed to occur in each time period. Intervening reproduction is by parthenogenesis.

Two alternative hypotheses were proposed to explain possible patterns of genetic change for populations living in seasonal environments (King, 1972). Under the model of incomplete genetic discontinuity, herein referred to as the 'Overlap Model', clonal groups emerge and are displaced from the pond at approximately the same times, but clone frequencies are adjusted by natural selection and vary from one season to the next (Figure 1). If adult stages occupy the lake for extended periods, two or more cycles of sexual reproduction may occur. By contrast, the model of complete genetic discontinuity of King, 1972, herein called the 'Nonoverlap Model', proposes that clones occupying the lake at different times differ in their times of emergence from resting eggs, periods of sexual reproduction, and displacement from the lake (Figure 2). Based on the significant differences found for the dynamics of seasonally divergent populations groups when measured under standard conditions, King (1972) suggested that his results were best interpreted as supporting the Nonoverlap Model. However, this interpretation was not based on measures of genetic variation and lacked definition of times of emergence, displacement, and sexual reproduction.

Field studies of seasonal succession in rotifers

Golf Course Pond

King (1977) applied polyacrylamide gel electrophoresis to analyze allozyme variation at the malic enzyme locus of *Asplanchna girodi* living in Golf Course Pond near Tampa, Florida. Samples were taken at weekly intervals from early April to July of 1976. Twenty-three clones were established from collections made in April and the first week of May; all were heterozygous for two alleles at the malic enzyme locus. However, 1 week later, in the second week of May, a new malic enzyme allele appeared and individuals homozygous for this allele totally dominated the population. In fact, only three of the 75 clones established from this collection were heterozygotes for the two alleles present earlier. Simultaneous with the appearance of a new malic enzyme genotype was a change in the alleles present at five other electrophoretic loci. Thus in a 7-day period there was an almost complete change in the genetic structure of the *A. girodi* population. These results provide strong support for the Nonoverlap Model.

To get more detailed information on this clonal substitution pattern in the following year, King and Snell (1980) made daily collections of the Golf Course Pond *A. girodi* during April and May, 1977. During this period 75–100 individuals from each collection were isolated and used to establish clones. Population growth rate under standard conditions and electrophoretic genotype at the malic enzyme locus were determined for each of the clones (King, 1980). While all three genotypes expected from the malic enzyme alleles found in the April and first week of May, 1976 were found, no individuals bearing the allele from the second week of May, 1976 were found in 1977.

Torreblanca Marsh

Gómez et al. (1995) studied *Brachionus* populations, classified at the time of the study as *B. plicatilis*, in a marsh on the Mediterranean coast of Spain. By screening four allozyme loci from July 1992 to November 1993, they found 11 composite genotypes in a small pond. These 11 genotypes could be clustered in three groups: one monomorphic, a second with five composite genotypes, and the third with four composite genotypes. Each group had unique alleles suggesting that there was no gene flow between groups. Mating experiments carried out in the laboratory confirmed within-group assortative mating and lack of hybridization (Gómez & Serra, 1995).

The three groups of genotypes were involved in a seasonal succession. Two groups, SS and SM, that are currently classified as *B. rotundiformis,* coexisted in July 1992. In September 1992 only SS individuals were detected. This group was replaced during October and November by another group, L, that is currently classified as *B. plicatilis*. Group L was collected in the marsh until early May, 1993 when it was replaced by SM and SS. The latter groups co-occurred from late spring through summer, but SS was the numerically dominant group during most of the summer period. This distribution pattern is correlated with salinity and temperature in the pond. Laboratory experiments confirmed the differential ecological specialization of the groups (Gómez et al., 1997). Gómez et al. (1995) also found evidence suggesting within-group temporal heterogeneity for *B. plicatilis*. These authors suggested that each group of genotypes is a biological species belonging to the same complex of sibling species.

Soda Lake

King & Zhao (1987) studied *B. plicatilis* from Soda Lake, a permanently stratified alkaline crater lake in northern Nevada. An electrophoretic analysis of clones established from different time periods revealed variation in six of the 11 isozyme systems studied. Three variable systems were chosen for further study and each of 138 clones was examined at the three loci. Each of these enzyme systems revealed genetic variation both within and between collections. During the course of the study, 38 of the 100 potential composite genotypes formed by combining observed electromorphs at the three loci were found.

In contrast to the results presented for Golf Course Pond and Torreblanca Marsh, no consistent seasonal presences or absences of specific clonal groups were found. Instead, different composite genotypes dominated different collections, but each had an extended period of presence in the lake and overlapped in its temporal distribution with a number of other clones. In laboratory studies, Zhao & King (1989) found significant differences among composite genotypes in their intrinsic rates of increase, net reproductive rates, and mean life spans. This suggests that at least some of the observed frequency variation among composite genotypes in Soda Lake was not neutral. King & Zhao therefore concluded that the B. plicatilis in Soda Lake constituted a single population that is best described by the Overlap Model. Further, if clone frequencies are determined by directional selection, that selection appears to be either too weak, or of too short a duration for the competitive exclusion of clones to occur. It is interesting to note that B. plicatilis has been present in all samples taken over a 4-year period from every month of the year, suggesting that its populations are permanent residents in Soda lake. B. plicatilis males or resting egg-bearing females have not been observed in any collection made from Soda Lake (although mictic females have been observed in Soda Lake clones in the laboratory.)

Tests of the predictions of the seasonal selection models

Quantity of variation

Both seasonal models predict that there will be significantly more genetic variation between populations separated in time than within clonal groups from the same collection. This prediction has been validated in each of the studies cited in the previous section. By using UPGMA cluster analysis, Gómez (1996) showed that maximum within-clonal group genetic distance was around 30% of the genetic distance among groups.

Competition between different seasonal specialists

The major distinction between the Overlap and Nonoverlap Models is based on the adaptive structure of the group of clones within which random mating is assumed to occur. As the clonal diversity of mating assemblages decreases, the potential for habitat specialization increases. At one extreme, if sexual reproduction were to occur only once per year, each of the clones in the population must be present and participate in the sexual process. In turn this requires that

each clone must have a broad niche so that it will be able to persist throughout the entire range of seasonal environments. At the opposite extreme, if each clone has a unique time of occurrence in the lake, niches are expected to be very narrow since each clone will occupy only a small portion of the entire seasonal spectrum.

These predictions have been examined with two distinct types of experiments. An indirect assessment of niche breadth can be conducted by testing variability of fitness characters among groups of similarly acclimated clones collected at the same and at different times. For example, King (1972) determined that Euchlanis dilatata collected from cool-water spring populations had relatively high rates of increase at 19°C and relatively low values of r at 27°C, whereas the opposite fitness relationships were found when testing warm-water clones. Similar studies yielding similar results have been conducted with B. plicatilis (Gómez et al., 1997; Zhao & King, 1979) and A. girodi (Snell & King, 1977). A note of caution should be introduced when considering such studies since each clone may have very different population-dynamic responses to different environmental conditions and, conversely, the same rate of increase may be obtained by quite different patterns of survivorship and fecundity (King, 1982).

Co-occurring clonal groups may also vary in their response to significant environmental features. Snell (1980) studied representatives of three genotypes at the malic enzyme locus of A. girodi collected from Golf Course Pond on 19 July 1977. The blue-green alga Anabaena flos-aquae isolated from the same pond was found to depress the reproductive rates of the rotifer. However, this depression was not the same for the three clones. Instead, one homozygote and the heterozygote were able to maintain positive reproductive rates at algal concentrations that were triple the concentration at which the other homozygote had a finite growth rate of zero. This sensitive genotype showed a sharp decrease in relative frequency during a subsequent Anabaena bloom in Golf Course Pond.

Because different clonal groups may share important ecological features, strong competition between clonal groups can be expected under the Nonoverlap Model. Snell (1977, 1979) has directly examined this prediction. In pairwise competition experiments between clones of A. brightwelli, rapid elimination of the inferior clone took place. Not surprisingly, competition was found to be frequency dependent. However, competitive exclusion between clones derived from spatially isolated habitats required longer periods to reach complete elimination of the inferior clone (approximately 16 generations) than when competition was between temporally separated clones from the same lake (five generations). Snell's results strongly support the potential role of interclonal competition as a process leading to reduced variability among co-occurring clones and the separation of clones into temporally discrete habitats when variability in competitive ability is high.

Temporal separation of male and resting egg production

To maintain clonal associations adapted to different sets of environmental conditions, mating between individuals belonging to different seasonal groups must not occur. This separation is made possible by temporal restrictions: under the Nonoverlap Model, only a fraction of the total number of clones in the lake is present during a given mictic period. If all clones were present whenever sexual reproduction occurred, recombination between different seasonal specialists would occur. This recombination would inhibit or preclude the development of discrete seasonal groups. The seasonal structures of the rotifers in Torreblanca Marsh and Golf Course Pond presented earlier therefore suggest the existence of a tight linkage between the timing of sexual reproduction and the adaptive states of the clonal groups occupying the habitat. Thus co-occurring clones having the same mixis period should have very similar dynamic characteristics. As a corollary we suggest that if two or more clones or groups of clones co-occur, yet have different population growth characteristics, they will also have different periods of sexual reproduction. As referenced above, the three clonal groups in Torreblanca Marsh are genetically discrete. Although this structure could arise by random mating followed by natural selection against hybrids, laboratory experiments provide a strong indication that gene flow does not occur between groups.

Similarly, the clones of *A. girodi* collected in the first and second weeks of May, 1976 from Golf Course Pond were genetically isolated in the laboratory. However, laboratory experiments demonstrated that random mating and recombination occur between individuals from the same collection (King, 1977) and between those from different collections that share the same malic enzyme alleles (King & Snell, 1977). At least for the abundant composite genotypes, the sea-

sonal succession in Soda Lake involved changes in frequency rather than the appearance of new alleles (King & Zhao, 1987). This is consistent with the expectation stated above from the Overlap Model.

Gómez et al. (1997) studied the mictic response of *Brachionus* strains collected in Torreblanca to salinity, temperature and density. The differential responses they observed would produce a 34% reduction in the number of encounters between males and females from different clonal groups. This figure is probably an underestimate as competition effects were not considered. Assortative mating behavior among these clonal groups (Gómez & Serra, 1995) can explain the lack of gene flow during the occasional overlap of mictic periods observed in the field (Gómez et al., 1995).

Studies with a number of species of *Brachionus* and *Asplanchna* have examined gametic isolation in the laboratory between clones of the same species collected from different sites. This topic is reviewed by Nogrady et al. (1993). Briefly, crosses between geographically separated strains of the same species are generally successful, however reciprocal crosses sometimes reveal asymmetries in both copulation frequencies and viability of the resultant resting eggs. The most likely source of the prezygotic barriers is a failure in the mate recognition system described by Snell and his co-workers (Rico-Martinez & Snell, 1995; Snell et al., 1988, 1995).

Temporal separation of resting egg hatching

Where populations are composed of narrow-niched specialists, it is of critical importance that resting egg hatching be synchronized with the environmental state. That is, a clone derived from a resting egg that hatches at the 'wrong' time will face either an inhospitable physical environment or an inhospitable biotic environment in which it is at a competitive disadvantage. Such clones would have low fitness and be subject to strong natural selection.

Unfortunately, we have less information on the biology of resting eggs than on any other part of the monogonont rotifer life history. The environmental cues that initiate completion of development and subsequent hatching of the resting egg are unknown. Clearly, however, the Nonoverlap Model is restricted to populations in which differential resting egg hatching occurs. We would also expect to see in these populations a broad temporal array rather than a brief concentration of hatching times focused in a single season. An indication that resting egg hatching does occur over an extended period in nature is provided by the work of Arndt (1991a,b) who studied estuarine rotifers of the Baltic Sea.

Succession of seasonally adapted populations and mixis patterns

Two factors affecting the dynamics of competition between taxonomically related clonal groups are seasonal specialization and mixis pattern. In this section we will consider both factors by modeling the dynamics of amictic and mictic females belonging to each of several competing groups. We will show that under certain conditions the occurrence of sexual reproduction alters the rate of competitive exclusion among clones having substantial niche overlap and significantly affects the possibility of achieving a nonoverlapping population structure.

The growth rates describing the dynamics for the ith population are

$$dA_i/dt = b_{i(N,\ t)}(1 - m_{i(t)})A_i - qA_i + \epsilon,$$
$$dM_i/dt = b_{i(N,\ t)}m_{i(t)}A_i - qM_i.$$

Index i ranges from 1 to s, where s is the total number of clonal groups in the lake. A_i and M_i are respectively the densities of amictic and mictic females of the ith clonal group; $m_{i(t)}$ is the mictic ratio (measured as the relative frequency of mictic daughters at time t); $b_{i(N,\ t)}$ is the birth rate of amictic females at a total density $N_t = \sum_i (A_{i,\ t} + M_{i,\ t})$; q is the mortality rate and ϵ is an immigration rate accounting for recruitment derived from resting egg hatching. Seasonal specialization and competition come into the model through $b_{i(N,\ t)}$, assumed to be

$$b_{i(N,t)} = b - ((b - q)K_{i(t)})N,$$
$$N = \sum(A_i + M_i),$$

where b is the intrinsic birth rate in the absence of intra- and interclonal competition, and K is the carrying capacity. The carrying capacity is assumed to be both time dependent and clone dependent, i.e., $K_{i(t)}$.

This model is consistent with an ecological scenario in which genetically diverse specialists consume the same resources, but vary in the efficiency of their physiological use of the resources because of genotype environment interactions. Individuals of each group have equivalent negative effects on the growth of all groups. In other words, we implicitly assume Lotka–Volterra competition coefficients equal to one, which

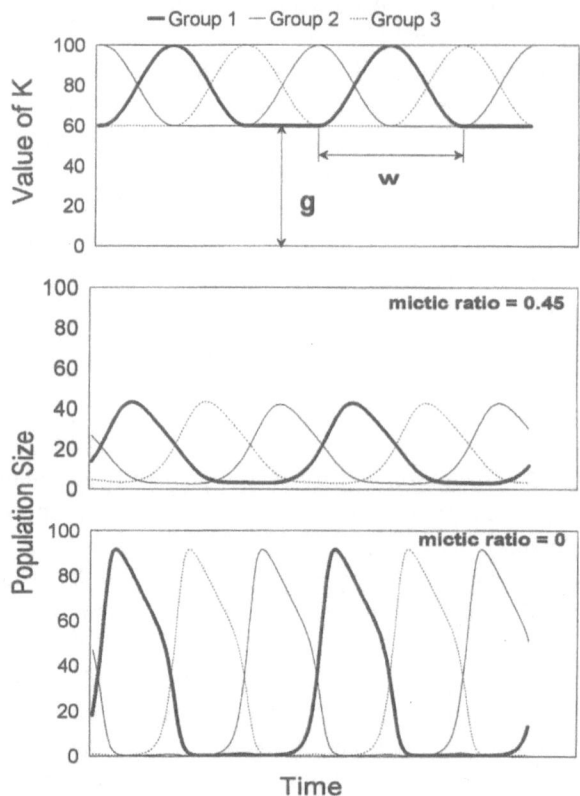

Figure 3. Seasonal succession of three rotifer groups: Variation in K values, and population densities with two different mictic ratios. (Assumed conditions: $b = 1.5$ day^{-1}, $q = 0.45$ day^{-1}, $\epsilon = 0.01$ female day^{-1}; K up to 100 females, $g = 60$ females, $w = 60$ days.)

is a convenient approach for competition between filter feeders having a very similar ecology.

$K_{i(t)}$ variation is based on modified sine functions having constant amplitude ($K_{max} - g$) and period (w) for a given simulation (see Figure 3, upper panel). Thus, carrying capacity variation is assumed to have the same shape for each group, but is time shifted so that each group has a unique temporal pattern. Parameter g represents the minimum K value in a particular run. The highest K value achieved in the oscillations ($K_{max} = 100$) is assumed to be the same in all simulations. The parameters g and w allowed us to modify the specialization degree between groups. The higher w and/or g, the lower the degree of specialization and the greater the potential interaction between groups. We indicate the degree of specialization by the overlap between K functions. Our coefficient of overlap between the K functions of groups i and j is

$$K_{ij}^{ov} = \frac{\sum(K_{i(t)} \cdot K_{j(t)})}{\sum(K_{j(t)})^2}.$$

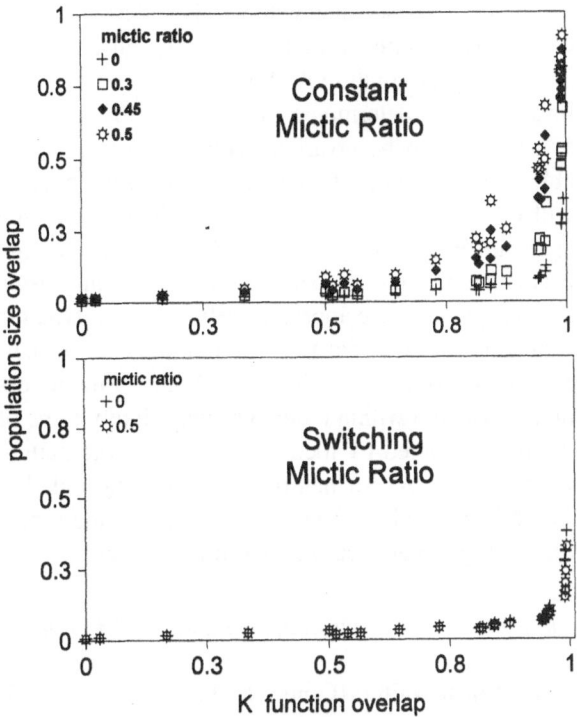

Figure 4. Relationships between the carrying capacity overlap over time and the overlap of the population sizes, evaluated for different mictic ratios. Upper panel, constant mictic ratio; lower panel, mictic ratio switching from 0 to 0.5 when the population growth is not positive. The dynamics of three rotifer groups was simulated. (Assumed conditions: $b = 1.5$ day^{-1}, $q = 0.45$ day^{-1}, $\epsilon = 0.01$ female day^{-1}; K up to 100 females; g and w varied among simulations from 0.01 to 80 females and from 30 to 90 days, respectively.)

This measure is analogous to the niche overlap coefficient of MacArthur & Levins (1967) and provides a measurement of the total overlap between groups i and j because of the assumed symmetries in the K values. A similar coefficient is used to evaluate the overlap between the time distribution of the populations, $K_{i(t)}$ values being replaced by population sizes.

The center and bottom panels of Figure 3 show population size variation under two different mictic ratios. Figure 4 (upper panel) presents relationships between the K overlap and the population size distribution overlap for different constant mictic ratios. When the overlap of the population size distributions was lower than 0.2, exclusion was complete except for very low population sizes maintained by constant immigration due to the hatching of resting eggs (see also in Figure 3, mictic ratio equal to zero). As expected in the presence of strong competition, competitive exclusion produces a time structure characterized by low overlap of successive groups. In the absence of mixis the only exceptions to such an extreme segrega-

tion between groups were found when the groups were assumed to be so similar that seasonal specialization was negligible (values of K overlap $\cong 1$). Of course, in the absence of seasonal specialization the competition that occurs is between equivalent phenotypes, and co-existence is possible.

The effect of mixis is to increase the temporal duration of overlap between groups; that is, mixis decreases the rate of competitive exclusion. This effect is negligible for low K overlap, but it becomes important as overlap increases. Note in Figure 4 (upper panel) that, if K overlap is close to 1, a mictic ratio of 0.5 produces a 3-fold increase in the overlap between the groups compared to the dynamics that occur in the absence of mixis.

These results are critically dependent on the assumption of a constant mictic ratio. We also examined competition among groups inducing mixis when their growth rates are not positive (Figure 4, lower panel). Permanent co-existence of the three groups is found when the coefficient of population-size overlap is higher than 0.14, with a few exceptions. In these simulations, mictic ratio did not have an important effect on the degree of population overlap, and in fact high mictic levels may produce a slight increase in the temporal segregation because mixis induction contributes to the population crash.

Discussion

A conceptual model of the genetic structure of seasonal clones

There are several interacting genetic elements that must occur if the population structure described by the Overlap Model is to maintain both high fitness in a given environment and genetic isolation between successive clonal groups. In order to clarify these interactions we have formulated a conceptual model that presents a hypothetical physical depiction of the system. First, there must be a receptor that responds to environmental cues to break diapause and reinitiate the genetic activity required for resting egg development and hatching. Second, there must be a mechanism to ensure that alleles conferring high fitness in the appropriate environment are transmitted as a block to the resting eggs that will produce the clones for the same environmental state in the future. Third, there must be a locus, or a group of loci, that responds to some environmental cue (or group of cues) to initiate

368

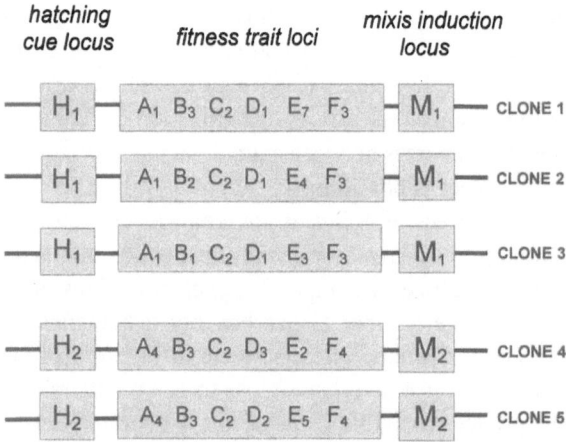

hatching cue locus | fitness trait loci | mixis induction locus

CLONE 1: H_1 — A_1 B_3 C_2 D_1 E_7 F_3 — M_1

CLONE 2: H_1 — A_1 B_2 C_2 D_1 E_4 F_3 — M_1

CLONE 3: H_1 — A_1 B_1 C_2 D_1 E_3 F_3 — M_1

CLONE 4: H_2 — A_4 B_3 C_2 D_3 E_2 F_4 — M_2

CLONE 5: H_2 — A_4 B_3 C_2 D_2 E_5 F_4 — M_2

Figure 5. Conceptual depiction of a gene complex controlling seasonal adaptation as envisioned under the Nonoverlap Model. Two loci, H and M, bearing alleles identified by temporal subscripts, determine the timing of resting egg hatching and mictic female production for seasonal groups 1 and 2. Loci A through F bear alleles influencing fitness in seasonal environments. See text for further description.

mixis and production of resting eggs. Although there are a number of possible chromosomal structures that would meet these requirements, the simplest model is that of the 'supergene' found in the land snail *Cepaea nemoralis* (Clarke et al., 1968) and a number of other species (Hedrick et al., 1978). Supergenes contain loci that influence related fitness traits. The tight linkage of these loci reduces the probability that important genetic correlations will be lost as a result of recombination.

Figure 5 presents a hypothetical genetic structure for two seasonal groups of clones. Patch 1 has three clones, patch 2 has two clones. If loci A–F were scored by allozyme electrophoresis, the investigator would recognize five composite genotypes.

As explained above, the Nonoverlap Model requires that clones hatch into an appropriate environment that will both support their physiological requirements and confer a competitive advantage. While the cues are likely to be environmental, the reinitiation of development requires gene action and locus H in Figure 5 is depicted as promoting differential resting egg hatching into seasonal environments 1 and 2. Similarly, mixis must occur late enough in a population cycle to permit the population to expand and produce mictic individuals, but not so late that the clonal group occupying a given environment is outcompeted by the members of a subsequent clonal group before resting egg production has occurred. The initiation of mixis is also expected to be under tight genetic control (locus

M, Figure 5). Associated with loci H and M is a series of loci determining fitness under a specific suite of environmental conditions. Some of these traits, for instance those determined by loci A and F in Figure 5, are likely to be invariant within groups but variable between groups. Others, exemplified by locus C, will be invariant in all groups. Additional loci, such as B and E, are likely to be variable both within and between clonal groups. The different clones in a group are expected to have approximately the same dynamic characteristics whether they are homozygotes or heterozygotes for fitness trait blocks. Some of the fitness block variation within clonal groups is likely to be of the type depicted by locus E, which has the pattern we would expect for neutral, or near-neutral alleles. These loci may have significant effects on the fitness of genotypes that occur in other time periods.

Characteristics of seasonally isolated clonal groups

In addition to rotifers (Gómez et al., 1995; King, 1977) seasonal succession is also common in cladocerans (e.g., Lynch, 1983), and may account for a significant part of the seasonal change in size and shape described by the process called cyclomorphosis (*sensu* Black & Slobodkin, 1987). The best described successions occur between genetically isolated clones or group of clones.

Clonal groups displaying seasonal succession are generally assumed to be members of a single species on the basis of classic taxonomic criteria. However, in the case of the *A. girodi* from Golf Course Pond, clonal groups established in the laboratory from the genetically distinct individuals taken in the first and second weeks of May, 1976 were each capable of within-group sexual crosses, but attempts to make between-group crosses were always unsuccessful. Heterozygotes indicating mating between the two groups were never observed in field collections. Similar observations have been made by Gómez et al. (1995) with the Brachionids from Torreblanca Marsh. It is therefore reasonable to suggest that the temporally isolated rotifers in both of these studies are better described as sibling species than as clonal variants of the same species. A broader evaluation of *Brachionus plicatilis* strains from a wide range of geographic regions by Gómez & Snell (1996) suggests that this complex is actually composed of more than three sibling species.

Genetic isolation between different groups may be based on simple temporal displacement (i.e., different groups have different bisexual periods). However, unless reproductive barriers have developed during the period of allopatry, the isolation is expected to be reversible. In other cases the members of different groups do not attempt copulation with each other suggesting that the genetic distance between the groups is much larger. In these cases the isolation may be irreversible. Research by Snell and his colleagues (Snell & Morris, 1993; Snell et al., 1988, 1995) has demonstrated the existence of a mate-recognition system in monogonont rotifers in which mating behavior is elicited by contact between a surface glycoprotein pheromone found on the female and specific receptors on the corona of the male. Even small changes in the biochemical configuration of these glycoproteins might serve to create reproductive isolation between closely related groups. A variety of experimental findings supports application of this mate-recognition system in the brachionids of Torreblanca Marsh (Gómez & Serra, 1995).

Other processes may also produce separation between clonal groups. A particularly intriguing example concerns the *Euchlanis dilatata* in Devil's Lake, Oregon (Walsh, 1995). Since this species is a littoral rotifer and is differentially associated with a variety of macrophytes, the possibility exists for spatial isolation between different segments of the population. Walsh (1992, and reported in Walsh & Zhang, 1992) found that two body-size morphs exist in this lake. The large morphotype, while conforming to the taxonomic descriptions of the species, was significantly larger (approximate mean adult length of 290 vs 230 μm) and was present throughout the year. The small morphotype was only present from May through July. Males were only observed in the small morphotype. Counts made by Walsh & Zhang (1992) demonstrated that males of the small morphotype had seven chromosomes and were haploid, females of the small morphotype were diploid, and females belonging to the large morphotype were triploid. It is interesting to note that the triploid form was found to have an extended temporal range in Devil's Lake. Both polyploid and aneuploid modifications of the genome have the capacity to create instant genetic isolation between clonal groups. Moreover, the parthenogenetic phase of the monogonont life cycle is an ideal vehicle for amplifying mutations, including chromosomal modifications that would block crosses between – but not necessarily within – different seasonal groups.

Competition between clonal groups

Mixis patterns are relevant to an understanding of clonal succession in several ways. First, as stressed in our presentation of the Overlap and Nonoverlap Models, in the absence of intrinsic barriers against hybridization, the timing of mixis induction will determine whether the genomes of different clonal groups fuse or remain genetically isolated. Second, the occurrence of mixis decreases population growth rate and can thus affect both the intensity and outcome of competition between clonal groups.

An important asymmetry among competing populations arises when they have different mictic ratios. This is a special version of the well-known 'cost of sex' argument (Maynard Smith, 1978; Williams, 1975). A difference in mictic ratio is produced if one clone induces mixis while others continue to reproduce parthenogenetically. Other things being equal, the clone with the higher mictic ratio has a decreased competitive ability because of the cost that mixis imposes on current growth rate. Therefore, when a clonal group starts to produce mictic females it is more likely to be excluded by other, non-mictic groups than if it had remained parthenogenetic.

The reduction in competitive ability accompanying sexual reproduction may have important consequences for the evolution of specialization. In a continuously changing environment, sexual reproduction by a clone should occur when the optimal environment is attained, for it is at that point that the number of amictic females, and hence the potential number of resting eggs, is at a maximum. If sexual reproduction is delayed the clone will loose fitness as its environment deteriorates. Population size will also be reduced as competition occurs with new clones that are reproducing exclusively by parthenogenesis. Hence, by initiating sexual reproduction in its optimal environment, the fitness loss associated with mixis – a reduction in competitive ability – is transferred to future environments in which the clone is unlikely to be present even if it does not initiate mixis. A corollary result would be a greater degree of temporal specialization.

A conclusion from our simulations is that sexual reproduction may facilitate coexistence of clones having different seasonal specializations if mixis is continuous and the mictic ratio is high. This result follows from the demographic effects of mixis in decreasing the growth rate which, in turn, increases the time required for displacement. We obtained this result even when we modeled strong competition, i.e., when the only difference between the two competing populations was in the temporal patterns of their carrying capacities. More complex models of resource use involving differential rates of resource exploitation could further increase the probability of coexistence (Rothhaup, 1988). By contrast, if mixis is not continuous or the mictic ratio is low, the outcome of competition is more likely to be a temporal exclusion of the competing populations.

Sibling species formation

A final consideration arising from our conclusions on optimal mictic patterns concerns the theoretical possibility of genetic divergence in sympatry. The question is whether rotifer populations can evolve from a pattern of clone frequency variation as described by the Overlap Model to a pattern of seasonal succession as described by the Nonoverlap Model. As suggested by Lynch (1985), divergence in the timing of bisexual reproduction may promote speciation in cyclical parthenogens. There are critical assumptions that must be met for this process to take place. First, an exclusively parthenogenetic growth phase facilitates the spread of mutations influencing the timing of mixis induction. Second, the bisexual reproductive period should be short enough to prevent mating between clones having different seasonal specializations. Under these assumptions, the two groups would attain sexual isolation by a combination of disruptive selection and assortative mating. At this stage divergence in mate recognition systems (Snell & Morris, 1993) might well reinforce the isolation between groups having different periods of mixis. Thus, while not the only source, we suggest that at least some of the growing number of examples of sibling species in rotifers may have their origin in the process of adaptation to seasonally varying environments.

Gaps in our understanding of rotifer life histories

Although our understanding of the population structure of rotifers living in seasonally varying habitats has come far in the past 25 years, there are still major gaps in our understanding of the relationship between the life cycle of monogonont rotifers and their population structure. Chief among these is the role that resting eggs play in determining repeatability of cycles from one year to the next. Several important questions remain virtually unexamined in this regard:

What are the cues that induce completion of development and hatching of resting eggs?

Pourriot & Snell stated in 1983 that the 'factors involved in the induction of resting egg development and hatching are poorly understood'. That statement remains an accurate assessment of our state of knowledge. Several environmental factors have been correlated with this process; for example, differential rates of resting egg hatching have been identified as a function of temperature (May, 1986), salinity (Gómez, personal communication) and light (Hagiwara & Hino, 1989). An approach emphasizing developmental physiology is needed to answer this question.

Do resting eggs from different clonal groups hatch at different times?

In introductory texts it is not unusual to find statements implying that rotifer resting eggs hatch only in the Spring. However, we know that there can be substantial variation in the duration of diapause and also in the timing of emergence. Arndt (1991a,b), for instance, studied six species of rotifers in a brackish estuary of the Baltic Sea. Because of the salinity, the shells of hatched resting eggs floated to the surface and provided a clear indication of the hatching period. Hatching of the resting eggs of most of the species occurred several times per year, and in two species hatching took place throughout the growing season.

Under the Overlap Model it is clearly required that the resting eggs giving rise to a particular clonal group hatch only when the environment is in a suitable state for physiological survival and probably for survival in the competitive environment as well. Hatchings that occur at other times would produce a substantial genetic load and loss of fitness for the parental group. Gómez (personal communication) has started to investigate this problem using clones from Torreblanca Marsh. In preliminary laboratory experiments she has found a differential response to salinity in the hatching of resting eggs from the SS, SM, and L strains. Her results are consistent with the general adaptive responses of the three clonal groups to salinity levels.

To explore this question it should be possible to place inverted funnel traps at a variety of locations of a lake or pond so that neonates from resting eggs could be gathered, cloned, and identified as to group by allozyme electrophoresis.

What is the distribution of resting egg viability?

Where clonal groups are associated with relatively narrow temporal periods, it seems likely that stochastic changes in the environment may create variability in reproductive success from one year to the next. Small changes in the mictic ratio or timing of mixis may produce significant changes in the production of resting eggs. This problem has particular importance for rotifers because in seasonal environments each of the clonal groups must be reconstituted each year. Resting eggs in sediments have the potential to act as a 'seed pool' from which populations can be reconstituted following reproductive failures. Various investigators have studied the viability of rotifer resting eggs isolated from sediment cores. For instance, Marcus et al. (1994) analyzed cores taken from the Pettaquamscutt estuary, Rhode Island and found viable resting eggs from an unstated species of *Brachionus* (identified as *B. plicatilis* by T. Snell, personal communication) that were >40 years old. The highest rate of hatching (77%) came from a sediment depth of 5–6 cm that had an approximate age of 13 years. Fu & Hirayama (personal communication) collected sediments from Lake Kaiike, Kagoshima, Japan, having an age of 100 years estimated using ^{210}Pb. Under laboratory conditions two resting eggs of *B. rotundiformis* from these sediments hatched and gave rise to viable amictic females. We look forward to more studies of this type in the future. Additional analyses of resting eggs from different sediment layers should permit us to gain an appreciation for the potential significance of resting eggs as a seed pool to reconstitute the population following the occasional failures that are likely to occur for a particular temporal group. The key to understanding the population structure of rotifers is to develop a dynamic understanding of the life cycle. It is not enough to consider cyclical parthenogens as sexual organisms with a minor modification, the ability to reproduce clonally. Instead, we must recognize that cyclical parthenogenesis may play a variety of roles in populations. These roles range from modification of ecological processes such as intra- and interspecific competition to fundamental evolutionary processes such as adaptation to temporally heterogeneous environments and speciation.

Acknowledgments

M. Serra was the recipient of a fellowship of the Ministerio Espa nol de Educación y Ciencia for a sabbatical visit to the Department of Zoology at Oregon State University. Support was also received from the Universitat de València for a subsequent visit by CEK to Valencia during which time this manuscript was prepared.

References

Arndt, H., 1991a. Population ecology of estuarine rotifers in an inlet of the Southern Baltic (Abstract). VI International Rotifer Symposium, Banyoles, Spain: 83.

Arndt, H., 1991b. Population dynamics and production of estuarine planktonic rotifers in the Southern Baltic: *Brachionus quadridentatus*. Acta Ichthiol. Piscatoria (Szczecin) 21: 7–15.

Black, R. W. II & L. B. Slobodkin, 1987. What is cyclomorphosis? Freshwat. Biology 18: 373–378.

Clarke, B., C. Diver & J. Murray, 1968. Studies on *Cepaea*. VI. The spatial and temporal distribution of phenotypes in a colony of *Cepaea nemoralis* (L.). Proc. r. Soc. B. 253: 521–548.

Gómez, A., 1996. Ecología genética y sistemas de reconocimiento de pareja en poblaciones simpátricas de rotíferos. Ph. D. Thesis. Universitat de València, Valencia.

Gómez, A., & M. Serra, 1995. Behavioral reproductive isolation among sympatric strains of the rotifer *Brachionus plicatilis* Müller 1786: insights into the status of this taxonomic species. Hydrobiologia 313/314: 111–119.

Gómez, A. & T. W. Snell, 1996. Sibling species and cryptic speciation in the *Brachionus plicatilis* species complex (Rotifera). J. Evol. Biol. 9: 953–964.

Gómez, A., M. Temprano & M. Serra, 1995. Ecological genetics of a cyclical parthenogen in temporary habitats. Journal of Evolutionary Biology 8: 601–622.

Gómez, A., M. J. Carmona & M. Serra, 1997. Ecological factors affecting mate system in the *Brachionus plicatilis* complex (rotifera). Oecologia 111: 350–356.

Hedrick, P. W., S. Jain & L. Holden, 1978. Multilocus systems in evolution. Evol. Biol. 11: 101–182.

King, C. E., 1972. Adaptation of rotifers to seasonal variation. Ecology 53: 408–418.

King, C. E., 1977. Genetics of reproduction, variation and adaptation in rotifers. Archiv für Hydrobiol. Beih. 8: 187–201.

King, C. E., 1980. The genetic structure of zooplankton populations. In Kerfoot, W. C. (ed.), Evolution and Ecology of Zooplankton communities. The University Press of New England, Hanover, NH, U.S.A.: 315–328.

King, C. E., 1982. The evolution of life span. In Dingle, H. & J. P. Hegman (eds), Evolution and Genetics of Life Histories. Springer-Verlag, New York: 121–138.

King, C. E., 1993. Random genetic drift during cyclical ameiotic parthenogenesis. Hydrobiologia 255/256: 205–212.

King, C. E. & T. W. Snell, 1977. Sexual recombination in rotifers. Heredity 39: 357–360.

King, C. E. & T. W. Snell, 1980. Density-dependent sexual reproduction in natural populations of the rotifer *Asplanchna girodi*. Hydrobiologia 73: 149–152.

King, C. E. & Y. Zhao, 1987. Coexistence of rotifer (*Brachionus plicatilis*) clones in Soda Lake, Nevada. Hydrobiologia 147: 57–64.

Lynch, M., 1983. Ecological genetics of *Daphnia pulex*. Evolution 37: 358–374.

Lynch, M., 1985. Speciation in Cladocera. Verh. int. Ver. Limnol. 22: 3116–3123.

Marcus, N. H., R. Luz, W. Burnett & P. Cable, 1994. Age, viability, and vertical distribution of zooplankton resting eggs from an anoxic basin: evidence of an egg bank. Limnol. Oceanogr. 39: 154–158.

May, L., 1987. Effect of incubation temperature on the hatching of rotifer resting eggs collected from sediment. Hydrobiologia 147: 335–338.

MacArthur, R. H. & R. Levins, 1967. The limiting similarity, convergence and divergence of coexisting species. American Naturalist 101: 377–385.

Maynard Smith, J., 1978. The Evolution of Sex. Cambridge University Press, Cambridge: 222 pp.

Nogrady, T., R. L. Wallace & T. W. Snell, 1993. Rotifera: Volume 1: Biology, Ecology, and Systematics. SPB Academic Publishing, The Hague: 137 pp.

Pourriot, R. & T. W. Snell, 1983. Resting eggs in rotifers. Hydrobiologia 104: 213–224.

Rico-Martinez, R. & T. W. Snell, 1995. Mating behavior and mate recognition pheromone blocking of male receptors in *Brachionus plicatilis* Müller (Rotifera). Hydrobiologia 313/314: 105–110.

Rothhaupt, K. O., 1988. Mechanistic resource competition theory applied to laboratory experiments with zooplankton. Nature 333: 660–662.

Serra, M. & C. E. King, Optimal rates of bisexual reproduction in cyclical parthenogens with density-dependent growth. J. Evol. Biol. in press.

Snell, T. W., 1979. Intraspecific competition and population structure in rotifers. Ecology 60: 494–502.

Snell, T. W., 1980. Blue-green algae and selection in rotifer populations. Oecologia 46: 343–346.

Snell, T. W. & E. M. Boyer, 1988. Thresholds for mictic female production in the rotifer *Brachionus plicatilis* (Müller). J. exp. mar. Biol. Ecol. 124: 73–85.

Snell, T. W. & B. L. Garman, 1986. Encounter probabilities between male and female rotifers. J. exp. mar. Biol. Ecol. 97: 221–230.

Snell, T. W. & F. H. Hoff, 1985. The effect of environmental factors on resting egg production in the rotifer *Brachionus plicatilis*. J. World Mar. Soc. 16: 484–497.

Snell, T. W. & C. E. King, 1977. Lifespan and fecundity patterns in rotifers: the cost of reproduction. Evolution 31: 882–890.

Snell, T. W. & P. D. Morris, 1993. Sexual communication in copepods and rotifers. Hydrobiologia 255/256: 109–116

Snell, T. W., M. J. Childress & B. C. Winkler, 1988. Characteristics of the mate recognition factor in the rotifer *Brachionus plicatilis*. Comp. Biochem. Physiol. 89: 481–485.

Snell, T. W., R. Rico-Martinez, L. N. Kelly & T. E. Battle, 1995. Identification of a sex pheromone from a rotifer. Mar. Biol. 123: 347–353.

Walsh, E. J., 1992. Ecological and genetic aspects of the population biology of the littoral rotifer *Euchlanis dilatata*. Ph.D. Dissertation, University of Nevada, Las Vegas.

Walsh, E. J., 1995. Habitat-specific predation susceptibilities of a littoral rotifer to two invertebrate predators. Hydrobiologia 313/314: 205–211.

Walsh, E. J. & L. Zhang, 1992. Polyploidy and body size variation in a natural population of the rotifer *Euchlanis dilatata*. J. Evol. Biol. 5: 345–353.

Williams, G. C., 1975. Sex and Evolution. Princeton University Press, Princeton, N.J.: 200 pp.

Zhao, Y. & C. E. King, 1989. Ecological genetics of the rotifer *Brachionus plicatilis* in Soda Lake, Nevada, USA. Hydrobiologia 185: 175–181.

Hydrobiologia **387/388**: 373–384, 1998.
E. Wurdak, R. Wallace & H. Segers (eds), Rotifera VIII: A Comparative Approach.
© *1998 Kluwer Academic Publishers.*

Review paper

Ecological genetics of *Brachionus* sympatric sibling species

Manuel Serra, África Gómez & María José Carmona
Departament de Microbiologia i Ecologia, Universitat de València, E-46100 Burjassot, Spain

Key words: Rotifera, zooplankton, cyclical parthenogens, coastal lagoons, environmental heterogeneity, allozymes, mating behavior, sexual reproduction, speciation, biodiversity, seasonal specialization

Abstract

In this paper we review previous studies on sympatric *Brachionus* populations in Torreblanca Marsh as a model of evolutionary and ecological relationships between closely related species. The marsh is a wetland on the Mediterranean coast of Spain with high spatial and temporal heterogeneity. Allozyme and morphometric analysis showed that *Brachionus* group *plicatilis* (formerly, *Brachionus plicatilis* and currently split into *B. plicatilis* and *B. rotundiformis*) was composed of three groups of genotypes with no evidence of gene flow between them (*B. plicatilis*, *B. rotundiformis* SM and *B. rotundiformis* SS). Correlations between seasonal and spatial distributions, on one hand, and temperature and salinity, on the other hand, were consistent with the results of experimental studies on population dynamics. Accordingly, *B. plicatilis* is a euryhaline, low temperature group; *B. rotundiformis* SM is adapted to high temperature and low salinity conditions; and *B. rotundiformis* SS is adapted to high temperature and high salinity conditions. The groups had different mictic responses to density, salinity and temperature, which can be explained to some extent as an adaptive escape response, given their different ecological preferences. These differences imply a partial ecological barrier to male–female encounter between groups. Mating experiments showed that most copulations occurred within a group. *B. plicatilis* has a mating recognition system different from those of either *B. rotundiformis* SM or SS, whereas the two *B. rotundiformis* groups had partially differentiated mating preferences. Cross-mating experiments performed in the laboratory failed to produce any detectable hybrids. We conclude that three sympatric sibling species inhabit Torreblanca Marsh. The remarkable association between genetic differences among clonal groups and their ecological preferences, mixis response and mating behavior is hypothesized to play a role in stabilizing sympatry, and gives insight into the evolution of genetic divergence and speciation in rotifers.

Introduction

Ecological genetics in monogonont rotifers is a research field whose origins can be traced back to the description of cyclomorphosis and the controversy over its proximate cause. Two possible processes – genetic substitution and phenotype plasticity – were recognized as possible causes of cyclomorphosis (e.g., Hutchinson, 1967), and a third possibility – that phenotypic changes were due to taxonomic species substitution – was also considered, providing an early link between ecological genetics and taxonomy in zooplankters. Modern ecological genetics of rotifers began with the introduction of demographic and biochemical techniques to population studies. King

(1972) analyzed *Euchlanis dilatata* populations by generating clonal lineages from females collected in a small lake on several dates. Life table experiments showed that genetic differences existed among individuals collected on different dates. Similarly, seasonal variation was found in *Asplanchna girodi* populations screened for allozyme variation (King, 1977a). Three important features in these two seminal works are: (1) parthenogenetic reproduction of rotifers was used to obtain clones for experimental and allozyme analysis; (2) they focused on seasonal population structure; (3) both allozyme and demographic techniques were used, which were later combined by King (1980). Moreover, King (1972, 1977a, 1980) put forward two conceptual models to describe the genetic structure of rotifer populations, which greatly influenced later research. These models were based on the degree of genetic discontinuity of rotifer populations and described patterns of seasonal genetic variation as well as the conditions giving rise to each pattern (see King & Serra, 1998).

Since the 1970s, technical and conceptual advances have provided new tools for the analysis of rotifer ecological genetics. Technical progress includes cost-effective isozyme screening (Gómez et al., 1995; King, 1972; Serra & Miracle, 1985; Snell & Winkler, 1984), new methods to promote resting-egg production and hatching (e.g., see reviews by Gilbert, 1974, and Pourriot & Snell, 1983; see also, Hagiwara et al., 1985, 1989, 1995; Minkoff et al., 1983; Snell & Hoff, 1987), which can be used to improve genetic interpretation of allozyme data, and standardized mating behavior tests (e.g., Snell, 1989; Snell & Hawkinson, 1983), which allows investigators to address a critical process involved in gene flow between genotypes and in species boundaries.

A theoretical framework for the interpretation of genetic data was provided by models of genetic seasonal succession (King, 1972, 1977a, 1980; see King & Serra, 1998). King (1977b, 1993, in press) also developed theoretical means to analyze population genetics in rotifers. He stressed the need for modifying standard population genetics theory to take into account particular features of monogonont rotifers such as the parthenogenetic phase of their life cycle. Thus, one of King's (1977b) conclusions is that the parthenogenetic phase accelerates selection, and affects genetic drift and neighborhood area (King, 1993; King & Murtaugh, 1997). Information on mate recognition in rotifers increased greatly in the last decade (Gilbert, 1963; Snell & Hawkinson, 1983; Snell & Morris, 1993; Snell et al., 1995), including the description of a surface glycoprotein that elicits male mating behavior. This allows a clearer interpretation of the level and causes of genetic divergence. Subsequently, species boundaries and sibling species occurrences are easier to detect (Snell, 1989; Serra et al., 1997). There is now a better understanding of the adaptive meaning of patterns in life cycle traits, such as dormancy (Gilbert, 1974), mixis induction (Carmona et al., 1993, 1995; King, 1980; Pourriot & Snell, 1983; Serra & Carmona, 1993; Serra & King, 1998; Snell, 1986, 1987; Snell & Boyer, 1988), sex allocation (Aparici et al., 1998), and mate choice (Gómez & Serra, 1996; Snell & Hawkinson, 1983), which helps to clarify the link between the process of adaptive divergence and genetic variation.

Figure 1. Map showing the location of Torreblanca Marsh. Temperature and salinity ranges (from July 1992 to November 1993) in the three ponds studied by Gómez et al. (1995) are also shown.

One of the aims of this review is to show how a set of approaches was combined to successfully elucidate the ecological genetics of a case study, i.e., sympatric populations of the former taxonomic species *Brachionus plicatilis* (hereafter, *Brachionus* group *plicatilis*) inhabiting Torreblanca Marsh. This is a wetland on the Mediterranean coast of Spain (Figure 1) which contains small ponds showing remarkable spatial differences in salinity as well as pronounced temporal variation in both salinity and temperature.

We will compare the findings of this case study with the results of other studies of *B.* group *plicatilis* strains. The first goal of this paper is to show the relevance of ecological genetics in rotifer taxonomy; in particular, how detailed studies of sympatric populations provide insight into taxonomy as compared to studies performed on allopatric strains. The second goal is to analyze relationships among genetic divergence, ecological specialization and features of sexual reproduction, i.e., mixis patterns and mating behavior. These relationships are relevant to understanding selective pressures acting on rotifers, and the processes of genetic divergence and speciation in these animals.

Genetic structure

Three small ponds in Torreblanca Marsh were sampled regularly (usually every 3 weeks) by Gómez et al. (1995) during the period July 1992–November 1993. These ponds showed a somewhat unpredictable character. For instance, Poza Sur (Figure 1) remained filled in summer 1992, but dried up in summer 1993. Moreover, the marsh may flood during some winters, which we observed in rainy years. *Brachionus* group *plicatilis* (at that time classified as *B. plicatilis*) was detected in 23 out of 56 samples. The group occurred in all samples collected in one of the ponds (Poza Sur). More than 1100 individuals were used to produce clones which were monitored for allozyme variation at four marker loci. The alleles found at each locus were: *Mdh-1*, four alleles; *Mdh-2*, three alleles (one of them being null); *Pgi*, five alleles; *Pgm*, four alleles.

A high correlation was found for the presence of specific alleles at different loci, and most of the possible multilocus genotypes were not detected. Thus, only 13 four-locus genotypes were observed, despite the high number of possible multilocus genotypes. This allowed the authors to classify the genotypes into three clonal groups (called SS, SM, and L), each one having specific alleles in most loci. No intergroup hybrids were found, although within each clonal group heterozygotes for all polymorphic loci were found. Accordingly, the authors concluded that gene flow between groups was absent. Figure 2 shows the results of a cluster analysis based on the estimated genetic distance between strains representative of all the genotypes found (Gómez, 1996). Genetic distance within clonal groups was about 30% the genetic distance among groups, which emphasizes the high divergence between these groups.

A biometric analysis showed that the groups found by Gómez et al. (1995) had different body size and shape (see Figure 3), L being the largest. L group had similar morphology to the so called L-morphotype (Hirayama, 1985; Ito et al., 1981), and is now identified as *B. plicatilis*, while SM and SS are similar to the S-morphotype and are now identified as *B. rotundiformis*. This correlation between morphological and allozyme variation has been reported in non-populational studies performed on *B.* group *plicatilis* (Fu et al., 1991a), and was one of the arguments used to split this taxon into two taxonomic species (i.e., *B. plicatilis* and *B. rotundiformis*; see Segers, 1995). Nevertheless, the findings by Gómez et al. (1995) strongly suggest that the number of sibling species belonging to *B.* group *plicatilis* is likely to be larger than two, since SS and SM groups – both being *B. rotundiformis* according to morphological description – did not share alleles in polymorphic loci, which indicates that gene flow between these groups did not exist. According to the new taxonomy, hereafter these clonal groups will be called *B. plicatilis*, *B. rotundiformis* SM and *B. rotundiformis* SS.

Groups found by Gómez et al. (1995) had different temporal and spatial distribution in Torreblanca Marsh. Very remarkably, the three groups were involved in a seasonal succession in one of the ponds

(Poza Sur). As the largest genotypes occurred in the fall–winter–early spring period, the intermediate in spring, and the smallest in the spring–summer–early fall period, taxon substitution looked like a typical cyclomorphosis in size.

Ecological specialization

Spatial and temporal distribution of clonal groups in Torreblanca Marsh indicated that clonal groups were ecologically specialized. Using factor analysis Gómez et al. (1995) showed that the relative frequency of the clonal groups was correlated to temperature and salinity. *B. plicatilis* was found under low temperature conditions, and tolerated a high range of salinity, *B. rotundiformis* SS was associated with high temperature–high salinity conditions, and *B. rotundiformis* SM occurred at low salinities and high temperatures. Although the results were suggestive on their own, experimental work was necessary to conclude that clonal group distribution was caused by a differential ecological response to temperature and salinity. Consequently, Gómez et al. (1997) studied the growth response of clonal groups to combined levels of salinity (9, 13 and 22 g/l) and temperature (10, 15, 20 and 25°C). Values of temperature and salinity were chosen to represent the conditions observed in the Marsh. Within the experimental range of temperatures, intrinsic rate of increase (r) increased with temperature, a common pattern in *Brachionus* group *plicatilis* species up to 27–35°C (Miracle & Serra, 1989). Different responses to temperature were found (Figure 4). Regarding the low temperature range, *B. rotundiformis* SS did not grow at 15°C. *B. rotundiformis* SM grew neither at 10°C nor at 15°C if salinity was high. By contrast *B. plicatilis* was able to grow at 10°C, at a low rate ($r = 0.1$ day^{-1}). The linear increase in growth rates with temperature had maximal slope for *Brachionus rotundiformis* SS, intermediate for *Brachionus rotundiformis* SM and low for *Brachionus plicatilis* (Gómez et al., 1997). Thus, a rise in temperature differentially increases the growth rates of the clonal groups. These results are consistent with the data reported in other works for *Brachionus* group *plicatilis* strains. For instance, Miracle & Serra (1989) using data from King & Miracle (1980) and from Serra (1987), found that the smaller the body size the steeper the slope relating growth rates to temperature. At high temperatures, growth rates of small strains (most probably, *B. rotun-*

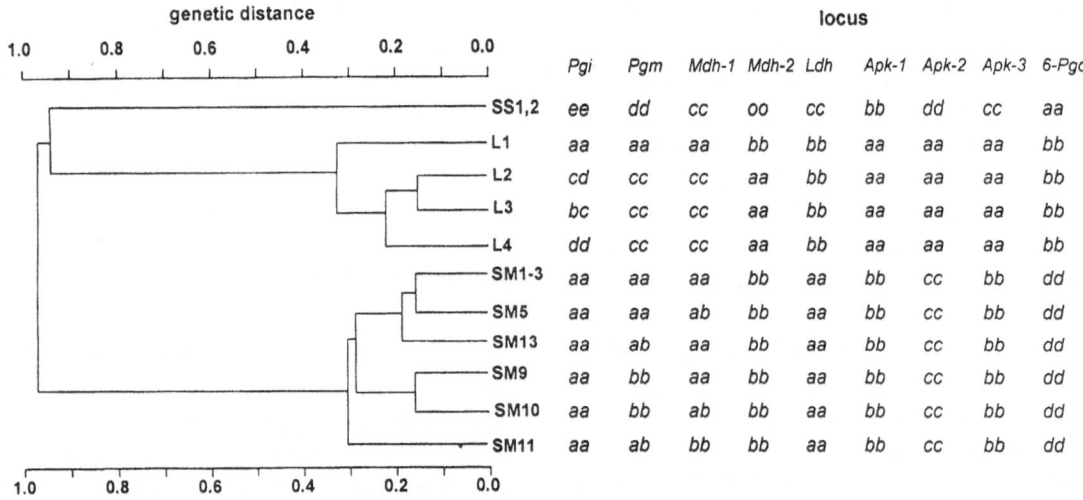

Figure 2. Dendrogram for the 13 genotypes of *Brachionus* group *plicatilis* found in Torreblanca Marsh based on Rogers' genetic distance. Genetic distance was computed from allozyme patterns shown on the right side (after Gómez, 1996).

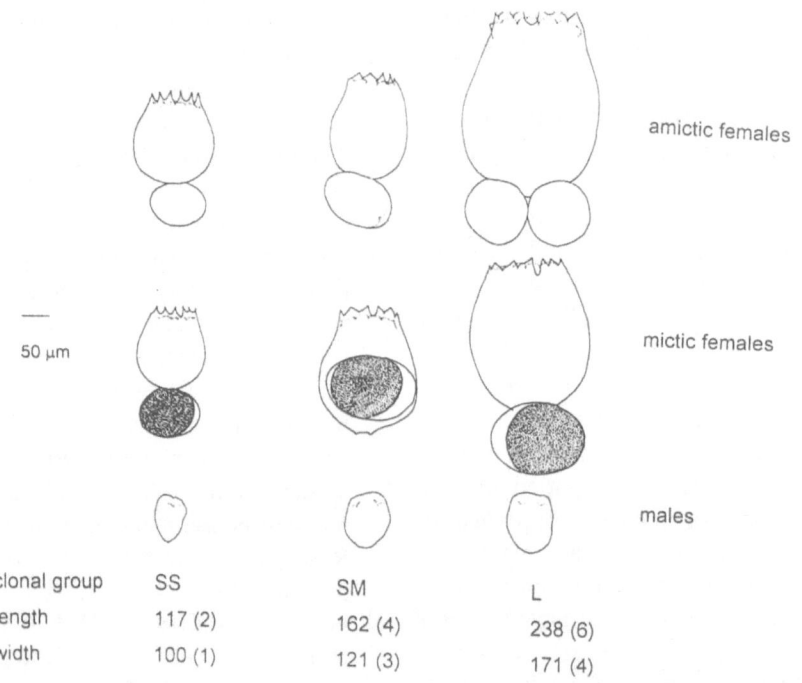

Figure 3. Drawing of individuals (amictic females, resting-egg-producing mictic females and males) belonging to the clonal groups SS, SM and L. Average (*n*=50) body length and width, and their standard errors (in parenthesis) are also shown (after Gómez, 1996).

diformis) were frequently higher than from those of large strains (e.g., Serra, 1987).

Response to salinity of clonal groups in Torreblanca Marsh was dependent on temperature (Figure 4). Growth rates tended to decrease with salinity, a tendency that is more remarkable for *B. rotundiformis* SM. Comparison of these results with data in the literature is difficult as we found differential responses to

salinity for the two clonal groups classified as the same taxonomic species (i.e., *B. rotundiformis*). In fact, optimal salinity values for strains likely belonging to *B. rotundiformis* are also scattered, ranging from 2 g/l (Pascual & Yúfera, 1983) to 40 g/l (Snell, 1986).

Growth rate responses to salinity and temperature of each clonal group are consistent with the correlations found in the field: *B. rotundiformis* SM is better

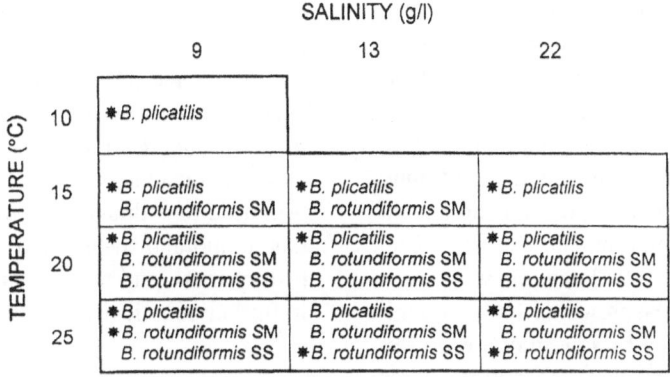

Figure 4. Tolerance to salinity and temperature of *B.* group *plicatilis* clonal groups inhabiting Torreblanca Marsh (Gómez et al., 1997). The asterisk indicates the clonal group(s) with highest population growth rate.

adapted to low salinity–high temperature than other clonal groups, *B. rotundiformis* SS is adapted to high salinity–high temperatures, and *B. plicatilis* is better adapted to low temperatures, showing quite high growth rates regardless of the salinity value. Differential ecological specialization of sympatric genotypes has been reported for *Euchlanis*, *Asplanchna* and *Brachionus* (King, 1972, 1980; King & Zhao, 1987; Snell, 1980), and it is assumed to have a role in the maintenance of diversity and in the seasonal genetic structure of rotifer populations (see King & Serra, 1998).

Mixis patterns

During the period July 1992–November 1993, Carmona et al. (1995) recorded population densities of *B.* group *plicatilis* in Torreblanca Marsh. Densities and mictic female frequency of *B. plicatilis* and *B. rotundiformis* were estimated from preserved samples. Additionally, mictic female frequency was also estimated by culturing females isolated from the field. *B. rotundiformis* SS and *B. rotundiformis* SM could not be distinguished in the preserved samples, but a large contribution of *B. rotundiformis* SS to the results for *B. rotundiformis* is likely, since according to allozyme data *B. rotundifromis* SS was much more frequent than *B. rotundiformis* SM. Reproduction was predominantly parthenogenetic in the clonal groups. Mictic females were found in most samples but proportions were lower than 29%. *B. plicatilis* had a relatively constant incidence of sexual reproduction throughout its presence in the ponds. In contrast, *B. rotundiformis* showed long periods with no or very few mictic females in the samples, and a relatively large mictic ratio at the end of the occurrence period. According to these observations the existence of alternative strategies – extended versus punctuated mixis patterns – in related species was suggested by Carmona et al. (1995).

Carmona et al. (1995) also detected an increase in mictic ratio with population density, which allowed them to estimate the threshold population density for mixis, termed mixis threshold. In accordance with the alternative mixis patterns, population density threshold for mixis in *B. plicatilis* was almost 3-fold higher than that for *B. rotundiformis* (23 vs 6.5 females/l). Differences between mixis thresholds of clonal groups inhabiting Torreblanca Marsh were also found in laboratory experiments (Gómez et al., 1997). *B. rotundiformis* SS had larger (around 8–10 times) mixis thresholds than *B. plicatilis*.

Laboratory experiments (Gómez et al., 1997) also showed a differential mictic response of clonal groups present in Torreblanca Marsh to salinity (9, 13 and 22 g/l) and temperature (15, 20 and 25°C). These two factors affected mixis levels both directly, affecting mixis thresholds, and indirectly, through the effect on growth rates and thus population density. *B. plicatilis* had its highest mictic ratio at low salinity. The highest mictic ratio of *B. rotundiformis* SS were observed at the combination of high salinity (13–22 g/l) and lowest growth temperature (20°C). *B. rotundiformis* SM showed its highest mictic ratios at combinations of low salinity (9–13 g/l) and intermediate temperature. In accordance with other studies, maximal mictic ratios were found in the salinity range 4–10 g/l for *B. plicatilis* (Hagiwara et al., 1989; Hino & Hirano, 1984, 1988; Lubzens et al., 1980, 1985; Pozuelo & Lubián, 1993) and in the range 25–32 g/l for *Brachionus rotundiformis* (Hagiwara & Lee, 1990; Hagiwara et al., 1989; Snell, 1986).

Adaptation to different habitats is expected to include evolution of different mixis patterns. The results reported for clonal groups inhabiting Torreblanca Marsh show that sympatric populations with different ecological niches also have different mictic responses. The alternative between extended and punctuated mixis patterns can be interpreted as an adaptation to different types of environments (Serra & King, in press). Thus, environments with random variation in the length of the growth season would favor an early onset of mixis accompanied by low mictic ratio (extended mixis pattern). In this way, the critical effects of a short growth season can be minimized because of early resting egg production, and when a long growth season occurs, a large population size can be achieved since parthenogenetic reproduction is not greatly reduced after mixis initiation. In contrast, a punctuated mixis pattern would be favored if the end of growth season is predictable. In this case, early investment in mixis has no advantages, and it slows down population growth. Accordingly, *B. plicatilis*, which showed an extended mixis pattern, occurred in late fall, winter and early spring, when salinity in the Marsh can suddenly decrease as a result of rain and floods. By contrast, *B. rotundiformis* SS, which showed a punctuated mixis pattern, seems to live in a more predictable environment, as the end of its growth season comes due to either a gradual increase in salinity, caused by evaporation in summer, or temperature decrease during early fall (Gómez, 1996). Mixis response to salinity and temperature is also genotype dependent, and an adaptive interpretation of this response can be hypothesized in some cases. *B. rotundiformis* SS showed a high mixis ratio at high salinity and at low temperatures, two conditions that anticipate its disappearance in the Marsh. Moreover, *B. plicatilis* increased mixis with decreased salinity, a cue for habitat deterioration in this group.

Mixis patterns, besides being an adaptation to adverse conditions, may play a role in segregating sexual periods and, as a result, in preventing gene flow (see King & Serra, 1998). The effects of mixis patterns on gene flow do not necessarily imply selection against hybridization, or reinforcement, which is a controversial topic in speciation theory (see Butlin, 1989; Paterson, 1978, 1982). These effects may be the byproduct of divergence in ecological preferences. Carmona et al. (1995) observed that sexual periods of *B. plicatilis* and *B. rotundiformis* may partially overlap. An overlap was observed in fall 1992, but not in 1993. Later, Gómez et al. (1997) estimated that the com-

bined effect of growth capability and mixis response to density, salinity and temperature would cause a minimum average 34% reduction in heterogamic sexual encounters between clonal groups. This value is likely to be an underestimation since competitive exclusion was not considered, and only the intrinsic growth rate was used to estimate growth capability. We conclude that temporal distribution of sex, as determined by life history traits, can reduce gene flow, which may result in further genetic divergence.

Mating behavior

Gómez & Serra (1995) studied mating behavior of the three clonal groups belonging to *Brachionus* group *plicatilis* that inhabit Torreblanca Marsh. They applied the mating tests developed by Snell & Hawkinson (1983) to eight strains of *Brachionus* group *plicatilis* isolated in ponds of Torreblanca Marsh, and to one strain of *Brachionus quadridentatus* as an outgroup. Their results depicted a clear pattern of associative mating. Thus, almost all copulations (90%, $n=87$) were observed within clonal groups. Most exceptions are males of *Brachionus rotundiformis* SM copulating with females of *Brachionus rotundiformis* SS. Mating initiation (male circling around female) showed a similar pattern. Despite the fact that males circled with some frequency around females of a different clonal group, percentages of inter-clonal group circling were significantly lower than those observed in homogamic tests. Again, the exceptions are *Brachionus rotundiformis* SM males, which did not demonstrate discrimination between *Brachionus rotundiformis* SS and *Brachionus rotundiformis* SM females. Interestingly, overlap of the bisexual periods of these two clonal groups has not been observed in the field. Some males seem to discriminate between females of their own strain and other strains belonging to the same clonal group, although the evidence for this behavior is not strong. Such a discrimination was not observed when strains belonging to the same group but isolated from different ponds were tested.

Given that copulation was observed between *B. rotundiformis* SM males and *B. rotundiformis* SS females, the possibility of hybridization between these two clonal groups could not be discounted, since the absence of hybrids in the field (Gómez et al., 1995) might have been due to either non-overlapping sexual periods or strong selection against hybrids. In order to determine if hybrids could be produced between *B.*

rotundiformis SM and SS, mating experiments using neonate males and females were conducted by Gómez (1996). Resting egg production was not found in any of four heterogamic tests despite the high number of young mictic females (e.g., 182 neonates) tested in some trials. By contrast, resting eggs were observed in all four homogamic tests. These results indicate that hybrids are not produced between *B. rotundiformis* SS and SM, and thus support the results derived from allozyme data.

Sympatric populations analysis vs. allopatric strains comparison

Studies on *Brachionus* group *plicatilis* populations inhabiting Torreblanca Marsh have shown that there are three sympatric groups of genotypes, with morphological differences in size and shape, and that the gene flow between genotypes is most likely completely absent. Each group is better suited to a different range of salinity and temperature, and the differences in mixis patterns can be interpreted adaptively as well. Ecological tolerance and differential mictic response cause differences in timing of mixis, which partially prevent intergroup sexual encounters. Moreover, if encounters occur, assortative mating and post-mating barriers tend to preclude hybrid formation. Thus the three clonal groups meet requirements to be considered different biological species according to the most common standards, as they are reproductively isolated (Mayr, 1970), they show genetic cohesion (Templeton, 1989), and, particularly significant from our point of view, they have different specific mate recognition systems (Paterson, 1985, 1993). Mate recognition systems are the core of the recognition species concept (Carson, 1994; Lambert et al., 1987; Paterson, 1985), a concept that is gaining considerable acceptance, and that can be successfully applied to rotifers. As a result, we conclude that more than two species are present in the complex of species *Brachionus* group *plicatilis*, although only two taxonomic species have been established until now (Segers, 1995).

Two types of approaches have been used recently to analyze genetic variation at low taxonomic levels of rotifers, analysis that has been focused particularly on the species complex *Brachionus plicatilis*. One approach is based on screening field populations for seasonal and/or spatial genetic variability, and the other is based on comparing strains isolated from several, sometimes distant places. Both approaches are complementary as the former facilitates ecological and population based interpretations and the latter provides insight into global patterns. In both approaches, some genotypes representative of the genetic diversity are usually analyzed for behavioral and/or ecologically significant traits. As far as we know, correlation between genetic differences and traits with potential adaptive significance has always been found regardless of the approach used (Fu et al., 1993; Gómez & Serra, 1995; Gómez & Snell, 1996; King, 1972, 1977a, 1980; King & Zhao, 1987; Miracle & Serra, 1989; Rico-Martínez & Snell, 1995a, b; Snell, 1980; Zhao & King, 1989).

Table 1 shows a comparison between the results found for clonal groups inhabiting Torreblanca Marsh and those from allopatric strain studies. Clusters of genotypes based on genetic distances are less clearly defined in allopatric strain studies. In these clusters, there is significant genetic heterogeneity within the main groups (see Fu et al., 1991a; Gómez & Snell, 1996). For instance, in the allozyme study performed by Fu et al. (1991a) on 67 strains with different geographic origins, two main groups, corresponding to what is now called *B. plicatilis* and *B. rotundiformis*, were found. Genetic distance within the same group can be very high, more than 80% of the genetic distance between groups (see Figure 2 for comparison). Gómez & Snell (1996), in their biometrical analysis, also found two clusters corresponding to *B. plicatilis* and *B. rotundiformis*, but with an intermediate strain and an important size variation within *B. plicatilis*. Diversity in optimal salinity of *B. rotundiformis* strains (Hagiwara et al., 1989; Ito et al., 1981; Pascual & Yúfera, 1983; Snell, 1986) also suggests important genetic diversity within this taxon.

According to studies performed in *Brachionus* group *plicatilis*, mating between strains of the same taxon (either *B. rotundiformis* or *B. plicatilis*) is more frequent than mating between strains belonging to different taxa. However, some important exceptions have been reported when allopatric strains have been tested, and accordingly the existence of either hybrid strains or more than two species has been suggested (Gómez & Snell, 1996; Rico-Martínez & Snell, 1995a,b). On the other hand, males often show strong homogamic preference even when compared to other allopatric strains belonging to their own taxon. This homogamic preference, if found, is quite weaker when sympatric clonal groups are compared. These differing results between approaches are not surprising as genetic dis-

Table 1. Genetic, ecological and behavioral diversity in strains of *Brachionus plicatilis* species complex

	Allopatric strains *B. plicatilis* (Bp) *B. rotundiformis* (Br)	Torreblanca Marsh populations[a] *B. plicatilis* (Bp) *B. rotundifromis* SM (Br SM) *B. rotundiformis* SS (Br SS)
Cluster analysis		
Within-group genetic heterogeneity	High (up to 80%)[b]	Low (up to 30%)
Biometry	High within-group diversity[c,d]; intermediate strains[c]	Low within-group diversity
Optimal growth[e]		
Temperature (°C)	Bp range: $27 - 30^+$ Br range: $30 - 35$	Bp: 25^+ Br SM: 25^+ Br SS: 25^+
Salinity (g/l)	Bp range: $2.5^- - 10$ Br range: $8^- - 40^+$	Bp: 9^- Br SM: 9^- Br SS: 13
Mating behavior		
Assortative mating and reproductive isolation between groups	Yes[f,g]	Yes
Heterogamic copulations	Yes[f]	Yes
Preference for heterogamic mates	Observed in a few cases[f]	Absent
Within-group homogamic preferences[h]	Frequently high[f]	Low

Results from studies on allopatric strains of *B. plicatilis* and *B. rotundiformis* and from the study on Torreblanca Marsh clonal groups (*B. plicatilis, B. rotundiformis* SM and *B. plicatilis* SS) are compared.
[a] Gómez & Serra (1995); Gómez et al. (1995, 1997).
[b] Fu et al. (1991a).
[c] Fu et al. (1991b).
[d] Gómez & Snell (1996).
[e] Range estimated from the optimal value in several studies (Hagiwara & Lee, 1990; Hagiwara et al., 1989; Ito et al., 1981; Lubzens et al., 1980, 1985; Pascual & Yúfera, 1983; Pozuelo & Lubián, 1993; Snell, 1986). The symbols − and + indicate that the highest growth rate was attained at the minimum or maximum values used for the environmental parameter.
[f] Gómez & Snell (1996); Rico-Martínez & Snell (1995a,b).
[g] Fu et al. (1993).
[h] Males show preferences for females of their own strain if compared to other strain belonging to the same group.

tance is *quite* shorter within sympatric clonal groups than within clusters of allopatric strains.

Taking into account this diversity, our conclusion that three sibling species belonging to *Brachionus* group *plicatilis* are present in Torreblanca Marsh is not weakened by data reported in allopatric strains studies. These latter studies clearly proved that *Brachionus plicatilis* strains do not belong to the same species that *Brachionus rotundiformis* strains (Gómez & Snell, 1996; Rico-Martínez & Snell, 1995a,b; Segers, 1995),

but it does not mean that only two species are present in *Brachionus* group *plicatilis*. Actually, both genetic distance analysis and mating behavior analysis of allopatric strains are consistent with the hypothesis that more than two species are present within these taxa.

Divergence processes in rotifer populations

Correlations between ecological preferences, life cycle traits, mating behavior and genetic differences have been found in populations of *B.* group *plicatilis* inhabiting Torreblanca Marsh, similarly to other studies on sympatric rotifer populations (King, 1972, 1980; King & Zhao, 1987; Snell, 1980). Genetic divergence, whether or not evolving in sympatry, should be associated to some degree of adaptive divergence in order to be stable in sympatry. If adaptive divergence does not occur, one of the following two outputs is expected in sympatry: (1) complete competitive exclusion, in the absence of gene flow, or (2) hybridization and genetic convergence to a single group, if gene flow exists. It is a matter for future research to address whether sympatry is a post-speciation phenomenon or whether populations may actually diverge in sympatry. Sympatric speciation is intrinsically difficult. However, we consider it more probable in rotifers than in bisexual animals with continuous reproduction, because seasonal variation and timing of sex may allow allochronic divergence (see Lynch, 1983; King & Serra, 1998), and because mutations affecting the mate recognition system may spread during the parthenogenetic phase, in which they are neutral, affecting a significant set of individuals before reproduction, and promoting their reproductive isolation.

To know how the above-mentioned correlations arise is a critical step in our understanding of rotifer evolution. Selection for divergence of timing of sex is expected after divergence of ecological preferences, as the optimal timing of sex is critically dependent on the ecological regime (Serra & Carmona, 1993; Serra & King, in press). Moreover, if specific population density is a cue for mixis initiation, shifting in timing of sex would be a pleiotropic effect of ecological divergence. The role of mating behavior and the mate recognition pheromone (Snell & Morris, 1993; Snell et al., 1995) in genetic divergence might be also important. Selection for divergence in mate recognition systems is unlikely if reproductive isolation between populations is not complete or almost complete (Butlin, 1989; Endler, 1989; Paterson, 1982). Nevertheless, that divergence would optimize gamete allocation after speciation if both populations co-occur. Alternatively, the evolution of divergence in mate recognition may be a side-effect of sexual selection, which may promote divergence and speciation (e.g., Lande, 1981). The evolution of mate recognition requires complementary changes in male and female. Thus, information on the recognition mechanisms would allow us to assess how likely such complementary changes are.

Given that mate recognition in rotifers is not visual, speciation may occur without important morpho-logical divergence (Gómez & Snell, 1996; Knowlton, 1993), which implies that sibling species may be relatively frequent in rotifers (Serra et al., 1997). Important morphological differences between some species belonging to *Brachionus* group *plicatilis* occur in body size. This trait is most probably a selectively relevant one (see Peters, 1983), as large body size has been related to advantages in growth rate at low temperature (Brooks & Dodson, 1965), which is consistent with seasonal distribution of clonal groups detected in Torreblanca Marsh. However, morphological differences between some biological species of rotifers, despite being statistically significant, are subtle (as those occurring between *B. rotundiformis* SS and *B. rotundiformis* SM), and a lack of correlation between size and mating preferences has been reported for some strains (Gómez & Snell, 1996). These findings suggest that if temperature or diet preferences regarding particle-size, leading to morphological divergence, are not involved in the ecological divergence between genotypes, then a particularly cryptic speciation may occur.

Final remarks

The studies on *Brachionus* populations in Torreblanca Marsh, as well as other literature in the field, suggest that sibling species may be a frequent phenomenon in monogonont rotifers, and point out the critical role that life cycle traits and mating behavior may play in genetic divergence. Moreover, these studies show that ecological divergence allows stable sympatry of sister species in heterogeneous habitats, favoring in this way the maintenance of biodiversity at lower taxonomic levels. Cryptic speciation and seasonal specialization pose some important points for future research. First, the extent of sibling species occurrence should be assessed, so that we can achieve an improved estimate of species biodiversity in Monogononta. Taxonomic species currently considered euryoic and/or with high morphological diversity are a priori good biological material for these types of studies. Second, since seasonal succession of clonal groups has been found, the likelihood of sympatric, allochronic speciation should be explicitly addressed. Evolutionary dynamics using ecological genetic models (e.g., Doebeli, 1996), taking into account the traits of the monogonont rotifer life cycle and the seasonality of the rotifer habitats, should also be addressed. Third, the relative importance of (1) temporal segregation of sex and (2) divergence in mate recognition system in speciation should be investigated, particularly how mate recognition system diverges. Fourth, phylogenetic trees for strains belonging to the same taxa in low taxonomic categories are required in order to improve the design and interpretation of mating tests and ecological analysis. Molecular genetic techniques give robust inference in this regard (Avise, 1994).

Our understanding of rotifer ecological genetics is mainly based on the study of active individuals living in a single pond. Information on the role of resting egg pool, gene flow between sites and spatial population structure is unfortunately scant, despite their importance for the study of speciation and adaptation. Some theoretical and empirical studies have been focused on these topics (e.g., Jenkins, 1995; King & Murtaugh, 1997), but more research is required to obtain a complete picture of how rotifer ecological and genetic diversity are structured.

Acknowledgments

This work was supported by Conselleria de Cultura Educació i Ciencia (Generalitat Valenciana; no- GV-2543/94) and Instituto Valenciano de Estudios e Investigación (Diputación Valenciana) to M.S. A.G. was also supported by fellowship of the Conselleria de Cultura Educació i Ciencia (Generalitat Valenciana). We thank two anonymous referees for critically reading the manuscript and for their language assistance.

References

Aparici, E., M. J. Carmona & M. Serra, 1998. Sex allocation in haplodiploid cyclical parthenogens with density-dependent proportion of males. Am. Nat. 152: 652–657.

Avise, J. C., 1994. Molecular Markers, Natural History and Evolution. Chapman and Hall, New York: 511 pp.

Brooks, J. L. & S. I. Dodson, 1965. Predation, body size and composition of the plankton. Science 150: 28–35.

Butlin, R., 1989. Reinforcement of premating isolation. In Otte, D. & J. A. Endler (eds), Speciation and its Consequences. Sinauer Associates, Sunderland, MA: 158–179.

Carmona, M. J., M. Serra & M. R. Miracle, 1993. Relationships between mixis in Brachionus plicatilis and preconditioning of the culture medium by crowding. Hydrobiologia 255/256 (Dev. Hydrobiol. 83): 145–152.

Carmona, M. J., A. Gómez & M. Serra, 1995. Mictic patterns of Brachionus plicatilis in small ponds. Hydrobiologia 313/314 (Dev. Hydrobiol. 109): 365–371.

Carson, H. L., 1994. Fitness and the sexual environment. In Lambert, D. M. & H. G. Spencer (eds), Speciation and the Recognition Concept: Theory and Application. The Johns Hopkins University Press, Baltimore: 126–137.

Doebeli, M., 1996. A quantitative genetic competition model for sympatric speciation. J. Evol. Biol. 9: 893–909.

Endler, J. A., 1989. Conceptual and other problems in speciation. In Otte, D. & J. A. Endler (eds), Speciation and its Consequences. Sinauer Associates, Sunderland, MA: 625–648.

Fu, Y., K. Hirayama & Y. Natsukari, 1991a. Genetic divergence between S and L type strains of the rotifer Brachionus plicatilis O.F. Müller. J. exp. mar. Biol. Ecol. 151: 43–56.

Fu, Y., K. Hirayama & Y. Natsukari, 1991b. Morphological differences between two types of the rotifer Brachionus plicatilis O. F. Müller. J. exp. mar. Biol. Ecol. 151: 29–41.

Fu, Y., A. Hagiwara & K. Hirayama, 1993. Crossing between seven strains of the rotifer Brachionus plicatilis. Nippon Suisan Gakkaishi 59: 2009–2016.

Gilbert, J. J., 1963. Contact chemorreception, mating behaviour, and sexual isolation in the rotifer genus Brachionus. J. exp. Biol. 40: 625–641.

Gilbert, J. J., 1974. Dormancy in rotifers. Trans. am. Micros. Soc. 93: 490–513.

Gómez, A., 1996. Ecología genética y sistemas de reconocimiento de pareja en poblaciones simpátricas de rotíferos. Ph.D. dissertation, University of Valencia, Valencia, Spain.

Gómez, A. & M. Serra, 1995. Behavioral reproductive isolation among sympatric strains of Brachionus plicatilis Müller, 1786: Insights into the status of this taxonomic species. Hydrobiologia 313/314 (Dev. Hydrobiol. 109): 111–119.

Gómez, A. & M. Serra, 1996. Mate choice in male Brachionus plicatilis rotifers. Funct. Ecol. 10: 681–687.

Gómez, A. & T. W. Snell, 1996. Sibling species in the Brachionus plicatilis species complex. J. Evol Biol. 9: 953–964.

Gómez, A., M. Temprano & M. Serra, 1995. Ecological genetics of a cyclical parthenogen in temporary habitats. J. Evol. Biol. 8: 601–622.

Gómez, A., M. J. Carmona & M. Serra, 1997. Ecological factors affecting gene flow in the Brachionus plicatilis complex (Rotifera). Oecologia 111: 350–356.

Hagiwara, A. & C. S. Lee, 1990. Resting egg formation of the L- and S-type rotifer Brachionus plicatilis under different water temperature. Nippon Suisan Gakkaishi 57: 1645–1650.

Hagiwara, A., A. Hino & R. Hirano, 1985. Combined effects of environmental conditions on the hatching of fertilized eggs of the rotifer Brachionus plicatilis collected from an outdoor pond. Bull. jpn. Soc. Sci. Fish. 51: 755–758.

Hagiwara, A., C. S. Lee, G. Miyamoto & A. Hino, 1989. Resting egg formation and hatching of the S-type rotifer Brachionus plicatilis at varying salinities. Mar. Biol. 103: 327–332.

Hagiwara, A., M. M. Jung, T. Sato & K. Hirayama, 1995. Interspecific relationships between marine rotifer Brachionus rotundiformis and zooplankton species contaminating in the rotifer mass culture tank. Fish. Sci. 61: 623–627.

Hino, A. & R. Hirano, 1984. Relationship between water temperature and bisexual reproduction rate in the rotifer Brachionus plicatilis. Bull. jpn. Soc. Sci. Fish. 50: 1481–1485.

Hino, A. & R. Hirano, 1988. Relationship between water chlorinity and bisexual reproduction rate in the rotifer Brachionus plicatilis. Nippon Suisan Gakkaishi 54: 1329–1332.

Hirayama, K., 1985. Biological aspects of the rotifer Brachionus plicatilis as a food organism for mass culture of seedling. Coll fr–jpn Oceanogr., Marseille 8: 41–50.

Hutchinson, G. E., 1967. A Treatise on Limnology, Vol. 2. J. Wiley, New York.

Ito, S., H. Sakamoto, M. Hori & K. Hirayama, 1981. Morphological characteristics and suitable temperature for the growth of several strains of the rotifer, *Brachionus plicatilis*. Bull. Fac. Fish., Nagasaki Univ. 51: 9–16.

Jenkins, D. G., 1995. Dispersal-limited zooplankton distribution and community composition in new ponds. Hydrobiologia 313/314 (Dev. Hydrobiol. 109): 15–20.

King, C. E., 1972. Adaptation of rotifers to seasonal variation. Ecology 53: 408–418.

King, C. E., 1977a. Genetics of reproduction, variation, and adaptation in rotifers. Arch. Hydrobiol. Beih. Ergebn. Limnol. 8: 187–201.

King, C. E., 1977b. Effects of cyclical ameiotic parthenogenesis on gene frequency population size. Arch. Hydrobiol. Beih. 8: 207–211.

King, C. E., 1980. The genetic structure of zooplankton populations. In Kerfoot, W. C. (ed.), Evolution and Ecology of Zooplankton Communities. University Press of New England, Hanover, NH: 315–328.

King, C. E., 1993. Random genetic drift during cyclical ameiotic parthenogenesis. Hydrobiologia 255/256 (Dev. Hydrobiol. 83): 205–212.

King, C. E. & M. R. Miracle, 1980. A perspective on aging in rotifers. Hydrobiologia 73 (Dev. Hydrobiol. 1): 13–19.

King, C. E. & P. Murtaugh, 1997. Effects of asexual reproduction on the neighborhood area of cyclical parthenogens. Hydrobiologia 358 (Dev. Hydrobiol. 124): 55–62.

King C. E. & M. Serra, 1998. Seasonal variation as a determinant of population structure in rotifers reproducing by cyclical parthenogenesis. Hydrobiologia 387/388 (Dev. Hydrobiol. 134): 361–372.

King, C. E. & Y. Zhao, 1987. Coexistence of rotifer (*Brachionus plicatilis*) clones in Soda Lake, Nevada. Hydrobiologia 147 (Dev. Hydrobiol. 42): 57–64.

Knowlton, N., 1993. Sibling species in the sea. Annu. Rev. Ecol. Syst. 24: 189–216.

Lambert, D. M, B. Michaux & C. S. White, 1987. Are species self defining? Syst. Zool. 36: 196–205.

Lande, R., 1981. Models of speciation by sexual selection on polygenic traits. Proc. natl. Acad. Sci. U.S.A. 78: 3721–3725.

Lubzens, E., R. Fishler & V. Berdugo-White, 1980. Induction of sexual reproduction and resting egg production in *Brachionus plicatilis* reared in seawater. Hydrobiologia 73 (Dev. Hydrobiol. 1): 55–58.

Lubzens, E., G. Minkoff & S. Marom, 1985. Salinity dependence of sexual and asexual reproduction in the rotifer *Brachionus plicatilis*. Mar. Biol. 85: 123–126.

Lynch, M., 1983. Ecological genetics of *Daphnia pulex*. Evolution 37: 358–374.

Mayr, E., 1970. Populations, Species and Evolution. Harvard University Press, Cambridge, MA: 453 pp.

Minkoff, G., E. Lubzens & D. Kahan, 1983. Environmental factors affecting hatching of rotifer (*Brachionus plicatilis*) resting eggs. Hydrobiologia 104 (Dev. Hydrobiol. 14): 61–69.

Miracle, M. R. & M. Serra, 1989. Salinity and temperature influence in rotifer life history characteristics. Hydrobiologia 186/187 (Dev. Hydrobiol. 52): 81–102.

Pascual, E. & M. Yúfera, 1983. Crecimiento en cultivo de una cepa de *Brachionus plicatilis* O.F. Müller en función de la temperatura y la salinidad. Invest. Pesquera 47: 151–159.

Paterson, H. E. H., 1978. More evidence against speciation by reinforcement. South Africa J. Sci. 74: 369–371.

Paterson, H. E. H., 1982. Perspective of speciation by reinforcement. South Africa J. Sci. 78: 53–57.

Paterson, H. E. H., 1985. The recognition concept of species. In Vrba, E. S. (ed.), Species and Speciation. Transvaal Museum Monograph No. 4, Pretoria: 21–29.

Paterson, H. E. H., 1993. Evolution and the Recognition Concept of Species. Collected writings. McEvey, S. F. (ed.). The Johns Hopkins University Press, Baltimore, MD, EEUU.

Peters, R. H., 1983. The Ecological Implications of Body Size. Cambridge University Press, New York.

Pourriot, R. & T. W. Snell, 1983. Resting eggs of rotifers. Hydrobiologia 104 (Dev. Hydrobiol. 14): 213–224.

Pozuelo, M. & L. M. Lubián, 1993. Asexual and sexual reproduction in the rotifer *Brachionus plicatilis* cultured at different salinities. Hydrobiologia 255/256 (Dev. Hydrobiol. 83): 139–143.

Rico-Martínez, R. & T. W. Snell, 1995a. Copulatory behavior and mate recognition pheromone blocking of male receptors in *Brachionus plicatilis* Müller (Rotifera). Hydrobiologia 313/314 (Dev. Hydrobiol. 109): 105–110.

Rico-Martínez, R. & T. W. Snell, 1995b. Male discrimination of female *Brachionus plicatilis* Müller and *Brachionus rotundiformis* Tchugunoff (Rotifera). J. exp. mar. Biol. Ecol. 190: 39–49.

Segers, H., 1995. Nomenclatural consequences of some recent studies on *Brachionus plicatilis* (Rotifera, Brachionidae). Hydrobiologia 313/314 (Dev. Hydrobiol. 109): 121–122.

Serra, M., 1987. Variación morfométrica, isoenzimática y demográfica en poblaciones de *Brachionus plicatilis*: diferenciación genética y plasticidad fenotípica. Ph.D. dissertation, University of Valencia, Valencia, Spain.

Serra, M. & M. J. Carmona, 1993. Mixis strategies and resting egg production of rotifers living in temporally-varying habitats. Hydrobiologia 255/256 (Dev. Hydrobiol. 83): 117–126.

Serra, M & C. E. King (in press). Optimal rates of bisexual reproduction in cyclical parthenogens with density dependent growth. J. Evol. Biol.

Serra, M. & M. R. Miracle, 1985. Enzyme polymorphism in *Brachionus plicatilis* populations from several Spanish lagoons. Verh. int. Verein. Limnol. 22: 2991–2996.

Serra, M., A. Galiana & A. Gómez, 1997. Speciation in monogonont rotifers. Hydrobiologia 358 (Dev. Hydrobiol. 124): 63–70.

Snell, T. W., 1980. Blue-green algae and selection in rotifer populations. Oecologia 46: 343–346.

Snell, T. W., 1986. Effect of temperature, salinity and food level on sexual and asexual reproduction in *Brachionus plicatilis* (Rotifera). Mar. Biol. 92: 157–162.

Snell, T. W., 1987. Sex, population dynamics and resting eggs production in rotifers. Hydrobiologia 144: 105–111.

Snell, T. W., 1989. Systematics, reproductive isolation and species boundaries in monogonont rotifers. Hydrobiologia 186/187 (Dev. Hydrobiol. 52): 299–310.

Snell, T. W. & E. M. Boyer, 1988. Thresholds for mictic-female production in the rotifer *Brachionus plicatilis* (Müller). J. exp. mar. Biol. Ecol. 124: 73–85.

Snell, T. W. & C. A. Hawkinson, 1983. Behavioral reproductive isolation among populations of the rotifer *Brachionus plicatilis*. Evolution 37: 1294–1305.

Snell, T. W. & F. J. Hoff, 1987. Fertilization and male fertility in the rotifer *Brachionus plicatilis*. Hydrobiologia 147 (Dev. Hydrobiol. 42): 329–334.

Snell, T. W. & P. D. Morris, 1993. Sexual communication in copepods and rotifers. Hydrobiologia 255/256 (Dev. Hydrobiol. 83): 109–116.

Snell, T. W. & B. C. Winkler, 1984. Isozyme analysis of rotifer proteins. Biochem. Syst. Ecol. 12: 199–202.

384

Snell, T. W., R. Rico-Martínez, L. N. Kelly & T. E. Battle, 1995. Identification of a sex pheromone from a rotifer. Mar. Biol. 123: 347–353.

Templeton, A. R., 1989. The meaning of species and speciation: a genetic perspective. In Otte, D. & J. A. Endler (eds), Speciation and its Consequences. Sinauer Associates, Sunderland, MA: 3–27.

Zhao, Y. & C. E. King, 1989. Ecological genetics of the rotifer *Brachionus plicatilis* in Soda Lake, Nevada, EEUU. Hydrobiologia 185: 175–181.

Hydrobiologia **387/388**: 385–393, 1998.
E. Wurdak, R. Wallace & H. Segers (eds), Rotifera VIII: A Comparative Approach.
© *1998 Kluwer Academic Publishers. Printed in the Netherlands.*

Allozyme electrophoresis: its application to rotifers

Africa Gómez
Departament de Microbiologia i Ecologia, Universitat de València, Burjassot, E-46100, Spain

Key words: protein electrophoresis, population genetics, cellulose acetate gel, alleles

Abstract

Allozyme electrophoresis is a well established technique for revealing genetic variation that has given useful results in a wide range of organisms. However, in rotifers it has been applied scarcely and only in a few species. In this paper the methods of acetate allozyme electrophoresis are introduced, including laboratory setup, equipment and staining recipes that have been successfully applied to brachionid rotifers. In addition, the literature published on allozyme electrophoresis in rotifers is reviewed and the main results and future prospects of the technique are discussed. We conclude that, despite the onset of DNA techniques, allozyme electrophoresis is promising and can yield important results in rotifer population genetics, ecology and systematics.

Introduction

Allozyme electrophoresis began to be used as a tool for the detection of genetic variation in populations three decades ago (Harris, 1966; Hubby & Lewontin, 1966; Lewontin & Hubby, 1966). Since then, it has been extensively used in taxonomic, phylogenetic and population genetic studies in a number of organisms (Avise, 1974; Ayala, 1976; Hebert & Beaton, 1989; Selander, 1976), and, due to the huge amount of information obtained, this technique has revolutionized population genetics and other evolutionary and ecological studies (Avise, 1994; Hedrick, 1985; May, 1992).

Protein electrophoresis is based on the fact that proteins in their native state with net charge differences migrate at different rates under the influence of an electric field (Richardson et al., 1986). The relative mobility of a protein depend on its electric charge (due to its primary structure) and on the shape and size of the molecule (which depends on its secondary and tertiary structure). Due to the number of factors involved (Freifelder, 1981), it is unusual that two types of protein show the same relative mobility in the pH range used in research (Richardson et al., 1986). If two proteins show different mobility after electrophoretic separation they will usually differ by at least one or more aminoacids. Enzymes, proteins with

catalytic function, can be visualized after electrophoretic separation using a specific staining protocol that incorporates the substrates and necessary cofactors of the enzyme being detected and a histochemical system that produces a visible product. This system often includes an oxidized salt, substrate of a final coupled reaction that yields a colored insoluble product. In some systems a linking enzyme is needed. The procedure often resolves alleles of the same locus. To improve resolution, the electrophoretic separation is made in a porous matrix (gel) embedded in an ionic buffer. Usually, the buffers used produce migration from negative to positive poles.

The advantages of this technique are numerous (Avise, 1994; Hedrick, 1985; Murphy et al., 1996), for example: (1) the data obtained are objective and easily measurable, on the basis of the mobility of different bands in gels; (2) the data are estimates of genetic variation (thus, they are not affected by phenotypic changes such as age or physiological changes) (3) the variants reveal independent Mendelian polymorphisms at several loci scattered around the genome, which provide molecular marker loci that are very useful for taxonomic studies, population ecology, mating systems, etc.; (4) a high number of individuals can be analyzed for numerous loci in nearly any organism, which allows a reliable estimate of allele frequencies and the accompanying measures of confidence; (5) the loci are generally co-dominant so that the heterozygotes can be identified; (76) there is a straightforward relationship to standard population genetic theory; (87) and last but not least the relatively cheap materials required and the short training time make it a very cost-effective technique. Of course, there are several potential problems when comparing allozyme electrophoresis to other molecular methods: (1) part of the genetic variation remains hidden, as not all nucleotide variants in a gene result in changes in the protein's net charge (Kreitman, 1983), therefore genetic variability measured using electrophoresis may underestimate the real variability by as much as 30% (Shaw, 1970); (2) only structural genes can be analyzed, that is, loci encoding for water-soluble proteins, therefore, the usual set of enzymes analyzed can represent a biased sample of the genome (Avise, 1994; Hedrick, 1985); (3) identification of loci or alleles in certain enzymes is sometimes difficult, and some variants are present only in certain tissues (Hedrick, 1985); (4) for small animals, it is necessary to sacrifice the individual to find the genotype; (5) allozyme variants may be under selection pressure or linked to selected loci and therefore constitute poor genetic markers (e.g. Carvalho, 1988). Nevertheless, most of these potential shortcomings do not present problems in the common uses of the technique in which genetic markers for ecological, systematics or population studies are needed.

Traditionally, starch, polyacrylamide and, to a lesser extent, agarose gels, have been used as a supporting medium for protein electrophoresis. However, recently several authors have recommended cellulose acetate gels, because of their versatility and ease of use, especially when working with small invertebrates (Hebert & Beaton, 1989; Wynne et al., 1992). There are two thorough manuals detailing the use of acetate electrophoresis (Hebert & Beaton, 1989; Richardson et al., 1986). Here we present a brief introduction to allozyme electrophoresis and its application to rotifers, but we strongly encourage any interested reader to refer to these books for further information. There are several advantages of the cellulose acetate support medium in comparison with other systems (Freifelder, 1981; Hebert & Beaton, 1989; Richardson et al., 1986; Rothe, 1994; Wynne et al., 1992), among them, economy in the reagents used in staining and running buffers, the low quantity of biological samples required (which is critical when we work with small organisms such as rotifers), high resolution and sensitivity for small samples, fast preparation and staining, quick electrophoretic separation (habitually 15–40 min, in comparison with more than 2 h for acrylamide), nontoxicity of gels, use of low voltage in electrophoresis (which reduces gel heating during electrophoretic separation) and ease of gel handling and conservation. Some drawbacks of this method include the cost of premade commercial gels, that gels cannot be sliced to stain with different enzyme systems and that staining may be difficult with low activity enzymes. An additional limitation of cellulose acetate is that proteins separate mainly on the basis of charge differences, with low or no separation by size, in contrast to starch and acrylamide (Richardson et al., 1986). This does not constitute a problem as allozymes rarely differ in mass or shape, and comparisons between different supporting media show an equal or superior resolution on cellulose acetate gels (Richardson et al., 1986). On the other hand, changes in electrophoretic conditions often reveal more allelic variation than changes in supporting media.

We will now introduce some useful terminology on allozyme analysis. The pattern of bands of a gel stained for an enzyme is called a *zymogram* (Richardson et al., 1986). When several bands of a given enzyme in a stained gel have different mobilities they are called *isozymes*. Isozyme variation can reflect genetic or non-genetic differences. Among genetic differences, enzyme variation can reflect the presence of one or more active loci in the tissue or organism studied, because of polyploidy, aneuploidy or gene duplication and specialization of each new locus. If isozymatic variation results from the presence of different alleles from the same locus, isozyme variants are called *allozymes*. In this case, the banding pattern of allelic variants depends on the quaternary structure of the enzyme (Richardson et al., 1986). As aminoacid substitutions that result in changes in banding patterns do not reflect all the variation in the locus, some authors prefer to call allozyme variants *electromorphs* (Hedrick, 1985). On the other hand, some isozymes reflect non-genetic changes in enzymes, such as posttranscriptional modifications, that may take place in vivo or in vitro (enzyme degradation during the sample extraction process through oxidization, deamination or acetilation, artifacts, etc.) that may affect resolution of the banding patterns after electrophoresis. Another useful term is *composite electromorph*, which is used to refer to a genotype for several allozyme loci.

Any study using isozyme data requires that zymograms be correctly interpreted, and this relies on a series of assumptions (Murphy et al., 1996). The basic assumption is that differences in mobility are due to genetic, heritable changes, that is, reflect DNA sequence differences between the genes encoding these proteins. Thus, care must be taken to ensure that the banding patterns we observe are due to either the presence of different alleles, or the presence of different enzyme loci. Codominance is also a common assumption. To meet this, a previous knowledge of the enzyme quaternary structure is often necessary. Other aspects of gel interpretation have been dealt with in great detail in Richardson et al. (1986).

Application to rotifers

Among zooplankters, rotifers offer several advantages in electrophoretic studies. First, in contrast to sexually reproducing zooplankters, such as copepods, rotifers are cyclical parthenogens and can be cloned in the laboratory. We can use as much biomass of genetically identical individuals as needed, and keep clones in culture for future isozyme analysis or other studies. Although the number of enzymes assayed per individual depends on animal size when using single individual extracts (Wolf, 1982) this is not a problem in rotifers, provided that culture is possible. Second, rotifers have short generation times and, consequently, high intrinsic growth rates, which minimizes the time necessary to breed enough individuals. In only a week period, it is possible to obtain ca 1000 individuals from a single *Brachionus plicatilis* female isolated from a field sample (personal observation).

Rotifers, however, have several disadvantages when compared with other zooplankters such as copepods or cladocerans. First of all, they have very small body sizes, which makes it difficult to use single individuals to determine allozyme patterns. Although strong staining enzyme systems (e.g., phosphoglucose isomerase (PGI)) can be analyzed in single individual extracts for large rotifers, when using small rotifers or low activity enzymes, cloning is required before attempting electrophoresis. This makes culture in the lab a necessary, and often labor-intensive, prior step of electrophoretic analyses, and makes difficult to study those species not easily cultured. In addition, mictic females and males can rarely be analyzed using this technique (Gómez et al., 1995). These drawbacks may be responsible for the scarcity of electrophoretic studies in rotifers (Walsh, 1993; see literature in Table 1), in comparison with cladocerans (see Hebert, 1987a, for a review). Nevertheless, many rotifer species have been cultured successfully (see Pourriot, 1977; Wal-

lace & Snell, 1991), and the ability of culturing rotifers should not be a limitation of the use of electrophoresis in rotifers, at least in the near future. A broader knowledge of the techniques and potential results of electrophoresis might contribute to encouraging rotifer researchers to use this technique.

In this paper I present a detailed protocol for cellulose acetate gel electrophoresis that has given good results with rotifers of the genus *Brachionus*, and revise the literature of electrophoresis in rotifers. A list of recipes of high activity and easily scorable enzyme systems is also provided.

Methodology

Sample preparation and protein extraction

An effective rotifer harvesting and concentration protocol is a critical step to obtain consistent results in allozyme electrophoresis. Rotifer clonal cultures obtained from single females are harvested for electrophoresis when they reach high densities (200–300 females per ml), usually 1 week for *Brachionus plicatilis*. Then, 10 ml of each culture are filtered through a small circular piece of Nytal mesh (mesh pore size, 30 or 50 μm, adjusted to clone size) using a plastic syringe and a 1.5-ml centrifuge tube, and washed sequentially with saline water and with 2 ml of chilled homogenizing buffer. After filtering and washing, rotifers are concentrated in a small drop of homogenizing buffer. Then the rotifer drop is transferred to 0.5-ml microcentrifuge tubes using a small spatula, and soluble proteins are extracted by homogenizing the rotifers in 50 μl of chilled 10% sucrose solution in 0.025 M Tris–glycine buffer, pH 8.5, with a small pestle made using Epoxy resin and a lancet. To pellet debris the samples are centrifuged at $15000 \times \boldsymbol{g}$ for 5 min at 4°C. After centrifuging, samples can be stored at −80°C until being used. As most proteins are temperature labile, sample preparation should be carried out either in a cold room or on ice.

For single female allozyme electrophoresis (suitable with rotifers longer than 200 μm and high activity enzymes such as PGI) a single female is washed in saline solution and then in homogenizing buffer. After this, it is placed in 10 μl of buffer and crushed in a 1-ml Eppendorf tube. From this point on, the protocol follows the one described below.

Table 1. Enzymes succesfully assayed in rotifers

	Enzyme	Species	Polymorphism	Loci	Alleles	Medium	Refs.
1.	AAT	*Brachionus plicatilis* (s.l.)	Yes	1	2	Starch	1
2.	ACP	*Asplanchna girodi,* *A. brightwelli*	Yes	?	?	Acrylam.	5
3.	ALP	*B. plicatilis* (s.l.)	Yes	1	2-4	Acrylam.	11, 13
4.	AO	*A. girodi*	No			Acrylam.	6
5.	APK	*B. plicatilis* (s.l.)	Yes	3	(2,5,4)	Acetate	3, 4
		M. quadricornifera	Yes	2	(2,1)	Acrylam.	9
6.	CHO	*B. plicatilis* (s.l.)	Yes	?	?	Acrylam.	8
7.	EST	*B. plicatilis* (s.l.)	Yes	2, 3	(5,2,4)	Acrylam.	11, 13
		A. brightwelli	No	?	?	Acrylam.	5
		A. girodi	Yes	?	?	Acrylam.	6
		M. quadricornifera	Yes	4, 2	7	Acrylam.	9
8.	FUM	*B. plicatilis* (s.l.)	No	?	?	Acrylam.	8
9.	GOT	*B. plicatilis* (s.l.)	Yes	1	3	Acrylam.	13
10.	G6PDH	*A. girodi, A. brightwelli*	Yes	?	?	Acrylam.	5, 6
11.	LAP	*M. quadricornifera*	Yes	1	3	Acrylam.	9
12.	LDH	*B. plicatilis* (s.l.)	Yes	1	9	Starch	1, 2
			No	?	?	Acrylam.	8, 11
			Yes	1	3	Acetate	3
		A. girodi	Yes	?	?	Acrylam.	5, 6
13.	MDH	*B. plicatilis* (s.l.)	Yes	3	(?,?,3)	Starch	1, 2
			Yes	1-2	(3,1-8)	Acrylam.	8, 11,13
			Yes	2	(3,3-5)	Acetate	3, 4
		A. girodi	Yes	?	?	Acrylam.	12
			No			Acrylam.	6
		A. brightwelli	Yes	?	?	Acrylam.	5
14.	ME	*B. plicatilis* (s.l.)	No			Acrylam.	8
			Yes	1-2	>3	Acetate	3
		A. girodi, *A. brightwelli*	Yes	1	2	Acrylam.	5, 6, 7
		M.quadricornifera	Yes	3	?	Acrylam.	9
15.	PGD	*B. plicatilis* (s.l.)	Yes	1	13	Starch	1, 2
			Yes	2	(4,4)	Acrylam.	13
			No	?	?	Acrylam.	8
			Yes	1	4	Acetate	3
16.	PGI	*B. plicatilis* (s.l.)	Yes	1	9	Starch	1, 2
			Yes	?	?	Acrylam.	8
			Yes	1	4-5	Acetate	3, 4
		A. girodi	Yes	?	?	Acrylam.	5, 6
		M. quadricornifera	Yes	1	4-6	Acrylam.	9, 10
17.	PGM	*B. plicatilis* (s.l.)	Yes	1	9	Starch	1, 2
			Yes	?	?	Acrylam.	8
			Yes	1	4	Acetate	3, 4
18.	SOD	*B. plicatilis* (s.l.)	Yes	1	4	Starch	1, 2
19.	TO	*B. plicatilis* (s.l.)	No	?	?	Acrylam.	8
		A. girodi	Yes	?	?	Acrylam.	5, 6

Refs.: (1) Fu et al., 1990a; (2) Fu et al., 1991; (3) Gómez et al., 1995; (4) Gómez & Snell, 1996; (5) King, 1977; (6) King, 1980; (7) King & Snell, 1977; (8) King & Zhao, 1987; (9) Pagani et al., 1991; (10) Ricci et al., 1989; (11) Serra & Miracle, 1985; (12) Snell, 1979; (13) Snell & Winkler, 1984.

Sample loading and electrophoretic separation

For electrophoresis we used the system provided by Helena Laboratories (Beaumont, Texas) for cellulose acetate gels. This system is composed of an electrophoresis chamber and an application kit (Super Z-12). Previous to the electrophoretic run, acetate gels are soaked in the running buffer (see below) for at least 20 min. Before loading the samples, the gel is blotted dry between sheets of filter paper. An aliquot of 6 μl of supernatant of the sample is loaded into each well of the sample plate using the applicator. Electrophoresis is carried out on cellulose acetate gels (Titan III, 76×76 mm, UK cat No. 3033) at 4°C, at a constant voltage (200 V) for 30 min, using 0.025 mM Tris–glycine, pH 8.5, as running buffer. Different buffers, or discontinuous buffer systems may be necessary depending on the enzymes tested. Current during the electrophoresis depends on the running buffer (3 mA/plate using 25 mM Tris–glycine, pH 8.5). Although refrigeration is not necessary, we usually run the electrophoresis in a refrigerator at 4°C. Clones with known zymograms are used as controls when needed.

Staining

Gels are stained in the dark using the agar overlay technique (Hebert & Beaton, 1989; see Appendices 1–3 for details on staining recipes and chemical preparation) immediately after the completion of the run. Staining procedures for a number of enzyme systems adapted for acetate electrophoresis can be found in Richardson et al. (1986) and Hebert & Beaton (1989). In Appendices 1–3 we provide detailed information and staining recipes for enzyme systems that have proved to give good results when used with rotifers (see Table 1).

Scoring

The genetic interpretation of zymograms can be checked through repeated surveys of clones maintained in the laboratory, as well as through agreement with the expected zymogram for the heterozygous locus with the standard enzyme quaternary structure. Another approach, more costly, but also more reliable, is to analyze the sexual offspring of presumptive heterozygous clones for a given allozyme locus to verify Mendelian segregation. Thus, resting eggs collected from the cultures, presumably the product of inbreeding, can be induced to hatch in the laboratory, and the zymograms of the resulting clones analyzed. When different protein mobilities are detected, they should be checked by running samples in contiguous wells.

If electrophoretic analysis, or at least sample preparation and freezing, can be done within a short time after the isolation of the females from the field samples, it can be ensured that parthenogenesis is the dominant mode of reproduction and that mutation, resting egg hatching, or other sources of genetic variation are negligible. No variation in zymogram patterns has been detected in clones maintained in our laboratory for long periods of time.

Final remarks

So far, electrophoretic studies in rotifers have been scant, and limited by the number of enzymes, species and sample size (Hebert, 1987b; Walsh, 1993). Since the introduction of the technique in rotifer studies by King (1977), the only species used have been *Asplanchna brigthwelli, A. girodi, Brachionus plicatilis* and *Macrotrachela quadricornifera*. Electrophoresis has been applied to these organisms to investigate population structure (Gómez et al., 1995; King, 1977, 1980; King & Zhao, 1987; Pagani et al., 1991; Ricci et al., 1989; Serra & Miracle, 1985), systematics (Fu et al., 1990a, 1991; Gómez et al., 1995; Serra & Miracle, 1985; Snell & Winkler, 1984), to obtain genetic markers for ecological experiments (Snell, 1979), to show the presence of sexual recombination (King & Snell, 1977), and to check for the trisomy or triploidy of abnormally large *B. rotundiformis* strains (Fu et al., 1990b). Some of these results have been rather surprising as they contradict widely held views. For example, rotifer populations have been shown to keep moderately high levels of genetic polymorphism (Gómez et al., 1995; King & Zhao, 1987), and the widely used species in aquaculture *B. plicatilis* is in fact a species complex (Gómez & Snell, 1996; Gómez et al., 1995). Some of the results have suggested the widespread occurrence of sibling species in monogonont rotifers (Gómez et al., 1995; King, 1977, 1980; see also Serra et al., 1997). The same studies challenge the preconception that cyclomorphosis in these organisms is mainly due to phenotypic plasticity.

In view of the results obtained so far, it seems that allozyme electrophoresis can still yield important results in four main fields of rotifer research:

(i) In studies of *population genetic structure*. This can include the study of the processes affecting allele frequencies: evidence for selection; gene flow; founder events; breeding structure; spatial and temporal population structure; hybridization phenomena; clonal structure in parthenogenetic species.

(ii) Within the field of *systematics*, this technique can aid in delineating species boundaries in sexual and parthenogenetic species (detecting sympatric or allopatric species); in the recognition of species complexes; recognition of hybrid zones; in phylogenetic reconstruction, especially for closely related species or genera (Avise, 1994; Buth, 1984). The technique is specially valuable when sibling species are involved (Gómez et al., 1995; King, 1980), as might be the case in many rotifer taxa (Gomez & Snell, 1996; Serra et al., 1997).

(iii) *in ecological studies*, including the growing field of ecological genetics: for example, in the association of certain genotypes with certain environments, to study clonal vertical migration; experiments on competition (Snell, 1979).

(iv) Finally, in *applied studies*, allozyme electrophoresis, due to its simplicity and low cost, can be used routinely for genetic identification, i.e., to check for the purity of stock clones employed in aquaculture or in laboratory research. The electrophoretic profiles of the clones used in the laboratory are established when they are obtained, and can be checked at regular intervals during the maintenance of these clones. This measure is advisable when several clones are kept together and, therefore, cross-contamination is possible.

In summary, protein electrophoresis has the potential to answer many questions in rotifer ecology, systematics and population structure, and, despite the breakthrough of DNA techniques, it is still a promising approach.

Acknowledgments

I am most grateful to Drs Manuel Serra, David Blesa, Paul Shaw and Lorenz Hauser for their many helpful suggestions and comments on previous versions of the manuscript.

Appendix 1: Recipes

Here we provide a number of enzyme staining recipes that have been adapted to give good results with rotifers using acetate gel electrophoresis and Tris–glycine as running buffer (see text). Quaternary structure and number of loci usually identified are also indicated. The following enzyme systems have been tried on rotifers and have given poor results: AK, GDH, HBDH, IDH, SDH, AO, AMY, HEX, CHO (see references in Table 1). In the recipes, >1 drop= is equivalent to 20 μl (see below for solution concentrations). Melted agar is added last, and the staining mixture immediately poured over the gel and kept in the dark at room temperature until the banding pattern appears.

Phosphoglucose isomerase (PGI; EC 5.3.1.9) dimer, 1 locus 0.75 ml Tris–HCl, pH 8.0; 1.5 ml NADP; 3 drops fructose-6-P; 3 drops MTT; 3 drops PMS**; 10 μl G6PDH**; 1.8 ml agar

Phosphoglucomutase (PGM; EC 2.7.5.1) monomer, 1 locus 0.75 ml Tris–HCl, pH 8.0; 1.5 ml NADP; 3 drops mgCl$_2$; 7 drops glucose-1-P; 5 drops MTT; 5 drops PMS**; 20 μl G6PDH**; 1.8 ml agar

Malate dehydrogenase (MDH; EC 1.1.1.37) dimer, 2-3 loci. 0.75 ml Tris–HCl, pH 8.0; 1.5 ml NAD; 3 drops mgCl$_2$; 12 drops malic substrate; 5 drops MTT; 5 drops PMS**; 1.8 ml agar

Malic enzyme (ME; EC 1.1.1.40) dimer? tetramer, 1 locus. 0.75 ml Tris–HCl, pH 8.0; 1.5 ml NADP; 2 drops mgCl$_2$; 12 drops malic substrate; 5 drops MTT; 5 drops PMS**; 1.8 ml agar

Lactate dehydrogenase (LDH; EC 1.1.1.27) tetramer, 1 locus. 0.75 ml Tris–HCl, pH 7.0; 1.5 ml NAD; 10 drops L-lactic acid; 5 drops MTT; 5 drops PMS**; 1.8 ml agar

Fumarate hydratase (FUM; EC 4.2.1.2) tetramer, 1 loci. 0.75 ml Tris–HCl, pH 8.0; 1.5 ml NAD; 12 drops fumaric acid (adjusted to pH 8.0); 5 drops MTT; 5 drops PMS**; 50 μl MDH**; 1.8 ml agar

Arginine phosphoquinase (APK; EC 2.7.3.3) monomer, 3 loci. 0.5 ml Tris–HCl, pH 8.0; 1.5 ml NADP; 5 drops mgCl$_2$; 100 μl phospho-L-arginine; 5 drops solution ADP+D-glucose; 5 drops MTT; 5 drops PMS**; 10 μl HEX**; 10 μl G6PDH**; 1.8 ml agar

Phosphogluconate dehydrogenase (PGD; EC 1.1.1.44) dimer, 1 locus. 0.75 ml Tris–HCl, pH 8.0; 1.5 ml NADP; 5 drops 6-phosphogluconic acid; 5 drops mgCl$_2$; 5 drops MTT; 5 drops PMS**; 1.8 ml agar.

** Add just after removing the plate from the electrophoresis tank. before adding the agar.

Appendix 2: Product preparation

Keep all solutions in the refrigerator on a suitable plastic tray for ready use. Use droppers and micropipettes to prepare staining solutions just before staining in small glass vials.

Unless otherwise stated, codes within parentheses after product names are Sigma catalog numbers.

Buffers

Tris–glycine, pH 8.5 (running buffer); 3 g/l Trizma base (T-1503) and 14.4 g/l glycine (G-7126). Check for pH. Prepare 2 l.

Tris–glycine, pH 8.5+10% (w/v) sucrose (homogenizing buffer). Prepare as above and add the sucrose.

Tris–HCl, pH 7.0 (staining buffer); 1.11 g Trizma base and 8.75 ml 1 M HCl. Adjust pH if needed and make up to 100 ml with distilled water.

Tris–HCl, pH 8.0 (staining buffer); 1.11 g Trizma base and 6.2 ml 1 M HCl. Adjust pH if needed and make up to 100 ml with distilled water.

Tris–HCl, pH 9.0 (staining buffer); 2.465 g Trizma base and 3.0 ml 1 M HCl. Adjust pH if needed and make up to 100 ml with distilled water.

General products

Agar (bacterial grade); 4.0 g agar in 250 ml distilled water. Melt it in a microwave oven (2–3 min) or in an autoclave (20 min, 1 atm). Store at 60EC between uses.

NADP (N-0505), 2 mg/ml. Prepare 100–200 ml.
NAD (N-7381), 2 mg/ml. Prepare 100–200 ml.
MgCl$_2$, 20 mg/ml. Prepare 15 ml in a dropper.
MTT (M-2128), 10 mg/ml. Prepare a stock solution (100 ml) and pour 15 ml in a dropper. (keep MTT and prepare solution at minimum light conditions).
PMS (P-9625), 2 mg/ml. Prepare a stock solution (100 ml) and pour 15 ml in a dropper (keep PMS and prepare solution at minimum light conditions).

Substrates, cofactors and other staining reagents

ADP (A-2754), 10 mg/ml. Prepare 15 ml and place them in a dropper.

DL-Lactic acid (L-1250). Place 15 ml in a dropper.

D-Glucose (G-8270), 315 mg/ml. To prepare for APK in an ADP solution.

Fructose-6-P (F-3627), 20 mg/ml. Prepare 15 ml and place them in a dropper.

Fumaric acid (F-1506), 100 mg/ml. Prepare 15 ml and place them in a dropper.

Glucose-1-P (grades III and VI in equal amounts; G-7000 and G-1259), 50 mg/ml. Prepare 15 ml and place them in a dropper.

Malic substrate. Stock solution is prepared with 180 ml water; 20 Tris–HCl, pH 9.0; 3.68 g sc l-malic acid (M-9138). Adjust to pH 8.0. Place 15 ml in a dropper.

Phospho-L-arginine (P-5139), 20 mg/ml. Prepare 15 ml and place them in a dropper.

6-Phosphogluconic acid (P-7877), 20 mg/ml. Prepare 15 ml and place them in a dropper.

Enzymes

G-6-PDH (G-6378), 200 units/ml.
MDH (M-7383), 2750 units/ml.
HEX (H-5500), 2700 units/ml.

Appendix 3

Laboratory equipment and supplies necessary or desirable for setting up an acetate gel allozyme electrophoresis laboratory.

A. Sample preparation and storage

Nytal mesh (30 and 50 Fm) and plastic syringes to filter rotifers; 1.5 and 0.5 ml centrifuge tubes; scissors, to cut Nytal mesh into 1 cm diameter filters; forceps, to handle mesh filters; parafilm[7] (to collect rotifer samples after filtering); small laboratory spatula (to place rotifer sample into Eppendorf tubes); homogenizers (made in the lab or commercially available); centrifuge (for pelleting debris), up to 13–15,000 g; ultra-cold freezer (only needed if samples are not processed immediately).

B. Gel preparation and electrophoretic separation

Bufferizers (commercially available, a plastic or glass 1 l container is also suitable) to soak the gels; application kit (Helena Electrophoresis, Super Z-12); gel electrophoresis tanks (platinum electrode); power supply; filter paper (for the wicks, blotting the gels after soaking and general use); applicator kit; cellulose acetate gels; millipore forceps (to handle acetate gels without damaging them).

392

C. Staining reagent preparation and scoring

Refrigerator (41C) with a freezing unit (−201C); analytical balance (0.1 mg–100 g); 15 ml glass vials (for stain mixtures); 15 ml plastic dropper bottles (to prepare staining solutions); pH meter; stir plate; pipette set and disposable pipette tips; disposable rubber gloves; microwave oven or autoclave (to melt agar and sterilize glassware); incubation oven (to keep agar melted and dry gels); light box (for scoring gels); dark boxes or drawers for gel incubation; general laboratory glassware (including petri dishes, Erlenmeyer flasks, 1 l bottles and volumetric flasks).

References

Avise, J. C., 1974. Systematic value of electrophoretic data. Syst. Zool. 23: 465–481.

Avise, J. C., 1994. Molecular Markers, Natural History and Evolution. Chapman and Hall, New York: 511 pp.

Ayala, F. J., 1976. Molecular Evolution. Sinauer, Sunderland, MA

Buth, D. G., 1984. The application of electrophoretic data in systematic studies. Ann. Rev. Ecol. Syst. 15: 501–522.

Carvalho, G. R., 1988. Differences in the frequency and fecundity of PGI-marked genotypes in a natural population of Daphnia magna (Crustacea: Cladocera). Func. Ecol. 2: 453–62.

Freifelder, D. 1981. Técnicas de Bioquímica y Biología Molecular. Editorial Reverté, Barcelona: 631 pp.

Fu, Y., K. Hirayama & Y. Natsukari, 1991. Genetic divergence between S and L type strains of the rotifer Brachionus plicatilis O.F. Müller. J. exp. mar. Biol. Ecol. 151: 43–56.

Fu, Y., Y. Natsukari & K. Hirayama, 1990a. A preliminary study on genetics of two types of the rotifer Brachionus plicatilis. In R.S. Svrjcek (ed.), Genetics in Aquaculture: Proceeedings of the Sixteenth US–Japan Meeting on Aquaculture, 20–21 October 1987, Charleston, South Carolina NOAA Technical report NMFS 92.

Fu Y., Y. Natsukari & K. Hirayama, 1990b. Strains of the rotifer Brachionus plicatilis having particular patterns of isozymes. In R. Hirano & I. Hanyu (eds), The Second Asian Fisheries Forum. Asian Fisheries Society, Manila, Philippines: 37–40.

Gómez, A. & T. W. Snell, 1996. Sibling species and cryptic speciation in the Brachionus plicatilis species complex (Rotifera). J. Evol. Biol. 9: 953–964.

Gómez, A., M. Temprano & M. Serra, 1995. Ecological genetics of a cyclical parthenogen in temporary habitats. J. Evol. Biol. 8: 601–622.

Harris, H., 1966. Enzyme polymorphism in man. Proc. r. Soc. Lond. B164: 298–310.

Hebert, P. D. N., 1987a. Genotypic characteristics of cyclic parthenogens and their asexual derivatives. In S.C. Stearns (ed.), The Evolution of Sex and its Consequences. Birkhauser Verlag, Basel: 175–195

Hebert, P. D. N., 1987b. Genotypic characteristics of the Cladocera. Hydrobiologia 145 (Dev. Hydrobiol. 35): 183–193.

Hebert, P. D. N. & M. J. Beaton, 1989. Methodologies for allozyme analysis using cellulose acetate electrophoresis: a practical handbook. Helena Laboratories, Beaumont, TX: 32 pp.

Hedrick, P. W., 1985. Genetics of Populations. Jones and Bartlett, Boston: 629 pp.

Hubby, J. L. & R. C. Lewontin, 1966. A molecular approach to the study of genic heterozygosity in natural populations. I. The number of alleles at different loci in Drosophila pseudoobscura. Genetics 54: 577–594.

King, C. E., 1977. Genetics of reproduction, variation and adaptation in rotifers. Arch. Hydrobiol. Ergeb. Limnol. 8: 187–201.

King, C. E., 1980. The genetic structure of zooplankton populations. In W. C. Kerfoot, (ed.), Evolution and Ecology of Zooplankton Communities. The University Press of New England, Hanover, (NH); 315–328.

King, C. E. & T. W. Snell, 1977. Sexual recombination in rotifers. Heredity 39: 357–360.

King, C. E. & Y. Zhao, 1987. Coexistence of rotifer (Brachionus plicatilis) clones in Soda Lake, Nevada. Hydrobiologia 147 (Dev. Hydrobiol. 42): 57–64.

Kreitman, M., 1983. Nucleotide polymorphism at the alcohol dehydrogenase locus of Drosophila melanogaster. Nature 304: 412–417.

Lewontin, R. C. & J. L. Hubby, 1966. A molecular approach to the study of genic heterozygosity in natural populations. II. Amount of variation and degree of heterozygosity on natural populations of Drosophila pseudoobscura. Genetics 54: 595–609.

May, B., 1992. Starch gel electrophoresis of allozymes. In A. R. Hoetzel (ed.), Molecular Genetic Analysis of Populations: a Practical Approach. IRL Press at Oxford University Press, Oxford: 1–27.

Murphy, R. W., J. W. Sites, D. G. Buth & C. H. Haufter, 1996. Proteins: isozyme electrophoresis. In D. M. Hillis, Moritz C. & Mable B. K. (eds), Molecular Systematics, 2nd edn. Sinauer Associates, Sunderland, MA, USA.

Pagani, M., C. Ricci & A. M. Bolzern, 1991. Comparison of five strains of a parthenogenetic species, Macrotrachela quadricornifera (Rotifera, Bdelloidea). II. Isoenzymatic patterns. Hydrobiologia 211: 157–163.

Pourriot, R., 1977. Food and feeding habits of Rotifera. Arch. Hydrobiol. Beib. Ergebn. Limnol. 8: 243–260.

Ricci, C., M. Pagani & A. M. Bolzern, 1989. Temporal analysis of clonal structure in a moss bdelloid population. Hydrobiologia 186/187 (Dev. Hydrobiol. 52): 145–152.

Richardson, B. J., P. R. Baverstock & M. Adams, 1986. Allozyme Electrophoresis. Academic Press, Sydney, Australia: 410 pp.

Rothe, G. M., 1994. Electrophoresis of Enzymes. Laboratory Methods. Springer Verlag, Berlin: 307 pp.

Selander, R. K., 1976. Genic variation in natural populations. In F. J. Ayala (ed.), Molecular Evolution. W. H. Freeman, San Francisco, CA: 284–332.

Serra, M. & M. R. Miracle, 1985. Enzyme polymorphism in Brachionus plicatilis populations from several Spanish lagoons. Verh. int. Verein. Limnol. 22: 2991–2996.

Serra, M., A. Galiana & A. Gómez, 1997. Speciation in monogonont rotifers. Hydrobiologia 358 (Dev. Hydrobiol. 124): 63–70.

Shaw, C. R., 1970. How many genes evolve? Biochem. Genet. 4: 275–283.

Snell, T. W., 1979. Intraspecific competition and population structure in rotifers. Ecology 60: 494–502.

Snell, T. W. & B. C. Winkler, 1984. Isozyme analysis of rotifer proteins. Biochem. Syst. Ecol. 12: 199–202.

Wallace, R. L. & T. W. Snell, 1991. Rotifera. In Ecology and Classification of North American Invertebrates. Academic Press, New York: 187–248.

Walsh, E. J., 1993. Rotifer genetics: integration of classic and modern techniques. Hydrobiologia 255/256 (Dev. Hydrobiol. 83): 193–204.

Wolf, H. G., 1982. A comparison of different electrophoretic techniques for the detection of isoenzymes in single Daphnids. Arch. Hydrobiol. 95: 521–531.

Wynne, I. R., H. D. Loxdale & C. F. Brookes, 1992. Use of a cellulose acetate system for allozyme electrophoresis. In Berry, R. J., T. J. Crawford & G. M. Hewitt (eds), Genes in Ecology. Blackwell Scientific Publications, Oxford: 534 pp.

Hydrobiologia **387/388**: 395–402, 1998.
E. Wurdak, R. Wallace & H. Segers (eds), Rotifera VIII: A Comparative Approach.
© *1998 Kluwer Academic Publishers.*

Measurements of the genome size of the monogonont rotifer *Brachionus plicatilis* and of the bdelloid rotifers *Philodina roseola* and *Habrotrocha constricta*

David B. Mark Welch & Matthew Meselson
Department of Molecular and Cellular Biology, Harvard University, Cambridge, MA 02138, U.S.A.

Key words: Rotifera, bdelloid, monogonont, genome size, DNA hydridization

Abstract

Genome size may be determined as the mass of genomic DNA per copy of a given sequence, multiplied by the number of copies of that sequence in the genome. Practical application of this relationship may be made by hybridizing a radiolabeled cloned segment of the genome to a known number of copies of the segment and to a known mass of genomic DNA separately immobilized on the same membrane. The ratio of the hybridization intensity per copy of the segment to the hybridization intensity per unit mass of genomic DNA is then taken to be the mass of genomic DNA per hybridizing sequence present in the genome. This ratio multiplied by the number of hybridizing sequences in the genome, determined by other means, is taken as the genome size. Employing this procedure with segments of the *hsp82* heat shock gene cloned from the monogonont rotifer *B. plicatilis* and from the bdelloid rotifers *P. roseola* and *H. constricta*, we estimate their genome sizes as 0.7, 2.2 and 1.0 pg, respectively.

Introduction

Rotifers of the Class Bdelloidea appear to reproduce only from unfertilized eggs (Donner, 1965; Ricci, 1987). Inhabiting stable and ephemeral fresh water habitats throughout the world, the class comprises some 350 named species, making it the largest and most diverse taxon for which mixis may be altogether lacking (Bell, 1987; Mayr, 1963). If bdelloids have evolved without sexual reproduction, they would offer a unique opportunity to identify the factors responsible for the otherwise near-universal maintenance of sexual reproduction in plants and animals.

As part of an investigation of the genomes of bdelloid rotifers, we have determined the genome sizes of two bdelloid species from different families, *Philodina roseola* Ehrenberg (Philodinidae) and *Habrotrocha constricta* Dujardin (Habrotrochidae), and of the facultatively sexual rotifer of the Class Monogononta, *Brachionus plicatilis* Müller.

In the general procedure followed, a series of aliquots of a cloned segment of the *hsp82* heat shock gene and of genomic DNA were spotted on the same nylon membrane and hybridized with an excess of the radiolabeled segment. The hybridization intensity was found to be linearly related to the number of segments or the mass of genomic DNA applied to each spot, and the slopes of the corresponding least-squares fitted lines were taken as the best estimates of hybridization intensity per cloned segment or per unit mass of genomic DNA. We then calculated genomic DNA content as the ratio of the hybridization intensity per cloned segment to the hybridization intensity per unit mass of genomic DNA, multiplied by the number of sequences in the genome that hybridize.

In work to be reported elsewhere, we have found two *hsp82* sequences in the monogonont *B. plicatilis* and, as expected for a sexually reproducing diploid, they are nearly identical, differing by only 0.3%. In contrast, we have not found closely similar *hsp82* sequences in either bdelloid species. Instead, we find four divergent sequences in *P. roseola*, designated *hsp82*-A, *hsp82*-A′, *hsp82*-B and *hsp82*-C, and three divergent sequences in *H. constricta*, designated *hsp82*-1, *hsp82*-2 and *hsp82*-3. The *hsp82*-A sequence differs from *hsp82*-B and *hsp82*-C by 13%, and these two sequences differ from each other by 2.7%; the *hsp82*-A and *hsp82*-A′ sequences differ from each other by 1.4%. The *hsp82*-1 sequence differs from *hsp82*-2 and *hsp82*-3 by 8%, which differ from each other by 2.6%.

Based on evidence cited in the discussion that each of the *hsp82* sequences occurs only once in the corresponding genome and that there are no other sequences in the genome to which they hybridize, we find the genomic DNA content of *B. plicatilis*, *P. roseola* and *H. constricta* to be 0.7, 2.2 and 1.0 pg, respectively.

Our method for determining genome size is essentially the reverse of the method used by Rivin et al. (1986) to determine the number of copies of a multicopy gene in a genome of known size.

Methods

Rotifer culture

We purchased *Philodina roseola* Ehrenberg from Carolina Biological Supply (Burlington, NC). One of us (MM) isolated *Habrotrocha constricta* from ground moss in Sandwich, Massachusetts. Dr Terry Snell provided resting eggs of *Brachionus plicatilis* Müller. Shortly after initial culturing, clonal cultures of each species were established from an individual rotifer that had been passed through at least 10 serial transfers. Final lines of *P. roseola* and *H. constricta* were derived from single eggs from these clonal isolates and have been cultured continuously since 1989 and 1993, respectively.

P. roseola and *H. constricta* are cultured in spring water (Belmont Springs, Lexington, MA), and *B. plicatilis* in artificial salt water (Snell et al., 1991). Media are sterilized by vacuum filtration through 0.22 μM cellulose acetate filters. Rotifers are cultured in 1–4 l glass beakers. Glass beakers are cleaned with detergent and Clorox bleach and then rinsed thoroughly; a magnetic stir bar is added, the opening is covered with six to eight layers of cheese cloth topped with a layer of aluminum foil, and the assemblage is sterilized by autoclaving. After sterilization the foil is removed and replaced with several layers of paper tissue.

Rotifers are fed a non-motile strain of *E. coli* grown with aeration by shaking at 37°C to OD_{600} 2.0–2.5 in 500 ml Luria–Bertani Medium (0.5% bacto-yeast extract, 1% bacto-tryptone, 0.5% NaCl, pH 7.0). Bacteria are concentrated by centrifugation and resuspended in 50 ml spring water, stored at 4°C, and used within 5 days. At each feeding *P. roseola* cultures are given ca 10 ml concentrated bacteria per liter; *H. constricta* and *B. plicatilis* cultures are given ca 3 ml/l. After feeding, cultures are magnetically stirred for 5 min, and cultures of *B. plicatilis* are further aerated by bubbling. Feeding is repeated when the bacterial concentration is considerably reduced as judged by reduced turbidity. Bdelloid cultures are kept at room temperature (17–25°C); *B. plicatilis* cultures are kept at 18°C on a 12L:12D light cycle.

Medium is changed every 7–10 days by passing the culture through ethanol-sterilized Nitex nylon screen (pore size 25 μM, catalogue number 3-25/16, Tetco Inc, Briarcliff Manor, NY) cut and folded to fit a sterile funnel. Rotifers collected on the screen are rinsed with 0.5–1 l of medium and washed into a beaker containing fresh medium.

Every 4–5 weeks bdelloid cultures are cleaned according to the following protocol: the rotifers are rinsed, resuspended in 500 ml 50 mM NaCl, 10 mM Tris (dispensed from 1 M stock, pH 8.0), 0.04 mg/ml lysozyme. After stirring for 5 min, sodium dodecyl sulfate (SDS, electrophoresis grade, BioRad, Richmond CA) is added to 0.1% and stirring is continued 1–2 min. The solution is then poured through ethanol-sterilized Nitex screen and the bdelloids on the screen are rinsed with 3–4 l of spring water. The cleaned rotifers are then washed into a beaker containing fresh spring water. Bdelloids survive this treatment; however, the use of a lower grade of SDS can greatly reduce survival.

Prior to DNA extraction, bdelloids are cleaned and washed into beakers containing fresh spring water, starved for 1–2 days, cleaned a second time, and washed into 50-ml polypropylene centrifuge tubes. After brief centrifugation the supernatant is quickly decanted and the rotifers are immediately frozen at −70°C. Dense 1 l cultures of *H. constricta* yield ca 0.7 g (wet weight), which includes rotifers, eggs and rotifer debris. Dense 3–4 l cultures of *P. roseola* yield at least 5 g.

Unlike the bdelloid rotifers, *B. plicatilis* does not survive treatment with SDS. Prior to collection for DNA extraction the rotifers, which have not been given food for 2 days, are rinsed and washed into artificial salt water in clean beakers, starved overnight, and rinsed again. Because these rotifers quickly resume swimming after centrifugation, they are not washed from the screen after final rinsing but are collected as a paste from the screen with a razor blade, placed in a 50-ml polypropylene centrifuge tube, compacted by brief centrifugation, and immediately frozen at

−70°C. Dense 1 l cultures of *B. plicatilis* yield ca 0.4 g (wet weight).

Cultures are periodically sampled and examined by phase-contrast microscopy or UV fluorescence microscopy in the presence of 1 μg/ml DAPI (4′, 6-diamidino-2-phenylindole). Contaminated cultures are destroyed.

Genomic DNA Preparation

Frozen rotifers are ground to a fine powder in a mortar and pestle under liquid nitrogen. The frozen powder is quickly poured into 50-ml polypropylene tubes containing 9.5 ml digestion buffer (100 mM NaCl, 50 mM EDTA, 50 mM Tris, pH 8.0), placing no more than 1.5 g in each tube. After gentle swirling and addition of 250 μl 20% SDS and 250 μl proteinase K solution (from a fresh stock solution of 2 mg/ml), the mixture is incubated at 55°C for 1.5–2 h with occasional gentle rocking. For *B. plicatilis*, the mixture is extracted once with an equal volume of 1:1 phenol:chloroform to remove material that otherwise precipitates at high CsCl concentration.

To each 10 ml of the digestion mixture is added 12.54 g CsCl and 300 μl of 1% ethidium bromide. After the CsCl is dissolved, the mixture is loaded into Quick-Seal Ultracentrifuge tubes (Beckman, Palo Alto, CA) by pouring through a syringe body fitted with a 16-gauge needle. The tubes are sealed and placed in a Beckman 80Ti rotor and centrifuged at 45 000 rpm (140 000 × g) for 55–66 h. After centrifugation the DNA band in each tube is visualized with long-wave UV light and removed after puncturing the tube with a 16-gauge needle fitted to a syringe. Ethidium bromide is extracted with isoamyl alcohol and CsCl is removed by dialysis against TE (10 mM Tris, 1 mM EDTA, pH 8.0). Approximately 300 μg of DNA is obtained per gram of frozen rotifer powder. DNA solutions are stored in TE at 4°C.

Preparation of hsp82 DNA segments and radiolabeled probe

We have cloned and sequenced *hsp82* heat shock genes from *B. plicatilis*, *P. roseola* and *H. constricta* (Mark Welch and Meselson, to be published). A ca 900-base pair region of each clone was amplified by PCR, and the amplified segments were electrophoretically separated on an agarose gel, isolated using GeneClean (Bio 101, La Jolla CA), and kept in TE at 4°C. Concentrations of purified segments were determined relative to DNA mass standards by comparisons of ethidium fluorescence of bands on electrophoretic gels.

Radiolabeled probes were made from *hsp82* segments by random primed DNA synthesis in the presence of [α^{32}P]dATP. Unincorporated nucleotides were removed by centrifugal gel-filtration through a column of G-25 Sephadex (Sigma, St Louis MO). Probe activity was above 1×10^9 cpm/μg.

Preparation of membranes

Dilutions of each *hsp82* segment were made in 6× SSC, 0.1 μg/ml bovine serum albumin and 1 μg/ml sonicated salmon sperm DNA, added as carrier (SSC is 0.15 M NaCl, 0.015 M sodium citrate). Pipet tips used in making dilutions were pre-soaked in the same solution. Genomic DNA was precipitated with ethanol, resuspended in 6× SSC and its concentration determined by UV spectrophotometry. Dilutions were then made as above.

Dilutions of *hsp82* segments and of genomic DNA were kept at 95°C for 5 min, chilled on ice, and applied in 1 ml aliquots to nylon membranes (BIO-TRANS, ICN Biomedicals, Costa Mesa, CA). In some cases, spots were built up from multiple aliquots, allowing time for drying between applications. After all applications had dried, membranes were kept 3 min on two layers of heavy filter paper soaked in 0.5 M NaOH, 1.5 M NaCl and 3 min on fresh filter paper soaked in the same solution. After 5 min on filter paper soaked in 0.5 M Tris, pH 8.0, 1.5 M NaCl, membranes were kept 2 min on filter paper soaked in 2× SSC and moved to dry filter paper. After drying, membranes were irradiated with UV light and baked 30 min at 80°C.

For each rotifer species, membranes were spotted with several different amounts of genomic DNA and with one or more series of spots containing different numbers of *hsp82* segment.

For *B. plicatilis* a series of spots containing the segment amplified from one of the two cloned *hsp82* sequences was applied to each of two membranes. In addition, two spots (10 and 1000 ng) of sonicated heat-denatured salmon sperm DNA were applied to one of the membranes.

For bdelloids, four separate series of spots containing *hsp82* segments were applied to each membrane: three series each containing segments of only one of three divergent sequences (*hsp82*-A, *hsp82*-B, and *hsp82*-C for *P. roseola*; *hsp82*-1, *hsp82*-2, and *hsp82*-3 for *H. constricta*) and a fourth series made up of spots which contained segments of all three sequences in equal amounts. In addition, one spot containing 1 μg of genomic DNA and 10^6 segments of hsp82-A or hsp82-1 was applied to the *P. roseola* and *H. constricta* membranes, respectively. One such membrane was prepared for *P. roseola* and two for *H. constricta*.

Hybridization and quantitation of hybridization intensity

Before hybridization, membranes were placed in 6× SET, 0.5% SDS, 5× Denhardt's (50× Denhardt's is 1% Ficoll, 1% polyvinylpyrrolidone, 1% bovine serum albumin; 20× SET is 3 M NaCl, 0.6 M Tris, pH 8.0, 0.02 M EDTA). After raising the temperature to 63°C, heat-denatured sonicated salmon sperm DNA was added to 0.1 mg/ml. After 2–4 h the solution was replaced with 6× SET, 0.1% SDS, 1× Denhardt's, pre-warmed to 63°C. Heat-denatured sonicated salmon sperm DNA was then added to 0.1 mg/ml, and heat-denatured probe to 1–2 ng/ml. Hybridization was conducted for 20 h at 63°C (standard stringency) or 68°C (high stringency). Membranes were washed at 65°C four times in 500 ml of either 2× SSC, 0.1% SDS (standard stringency) or 0.33× SSC, 0.1% SDS (high stringency), rinsed in 2× SSC and wrapped in plastic wrap.

B. plicatilis membranes were hybridized at standard stringency with probe made from the same *B. plicatilis* hsp82 segment used for spotting. After measurement of hybridization intensity, probe was completely removed from one membrane by washing twice in 2 l of 0.1% SDS at 85°C for 20 min. The membrane was then hybridized and washed at high stringency. The *P. roseola* membrane was hybridized at high stringency with hsp82-A probe and at standard stringency with a mixture of hsp82-A and hsp82-B. The *H. constricta* membranes were hybridized at standard stringency with a mixture of hsp82-1 and hsp82-2 which had been labeled under identical conditions. At standard stringency the hsp82-2 probe hybridizes with equal efficiency to itself and to hsp82-3 but less efficiently to hsp82-1. At high stringency hsp82-1 probe cross-hybridizes appreciably to hsp82-2 and hsp82-3.

Membranes were placed on a Fuji phosphor-imaging plate and hybridization intensities were measured in a Fuji BAS 2000 Bio-Imaging Analyzer. The integrated intensity of hybridization over each spot, corrected for background, was determined for two exposures of each hybridization, using the manufacturer's software.

Results

There was no evidence of hybridization of hsp82 probe to spots containing only salmon sperm DNA, indicating that the carrier DNA does not contribute to the hybridization intensity. Also, the hybridization intensity from spots containing genomic DNA plus a hsp82 segment equaled the sum of the hybridization intensities for the two DNA samples applied as separate spots, showing that hsp82 sequences in the genomic spots hybridize with the same efficiency as those in the segment spots.

Brachionus plicatilis

Figure 1a shows the intensity of hybridization by the *B. plicatilis* hsp82 probe plotted against the number of *B. plicatilis* hsp82 segments applied to the membrane hybridized and washed at high stringency. Figure 1b shows hybridization intensity in the same units plotted against the applied mass of genomic DNA for the same exposure of the membrane. The respective slopes are 6.5 hybridization units/10^6 hsp82 segments and 18.5 hybridization units/μg of genomic DNA, giving a ratio of 0.35 pg of genomic DNA per hsp82 segment. Measurements from an independent exposure of the membrane from this hybridization gave a ratio of 0.36. Two exposures of this membrane hybridized at standard stringency gave ratios of 0.35 and 0.34, and two exposures of the other membrane hybridized at standard stringency gave ratios of 0.38 and 0.35. The overall average of these values is 0.36 pg of genomic DNA per hsp82 segment.

Philodina roseola

Figure 2a shows the intensity of hybridization by the hsp82-A probe plotted against the number of hsp82-A segments present in spots of hsp82-A alone and in spots containing three *P. roseola* hsp82 sequences, measured on a membrane hybridized and washed at high stringency. The slopes of the resulting lines are essentially the same: 6.1 and 6.0 hybridization units per 10^6 segments for the hsp82-A series and for the series containing all three hsp82 sequences, respectively. No hybridization was detected to the series of spots containing only hsp82-B or hsp82-C.

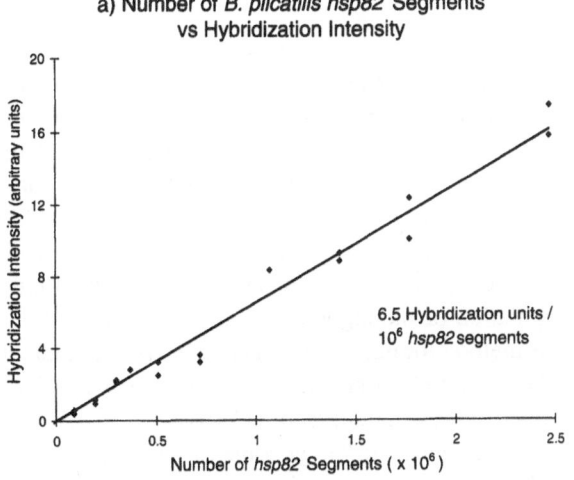

a) Number of *B. plicatilis hsp82* Segments
vs Hybridization Intensity

6.5 Hybridization units /
10^6 *hsp82* segments

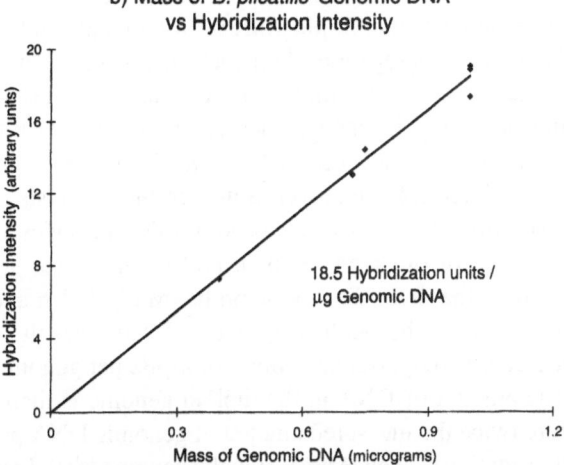

b) Mass of *B. plicatilis* Genomic DNA
vs Hybridization Intensity

18.5 Hybridization units /
μg Genomic DNA

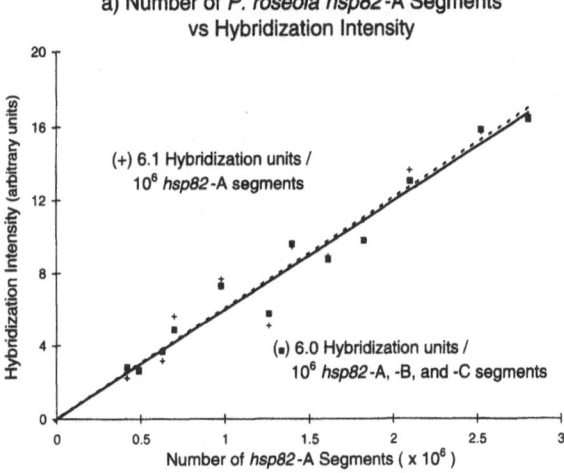

a) Number of *P. roseola hsp82*-A Segments
vs Hybridization Intensity

(+) 6.1 Hybridization units /
10^6 *hsp82*-A segments

(■) 6.0 Hybridization units /
10^6 *hsp82*-A, -B, and -C segments

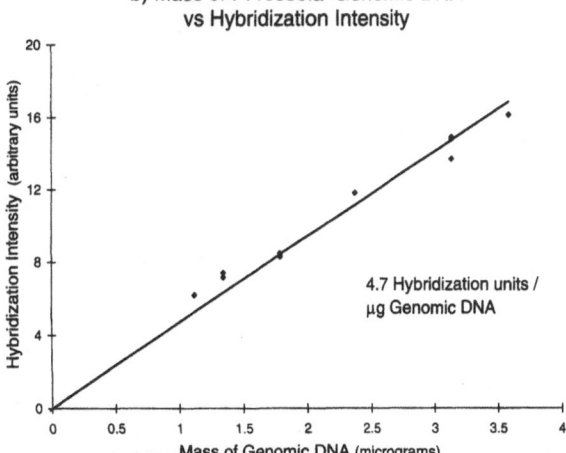

b) Mass of *P. roseola* Genomic DNA
vs Hybridization Intensity

4.7 Hybridization units /
μg Genomic DNA

Figure 1. Hybridization intensity of spots containing (a) *B. plicatilis hsp82* segments and (b) *B. plicatilis* genomic DNA, from a membrane hybridized at high stringency with *B. plicatilis hsp82* probe, measured in arbitrary units. The least-squares linear regression is shown on each plot with the slope of the line expressed as the number of hybridization units per unit of DNA.

Figure 2. Hybridization intensity of spots containing (a) *P. roseola hsp82*-A segments and (b) *P. roseola* genomic DNA, from a membrane hybridized at high stringency with a probe made from *hsp82*-A, measured in arbitrary units. (a) The hybridization intensity for the series of spots containing *hsp82*-A alone (+) and the hybridization intensity of spots containing segments of three sequences in equal amounts (*hsp82*-A, -B and -C, ■). No signal above background was detected for any of the spots containing only *hsp82*-B or *hsp82*-C segments. The least-squares linear regression is shown on each plot (dashed for the *hsp82*-A series) with the slope of each line expressed as the number of hybridization units per unit of DNA.

Figure 2b shows hybridization intensity in the same units plotted against the applied mass of genomic DNA for the same exposure of the membrane. The slope is 4.7 hybridization units/μg genomic DNA, giving a ratio of 1.30 pg of genomic DNA per *hsp82*-A segment. Repeating these measurements for two additional exposures of the membrane gave ratios of 1.43 and 1.42. The overall average of these values is 1.4 pg of genomic DNA per *hsp82*-A segment.

Habrotrocha constricta

Figure 3a shows the intensity of hybridization by a mixture of the *hsp82*-1 and *hsp82*-2 probes plotted against the number of *hsp82* segments present in spots

containing *hsp82*-1, *hsp82*-2 or *hsp82*-3, measured on a membrane hybridized and washed at standard stringency. It is seen that the slopes of the three resulting lines are nearly the same, with an average value of 19.2 hybridization units/10^6 *hsp82* segments.

Figure 3b shows hybridization intensity in the same units plotted against the applied mass of genomic DNA for the same exposure of the membrane. The slope of the resulting line is 54.1 hybridization units/μg genomic DNA, giving a ratio of 0.35 pg

a) Number of *H. constricta* *hsp82*-1, -2, and -3 Segments vs Hybridization Intensity

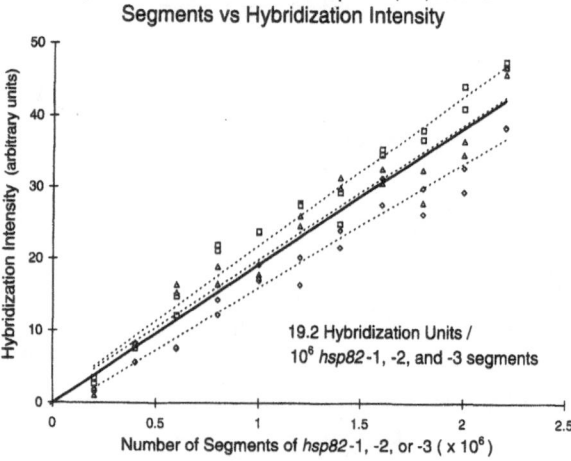

19.2 Hybridization Units /
10^6 *hsp82*-1, -2, and -3 segments

b) Mass of *H. constricta* Genomic DNA vs Hybridization Intensity

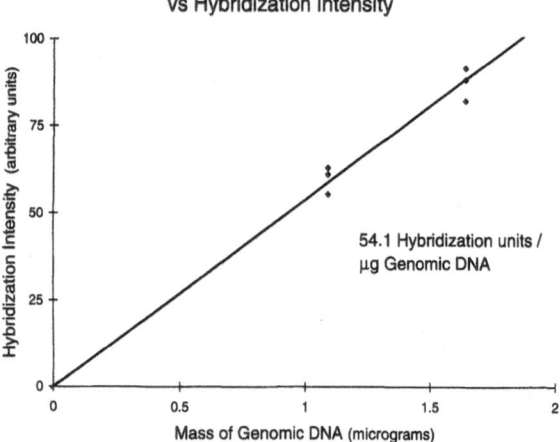

54.1 Hybridization units /
μg Genomic DNA

Figure 3. Hybridization intensity of spots containing (a) *H. constricta hsp82*-1, -2 or -3 segments and (b) *H. constricta* genomic DNA, from a membrane hybridized at standard stringency with an equal mixture of hsp82-1 and hsp82-2 as probes, measured in arbitrary units. (a) The relationship of hybridization intensity to number of segments applied is shown for *hsp82*-1 (open diamonds), *hsp82*-2 (open triangles), and *hsp82*-3 (open squares), with dashed lines for each least-squares linear regression. The least-squares linear regression of the average hybridization intensity of the three sequences for each amount of segment applied is shown as a solid line, with the slope of the line expressed as the number of hybridization units per unit of DNA.

Discussion

Determinations of genome size based on fluorometric or photometric measurements of nuclear DNA content require knowledge of the cell cycle stage and of the level of ploidy or polyteny. Both bdelloids and monogononts are known to have cells of differing nuclear DNA content (Jones & Gilbert, 1977; Pagani et al., 1993) and ploidy variation has been reported within a population of monogononts (Walsh & Zhang, 1992). Our method for determining genome size is independent of the number of copies of the genome in a nucleus. It does, however, require knowledge of the copy number of sequences in the genome that hybridize to the probe. Karyotype analysis (Rumengan et al., 1991) and the results of genetic crosses (Snell & Winkler, 1984) indicate that *B. plicatilis* is diploid and exhibits Mendelian segregation. As expected for alleles in a sexually reproducing diploid, sequencing of numerous independently isolated clones reveals only two different *hsp82* sequences in *B. plicatilis* and these are nearly identical. Moreover, Southern blot analysis of restriction digests of whole genome DNA provides no evidence of other copies of *hsp82* or of related sequences that hybridize to the probe (Mark Welch and Meselson, to be published). We conclude that there are only two hybridizing copies of *hsp82* per genome. The amount of DNA in the diploid genome is therefore twice the measured amount of genomic DNA per hybridizing sequence present in the genome, or 0.7 pg.

In the bdelloids *P. roseola* and *H. constricta*, sequencing of numerous independently isolated clones reveals four copies of *hsp82* in *P. roseola* and three copies in *H. constricta*. As with *B. plicatilis*, Southern blot analysis provides no evidence of additional copies of *hsp82* or of related sequences that hybridize to the probe (Mark Welch and Meselson, to be published).

The probe used for *P. roseola*, *hsp82*-A, does not hybridize to *hsp82*-B and *hsp82*-C at the higher stringency we used, so that there are two hybridizing sequences per genome. Therefore the amount of DNA per genome is twice the measured amount of genomic DNA per sequence that hybridizes to the probe, or 2.75 pg. Measurements at standard stringency with a mixture of *hsp82*-A and *hsp82*-B as probe gave an average value for the amount of DNA per genome of 1.72 pg. The overall average from all measurements is 2.2 pg.

In *H. constricta*, the three different *hsp82* copies are less divergent than those in *P. roseola* and we have not found hybridization conditions that allow their clear discrimination. We therefore used a probe

of genomic DNA per copy of each of the *hsp82* sequences. An additional exposure of this membrane and two independent exposures of a separate membrane gave ratios of 0.30, 0.28 and 0.33. The overall average of these values is 0.32 pg of genomic DNA per copy of each of the three *hsp82* sequences.

401

Table 1. Total nuclear DNA content reported in some vermiform taxa

Taxon[1]	Total nuclear DNA content (pg)	Reference
Rotifera		
Monogononta		
Brachionus plicatilis	0.7	This study
Bdelloidea		
Philodinidae		
Philodina roseola	2.2	This study
Macrotrachela quadricornifera	1.5	Pagani et al., 1993
Habrotrochidae		
Habrotrocha constricta	1.0	This study
Gastrotricha (8)[2]	0.3–1.3	Balsamo & Manicardi, 1995
Priapulida (*Halicryptus spinulosus*)	1.1	Schreiber, 1994
Nematoda (7)	0.2–1.3	Wood, 1988
Platyhelminthes: Turbellaria (4)	1.0–1.9	Martens et al., 1989
Tardigrada (15)	0.2–1.6	Garagna et al., 1996
Annelida: Polychaeta (11)[3]	0.8–2.3	Sella et al., 1993

[1] The number of species examined is indicated in parentheses where species are not specified.
[2] Including only those species for which both haploid and somatic DNA content was measured; seven of the eight species examined have genome sizes between 0.3 and 0.6 pg.
[3] Nine of 11 species examined have genome sizes between 0.8 and 0.9 pg.

containing *hsp82*-1, which hybridizes efficiently only to itself, and *hsp82*-2, which hybridizes efficiently both to itself and to *hsp82*-3. This makes three copies of the *hsp82* sequence per genome that hybridize to the mixed probe. Accordingly, the amount of DNA per genome is three times the measured amount of genomic DNA per hybridizing sequence, or 1.0 pg.

The only published estimate of genomic DNA content in rotifers of which we are aware is that of Pagani et al. (1993), who report a total G1 nuclear DNA content of 1.5 pg for *Macrotrachela quadricornifera*, based on feulgen staining of whole nuclei, similar to our estimate of 2.2 pg for *P. roseola*, a member of the same bdelloid family.

It may be of interest to compare rotifer genome sizes with reported 2C values for other taxa which may be loosely designated Vermes (e.g., Hyman, 1940; Kükenthal & Krumbach, 1931; Linnaeus, 1735). As may be seen in Table 1, the values we find for the monogonont and both of the bdelloids we have examined fall within the range reported for the other taxa.

Our calculations of the genome sizes of *P. roseola* and of *H. constricta* are based on the assumption that each of the *hsp82* sequences cloned from each of these species is present only once per genome. Direct measurements of total G1 nuclear DNA content giving values in agreement with the genome sizes we report here would provide confirmation for our evidence that bdelloids, unlike sexually reproducing diploids, do not posses allele pairs.

References

Balsamo, M. & G. C. Manicardi, 1995. Nuclear content in Gastrotricha. Experientia 51: 356–359.
Bell, G., 1987. The Masterpiece of Nature: the Evolution and Genetics of Sexuality. University of California Press, Berkeley: 635 pp.

Donner, J., 1965. Ordnung Bdelloidea. Akademie Verlag, Berlin: 297 pp.

Jones, P. A. & J. J. Gilbert, 1977. Polymorphism and polyploidy in the rotifer *Asplanchna sieboldi*: relative DNA contents in tissues of saccate and campanulate females. J. exp. Zool. 201: 163–168.

Garagna, S., L. Rebecchi & A. Guidi, 1996. Genome size variation in Tardigrada. J. linn. Soc., Zool. 116: 115–121.

Hyman, L., 1940. The Invertebrates, Vol. 1. McGraw-Hill, New York: 726 pp.

Kükenthal, W. & T. Krumbach, 1931. Handbuch der Zoologie, Vol. 2. Walter de Gruyter, Berlin.

Linnaeus, C., 1735. Systema Naturae. In Engel-Ledeboer, M. S. J. & H. Engel (eds), 1964. Dutch Classics on the History of Science, vol. 8. B. de Graaf, Nieuwkoop, The Netherlands: 47 pp.

Mayr, E., 1963. Animal Species and Evolution. Harvard University Press, Cambridge: 797 pp.

Martens, P. M., M. C. Curini-Galletti & P. Van Oostveldt, 1989. Polyploidy in Proseriata (Platyhelminthes) and its phylogenetic implications. Evolution 43: 900–907.

Pagani, M., C. Ricci & C. A. Redi, 1993. Oogenesis in *Macrotrachela quadricornifera* (Rotifera, Bdelloidea) I. Germarium eutely, karyotype and DNA content. Hydrobiologia 255/256: 225–230.

Ricci, C., 1987. The ecology of bdelloids: how to be successful. Hydrobiologia 147: 117–127.

Rivin, C. J., C. A. Cullis, & V. Walbot, 1986. Evaluating the quantitative variation in the genome of *Zea mays*. Genetics 113: 1009–1019.

Rumengan, I. F. M., H. Kayano & K. Hirayama, 1991. Karyotypes of S and L type rotifers *Brachionus plicatilis* O.F. Müller. J. exp. Mar. Biol. Ecol. 154: 171–176.

Schreiber, A., 1994. DNA content in blood cells of *Halicryptus spinulosus*, a species of the phylum Priapulida. Naturwissenschaften 80: 455–456.

Sella, G., C. A. Redi, L. Ramella, R. Soldi & M. C. Premoli, 1993. Genome size and karyotype length in some interstitial polychaete species of the genus Ophryotrocha (Dorvilleidae). Genome 36: 652–657.

Snell, T. & B. C. Winkler, 1984. Isozyme analysis of rotifer proteins. Biochem. Syst. Ecol. 12: 199–202.

Snell, T., B. D. Moffat, C. Janssen & G. Persoone, 1991. Acute toxicity tests using rotifers. III. Effects of temperature, strain, and exposure time on the sensitivity of *Brachionus plicatilis*. Envir. Tox. Wat. Qual. 6: 63–75.

Walsh, E. J. & L. Zhang, 1992. Polyploidy and body size variation in a natural population of the rotifer *Euchlanis dilatata*. J. evol. Biol. 5: 345–353.

Wood, W. B. (ed.), 1988. The Nematode *Caenorhabditis elegans*. Cold Spring Harbor Press, Cold Spring Harbor, New York: 667 pp.

Hydrobiologia **387/388**: 403–407, 1998.
E. Wurdak, R. Wallace & H. Segers (eds), Rotifera VIII: A Comparative Approach.
© *1998 Kluwer Academic Publishers.*

Karyotypes of bdelloid rotifers from three families

Jessica L. Mark Welch & Matthew Meselson
Department of Molecular and Cellular Biology, Harvard University, 7 Divinity Avenue, Cambridge,
MA 02138, U.S.A.

Key words: rotifer, bdelloid, chromosome number, karyotype

Abstract

We have determined the chromosome numbers in embryo nuclei of four species of bdelloid rotifers from three families. In agreement with Hsu (La Cellule 57: 283–296, 1956), we find that *Philodina roseola* has 13 chromosomes. We find that *Macrotrachela quadricornifera* has 10 chromosomes and that *Habrotrocha constricta* and *Adineta vaga* both have 12. This is the first report of the chromosome number of a member of the family Adinetidae.

Introduction

Bdelloid rotifers appear to be completely asexual, reproducing parthenogenetically from mitotically produced eggs. In an organism that never undergoes meiosis, selection against the accumulation of chromosomal translocations in the germ line should be greatly reduced. Fossils found in amber suggest that bdelloids have existed for tens of millions of years (Poinar & Ricci, 1992; Waggoner & Poinar, 1993). If they have been reproducing without meiosis for this length of time, bdelloids may have accumulated extensive chromosomal rearrangements and lost morphologically homologous chromosome pairs.

In the first published examination of the karyotype of a bdelloid rotifer, Hsu (1956a) reported the bdelloid *Philodina roseola* to have 13 chromosomes, including three that have no apparent homolog: one long chromosome and two dot chromosomes of unequal size. Since this observation, the question of chromosomal homology in bdelloids has continued to attract interest (Pagani et al., 1993).

As a further step toward determining whether bdelloids have homologous chromosome pairs, we have examined the chromosomes of four bdelloid species from three families.

Materials and methods

Rotifers

Four species were examined in this study:*Habrotrocha constricta* Dujardin, isolated in 1993 from ground moss in Sandwich, Barnstable County, Massachusetts; *Philodina roseola* Ehrenberg, purchased in 1989 from Carolina Biological Supply Co., Burlington, North Carolina; and *Macrotrachela quadricornifera* Milne and *Adineta vaga* Davis, both provided by C. Ricci.

Culturing of rotifers

Isolation of single rotifers

Shortly after the initial culture of each species of rotifer was established in our laboratory, single animals for further culture were isolated by serial passage as follows. A single animal was removed from the mass culture with a micropipette and deposited into sterile spring water in a 24-well plastic cell culture dish. Using a fresh pipette tip each time, the single animal was transferred into a fresh well at least 10 times successively to ensure that only one rotifer was used to start each clonal culture. In the cases of *H. constricta* and *P. roseola*, a single egg from the clonal culture was used to start our final laboratory culture.

Food and culture medium

M. quadricornifera is cultured in $1 \times$ freshwater medium ($10 \times$ medium contains 0.96 g $NaHCO_3$, 0.1 g

$CaCl_2 \cdot 2H_2O$, 0.6 g $MgSO_4 \cdot 7H_2O$, and 0.04 g KCl per liter, sterilized by filtration). All other bdelloid rotifers in our laboratory are cultured in commercial spring water (Belmont Springs Water Co., Lexington, MA), sterilized by filtration through an 0.22 μm cellulose acetate filter (Corning Costar Corp., Cambridge, MA).

P. roseola, *H. constricta*, and *A. vaga* are fed *Escherichia coli*. *M. quadricornifera* are fed *Saccharomyces cerevisiae*.

Preparation of chromosome squashes from embryos

To prepare embryos for chromosome squashes, 1–2 ml of a dense culture of bdelloids is treated with 10 ml of a solution of 10% Clorox (commercial bleach), for a final concentration of approximately 0.5% sodium hypochlorite. This treatment dissolves much of the debris in the culture; embryos sink to the bottom. After 2–5 minutes the supernatant is removed by pipetting and replaced with sterile distilled water. The distilled water wash is repeated 4–5 times to remove bleach.

At least 100 cleaned embryos are pipetted onto a microscope slide and prepared by a modification of the method of Pagani et al. (1993). A cover slip is placed over the embryos, the slide is inverted onto 4–5 layers of filter paper, and the embryos are squashed by gentle pressure on the slide. The slide is then placed immediately onto dry ice for at least ten minutes or into liquid nitrogen for about ten seconds, the cover slip is pried off using a razor blade, and the slide is placed into a fixative of 75% methanol, 25% glacial acetic acid at 4 °C for at least 30 minutes. After fixation the slide is dehydrated through an ethanol series (5 minutes each in ice-cold 70%, 95%, and 100% ethanol), air-dried, and stored over silica gel in a desiccator at 18 °C.

Staining, mounting, and fluorescence microscopy

For fluorescence microscopy the slides are stained for 5 minutes with 1 μg/ml DAPI (4′,6-diamidino-2-phenylindole) in 4× SSC (20× SSC is 175.3 g NaCl, 88.2 g sodium citrate per liter, pH adjusted to 7 with HCl), rinsed twice in 4× SSC, and mounted under a 24 × 60 mm cover slip in ProLong antifade solution (Molecular Probes, Eugene, OR). Slides are viewed with a Zeiss Axiophot microscope equipped with a mercury arc lamp and epifluorescent filters.

The best-separated chromosomes occur in young embryos, generally those with fewer than 60 nuclei. Because the majority of embryos in each preparation are older than this, it is necessary to scan at least 1,000 embryos to find several nuclei with well-separated chromosomes all in a single focal plane. Alternatively, one can collect younger embryos selectively, by transferring adults into a clean dish, waiting approximately 6 hours, and collecting the embryos that were produced.

Images are recorded using a CCD camera with intensifier (Videoscope, Herndon, VA) and a Macintosh computer using the public domain software NIH Image (available on the World Wide Web at http://rsb.info.nih.gov/nih-image/). Each karyotype is prepared from an image of a single nucleus using Adobe Photoshop software.

Results

Two representative nuclei of each species are shown in Figures 1–4. Nuclei of *Philodina roseola* embryos have 13 chromosomes (Figure 1), including two dot-shaped chromosomes less than 1 μm long, ten chromosomes of 2–3 μm in length, and one chromosome approximately 4 μm long. Embryo nuclei of *Macrotrachela quadricornifera* have 10 chromosomes ranging in size from approximately 2.5 to 4 μm (Figure 2). Embryo nuclei of *Habrotrocha constricta* have 12 chromosomes approximately 2–4 μm long (Figure 3), and embryo nuclei of *Adineta vaga* have 12 chromosomes ranging in size from approximately 1 to 3 μm (Figure 4). These size ranges are approximate because the length of the chromosomes changes with their degree of condensation as they progress through mitosis. Their apparent length in photographs also depends on whether they are lying flat in the plane of the photograph or at an angle to it, as occurs in the second nucleus in Figure 4.

Discussion

Our finding of 13 chromosomes in *P. roseola*, including one long and two dot chromosomes, exactly matches the description in Hsu (1956a). This is the only one of the four bdelloid species we examined in which distinctive individual chromosomes were observed. In *Macrotrachela quadricornifera*, we find 10 chromosomes but Pagani et al. (1993) report only 5. The photograph in Pagani et al. shows 10 chromosomes, which the authors interpreted as a nucleus in anaphase. In our preparations, we have never seen a nucleus with only five chromosomes. Therefore

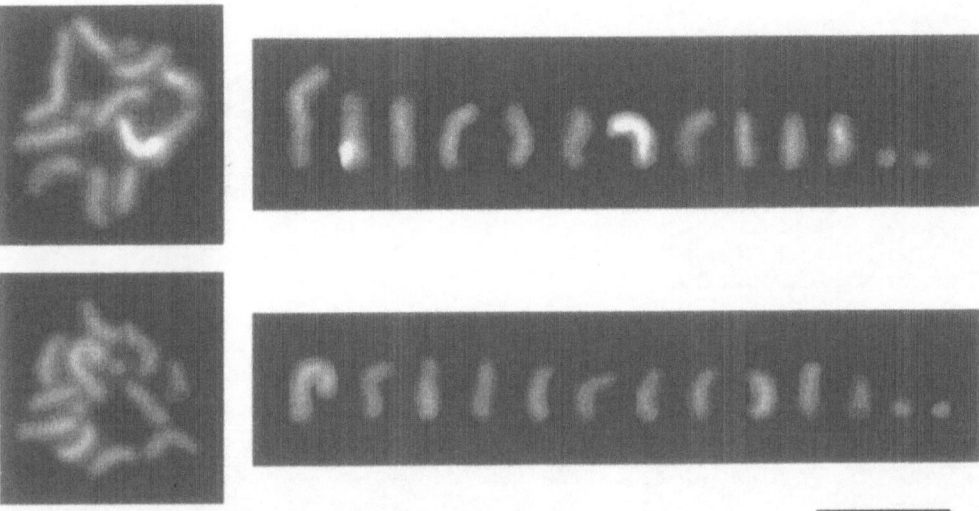

Figure 1. Left: two mitotic nuclei from early embryos of *Philodina roseola*. Right: the chromosomes from these nuclei rearranged to display a karyotype. Chromosome number = 13. Scale bar = 5 μm.

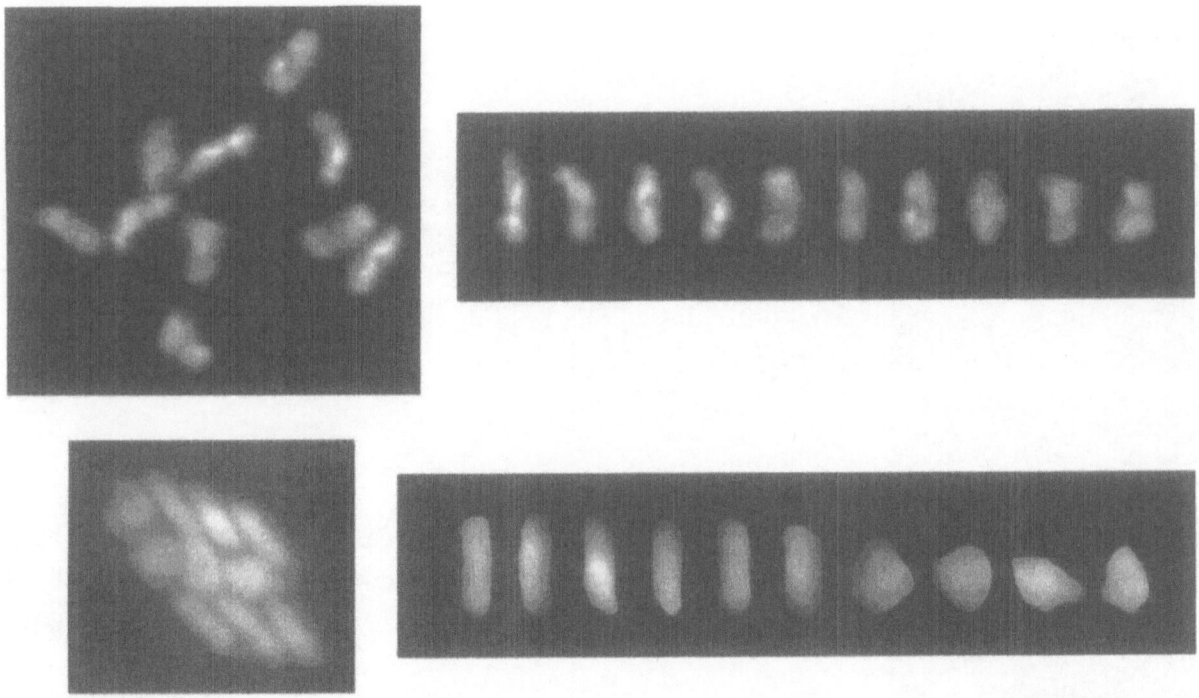

Figure 2. Left: two mitotic nuclei from early embryos of *Macrotrachela quadricornifera*. Right: the chromosomes from these nuclei rearranged to display a karyotype. Chromosome number = 10. Scale bar = 5 μm.

406

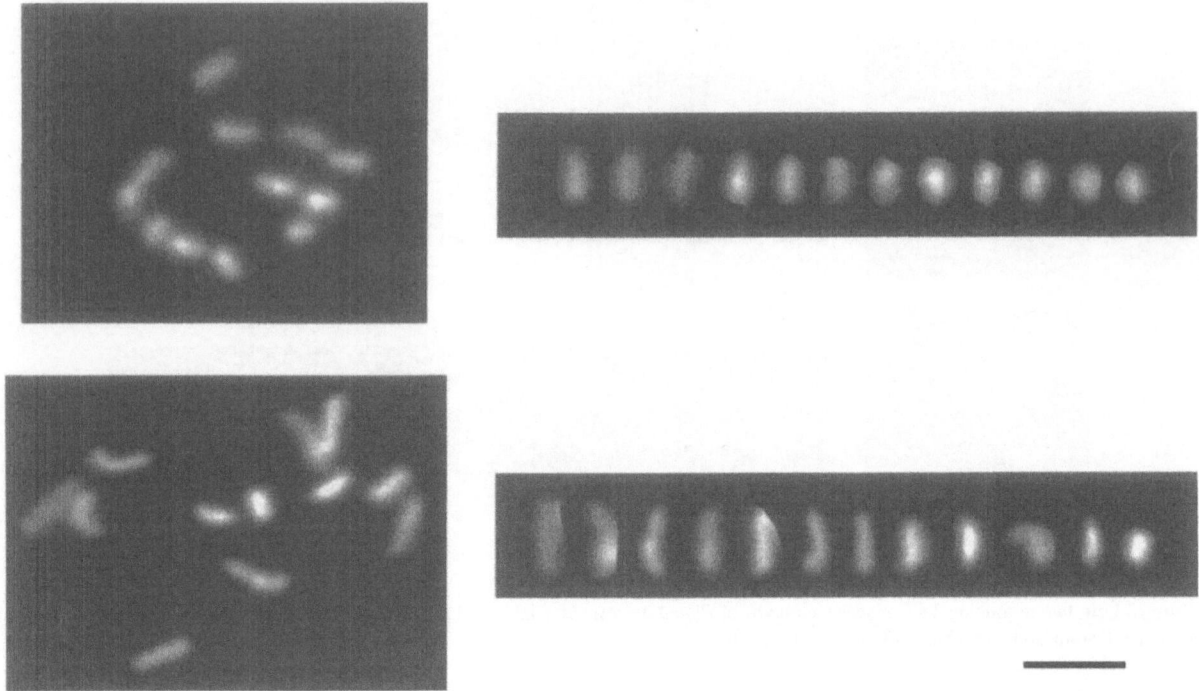

Figure 3. Left: two mitotic nuclei from early embryos of *Habrotrocha constricta*. Right: the chromosomes from these nuclei rearranged to display a karyotype. Chromosome number = 12. Scale bar = 5 μm.

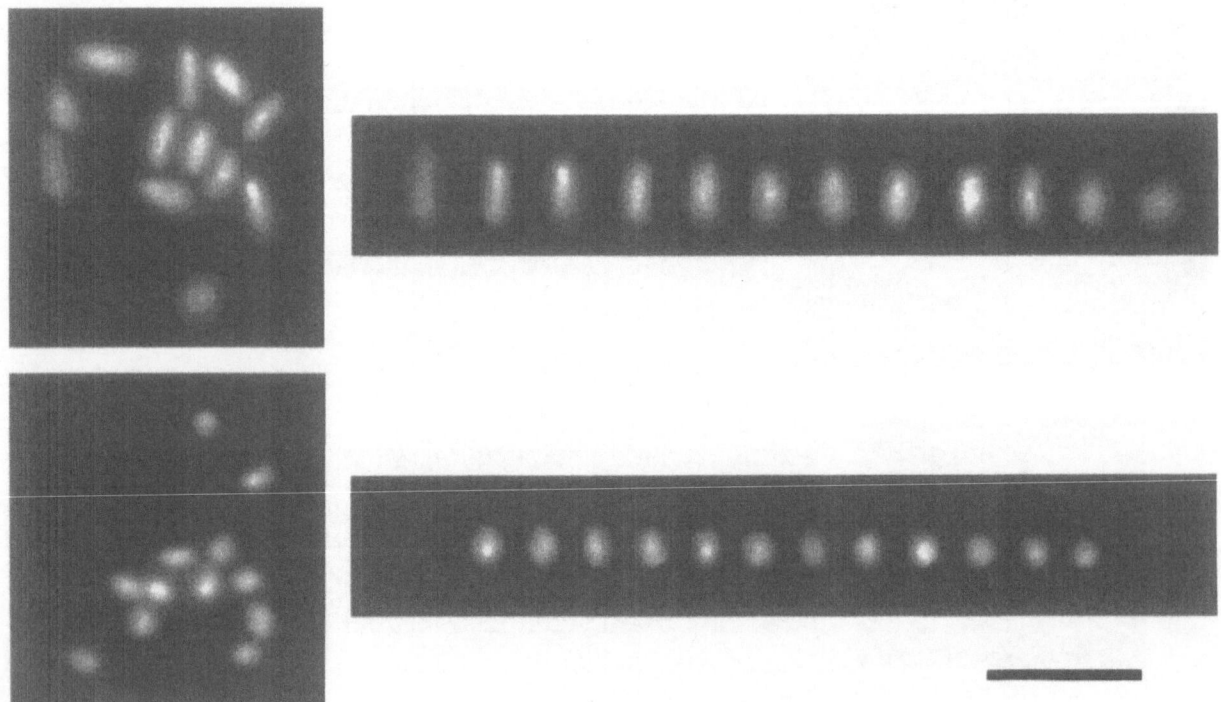

Figure 4. Left: two mitotic nuclei from early embryos of *Adineta vaga*. Each photo in Figure 4 is a montage of two focal planes, in order to bring each chromosome into focus. Right: the chromosomes from these nuclei rearranged to display a karyotype. Chromosome number = 12. Scale bar = 5 μm.

we believe that the true chromosome number in *M. quadricornifera* is 10.

There are two other bdelloids for which chromosome numbers are published: *Habrotrocha tridens* Milne, in which Hsu (1956b) reported 13 chromosomes, two of them dot-shaped, and *Habrotrocha rosa* Donner, in which Plasota and Plasota (1980) reported 14 chromosomes. Thus, including the present study, published observations of bdelloid chromosome numbers range from 10 to 14. Pagani et al. (1993) cite reports of bdelloid chromosome numbers ranging from 8 to 16; this range includes unpublished work.

Do bdelloids have pairs of homologous chromosomes? Although there are size differences among the chromosomes in each karyotype, the variation in *M. quadricornifera*, *H. constricta*, and *A. vaga* appears continuous and no chromosome appears to possess distinguishing characteristics that would allow it to be readily identified in a karyotype. Only in *P. roseola* are there single chromosomes that can be readily distinguished: the longest chromosome and the two unequal dots. However, there are numerous cases in other taxa in which chromosomes that differ in size nonetheless pair at meiosis, including cases in which a single chromosome pairs with two or more partners (White, 1973). Thus the *P. roseola* karyotype does not prove that bdelloids lack meiosis.

Conversely, the karyotypes shown here provide no evidence that meiosis or chromosomal pairing do occur. There is no clear morphological evidence for homologous chromosome pairs. Because bdelloid chromosomes are so small, simple inspection of their shape and size can provide little information about the presence or absence of chromosomal homology.

Further study, using more advanced techniques to detect the location of homologous sequences on the chromosomes, will be necessary to settle this question.

Acknowledgements

We thank Claudia Ricci for cultures of *Adineta vaga* and *Macrotrachela quadricornifera*, and we are grateful to Claudia Ricci and Aydin Örstan for assistance in verifying the identity of *Habrotrocha constricta*. We thank Nancy Kleckner for the use of her fluorescence microscope. This research was supported by a National Science Foundation grant to M. M. Additional support for J. M. W. was provided by an NIH Genetics Training Grant to Harvard University.

References

Hsu, W. S., 1956a. Oogenesis in the Bdelloidea rotifer *Philodina roseola* Ehrenberg. La Cellule 57: 283–296.

Hsu, W. S., 1956b. Oogenesis in *Habrotrocha tridens* (Milne). Biol. Bull. 111: 364–374.

Pagani, M., C. Ricci & C. A. Redi, 1993. Oogenesis in *Macrotrachela quadricornifera* (Rotifera, Bdelloidea) I. Germarium eutely, karyotype and DNA content. Hydrobiologia 255/256: 225–230.

Plasota, K. & M. Plasota, 1980. The determination of the chromosome number of *Habrotrocha rosa* Donner, 1949. Hydrobiologia 73: 43–44.

Poinar, G. O. Jr. & C. Ricci, 1992. Bdelloid rotifers in Dominican amber: Evidence for parthenogenetic continuity. Experientia 48: 408–410.

Waggoner, B. M. & G. O. Poinar Jr., 1993. Fossil habrotrochid rotifers in Dominican amber. Experientia 49: 354–357.

White, M. J. D., 1973. Animal Cytology and Evolution, 3rd edn. Cambridge University Press, London: 961 pp.

Hydrobiologia **387/388**: 409–419, 1998.
E. Wurdak, R. Wallace & H. Segers (eds), Rotifera VIII: A Comparative Approach.
© *1998 Kluwer Academic Publishers.*

Does the evasive behavior of *Hexarthra* influence its competition with cladocerans?

Anupama Kak & T. Ramakrishna Rao
Department of Zoology, University of Delhi, Delhi-110 007, India

Key words: Cladocera, escape behavior, food level, interference competition, Rotifera

Abstract

We tested the hypothesis that evasive movements by *Hexarthra mira* reduce adverse effects of interference competition with cladocerans permitting coexistence. To do this we studied the population growth of *Hexarthra mira* and two non-evasive *Brachionus* species in the presence of one of either two cladocerans (*Daphnia similoides*, *Ceriodaphnia cornuta*) at three food (*Chlorella*) levels (0.5, 2 and 4×10^6 cells ml^{-1}) at 25°C. The non-evasive, but larger-sized *B. calyciflorus* was suppressed by *D. similoides* at all food levels tested, and by *C. cornuta* at high food levels only. The smaller *B. angularis* showed similar trends with *D. similoides*, but with *C. cornuta* it persisted and increased in population size at the medium and high food levels. *Hexarthra* was able to coexist with both the cladocerans regardless of food level. However, population growth rate of *Hexarthra* was affected significantly in presence of *D. similoides*, but not in the presence of *C. cornuta*. We suggest that evasive behavior of *Hexarthra* helps it coexist with large cladocerans by reducing the frequency of its being drawn into their branchial chambers.

Introduction

In many freshwater ecosystems, the food niches of cladocerans and rotifers overlap considerably, leading to competition in which rotifers are often suppressed by large cladocerans (Gilbert, 1988a). The inverse relationship often observed between the relative abundances of cladocerans and rotifers in nature has been cited as evidence for their competitive interactions (Gilbert, 1988a; Lampert & Rothhaupt, 1991; May & Jones 1989; Vanni, 1986). Rotifers have been shown to be adversely affected by both exploitative and interference forms of competition with cladocerans (Burns & Gilbert, 1986; Fradkin, 1995; Gilbert, 1985; Gilbert & Stemberger, 1985). The mechanisms of interference competition include inhibition of suspension feeding and injuries or even incidental carnivory resulting from the rotifers being swept into the branchial chamber of the filter-feeding cladocerans (Burns & Gilbert, 1986; MacIssac & Gilbert, 1991; Schneider, 1990). Because exploitative and interference competition are influenced by ambient food concentrations, food levels could affect the outcome of competition between rotifers and cladocerans (Bengtsson, 1987; Romanovsky

& Feniova, 1985). The body size of the competing rotifer species, or perhaps more importantly the relative size of the cladoceran and rotifer, is also a determinant of the competitive outcome, smaller rotifers being generally more susceptible (MacIssac & Gilbert, 1989). Among the many predator-deterrent morphological and behavioral attributes evolved by rotifers are the sudden escape responses of certain species such as *Polyarthra*, *Hexarthra* and *Filinia*, which are effective against *Asplanchna* predation (Gilbert, 1985; Gilbert & Williamson, 1978; Iyer & Rao, 1996). These escape responses also may mitigate the adverse effects of interference competition with cladocerans through a reduction in the frequency of potentially harmful close encounters (Gilbert, 1987; Wickham & Gilbert, 1991).

In many eutrophic ponds and lakes we often observed that *Hexarthra mira* coexists, at times in fairly large numbers, with *Daphnia similoides* and *Ceriodaphnia cornuta*. Thus we conducted a laboratory study using three food levels, to test the hypothesis that the evasive, escape responses of *H. mira* reduce intensity of interference competition with cladocerans and facilitate its coexistence with them. We then compared the population growth responses of *H. mira* in competition with *D. similoides* and *C. cornuta* to those of two different-sized, non-evasive, rotifer species (*B. angularis*, *B. calyciflorus*) under similar conditions.

Materials and methods

Six experiments were conducted in total (two cladocerans × three rotifer species) to study the competitive interactions of the rotifers *Hexarthra mira*, *Brachionus angularis* and *B. calyciflorus* individually with each of the two cladocerans, *Daphnia similoides* and *Ceriodaphnia cornuta*. All experimental animals used in the tests were taken from laboratory cultures maintained continuously for at least a year. The cultures of these species were clones, developed in each case from a single parthenogenetic female isolated from local pond plankton. Adult body sizes (μm±SE) of all test species used are as follows: Rotifers (*Hexarthra mira* (170±4.2), *Brachionus calyciflorus* (179±4.3) and *Brachionus angularis* (75±1.8)); Cladocerans (*Daphnia similoides* (2134±42.5) and *Ceriodaphnia cornuta* (528±11)). All cultures were maintained on a diet of laboratory cultured *Chlorella* ($1–3 \times 10^6$ cells ml^{-1}) which was also used in the experiments. Mass cultures of *Chlorella* were developed using a vitamin-enriched medium (Kuhl & Lorenzen, 1964). Based on preliminary tests with all the test species, three food levels ($0.5, 2.0$ and 4.0×10^6 cells ml^{-1} of *Chlorella*) were used. The experimental design included, at each food level, six glass beakers containing replicated sets of cladoceran alone, rotifer alone and both together. Each beaker contained 200 ml medium of autoclaved tap water with *Chlorella*, concentrated from mass cultures and re-suspended to obtain the required concentration. Into each of the experimental vessels we introduced 10 neonate individuals of the cladoceran and 150 individuals of the rotifer, either as separate species or together. All the 18 test beakers (three food levels × three treatments × two replicates) were covered with foil and placed in an unilluminated BOD incubator at 25±1.5 °C. The beakers were gently shaken periodically to minimize sedimentation of *Chlorella*. At 48±4 h intervals, the animals in each beakers were counted and all the live individuals were transferred into a vessel with fresh medium and food. The experimental duration ranged from 10 to 16 days. For rotifers the populations were counted in total when densities were low (<800 animals) or in three aliquot sample estimates when the densities were higher. The cladocerans were counted in total.

The population growth rates (r) for each species alone and in combination with the other competing species, were calculated using the exponential growth equation. In cases where the population growth curves showed crests and troughs we used the average of r values calculated using different intervals of t over a period of 6–12 days during which a stable age distribution was assumed to have been attained. Effects of competition were evaluated in terms of the percent change in population growth rate of the test species relative to that achieved in the single species control. The cladoceran-imposed mortality rate on the rotifer was calculated as the difference between rotifer population growth rates with and without cladocerans (Gilbert, 1988b). The statistical significance of the differences in the single and mixed species groups in relation to food level were tested using a two-way ANOVA.

Results

Fate of rotifers

Hexarthra mira was able to coexist, although at densities lower than in controls, with both *C. cornuta* (Figure 1) and *D. similoides* (Figure 2) at all the food levels tested. The population growth rate of the evasive rotifer was affected significantly in the presence of *D. similoides* ($P<0.001$), but not in the presence of the smaller cladoceran *C. cornuta* ($P>0.5$). Food level effects on its growth rate were significant ($P<0.01$), as were its interaction effects ($P<0.05$), with either *Daphnia* or *Ceriodaphnia*, the latter indicating that the cladoceran effects on *H. mira* growth rate were food level-dependent.

The non-evasive and considerably smaller-sized *Brachionus angularis* was able to coexist with *C. cornuta* only at medium and high food levels (Figure 3). At low and medium food levels, the growth rates of the rotifer were actually higher in presence of the cladoceran than in the control, although the differences were statistically significant only at the medium food level ($P<0.001$). In the presence of *Daphnia*, the rotifer was adversely affected regardless of the food level (Figure 4). Its growth rates were significantly lower ($P<0.01$) in the mixed species treatment than in controls. However, *D. similoides* could not suppress the rotifer completely during the experimental period. With either of the cladoceran species food effects were significant ($P<0.01$) but the interaction effect was significant only with *C. cornuta*.

Ceriodaphnia cornuta   *Hexarthra mira*

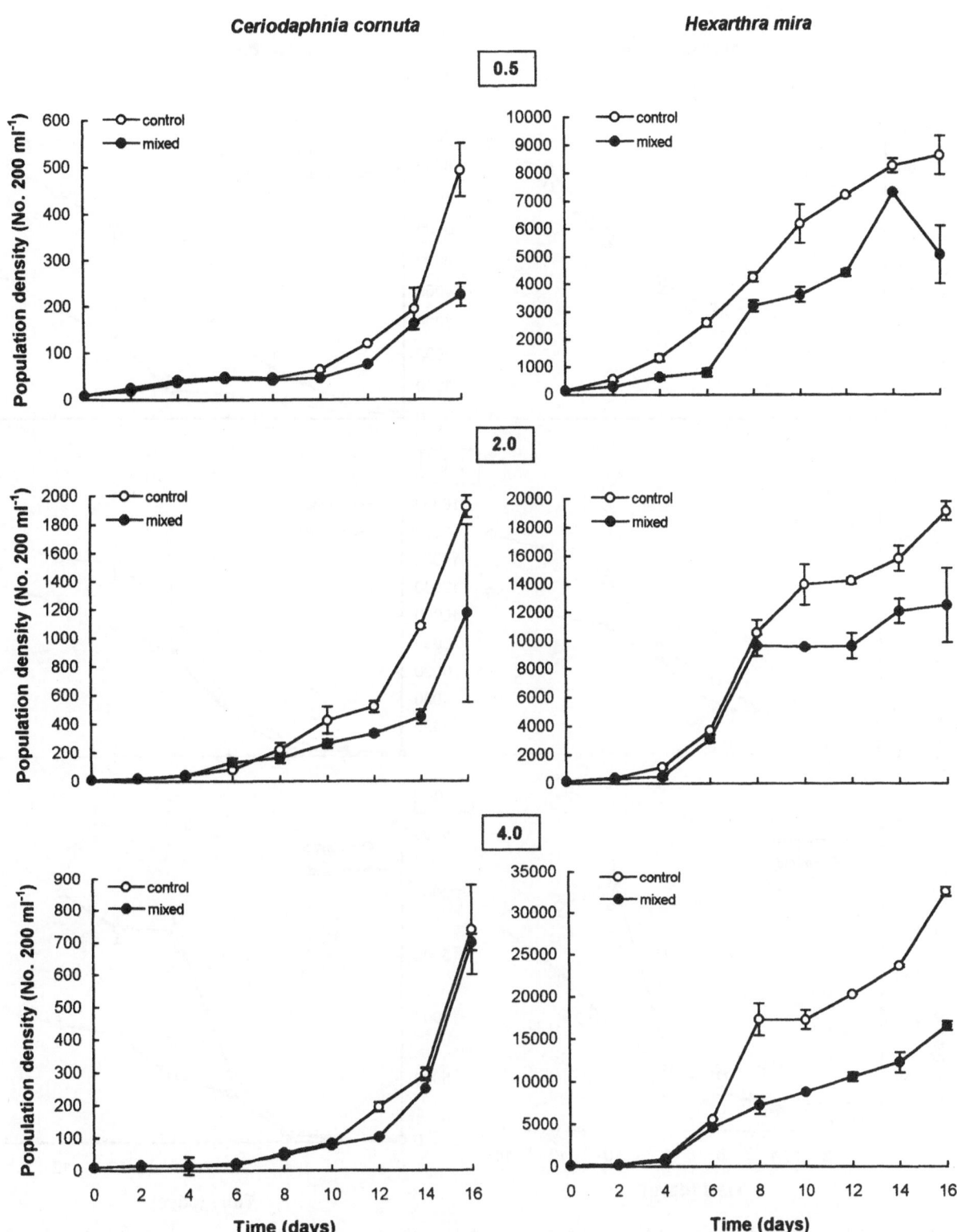

Figure 1. Population growth trajectories of *Hexarthra mira* and *Ceriodaphnia cornuta* in controls and mixed-species treatments at three food levels. Values are mean±SE.

412

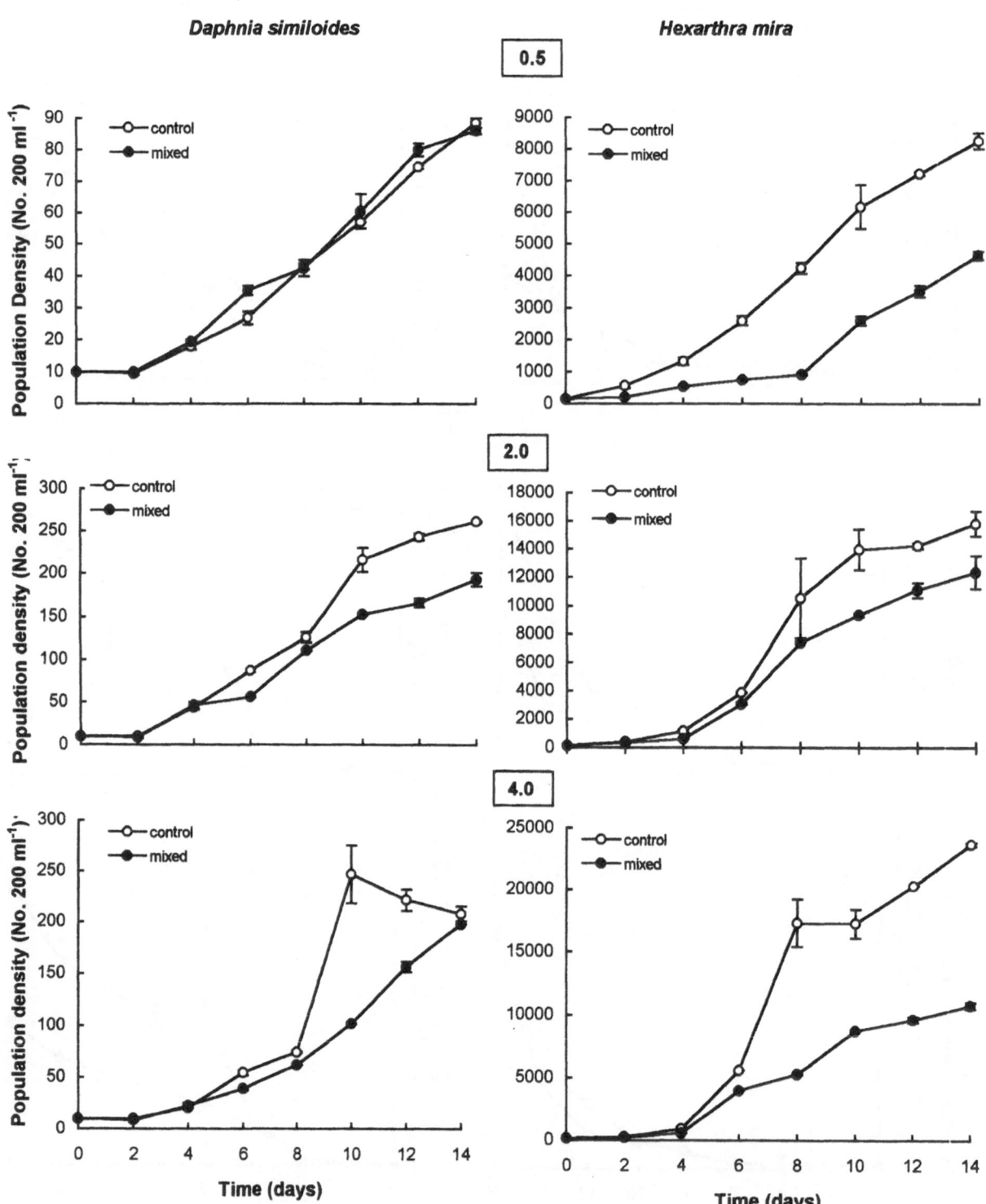

Figure 2. Population growth trajectories of *Hexarthra mira* and *Daphnia similoides* in controls and mixed-species treatments at three food levels. Values are mean±SE.

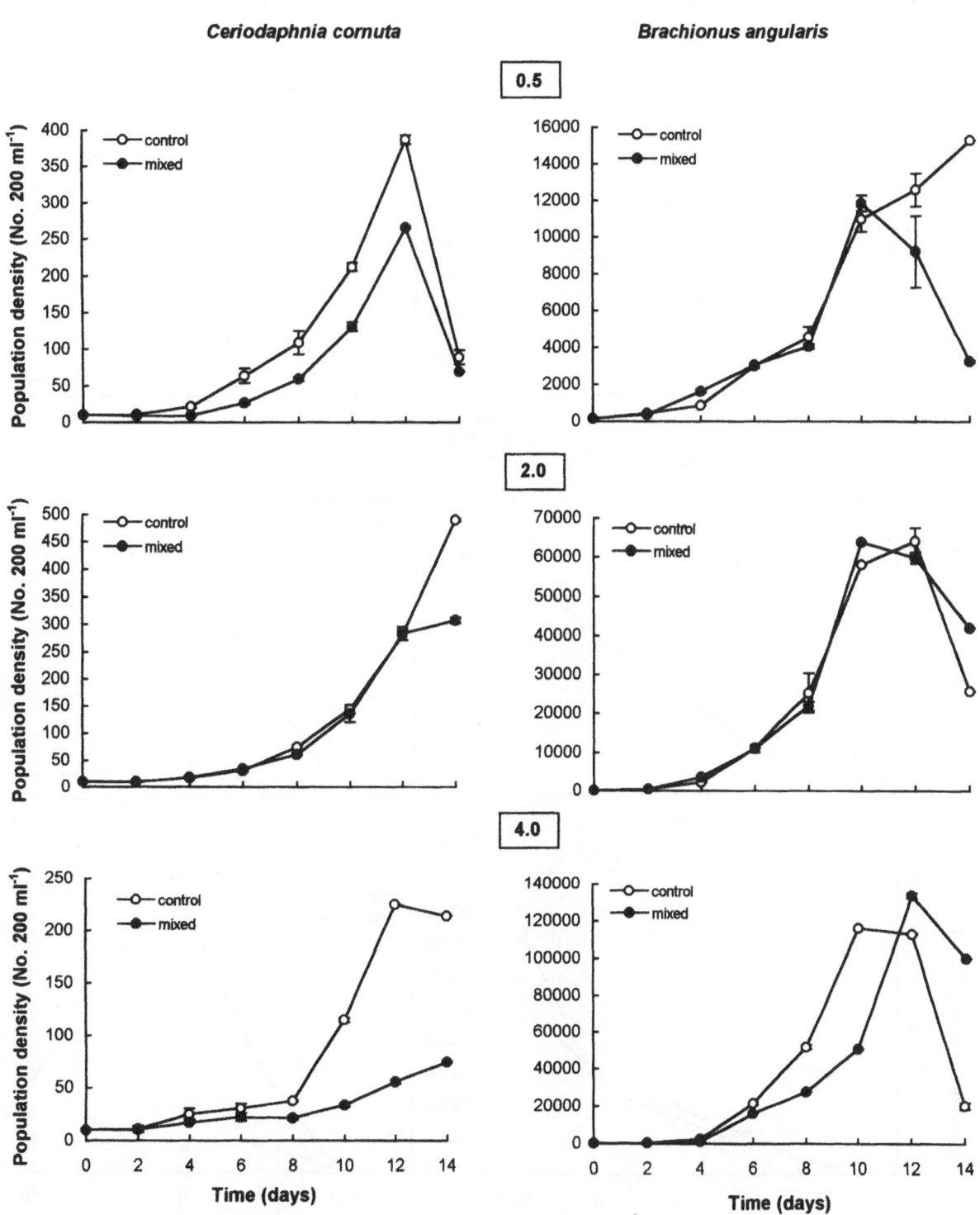

Figure 3. Population growth trajectories of *Brachionus angularis* and *Ceriodaphnia cornuta* in controls and mixed-species treatments at three food levels. Values are mean±SE.

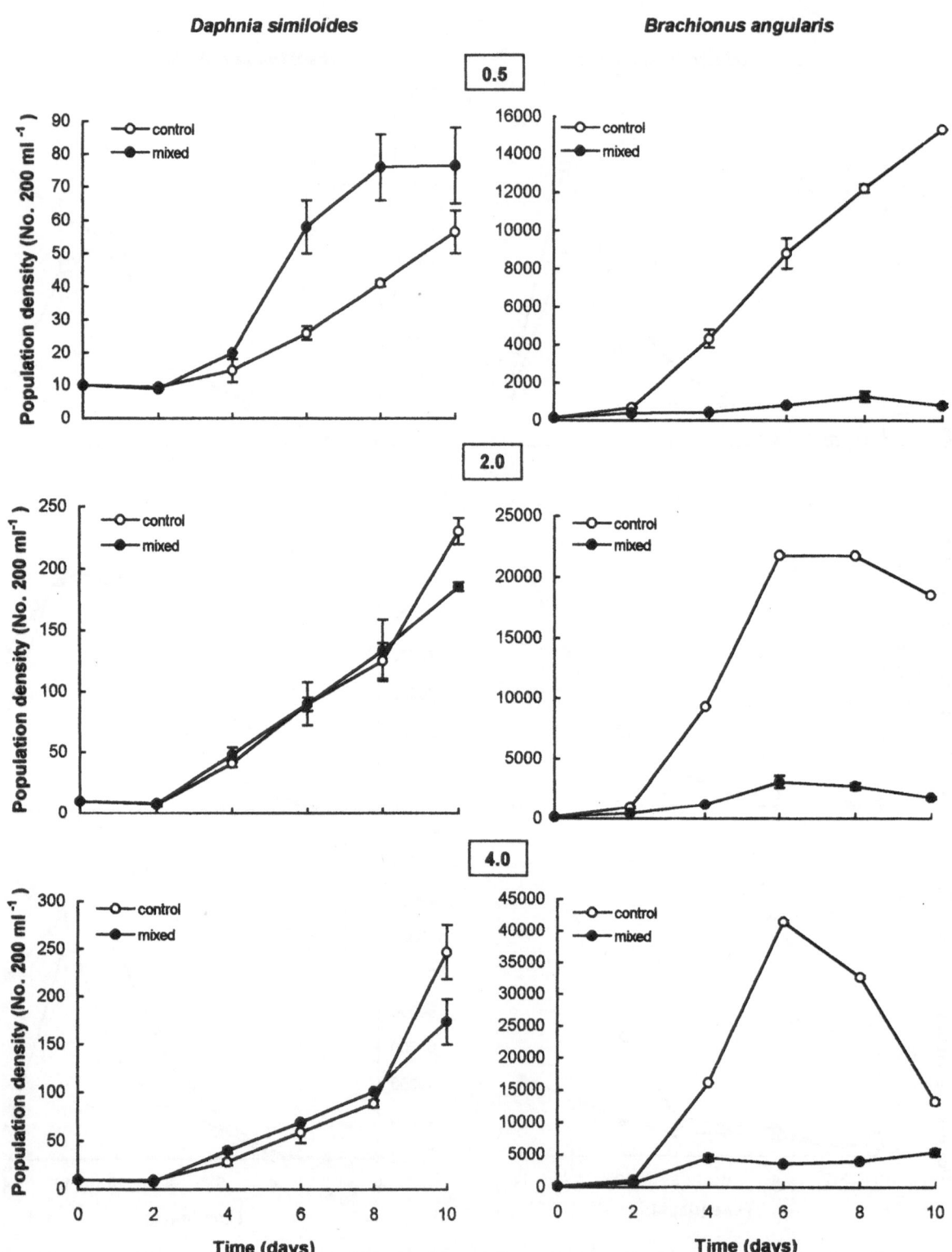

Figure 4. Population growth trajectories of of *Brachionus angularis* and *Daphnia similoides* in controls and mixed-species treatments at three food levels. Values are mean±SE.

Brachionus calyciflorus, non-evasive but with body size close to that of *H. mira*, was adversely affected by the cladocerans, less so by *Ceriodaphnia* (Figure 5) than by *Daphnia* (Figure 6). In competition with *Daphnia*, *B. calyciflorus* was completely eliminated at all food levels (Figure 6), with *r* values significantly lower ($P<0.001$) than in the controls. The *r* value of the rotifer in mixed-species cultures was, in fact, negative at both low and medium food levels. With the smaller-sized *C. cornuta*, *B. calyciflorus* population growth was strongly inhibited only at the high food levels (Figure 5). Although the rotifer maintained itself during the experimental period, its eventual extinction was predicted by the negative *r* value. At medium food levels, *B. calyciflorus* was only slightly inhibited by the cladoceran.

Fate of cladocerans

Ceriodaphnia and *Daphnia* also were affected by the presence of rotifers. *Ceriodaphnia* was adversely affected by presence of *B. angularis* (Figure 3), but not significantly so by that of *B. calyciflorus* (Figure 5) or *H. mira* (Figure 1). The *r* values of *Ceriodaphnia* in mixed treatments with either of these two were not significantly different from those in the controls ($P>0.25$). With *B. angularis* as the competitor in the medium, *C. cornuta* reached densities lower than in controls, particularly at high food levels (Figure 3). At any food level, *r* values of the cladoceran were significantly lower ($P<0.001$) in the presence of *B. angularis* than in the controls.

Population growth of *Daphnia* was affected more by *H. mira* (Figure 2) and *B. calyciflorus* (Figure 6) than by *B. angularis* (Figure 4). At medium and high food levels, its *r* values were significantly lower with *H. mira* than in controls ($P<0.001$). Densities achieved by *Daphnia* were actually higher in presence of *B. angularis* than in controls at low food level, though the difference in *r* value was not significant $P>0.25$). At medium and high food levels *r* values of *D. similoides* were significantly lower ($P<0.05$) than the controls.

Discussion

Among the three rotifers tested, *Hexarthra mira* suffered the least amount of cladoceran-inflicted mortality with either *Daphnia* or *Ceriodaphnia* (Figure 7). Although its growth rates were less in the presence

of *Daphnia*, *Hexarthra* coexisted with both cladocerans for the duration of the experiment. Among factors influencing the outcome of competition are body size and vulnerability of the rotifer, body size and concentration of the cladoceran, and food level (Gilbert, 1988a). In terms of its body size in relation to that of its competitor, *Hexarthra* is close to *B. calyciflorus* which was suppressed by *D. similoides* at all food levels. From a comparison of cladoceran-imposed death rates, *B. calyciflorus* suffered six (high food levels) to 15 (low food levels) times more than *H. mira* in the presence of *Daphnia*. Susceptibility of *B. calyciflorus* to interference competition from *Daphnia* has been recorded in another study as well (Gilbert, 1985). The mechanism that helps *Hexarthra* reduce the adverse effects of interference competition with *Daphnia* is most probably very similar to that elucidated for *Polyarthra*, another rotifer capable of sudden escape responses (Gilbert, 1985, 1987; Kirk & Gilbert, 1988). While its sudden escape responses may mitigate adverse effects of interference competition, *Hexarthra* still experiences the effects of exploitation competition, reflected in its lowered growth rates in the presence of *Daphnia*. The density of *Daphnia* used in our study ($50 \, 1^{-1}$) is not uncommon in local eutrophic waters; other studies (Burns & Gilbert, 1986; Fradkin, 1995) indicate that at such high cladoceran densities even the less susceptible rotifers could be adversely affected by interference competition. In an enclosure study, Fradkin (1995) observed that the number of *Polyarthra* also declines at higher densities of *Daphnia pulicaria*.

The small, non-evasive *B. angularis*, has a low body size ratio (BSR) compared to *D. similoides* and *C. cornuta* (BSR=0.04 and 0.14, respectively). It could not be suppressed completely by *Daphnia* during the test period, although the mortalities imposed on it were very high, particularly at low and medium food levels. With smaller-sized *Ceriodaphnia*, *B. angularis* showed no significant reduction in its growth rate. This brachionid had the highest *r* values (nearly three times that of *B. calyciflorus* at the highest food level tested), and was able to compensate for mortality imposed by the cladoceran. Where exploitative competition is the dominant form of competition, smaller rotifers such as *B. angularis* probably have an advantage because of their low food requirements (MacIsaac & Gilbert, 1991).

The consequences of competition between rotifers and cladocerans have received relatively little attention. Small cladocerans are more susceptible to com-

416

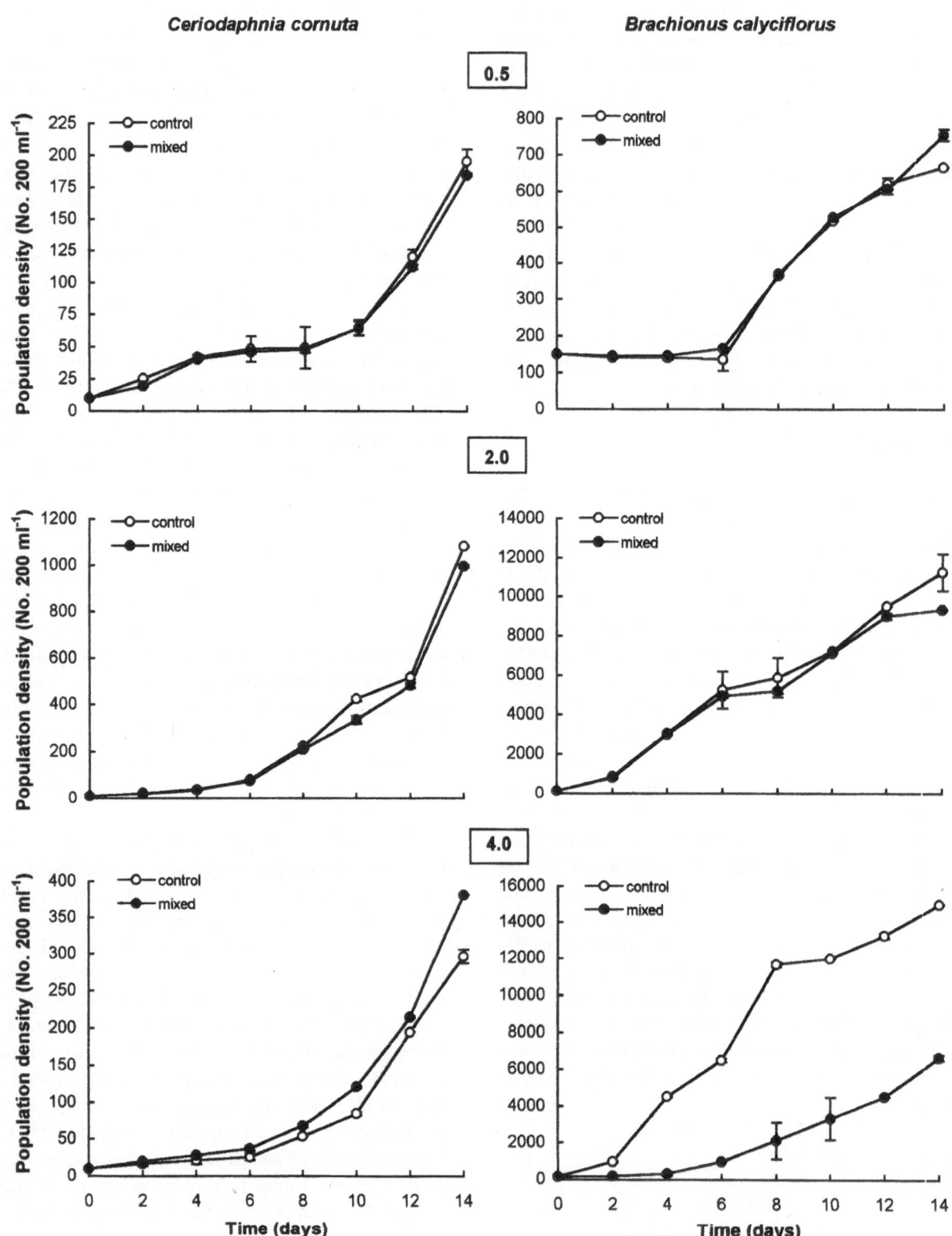

Figure 5. Population growth trajectories of *Brachionus calyciflorus* and *Ceriodaphnia cornuta* in controls and mixed-species treatments at three food levels. Values are mean±SE.

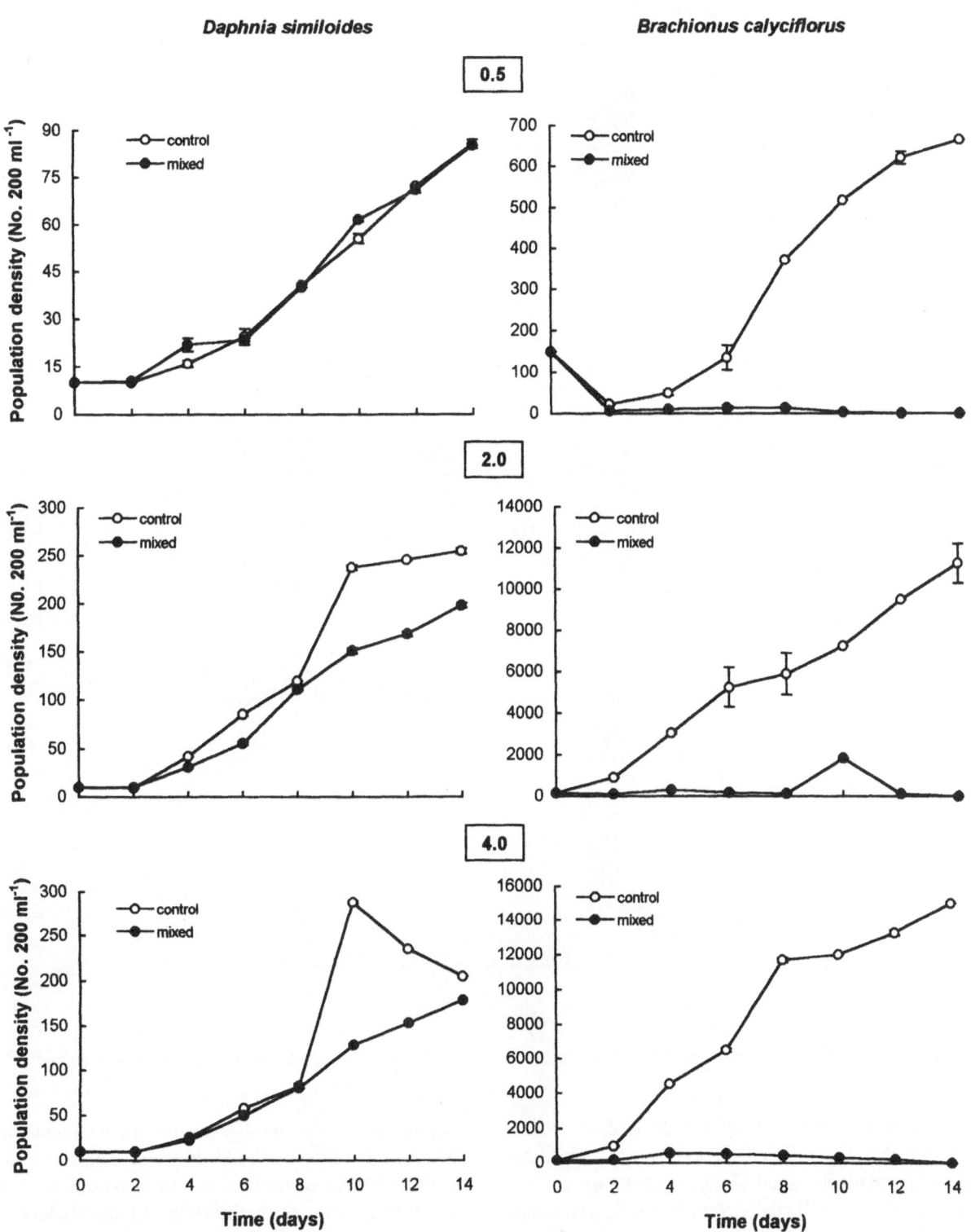

Figure 6. Population growth trajectories of *Brachionus calyciflorus* and *Daphnia similoides* in controls and mixed-species treatments at three food levels. Values are mean±SE.

418

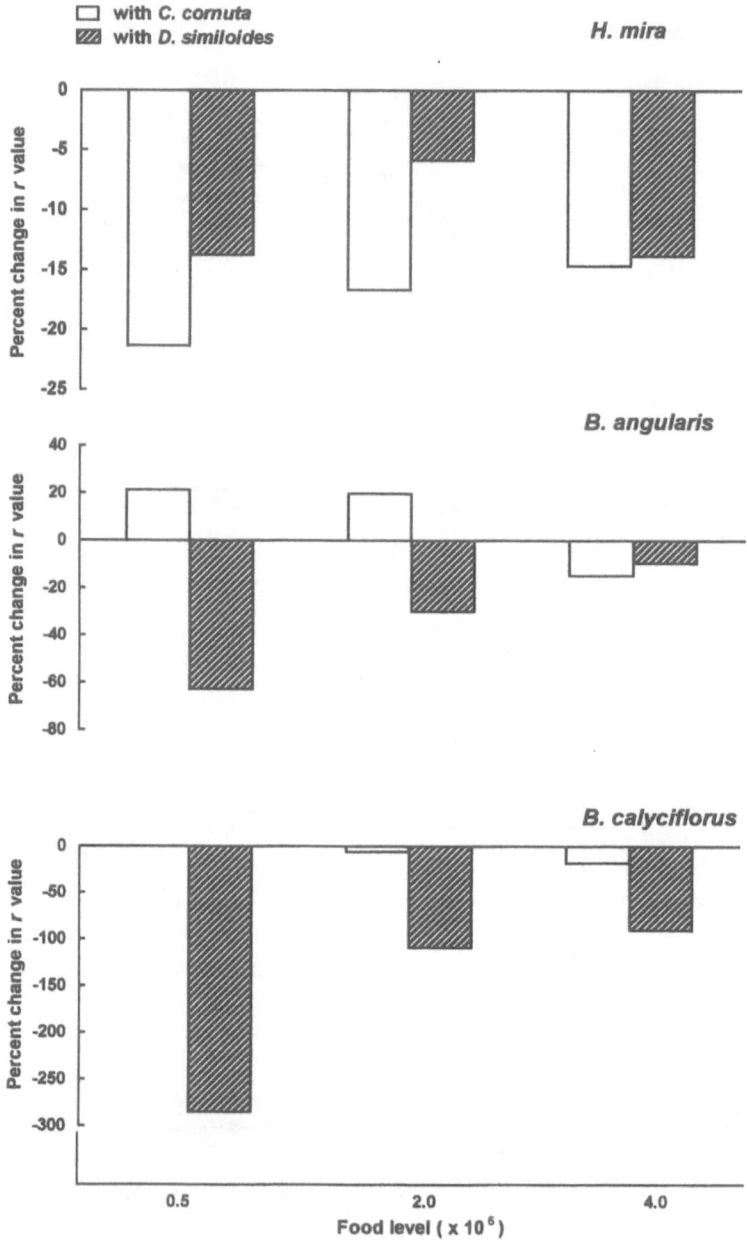

Figure 7. Percent change relative to controls in the population growth rates (*r*) of the three rotifer species in the presence of cladocean.

petition with rotifers. MacIsaac & Gilbert (1989) reported that *Keratella cochlearis* was able to eliminate the relatively small *Daphnia ambigua*. On the other hand, the small cladoceran, *Scapholeberis kingi*, while being able to suppress soft-bodied *Synchaeta oblonga*, was itself unaffected by the rotifer (Gilbert & MacIsaac, 1989). In our study, *Daphnia*, but not *Ceriodaphnia*, was adversely affected by *H. mira* and *B. calyciflorus*, whereas the opposite was true with *B.*

angularis. At low food level, the higher population growth rate of *D. similoides* in the presence of *B. angularis* than in controls is due to increased levels of incidental carnivory on the rotifer by the cladoceran. In a separate short-duration experiment using tethered *Daphnia*, we observed that its ingestion rates with *B. angularis* increased with decreasing levels of algal concentration and increasing densities of the rotifer. We further found that *D. similoides* could not only survive but actually produce broods in a medium containing moderate to high densities of *B. angularis* but devoid of any phytoplankton.

B. angularis should be more susceptible to interference (because of its small size) and less so to exploitation (probably because of a lower threshold food concentration). Our results however indicate that the relative importance of interference and exploitation to this species is dependent upon the ambient food levels; interference is of greater importance at low food levels, but exploitation at higher food levels. At high food levels, *B. angularis* is able to compensate the mortality imposed by interference competition with its very high growth rates and coexist with the cladoceran. Although *B. calyciflorus* and *H. mira*, by virtue of their relatively large size, should be more susceptible to exploitation than interference, our study indicates otherwise. Interference was probably of greater importance to *B. calyciflorus*, as it was strongly affected by the presence of *Daphnia* at any food level (Figure 6). The growth rates of *H. mira* were not too different from those of *B. calyciflorus* at any food level, in the absence of the competing cladoceran. This suggests that *H. mira* did not suffer the same fate as *B. calyciflorus* only due to its evasive behavior which reduces its susceptibility to interference.

Acknowledgements

We benefited immensely from constructive criticism of an earlier version by John Gilbert, to whom we are thankful. Financial support provided by the Department of Science and Technology, Government of India, is gratefully acknowledged.

References

Bengtsson, J., 1987. Competitive dominance among Cladocera: are single factors enough? Hydrobiologia 145: 245–250.

Burns, C. W. & J. J. Gilbert, 1986. Effects of daphnid size and density on interference between *Daphnia* and *Keratella cochlearis*. Limnol. Oceanogr. 31: 848–858.

Fradkin, S. C., 1995. Effects of interference and exploitative competition from large-bodied cladocerans on rotifer community structure. Hydrobiologia 313/314: 387–393.

Gilbert, J. J., 1985. Competition between rotifers and *Daphnia*. Ecology 66: 1943–1950.

Gilbert, J. J., 1987. The *Polyarthra* escape response: defence against interference from *Daphnia*. Hydrobiologia 147: 235–238.

Gilbert, J. J., 1988a. Suppression of rotifer populations by *Daphnia*: a review of the evidence, the mechanisms, and the effects on zooplankton community structure. Limnol. Oceanogr. 33: 1286–1303.

Gilbert, J. J., 1988b. Susceptibilities of ten rotifer species to interference from *Daphnia pulex*. Ecology 69: 1826–1838.

Gilbert, J. J., 1989. Competitive interactions between the rotifer *Syncheata oblonga* and the cladoceran *Scapholeberis kingi* Sars. Hydrobiologia 186/186: 75–80.

Gilbert, J. J. & C. E. Williamson, 1978. Predator–prey behaviour and its effect on rotifer survival in associations of *Mesocyclops edax, Asplanchna girodi, Polyarthra vulgaris* and *Keratella cochlearis*. Oecologia 37: 13–22.

Gilbert, J. J. & R. S. Stemberger, 1985. Control of *Keratella* populations by interference competition from *Daphnia*. Limnol. Oceanogr. 30: 180–188.

Iyer, N. & T. R. Rao, 1996. Responses of the predatory rotifer *Asplanchna intermedia* to prey species differing in vulnerability: laboratory and field studies. Freshwat. Biol. 36: 521–533.

Kirk, K. L. & J. J. Gilbert, 1988. Escape behavior of *Polyarthra* in response to artificial flow stimuli. Bull. Mar. Sci. 43: 551–560.

Kuhl, A. & H. Lorenzen, 1964. Handling and culturing of *Chlorella*. In Prescott, D. M. (ed.), Methods in Cell Physiology, Vol I. Academic Press, New York, London: 159–187.

Lampert, W. & K. O. Rothhaupt, 1991. Alternating dynamics of rotifers and *Daphnia magna* in a shallow lake. Arch. Hydrobiol. 120: 447–456.

MacIssac, H. J. & J. J. Gilbert, 1989. Competition between rotifers and cladocerans of different body sizes. Oecologia 81: 295–301.

MacIssac, H. J. & J. J. Gilbert, 1991. Discrimination between exploitative and interference competition between Cladocera and *Keratella cochlearis*. Ecology 72: 924–937.

May, L. & D. H. Jones, 1989. Does interference competition from *Daphnia* affect populations of *Keratella cochlearis* in Loch Leven, Scotland? J. Plankton Res. 11: 445–461.

Romanovsky, Y. E. & I. Y. Feniova, 1985. Competition among Cladocera: effect of different levels of food supply. Oikos 44: 243–252.

Schneider, D. W., 1990. Direct assessment of the independent effects of exploitative and interference competition between *Daphnia* and rotifers. Limnol. Oceanogr. 35: 916–922.

Wickham, S. A. & J. J. Gilbert, 1991. Relative vulnerabilities of natural rotifer and ciliate communities to cladocerans: laboratory and field experiments. Freshwat. Biol. 26: 77–86.

Vanni, M. J., 1986. Competition in zooplankton communities: suppression of small species by *Daphnia pulex*. Limnol. Oceanogr. 31: 1039–1056.

Hydrobiologia **387/388**: 421–425, 1998.
E. Wurdak, R. Wallace & H. Segers (eds), Rotifera VIII: A Comparative Approach.
© *1998 Kluwer Academic Publishers.*

Colony size in *Conochilus hippocrepis*: defensive adaptation to predator size

Maria Diéguez & Esteban Balseiro
Dept. Ecology, Centro Regional Universitario Bariloche, Unidad Postal Universidad, 8400 Bariloche, Argentina

Key words: Conochilus, coloniality, copepod predation, copepod development

Abstract

Conochilus hippocrepis colonies were analysed in relation to the presence and size of the predaceous calanoid copepod *Parabroteas sarsi*. *Conochilus* colonies increase in size throughout the season from May to August and then disappear from the lake. Simultaneously, *Parabroteas* developed from CI to CV and adults. We observed that when the predaceous copepod begins to prey on *Conochilus*, colony size increases in relation to maxilliped length of the predator. Our results show that the increasing size of the colony of *Conochilus* is an effective defense against *Parabroteas* predation.

Introduction

Planktonic rotifers are an important component of the diet of aquatic predators, especially of inverteb-rate predators, who prefer smaller prey individuals such as rotifers (Williamson, 1983). Since carnivorous planktonic invertebrates are size-dependent predators (Zaret, 1978; Gliwicz & Pijanowska, 1989), changes in the body shape and size of their prey may be effect-ive antipredator responses (Sih, 1987; Dodson, 1989). Among the different defenses of planktonic rotifers against predators, coloniality increases effective size and therefore may protect individuals in the colony from predation (Stemberger & Gilbert, 1987; Wallace, 1987).

All species of the genus *Conochilus* form colonies of different sizes. This feature may provide protection from predation by copepods, *Asplanchna* and *Lepto-dora* (Gilbert, 1980; Williamson, 1983; Edmondson & Litt, 1987; Wallace, 1987). *Conochilus* colonies often persist in Northern Hemisphere lakes in the presence of copepod predators like *Epischura* and *Diacyclops* (Stemberger & Evans, 1984; Stemberger & Gilbert, 1987). Similarly, in a small fishless pond of the South-ern Andes (41° S) *Conochilus hippocrepis* (Schrank) coexists with the large predaceous calanoid copepod *Parabroteas sarsi* Daday. Early studies revealed that

total prey length of *Daphnia* is a good predictor of susceptibility to *P. sarsi* (Balseiro & Vega, 1994).

In this study we analysed by indirect evidence the effectiveness of the coloniality of *Conochilus hippo-crepis* in permitting the coexistence with the predator *Parabroteas sarsi*.

Methods

The study was carried out in Laguna Fantasma, a small, fishless, temporary pond near Nahuel Huapi lake (41° S, 72° W, Argentina). The hydroperiod of the pond extends from autumn (April) to early sum-mer (December). From mid summer to early autumn, the pond remains dry and often freezes during winter (July). In particular, in 1994 the ice remained only for one week. A maximum depth of 1.3 m was recor-ded during June and July. Based on our unpublished observations of samples taken in early autumn, we know that *Conochilus hippocrepis* population devel-ops from resting eggs that hatch just after the first autumn rainfalls fill the pond. The pond was sampled for *C. hippocrepis* approximately every 10 days (range = 6–20 days) over a three month interval (May 31, 1994 to August 31, 1994). We both took qualitative and quantitative samples of the zooplankton of Laguna Fantasma to monitor the population dynamics and an-imals sizes of *C. hippocrepis* and *P. sarsi*. Qualitative

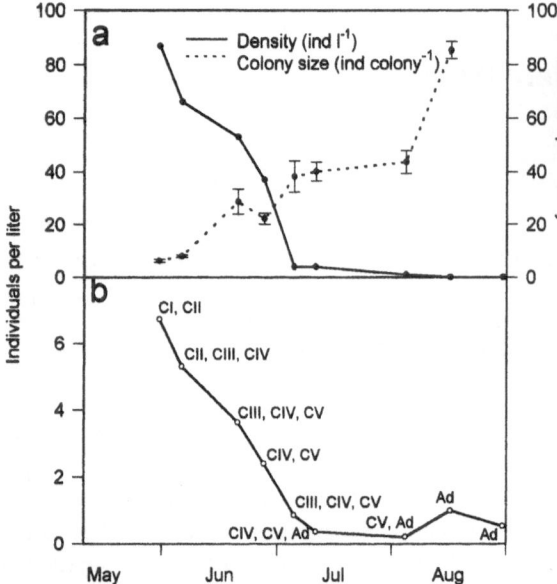

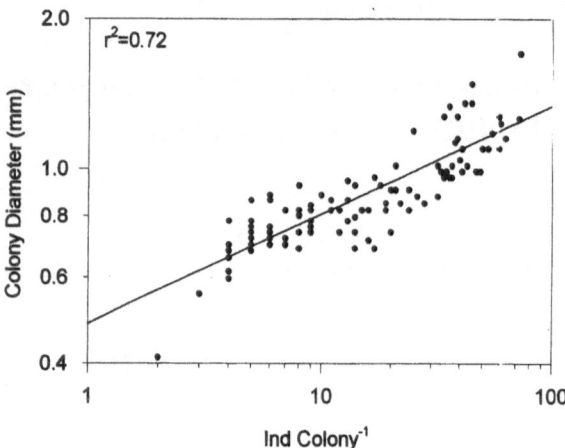

Figure 2. Colony size (ind colony^{-1}) and colony diameter (mm) of *Conochilus hippocrepis* in Laguna Fantasma, autumn–winter 1994.

Figure 1. Conochilus hippocrepis and *Parebroteas sarsi* densities in Laguna Fantasma during autumn-winter 1994. a: *Conochilus hippocrepis* density (ind l^{-1}) and colony size (ind colony^{-1}). b: *Parabroteas sarsi* density (ind l^{-1}) indicating the developmental stage present in each sampling date.

samples were obtained by horizontal trawls with a 35 μm mesh net from the surface to near the bottom. To obtain quantitative samples 30 litres of pond water were collected using a Schindler-Patalas trap and filtered through a 35 μm mesh net.

On each sampling date, living colonies of *Conochilus* were counted and measured, and the number of individuals per colony was determined. Colony diameter was estimated as the greater axial diameter. The density of each copepodid instar of the predator was estimated, and body size and length of the maxilliped were measured. For the analysis of the data, a weighted mean of the maxilliped length for each sampling date was obtained as the sum of the products of the density of the instar and the maxilliped length divided by copepod density. This estimation was called weighted maxilliped length (WML).

$$WML = \frac{\sum_{i=1}^{6} d_i \cdot m_i}{\sum_{i=1}^{6} d_i},$$

where i is the copepodid instar from CI to CVI (adults), d_i the instar density and m_i the maxilliped length of the instar.

Results

During the winter period the temperature of Laguna Fantasma remained very low, with values below 7 °C. *Conochilus hippocrepis* started the season with high densities of more than 80 individuals l^{-1}. *Conochilus* density declined to 4 ind l^{-1} in July and remained low until early August. The rotifer was not detected in the quantitative samples in late August (Figure 1a). On the other hand, colony size (individual per colony) increased throughout the population cycle from 8 up to 80 ind colony^{-1} (Figure 1a). As a consequence, colony diameter showed a similar trend. However, this increment is not linear with the number of individuals per colony. Recruitment of the first 10 individuals to the colony results in a two-fold increment in diameter, and another two-fold increase is reached with a ten-fold increase in colony size, fitting a power model ($R^2 = 0.72$, $P < 0.05$) (Figure 2).

The population of the predator *Parabroteas sarsi* (first instar copepodid) was also high at the beginning of the study period, with a density of 7 ind l^{-1} (Figure 1b). *P. sarsi* has a single cohort in each hydroperiod and, as this cohort develops, copepodid instars grow from CI in autumn to adults in late winter and spring (Figure 1b). This implies that the predator changes its size from 1.5 mm total length for CI to 4.5 mm for adults. The development of the cohort is rather synchronised, so in each date only two or three instars were found together (Figure 1b). Maxilliped length also showed a monotonical increase through the sampling period, from 0.6 mm in CI to 1.6 mm in CV and adults ($R^2 = 0.95$, $P < 0.05$) (Figure 3). From CI to the beginning of adulthood the copepod

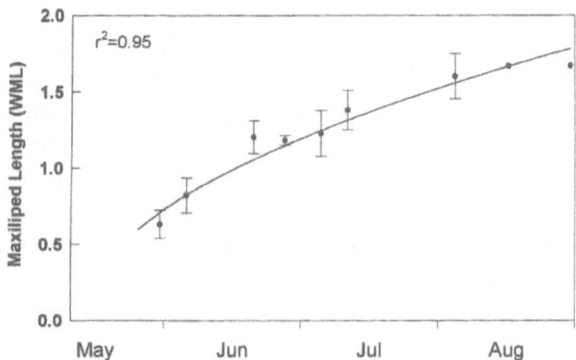

Figure 3. Change of weighted maxilliped length (WML) of the predaceous calanoid copepod *Parabroteas sarsi* in Laguna Fantasma, autumn–winter 1994.

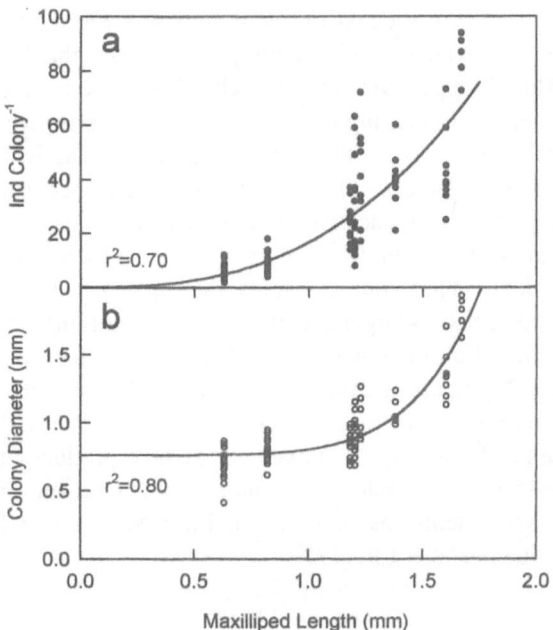

Figure 4. Relationship between colony size and predator size (weighted maxilliped length). a: Colony size expressed as ind per colony. b: colony size expressed as diameter.

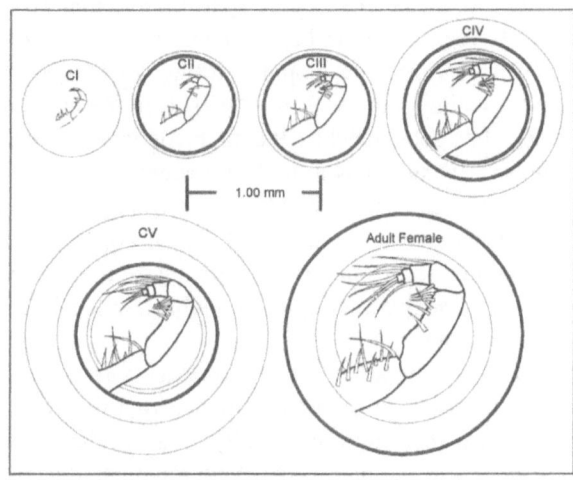

Figure 5. Schematic representation of coexisting relative sizes of maxilliped of *P. sarsi* and colonies of *Conochilus hippocrepis* (circles). Thick circles represent mean colony size when the particular stage of the predator was dominant. Thinner circles represent also the mean colony size when the particular stage of the predator was present but not as dominant.

that there is no colony of a predictable diameter less than 0.7 mm, which is the size of a colony with two individuals (Figure 4b). Colonies of this size coexist with the copepodid I (CI) stage, which apparently can not ingest *Conochilus* (Vega, 1995). Colonies begin to increase in size when larger copepodids, with larger maxillipeds (Figure 5), appear in the lake. It is remarkable that the maxilliped size of the dominant copepodid stage (measured unextended) was always less than the mean colony diameter (Figure 5).

Discussion

Coloniality is not a common feature within planktonic rotifers (Wallace, 1987). There are two main hypotheses that may explain the possible advantage of building colonies. An energetic advantage may be achieved if the clearance rate per individual is higher in colonies than in solitary individuals or, if there is an increase in filtering efficiency (Wallace, 1987). However, Wallace (1987) concluded that no increase in clearance rate was associated with colony size in different colonial rotifers. Coloniality can involve an antipredator defense as bigger colonies would be more difficult to manipulate by invertebrate predators such as cyclopoid copepods and *Asplanchna* (Stemberger & Gilbert, 1987). *Parabroteas sarsi* is a very large and voracious calanoid copepod that, as an adult, can

coexists with *Conochilus* and after the rotifer disappearance the large *Daphnia middendorifiana* starts its population cycle during spring.

Positive relationships were found between the length of the maxilliped and both individuals per colony ($R^2 = 0.70$, $P < 0.05$) and colony diameter ($R^2 = 0.80$, $P < 0.05$) (Figure 4a, b). In both cases the relationships were of the form $y = a + bx^c$, where y is either colony diameter or individuals per colony and x the maxilliped length (WML). This function relates colony diameter and WML, and it indicates easily capture and ingest prey up to 1.6 mm long (Balseiro & Vega, 1994). In Laguna Fantasma, *Conochilus* coexists with each developmental stage of *P. sarsi*. In the beginning of the season, *P. sarsi* is in the naupliar stage, but soon reaches the first copepodid instar. Although *P. sarsi* has a marked tendency for macrophagy during this time, it only consumes large diatoms and thecamoebae (Vega, 1995). From copepodid II to adulthood, *P. sarsi* is able to ingest rotifers including *Conochilus* (Diéguez, unpublished laboratory observations).

An increase in prey size decreases vulnerability to an invertebrate predator (Gliwicz & Pijanowska, 1989), but there may be a bottleneck, as the prey's offspring can be much smaller than adults. Balseiro & Vega (1994) showed that *Daphnia middendorffiana* solves this bottleneck by the production of a long tail spine in the juveniles. This spine diminishes the predation rate of *P. sarsi*. *Conochilus*, which in Laguna Fantasma coexists with this same predator, has an individual adult size within the size range vulnerable to *P. sarsi*, so not only are the offspring vulnerable but the adults too.

Development of colonies has been observed to protect *Conochilus unicornis* from the effect of interference competition with *Daphnia pulex* (Gilbert, 1988). The advantage of creating a colony is that rotifers may increase their effective size without increasing individual size. Besides, offspring are immediately included in the colony and thus, protected against predators in spite of their smaller size. An increase in predator size may induce an increase in colony size maintaining a similar level of protection. The size of the colony would be determined by the ability of the predator to capture and ingest it. However, it is not necessary to assume that *Parabroteas* chemically induces larger colonies in *Conochilus*. Such an increase in colony size may be a consequence of selective removal of smaller colonies by the predator. As is shown in Figure 5, *Conochilus* colonies are always larger than the mouthpiece of its predator, perhaps indirect evidence that smaller colonies have been removed by predation. Indeed, Vega (1995) found *Conochilus* trophi in CIII CIV and CV gut contents of *P. sarsi*.

The capability of an invertebrate predator to capture and ingest a given prey size should be directly related to the size of the mouthparts of the predator. Obviously, the predator's mouthparts will increase in size during postembryonic development. In a predator with a synchronous cohort development, as observed for *P. sarsi* in Laguna Fantasma, the vulnerable size of prey will change monotonically with time as the predator develops. Thus, if the prey population is affected by the change in the predator size, it would be expected to show a monotonic change in its size. Indeed we observed that the colony size, and the number of individuals per colony, increase with time during the season. The size of *Conochilus* colonies at the beginning is about 700 μm and then increases rapidly. This increase starts when *Parabroteas* maxillipeds reach the size that allow it to capture colonies with two or three individuals. Then the colony size goes on increasing according to the increase in maxilliped size of the predator (Figure 4b).

In natural populations of *C. hippocrepis* we did not observe colonies larger than 1.8 mm in diameter, about 80 ind. colony^{-1}. Colonies of this size were only observed coexisting with adults of *Parabroteas* in late winter and then disappeared from the lake. This disappearance may be related to some negative effect of increasing colony size. For example, when colonies are large, the filtering efficiency of individuals may decrease. This would imply that increasing size would enhance antipredation defense, but simultaneously would decrease energy uptake by the individuals. Thus, one can think that there is a critical size. Colonies larger than this one would not survive in a food limiting environment, but smaller ones would be preyed on by *Parabroteas* adults. However, if increasing colony size reduces filtering efficiency, the upper limit may be influenced by lake productivity, with larger colonies being found in less food-limiting environments, as observed in Lake Washington by Edmondson & Litt (1987).

Acknowledgements

We want to acknowledge Dr J. J. Gilbert and an anonymous referee for the helpful suggestions that certainly improved the manuscript. We also thank Dr R. L. Wallace for valuable discussions during the meeting. This work was supported by ANPCyT PMT-PICT0389 to EGB and UNC Grant B/701 to B. Modenutti. M.C. Diéguez was supported by Fundación Antorchas Grant A-13393/1-4.

References

Balseiro, E. G. & M. Vega, 1994. Vulnerability of *Daphnia middendorffiana* to *Parabroteas sarsi* predation: the role of the tail spine. J. Plankton Res. 16: 783–793.

Dodson, S. I., 1989. Predator-induced reaction norms. BioScience 39: 447–452.

Edmondson, W. T. & A. H. Litt, 1987. *Conochilus* in Lake Washington. Hydrobiologia 147: 157–162.

Gilbert, J. J., 1980. Feeding in the rotifer *Asplanchna*: behavior, cannibalism selectivity, prey defenses, and impact on rotifer community. In W. C. Kerfoot (ed.), Evolution and Ecology of Zooplankton Communities. University Press of New England, Hanover: 158–172.

Gilbert, J. J., 1988. Susceptibilities of ten rotifer species to interference from *Daphnia pulex*. Ecology 69: 1826–1838.

Gliwicz, Z. M. & J. Pijanowska, 1989. The role of predation in zooplankton succession. In W. C. Kerfoot & A. Sih (eds), Predation: Direct and Indirect Impacts on Aquatic Communities. University Press of New England, Hanover: 253–296.

Sih, A., 1987. Predators and prey lifestyles: an evolutionary and ecological overview. In W. C. Kerfoot & A. Sih (eds), Predation: Direct and Indirect Impacts on Aquatic Communities. University Press of New England, Hanover: 203–224.

Stemberger, R. S. & M. S. Evans, 1984. Rotifer seasonal succession and copepod predation in Lake Michigan. J. Great Lakes Res. 10: 417–428.

Stemberger, R. S. & J. J. Gilbert, 1987. Defenses of planktonic rotifers against predators. In W. C. Kerfoot & A. Sih (eds), Predation: Direct and Indirect Impacts on Aquatic Communities. University Press of New England, Hanover: 227–239.

Vega, M., 1995. La depredación intrazooplanctónica: un estudio sobre *Parabroteas sarsi*. Doctoral Dissertation, University of Comahue, 211 pp.

Wallace, R. L., 1987. Coloniality in the phylum Rotifera. Hydrobiologia 147: 141–155.

Williamson, C. E., 1983. Invertebrate predation on planktonic rotifers. Hydrobiologia 104: 385–396.

Zaret, T. M., 1978. A predation model of zooplankton community structure. Verh. int. Ver. Limnol. 20: 2496–2500.

Hydrobiologia **387/388**: 427–436, 1998.
E. Wurdak, R. Wallace & H. Segers (eds), Rotifera VIII: A Comparative Approach.
© *1998 Kluwer Academic Publishers.*

Spatial segregation between rotifers and cladocerans mediated by *Chaoborus*

María J. González
Department of Biological Sciences, Wright State University, Dayton, OH 45435, U.S.A.
Tel.: [+1] 937 775-2301; Fax: [+1] 937 775-3320; E-mail: maria.gonzalez@wright.edu

Key words: Chaoborus, vertical migration, spatial overlap, *Daphnia*

Abstract

In a field experiment I examined the effect of *Chaoborus* spp, on the vertical distribution of three rotifer species, *Kellicottia longispina, Keratella cochlearis* and *Polyarthra* sp. and on the spatial overlap of these rotifer species with three *Daphnia* species (*D. pulicaria, D. rosea* and *D. retrocurva*). In the presence of *Chaoborus*, total rotifer abundance increased, while total cladoceran abundance decreased. Patterns of migratory behavior varied among rotifer species. *Kellicottia longispina* and *Polyarthra* sp. showed vertical migration, while *K. cochlearis* did not. *Kellicottia longispina* mean depth was deeper during the day than during the night. The presence of *Chaoborus* had no significant effect on its vertical distribution. *Polyarthra* mean depth was significantly shallower during the day than during the night, but a marginally significant interaction suggests that day–night differences occurred only in the absence of *Chaoborus*. No vertical migration was observed in any *Daphnia* species in the absence of *Chaoborus*. *D. pulicaria* mean depth was significantly shallower in the presence of *Chaoborus*, and a marginally significant *Chaoborus*×time interaction suggests that *D. pulicaria* migrate upward during the night. The spatial overlap of *K. longispina* with each *Daphnia* species was not affected by *Chaoborus*. *Keratella cochlearis* was spatially segregated from *D. pulicaria* in the absence of *Chaoborus*, but the spatial overlap between these two species significantly increased in the *Chaoborus* treatment. Spatial segregation occurred between *Polyarthra* and *D. pulicaria* in absence of *Chaoborus*, however a significant *Chaoborus*×time interaction indicated that the spatial segregation occurred only during the day. These results suggest that *Chaoborus* could have complex indirect effects on rotifer–*Daphnia* interactions. Rotifer populations could be released from competition due to *Chaoborus* predation on *Daphnia*. *Chaoborus* presence, however, could intensify rotifer–*Daphnia* competitive interactions by increasing their spatial overlap.

Introduction

Diel vertical migration of zooplankton is a widespread phenomenon (Hays et al., 1994; Hutchinson, 1967). Most of the studies on diel vertical migration have focused on crustaceans (Dini & Carpenter, 1992; González & Tessier, 1997; Lampert, 1987; Neill, 1990; Ohman et al., 1983; Stich & Lampert, 1981) while few studies have documented this behavior in rotifers (Cruz-Pizarro, 1978; Dumont, 1968; George & Fernando, 1970; Magnien & Gilbert, 1983; Pivoda, 1977; Rutter-Kolisko, 1975). Predator–prey interactions are strongly dependent on the spatial distribution of predator and prey (Sih, 1987; Williamson, 1993). Diel vertical migration therefore can represent

an effective mechanism to minimize predation risk by visual and tactile predators (Gliwicz, 1986; Lampert, 1989; Stich & Lampert, 1981; Zaret & Suffern, 1976). Laboratory and field experiments have shown that cladocerans and copepods can alter their habitat use in the water column in response to the presence of fish and invertebrate predators (Dini & Carpenter, 1992; González & Tessier, 1997; Leibold, 1990; Neill, 1990; Wright & Shapiro, 1990). However, the effect of predators on rotifer vertical migration has not been experimentally tested.

Competition is an important factor regulating the composition and structure of rotifer communities (De-Mott, 1989). There is generally an inverse relationship between the presence of large cladocerans and the abundance of rotifers (Fradkin, 1995; Gilbert, 1988a; Lampert & Rothhaupt, 1991). Furthermore there is considerable evidence from in situ (Fradkin, 1995; Neill, 1984; Vanni, 1987) and laboratory experiments (Burns & Gilbert, 1986a,b; Gilbert, 1985a, 1988b; Gilbert & Stemberger, 1985; Stemberger & Gilbert, 1987) that rotifers are outcompeted by large zooplankton, in particularly *Daphnia* ≥ 1.2 mm. Competition occurs through both exploitative and interference competition. Habitat segregation is an important mechanism allowing coexistence of potential competitors (Leibold, 1991; Schoener, 1974). Therefore changes in vertical diel migration by rotifers and/or cladocerans as a response to the presence of a predator may affect the competitive interactions between these two taxa by altering their spatial overlap.

The objectives of this study were: (1) to experimentally examine the effect of an invertebrate predator, *Chaoborus* spp, on the vertical distribution of three rotifer species, *Kellicottia longispina*, *Keratella cochlearis* and *Polyarthra* sp. and (2) to document the effect of *Chaoborus* on the spatial overlap of these rotifer species with three *Daphnia* species.

Study site

The study site was Warner Lake, a small (26 ha), oligotrophic, hardwater (pH 8.5, alkalinity 100 meq/l, R. Bachmann unpublished data) kettle lake, located in Barry County, MI. The lake is relatively deep ($Z_{max}=16$ m) and remains thermally stratified throughout summer; the epilimnion extends to nearly 4 m. Warner Lake is typical of other lakes in the region in that larvae of two species of the invertebrate predator *Chaoborus* (*C. punctipennis* Say and *C. flavicans* Meigen) co-occur in the lake throughout the year, reaching total densities of up to 1 ind. l^{-1} in the summer (K. Geedey unpublished data, M. González personal observation).

Both *Chaoborus* species forage on zooplankton, but differ in their ability to handle larger prey items. *Chaoborus flavicans* is larger than *C. punctipennis* and is believed to be a more efficient predator of larger zooplankton (Moore, 1988; Soranno et al., 1993). Instars III and IV of both species, however, can prey on small species of crustacean zooplankton or juvenile stages of larger species, while instars I and II prey more effectively on rotifers (Elser et al., 1987; Mackay et al., 1990; Moore, 1988).

The rotifer assemblage of Warner Lake was dominated by *Kellicottia longispina*, *K. cochlearis* and *Polyarthra* sp. during the study period. Three species of *Daphnia* (*D. pulicaria*, *D. rosea* and *D. retrocurva*) coexist during summer in Warner Lake and are the major herbivorous zooplankton in the lake.

The three *Daphnia* species differ in body size and vertical habitat use. *Daphnia pulicaria* is the largest of the three species (mean size, 1.6 mm), *D. rosea* intermediate (mean size, 1.1 mm) and *D. retrocurva* the smallest (mean size, 0.80 mm). The contrast in *Daphnia* body size indicates the potential competitive interaction with rotifers. Among these three species, I expected *D. pulicaria* to be the strongest competitor with rotifers. In contrast, *D. retrocurva* should be the weakest competitor with rotifers. Late instars of *Chaoborus* depressed *Daphnia* abundance relative to predator-free controls (González & Tessier, 1997). Therefore, I expected that rotifers would be more strongly affected by the indirect effect of *Chaoborus* predation on the *Daphnia* assemblages than by the direct effect of *Chaoborus* predation on the rotifers.

Methods

Enclosure experiment

This study was part of a field experiment conducted during late July 1992 to assess the interactive effects of fish and *Chaoborus* on the abundance and habitat use of the *Daphnia* assemblage in Warner Lake. Here I present the experimental design and results of two treatments (presence and absence of *Chaoborus*). Details of the experimental design, and the effect of fish are described elsewhere (González & Tessier, 1997). The enclosures were cylindrical polyethylene bags sealed at the bottom, and suspended from a wooden frame buoyed with Styrofoam floats. Enclosures had a surface area of 1 m^2 and were 10 m deep, with a volume of 10 000 l.

The enclosures were filled with water taken from a depth of 2–8 m pumped through an 80-μm mesh, enabling phytoplankton and small rotifers to pass into the bags, but excluding larger zooplankton such as *Daphnia* and *Chaoborus*. Two days after filling, I stocked the bags with Warner lake densities of large zooplankton collected from Warner Lake using an 80-μm mesh plankton net. To avoid the addition of *Chaoborus* to the bags, zooplankton were collected by vertical hauls taken from 9 m to the surface during the day, when *Chaoborus* was in the hypolimnion. Predation treatments began 2 days after prey zooplankton were stocked.

Chaoborus (instars III and IV) were obtained from nearby Wintergreen Lake, where they were numerous and the relative abundance of *C. punctipennis* and *C. flavicans* was similar to that in Warner Lake (40:60, respectively). I separated *Chaoborus* from other zooplankton in the laboratory and added them randomly to three enclosures at the density observed in Warner Lake (0.8 ind. l^{-1}).

Temperature and oxygen conditions in the bags were very similar to those in the lake and were also similar among the bags. Temperature ranged from 21°C at the surface to 7.5°C at 10 m, with a thermocline at 4 m. The oxygen concentration ranged from 9 mg l^{-1} at the surface to 6.6 mg l^{-1} at 10 m.

Abundance patterns

Initial and final densities of *Chaoborus* were calculated from vertical hauls taken with an 80-μm mesh plankton net (diameter, 30 cm). The abundance of rotifers and cladocerans were calculated from zooplankton samples taken once during the day and once during the night at different depths (0, 2, 4, 6, 8 and 10 m), using a 20-l Schindler–Patalas trap equipped with an 80-μm mesh net. Since this study was part of a larger field experiment (four treatments, 12 bags), and much time was required to sample each bag at all depths, we randomly selected two bags from each treatment for the depth specific sampling. Samples were preserved in cold 5% sugar–Formalin solution.

The use of a 80-μm mesh to sample the bags probably excluded small rotifers from the collections. Therefore our samples may not provide a precise quantitative estimate of abundance patterns and species composition of the rotifer community in the different treatments. Analysis of zooplankton samples taken with a 63-μm mesh plankton net during early July, however, revealed that similar to my enclosures, *K. longispina*, *K. cochlearis* and *Polyarthra* sp. were the dominant rotifer species in Warner Lake. Furthermore, I focused my analysis on the relative treatment differences in migratory patterns and spatial overlap among these three rotifers and *Daphnia*. Although such parameters were certainly affected by the sampling methodology, I expected that the effect was similar in all treatments and relative differences would not be affected.

To evaluate the effect of *Chaoborus* on the density of rotifers and *Daphnia*, I calculated a mean density among the depth specific samples in each bag. I compared the mean density of each prey species in enclosures with and without *Chaoborus* using a *t*-test. Data were log transformed prior to analysis to meet assumptions of variance homogeneity (Sokal & Rohlf, 1988).

Vertical distribution patterns

I calculated the proportion of *Chaoborus* and each rotifer in the depth-specific samples during day and night. I also calculated a mean depth for each rotifer and *Daphnia* species during day and night. To test the effect of *Chaoborus* on the vertical distribution of rotifer and *Daphnia* species, I compared their mean depths in presence and absence of *Chaoborus* using a repeated-measures ANOVA (SAS Institute, 1990).

Overlap between rotifers and Daphnia

I calculated the overlap index between rotifers and cladocerans using the equation described by Williamson & Stoeckel (1990):

$$O_{ij} = \frac{\sum\limits_{z=1}^{m} (N_{jz} n_{iz})m}{\sum\limits_{z=1}^{m}(N_{jz}) \, (\sum\limits_{z=1}^{m} n_{jz})},$$

where z represents depth, m is number of depth points sampled, N_{jz} is the density of *Daphnia* at a given depth and n_{iz} is the density of rotifers at a given depth. Overlap index <1 indicates spatial segregation between the rotifer and *Daphnia* species. Overlap index$=1$ indicates that rotifer and/or *Daphnia* species are uniformly distributed in the water column. Overall index >1 indicates aggregation of the rotifer and *Daphnia* species in certain strata of the water column. We compared the overlap index (O_{ij}) among each rotifer and *Daphnia* species in enclosures with and without *Chaoborus* using a repeated measures ANOVA (SAS Institute, 1990).

430

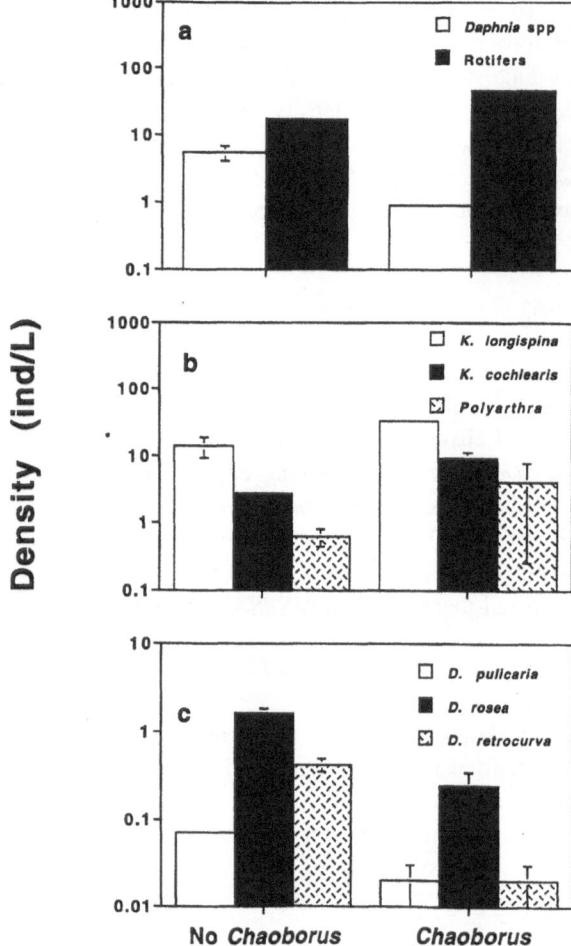

Figure 1. Total mean density (±SE) of dominant rotifers and *Daphnia* species (a), Mean density (±SE) of *K. longispina*, *K. cochlearis* and *Polyarthra* spp. (b) and *D. pulicaria*, *D. rosea* and *D. retrocurva* (c) in the absence and presence of *Chaoborus*. Error bars that are not apparent are smaller than the symbols.

Results

Abundance patterns

In the *Chaoborus* treatment, mean density of *Chaoborus* declined during the 2-week period of the experiment. The initial density was 0.90±0.27 ind./l 2 days after the initial addition, while the final mean density was 0.50±0.06 ind./l. This decline however was not statistically significant (df=3, t=1.69, P= 0.20). *Chaoborus* species composition significantly changed throughout the experiment (df=3, t=3.83, P=0.03). Initially the relative proportions of *C. punctipennis* and *C. flavicans* were similar (0.53 and 0.47, respectively), but *C. punctipennis* relative proportion declined through time (0.27 and 0.73, respectively,

González & Tessier, 1997). The decline could be attributed to the earlier pupation of *C. punctipennis* compared to *C. flavicans* (von Ende, 1982). Similar changes in *Chaoborus* species composition were observed in Warner and Wintergreen lakes during the experiment (M. González, personal observation).

Total rotifer abundance increased in the presence of *Chaoborus* (Figure 1a; df=2, t=24.18, P=0.04). Abundances of *K. longispina* and *K. cochlearis* were significantly higher in the *Chaoborus* treatment (Figure 1b; df=2, t=3.56 and 7.15, P<0.05). However, *Polyarthra* abundance was not statistically different between treatments (Figure 1a; df=2, t=1.79, P=0.22). Total *Daphnia* density significantly decreased in the presence of *Chaoborus* (Figure 1a; df=2, t=6.13, P=0.02). The abundance of *D. rosea* and *D. retrocurva* were significantly lower in the *Chaoborus* enclosures (Figure 1c; df=2, t=6.09 and 4.30, P<0.05), while *D. pulicaria* abundance was similar in presence and absence of *Chaoborus* (Figure 1c; df=2, t=0.27, P=0.81).

Vertical distribution patterns

Chaoborus showed a strong diel migration pattern. During the day *Chaoborus* stayed on the bottom of the bags, and migrated up in the water column at night. *Chaoborus* was most abundant at 2 m at night and its mean depth was 4.72±0.29 m (González & Tessier, 1997).

Kellicottia longispina showed a distinct pattern of vertical migration. Its mean depth was significantly deeper during the day than during the night (Figure 2a, df=1, 2, F=30.80, P=0.03). *Chaoborus*, however, had no significant effect on *K. longispina* vertical distribution (Figure 3d, df=1, 2, F=1.30, P=0.37). This rotifer was concentrated between 4 and 8 m during the day and it was more uniformly distributed during the night (Figure 3a,d). *Keratella cochlearis* showed no diel migration pattern (df=1, 2, F=3.9 P=0.18) and its mean depth was similar in both treatments (Figure 2b, df=1, 2, F=6.18 P=0.13). During the day the highest abundance of *K. cochlearis* occurred at 10 m, while at night this rotifer was uniformly distributed (Figure 3b,e). *Polyarthra* sp. distribution was more complex. Its mean depth was significantly shallower during the day than during the night (Figure 2c, df=1, 2, F=19.3, P=0.05), but a marginally significant interaction (df=1, 2, F=10.55, P=0.06) suggests that day–night differences occurred only in the absence of *Chaoborus*. *Polyarthra* seemed to avoid the metalim-

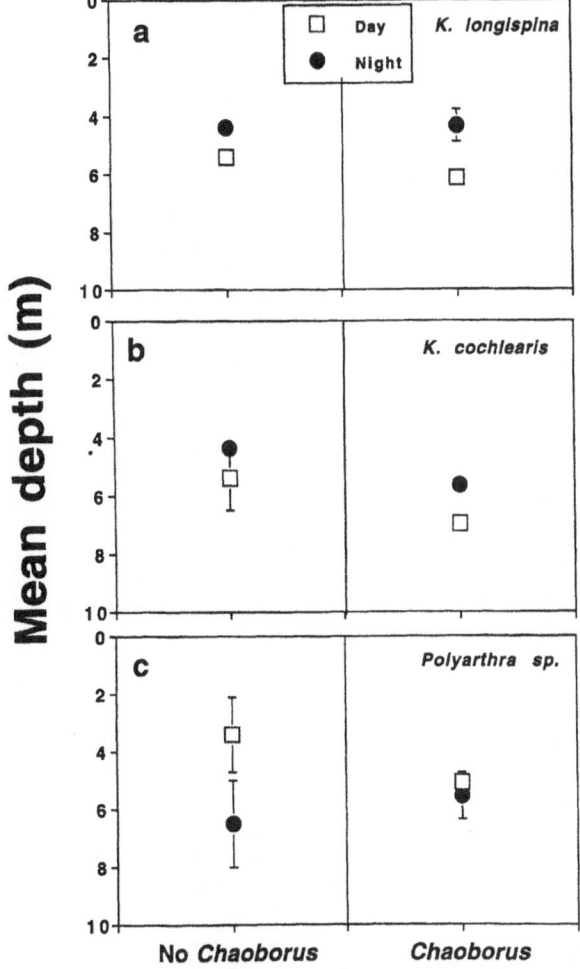

Figure 2. Mean depth (±SE) of (a) *K. longispina*, (b) *K. cochlearis* and (c) *Polyarthra* sp. during the day and night in the absence and presence of *Chaoborus*. Error bars that are not apparent are smaller than the symbols.

nion (6 m) in the absence of *Chaoborus*, but it was more evenly distributed in the water column in the presence of *Chaoborus* (Figure 3c,f).

In the absence of *Chaoborus* the vertical distribution of *Daphnia* was similar during day and night. *Daphnia pulicaria* remained in the metalimnion (Figure 4a), while *D. rosea* and *D. retrocurva* remained in the epilimnion (Figure 4b,c). In the presence of *Chaoborus*, only *D. pulicaria* mean depth was significantly shallower (Figure 4d; df=1, 2, F=24.7, P=0.03). A marginally significant interaction (df=1, 2, F=12.3, P=0.07) suggests that *Chaoborus* presence affected *D. pulicaria* only during the night (Figure 4d). For *D. rosea* and *D. retrocurva* a greater proportion of individuals were observed at 0 m in the presence of *Chaoborus* (Figure 4e,f), however, no significant

differences were detected (df=1, 2, F=0.44 and 0.88, $P > 0.1$).

Overlap between rotifers and Daphnia

The overlap of *K. longispina* with each *Daphnia* species was similar between treatments (Figure 5a–c; Table 1). This rotifer was not significantly segregated from *Daphnia* species (Table 1).

Chaoborus significantly affected the overlap of *K. cochlearis* and *Polyarthra* with *D. pulicaria*. *Keratella cochlearis* was spatially segregated from *D. pulicaria* in the absence of *Chaoborus* but not in its presence (Figure 5d, Table 1). I observed similar behaviour between *Polyarthra* and *D. pulicaria* (Figure 5g, Table 1), however a significant *Chaoborus*×time interaction (Table 1) indicated that the spatial segregation occurred only during the day.

Discussion

Patterns of migratory behavior were species specific. *Keratella longispina* and *Polyarthra* sp. showed vertical migration, while *K. cochlearis* did not show migratory behavior. Furthermore, the habitat use was opposite between the migratory species. *Kellicottia longispina* remained deeper in the water column during the day, while *Polyarthra* sp. remained deeper in the water column during the night. Spatial segregation among rotifer species has been documented in previous studies (Cruz-Pizarro, 1978; Miracle, 1977; Pivoda, 1977; Ruttner-Kolisko, 1975).

No previous information exists on the migratory behavior of *K. longispina*. For *K. cochlearis* and *Polyarthra* sp., however, my results confirm previous findings of great variability in the migratory pattern in a particular species among lakes (Miracle, 1977). The lack of migratory behavior of *K. cochlearis* has been previously reported in deep lakes (>8 m, Begg, 1976; Pivoda, 1977; Rutter-Kolisko, 1975), but Williamson & Magnien (1982) and Cruz-Pizarro (1978) observed that *K. cochlearis* vertically migrated in shallow lakes (<5 m). The vertical migration pattern of *Polyarthra* sp. observed in this study was similar to the pattern reported by Rutter-Kolisko (1975) for *P. dolicoptera*. In contrast, Pivoda (1977) observed no vertical migration for this species.

The higher abundance of *K. longispina* and *K. cochlearis* in the presence, as opposed to in the absence, of *Chaoborus* suggests a strong positive effect of this predator on rotifers by reducing *Daphnia*

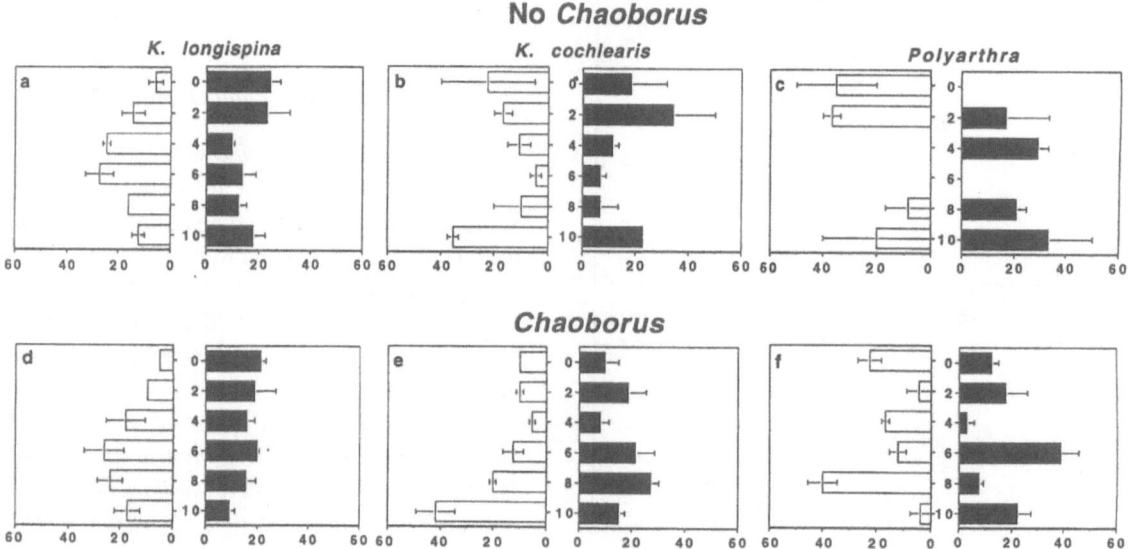

Figure 3. Vertical distribution (±SE) of (a,d) *K. longispina*, (b,e) *K. cochlearis* and (c,f) *Polyarthra* sp. during the day (open bars) and night (dark bars) in the absence and presence of *Chaoborus*. Error bars that are not apparent are smaller than the symbols.

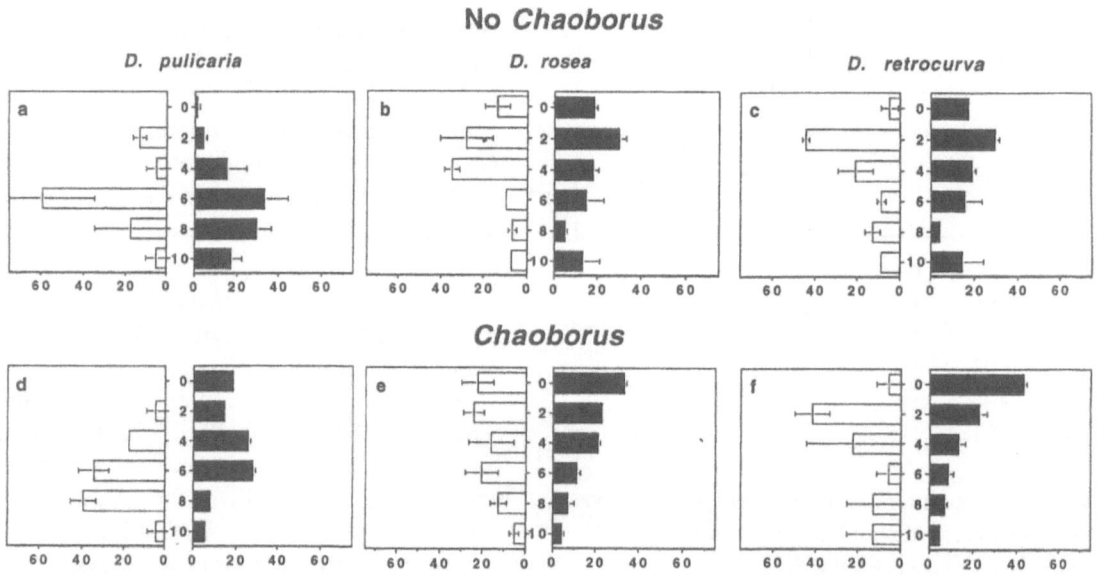

Figure 4. Vertical distribution (±SE) of (a,d) *D. pulicaria*, (b,e) *D. rosea* and (c,f) *D. retrocurva* during the day (open bars) and night (dark bars) in the absence and presence of *Chaoborus*. Error bars that are not apparent are smaller than the symbols.

abundance, thereby releasing rotifers from competition. This increase in rotifer abundance probably was caused by the higher proportion of late *Chaoborus* instars (III and IV) in my enclosures which prey probably on *Daphnia*.

Behavioral responses of rotifers to the presence of *Chaoborus* were less dramatic than the behavioral response of *Daphnia* (Dini & Carpenter, 1992; González and Tessier, 1997). The lack of behavioral responses

of *K. longispina* to *Chaoborus* presence suggests that other factors instead of *Chaoborus* predation are responsible for its diel vertical migration.

Changes observed in the overlap of rotifers and *D. pulicaria* as a response to *Chaoborus* suggest that this predator could have a negative effect on rotifers by increasing the spatial overlap with strong competitors such as *D. pulicaria*. The changes in the *K. cochlearis* –*D. pulicaria* overlap were caused by the strong mi-

433

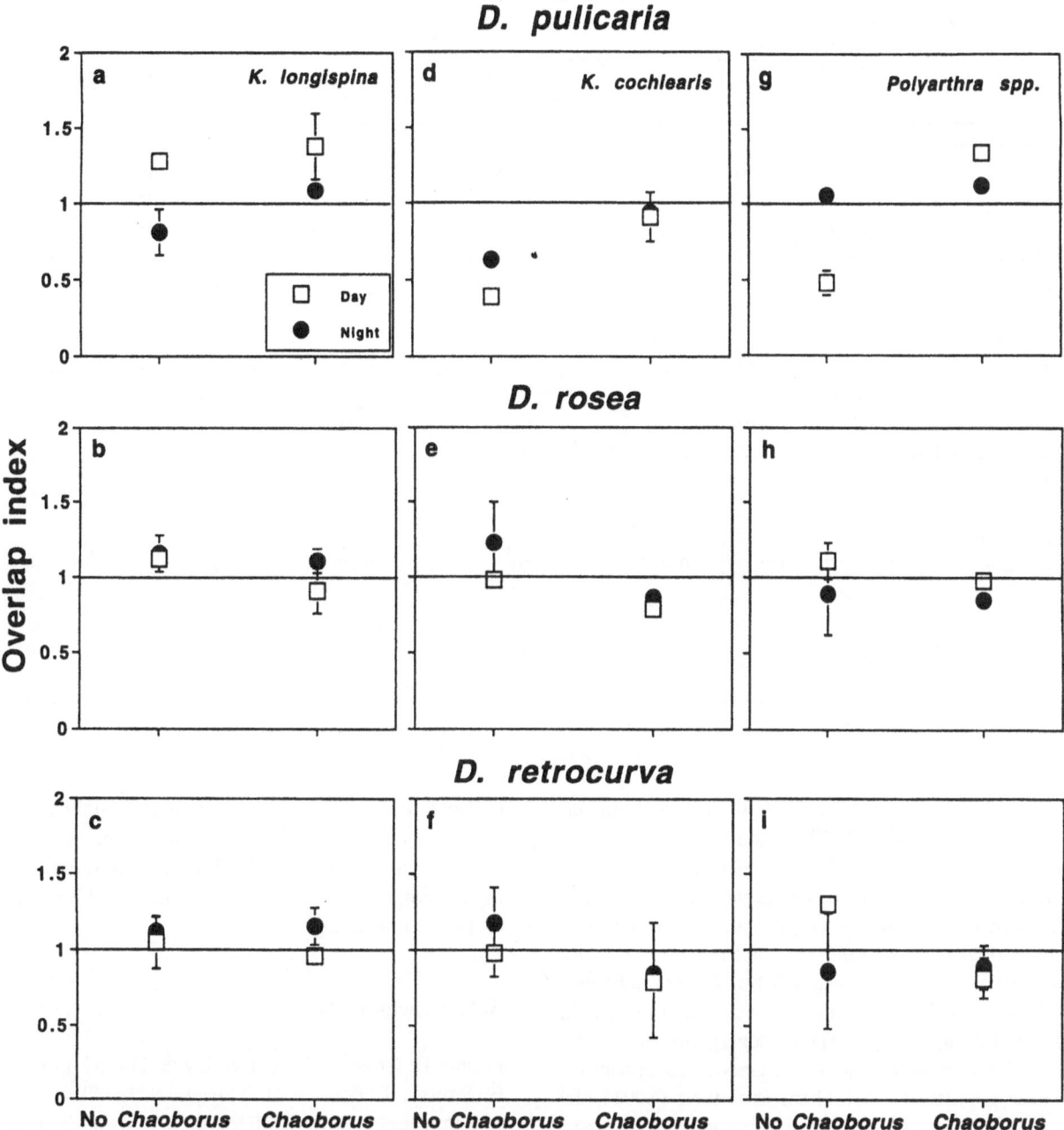

Figure 5. Overlap values (±SE) of *K. longispina* (a,b,c), *K. cochlearis* (d,e,f) and *Polyarthra* sp. (g,h,i) with *D. pulicaria*, *D. galeata* and *D. retrocurva* during the day (open squares) and night (dark circles) in the absence and presence of *Chaoborus*. Error bars that are not apparent are smaller than the symbols.

gratory behavior response of *Daphnia* to *Chaoborus*. *Daphnia pulicaria* showed no migratory behavior in the absence of *Chaoborus*, remaining in the metalimnion (6 m), while in the presence of *Chaoborus* this species migrated upwards during the night. Changes in the migratory behavior of *Polyarthra* sp., however, contributed to the greater overlap with *D. pulicaria* in

the *Chaoborus* enclosures. In absence of *Chaoborus*, this rotifer avoided the metalimnion where *D. pulicaria* was concentrated, while in the presence of *Chaoborus*, *Polyarthra* sp. showed a more uniform vertical distribution.

Polyarthra spp. exhibit an escape response which provides an effective defense against some inverte-

Table 1. Repeated-measures ANOVA on rotifer–*Daphnia* overlap (df=1, 2 for all rotifer species)

| | Overlap | | | | | | | | |
| | D. pulicaria | | | D. rosea | | | D. retrocurva | | |
	MS	F value	P value	MS	F value	P value	MS	F value	P value
K. longispina									
Chaoborus	0.07	1.904	0.30	0.033	8.23	0.10	0.002	0.072	0.81
Error	0.037			0.004			0.021		
Time	0.286	8.871	0.10	0.027	0.624	0.51	0.034	0.951	0.432
Chaoborus × Time	0.05	0.47	0.56	0.013	0.311	0.63	0.008	0.218	0.69
Error	0.032			0.043			0.035		
K. cochlearis									
Chaoborus	0.34	61.73	0.02*	0.149	5.821	0.13	0.128	1.447	0.35
Error	0.006			0/026			0.088		
Time	0.033	1.03	0.41	0.056	1.09	0.40	0.033	0.249	0.67
Chaoborus × Time	0.021	0.666	0.500	0.015	0.298	0.64	0.011	0.087	0.79
Error	0.032			0.051			0.132		
Polyarthra sp.									
Chaoborus	0.437	193.2	0.005*	0.002	0.179	0.713	0.099	1.528	0.347
Error	0.002			0.013			0.065		
Time	0.063	6.59	0.12	0.002	0.209	0.69	0.07	0.633	0.52
Chaoborus × Time	0.316	33.04	0.03*	0.019	1.732	0.318	0.138	1.24	0.39
Error	0.01			0.011			0.111		

brate predators (Gilbert, 1985b; Gilbert & Williamson, 1978, Moore & Gilbert, 1987) and interference competition from *Daphnia* (Gilbert, 1987; Kirk & Gilbert, 1988). The drastic differences in vertical distribution between *Polyarthra* and *D. pulicaria* in the absence of *Chaoborus* suggest that spatial segregation may be another mechanism by which *Polyarthra* could be less vulnerable to *Daphnia* interference competition. This result also suggests that spatial segregation should be considered in the evaluation of cladoceran impact on rotifer populations. Wickham & Gilbert (1991) observed that *Polyarthra vulgaris* growth rate declined in the presence of *D. pulex* in 3-l enclosures, but no decline was observed in larger field enclosures. Differences in spatial segregation in the large enclosures may explain the reduced vulnerability in these enclosures.

In summary, this study provides evidence on how competitive interactions between rotifers and large *Daphnia* could be affected by *Chaoborus*. Rotifer populations could be released from competition due to *Chaoborus* predation on *Daphnia*. On the other hand, alterations in habitat segregation mediated by *Chaoborus* could intensify competitive interactions between these taxa by increasing their spatial overlap. These results suggest the importance of incorporating diel vertical migration and habitat use in understanding the combined effects of competition and predation on rotifer assemblages.

Acknowledgments

I thank D. Crossen, C. K. Geedey, S. Hu, M. Leibold, E. Smiley, J. Tsao, A. J. Tessier, A. M. Turner, M. J. Vanni and P. Woodruff for their assistance in the field and laboratory. I also thank Lyle Champion for kindly providing access to Warner Lake and J. J. Gilbert for inspiring this work. I was supported by a Michigan State University Postdoctoral Fellowship during the performance of the field experiment and by Ohio Board of Regents/Research Challenge Grant during the sample processing and manuscript preparation. This is contribution number 839 of the W. K. Kellogg Biological Station.

References

Begg, G. W., 1976. The relationship between the diurnal movements of some of the zooplankton and the sardine *Limnothrisa miodon* in Lake Kariba, Rhodesia. Limnol. Oceanogr. 21: 529–539.

Burns, C. W. & J. J. Gilbert, 1986a. Effects of dapnid size and density on interference between *Daphnia* and *Keratella cochlearis*. Limnol. Oceanogr. 31: 848–858.

Burns, C. W. & J. J. Gilbert, 1986b. Direct observation of the mechanism of interference between *Daphnia* and *Keratella cochlearis*. Limnol. Oceanogr. 31: 859–866.

Cruz-Pizarro, L., 1978. Comparative vertical zonation and diurnal migration among Crustacea and Rotifera in the small high mountain lake La Caldera (Granada, Spain). Verh. int. Ver. Limnol. 20: 1026–1032.

DeMott, W. R., 1989. The role of competition in zooplankton succession. In U. Sommer (ed.), Plankton Ecology. Springer-Verlag, Berlin: 195–252.

Dini, M. L. & S. R. Carpenter, 1992. Fish predators, food availability and diel vertical migration in *Daphnia*. J. Plankton Res. 14: 359–377.

Dumont, H. J., 1968. A study of a man-made freshwater reservoir in eastern Flanders (Belgium) with special reference to the vertical migration of the zooplankton. Hydrobiologia 32: 97–130.

Elser, M. M., C. N. von Ende, P. A. Sorrano & S. R. Carpenter, 1987. *Chaoborus* populations: response to food web manipulation and potential effects on zooplankton communities. Can. J. Zool. 65: 2846–2852.

Fradkin, S. C., 1995. Effects of interference and exploitative competition from large-bodied cladocerans on rotifer community structure. Hydrobiologia 313: 387–393.

George, M. G. & C. H. Fernando, 1970. Diurnal migration of three rotifers in Sunfish Lake, Ontario. Limnol. Oceanogr. 15: 218–223.

Gilbert, J. J., 1985a. Competition between rotifers and *Daphnia*. Ecology 66: 1945–1950.

Gilbert, J. J., 1985b. Escape responses of the rotifer *Polyarthra*: a high-speed cinematographic analysis. Oecologia 66: 322–331

Gilbert, J. J., 1987. The *Polyarthra* escape response: defense against interference from *Daphnia*. Hydrobiologia 147 (Dev. Hydrobiol. 42): 235–238.

Gilbert, J. J., 1988a. Suppression of rotifer populations by *Daphnia*: a review of the evidence, the mechanisms and the effects on zooplankton community structure. Limnol. Oceanogr. 33: 1286–1303.

Gilbert, J. J., 1988b. Susceptibilities of ten rotifer species to interference from *Daphnia pulex*. Ecology 69: 1826–1838.

Gilbert, J. J. & R. S. Stemberger, 1985. Control of *Keratella* populations by interference competition from *Daphnia*. Limnol. Oceanogr. 30: 180–188.

Gilbert, J. J. & C. E. Williamson, 1978. Predator–prey behavior and its effect on rotifer survival in associations of *Mesocyclops edax*, *Asplanhna girodi*, *Polyarthra vulgaris* and *Keratella cochlearis*. Oecologia 37: 13–22.

Gliwicz, M. Z., 1986. Predation and the evolution of vertical migration in zooplankton. Nature 320: 746–748.

González, M. J. & A. J. Tessier, 1997. Habitat segregation and interactive effects of multiple predators on a prey assemblage. Freshwat. Biol. 37: 179–191.

Hays, G. C., C. A. Proctor, W. G. John, & A. J. Warner, 1994. Interspecific differences in the diel vertical migration of marine copepods: The implications of size, color, and morphology. Limnol. Oceanogr. 39: 1621–1629.

Hutchinson, G. E., 1967. A Treatise on Limnology. Vol 2. Wiley, New York.

Kerfoot, W. C., 1987. Cascading effects and indirect pathways. In: Kerfoot, W. C. & A. Sih (eds), Predation: Direct and Indirect Impacts on Aquatic Communities. University Press: 57–70.

Kirk, K. & J. J. Gilbert, 1988. The escape behavior of *Polyarthra* in response to artificial flow stimuli. Bull. mar. Sci. 43: 551–560.

Lampert, W., 1987. Vertical migration of freshwater zooplankton: indirect effect of vertebrate predators on algal communities. In: Kerfoot, W. C. & A. Sih (eds), Predation: Direct and Indirect Impacts on Aquatic Communities. University Press: 203–224.

Lampert, W., 1989. The adaptive significance of diel vertical migration of zooplankton. Funct. Ecol. 3: 21–27.

Lampert W. & K. O. Rothhaupt, 1991. Alternating dynamics of rotifers and *Daphnia magna* in a shallow lake. Arch. Hydrobiol. 120: 447–456.

Leibold, M. A., 1990. Resources and predation can affect the vertical distribution of zooplankton. Limnol. Oceanogr. 35: 938–944.

Leibold, M. A., 1991. Trophic interactions and habitat segregation between competing *Daphnia* species. Oecologia 86: 510–520.

MacKay, N. A., S. R. Carpenter, P. A. Soranno & Vanni M. J., 1990. The impact of two *Chaoborus* species on a zooplankton community. Can. J. Zool. 68: 981–985.

Magnien, R. E. & J. J. Gilbert, 1983. Diel cycles of reproduction and vertical migration in the rotifer *Keratella crassa* and their influence on the estimation of population dynamics. Limnol. Oceanogr. 28: 957–969.

Miracle, M. R., 1977. Migration, patchiness and distribution in time and space of planktonic rotifers. Arch. Hydrobiol. Beih. Ergebn. Limnol. 8: 19–37.

Moore, M. V., 1988. Differential use of food resources by the instars of *Chaoborus punctipennis*. Freshwat. Biol. 19: 249–268.

Moore, M. V. & J. J. Gilbert, 1987. Age-specific *Chaoborus* predation on rotifer prey. Freshwat. Biol. 17: 223–236.

Neill, W. E., 1984. Regulation of rotifer densities by crustaceans zooplankton in an oligotrophic montane lake in British Columbia. Oecologia 61: 175–181.

Neill, W. E., 1990. Induced vertical migration in copepods as a defence against invertebrate predation. Nature 345: 524–526.

Ohman, M. D., B. Froat & E. Cohen, 1983. Reverse diel vertical migration: an escape response from invertebrate predators. Science 220: 1404–1407.

Pivoda, B., 1977. Migration of planktonic rotifers in Lunzer Obersee (Austria). Arch. Hydrobiol. Beih. Ergebn. Limnol. 8: 50–52.

Ruttner-Kolisko, A., 1975. The vertical distribtion of plankton rotifers in a small alpine lake with a sharp oxygen depletion (Lunzer Obersee) Verh. int. Ver. Limnol. 19: 1286–1294.

Statistical Analysis System Institute, 1990. SAS/STAT User's Guide, Version 6. 4th edn. SAS Institute, Cary, NC, U.S.A.

Schoener, T., 1974. Resource partitioning in ecological communities. Science 185: 27–39.

Sih, A., 1987. Predator and prey lifestyles: an evolutionary and ecological overview. In: Kerfoot, W. C. & A. Sih (eds), Predation: Direct and Indirect Impacts on Aquatic Ccommunities. University Press: 203–224.

Sokal, R. R. & F. R. Rohlf, 1988. Biometry. W. H. Freeman, N.Y.

Soranno, P. A., S. R. Carpenter & S. M. Moegenburg, 1993. Dynamics of the phanton midge: implications for zooplankton. In: Carpenter, S. R. & J. F. Kitchell (eds), The Trophic Cascade in Lakes. Cambridge University Press: 103–115.

Stemberger, R. S. & J. J. Gilbert, 1987. Rotifer threshold food concentrations and the size-efficiency hypothesis. Ecology 68: 181–187.

436

Stich, H. & W. Lampert, 1981. Predator evasion as an explanation of diurnal vertical migration by zooplankton. Nature: 293: 396–398.

Tessier, A. J. & M. A. Leibold, 1997. Habitat use and ecological specialization within lake *Daphnia* populations. Oecologia 109: 561–570.

Vanni, M. J., 1987. Food availability, fish predation, and the dynamics of a zooplankton community coexisting with planktivorous fish. Ecol. Monogr. 57: 61–88.

von Ende, C. N., 1982. Phenology of four *Chaoborus* species. Envir. Ent. 11: 9 –16.

Wickham, S. E. & J. J. Gilbert, 1991. Relative vulnerabilities of natural rotifer and ciliate communities to cladocerans: laboratory and field experiments. Freshwat. Biol. 26: 77–86.

Williamson, C. E., 1993. Linking predation risk models with behavioral mechanisms: Identifying population bottle-necks. Ecology 74: 320–331.

Williamson, C. E. & R. E. Magnien, 1982. Diel vertical migration in *Mesocyclops edax*: Implications for predation rate estimates. J. Plankton Res. 4: 329–339.

Williamson, C. E. & E. Stoeckel Mark, 1990. Estimating predation risk in zooplankton communities: the importance of vertical overlap. Hydrobiologia 198 (Dev. Hydrobiol. 60): 125–131.

Wright, D. & J. Shapiro, 1990. Refuge availability: a key to understanding the summer disappearance of *Daphnia*. Freshwat. Biol. 24: 43–62.

Zaret, T. M. & J. S. Suffern, 1976. Vertical migration in zooplankton as a predator avoidance mechanism. Limnol. Oceanogr. 21: 804–813.

Hydrobiologia **387/388**: 437–444, 1998.
E. Wurdak, R. Wallace & H. Segers (eds), Rotifera VIII: A Comparative Approach.
© *1998 Kluwer Academic Publishers.* .

Morphotype-specific predation in the trimorphic rotifer *Asplanchna silvestrii*

Stephanie E. Hampton

Department of Biological Sciences, Dartmouth College, 6044 Gilman, Hanover, NH 03755-3576, U.S.A.

Key words: predation, polymorphism, cannibalism, *Asplanchna*, rotifers, zooplankton

Abstract

Several species of the predatory rotifer *Asplanchna* exhibit dramatic diet-induced trimorphism. The three morphotypes differ greatly in body shape and size, attributes that should affect predation ability. It has been hypothesized that these morphotypes evolved to exploit different prey assemblages in different environments. Here I compare the predatory behavior of the campanulate morphotype (the largest) to that of the cruciform morphotype (the intermediate) using crustacean and conspecific prey. These prey are known to induce production of the greatest proportion of campanulates. I hypothesize that the campanulate is better able to exploit these relatively large prey than are cruciforms. The campanulates did have higher ingestion rates with conspecific prey, but the ingestion rates of the morphotypes were not different with the crustacean prey, due to the campanulate's relatively low probability of attacking the crustacean prey. The campanulate attack probability is higher with both conspecific and crustacean prey than has been previously reported for campanulate *A. silvestrii* with smaller rotifer prey. While the campanulate handles both relatively large prey with comparative ease, and is more likely to attack these prey than smaller rotifer prey, the campanulate morphotype seems most effective at cannibalism due to its high preference for congeneric prey.

Introduction

The rotifer *Asplanchna* is a common predator in freshwater zooplankton communities. Several species exhibit marked diet-induced polymorphism (Gilbert, 1980a). While the feeding behavior of *Asplanchna* has attracted much attention (Hurlbert et al., 1971; Green & Lan, 1974; Salt, 1977; Gilbert, 1978; Gilbert & Williamson, 1978; Commins & Salt, 1988; Starkweather & Walsh, 1989; Urabe, 1992; Conde-Porcuna et al., 1993; Gilbert & Jack, 1993; Sarma, 1993; Conde-Porcuna & Sarma, 1995; Iyer & Rao, 1996), the ability of specific morphotypes to feed on various prey types has not been thoroughly addressed. Here I assess the potential of two morphotypes of *A. silvestrii* to feed on two relatively large prey types, crustacean and congeneric prey.

The polymorphism of *A. silvestrii*, though it has not been directly experimentally addressed, appears to be quite similar to that of *A. sieboldi* (Gilbert &

Confer, 1986). *Asplanchna* emerges from diapause in the small saccate form (0.5–0.8 mm in *A. silvestrii*), but a diet containing a specific molecular form of vitamin E (α-tocopherol) induces production of the larger morphotypes (Figure 1) in subsequent generations. The saccate is probably a relatively ephemeral morphotype in natural populations, since α-tocopherol is produced by photosynthetic organisms and consequently present in algivorous prey. The cruciform has both a larger body size (typically 1.0–1.9 mm in *A. silvestrii*) and body wall outgrowths known to protect it from cannibalism (Gilbert, 1973). The largest morphotype, the campanulate, has a still larger body size (typically 1.6–2.0 mm in *A. silvestrii*), no body wall outgrowths, and a disproportionately broad corona. Campanulates appear to be produced at the highest frequency when cultured on congeners or crustaceans (Powers, 1912; Gilbert, 1975).

Hurlbert et al. (1972) first speculated that polymorphism may have evolved in these species to allow

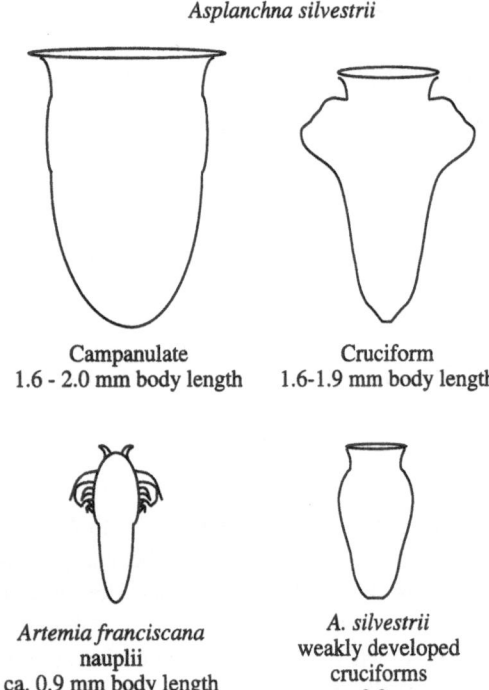

Asplanchna silvestrii

Campanulate
1.6 - 2.0 mm body length

Cruciform
1.6-1.9 mm body length

Artemia franciscana
nauplii
ca. 0.9 mm body length

A. silvestrii
weakly developed
cruciforms
ca. 0.9 mm

Figure 1. Predator morphotypes (campanulate and cruciform) and prey items (*A. franciscana* and small *A. silvestrii*) used in experiments, with approximate body lengths. Body wall outgrowths on the top right cruciform shape are shown fully extended, although these outgrowths are usually collapsed against the body unless the cruciform is attacked.

exploitation of different prey assemblages in a variable environment. The size of prey *Asplanchna* can ingest appears to be a direct function of the corona size. Thus it would be adaptive for larger morphotypes to be produced when larger prey are more available, and the campanulate's disproportionately large corona should provide an advantage over the cruciform in feeding on large prey items. Gut content analysis of individuals from natural populations of trimorphic *Asplanchna* has shown that diet does vary among the morphotypes (Hurlbert et al., 1972); campanulates ate more copepodites, cladocerans, small insects, and conspecific prey than the other morphotypes.

Gilbert has described the propensity for cannibalism in other *Asplanchna* species (reviewed in 1980b). He found the largest phenotypes in all the polymorphic species to have high preference and ability for cannibalism, while the only non-polymorphic species he studied did not attack conspecific prey. The high campanulate potential for cannibalism, coupled with observations that campanulate production is higher on congeneric than other rotifer diets, suggests that can-

nibalism is an adaptive function of the campanulate morphotype.

A comparison of the predatory behavior of *Asplanchna* with congeners and crustaceans is of acute interest since crustacean prey is the only other prey type known to induce high levels of campanulate production in cultures. If trimorphism is an adaptation facilitating exploitation of a wider range of prey, one would expect greater production of certain morphotypes in response to a given prey to correspond with greater potential to ingest that prey. Accordingly, recent work (Hampton & Starkweather, in press) demonstrated that cruciform *A. silvestrii* were generally more effective predators on rotifer prey than were campanulates, due to low probability of campanulates attacking rotifer prey after encounter.

In an effort to explain the differentiation in food niches that may have contributed to the evolution of trimorphism, I have conducted observational experiments on *A. silvestrii* cruciforms and campanulates with conspecific and crustacean prey. Here I test the hypothesis that campanulates have higher feeding rates than cruciforms on these relatively large prey. In addition, I address the expectation that campanulates demonstrate higher preference for conspecific and crustacean prey than has been reported for smaller rotifer prey.

Materials and methods

I cultured a mixture of clones of *A. silvestrii* from Little Fish Lake, Nevada, U.S.A., individually in 2 ml tissue culture well plates in glass-fiber-filtered lake water (Storr's Pond, New Hampshire, U.S.A.). All rotifer and algal cultures were kept at 19 °C on a photoperiod (LD 16:8). Large cruciforms and campanulates (Figure 1) ate neonatal *Artemia franciscana* (0.5–0.6 mm body length) hatched from cysts (Argent Chemical Laboratories) at 20–22 °C in filtered lake water treated with 1 gNaCl l^{-1}. Smaller cruciforms (Figure 1) ate *B. calyciflorus* that had been reared on *Cryptomonas* spp. Cryptomonad algae were cultured on a modified MBL medium (Stemberger, 1981). *Artemia* individuals used as prey in experiments (Figure 1) were approximately two days old and larger than those used to culture *A. silvestrii*.

To determine the relative abilities of cruciforms and campanulates to ingest *A. franciscana* and smaller *A. silvestrii*, I conducted observational experiments that allowed me to analyze behavioral interactions between predator morphotypes and prey types. I carried out 50 experiments with each prey type, equally divided between cruciforms and campanulates. My methods closely resembled those of similar studies (Gilbert, 1978; Gilbert & Williamson, 1978; Starkweather & Walsh, 1989; Sarma, 1993; Iyer & Rao, 1996).

Before the experiments, I isolated predators from food for 4 to 6 hours and recorded body length to the nearest 0.1 mm. The predators in the experiments were all adults between 2 and 5 days old. Although many intermediate forms occurred in cultures, I chose only those individuals which most strongly resembled the cruciform and campanulate morphotypes (Figure 1). All cruciforms had maximally developed body wall outgrowths (Birky & Power, 1969), and campanulates all had characteristically broad coronae and no body wall outgrowths (Gilbert, 1980a).

I pipetted 30 similarly sized individuals of a species (Figure 1) into 1 ml of filtered lake water in a glass depression plate. After placing one 'starved' predator into the depression, I observed the interactions under 12× magnification until one prey had been ingested or 10 minutes had passed with no ingestion. Events noted were encounter (prey contacting *A. silvestrii*'s corona), attack (orientation of the corona toward prey followed by opening the mouth to draw in prey), time of both successful and unsuccessful captures (capture defined here as partial or complete entrapment of the prey in the mastax), and time of lost capture or ingestion (ingestion defined as movement of the prey past the trophi). If no ingestion occurred, I exposed the predator to a high density of another, presumably 'easier', prey item (*B. calyciflorus* after *Artemia* prey experiments, and neonatal *Artemia* after conspecific prey experiments). If no ingestion of the second item occurred within 20 minutes, the predator was assumed unhealthy or otherwise unusual and the experimental results obtained from that predator were discarded. Such exclusion was rare and normally coupled with notably low or zero attack probability and unusual behavior such as frequent contraction of the corona. After termination of each experiment, I returned the predator and prey to cultures to contribute offspring for later use, but did not reuse individuals in experiments.

I used the Kruskal-Wallis ranks test to analyze the effect of morphotype and prey type on encounter rate, proportion of encounters followed by attack, proportion of attacks followed by successful capture, handling time (time from successful capture to ingestion), and ingestion rate.

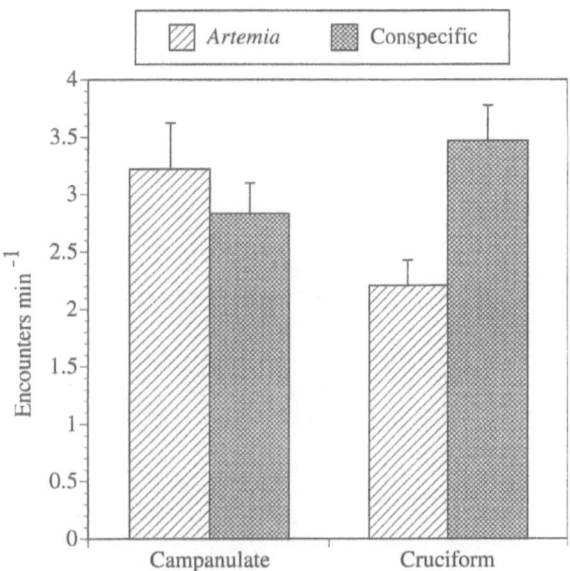

Figure 2. Encounter rates between the campanulate and cruciform morphotypes of *A. silvestrii* and two prey types, *A. franciscana* and small *A. silvestrii*. Error bars are standard error.

Results

Encounter rates

The encounter rates of campanulates and cruciforms with prey (Figure 2) were not significantly different ($p = 0.5955$). While campanulate encounter rates with *Artemia* and conspecific prey were similar ($p = 0.8461$), cruciforms had a significantly higher encounter rate with conspecifics than with *Artemia* ($p = 0.0046$). Campanulate mean encounter rate was higher than that of cruciforms with *Artemia*, although not significantly different ($p = 0.0914$). However, cruciforms had a slightly higher rate of encounter than campanulates with conspecifics, also not significant ($p = 0.1682$).

Attack after encounter

Both morphotypes expressed a similar attack probability when both prey types were pooled ($p = 0.8246$), but the behavior of the campanulates was dramatically different between the prey types (Figure 3). Although cruciform attack probability did not differ

440

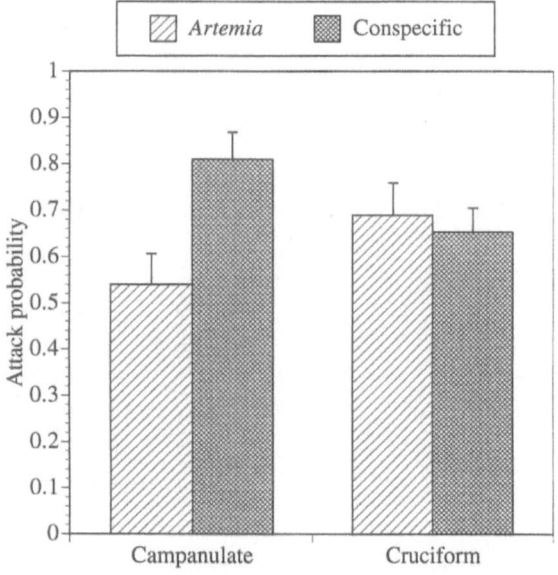

Figure 3. Probabilities of attack (proportion of encounters followed by attack) by the campanulate and cruciform morphotypes of *A. silvestrii* on two prey types, *A. franciscana* and small *A. silvestrii*. Error bars are standard error.

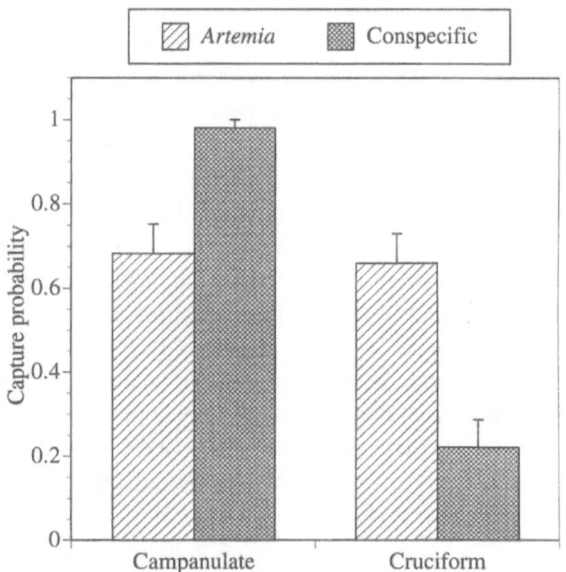

Figure 4. Probabilities of capture (proportion of attacks followed by successful captures) for the campanulate and cruciform morphotypes of *A. silvestrii* with two prey types, *A. franciscana* and small *A. silvestrii*. Error bars are standard error.

with either prey (p = 0.4945), the campanulates attacked conspecific prey significantly more frequently than *Artemia* (p = 0.0046). Campanulates were also significantly more likely to attack conspecific prey than were cruciforms (p = 0.037).

Capture after attack

Campanulates and cruciforms did not differ in probability of capture after attack (p = 0.2934), but both predators showed different capture probabilities with each prey type. The campanulate capture probability was significantly higher with conspecifics than *Artemia* (p = 0.0004), and the cruciform capture probability was significantly lower with conspecifics than *Artemia* (p = 0.0000). While campanulate and cruciform capture probability was very similar with *Artemia* (p = 0.814), the difference in probability of campanulates and cruciforms capturing conspecific prey was highly significant (p = 0.0000). Campanulates had almost complete success in capturing conspecifics after attack while cruciforms had relatively low probability of capture after attack (Figure 4). Ten of the twenty-five cruciforms presented with conspecific prey were entirely unable to successfully capture one prey item within the ten-minute experimental period.

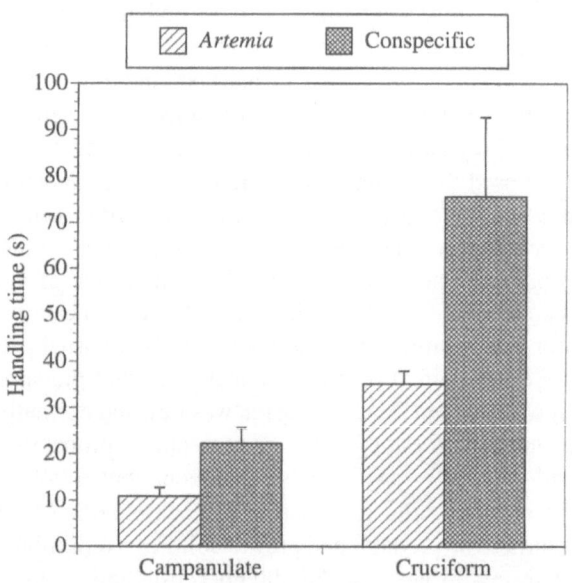

Figure 5. Handling times (time from successful capture to ingestion) for the campanulate and cruciform morphotypes of *A. silvestrii* with two prey types, *A. franciscana* and small *A. silvestrii*. These data do not include the ten cruciforms that were unable to ingest conspecific prey. Error bars are standard error.

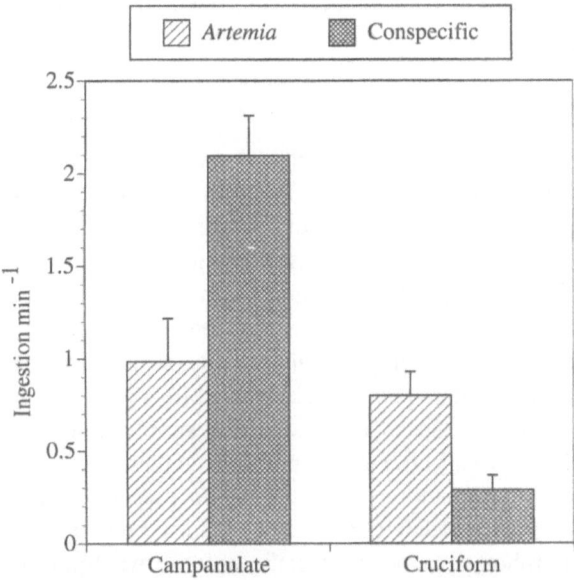

Figure 6. Ingestion rates for the campanulate and cruciform morphotypes of *A. silvestrii* with two prey types, *A. franciscana* and small *A. silvestrii*. Error bars are standard error.

Handling time

The handling time values (Figure 5) include only those predators which successfully ingested prey, so the cruciform mean handling time with conspecifics reflects the fifteen cruciforms that successfully ingested conspecifics. Campanulates were better able to ingest both prey types after capture than cruciforms ($p<0.0001$). Conspecific prey took significantly longer to ingest after capture than *Artemia* in general ($p = 0.0503$), and within each morphotype (campanulates, $p = 0.0015$; cruciforms, $p = 0.0055$). Cruciforms required significantly more time to handle both *Artemia* and conspecific prey than did campanulates ($p < 0.0001$ for each prey).

Ingestion rate

Campanulates demonstrated greater ability than cruciforms to ingest prey ($p = 0.0000$). However, campanulate ability differed quite significantly between prey types ($p = 0.0000$), showing a much higher rate of ingestion with conspecific prey than *Artemia* (Figure 6). Cruciforms, in contrast, ingested *Artemia* at a significantly higher rate than conspecific prey ($p = 0.0004$). There was no significant difference in ingestion rates of campanulates and cruciforms with *Artemia* ($p = 0.7124$), but campanulates showed a

considerably greater ability to ingest conspecifics than cruciforms ($p = 0.0000$).

Discussion

Cannibalism appears to have been a very strong agent of selection in the trimorphic *Asplanchna* species. Cruciforms and males are heavily defended against cannibalism by body wall outgrowths (Gilbert, 1973), and campanulates in all known trimorphic species have demonstrated great potential for cannibalism (*A. intermedia* and *A. sieboldi*, Gilbert, 1980b). The *A. silvestrii* campanulate demonstrated a strong preference for conspecific prey in this study and a greater ability to capture and ingest conspecific prey than the cruciform. Due to its comparatively low preference for the crustacean prey, the campanulate mean ingestion rate of *Artemia* was similar to cruciforms, despite the campanulate's greater ability to handle this prey. Each aspect of the observed predation process is discussed below in further detail.

Encounter rates

Based on swimming speeds and corona sizes of *Asplanchna* campanulates and saccates (the smallest morphotype), Gilbert & Stemberger (1985) calculated that one of the benefits of the giant *Asplanchna* campanulate morphotype would be a higher encounter rate with prey due to the disproportionately large corona. Previous work with *A. silvestrii* (Hampton & Starkweather, in press) supports this hypothesis with campanulate and saccate morphotypes. However, the authors did not find that cruciforms and campanulates differed in their encounter rates. These findings are consistent with the results of the present study which show no difference in campanulate and cruciform encounter rate.

It is unclear why conspecific prey were encountered more frequently by cruciforms than were *Artemia*. The search behavior of *Asplanchna* is generally thought to be independent of prey before encounter (Gilbert, 1980b; Hampton & Starkweather, in press), so it is unlikely that cruciforms changed search behavior in the presence of either prey, particularly since campanulates did not have significantly different encounter rates between prey. It is possible that the campanulate encounter rate with conspecific prey is not representative of the potential campanulate encounter rate, due to the high probability of a

442

campanulate attacking its first or second encountered prey. Consequently, most of the observed time in these treatments was spent in handling rather than searching. In contrast, due to the low probability of capturing attacked conspecific prey, cruciforms occupied more time with encountering and attacking prey.

Attack after encounter

Asplanchna is known to be a very selective predator (reviewed in Gilbert, 1980b). Recognition of the prey item occurs chemically during the encounter, and the decision whether or not to attempt to eat a prey item is made in this initial recognition, rather than after attack and capture. Thus *Asplanchna*'s preferences are often inferred from probability of attack after encounter. Campanulates showed a distinctly greater preference for conspecific prey than the crustacean prey, while cruciforms attacked either prey with similar frequency.

The mean probability of campanulate attack on both relatively large prey in this study (0.54) is higher than attack probabilities noted for *A. silvestrii* campanulates with smaller rotifer prey; Hampton & Starkweather (in press) reported mean attack probabilities ranging from 0.06–0.28 for campanulates with rotifers sized 250–450 mm long. A strong campanulate preference for large prey may save energy in foraging. Gilbert & Stemberger (1985) calculated that campanulate *A. silvestrii* would have to eat 20 smaller rotifers (*Synchaeta oblonga*) to maintain the same reproductive rate that could be maintained by 1.5 ingestions of the large prey *A. brightwelli*. The authors suggested that prey capture and ingestion may be an energetically costly process in *Asplanchna*. If so, campanulates may benefit energetically from ignoring small prey and ingesting only large prey when encountered. Since the campanulate morphotype is induced at the highest frequency by the ingestion of conspecific and crustacean prey, the decision to ignore small prey in favor of larger prey which may be encountered later is probably not so risky. A period of starvation will make *A. silvestrii* less discriminating (Gilbert, 1980b), such that it is unlikely to starve if only surrounded by small prey.

Genetic relationships among predators and conspecific prey were not noted. Some variation in the attack response of cannibals may have been caused by recognition of clonemates. Species recognition and discrimination among conspecific prey morphotypes are known in *A. intermedia* (formerly known as *A. sieboldi* clone 12C1; Gilbert, 1976, 1980b), but clonal recognition has not been confirmed in *Asplanchna* cannibals.

Capture after attack

When the results from both prey types were pooled, campanulates were more likely to capture prey after attack than cruciforms, due to the higher campanulate probability of successful capture of conspecifics (Figure 4). Campanulate attacks on *Artemia* may have been less successful than those on conspecific prey for several reasons. *Artemia* actively swam away from campanulates after attack and while partially captured, but the small *A. silvestrii* simply contracted their coronae after attack, pushing out their slight body wall outgrowths. While *Artemia*'s escape behavior afforded moderate protection against both predator morphotypes, the conspecific defense was quite effective against cruciforms but not campanulates. In addition, *Artemia* may not have elicited as strong an attack response from campanulates as did conspecific prey (see Gilbert, 1978).

Handling time

Campanulates required less time for handling both prey than did cruciforms. This difference was particularly pronounced with conspecific prey. While handling conspecific prey was significantly more time-consuming for both morphotypes, the difference between prey types was more moderate for campanulates than cruciforms (Figure 5). In addition, 40% of the cruciforms observed were entirely incapable of ingesting conspecific prey. Campanulates clearly experience a benefit over cruciforms in handling these large prey.

Ingestion rate

The campanulate morphotype had a higher ingestion rate than the cruciform with these prey overall, but the difference was driven mainly by its very high relative ingestion rate of conspecific prey (Figure 6). Despite the campanulates' comparative ease of handling *Artemia*, the campanulate ingestion rate with *Artemia* was similar to that of cruciforms, probably due to the somewhat lower probability of campanulate attack on the crustacean prey. Thus campanulates and cruciforms appear to be equally able to exploit the crustacean prey *Artemia*, but the campanulate morphotype was clearly more proficient in cannibalism. This very high potential for cannibalism in *A. silvestrii* is similar to the voracious cannibalism described in campanulate morphotypes of the trimorphic species *A. intermedia* and *A. sieboldi* (Gilbert, 1980b).

While campanulate ingestion rate of the crustacean prey was not higher than that of the cruciform as predicted, it should be noted that campanulates gained the same amount of prey as the cruciforms for less effort. With fewer attacks and shorter handling time, campanulates achieved the same ingestion rates as cruciforms. Attack and capture may be costly activities for *Asplanchna*, but campanulates also have to maintain higher body mass than cruciforms such that reduced energy expenditure may not necessarily result in higher campanulate fitness with this prey.

Currently, evidence supports the hypothesis that campanulates have evolved mainly to exploit congeneric prey, but more work is necessary to unravel the complicated relationship between *Asplanchna* polymorphism and feeding behavior. The following research would help elucidate this relationship: (1) Observational experiments with other crustacean prey are necessary to determine whether preference is similarly relatively low, particularly since *Asplanchna* and *Artemia* are not known to co-occur. (2) Larger crustacean prey should also be tested, since the campanulate can probably ingest some larger crustaceans than cruciforms. Consequently, the campanulate morphotype would be adaptive in a zooplankton community composed mainly of larger crustaceans, even if attack probability remained lower than that of cruciforms. (3) It is known that production of campanulates in the trimorphic species increases, relative to cruciforms, when the diet includes congeneric or crustacean prey, but it is necessary to quantify this production. It would be interesting to know how closely the proportional production of campanulates on different diets corresponds to preference for and ability to ingest prey types. It seems crucial to address these issues in order to understand the dynamics of this continuous feedback loop of selective feeding and polymorphism.

Acknowledgments

I thank John Gilbert for the many constructive discussions that shaped this study, Steve Fradkin for practical advice throughout the project and extensive comments on the manuscript, and an anonymous reviewer whose suggestions improved the clarity of the manuscript. Kevin Kirk and Peter Starkweather generously provided *Asplanchna* and *Artemia*.

References

Birky, C. W. & J. A. Power, 1969. The developmental genetics of polymorphism in the rotifer *Asplanchna*. J. exp. Zool. 170: 157–168.

Commins, M. L. & G. W. Salt, 1988. Some patterns of predation and prey selection by the rotifer *Asplanchna girodi* in replicated outdoor tanks. Verh. int. Ver. Limnol. 23: 2028–2032.

Conde-Porcuna, J. M., R. Morales-Baquero & L. Cruz-Pizarro, 1993. Effectiveness of the caudal spine as a defense mechanism in *Keratella cochlearis*. Hydrobiologia 255/256: 283–287.

Conde-Porcuna, J. M., & S. S. S. Sarma, 1995. Prey selection by *Asplanchna girodi* (Rotifera): the importance of prey defence mechanisms. Freshwat. Biol. 33: 341–348.

Gilbert, J. J., 1973. The induction and ecological significance of gigantism in the rotifer *Asplanchna sieboldi*. Science 181: 64–66.

Gilbert, J. J., 1975. Polymorphism in the rotifer *Asplanchna sieboldi*: Variability in the body-wall-outgrowth response to dietary tocopherol. Physiol. Zool. 48: 409–419.

Gilbert, J. J., 1976. Selective cannibalism in the rotifer *Asplanchna sieboldi*: contact recognition of morphotype and clone. Proc. natl. Acad. Sci. U.S.A. 73: 3233–3237.

Gilbert, J. J., 1978. Selective feeding and its effect on polymorphism and sexuality in the rotifer *Asplanchna sieboldi*. Freshwat. Biol. 8: 43–50.

Gilbert, J. J., 1980a. Developmental polymorphism in the rotifer *Asplanchna sieboldi*. Am. Sci. 68: 636–646.

Gilbert, J. J., 1980b. Feeding in the rotifer *Asplanchna*: behavior, cannibalism, selectivity, prey defenses, and impact on rotifer communities. In W. C. Kerfoot (ed.), Evolution and Ecology in Zooplankton Communities. The University Press of New England, Hanover (NH): 158–172.

Gilbert, J. J. & J. L. Confer, 1986. Gigantism and the potential for interference competition in the rotifer genus *Asplanchna*. Oecologia 70: 549–554.

Gilbert, J. J. & J. D. Jack, 1993. Rotifers as predators on small ciliates. Hydrobiologia 255/256: 247–253.

Gilbert, J. J. & R. S. Stemberger, 1985. The costs and benefits of gigantism in polymorphic species of the rotifer *Asplanchna*. Arch. Hydrobiol. 21: 185–192.

Gilbert, J. J. & C. E. Williamson, 1978. Predator-prey behavior and its effect on rotifer survival in associations of *Mesocyclops edax*, *Asplanchna girodi*, *Polyarthra vulgaris*, and *Keratella cochlearis*. Oecologia 37: 13–22.

Green, J. & O. B. Lan, 1974. *Asplanchna* and the spines of *Brachionus calyciflorus* in two Javanese sewage ponds. Freshwat. Biol. 4: 223–226.

Hampton, S. E. & P. L. Starkweather, in press. Differences in predation among morphotypes of the rotifer *Asplanchna silvestrii*. Freshwat. Biol.

Hurlbert, S. H., M. S. Mulla & H. R. Willson, 1972. Effects of an organophosphorus insecticide on the phytoplankton, zooplankton, and insect populations of fresh-water ponds. Ecol. Monogr. 42: 269–299.

Iyer, N. & T. R. Rao, 1996. Responses of the predatory rotifer *Asplanchna intermedia* to prey species differing in vulnerability: laboratory and field studies. Freshwat. Biol. 36: 521–533.

Powers, J. H., 1912. A case of polymorphism in *Asplanchna* simulating a mutation. Am. Nat. 46: 441–462, 526–552.

Salt, G. W., 1977. An analysis of the diets of five sympatric species of *Asplanchna*. Arch. Hydrobiol. 8: 123–125.

Sarma, S. S. S., 1993. Feeding responses of *Asplanchna brightwelli* (Rotifera): laboratory and field studies. Hydrobiologia 255/256: 275–282.

444

Starkweather, P. L. & E. J. Walsh, 1989. Influence of cyanobacterial diet on *Asplanchna* predation risk in *Brachionus calyciflorus*. Hydrobiologia, 186/187: 35–38.

Stemberger, R. S., 1981. A general approach to the culture of planktonic rotifers. Can. J. Fish. aquat. Sci. 38: 721–724.

Urabe, J., 1992. Midsummer succession of rotifer plankton in a shallow eutrophic pond. J. Plankton Res. 14: 851–866.

Hydrobiologia **387/388**: 445–449, 1998.
E. Wurdak, R. Wallace & H. Segers (eds), Rotifera VIII: A Comparative Approach.
© *1998 Kluwer Academic Publishers. Printed in the Netherlands.*

Uptake of latex beads as size-model for food of planktonic rotifers

Diethelm Ronneberger
Institute of Freshwater Ecology and Inland Fisheries, Department of Stratified Lakes,
D-16775 Neuglobsow, Germany

Key words: clearance rate, latex beads, size frequency distribution

Abstract

Selection of particles ingested by planktonic rotifers was investigated using different-sized fluorescent latex beads (ca 0.05–10 μm diameter). *Keratella cochlearis*, *Keratella quadrata*, and *Polyarthra dolichoptera* preferred 1–2-μm diameter beads, whereas *Synchaeta pectinata* selected for 2–3-μm particles. However, the size-frequency distribution of beads ingested indicated great variability. The 'incipient limiting level' of particle concentration (ILL) for rotifers tested with 2-μm beads was high. *Keratella cochlearis* had the lowest ILL with 5×10^5 beads ml^{-1} and *Synchaeta pectinata* the highest with 1×10^6 beads ml^{-1}. The clearance rate (CR) varied for all species between 0.1 and 0.4 μl ind.$^{-1}$ h^{-1}. *Keratella cochlearis* had the highest CR with 0.4 μl ind.$^{-1}$ h^{-1} at food concentration of 5×10^5 beads ml^{-1}. *S. pectinata* and *P. dolichoptera* had their maximum CR with 0.34 and 0.20 μl ind.$^{-1}$ h^{-1}, respectively, at 7.5×10^5 beads ml^{-1}.

Introduction

Investigations of food uptake by rotifers are essential for a better understanding of interactions in zooplankton. In a review Pourriot (1965) reported that rotifers prefer algal food, whereas bacteria played a minor role in their diet. He described detritus as an additional food resource for *Keratella cochlearis*. Intensive investigations on the nutrition biology of rotifers were carried out in the 1980s by Bogdan & Gilbert (1982, 1984, 1987), Gilbert & Durand (1990), Gilbert & Bogdan (1981), and Starkweather & Bogdan (1980). Investigations using fluorescent-labeled particles (fluorescent-labeled beads, bacteria (FLB), or algae (FLA) as model food for rotifers are rather new (Ooms-Wilms 1993, 1995; Telesh et al., 1995; Vadstein et al., 1993). Moreover, the majority of these investigations were carried out on cultured rotifers (mostly *Brachionus*) or single species, such as *Filinia* and *Keratella*. The clearance rate of *Filinia terminalis* and *Keratella cochlearis* was estimated using FLB and FLA, respectively (Ooms-Wilms, 1993; Telesh et al., 1995).

The goal of this investigation was to investigate feeding by *Keratella cochlearis*, *Keratella quadrata*, *Polyarthra dolichoptera* and *Synchaeta pectinata* using different-sized fluorescent latex beads.

Materials and methods

Rotifers were sampled on five occasions during autumn 1996 and spring 1997 in the oligotrophic Lake Stechlin using a 55-μm plankton net. Crustaceans were removed using a 100-μm sieve. Temperature at the sampling site ranged between 4 and 7.5°C; thus, before the experiments were performed, the rotifers were adapted to room temperature (20°C) for 4 h. For the experiments 50 individuals were picked and transferred to small vials containing 2 ml of prefiltered (0.2 μm) lake water and the desired number of fluorescent beads of different sizes. The reported diameters of these beads were as follows: 0.04, 0.1, 0.43, 1.01, 2.16, 2.97, 6 and 10 μm. Experiments for each species were run in replicate. After an incubation period of 5 min, rotifers were fixed with formalin, separated from the beads by sieving through a 55-μm net, and washed. Then the number of beads ingested was counted using inverted fluorescence microscopy. All rotifers from both replicates were pooled (i.e., 100 individuals) and examined for the number of beads ingested. In order to estimate mean ingestion and clearance rate all animals were considered, including those with no beads in their guts. Only those beads found in the mastax were counted. The ingestion rate was determined using the calculated mean values. The clearance rate was

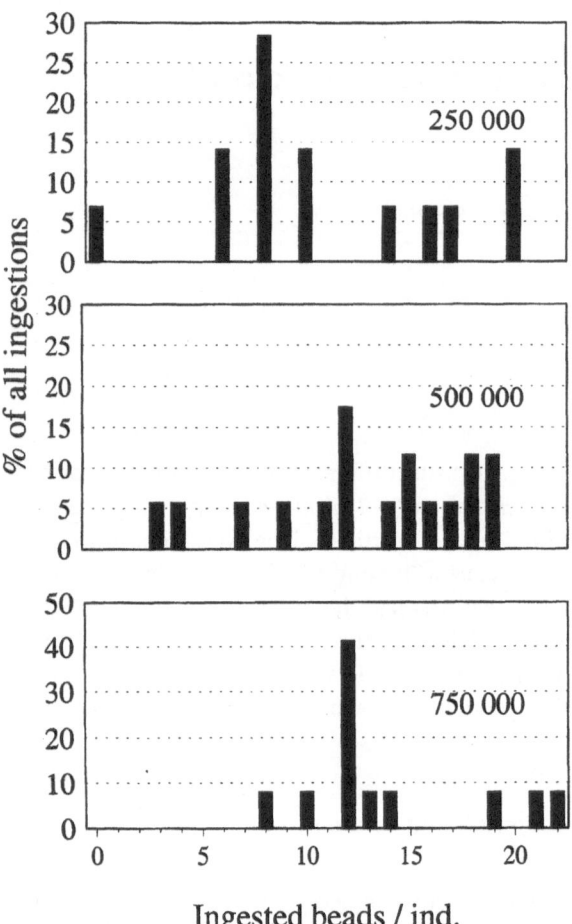

Figure 1. Frequency distribution of ingestion by *Keratella cochlears* at different bead concentrations of 1-μm beads.

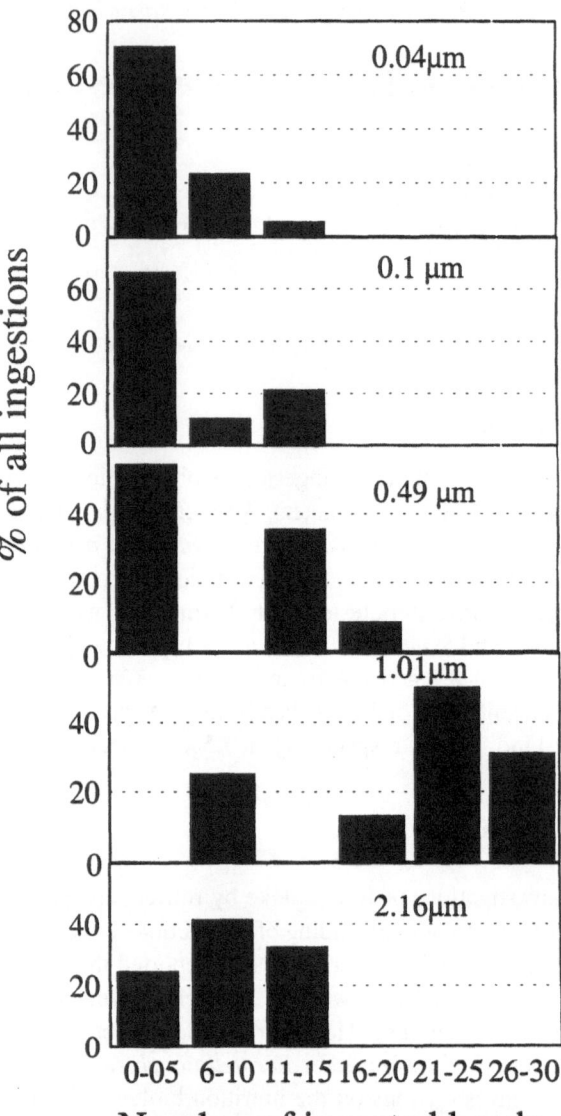

Figure 2. Frequency distribution of ingestion by *Keratella cochlears* at different bead sizes. Bead concentration, 250 000 beads ml⁻¹; mean bead sizes are indicated in each panel.

calculated according to the equation of Bogdan and Gilbert (1984): CR=IR/TC, where CR is the clearance rate in μl ind.$^{-1}$ h^{-1}, IR is the ingestion rate in beads ind.$^{-1}$ h^{-1}, and TC is the concentration of tracer cells in numbers ml^{-1} in the experimental vials.

Results

The number of beads ingested by *K. cochlearis* at different food concentrations is shown in Figure 1. The lowest number of beads found in an animal was 0, the highest was 22. Concerning the number of beads found in the mastax no typical frequency distribution was discovered. The mean number of particles ingested did not depend on the ambient concentration.

Figure 2 shows the frequency distribution of the number of beads ingested by *K. cochlearis* as influenced by particle size. There seemed to be a tendency to ingest a higher number of particles as the particle

size increased up to 1 μm. At 2 μm diameter the number of particles found inside the animals declined.

The ingestion of different-sized particles in other rotifers such as *K. quadrata*, *P. dolichoptera*, and *S. pectinata* was similar in that all of them revealed a unimodal distribution (Figure 3). This was true for the mean as well as for the maximum number of particles ingested. Although the frequency distribution of different-sized particles found in the intestine of the species was rather broad and overlapped, it seems that they concentrate on slightly different sizes. The greatest difference was found between *Keratella*

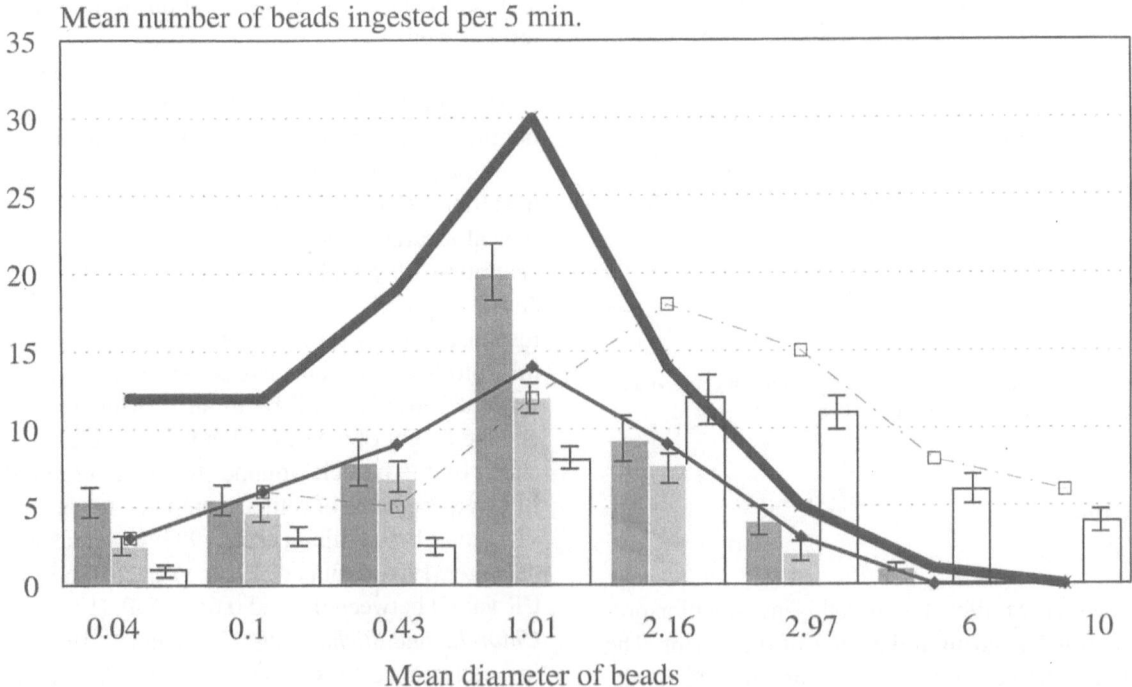

Figure 3. Mean ingestion of different bead sizes by three rotifers. Lines show the maximal ingestion for each species. Symbols are as follows: (■ and thick black line) *Keratella quadrata*; (▨, and thin black line) *Polyarthra dolichoptera*; (□ and dotted line) *Synchaeta pectinata*. Bars are standard deviations.

cochlearis, preferring particles of approximately 0.5–2 μm, and *Synchaeta pectinata* selecting beads having a diameter of 1–3 μm. Moreover, the latter species was able to ingest beads as big as 6 μm. *Polyarthra dolichoptera* was found to behave intermediate.

The ingestion rate as influenced by the particle concentration is shown in Figure 4. The overall picture reveals differences between the species investigated. *K. cochlearis* and *K. quadrata* had increasing ingestion rates up to a particle concentration of 5×10^5 beads ml^{-1} reaching a saturation level at higher densities. *P. dolichoptera* and *S. pectinata* had invariant ingestion rates at particle concentrations ranging from $2.5–5 \times 10^5$ beads. Increasing ingestion rates were found at $5–7.5 \times 10^5$ beads ml^{-1}. At higher particle concentrations, a saturation level was reached. From such data one may estimate the Incipient Limiting Level (ILL); this is the concentration of particles above which there is no significant increase in IR. In this study I found the ILL to vary among species; for *K. cochlearis* and *K. quadrata* it was found at 5×10^5 beads ml^{-1}; for *P.dolichoptera* at 7.5×10^5 beads ml^{-1}. In *Synchaeta pectinata* there was neither a distinct ILL nor a clear saturation level.

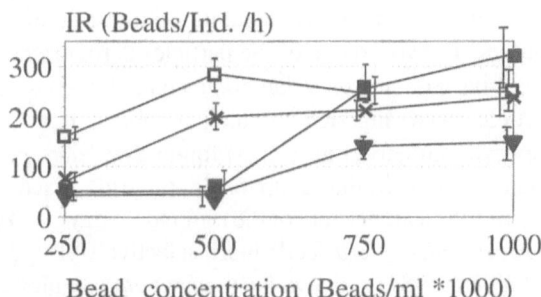

Figure 4. Ingestion rate (beads ind.$^{-1}$ h^{-1}) of different bead concentrations for 2-μm beads of *Keratella cochlearis* (⋆), *Keratella quadrata* (–□–), *Polyarthra dolochoptera* (–▼–) and *Synchaeta pectinata* –■–. Bars are standard deviation.

As a consequence of differences in IR the CR also was different. In relation to particle concentration the values varied between 0.01 and 0.35 μl ind.$^{-1}$ (Figure 5). *K. cochlearis* and *K. quadrata* had the highest CR at lower concentration of beads, *P. dolichoptera* and *S. pectinata* increased their CR at higher bead concentrations.

448

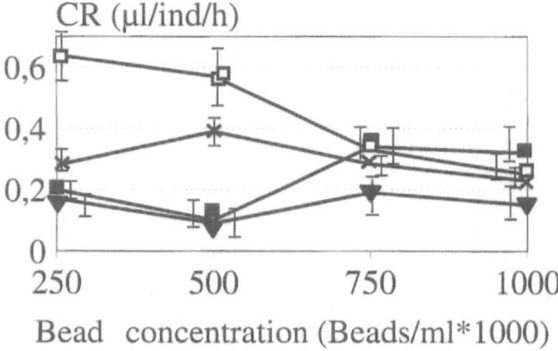

Figure 5. Clearance rate (μl ind^{-1} . h^{-1}) of different concentration for four rotifer species. For details see Figure 4.

Discussion

The planktonic rotifers tested in this investigation preferred particles having a diameter of 0.5–3 μm. The differences found between species are most likely the result of differences in corona size. *K. cochlearis* has the smallest corona, while *S. pectinata* possesses the largest. The corona of *P. dolichoptera* is intermediate in size (Bogdan & Gilbert, 1987; Telesh et al., 1995). These differences are reflected in the size–frequency distribution of the beads found in the intestine of the rotifers. The size range of the particles is representative of bacteria and picoplanktonic algae. Larger-sized particles were ingested in small numbers, if at all. However, Gilbert et al. (1993) found *Synchaeta pectinata* to ingest ciliates up to 45–60 μm, reaching maximum clearance rate of 0.81 ml ind.$^{-1}$ day^{-1}. *Polyarthra dolichoptera* feeds more selectively (Bogdan & Gilbert, 1987) and *Synchaeta pectinata* can play an important role in the microbial food web by grazing on protozoans (Gilbert et al., 1993).

Surprisingly, large individual differences in IR, ranging from 0 to 30 beads ind.$^{-1}$ 5 min^{-1} were observed. The reasons for this are unclear. A possibility would be temperature stress due to the differences between sampling site and laboratory. Nevertheless, Ooms-Wilms (1993) applying the same methodological approach found the same effects with even longer adaptation times. It is known from studies of planktonic crustaceans that starved animals have much higher ingestion rates than well-fed ones (Geller, 1975). The significance of this factor is hard to judge because the ambient particle concentration at the sampling site has not been estimated during this investigation. However, temperature adaptation for 4

h should lead to some equilibration in the starvation level as well.

Concerning IR and CR of filter-feeding planktonic crustaceans there is a typical kinetics found in a large number of experiments (saturation, exponential decrease, e.g., Rigler, 1971). This seems to be different in rotifers. *K. cochlearis* and *K. quadrata* exhibited a typical saturation curve reaching a plateau at a concentration of 2.5×10^5 beads ml^{-1}. *P. dolichoptera* and *S. pectinata*, on the other hand, showed a logistic response curve with constant IR up to 5×10^5 beads ml^{-1} followed by an increase. The saturation level was reached at 7.5×10^5 beads ml^{-1}. Although the ILL and the height of the saturation level are clearly influenced by the particle volume (Kasprzak et al., 1986) food requirements of rotifers compared to cladocerans seem to be high (Gilbert et al., 1993; Kasprzak & Ronneberger, 1985; Lampert, 1977; Vadstein et al., 1993). CR varied between 0.1 and 0.65 μl ind.$^{-1}$ h^{-1}. Using *Chlorella* and *Stichococcus* Telesh et al. (1995) found lower clearance rates as compared to smaller particles.

Many crustaceans, especially calanoid copepods, select their food according to taste (DeMott, 1988, 1989). At present, there is no indication of a preference for flavoured beads in rotifers (DeMott, 1988; Ooms-Wilms, 1993; Rothhaupt, 1990). Ooms-Wilms (1993) found *Filinia longiseta* to prefer unflavoured beads over bacteria (ca 0.5 μm). Differences in the ingestion of live and dead food particles were observed in *Keratella* (Starkweather & Bogdan, 1980). It was observed in our experiments that *Polyarthra dolichoptera* has lower ingestion rates than the smaller *Keratella cochlearis*. This may be due to the greater ability of *Polyarthra dolichoptera* to select its food (Gilbert & Bogdan, 1980).

In conclusion, the planktonic rotifer species *K. cochlearis*, *K. quadrata*, *P. dolichoptera* and *S. pectinata* prefer bacteria- and picoplankton-sized particles at least under the conditions chosen in these experiments. According to my results *Polyarthra dolichoptera* is likely to feed more selectively on bacteria and chlorococcal algae. *Synchaeta pectinata* may play another role by grazing on protozoans (Gilbert et al., 1993). Further investigations should address the question as to whether particle flavour is of importance in food selection and grazing by rotifers.

Acknowledgments

I thank P. Kasprzak and Brook Lemma who provided many helpful comments and critically read the manuscript and assisted in translating it into English. Liz Wurdak has corrected for English usage.

References

Bogdan K. G. & J. J. Gilbert, 1982. Seasonal pattern of feeding by natural populations of *Keratella*, *Polyarthra* and *Bosmina*: clearance rates, selectivities, and contributions to community grazing. Limnol. Oceanogr. 27: 918–934.

Bogdan K. G. & J. J. Gilbert, 1984. Body size and food size in freshwater zooplankton. Proc. Nat. Acad. Sci. U.S.A. 81: 6427–6431.

Bogdan K. G. & J. J. Gilbert, 1987. Quantitative comparison of food niches in some freshwater zooplankton. Oecologia 72: 331–340.

DeMott, W. R., 1986. The role of taste in food selection by freshwater zooplankton. Oecologia 69: 334–340.

DeMott, W. R., 1988. Discrimination between algae and artificial particles by freshwater and marine copepods. Limnol. Oceanogr. 33: 397–408.

DeMott, W. R., 1989. Optimal foraging theory as a predictor of chemically mediated food selection by selection by suspension-feeding copepods. Limnol. Oceanogr. 34: 140–154.

Geller, W., 1975. Die Nahrungsaufnahme von *Daphnia pulex* in Abhängigkeit von der Futterkonzentraton, der Temperatur, der Körpergröße und dem Hungerzustand der Tiere. Arch. Hydrobiol. Suppl. 48: 47–107.

Gilbert J. J. & K. G. Bogdan, 1981. Selectivity of *Polyarthra* and *Keratella* for flagellate and aflagellate cells. Verh. int. Ver. Limnol. 21: 1515–1521.

Gilbert, J. J. & M. W. Durand, 1990. Effect of *Anabaena flos-aquae* on the abilities of *Daphnia* and *Keratella* to feed and reproduce on unicellular algae. Freshwat. Biol. 24: 577–596.

Gilbert, J. J. & J. D. Jack, 1993. Rotifers as predators on small ciliates. Hydrobiologia 255/256: 247–253.

Kasprzak, P. & D. Ronneberger, 1985. The secondary production. In Casper (ed.), Lake Stechlin. A Temperate Oligotrophic Lake. Dr W. Junk Publishers, Dordrecht: 323–345.

Kasprzak, P., V. Vyhnálek & M. Straskraba, 1986. Feeding and food selection in *Daphnia pulicaria*. Limnologica 17: 309–323.

Lampert, W., 1977. Studies on the carbon balance of *Daphnia pulex* DE GEER as related to environmental conditions. II. The dependence of carbon assimilation on animal size, temperature, food concentration and diet species. Arch. Hydrobiol. Suppl. 48: 361–368.

Ooms-Wilms, A., G. Postema & R. D. Gulati, 1993. Clearance rates by the rotifer *Filinia longiseta* (Ehrb.) measured using three tracers. Hydrobiologia 255/256: 255–260.

Ooms-Wilms, A. L., G. Postema & R. D. Gulati, 1995. Evaluation of bacterivory of Rotifera based on measurements of in situ ingestion of fluorescent particles, including some comparisons with Cladocera. J. Plankton Res. 17: 1057–1077.

Pourriot, R., 1965. Recherches sur l'écologie des rotifères. Vie Milieu: (Suppl. 21) 224 pp.

Rigler, F. H., 1991. Methods for the measurement of assimilation of food by zooplankton. In Edmondson, W. T. & G. G. Winberg (eds), Secondary Productivity in Fresh Waters, IBP Handbook No. 17, Blackwell, Oxford and London: 264–269.

Rothhaupt, K. O., 1990. Differences in particle size-dependent feeding efficiencies of closely related rotifer species. Limnol. Oceanogr. 35: 16–23.

Starkweather, P. L. & K. G. Bogdan, 1980. Detrital feeding in natural zooplankton communities: discrimination between live and dead algal foods. Hydrobiologia 73: 83–85.

Telesh, I. V., A. Ooms-Wilms & R. D. Gulati, 1995. Use of fluorescent labelled Algae to measure the clearance rate of the rotifer *Keratella cochlearis*. Freshwat. Biol. 33: 349–355.

Vadstein, O., O. Gunvor & Y. Olsen, 1993. Particle size dependent feeding by the rotifer *Brachionus plicatilis*. Hydrobiologia 255/256: 261–267.

Hydrobiologia **387/388**: 451–457, 1998.
E. Wurdak, R. Wallace & H. Segers (eds), Rotifera VIII: A Comparative Approach.
© *1998 Kluwer Academic Publishers.*

Ecology versus taxonomy: is there a middle ground?

D. L. Nielsen*, R. J. Shiel & F. J. Smith
Cooperative Research Centre for Freshwater Ecology, P.O. Box 921, Albury 2640, NSW, Australia
(*author for correspondence)

Key words: taxonomic resolution, ecological studies, flooding, Rotifera

Abstract

In many long-term, intensive experimental and field studies there often arises a need to trade off taxonomic resolution for ecological answers. Compounding this problem is a taxonomic impediment, the lack of experienced taxonomists capable of processing large numbers of samples to species resolution, especially in groups such as the Rotifera. This paper has two aims: (1) To investigate the level of taxonomic resolution required to determine the impact of a disturbance, in the form of a flood event; (2) to compare the impact of different taxonomic resolutions in assessing biodiversity. Results suggest both family and generic resolution can be used to determine the impact of a flood event and that these levels have some applicability to biodiversity studies. Relatively inexperienced taxonomists who can identify the common rotifers to generic level, can be relied upon to detect disturbance to community structure but their data become unreliable when assessing biodiversity.

Introduction

The responses of macroinvertebrate communities to disturbance, in particular impacts caused by external forces such as mining or pollution, are well documented. The use of multivariate methods such as multi-dimensional scaling (MDS) is widespread in determining effects that disturbance may have (Clarke, 1993; Clarke & Ainsworth, 1993). Multivariate methods also have been used to classify different systems in terms of their community structure (Wright et al., 1984; Pontin & Langley, 1993). Many of these studies have examined changes based on taxonomic resolution at the species level but more recently marine studies have shown that ecologically significant results can be obtained at either generic or family level of taxonomic resolution (Clarke & Warwick, 1994).

For many large-scale ecological studies the experience and resources required for species resolution are often not available. In many groups the taxonomy is relatively unknown or unreliable, hence community studies at taxonomic levels other than species may need to be undertaken. To overcome the shortage of professional taxonomists the use of 'biodiversity technicians' has been suggested, but results often are unreliable due to incorrect identification of individu-

als. The quality of results may also vary considerably among taxonomic groups that are considered and with the ability of individual technicians (Cranston & Hillman, 1992; Hammond, 1994).

Measurement of biodiversity is sometimes thought to require species resolution but can be measured in many different ways and at different levels (May, 1994). In many systems, such as the neotropics (Prance, 1994) or Australian ant communities (Andersen, 1995), lower taxonomic resolutions have been shown to be inadequate in assessing biodiversity, but in some macroinvertebrate studies there have been high correlations between the number of families present and the total number of species (Growns & Growns, 1996). Indicator or focal groups have also been suggested as means of estimating biodiversity (Hammond, 1994).

In this paper we will compare the use of family, genus and species levels of resolution in a study assessing the impact of flooding on rotifer communities, and their applicability to biodiversity measurements. The samples were examined by two taxonomists. Taxonomist 'A' had a honours degree in biology plus eighteen months experience in rotifer taxonomy. Taxonomic resolution was to the generic level. Taxonomist 'B' was a professional rotifer systematist with post-

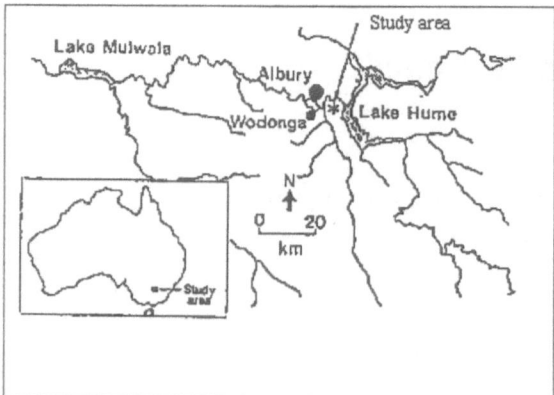

Figure 1. Location map of study area on River Murray Floodplain in northern Victoria, Australia.

graduate qualifications and over 20 years experience. Taxonomic resolution was to species.

Study site

To examine the effect of changed patterns of inundation on the microfauna of Australian billabongs (oxbow lakes), sixteen artificial billabongs were constructed four kilometres south-east of Albury, New South Wales, Australia (Figure 1). The sixteen billabongs were divided into four treatments (each with four replicates). In this paper we will only be examining rotifer communities in the unflooded billabongs and the impact of inundation on rotifer communities in the winter/spring treatment billabongs over a flood event in September 1996. Each billabong was 4.5 m in diameter and with a shallow ephemeral area and a deep section with a maximum depth of approximately 0.90 m. They were planted initially with both *Myriophyllum papillosum* Orch. and *Vallisneria* sp.

Methods

Zooplankton samples were collected using a Jabsco pump (14 L/min.). The inlet pipe was moved through the water column sampling both deep and shallow areas for a period of two min. The sampled volume was filtered through a 35 μm mesh net and the filtrate was returned to the pond. The collected sample volume was recorded and a 10 ml subsample extracted and preserved in 70% ethanol. The remaining organisms were returned live to the ponds.

In September 1996 the winter treatment billabongs were flooded, resulting in a two-fold increase in volume. Four samples were collected every second day prior to flooding (samples 1–4), immediately after flooding on the 03.IX.96 (sample 5) and then at 12 hours, 24 hours, 48 hours, etc. until the 25.IX.96 (samples 6–12).

The rotifer component of these samples was then counted and identified by the two taxonomists. Taxonomist 'A' counted and identified successive 1 ml aliquots taken from the 10 ml subsample until a minimum of 200 individuals had been counted. Animals were counted in a Sedgewick-Rafter counting chamber and identified using darkfield microscopy. Generic identification was based on the taxonomic key of Shiel (1995). Taxonomist 'B' counted a single 1 ml aliquot from the 10 ml subsample. Animals were counted in a Sedgewick-Rafter counting chamber and identified using darkfield microscopy or compound microscopy (Nomarski optics) after erosion of trophi using sodium hypochlorite. In both cases the resulting data were expressed as animals per litre.

Statistical analysis

To determine if there were changes in community structure at the different taxonomic levels identified by each taxonomist, the data were analysed using multivariate techniques developed at Plymouth Marine Laboratories, England, on data that had undergone a fourth root transformation (Clarke, 1993; Clarke & Warwick, 1994). These techniques require the computation of a similarity matrix. The most common and robust matrix currently used in ecological studies is derived from Bray-Curtis coefficient of similarity (Faith et al., 1991).

Specific techniques used were:

1. Non-metric multi-dimensional scaling (MDS). This method is relatively free of restrictive assumptions and provides a robust means of presenting multivariate data through ordination. To obtain a meaningful stress level (<0.20) the MDS was calculated from the means of the four replicates at each sampling time.

2. Two-way nested analysis of similarity (ANOSIM). This allows for formal hypothesis testing by estimating the probability that dissimilarities calculated between groups identified a priori could occur by chance. This is analogous to the univariate analysis of variance.

Table 1. Total number of taxa recognised by each taxonomist, at each level of discrimination, from the flooded and unflooded billabongs for the sampling period

	Taxonomist 'A'		Taxonomist 'B'	
	Flooded	Unflooded	Flooded	Unflooded
Families	12	13	27	25
Genera	20	19	49	40
Species			96	88
Density Mean	2610	3574	2062	2744
Standard error	359	296	315	238

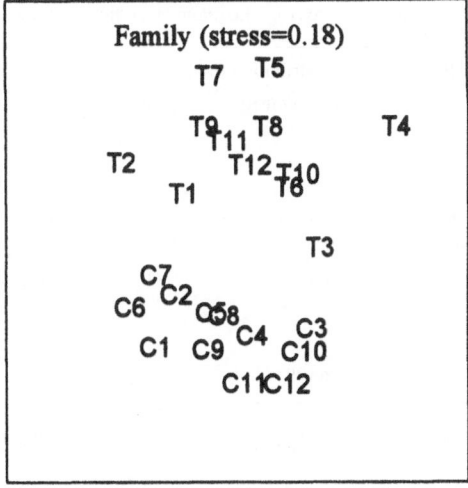

3. Similarity percentages (Simper). This allows determination of which species are contributing to the dissimilarities between groups and their percentage contribution to the dissimilarity.

The statistical analysis program SYSTAT® (Ver. 6.0.1.) was used for univariate analysis of variance (ANOVA) to determine differences in the numbers and types of animals identified by the two taxonomists, and linear regressions were used to compare the number of taxa at each taxonomic level from each sample identified by taxonomist 'B', to test for any relationship that may assist in the use of lower levels of taxonomic resolution in determination of biodiversity. Rotifer densities were fourth root transformed to remove any heterogeneity of variances prior to analysis (Underwood, 1997).

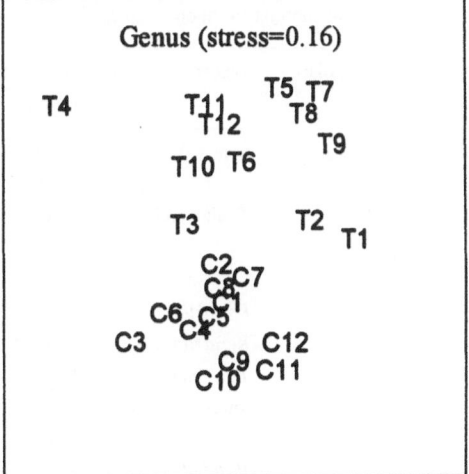

Results

Taxonomist 'B' identified significantly more taxa than Taxonomist 'A' ($P<0.001$) at both the family and genus levels ($P<0.001$). There was no significant difference in numbers of taxa collected between the flooded and unflooded treatments by either taxonomist. There was no significant difference between the two taxonomists when density (animals per litre) was calculated ($P=0.105$) (Table 1).

The data from the flooded billabongs compiled by both taxonomists were further analysed to determine the contribution of individual taxa to the density, at each taxonomic level (Table 2). Simper analysis on the data from both taxonomists indicates that these dominant genera contribute a significant proportion to the dissimilarity between the pre- and post-flood groups within the treatment samples (Table 3).

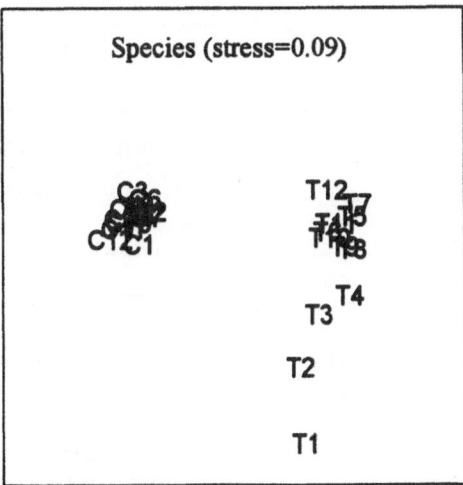

Figure 2. MDS plots for the family, genus and species resolution from Taxonomist 'B'. (C = unflooded, T = treatment. 1–4 represent pre-flood samples, 5–12 post-flood samples).

Table 2. Contribution of dominant taxa at each taxonomic level to total density within group. Data from taxonomist 'B' flood samples. Numbers in brackets are from Taxonomist 'A' flood samples

Family	Contribution to total density (%)	No. of genera present	Dominant genus	Contribution to total density within family (%)	No. species present in each genus	Dominant species within each genus (*)	Contribution to total density within genus (%)
Synchaetidae	39 (39)	2 (2)	*Polyarthra*	98 (99)	2	*P. dolichoptera*	91
Brachionidae	47 (45)	3 (3)	*Keratella*	99 (96)	2	*K. procurva*	99
Trichocercidae	10 (14)	1 (1)	*Trichocerca*	100 (100)	12	*T. dixon-nuttalli*	35
						T. similis	46
						T. brachyura	10
Total	96 (98)						

* Taxonomist 'A' was only required to identify to the generic level.

Table 3. Contribution to dissimilarity between treatment groups pre- and post-flood. Data from Taxonomist 'B'. Numbers in brackets are from Taxonomist 'A' flood samples

Resolution	Dissimilarity index	Taxon	Contribution to dissimilarity (%)
Family	34 (29)	Synchaetidae	15 (17)
		Brachionidae	12 (16)
		Trichocercidae	9 (12)
		total	36 (45)
Genus	44 (35)	*Polyarthra*	10 (13)
		Keratella	9 (12)
		Trichocerca	6 (9)
		total	25 (34)
Species	59	*P. dolichoptera*	9 (*)
		K. procurva	5
		T. dixon-nuttalli	5
		T. similis	5
		T. brachyura	3
		total	27

*Taxonomist 'A' was only required to identify to the generic level.

Table 4. Two-way nested ANOSIM *R* and *P* values derived from pre- and post- flood groups selected *a priori*

Taxonomist	Resolution	Flooded		Unflooded	
		R value	*P* value	*R* value	*P* value
'A'	Family	0.501	<0.05	0.127	0.186
	Genus	0.318	<0.05	0.108	0.226
'B'	Family	0.654	<0.05	0.090	0.297
	Genus	0.664	<0.05	-0.024	0.523
	Species	0.678	<0.05	0.129	0.226

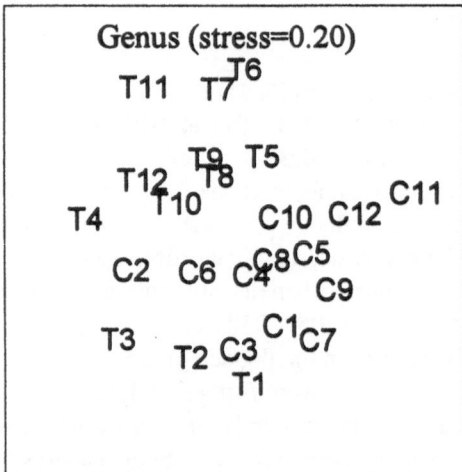

Figure 3. MDS plots for the family and genus resolution from Taxonomist 'A'. (C = unflooded, T = treatment. 1–4 represent pre-flood samples, 5–12 post-flood samples).

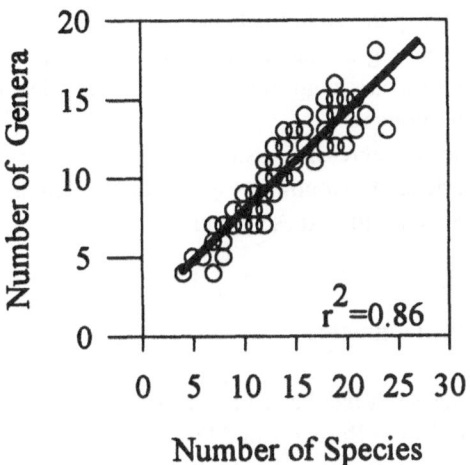

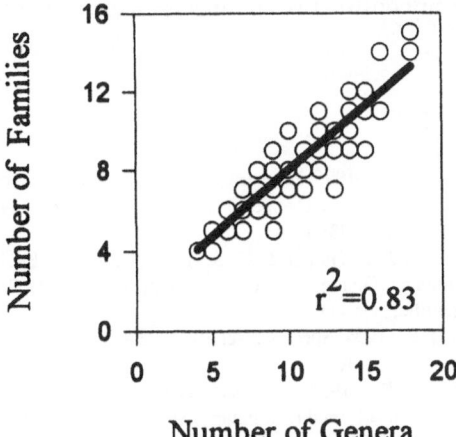

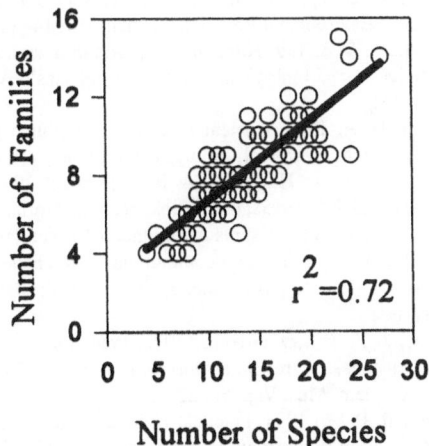

Figure 4. Linear regression for the three combinations of taxonomic resolution using data from both treatments identified by Taxonomist 'B' (*n* = 96).

Ordination plots were graphed for each taxonomic level for Taxonomist 'B' (Figure 2) and for Taxonomist 'A' (Figure 3). ANOSIM indicates that both taxonomist were detecting an effect on community structure as a result of flooding (Table 4).

Linear regression comparing the numbers of taxa identified at each taxonomic level by taxonomist 'B', using the three combinations of resolution possible, indicates that a strong relationship exists between the different resolutions (Figure 4). This suggests that the use of lower levels of resolution such as family and genera could be used to estimate biodiversity.

Discussion

Significant differences in the number of taxa identified by the two taxonomists were recorded at the two levels of discrimination comparable between the two taxonomists (Table 1), but there was no difference in density calculated between the two. This suggests that the difference in the number of taxa was due to the occurrence of rare or difficult to identify taxa that were incorrectly identified by 'A'. This has occurred even though 'A' examined more of the sample than 'B'.

The systems were dominated by a small proportion of the total number of taxa (Table 2) which may be indicative of an unstable environment (Begon et al., 1990, but see Dumont, 1994). Although this may be an artefact of the experimental units and it should be investigated in natural environments it has been reported elsewhere. For example within the marine environment, which is relatively stable, there are 34 phyla, of which 8 phyla contain more than 90% of all the known marine species. In terrestrial environments, which can be considered to be more unstable than marine environments, there are fewer phyla, but more species, of which more than 90% belong to the phylum Arthropoda (Briggs, 1994). However, it should be recognised that there exists a continuum, with stability and instability at opposite extremes, both having low species diversity: the greatest diversity existing at an intermediate between the two (Petraitis et al., 1989). The effect of a disturbance may appear in only a few species which may change over time and vary between systems. Severe disturbance may also change the dominant species.

The level of taxonomic discrimination in this study is important in determining the effects of disturbance on community structure, with the strongest effect occurring at the species level. Significant changes were detected at both generic and family levels but with a corresponding loss of information. The increase in R value as taxonomic discrimination increases indicates an improvement in discrimination between groups. This is contrary to Clarke & Warwick (1994) who suggest that lower discrimination of taxonomy reflected pollution gradients better than species level.

At each taxonomic level there was a significant effect of inundation on the rotifer community, which did not occur in the unflooded treatment. Even though Taxonomist 'A' identified fewer taxa there was still a significant effect, however the ANOSIM test statistic 'R' was reduced and there was a stronger result at the family level. This change was in all likelihood due to the smaller number of taxa recognised. However the key taxa recognised by both taxonomists, even though they compose only a small proportion of the available taxa, are the major contributors to the dissimilarity index at all taxonomic levels (Table 3). As taxonomic resolution is increased, the dissimilarity between the pre- and post-flood groups increase indicating a higher power of resolution. The misidentification of taxa by Taxonomist 'A' and the resulting smaller taxa list, resulted in the dissimilarity between groups becoming weaker when compared to Taxonomist 'B'.

In all cases there was a strong linear relationship between the different taxonomic levels with a correlation value (r^2) that explained more than 70% of the variation. This suggests that lower taxonomic resolutions can be used for rotifers in biodiversity studies, however it will not be applicable in studies using biodiversity technicians where not all families and genera are being recorded. This would lead to an underestimation of the existing biodiversity.

Species resolution is and always will be the ideal. However, given time constraints and the lack of highly skilled taxonomists specialising in groups such as rotifers, biodiversity technicians can be used when determining the effect of inundation on rotifer communities. Estimates of biodiversity using their data would under represent the actual biodiversity.

Conclusions

1. Lower levels of taxonomic discrimination (genus and family) are applicable when examining the effects of disturbance on planktonic communities.
2. Biodiversity technicians can be used when examining disturbance on planktonic communities

provided that they are given enough training to identify the common taxa.
3. Linear regression between levels of taxonomic resolution are high for biodiversity studies within billabongs.
4. Linear regression using data obtained by inexperienced taxonomists would underestimate biodiversity within these systems.

Acknowledgments

This research has been funded by the Land and Water Resources Research and Development Corporation and the Co-operative Research Centre for Freshwater Ecology. The authors would also like to thank Dr Jane Growns, Dr John Whittington and Dr Terry Hillman for their constructive comments on the manuscript.

References

Anderson, A. N., 1995. Measuring more of biodiversity: Genus richness as a surrogate for species richness in Australian ant faunas. Biol. Conserv. 73: 39–43.

Begon, M., J. L. Harper & C. R. Townsend, 1990. Ecology: individuals, populations and communities (2nd edn). Blackwell Scientific, London.

Briggs, J. C., 1994. Species diversity: land and sea compared. Syst. Biol. 43: 130–135.

Clarke, K. R., 1993. Non-parametric multivariate analysis of changes in community structure. Aust. J. Ecol. 18: 117–143.

Clarke, K. R. & M. Ainsworth, 1993. A method of linking multivariate community structure to environmental variables. Mar. Ecol. Prog. Ser. 92: 205–219.

Clarke, K. R. & R. M. Warwick, 1994. Change in Marine Communities: an approach to statistical analysis and interpretation. Natural Environment Research Council, UK, 144 pp.

Cranston, P. S. & T. J. Hillman, 1992. Rapid assessment of biodiversity using 'biological diversity technicians'. Aust. Biol. 5: 144–155.

Dumont, H. J., 1994. Ancient lakes have simplified pelagic food webs., Arch. Hydrobiol. Beih. Ergebn. Limnol. 44: 223–234.

Faith, D. P., C. L. Humphrey & P. L. Dostine, 1991. Statistical power and BACI designs in biological monitoring: Comparative evaluation of measures of community dissimilarity based on benthic macroinvertebrate communities in Rochelle Mine creek, Northern Territory, Australia. Aust. J. mar. Freshwat. Res. 42: 589–602.

Growns, J. E. & I. O. Growns, 1996. Predicting species richness for Australasian freshwater macroinvertebrates: Do we want to know? Mem. Mus. Vict. 56: 223–230.

Hammond, P. M., 1994. Practical approaches to the estimation of the extent of biodiversity in speciose groups. Phil. Trans. r. Soc. Lond. B 345: 101–118.

May, R. M., 1994. Conceptual aspects of the quantification of the extent of biological diversity. Phil. Trans. r. Soc. Lond. B 345: 13–20.

Petraitis, P. S., R. E. Latham & R. A. Niesenbaum, 1989. The maintenance of species diversity by disturbance. Q. Rev. Biol. 64: 393–418.

Pontin, R. M. & J. M. Langley, 1993. The use of rotifer communities to provide a preliminary national classification of small water bodies in England. Hydrobiologia 255/256: 411–419.

Prance, G. T., 1994. A comparison of the efficacy of higher taxa and species numbers in the assessment of biodiversity in the neotropics. Phil. Trans. r. Soc. Lond. B 345: 89–99.

Shiel, R. J., 1995. A guide to identification of rotifers, cladocerans and copepods from Australian inland waters. Identification guide No. 3. Co-operative Research Centre for Freshwater Ecology, 144 pp.

Underwood, A. J., 1997. Experiments in ecology: Their logical design and interpretation using analysis of variance. Cambridge University Press, United Kingdom.

Wright, J. F., D. Moss, P. D. Armitage & M. T. Furse, 1984. A preliminary classification of running-water sites in Great Britain based on macro-invertebrate species and prediction of community type using environmental data. Freshwat. Biol. 14: 221–256.

Hydrobiologia **387/388**: 459–467, 1998.
E. Wurdak, R. Wallace & H. Segers (eds), Rotifera VIII: A Comparative Approach.
© *1998 Kluwer Academic Publishers.*

Ecological water quality assessment of the Bütgenbach lake (Belgium) and its impact on the River Warche using rotifers as bioindicators

Yves Marneffe, Sophie Comblin & Jean-Pierre Thomé
Université de Liège, Laboratoire d'Ecologie Animale et d'Ecotoxicologie, Institut de Zoologie,
Quai Van Beneden 22, 4020 Liège, Belgium
E-mail: y.marneffe@ulg.ac.be

Key words: Bütgenbach lake, water quality, bioindicator(s), biomonitoring, Rotifera, ecotoxicity

Abstract

Results are presented on a study of the zooplankton of Bütgenbach reservoir and from the Warche and Holzwarche rivers which feed the reservoir (March and October 1996). The zooplankton was dominated by rotifers in spring and by crustaceans (cladocerans and copepods) in summer and autumn. A temperature gradient developed during summer and a drastic depletion of oxygen and increase in ammonia concentrations was observed below 7 m depth. The water quality of the River Warche was compared upstream and downstream of the lake using bioindication by rotifers and by reproduction ecotoxicological tests on *Brachionus calyciflorus* on the other hand. The bioindicators reveal an overall improvement in water quality of the Warche downstream of the reservoir, whereas the toxicity assays show a decline in water quality downstream of the lake during the stratification period, due to the release of hypolimnetic water from the dam. So, under special conditions, ecotoxicity assays appear to be more sensitive than bioindication using rotifers saprobic valences.

Introduction

Most of the Belgian hydrographic network is polluted by industrial, domestic and agricultural effluent. Biotic indices using macroinvertebrates reveal a marked difference between the north (with generally low water quality) and the south of the country (where the watercourses are generally only slightly polluted) (IHE, 1990; Ministère de la Région Wallonne, 1994). The Hautes-Fagnes/Eifel area, which is located far from the main industrial centres, so far appeared protected from pollution sources. More recently, however, groundwater, springs, lakes and rivers in this area, formerly famous for their high water quality, have become contaminated with a range of pollutants. Agriculture, sylviculture, animal breeding, and rural industries (e.g., dairy industries, sawmills, paper mills, tanneries) now strongly alter the physical, chemical and biological characteristics of these waters. Previous studies have shown a rapid decrease of the water quality of the River Warche a few kilometres downstream of its spring, due to the use of chemical fertilizers and the decomposition of organic sewage in the river

(Fabri, 1977; Fabri & Leclercq, 1977; Descy et al., 1981; Marneffe et al., 1995, 1997). This river, together with the River Holzwarche, feed the Bütgenbach lake bringing these pollutants into the reservoir. In addition to this, an important tourist centre has been built near the reservoir which discharges concentrated sewage effluent into the lake.

Zooplankton species successions and spatial distributions result from differences in ecological tolerance to various abiotic and biotic environmental parameters. Despite the fact that factors such as predation (Gulati, 1990; Seda & Duncan, 1994; Taleb et al., 1994; Visman et al., 1994) or competition (Gilbert, 1988; Lampert & Rothhaupt, 1991; Dohet & Hoffmann, 1995) may alternately induce changes in zooplankton communities favouring one species over another (Arnoth & Vanni, 1993), rotifers are usually considered to be useful indicators of water quality and of trophic status (Gannon & Stemberger, 1978; Sladecek, 1983; Blancher, 1984; Marneffe et al., 1995, 1997). Indeed, due to high population turnover rates, rotifers are particularly sensitive to changes in water

460

quality (Sládecek, 1983). Their community structure not only allows estimates of the level of pollution, but also can indicate the trend of general conditions over time.

Although the use of bioindicators is relatively simple, fast, cheap, reliable and achievable, it is now well established that chemical measurements and ecotoxicity assays must also be used as complementary tools for the assessment of river water quality. For many years, rotifers have been used for chronic toxicity tests based on demographic parameters (Rao & Sarma, 1986; Snell & Moffat, 1992; Fernández-Casalderrey et al., 1993; Snell & Carmona, 1995), on swimming behaviour (Janssen et al., 1994; Charoy et al., 1995), on feeding behaviour (Ferrando et al., 1993) or on esterase biomarkers (Burbank & Snell, 1994).

This study was designed to determine the physical, chemical and biological characteristics of the water in the Bütgenbach lake and to assess the River Warche water quality using the complementary tools mentioned above.

Description of studied sites and sampling periodicity

The Bütgenbach dam was built in 1932 on the upper reaches of the River Warche. Its purpose was to provide hydroelectric power and regulate the volume of the downstream reservoir of Robertville. These activities have resulted in marked variations in water level in Bütgenbach reservoir throughout the year. Two rivers, Warche and Holzwarche, feed the lake which is located at an elevation of 550 m. Its catchment covers an area of 72 km², which includes about 80% pastures and a population density of about 57 persons km^{-2}. During summer, this density generally increases by more than 40% due to tourism. The reservoir's maximum volume reaches 11 10^6 m^3 and it has a surface of 120 ha. The maximum depth amounts to 23 m and the mean depth is about 9 m.

From April to October 1995, samples were collected twice a month with a 5-l Van Dorn bottle for physical, chemical and biological analyses at three stations (Figure 1): station A located near the River Warche mouth (sampling depths: surface, 1, 2, 3 or 4 m), station B located in the central part of the lake (sampling depths: surface, 1, 2, 5 and 8 m) and station C located near the dam (sampling depths: surface, 1, 2, 5, 10 and 15 m). Additional samples were collected upstream of the lake in the Warche and Holzwarche

Rivers and downstream of the lake (200 m and 5 km downstream) at two stations located on the River Warche.

Material and methods

Zooplankton analysis and use of the rotifer communities as bioindicators

Zooplankton was concentrated by filtering 2 l of lake water through a 37 μm mesh net fitted in a special system to keep the plankton submerged and prevent damage to the fragile specimens. This system was tested to ensure that all of the organisms entering the system were collected on the net. Rotifers, cladocerans and copepods were preserved in 4% formalin solution after anaesthesia with mineral water saturated with CO_2, identified to genus or species level, and counted. Several identification keys were used for these determinations (Dussart, 1969; Harding & Smith, 1974; Pontin, 1978; Amoros, 1983; Pourriot & Francez, 1986). Rotifers were classified in different classes according to their saprobic valences as determined by Sládecek (1983). The zooplankton of the River Warche was concentrated by filtering 50 l of water through a 37 μm mesh net.

Physical and chemical variables

Ammonium, nitrites, nitrates, phosphates and silicon were analysed by standard photometric methods with an SQ 118 photometer and Spectroquant$^{(R)}$ kits (Merck, Darmstadt, Germany). Ammonium ions were assayed as indophenol blue at 690 nm (Bolleter et al., 1961). Nitrates (after reduction) and nitrites were assayed with sulphanilic acid and 1-naphtylamine, the absorbance being measured at 526 nm (Bratton et al., 1939). Orthophosphate ions were reduced to phosphomolybdenum blue (PMB) whose absorbance was measured at 712 nm. Silicates were reduced to silicomolybdenum blue whose absorbance was measured at 650 nm (Rodier, 1965). Temperatures and oxygen concentrations were measured by oxymeter (OXI 96 WTW, Weilheim, Germany) at 1 m depth intervals.

Reproduction toxicity tests

Reproduction toxicity tests were conducted using the freshwater rotifer *Brachionus calyciflorus*. Water to be tested was prepared by filtering field samples

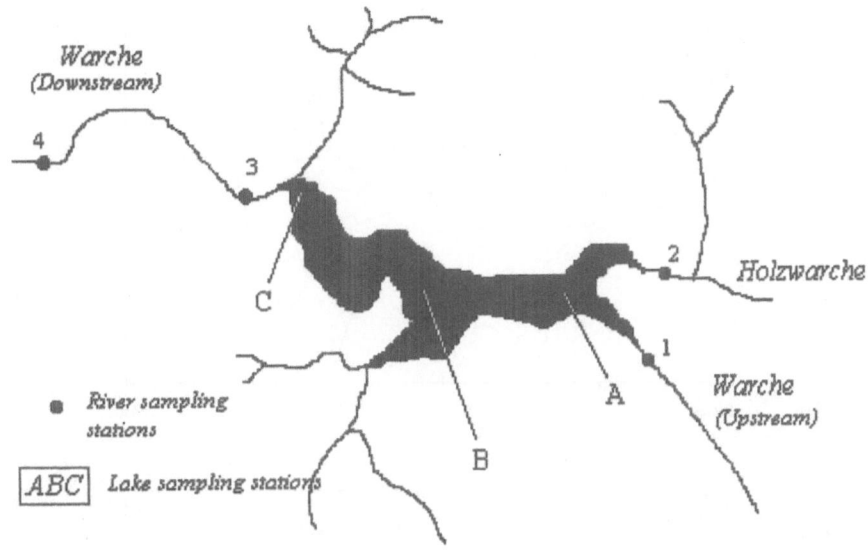

Figure 1. Map of Bütgenbach reservoir showing the location of the lake sampling stations (A–C) and river sampling stations (1–4).

Table 1. Mean values of the main physical and chemical parameters measured in the River Warche and Bütgenbach reservoir (upper 2 m of water column)

	Warche upstream		Holzwarche		A		B		C		Warche downstr.	
	$\bar{x}$	SE	$\bar{x}$	SE	$\bar{x}$	SE	$\bar{x}$	SE	$\bar{x}$	SE	$\bar{x}$	SE
Temperature (°C)	12.7	3.2	13.2	3.6	14.7	4.9	14.7	4.9	14.6	5.0	9.4	2.9
pH	7.4	0.4	7.5	0.2	7.4	0.6	7.5	0.6	7.6	0.6	7.3	0.4
Oxygen (mg l^{-1})	9.5	1.6	9.6	1.0	10.1	1.1	10.4	1.3	10.7	1.7	9.1	2.2
Oxygen (% sat.)	96	9	98	9	107	12	109	13	112	13	84	17
N–NO$_3^-$ (mg l^{-1})	3.7	1.0	3.1	1.2	2.7	0.7	2.9	0.7	3.3	0.7	2.6	0.5
N–NO$_2^-$ (mg l^{-1})	0.04	0.03	0.02	0.01	0.02	0.01	0.02	0.01	0.02	0.01	0.05	0.04
N–NH$_4^+$ (mg l^{-1})	0.09	0.08	0.10	0.07	0.02	0.03	0.03	0.04	0.04	0.04	0.56	0.41
Si (mg l^{-1})	2.5	0.5	2.0	0.7	0.8	0.5	0.6	0.4	0.4	0.3	1.4	0.7
PO$_4^{3-}$ (mg l^{-1})	0.34	0.32	0.19	0.20	0.02	0.03	0.01	0.01	0.01	0.02	0.13	0.21

through 0.22 μm filter units (Sterivex^{TM}-GS, Millipore, Bedford, MA, U.S.A.). Test animals were obtained by hatching commercially available cysts (Creasel, Deinze, Belgium) and fed a suspension of *Dictyosphaerium ehrenbergianum* at a final concentration of 300 000 cells ml^{-1}. Five newborn rotifers were placed in triplicate wells (sterile 24-well NUNC$^{(R)}$ plates) with 2 ml of water in each well. The culture wells were observed twice daily, during which time their offspring were removed and the medium was renewed. Adults, new-borns and eggs were counted.

Results

Physical and chemical quality of the lake water and impact on the River Warche

The mean values of the main physical and chemical parameters measured are shown in Table 1. The lake values refer to the upper 2 m of water column. The spatial and temporal variations in water temperature, percentage oxygen saturation and ammonium concentrations at Station C are shown in Figure 2 (a, b, c, respectively). The temperature of the lake water follows a standard pattern of evolution in eutrophic temperate lakes with a marked thermocline appearing during summer. Between April and October 1995, the

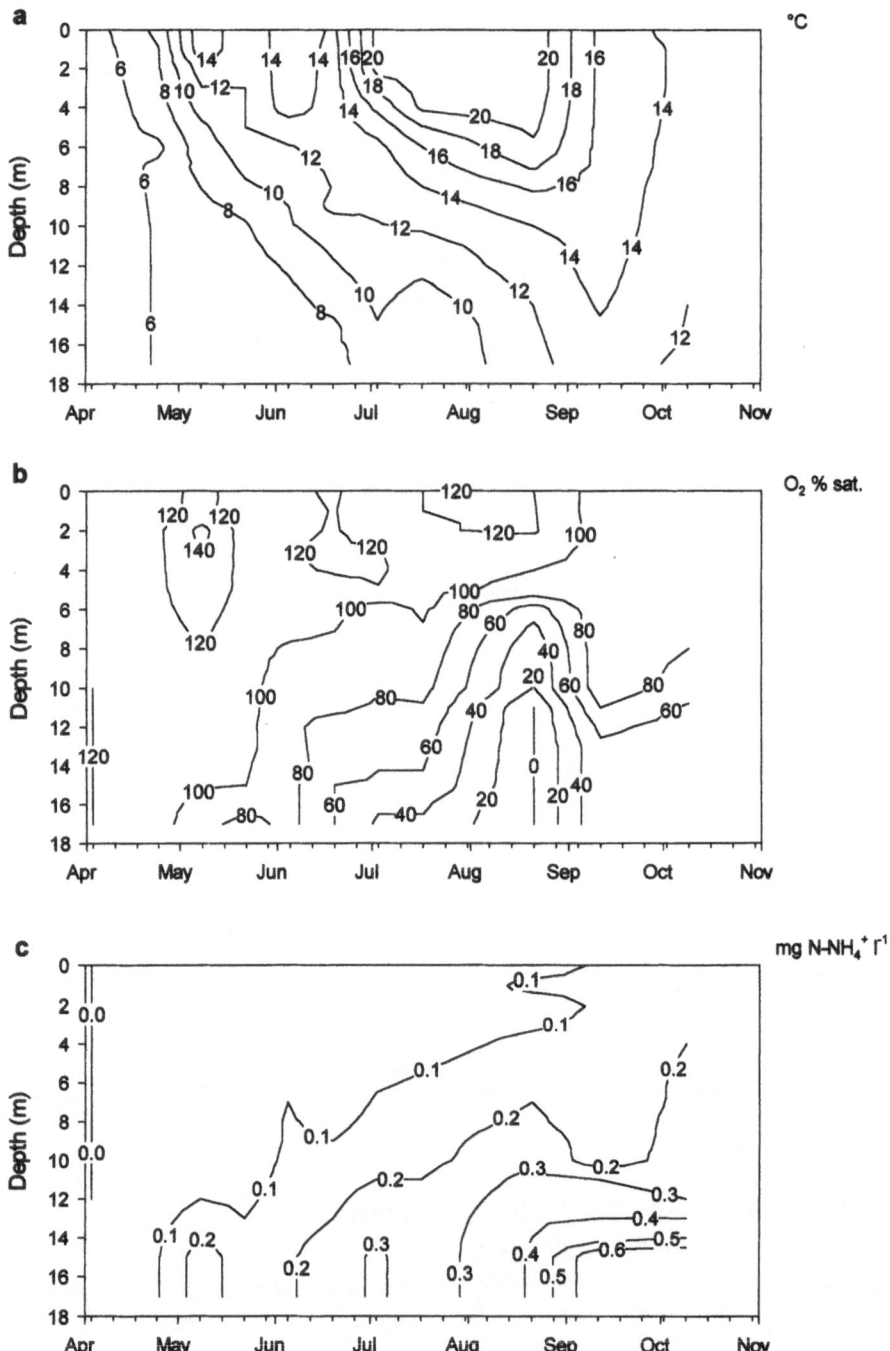

Figure 2. Spatial and temporal variations in water temperature (°C) (a), percentage oxygen saturation (b) and ammonium concentrations (mg N–NH$_4^+$ l^{-1}) (c) at Station C.

water temperature ranged between 5 and 22 °C. A vertical temperature gradient progressively developed between May and September which disappeared again during autumn. A severe oxygen deficit occurred near the bottom during summer (minimum value: 0 mg l^{-1}

at 12 m depth in August 1995). Ammonium concentrations did not significantly increase from Station A to Station C (ANOVA; $P \leq 0.001$), but increased from the surface to the bottom during summer at Station C (Figure 2c). In contrast, silicon concentrations (Table

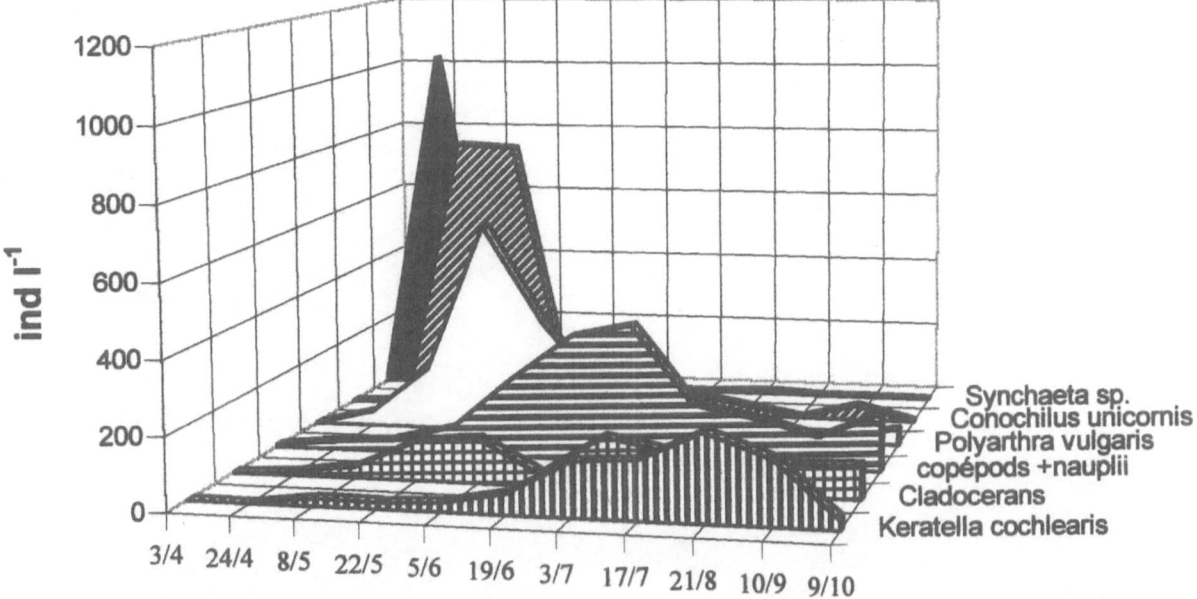

Figure 3. Temporal variation of zooplankton community at the Station C (mean densities in the upper 5 metres of the water column).

1) were significantly higher in the upstream part of the reservoir (Station A) than near the dam (Station C) (ANOVA; $P \leq 0.001$). Of the other nutrient studied, phosphate concentrations were relatively low ranging from 0 to 0.14 mg l^{-1} (mean annual value 0.01 mg l^{-1}) and increasing slightly near the bottom of the reservoir at Station C (mean annual value 0.03 mg l^{-1}). Nitrate concentrations ranged from 1.4 to 5.4 mg $N–NO_3^- \ l^{-1}$ and nitrite concentrations ranged from 0 to 0.04 mg $N–NO_2^- \ l^{-1}$. For both nitrogen compounds there was no significant differences between sampling stations A, B and C (ANOVA; $P \leq 0.001$). The impact of the lake on the physical and chemical quality of the River Warche during spring, summer and autumn–winter periods has been discussed elsewhere (Marneffe et al., 1997).

Composition of the zooplankton

The zooplankton of Bütgenbach reservoir was clearly dominated by rotifers; crustaceans (cladocerans and copepods) were less abundant. However, crustaceans represented an important part of the total zooplankton biomass due to their high individual weights. A list of taxa found in the lake during the study is shown in Table 2, together with an indication of their respective abundances. Over the course of the year, there was a succession of species in the lake. Figure 3 shows the

zooplankton succession at Station C. Rotifers dominated in Spring (*Synchaeta* spp. *Conochilus unicornis* and *Polyarthra vulgaris*). Copepods and cladocerans were most abundant during summer. *Keratella cochlearis* was abundant from July to October. Moreover, the faunal spectrum and species densities varied with sampling location.

Use of the rotifer communities as bioindicators

Mean annual densities were calculated for each rotifer species collected at the different sampling stations (upper 5 m of the water column). These different species were classified in different classes according to their saprobic valences as determined by Sládecek (1983). Figures 4 and 5 show the proportions of these different classes at each sampling station. The proportion of oligosaprobic species increased from A to C whereas the proportion of β–α mesosaprobic species decreased (Figure 4). This suggested an improvement in the lake surface water quality from the upstream site to the downstream site. The water quality of the River Warche similarly improved downstream of the lake (Figure 5). This observation is in good agreement with the determination (at the same sampling stations) of biotic indices based on macroinvertebrates as discussed elsewhere (Marneffe et al., 1997).

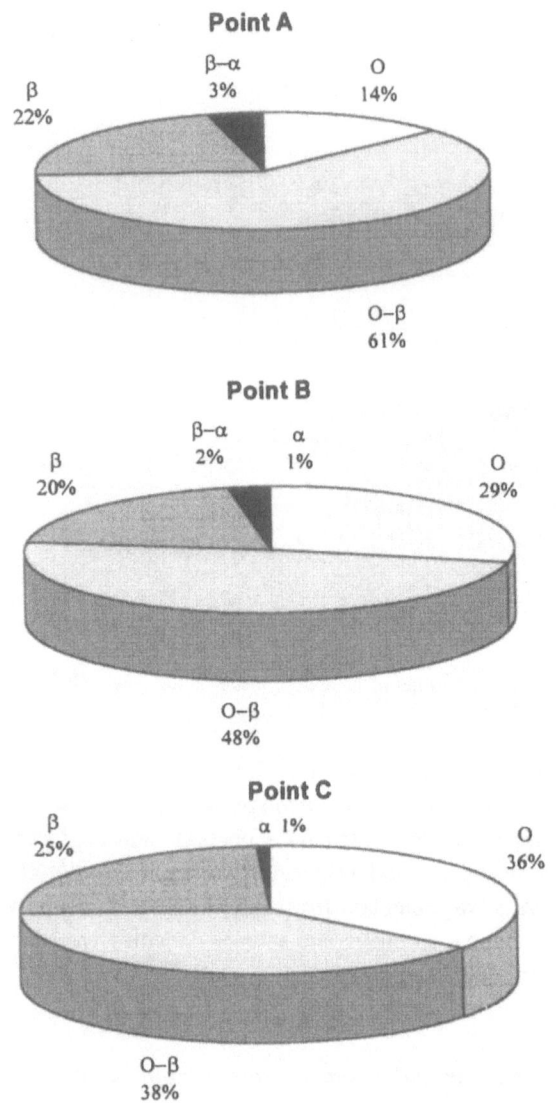

Point A

β–α 3%
O 14%
β 22%
O–β 61%

Point B

β–α 2%
α 1%
β 20%
O 29%
O–β 48%

Point C

β 25%
α 1%
O 36%
O–β 38%

Figure 4. Proportions of the different rotifer species of the Bütgenbach lake classified according to their saprobic valences.

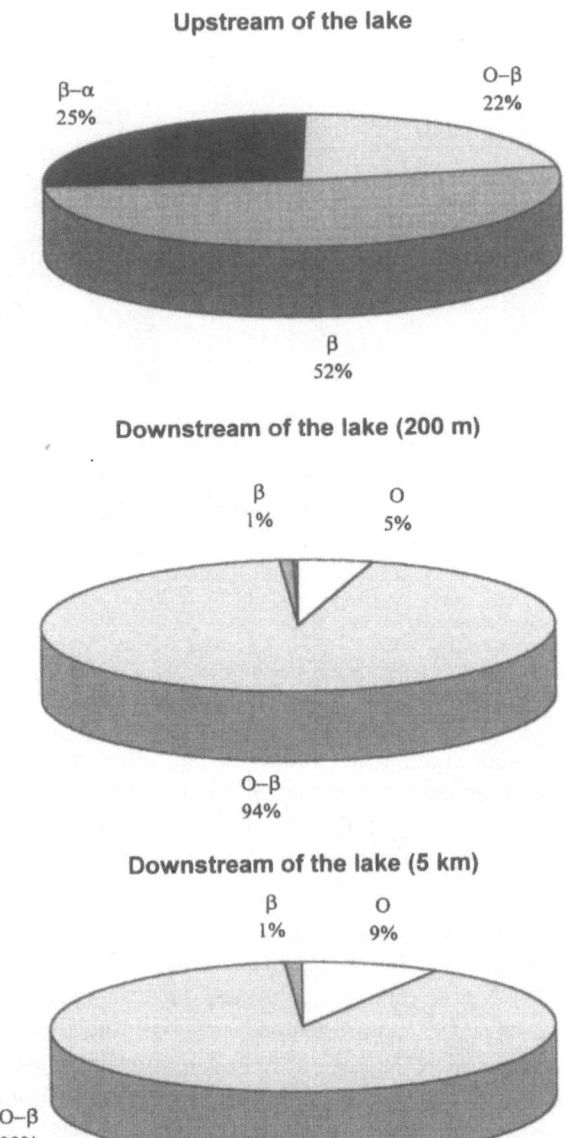

Upstream of the lake

β–α 25%
O–β 22%
β 52%

Downstream of the lake (200 m)

β 1%
O 5%
O–β 94%

Downstream of the lake (5 km)

β 1%
O 9%
O–β 90%

Figure 5. Proportions of the different rotifer species of the River Warche classified according to their saprobic valences.

Reproduction toxicity tests

Figure 6 shows the cumulative evolution of the number of young females produced per rotifer under standardized conditions in water originating from different stations located upstream and downstream of the lake. The reproduction ability was lower in the water sampled just downstream of the lake than in water collected upstream, but it increased again 5 km downstream, confirming the improvement in water quality suggested by the bioindicators at this station. The water tested in this experiment was sampled during the lake stratification period when ammonium concentra-

tions were high near the bottom of the lake and in the River Warche just downstream of the reservoir (1.1 mg N–NH_4^+ l^{-1}). Therefore, this observation does not contradict the improvement in water quality observed at this station on the basis of macroinvertebrate biotic indices or of rotifer communities used as bioindicators (integration of the quality over a longer period of time), as water containing high ammonium concentrations is released from the bottom of the lake into

Table 2. List of species or genera found in the Bütgenbach lake along with their maximum densities at the different sampling stations

	Maximum densities (ind l^{-1})		
	Point a	Point B	Point C
ROTIFERS			
Asplanchna priodonta (Gosse, 1850)	171	93	54
Brachionus angularis (Gosse, 1851)	10	10	0
Brachionus calyciflorus (Pallas, 1766)	0	0	3
Conochiloides natans (Seligo, 1900)	9	17	25
Conochilus unicornis (Rousselet, 1892)	280	1160	1080
Conochilus hippocrepis (Schrank, 1830)	10	30	25
Epiphanes senta (O. F. Müller, 1773)	4	0	0
Euchlanis dilatata (Ehrenberg, 1832)	1	0	0
Filania sp.	22	9	9
Filania longiseta (Ehrenberg, 1834)	30	15	12
Gastropus hyptopus (Ehrenberg, 1838)	0	2	2
Kellicottia longispina (Kellicott, 1879)	10	28	37
Keratella cochlearis (Goss, 1851)	973	600	586
Keratella quadrata (O. F. Müller, 1786)	49	150	91
Lecane sp.	5	3	6
Lepadella sp.	8	0	6
Notholca squamula (O. F. Müller 1786)	2	0	0
Ploesoma hudsoni (Imhof, 1891)	9	10	0
Polyarthra major (Burckhart, 1900)	44	25	10
Polyarthra sp.	150	57	52
Polyarthra vulgaris (Carlin, 1943)	803	513	800
Pompholyx sulcata (Hudson, 1845)	8	2	5
Synchaeta pectinata (Ehrenberg, 1832)	82	263	489
Synchaeta sp.	224	942	1720
Trichocerca similis (Wierzejski, 1893)	4	6	3
Trichocerca longiseta (Schrank, 1802)	5	10	9
Trichocerca sp.	4	3	1
COPEPODS			
Calanoids	42	63	56
Cyclopoids	113	150	158
Nauplii	308	342	597
CLADOCERANS			
Bosmina coregoni (Baird, 1857)	528	420	481
Bosmina longirostris (O. F. Müller 1785)	161	370	102
Chydoridae	11	6	0
Daphnia cucullata (Sars, 1862)	269	229	249
Daphnia hyalina (Leydig, 1860)	10	29	8
Daphnia longispina (O. F. Müller, 1785)	25	29	15
Daphnia obtusa (Kurz, 1874)	6	10	9
Diaphanosoma brachyurum (Liévin, 1848)	388	143	56
Leptodora kindtii (Focke, 1844)	1	1	1
Sididae	3	2	3
Simocephalus sp.	9	0	0

the River Warche. This could have a serious negative effect on animal communities downstream.

Discussion

Along the first ten kilometres of the River Warche, a decrease in physical, chemical and biological water quality is induced by the input of nitrogen compounds from the main tributaries and effluents where high nutrient concentrations have been measured (Fabri & Leclercq, 1977; Marneffe et al., 1997; Descy et al., 1981). As a result, both chemical and biological quality decreases downstream of Büllingen, which is just upstream of the lake. Verbanck & Versaen (1996) showed that the River Warche is the source of 82% of the total phosphorus (P) load (mainly constituted of ortho phosphates) to the lake. They estimated this total load to be 2.5 tonnes of phosphorus per year. These authors also calculated that 2 tonnes of this incoming P are retained each year in bottom sediments. This study and our results show that nitrogen compounds also originate mainly from the rivers Warche and Holzwarche. The high nutrient concentrations near the reservoir inlet are a well-known phenomenon that has a direct impact on phyto- and zooplankton communities. In the same way, biological developments act on nutrient concentrations; e.g., silicon concentrations that decrease from upstream to downstream probably as a consequence of diatom uptake. From the classification of the different rotifer species according to their saprobic valences (Sládecek, 1983) it appears that the oligotrophic species proportion increases from the reservoir inlet to the dam (annual mean on the five first meters of depth) whereas oligo-beta mesosaprobic and beta-alpha mesosaprobic species proportion decreases. These observations are in good agreement with the decrease of nutrient concentrations from upstream to downstream. The use of rotifers as bioindicators also emphasises an improvement of the overall water quality in the River Warche which agrees with the improvement observed using biotic indices based on macroinvertebrates (Marneffe et al., 1997). Thus, the lake seems to function like a large purification system. However, the situation is not as beneficial as it seems. A vertical gradient of physical, chemical and biological parameters appears during the summer period. This vertical heterogeneity was not observed at Station A due to the shallowness of the lake and to the effect of influent river at this station. On the contrary, at Station C, located near the dam, phyto- and zooplankton are concentrated on the upper five metres during summer, and anoxic conditions occur below 10 m of depth in August. During this period, dissolved oxygen can be rapidly reduced by intense microbial

466

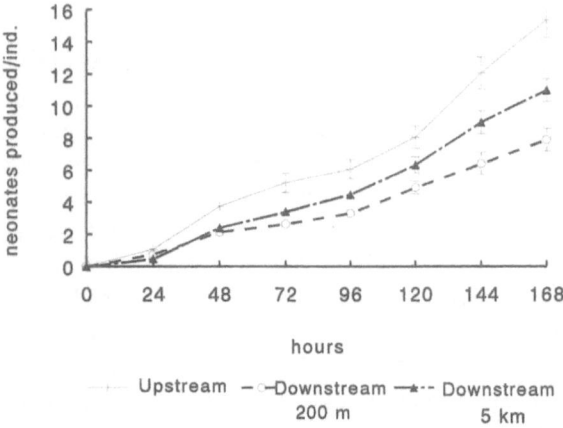

Figure 6. Cumulative evolution of the number of neonates produced per rotifer placed in standardised conditions in the water originating from different sampling stations upstream and downstream of the lake.

activities occurring at the sediment/water interface and can induce denitrification processes (Wetzel, 1983). As a consequence of organic matter decomposition and of excretion, ammonium concentrations increase markedly near the bottom of the lake. At the end of the summer period, bluegreen algae blooms occur near the surface inducing important consumption of carbon dioxide and consequently, surface pH values can reach 10. All these perturbations have important consequences on the water quality of the Warche since the water that feeds this river is released from the base of the hydroelectric dam. During summer very high phosphate and ammonium concentrations also are observed in the River Warche downstream of the lake, whereas silicon concentrations decrease. Reproduction toxicity tests using *Brachionus calyciflorus* revealed a decrease of the reproduction capacity of *Brachionus calyciflorus* placed in the water sampled just downstream of the lake, and a partial restoration of this capacity for rotifers placed in water sampled 5 km downstream of the dam. The deterioration of the water quality near the bottom of the lake during the stratification period could therefore have a negative impact on animal communities downstream. Under conditions of stratification ammonium concentrations were high $(1.1 \text{ mg N–NH}_4^+ \text{ l}^{-1} \text{ i.e., } 1.4 \text{ mg NH}_4^+ \text{ l}^{-1})$ in the River Warche water just downstream of the reservoir and it is worth noticing that the LC 50 of free ammonia tested on *Brachionus calyciflorus* ranges from 3.2 to 4.6 mg l^{-1} (Snell & Janssen, 1995).

Conclusion

Physical and chemical measurements, bioindicator determinations, and ecotoxicity tests used as complementary tools to assess the water quality of the Bütgenbach lake and its impact on the River Warche show generally the same trends in the longitudinal evolution of water quality from upstream to downstream. A strong deterioration of water quality upstream of the lake due to sewage effluent from Büllingen is detected by measurements of physical, chemical and biological parameters. The presence of the lake on the watercourse seems to result in an overall improvement in water quality downstream. However, toxicity tests on *Brachionus calyciflorus* show that, when the reservoir is stratified, the water released from the base of the dam could have a negative effect on the water quality in the River Warche.

Acknowledgements

This research was supported by a grant from the 'Fonds National de la Recherche Scientifique'.

References

Amoros, C., 1983. Cladocères. Extrait du Bulletin de la Société Linéenne de Lyon 53ème année 3/4, 63 pp.

Arnott, S. E. & M. J. Vanni, 1993. Zooplankton assemblages in fishless bog lakes: influence of biotic and abiotic factors. Ecology 74(8): 2361–2380.

Blancher, E. C., 1984. Zooplankton-trophic state relationship in some north and central Florida lakes. Hydrobiologia 109: 251–263.

Bolleter, W. T., C. J. Bushman & P. W. Tidwell, 1961. Ammonium Analysis. Analyt. Chem. 33: 592–594.

Bratton, A. C., E. K. Marshall, D. Babitt & A. R. Hendrickson, 1939. Nitrite Determination in Fresh Water. J. Biol. Chim. 128: 537–550.

Burbank, S. E. & T. W. Snell, 1994. Rapid toxicity assessment using esterase biomarkers in *Brachionus calyciflorus* (Rotifera). Envir. Toxicol. Wat. Qual. 9: 171–178.

Charoy, C. P., C. R. Janssen, G. Persoone & P. Clement, 1995. The swimming behaviour of *Brachionus calyciflorus* (rotifer) under toxic stress. I. The use of automated trajectometry for determining sublethal effects of chemicals. Aquat. Toxicol. 32: 271–282.

Descy, J. P., A. Empain & J. Lambinon, 1981. La qualité des eaux courantes en Wallonie-Bassin de la Meuse. Secrétariat d'état à l'environnement, à l'aménagement du territoire et à l'eau pour la Wallonie. 18 pp.

Dohet, A. & L. Hoffmann, 1995. Seasonnal succession and spatial distribution of zooplankton community in the reservoir of Esh-sur-Sûr (Luxembourg). Belg. J. Zool. 125(1): 109–123.

Dussart, B., 1969. Les copépodes des eaux continentales d'Europe Occidentale. Tome II: Cyclopoïdes et biologie. Ed. Boubée et Cie, 292 pp.

Fabri, R., 1977. Végétation, production primaire et caractéristiques physico-chimiques d' une rivière de Haute Ardenne (Belgique): la Warche supérieure. Lejeunia, nouvelle série 87: 1–43.

Fabri, R. & L. Leclercq, 1977. Végétation et caractéristiques physico-chimiques des eaux de trois rivières de Haute Ardenne (Belgique): la Helle, la Roer et la Warche. Bull. Soc. r. Bot. Belg. 110: 202–216.

Fernández-Casalderrey, A., M. D. Ferrando & E. Andreu-Moliner, 1993. Chronic toxicity of methylparathion to the rotifer *Brachionus calyciflorus* fed on *Nannochloris oculata* and *Chlorella pyrenoidosa*. Hydrobiologia 255/256: 41–49.

Ferrando, M. D., C. R. Janssen, E. Andreu & G. Persoone, 1993. Ecotoxicological studies with the freshwater rotifer *Brachionus calyciflorus*. III. The effects of chemicals on the feeding behavior. Ecotox. Envir. Saf. 26: 1–9.

Gannon, J. E. & R. S. Stemberger, 1978. Zooplankton (especially crustaceans and rotifers) as indicators of water quality. Trans. am. micros. Soc. 97: 16–35.

Gilbert, J. J., 1988. Suppression of rotifer populations by *Daphnia*: a review of the evidence, the mechanisms and the effects on zooplankton community structure. Limnol. Oceanogr. 33: 1286–1303.

Gulati, R. D., 1990. Zooplankton structure in the Loosdrecht Lakes in relation to trophic status and recent restoration measures. Hydrobiologia 191: 173–188.

Harding, J. P. & W. A. Smith, 1974. A key to the British Freshwater cyclopid and calanoid copepods. Freshwater Biological Association, 56 pp.

IHE, 1990. Carte de la qualité des cours d'eau de Belgique.

Janssen, C. R., M. D. Ferrando & G. Persoone, 1994. Ecotoxicological studies with the freshwater rotifer *Brachionus calyciflorus*. IV. Rotifer behavior as a sensible and rapid sublethal test criterion. Ecotox. Envir. Saf. 28: 244–255.

Lampert, W. & K. O. Rothhaupt, 1991. Alternating dynamics of rotifers and *Daphnia magna* in a shallow lake. Arch. Hydrobiol. 120: 391–392.

Marneffe, Y., J. C. Bussers, M. Louvet & J. P. Thomé, 1995. La qualité physico-chimique et biologique de la Warche et son incidence sur l'Amblève. Bull. Soc. r. Sci. Liège 64: 77–103.

Marneffe, Y., S. Comblin, J. C. Bussers & J. P. Thomé, 1997. Biomonitoring of the water quality in the River Warche (Belgium): Impact of tributaries and sewage effluent. Neth. J. Zool. 47: 111–124.

Ministére de la Région Wallonne, 1994. Carte de la qualité biologique des eaux de surface.

Pontin, R. M., 1978. A Key of British Freshwater Planktonic Rotifera. Freshwater Biological Association, 179 pp.

Pourriot, R. & A. J. Francez, 1986. Rotifères. Extrait du Bulletin de la Société Linéenne de Lyon 55ème année 5, 37 pp.

Rao, T. R. & S. S. S. Sarma, 1986. Demographic parameters of *Brachionus calyciflorus* Muller (Rotifera) exposed to sublethal DDT concentrations at low and high food levels. Hydrobiologia 139: 193–200.

Rodier, J., 1965. L'analyse chimique et physico-chimique de l'eau. Eaux naturelles et eaux usées, Troisième édition. Dunod, Paris, 412 pp.

Seda, J. & A. Duncan, 1994. Low fish predation pressure in London reservoirs: II. Consequences to zooplankton community structure. Hydrobiologia. 291: 179–191.

Sládecek, V., 1983. Rotifers as indicators of water quality. Hydrobiologia 100: 169–201.

Snell, T. W. & M. J. Carmona, 1995. Comparative toxicant sensitivity of sexual and asexual reproduction in the rotifer *Brachionus calyciflorus*. Envir. Toxicol. Chem. 14(3): 415–420.

Snell, T. W. & C. R. Janssen, 1995. Rotifers in ecotoxicology: a review. Hydrobiologia 313/314: 231–247.

Snell, T. W. & B. Moffat, 1992. A 2-d life cycle test with the rotifer *Brachionus calyciflorus*. Envir. Toxicol. Chem. 11: 1249–1257.

Taleb, H., P. Reyes-Marchant & N. Lair, 1994. Effect of vertebrate predation on the spatio-temporal distribution of cladocerans in a eutrophic lake. Hydrobiologia 294: 117–128.

Verbank, M. & M. Versaen, 1996. Etude des processus d'eutrophisation et des incidents de pollution dans le lac de Bütgenbach. Rapport d'étude, 26 pp.

Visman V., D. J. McQueen & E. Demers, 1994. Zooplankton spatial patterns in two lakes with contrasting fish community structure. Hydrobiologia 284: 177–191.

Wetzel, R. G., 1983. Limnology, 2nd edn. Saunders, New York, 767 pp.

Hydrobiologia **387/388**: 469–476, 1998.
E. Wurdak, R. Wallace & H. Segers (eds), Rotifera VIII: A Comparative Approach.
© *1998 Kluwer Academic Publishers.*

Size-structure dynamics of the rotifer chemostat: a simple physiologically structured model

James N. McNair[1], Martin E. Boraas[2] & Dianne B. Seale[2]
[1]*Patrick Center for Environmental Research, The Academy of Natural Sciences of Philadelphia, 1900 Benjamin Franklin Parkway, Philadelphia, PA 19103, U.S.A.*
[2]*Department of Biological Sciences, University of Wisconsin at Milwaukee, P.O. Box 413, Milwaukee, WI 53201, U.S.A.*

Key words: population dynamics, size structure, dilution rate

Abstract

The classical chemostat models of Monod and others were designed for unicellular organisms. We summarize evidence that these models are not adequate for the rotifer chemostat, then propose a new, physiologically structured model that resolves some of their key problems yet remains biologically simple. The new model includes separate ontogenetic stages for eggs and free-swimming rotifers, with generalized age structure in the egg stage and body mass structure in the free-swimming stage. We present several numerical examples to illustrate the model's behavior, and we compare these in a preliminary way with experimental evidence from the literature.

Introduction

Current knowledge of the principles of population dynamics is unsatisfactory on both empirical and theoretical grounds. The main problems are a weak empirical base, overly simple models with but vague links to real systems, and little effective interplay between theory and experiment. We therefore believe there is a need for renewed emphasis on carefully controlled laboratory studies of population dynamics, and on the development of structured population models that can be related clearly, directly, and convincingly to real experimental systems.

The rotifer chemostat is an excellent system for conducting laboratory population-dynamics studies. Relevant experimental techniques, concepts, and sample data are outlined in a companion paper (Boraas et al., 1998). Here we summarize the basic ideas underlying classical models of the chemostat, show that these models are inadequate for rotifers, and outline a new model that we believe has the potential to provide a more satisfactory theory.

Classical chemostat models

The main component of the rotifer chemostat is an enclosed culture vessel containing a suspension of rotifers and algae (Figure 1A). The culture is kept well mixed to maintain spatial homogeneity. Fresh algal suspension (from an algal chemostat) is pumped into the culture at a continuous, regulated rate. Since the culture volume is fixed, outflow exactly balances inflow (dimensions: volume·time^{-1}). The culture vessel is kept in the dark (to minimize algal growth and division) under tightly controlled physical conditions and is monitored regularly. (See Boraas et al., 1998, for additional details.)

In attempting to characterize the rotifer chemostat mathematically, previous studies have employed models originally developed for unicellular organisms, including Monod's (1950) model and a few straightforward variants. We refer to these collectively as classical chemostat models. A typical example is the Monod-Herbert model (Herbert, 1958), which differs from Monod's model simply by permitting loss of biomass via catabolism.

All classical chemostat models assume that the state of a rotifer population at any time t can be ad-

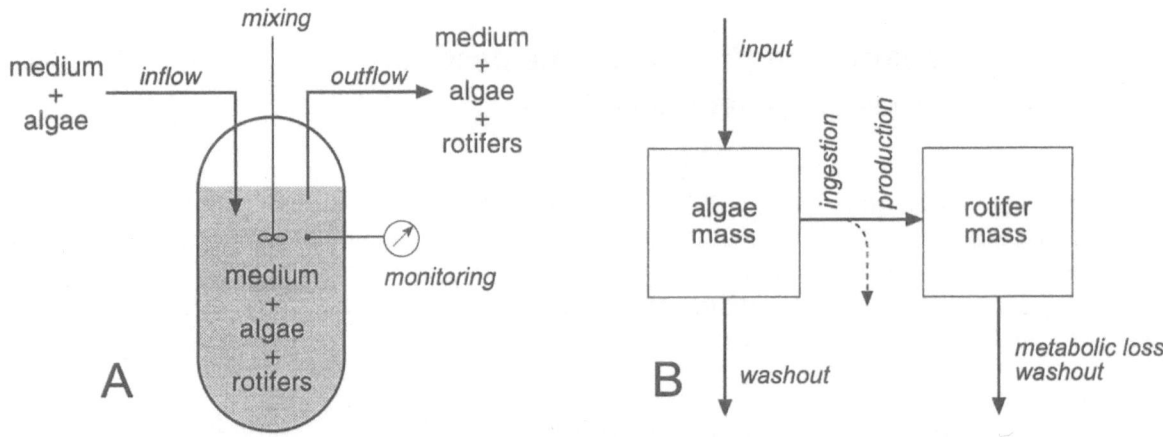

Figure 1. A—Basic components of the rotifer chemostat (highly schematic). See text for description, and Boraas et al. (1998) for additional details. B—Schematic of the Monod-Herbert model. See text for description.

equately characterized by a single number $M(t)$, which usually is some measure of the total mass of the population (e.g. total carbon). The dynamics of $M(t)$ are assumed to be determined by the difference between the rate at which total rotifer biomass grows (as a result of ingesting algae and converting it into rotifer biomass) and the rates at which total rotifer biomass is diminished by catabolism and by being washed out of the chemostat. Similarly, it is assumed that the state of the algae population can be adequately characterized by a single number representing its total mass, whose dynamics are determined by the difference between the rate at which algal biomass is fed into the chemostat and the rates at which it is removed by rotifer ingestion and by being washed out of the system.

Equations governing the states of the algae and rotifer populations are usually stated in terms of the respective volume-normalized mass concentrations $A(t)$ and $R(t)$ (dimensions: mass·volume^{-1}); e.g. $R(t) = M(t)/V$, where V is the (constant) culture volume. The Monod-Herbert model, for example, can be written in the form:

$$\frac{dA}{dt} = A_0 D - AD - \frac{I_{sup} A R}{K_h + A},$$

$$\frac{dR}{dt} = Y \frac{I_{sup} A R}{K_h + A} - \rho R - DR, \qquad (1)$$

where the first and second equations specify the instantaneous rates of change in mass concentrations of

algae and rotifers, respectively, and

$A_0 =$ algal mass concentration in the feed;

$D =$ dilution rate (inflow divided by V);

$I_{sup} =$ asymptotic mass-specific rate of ingestion by rotifers;

$K_h =$ half-saturation constant for ingestion;

$Y =$ yield coefficient (rotifer biomass produced per unit algal biomass ingested);

$\rho =$ mass-specific rate of metabolic loss of rotifer biomass.

The three terms on the right side of the algae equation correspond to algal input, washout, and ingestion by rotifers. The three terms on the right side of the rotifer equation correspond to production of rotifer biomass from ingested algae, metabolic loss of biomass, and washout. The model is shown schematically in Figure 1B.

It is important to note that equation (1) asserts that the dynamics of algal and rotifer mass can be understood and predicted with no knowledge of the internal structure of either population (e.g. age or size structure). Thus, according to (1), if we set up a chemostat with fixed initial masses of algae and rotifers, it should make no difference to the system's short-term dynamics whether the rotifer population consists entirely of eggs (which would not ingest algae, and would neither reproduce nor increase in dry mass), entirely of gravid females (which would ingest algae, reproduce, and possibly increase in dry mass, as well), or of some

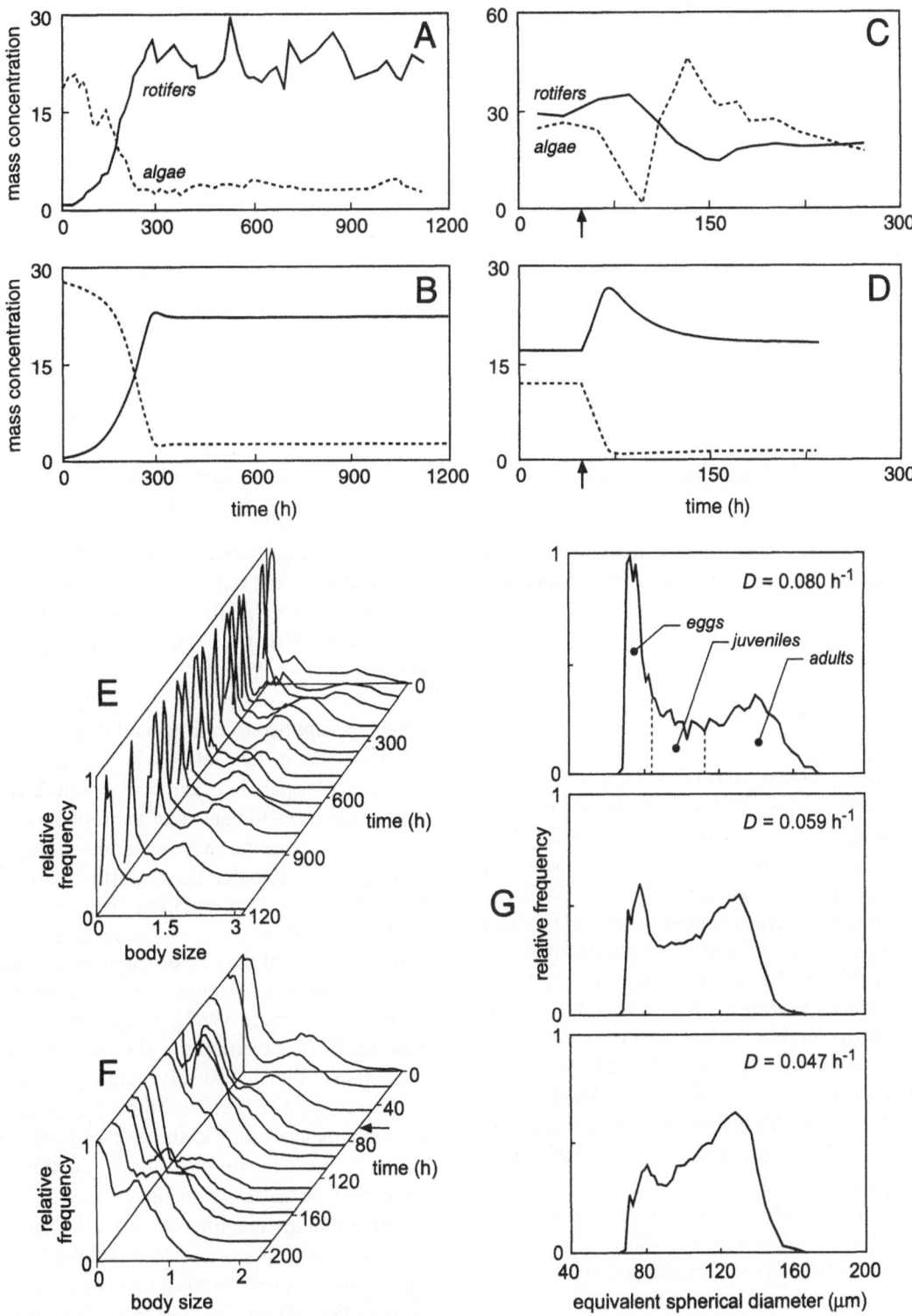

Figure 2. Examples of rotifer chemostat dynamics in laboratory experiments. All examples apply to *Brachionus calyciflorus*. A—Approach to quasi-steady state. Mass concentration units: μg·mL^{-1} for rotifers, μg·mL^{-1} × 5.7 for algae. B—Dynamics predicted by the Monod-Herbert model for data-series A. Units: same as in A. C—Effect of a downward shift in dilution rate (arrow) following the end of data-series A. Units: μg·mL^{-1} for rotifers and algae. D—Dynamics predicted by the Monod-Herbert model for data-series C. Units: same as in C. E, F—Observed dynamics of the rotifer size distribution in data-series A and C. Size distributions are normalized to achieve a constant egg-peak height; non-normalized distributions are shown in Figure 5 of Boraas et al. (this volume). Body size units: μm^3 × 10^6. G—Steady-state rotifer size distributions at three different dilution rates. Top distribution shows approximate sizes of eggs, juveniles, and adults. Data: A—F: Boraas (1983), G: Bennett & Boraas (1989).

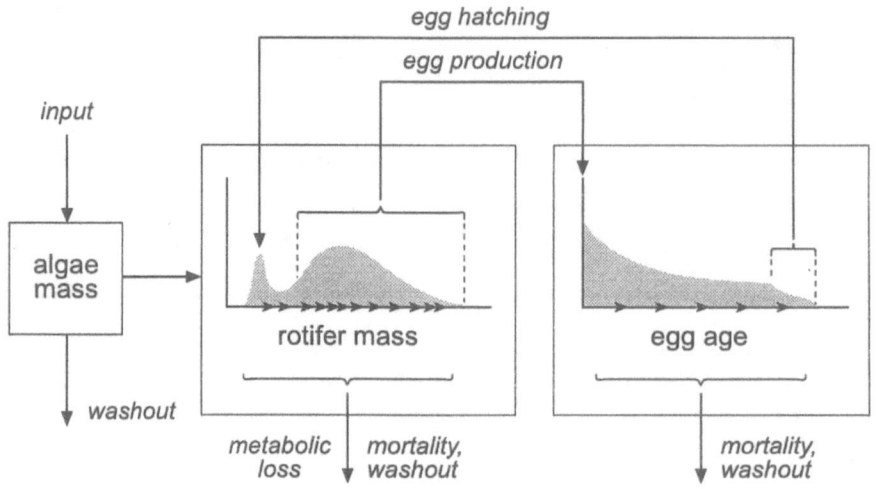

Figure 3. Schematic of the proposed physiologically structured chemostat model. See text for description.

Key experimental evidence

We now show results of several experiments that permit an assessment of classical chemostat models, here represented by the Monod-Herbert model. Figure 2A shows the results of an experiment conducted by Boraas (1983) in which *Brachionus calyciflorus* was introduced into a chemostat at low abundance and allowed to grow. Figure 2B shows the dynamics predicted by a calibrated Monod-Herbert model. Note the contrast between the smooth approach to steady state predicted by the model and the pronounced and persistent fluctuation observed. This result is typical of such experiments (e.g. Rothaupt, 1993; Walz, 1993) and reveals that the Monod-Herbert model exhibits too little tendency toward oscillation, compared to real rotifer populations.

A more striking discrepancy is demonstrated in panels C and D of Figure 2, which show the results of a downward D-shift experiment conducted by Boraas (1983). The chemostat was allowed to run for roughly 1000 h at $D = 0.045$ h^{-1}, after which D was abruptly decreased to 0.0135 h^{-1} (by reducing the pump speed). As the figure shows, the observed transient dynamics following the downward shift in D (panel C) bear little resemblance to the behavior predicted by the calibrated Monod-Herbert model (panel D). Most

notably, the algae showed a dramatic resurgence following their initial crash, whereas the model predicted essentially none. A similarly gross discrepancy was observed by Walz (1993) in a D-shift experiment with *Brachionus angularis*.

Another serious problem with classical chemostat models as applied to rotifers is that they are unable to address phenomena dealing with population structure. For example, panels E and F of Figure 2 (from Boraas, 1983) show dynamics of the rotifer size structure corresponding to the mass dynamics in panels A and C. In panel F, note that the D shift is followed by loss of the egg peak, accumulation of juveniles and small adults, then return of the egg peak (egg, juvenile, and adult segments of the size distribution are identified in panel G). Another example appears in panel G (from Bennett & Boraas, 1989), which shows the steady-state rotifer size distribution at three different dilution rates. Note that the steady-state egg peak is taller relative to the adult peak at higher dilution rates. Size-structure patterns such as these provide valuable clues about the mechanisms of rotifer population regulation but cannot be addressed using classical models.

Based on the inability of classical chemostat models to account for observed transient dynamics of total mass, and on their inability to address observed patterns in population structure (and also for theoretical reasons beyond the scope of this paper), we believe that these models are inadequate tools for studying the rotifer chemostat. In the next section, we propose a new model that resolves some of these problems.

mix of life stages. If this property were to hold for real rotifer chemostats, it would be a most remarkable one, indeed.

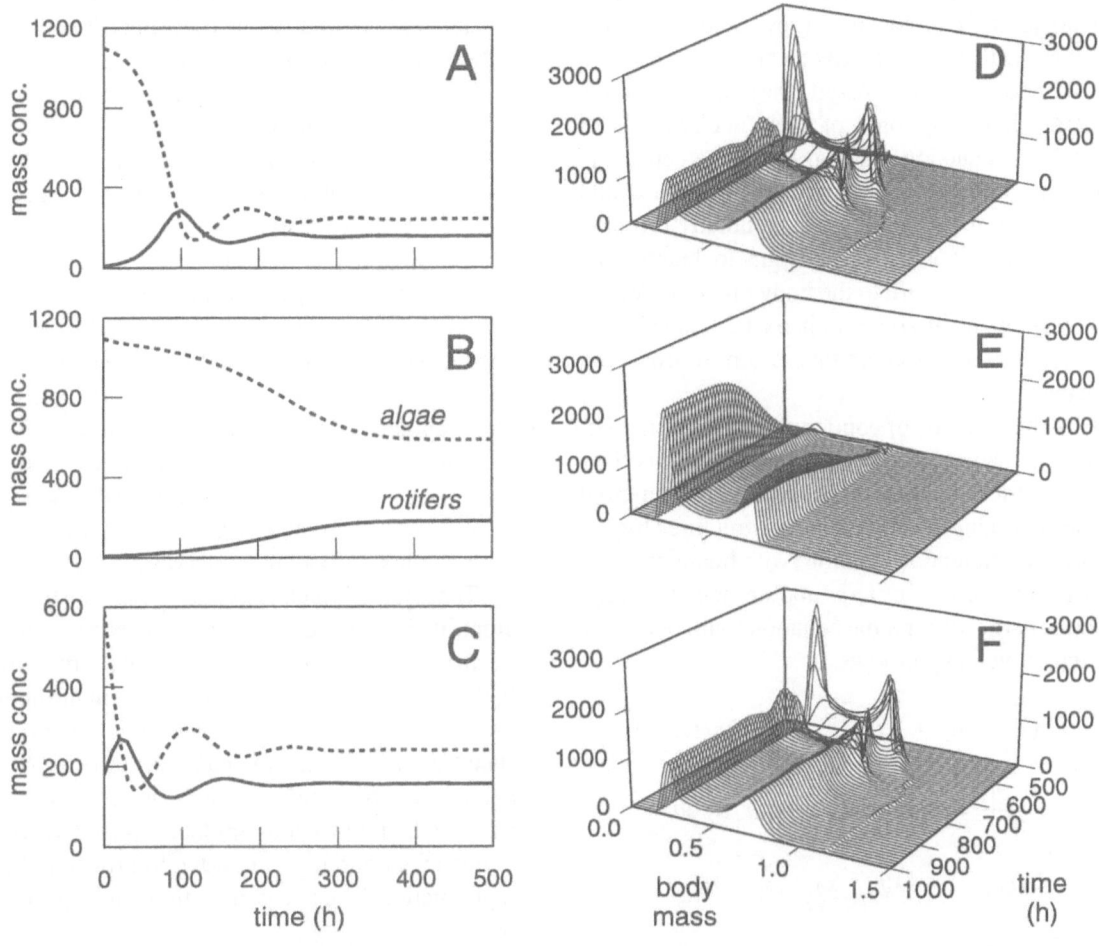

Figure 4. Examples of model dynamics. A—Total algal and rotifer masses at $D = 0.02$ h^{-1}. B—Total algal and rotifer masses at $D = 0.05$ h^{-1}. C—Continuation of B after a step-change in dilution rate to $D = 0.02$ h^{-1}. D—Rotifer size distributions corresponding to A. Vertical axis is population density. E—Rotifer size distributions corresponding to B. F—Rotifer size distributions corresponding to C.

A simple physiologically structured model

Our objective in developing a new model of the rotifer chemostat was to add the minimum biological detail necessary to account for currently known phenomena, and to increase the correspondence between model components and experimentally measurable properties. We therefore decided to (a) add only a few key components of structure to the rotifer population, (b) focus on structural components that are experimentally measurable and directly related to basic physiological mechanisms, (c) keep the model simple enough so it is both computationally tractable and reasonably transparent to underlying principles, and (d) pose the model in a form that can be reduced to a classical model via specializing assumptions (so the reasons for differences in model properties can be clearly identified).

The basic idea behind the model is illustrated in Figure 3. Recall from Figure 1B that classical chemostat models can be diagramed as two (connected) boxes with no internal structure. In the new model, the algae box of the classical model is retained, but two types of structure are added to the rotifer box. First, the population is divided into two discrete stages: eggs and free-swimming rotifers (hereafter, rotifers). Second, each stage is given a different type of continuous internal structure: generalized age for eggs (e.g. degree days) and body mass for rotifers. Thus, the new model is only slightly more complex than classical models and remains biologically quite simple.

The model works as follows. Rotifers continuously ingest algae and consequently grow along the body mass axis (with negative growth allowed). Once a threshold body mass is crossed, they become adults and begin allocating some of their net assimilated mass to egg production. The eggs produced enter the age axis of the egg stage at generalized age 0 and begin a process of development, eventually crossing a threshold age at which they begin to hatch. The neonates produced return to the body mass axis of the rotifer stage, with variability in neonatal body size allowed. The newborn rotifers then begin to grow, and the life cycle is complete.

Using the methods of continuum transport modeling, the schematic of Figure 3 is easily translated into a set of equations comprising an ordinary differential equation governing the algae, a mass-structured hyperbolic partial differential equation (with boundary condition) governing the rotifers, and an age-structured hyperbolic partial differential equation (with boundary condition) governing the eggs:

$$\frac{dA}{dt} = (A_0 - A)D - \varepsilon \int_{x_0}^{\infty} F(A, x) n(t, x)\, dx$$

$$\frac{\partial n}{\partial t} + \frac{\partial (\gamma n)}{\partial x} = b(x) \int_0^{\infty} \nu(a) n_e(t, a)\, da$$

$$- [\mu(A, x) + D] n(t, x), \quad x > x_0$$

subject to $n(t, x_0) = 0$ when $\gamma(A, x_0) > 0$

$$\frac{\partial n_e}{\partial t} + u \frac{\partial n}{\partial a} = -[\nu(a) + \mu_e(a) + D] n_e(t, a),$$

$$a > 0$$

subject to $u n_e(t, 0) = \int_{x_0}^{\infty} \beta(A, x) n(t, x)\, dx,$

(2)

where:

$A(t)$ = total mass of algae at time t, per unit volume of culture;

$n(t, x)\, dx$ = number of rotifers with body mass between x and $x + dx$ at time t, per unit volume of culture;

$n_e(t, a)\, da$ = number of eggs with generalized age between a and $a + da$ at time t, per unit volume of culture;

$\gamma(A, x)$ = rate of growth in rotifer body mass at body mass x;

$b(x)$ = probability density function for the mass of a rotifer egg;

$\nu(a)$ = egg maturation rate at generalized age a;

$\mu(A, x)$ = mortality rate of rotifers at body mass x;

D = dilution rate;

u = temperature-dependent rate of generalized aging in eggs;

$\mu_e(a)$ = egg mortality rate at generalized age a;

A_0 = input mass concentration of algae;

ε = algal cell mass;

$F(A, x)$ = rotifer ingestion rate at body mass x (cells per time per rotifer);

x_0 = lower limit of body mass for a rotifer;

$\beta(A, x)$ = rotifer egg production rate at body mass x.

Other choices of the boundary condition for the rotifer equation are plausible. All quantities except γ are restricted to being nonnegative, and we assume $b(x) > 0$ for $x_1 < x < x_2$, and $b(x) = 0$ otherwise (so all egg masses fall within a finite interval).

The rates at which eggs and rotifers are transported along their respective structural axes are determined by the rates of generalized aging and of growth in body mass. For simplicity, we assume in the numerical examples below that generalized aging occurs at a constant rate $u = 1$, which is equivalent to chronological aging. Growth in body mass is assumed to be given by the amount of net assimilation (gross assimilation minus metabolic loss) allocated to growth rather than reproduction. The net assimilation rate $\alpha(A, x)$ is given by

$$\alpha(A, x) = [\epsilon p(A, x) - \xi(A, x)] F(A, x) - \rho x^{\theta}.$$

where $p(A, x)$ = assimilation fraction, $\xi(A, x)$ = assimilate spent on acquisition and processing, per ingested cell, and ρx^{θ} = resting metabolic loss rate. Growth and fecundity are then given by

$$\gamma(A, x) = \alpha(A, x) \phi_s(\alpha, x),$$

$$\beta(A, x) = \frac{\alpha(A, x) \phi_r(\alpha, x)}{\int y b(y)\, dy},$$

where $\phi_s(\alpha, x)$ and $\phi_r(\alpha, x)$ are the proportional allocations to somatic growth and reproduction, and the denominator in the formula for $\beta(A, x)$ is the average mass of an egg (which converts reproduction from mass of eggs per time to number of eggs per time). Proportional allocation to somatic growth and reproduction are related by

$$\phi_r(\alpha, x) = 1 - \phi_s(\alpha, x),$$

so it suffices to specify only $\phi_r(\alpha, x)$.

In the numerical examples below, $\phi_r(\alpha, x)$ is assumed to be the product of a component depending only on x and a component depending only on α, with allocation working as follows. No net assimilate is allocated to reproduction if body mass is too small (in which case the size-dependent component is zero) or if net assimilation is too small (in which case the assimilation-dependent component is zero). The size-dependent component is constant except in a certain interval on the body-mass axis (the maturation window) over which it increases from the juvenile value (= 0) to the adult value. The assimilation-dependent component is zero for negative net assimilation and increases toward a positive asymptote with increasing positive net assimilation. These assumptions are for purposes of illustration and probably will require adjustment in actual applications.

Numerical examples

We now illustrate the behavior of model (2). Our purpose is merely to show a few types of behavior the model can exhibit, and to compare these with available data in a qualitative and preliminary way. Thorough exploration of the model's properties and accurate estimation of parameter values are major tasks and are beyond the scope of this paper.

Panels A–C in Figure 4 show examples of total-mass dynamics. Panels A and B show the initial behavior and approach to steady state with $D = 0.02$ h^{-1} and $D = 0.05$ h^{-1}, respectively. Transient oscillation is evident in A but effectively disappears at high dilution rates, as in B. The dilution rate of example B was shifted downward to $D = 0.02$ h^{-1} at $t = 500$ h, and the subsequent dynamics are shown in panel C. Note that the initial crash in algal mass is followed by a pronounced resurgence, which is roughly the behavior observed by Boraas (1983) and Walz (1993) in their laboratory experiments (e.g. Figure 2C). The assimilation-dependent component of the allocation function plays an important role in determining how pronounced this resurgence is.

Panels D–F in Figure 4 show dynamics of the rotifer size structure during the same numerical experiments whose mass dynamics are shown in panels A–C. The size distribution shows transient wave-like oscillations in D, but these disappear at high dilution rates, as in E. Similar oscillations are set off by a downward shift in dilution rate, as shown in panel F. Note in particular the initial loss of the egg peak (as the crash in residual algae causes adults to divert net assimilation away from reproduction), the accumulation of juven-

iles and small adults, then return of the egg peak at a lower height. This is basically the pattern observed by Boraas (1983), shown in Figure 2F above and in Figure 5 of Boraas et al. (this volume). Also note that the steady-state egg peak is higher relative to the adult peak when $D = 0.05$ h^{-1} (panel E) than when $D = 0.02$ h^{-1} (panels D, F). This result is consistent with the pattern observed by Bennett & Boraas (1989), shown in Figure 2G.

Discussion

Our preliminary numerical results suggest that the new model is promising, but final judgement on its value awaits rigorous empirical tests. We expect laboratory experiments to indicate that adjustments in the model are necessary, and we note in this connection that various modifications and embellishments can be made without destroying the model's tractability. For example, accounting for observed chemostat dynamics at low dilution rates may require modeling the accumulation of fecal debris in the culture, since ingestion of this material is likely to alter gross assimilation by rotifers. Or it might be necessary to allow egg size to vary with factors such as maternal body mass or net assimilation rate. Generalized or chronological aging can also be incorporated in the rotifer equation (e.g. if maturation to adulthood must be allowed to depend in part on age, or if senescence is important), though we note that there is no adequate way to measure age in a chemostat.

If experiments indicate that substantially greater biological detail must be incorporated into the model, then it probably will be necessary to abandon the continuum transport modeling framework in favor of a stochastic, individual-based, computer simulation model where the fate of each individual in the rotifer population is tracked separately through time. Such models have fewer computational constraints than continuum transport models, mainly because they avoid the realm of multidimensional partial differential equations and their attendant numerical difficulties. It must be remembered, however, that regardless of the modeling framework employed, including a high degree of biological detail will produce a model whose behavior is largely incomprehensible. Too much detail is therefore as bad as too little in research aimed at elucidating basic principles.

476

Acknowledgement

This work was partially funded by a grant to JNM from the Environmental Associates of the Academy of Natural Sciences of Philadelphia.

References

Bennett, W. N. & M. E. Boraas, 1989. Comparison of population dynamics between slow- and fast-growing strains of the rotifer *Brachionus calyciflorus* Pallas in continuous culture. Oecologia 81: 494–500.

Boraas, M. E., 1979. Dynamics of Nitrate, Algae, and Rotifers in Continuous Culture: Experiments and Model Simulations. Ph.D. thesis, Pennsylvania State University.

Boraas, M. E., 1983. Population dynamics of food-limited rotifers in two-stage chemostat culture. Limnol. Oceanogr. 28: 546–563.

Boraas, M. E., D. B. Seale, J. E. Boxhorn & J. N. McNair, 1998. Rotifer size distribution changes during transient phases in open cultures. Hydrobiologia 387/388 (Dev. Hydrobiol. 134): 477–482.

Herbert, D., 1958. Some principles of continuous culture. In G. Tunevall (ed.), Recent Progress in Microbiology. Blackwell, London: 381–396.

Monod, J., 1950. La technique de culture continue; theorie et applications. Ann. Inst. Pasteur, Lille 79: 390–410.

Rothaupt, K. O., 1993. Critical consideration of chemostat experiments. In N. Walz (ed.), Plankton Regulation Dynamics. Springer-Verlag, New York: 217–225.

Walz, N., 1993. Model simulations of continuous rotifer cultures. Hydrobiologia 255/256: 165–170.

Hydrobiologia **387/388**: 477–482, 1998.
E. Wurdak, R. Wallace & H. Segers (eds), Rotifera VIII: A Comparative Approach.
© *1998 Kluwer Academic Publishers.*

Rotifer size distribution changes during transient phases in open cultures

Martin E. Boraas[1], Dianne B. Seale[1], Joseph E. Boxhorn[1] & James N. McNair[2]
[1]*Department of Biological Sciences, University of Wisconsin-Milwaukee, Milwaukee, WI 53201, U.S.A.*
[2]*Patrick Center for Environmental Research, The Academy of Natural Sciences, 1900 Benjamin Franklin Parkway, Philadelphia, PA 19103-1195, U.S.A.*

Key words: Rotifera, *Brachionus ruhens*, *Brachionus calyciflorus*, population dynamics, steady-state growth, continuous culture, size distributions

Abstract

In laboratory studies, rotifers (*Brachionus calyciflorus*) were monitored under well-defined environmental conditions at different supply rates of a unicellular algal food (*Chlorella vulgaris*). Rotifer size frequency distributions are described for conditions of steady-state growth, exponential increase, and starvation. Temporal fluctuations in size-age structure are described for cultures during transient conditions during the approach to a steady state and following step changes in food supply rate. The size structures of the populations displayed definite and reproducible shifts among typical patterns during transient conditions, reflecting the physiological and other dynamic processes that underlay the population dynamics. Size structure probably is a key variable that should be included in models for predicting growth dynamics during transient growth conditions.

Introduction

The goal of our research is to use a combination of experimental and modeling approaches to evaluate the mechanisms determining rotifer growth. For population models to have practical use, they must be evaluated with comparisons to experimental data. The success of this approach depends on congruence between the model and the system it attempts to describe, or there is a risk of attempting to model 'noise' rather than the relevant biological information. For this reason, we use a variety of rotifer continuous cultures for our experimental studies (Boraas, 1983, 1993b; Boraas & Bennett, 1988). In any continuous culture, the steady-state population provides the ideal starting condition for evaluating the magnitude and duration of transient fluctuations in physiological conditions and demographic structure. If one begins with a well-defined steady-state size structure, it becomes possible to (1) describe and classify deviations from this defined structure, and (2) describe precisely the conditions that lead to these deviations.

There are three types of transient responses that can be evaluated in continuous cultures: step changes, pulse changes, and the approach to steady state from an inoculum. The step change is potentially most revealing for pertinent biological phenomena since it can involve either an increase or decrease in the value of a dependent variable. In a step change, once the population has entered a steady state, the value of the dependent variable, such as the food supply rate, is changed. The population then responds to this change and can be monitored as it approaches a new steady state. In a pulse or spike change, a dependent variable is changed for a brief time; for example, a pulse of extra food can be injected into the culture. In this case, the population can be monitored as it responds and then returns to its former steady state. In the third case, an inoculum of rotifers is added to a continuous culture in which algae are at maximum densities. Initially, the rotifers experience a condition of food saturation. Consequently, the population enters a state of exponential increase at its biological maximum specific growth rate, r_{max}, for the given environmental conditions. As the population grows, it consumes food, causing the residual algal concentrations to decline within the culture. The population specific growth rate, r, slows with declining food; the population and its residual food supply can be monitored as they enter the steady state.

478

In this paper, we describe the size distributions of rotifers (*Brachionus calyciflorus*) from our cultures, grown under well-defined environmental conditions, at different rates of supply of food (the unicellular alga, *Chlorella vulgaris*). This paper describes rotifer size distributions under three distinct environmental conditions: at the steady state, during the approach to the steady state, and after a step change in the food supply rate. Size distributions following a pulse change are described elsewhere (Boraas, 1993a).

Materials and methods

We have used a variety of continuous and semi-continuous culture systems, described in previous publications (Boraas, 1983, 1993a, b; Boraas & Bennett, 1988; Bennett & Boraas, 1989; Boraas et al., 1990). Rotifers (*Brachionus calyciflorus* Pallas) were obtained from J. Gilbert (chemostat cultures) or were isolated from the Milwaukee Harbor (semi-continuous cultures: Boraas & Bennett, 1988). The rotifer studies described here were conducted in a two-stage system consisting of a first-stage algal chemostat (in the light) and a second-stage rotifer culture vessel (in the dark, to eliminate algal growth). The first stage can be considered a separate culture, from which algae (*Chlorella vulgaris*: UTEX 26) are harvested to feed the rotifers in the second stage (Boraas, 1983, 1993b). The semi-continuous culture vessels were 1- or 2-l Erlenmeyer flasks containing 500 ml and mounted on a shaker table (Boraas et al., 1993). The algal culture was mixed using charcoal-filtered, hydrated, filter-sterilized air. Temperature was maintained at 25 ± 0.05 °C for chemostat cultures and 23 ± 1 °C for semi-continuous cultures. For all cultures, algae in the first stage were maintained at a constant dilution rate, to maintain the algae in a constant physiological condition. In these culture systems, the rotifer specific growth rate is determined by the culture's dilution rate, D (D = flow rate/culture volume) (Boraas, 1983, 1993b).

The second-stage rotifer culture vessel consisted of a constant-volume culture vessel and an outflow flask. Sterile-filtered air was supplied to the rotifer stage, minimizing settling of algae in the pump lines and mixing the rotifer stage. For chemostat cultures, a peristaltic pump delivered the algal food to the rotifer culture for the selected value of D. For semi-continuous cultures, a constant volume of the rotifer

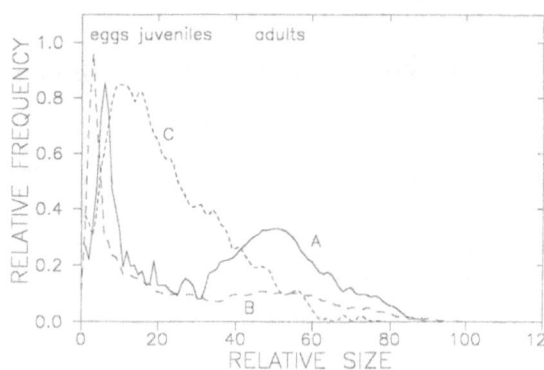

Figure 1. Characteristic rotifer size distributions in continuous culture. Distribution A (solid line) indicates rotifers from steady-state cultures. The peaks and valley correspond to the age categories indicated on the figure. Distribution B (long dashes) indicates a population growing at or near r_{max}. The scale of the graph is compressed somewhat relative to A. Distribution C (short dashes) represents a population in decline, i.e. undergoing starvation.

culture was daily removed and replaced with algal culture (after Boraas, 1993b; Boraas et al., 1990).

Both the rotifer and algal populations were quantified routinely with electronic particle counters. Traditional light microscopy was used to classify the corresponding size-age categories. The chemostat data were collected with a Model ZBI Coulter Counter and Model 100 Channelyzer (see Boraas, 1983). The semi-continuous data were collected with a Particle Data Elzone 280-PC. Three-dimensional plots of time-course data were produced by the program SYSTAT.

Samples were removed under vacuum from the culture vessel through a sterilized silicone rubber port with a 25-gauge needle. When passed through the needle, rotifers were stripped of their eggs. Rotifers and eggs were enumerated and sized within 0.5 h of sampling. This kind of automated counting system provides a precise, rapid method for obtaining counts, size distributions, and biovolume estimates of plankton samples (Boraas, 1983). This system makes it possible for replicated counts of 100 to 128 size classes of rotifers to be collected, analyzed and saved to a disk file in about 7.5 min.

Results

Rotifer size distributions. The size distributions of the rotifers we have observed in our cultures can be characterized as falling into three representative categories.

1. Steady-state rotifer cultures. The steady-state rotifer populations at intermediate values of D (about 0.7 d^{-1}) typically have a distinct size-frequency distribution (Distribution A, Figure 1). The largest peak in the distribution is associated with the rotifer eggs. A second peak is associated with adults. The intermediate size category (the juvenile category) typically is less abundant than either eggs or adults in the steady-state culture.

2. Maximum population increase. When the residual food concentration is very high, e.g. immediately after the culture vessel is filled with algal suspension and the rotifers are inoculated, the rotifers increase exponentially at r_{max} for the culture conditions. In a continuous culture, this is necessarily a transient phase since r must equal D in the steady state. These population size-age structures are quite distinctive (Distribution B, Figure 1). The egg peak of the rotifers is very pronounced while the proportions of the numbers in the adult and juvenile size-age classes are similar. The maximum sizes of adults are usually slightly larger than during a steady state, largely due to internal developing eggs.

3. During starvation the larger size categories and eggs almost vanish (Distribution C, Figure 1). Eggs and juveniles are rare or absent at this time. Empty loricas and other debris are common. Living rotifers are thin, show little internal structure and move slowly. Evidently, with a sufficiently low food supply, the animals lose volume both by stopping egg production and by cell shrinkage. In continuous cultures, this pattern can be seen at very low dilution rates (e.g. Bennett & Boraas, 1988) and after step-down changes in D (this paper).

As the value of D can be any value that is physiologically possible for the organisms, intergradations among these 'typical' size distributions are often seen.

Temporal size-age structure fluctuations in a two-stage rotifer culture. The time course of the approach to a steady state in a chemostat culture ($D = 0.83$ d^{-1}) is shown in Figure 2. After an initial lag, the population entered a period of exponential increase and exhibited the size-distribution pattern described in Figure 1B. Then, as the algal density declined, both the rotifers and their food entered a steady state (Figure 2). Although the population was growing at its physiological maximum, the transition to a steady state was gradual

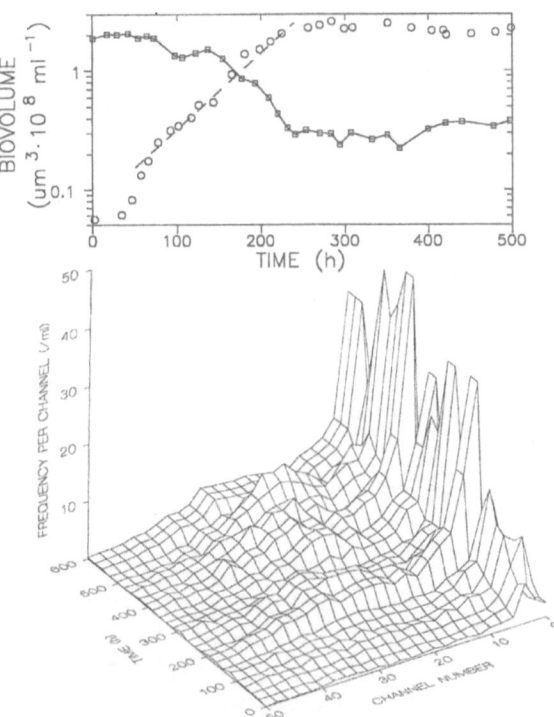

Figure 2. The time course of the components of a culture approaching and entering a steady state. In the upper panel, the circles indicate total rotifer biovolume. The dashed line indicates the period of exponential increase. The squares and solid line indicate the residual algae, i.e. the algae not consumed in the second stage. The lower panel indicates size-frequency distributions as rotifer frequency per ml per channel.

as shown by the size distribution changes (Figure 2, lower panel).

Following a step reduction in dilution rate in a chemostat culture, from 1.1 to 0.32 d^{-1}, the rotifer and algal populations first experienced transients and then entered a new steady state (Figure 3). Immediately after the step change, egg production (as eggs/adult) declined, and then rebounded, eventually becoming stable (Figure 3). Within 24 h of the step change, the size category associated with eggs dramatically declined, and the body size of rotifers declined precipitously (Figure 3), resembling the pattern for starvation (Figure 1C). Within 72 h, a new 'steady-state' size structure was established, which differed from the structure seen before the step change: the egg peak was present, but less pronounced, and rotifers in the 'adult' peak were at an intermediate body size that was much smaller than seen in the previous steady state.

480

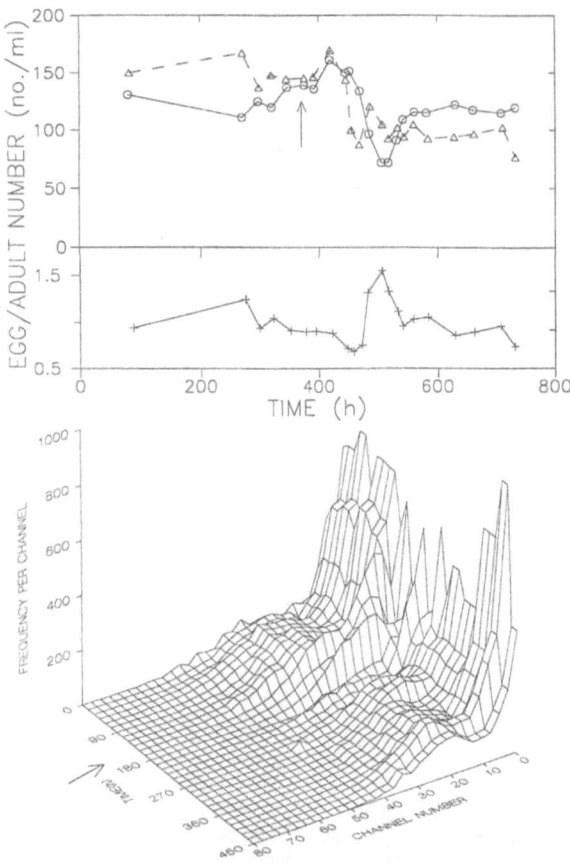

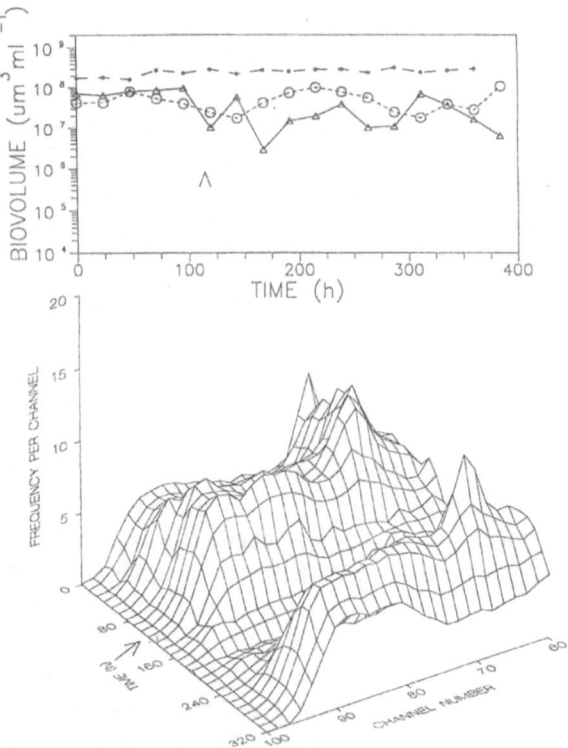

Figure 3. The upper panel shows the time course of a two-stage rotifer culture undergoing a step change in dilution rate, indicated by the arrow. Panel A shows adult rotifer numbers as circles and a solid line and numbers of eggs as triangles and a dashed line. Panel B shows the ratio of eggs/adults in Panel A. The lower panel shows size distributions of rotifers in the same culture. Note the direction of the time axis in this and following figures. The arrow indicates the step change. Further discussion in text.

Figure 4. The upper panel shows the time course of a semi-continuous two-stage rotifer culture. Total rotifer biovolume is indicated by open circles and a dotted line. Input algal concentrations are indicated by closed circles and a dashed line. Residual algal concentrations are indicated by triangles and a solid line. The lower panel shows size distributions of rotifers for the same culture. The step change is indicated by the arrows. Further discussion in text.

Discussion

General approach. The study of population dynamics endeavors to disentangle the relative importance of the many factors that interact to determine population growth and to recombine them in reasonable predictive models. For convenience, we can classify them either as physical environmental factors, such as temperature, that are not influenced by the populations themselves, and biological factors, such as food levels, that are modified by the growth of the populations. Models needed to describe and predict dynamics of steady-state populations can be much simpler than for other populations. In contrast, models that cannot account for steady-state populations, or populations perturbed from steady states, are unlikely to be useful for understanding population growth in more complex environments.

Following a reduction in D in a semi-continuous culture from 0.45 d^{-1} to 0.20 d^{-1}, the rotifer biovolume fluctuated smoothly and the residual algal density (also as biovolume) declined by about a factor of 5 (Figure 4). The rotifer size distributions showed a dramatic change, with egg production declining immediately after the shift with a small egg peak during the decline (Figure 4), similar to the 'starvation' pattern (Figure 1C). Next, the size distribution returned to the shape before the shift; but then the egg peak and the population density declined again. It should be noted that the semi-continuous samples were always taken at the daily dilution event, which may have synchronized egg production to some extent.

In our approach, first, we eliminate the need to include physical factors as variables in our population models by using experimental systems that hold these factors constant. In this manner, we can reduce (but not eliminate) many sources of variability and ambiguity, so that the biological factors affecting population growth can be emphasized, rather than environmental noise.

Secondly, we use open systems for all our cultures. We provide a constant supply of food, which is as invariant as possible with respect to composition, nutritional value, and input concentration. Thus, external factors that might alter food availability are eliminated as potential variables. Within the experimental system, the populations use this food for growth. We can then evaluate the dynamics of the populations and their food supply.

Thirdly, we develop models in a step-wise fashion. We begin with simple models based on a fundamental axiom for population growth: populations must have a food resource for growth. The growth of any population necessarily converts some food into organism biomass. We use simple resource-based models to predict population growth on a resource. Ideally, experimental data can be compared unambiguously to the model predictions to reveal where the models fall short. Areas of discrepancy between model and data are then crucial for revealing where modifications of the models are needed.

Application of models to rotifer population growth. The most commonly used models for population growth are not adequate to describe rotifer population growth in these cultures. For example, the logistic equation does not explicitly describe variations in population age structure with changing environmental conditions.

Continuous cultures have been used since Monod (1950) and Novick & Szilard (1950) independently published the procedure as a method for growing bacteria in steady states, and since then have been applied to many problems (Pirt, 1975; Tilman, 1982; Walz, 1993). The mathematics of single-species chemostat and semi-continuous growth of unicellular organisms are reasonably well understood (e.g. Herbert et al., 1956; Williams, 1972; Pirt, 1975). In previous papers, we have described application of these models to our rotifer cultures (Boraas, 1983, 1993a, b; Boraas & Bennett, 1989). In spite of the obvious differences between unicells and metazoans, these mathematical models are sufficient to describe many aspects of steady-state growth dynamics of rotifers. For our purposes, they are far superior to the logistic growth model, often used to model animal population growth. The logistic equation ignores the essential requirement for population growth: organisms require some food supply for growth. If we estimate the carrying capacity for rotifers in one culture system, we have no explicit means for using that information to predict rotifer population dynamics in a new environment, e.g. one with a different food level. The logistic equation cannot be easily modified to include variations in population age structure with changing environmental conditions.

Our previous studies have shown that our cultures satisfy an important assumption of the Monod-Herbert model: food is the rate-limiting factor for population growth. Monod (1942) first demonstrated that the specific growth rate of bacteria is a hyperbolic function of the limiting resource concentration, in which the maximum rate is approached as an asymptote. In our culture systems, similarly the rotifer specific growth rate is a hyperbolic function of food level (i.e. algal density) (Boraas, 1983). Data show a good fit to the Monod (1942) model for microbial growth:

$$r = r_{max}S/[K_s + S],$$

where r is growth rate, r_{max} is the maximum specific growth rate, S is substrate concentration (i.e. algal density), and K_s is the concentration of S where r is one-half of r_{max} (Boraas, 1983; Walz, 1993).

The Monod-Herbert model has an important limitation on its ability to predict population growth dynamics: it assumes a steady-state population. In its simplest form, the Monod-Herbert model assumes either that all individuals in the population are identical or that their size-age structure does not change in time. Obviously, the unmodified Monod-Herbert model cannot be used to predict any of the shifts in size structure we have observed at different values of D, nor can it predict some key aspects of transient growth conditions, when a population is not in a steady state.

The results in this paper indicate that additional detail is needed to account for animal population growth dynamics under transient conditions. In transient growth conditions, rotifer size structure demonstrates characteristic size structures that seem to reflect the physiological condition of the rotifers. In order to take account of these fluctuations, the basic model must be modified to allow for shifts in fecundity, resource allocation within each individual, and size-dependent metabolic rates that are functions of changes in algal supply rates (McNair et al., 1998).

References

Bennett, W. N. & M. E. Boraas, 1989. Comparison of population dynamics between slow- and fast-growing strains of the rotifer *Brachionus calyciflorus* in continuous culture. Oecologia 81: 494–500.

Boraas, M. E., 1983. Population dynamics of food-limited rotifers in two-stage chemostat culture. Limnol. Oceanogr. 28: 546–563.

Boraas, M. E., 1993a. Single-stage predator–prey algal-rotifer chemostat culture. In Walz, N. (ed.), Plankton Regulation Dynamics: Experiments and Models in Rotifer Continuous Culture. Springer-Verlag, Berlin: 51–61.

Boraas, M. E., 1993b. The growth of *Brachionus rubens* in semi-continuous culture. In Walz, N. (ed.), Plankton Regulation Dynamics: Experiments and Models in Rotifer Continuous Culture. Springer-Verlag, Berlin: 151–158.

Boraas, M. E. & W. N. Bennett, 1988. Steady-state rotifer growth in a two-stage, computer controlled turbidostat. J. Plankton Res. 10: 1023–1038.

Boraas, M. E., D. B. Seale & J. B. Horton, 1990. Resource competition between two rotifer species (*Brachionus rubens* and *B. calyciflorus*): an experimental test of a mechanistic model. J. Plankton Res. 12: 77–87.

Herbert, D., 1958. Some principles of continuous culture. In Tunevall, G. (ed.), Recent Progress in Microbiology. Blackwell, Oxford: 181–196.

Herbert, D., R. Elsworth & R. C. Telling, 1956. The continuous culture of bacteria: a theoretical and experimental study. J. Gen. Microbiol. 14: 601–622.

McNair, J. N., M. E. Boraas & D. B. Seale, 1998. Size-structure dynamics of the rotifer chemostat: a simple physiologically structured model. Hydrobiologia 387/388 (Dev. Hydrobiol. 134): 469–476.

Monod, J., 1942. Recherches sur la croissance des cultures bacteriennes. Hermann et Cie, Paris. 210 pp.

Monod, J., 1950. La technique de culture continue: théorie et applications. Ann. Inst. Pasteur Paris 79: 390–410.

Novick, A. & L. Szilard, 1950. Description of the chemostat. Science 112: 715–716.

Pirt, S. J., 1975. Principles of Microbe and Cell Cultivation. Blackwell, Oxford. 174 pp.

Tilman, D., 1982. Resource Competition and Community Structure. Princeton University Press, Princeton. 269 pp.

Walz, N., 1993. Regulation models in rotifer chemostats. In Walz, N. (ed.), Plankton Regulation Dynamics: Experiments and Models in Rotifer Continuous Culture. Springer-Verlag, Berlin: 135–150.

Williams, F. M., 1972. Mathematics of microbial populations, with emphasis on open systems. In Deevy, E. S. (ed.), Growth by Intussusception: Ecological Essays in Honor of G. Evelyn Hutchinson. Trans. Conn. Acad. Arts Sci., New Haven, Vol. 44, Archon Books, Hamden, Conn.: 396–426.

Hydrobiologia **387/388**: 483–487, 1998.
E. Wurdak, R. Wallace & H. Segers (eds), Rotifera VIII: A Comparative Approach.
© 1998 Kluwer Academic Publishers.

Influence of the food ration and individual density on production efficiency of semicontinuous cultures of *Brachionus*-fed microalgae dry powder

N. Navarro[1] & M. Yúfera[2,*]

[1]*Cultivos Piscícolas Marinos S. A. Apartado 119, 11100 San Fernando, Cádiz, Spain*
[2]*Instituto de Ciencias Marinas de Andalucia (CSIC), Apartado Oficial, 11510 Puerto Real, Cádiz, Spain*
(*author for correspondence)

Key words: Rotifera, semicontinuous culture, dried microalgae, *Brachionus*

Abstract

In the present study we examined the effect of food ration and individual density at the time of starting the harvesting stage, in determining the demographic parameters in a semicontinuous culture of the rotifers *Brachionus plicatilis* and *B. rotundiformis* fed freeze dried *Nannochloropsis oculata*. Three daily food rations (25, 50 and 100 mg l^{-1}) and three rotifer densities (250, 500 and 1000 rotifers ml^{-1}) have been tested, in trying to maintain a constant availability of microalgal cells of 100 ng per individual. Results showed that with 10% daily dilution, the mean rotifer density during the steady state remained at values similar to those pre-fixed at the beginning of the harvesting. The production (rotifers $l^{-1} d^{-1}$) increased with the food ration, but the system efficiency (mg-rotifers produced per mg-microalgae) was the same irrespective of the ration.

Introduction

In recent years, the use of concentrated algae cells in the mass culture of the rotifer *Brachionus* has opened new horizons for mass production and research on this organism (Maruyama & Hirayama, 1993; Lubzens et al., 1995; Yúfera & Navarro, 1995; Yoshimura et al., 1996). This is due not only to the evident reduction of the total volume required in order to obtain the final product in mass culture, but also to the possibility of testing food densities higher than can be obtained with unprocessed microalgal suspensions. Another aspect contributing to the enhancement of the rotifer production has been the improvement of semicontinuous culture techniques. This system can be assimilated to a chemostat when the dilution rate is constant and frequent, and the fresh medium contains a constant amount of food (Schlüter et al., 1987; Boraas, 1993). When the quasi-steady state is attained, the growth rate compensates the dilution rate, and the population maintains its demographic characteristics for long periods. While the growth rate during this stage is determined exclusively by the dilution rate, other parameters such as rotifer density and production are also determined by the food ra-

tion (considering the physical and chemical culture conditions such as temperature or salinity as fixed).

In a previous study (Yúfera & Navarro, 1995), we showed that freeze-dried microalgae powder, as a sole food source, results in good population growth of rotifers in batch culture with a daily addition of food. In order to optimize the biomass production in semi-continuous culture system using microalgae dry powder, we examined the effect of the food ration and individual density at the onset of the harvesting stage, in determining the demographic parameters in two closely-related species *Brachionus plicatilis* and *Brachionus rotundiformis*.

Material and methods

Biometric and biomass characteristics of rotifer strains of *Brachionus plicatilis* (strain S 1) and *B. rotundiformis* (strain Bs) have been treated in previous papers (Yúfera, 1982; Yúfera et al., 1993). Microalgae (*Nannochloropsis oculata*; Eustigmatophyceae) dry powder was obtained as follows. Microalgal dry cells were collected by centrifugation from cultures in the growth phase. The cells were subsequently washed with an isotonic medium, frozen at $-80\,^\circ$C and freeze-dried. Rotifers were cultured in 1-liter flasks with gentle aeration, kept in a thermoregulated chamber at 25 °C and salinity 18. Three daily rations were tested: 25, 50 and 100 mg l^{-1}. Cultures started at low rotifer densities (25–50 rotifers ml^{-1}) and harvesting began before exponential growth ceased, when the rotifer density reached 250, 500 and 1000 rotifers ml^{-1} for the three rations, respectively (Yúfera & Navarro, 1995); at such preestablished individual densities and food rations, the amount of microalgal cells available per rotifer is 100 ng in all cases. The rotifer cultures were harvested and diluted once a day. The dilution rate was 10% of the culture volume per day.

All experiments were performed in triplicate. Each day, the number of females and eggs were counted in 3 ml samples. Daily rotifer density was calculated as $\sqrt{(D_t \cdot D_i)}$, D_t being the rotifer density just before the dilution and D_i the rotifer density just after the previous dilution (24 h before). Daily population growth rate was calculated as $\mathrm{Ln}\,D_t - \mathrm{Ln}\,D_i$. Taking into account the differences in the duration of the steady state among replicates, mean daily rotifer density and mean daily growth rate have been considered to characterize this period under the different food conditions. In addition, weekly growth rates (mean of the daily growth rates for each week) were determined for a better monitoring of the cultures. We considered that the steady state ended when the weekly growth rate decreased strongly in comparison with the previous ones. Daily production during the steady state (P) was calculated as the number of harvested rotifers divided by total number of days in the period. System efficiency (E) defined as the percentage of dry microalgae transformed into dry rotifer biomass in each treatment was calculated as

$$E = P \cdot W \cdot 100 \cdot R^{-1},$$

where P is the daily production (rotifers l^{-1} d^{-1}); W is the rotifer dry weight; and R is the corresponding daily dry microalgal ration. Dry weight values of both rotifer strains are taken from Yúfera et al. (1993).

Results

The rotifers fed *N. oculata* dry powder in semicontinuous culture grew well under all food rations, and after starting the daily harvesting, the population remained stable for several weeks in both *Brachionus* species, although some variability among replicates was evident. After this quasi-steady state period, growth rate

and female density decreased gradually. An example of these cultures is shown in Figure 1. As expected, the growth rates were close to 0.1 d^{-1}. The mean rotifer density and the amount of microalgae cells per rotifer during the steady-state remained at values similar to those pre-established for the beginning of the harvesting period, although in *B. rotundiformis* fed 100 mg l^{-1} d^{-1}, the available cells per rotifer resulted significantly higher than pre-fixed (Table 1).

Daily production during the steady-state phase increased significantly ($P < 0.01$) according to the average rotifer density (Figure 2). The maximum values obtained with 100 mg l^{-1} d^{-1} dry microalgae powder were about 103 000 individuals l^{-1} d^{-1} in the case of *B. plicatilis*, and 83 000 individuals l^{-1} d^{-1} in that of *B. rotundiformis*. Expressed in dry weight, the production ranged from 11 to 46 and from 6 to 19 mg l^{-1} d^{-1} for *B. plicatilis* and *B. rotundiformis*, respectively.

For each *Brachionus* species, the system efficiency during the steady-state phase (percentage of the supplied dry microalgae converted into dry rotifer biomass) was quite similar irrespective of the food ration ($P > 0.05$). The system efficiency was higher in *B. plicatilis* ($\approx$45%) than in *B. rotundiformis* ($\approx$20%).

Discussion

Our results confirm that the dilution rate and the amount of food supplied determine the dynamics of the system. For a given dilution rate, the rotifer population stabilizes at higher individual densities with increased food ration, maintaining a relation of approximately 100 ng algal dry weight per individual per day. These values of daily algal cell availability are the minimum required to maintain the rotifer populations growing at a rate of 0.1, and therefore with a given age structure and fecundity. The rotifer density averages obtained in this study are similar to those at which the populations finished the exponential growth when the cultures are not diluted (Yúfera & Navarro, 1995). Under such growth conditions, the populations are primarily constituted of old individuals with a relatively low fecundity in which a great part of the consumed biomass is allocated to maintenance. This situation changes to progressively younger populations with higher fecundity when the dilution rate is increased (Navarro, 1996).

The food was consumed with the same efficiency under the three tested rations. Therefore, any of

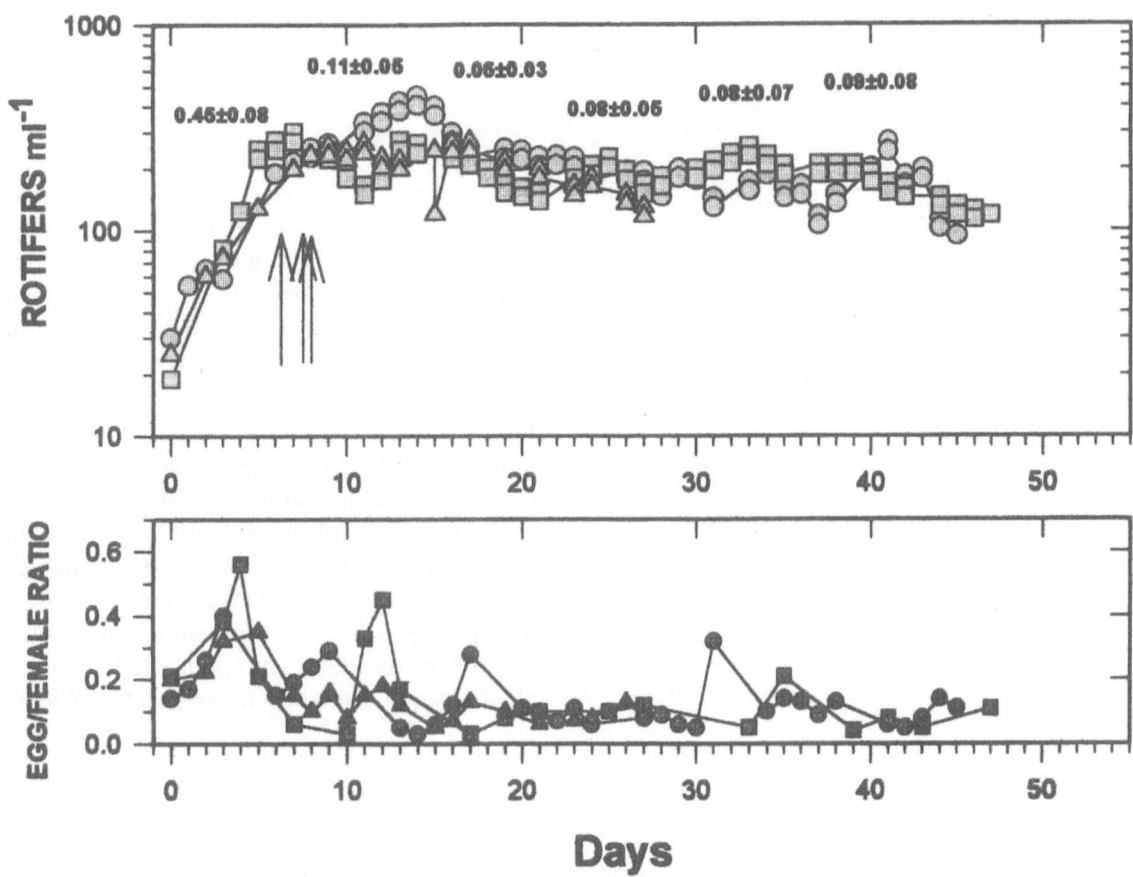

Figure 1. *B. plicatilis* growth curves and the corresponding egg/female ratio in semicontinuous culture method with a daily supply of 25 mg l^{-1} of dry microalgae. Numbers indicate weekly population growth rate. Arrows indicate the start of harvesting.

Table 1. Individual density (*D*) and the amount of microalgal powder available per rotifer (*A*) reached during the steady state stage (mean and standard deviations of three replicates). Pre-established values at the beginning of harvesting (D_0 and A_0) were also included for comparison

	Ration (mg l^{-1} d^{-1})	D_0 (ind ml^{-1})	D (ind ml^{-1})	A_0 (ng ind^{-1})	A (ng ind^{-1})
B. plicatilis	25	250	222 ± 59	100	112 ± 22
	50	500	521 ± 124	100	96 ± 18
	100	1000	1033 ± 280	100	97 ± 21
B. rotundiformis	25	250	221 ± 49	100	113 ± 20
	50	500	416 ± 69	100	120 ± 17
	100	1000	818 ± 64	100	122 ± 9*

Value with asterisk is significantly ($P < 0.05$) different from 100.

486

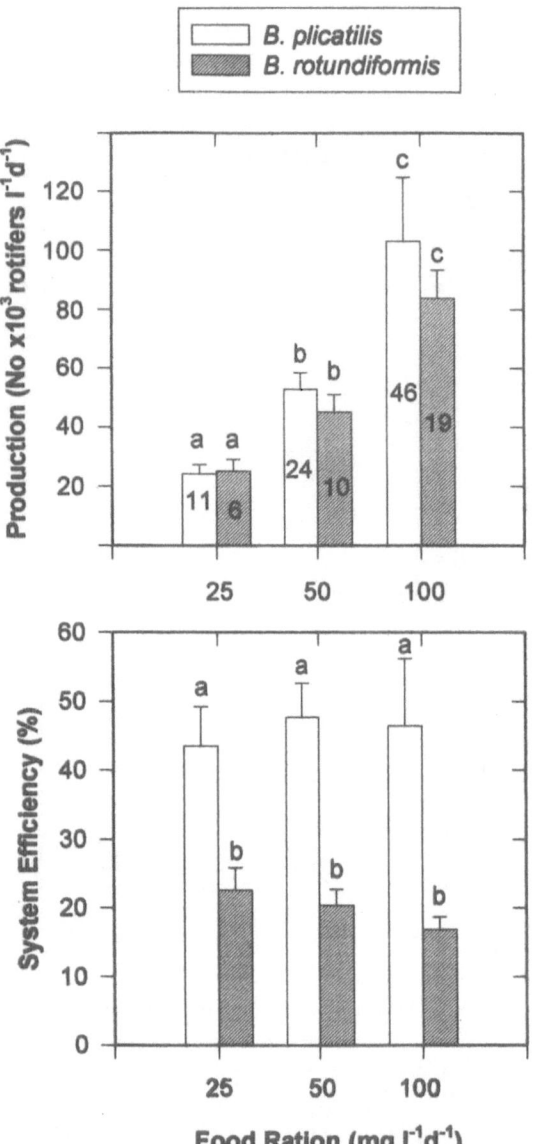

Figure 2. Rotifer production (individuals l^{-1} d^{-1}) and system efficiency (percentage of dry microalgae transformed into dry rotifer biomass) in semicontinuous culture methods under three food rations. Vertical bars indicate the standard deviation. Numbers within columns indicate the corresponding value of dry matter produced (mg l^{-1} d^{-1}). Columns with the same letters are not significantly ($P < 0.05$) different.

these food rations can be appropriate for rotifer mass culture. The choice will depend on the production demanded. This constancy of the food conversion efficiency with increasing food concentration supports the feasibility of obtaining high density and high productive intensive mass cultures as those obtained with concentrated live algal cells (Yoshimura et al., 1996).

An aspect not considered in this study is the bacteria load generated by the microalgal dry cells in the culture which can constitute a noteworthy contribution of food to rotifers (Aoki & Hino, 1996). Anyway, there are no indications that such contribution affects the efficiency obtained under different microalgae rations during the steady state, although it could affect the stability.

Concentrated and frozen *N. oculata* has been used efficiently in large scale cultures (Hagiwara et al., 1993; Hamada et al., 1993; Balompapueng et al., 1997). Furthermore, mass production of rotifers with microalgal powder has been studied under different culture conditions and using other microalgal species (i.e., Hirayama & Nakamura, 1976; Gatesoupe & Robin, 1981). The present study confirms the usefulness of such type of preservation in *N. oculata*, a species of high quality – via rotifers – for larval rearing of marine fish, and for establishing highly productive *Brachionus* culture systems. Although this study has been performed in one litre flasks, we tested the viability of dry *Nannochloropsis* in 30-litre cultures (Navarro, 1996). The optimization of the system efficiency will depend exclusively on the dilution rate.

Acknowledgments

This work was supported by the Comisión Interministerial de Ciencia y Tecnología, Spain (CICYT Project AGF-94-0756). We thank J. M. Espigares, and J. A. Miquel for their helpful technical assistance.

References

Aoki, S. & A. Hino, 1996. Nitrogen flow in a chemostat culture of the rotifer *Brachionus plicatilis*. Fish. Sci. 62: 8–14.

Balompapueng, M. D., A. Hagiwara, A. Nishi, K. Imaizumi & K. Hirayama, 1997. Resting egg formation of the rotifer *Brachionus plicatilis* using a semicontinuous culture method. Fish. Sci. 63: 236–241.

Boraas, M. E., 1993. Semicontinuous culture methods. In N. Walz (ed.), Plankton Regulation Dynamics. Experiments and Models in Rotifer Continuous Cultures. Springer-Verlag, Berlin, Ecol. Studies 98: 13–20.

Gatesoupe, F. J. & J. H. Robin, 1981. Commercial single-cell protein either as a sole food source or in formulate diets for intensive and continuous production of rotifers (*Brachionus plicatilis*). Aquaculture 25: 1–15.

Hagiwara, A., K. Hamada, A. Nishi, K. Imaizumi & K Hirayama, 1993. Mass production of rotifer resting eggs in 50 m^3 tanks. Nippon Suisan Gakkaishi 59: 93–98.

Hamada, K., A. Hagiwara & K. Hirayama, 1993. Use of preserved diet for rotifer *Brachionus plicatilis* resting egg formation. Nippon Suisan Gakkaishi 59: 85–91.

Hirayama, K. & K. Nakamura, 1976. Fundamental studies on physiology of rotifers in mass culture. V. Dry *Chlorella* powder as food for rotifers. Aquaculture 22: 149–163.

Lubzens, E., O. Gibson, O. Zmora & A. Sukenik, 1995. Potential advantages of frozen algae (*Nannochloropsis* sp.) for rotifer (*Brachionus plicatilis*) culture. Aquaculture 13(3): 295–309.

Maruyama, I. & K. Hirayama, 1993. The culture of the rotifer *Brachionus plicatilis* with *Chlorella vulgaris* containing vitamin b 12 in its cells. J. World Aquacult. Soc. 24: 194–198.

Navarro, N., 1996. Cultivo del rotifero *Brachionus plicatilis* OF. Müfler alimentado con microalgas, liofilizadas. Dinámica de la población y calidad nutritiva como, alimento larvario de peces marinos. Tesis Doctoral, Universidad de Cádiz, 257 pp.

Schlüter, M., C. J. Soeder & J. Groeneweg, 1987. Growth and food conversion of *Brachionus rubens* in continuous culture. J. Plankton Res. 9: 761–783.

Yoshimura, K., A. Hagiwara, T. Yoshimatsu & C. Kitajima, 1996. Culture technology of marine rotifers and the implications for intensive culture of marine fish in Japan. Mar. Freshwat. Res. 1996: 217–222.

Yúfera, M., 1982. Morphometric characterization of a small-sized strain of *Brachionus plicatilis* in culture. Aquaculture 27: 56–61.

Yúfera, M. & N. Navarro, 1995. Population growth dynamics of the rotifer *Brachionus plicatilis* cultured in non-limiting food condition. Hydrobiologia 313/314: 399–405.

Yúfera, M., E. Pascual & J. Guinea, 1993. Some factors influencing the biomass of the rotifer *Brachionus plicatilis* in culture. Hydrobiologia 255/256: 159–164.

Hydrobiologia **387/388**: 489–494, 1998.
E. Wurdak, R. Wallace & H. Segers (eds), Rotifera VIII: A Comparative Approach.
© *1998 Kluwer Academic Publishers.*

Effect of water viscosity on the population growth of the rotifer *Brachionus plicatilis* Müller

Atsushi Hagiwara, Nozomu Yamamiya & Adriana Belem de Araujo
Faculty of Fisheries, Nagasaki University, Bunkyo, Nagasaki 852-8521, Japan

Key words: viscosity, population growth, *Brachionus plicatilis*

Abstract

We measured the relative viscosity of water taken from a 500 ml rotifer batch culture against natural sea water using a capillary type viscometer (Ostwald type) during 30 days. The relative viscosity ranged from 1.054 to 1.148. We observed a negative correlation between population growth rate and viscosity. We experimentally tested the effect of viscosity on reproduction, swimming activity and ingestion rate of *Brachionus plicatilis*. By dissolving methyl cellulose, the relative viscosity of natural sea water (salinity 22 ppt) was increased from 1.0 (control group) to 1.022, 1.031, 1.078 and 1.169. At higher viscosity, individually cultured rotifers at 25 °C that were fed *Nannochloropsis oculata* showed slower swimming, lower ingestion rate and lower fecundity than control. By increasing the relative viscosity from 1 to 1.169, the swimming activity index decreased from 28.3 to 4.9 (1 mm^2 square grids/30 s), the ingestion rate decreased from 1.13 to 0.34 ($\times 10^3$ cells ind^{-1} h^{-1}), and the number of offspring per amictic female decreased from 20.8 ± 3.0 to 8.1 ± 3.7 (mean ± SD).

Introduction

Viscous forces predominate over inertial forces for microorganisms with a low Reynolds number. For microorganisms, moving in water is like swimming in a pool of molasses for a human (Mann & Lazier, 1991). Theoretical mechanisms of particle retention in fluid dynamic environments have been useful in clarifying feeding mechanisms of copepods (Koehl & Strickler, 1971) and rotifers (Rubenstein & Koehl, 1977; Starkweather, 1995).

Monogonont rotifers inhabit eutrophic waters where the viscosity of water is greater than that of oligotrophic waters. The euryhaline rotifers, *Brachionus plicatilis* and *B. rotundiformis*, have become popular as models for investigating fundamental principles of aquatic ecology, as live food in aquaculture and as test animals for eco-toxicological research. Population growth of rotifers can decrease due to an increase in free ammonia in the culture (Yu & Hirayama, 1986) or an increase in certain bacteria and protozoa (Maeda & Hino, 1991). It is theoretically predicted that food ingestion in rotifers could be promoted by water turbulence due to a decrease of the boundary layer around the body surface of rotifers. On the other hand, rotifer

swimming or food ingestion would be suppressed by an increase in water viscosity. Information is scarce, however, on the effect of such an environment on the physiological state of cultured rotifers. In this study, we examined (1) the relation between growth rate of rotifers in a laboratory culture and viscosity of the culture and (2) the effect of changes in water viscosity on growth, reproduction, feeding and swimming behavior of the rotifer *Brachionus plicatilis*.

Materials and methods

B. plicatilis (Tokyo strain) used in these experiments was cultured under steady-state conditions. *Nannochloropsis oculata* grown in modified Erd–Schreiber medium (Hagiwara et al., 1994) was used as food. Algal density in rotifer cultures was adjusted daily to replenish 7×10^6 cells ml^{-1}. Culture temperature was 25 ± 1 °C, salinity was 22 ppt and photoperiod at 0 L: 24 D. Salinity was adjusted by diluting natural sea water collected at Nomo Fisheries Station, Nagasaki University with distilled water.

The effect of viscosity on rotifer reproduction in a batch culture

B. plicatilis was mass cultured in a 500 ml glass beaker for 29 days. The culture was initiated with 4 rotifers /ml. A 20 ml water sample for viscosity measurement was collected daily, and frozen at −30 °C after GF/C (Whatman) filtration. A 5 ml water sample was also collected to determine rotifer density under a stereo-microscope. Water viscosity was estimated by using a capillary type viscometer (Ostwald type, 0.5 mm tube diameter, Vidrex, Fukuoka, Japan). The viscometer was placed in a water bath (25 ± 1 °C) for viscosity measurement (Yamato Kagaku Co., Tokyo, Japan). A 5 ml GF/C (Whatman) filtered sample was added to the viscometer capillary and kept for 10 min. After the tested sample was acclimated to 25 °C, the water sample was released inside the capillary tube and the time required for the water mass to fall between two menisci (about 3 ml volume) was monitored. This procedure was repeated ten times for each sample. Between measurements of each sample, the viscometer was washed and rinsed with distilled water and 99% ethanol. Falling time (seconds) in the capillary tube was 106.65 ± 0.18 and 112.57 ± 0.33 (mean ± SD) for distilled water and natural sea water (diluted to 22 ppt salinity by distilled water), respectively. The relative viscosity of culture water sample was calculated as a ratio of the falling time of the culture sample against that of 22 ppt sea water.

Regression analysis was conducted to examine the relation between culture water viscosity and daily population growth ratio of rotifers.

Effects of viscosity on rotifer performance

We tested the effect of viscosity changes on the reproduction, swimming speed, ratio of swimming to attached individuals and ingestion rate of B. plicatilis.

Viscosity of the experimental sea water was adjusted by the addition of methyl cellulose (13–18 mPa s of a 2% aqueous solution at 20 °C, Wako Chemical Co, Tokyo, Japan) to 0.0125, 0.025, 0.05 and 0.1%. The relative viscosity against 22 ppt diluted sea water was 1 (control, no addition of methyl cellulose), 1.022, 1.031, 1.078 and 1.169. These media were used to determine the effect of sea water viscosity on rotifer fecundity, swimming speed, and ingestion rates. The relative viscosity was measured by a viscometer (see previous section).

Amictic females carrying first laid eggs were prepared following the protocol described by Satuito &

Hirayama (1986). Rotifers were individually transferred to 2 ml N. oculata suspension medium in multi-well plates (Iwaki 24 well tissue culture plate, Tokyo, Japan). Ten rotifers were tested at each viscosity. Maternal females were transferred to newly prepared N. oculata suspension medium once a day. After transfer, neonates were counted and removed. The number of offspring produced during the lifetime as well as an intrinsic rate of natural increase (Birch, 1948) was calculated. The swimming activity at different viscosities was determined in 24- and 96-hour-old rotifers by the method reported by Snell et al. (1987). One minute after the experimental rotifer was transferred to the test medium, the number of 1 mm^2 grids trespassed by a rotifer during a 30 s observation period was monitored and used as an index of rotifer swimming activity.

The ratio of swimming to attached rotifers was determined 24 h after the inoculation of 20 rotifers into 10 ml medium at the varying viscosities. Three replicates for each treatment were prepared with newborn and egg-bearing females.

Ingestion rates (cells rotifer^{-1} h^{-1}) under different viscosity conditions were determined directly on batch cultures. From an exponentially grown 500 ml rotifer culture, 1000 rotifers were pipetted into screw cap glass vials containing 5 ml medium that was adjusted to the tested viscosities. N. oculata was added to adjust the cell density to 7×10^6 cells/ml. Three replicates were prepared, as well as controls that contained N. oculata alone. Cultures were kept under total darkness. After one hour, 0.5 ml 10% formaldehyde solution was added and the algal cell density was counted in a hemacytometer. The ingestion rate was computed using the following equation (Fuller & Clarke, 1936; Nimura, 1963):

$$G_t = V(C_t - C_0) \frac{\log C_t' - \log C_t}{\log C_t - \log C_0},$$

where, G_t is food uptake per animal during time t, C_0 and C_t are food densities of rotifer cultures at time t and 0, respectively, and C_t' is the food density of control cultures (without rotifers) at time t.

One-way analysis of variance (ANOVA) using Systat 7.0 (SPSS Inc.) was conducted to identify significant effects of culture water viscosity on reproduction, swimming activity and ingestion rate. Multiple comparison was conducted using Tukey HSD to determine which treatments were significantly different.

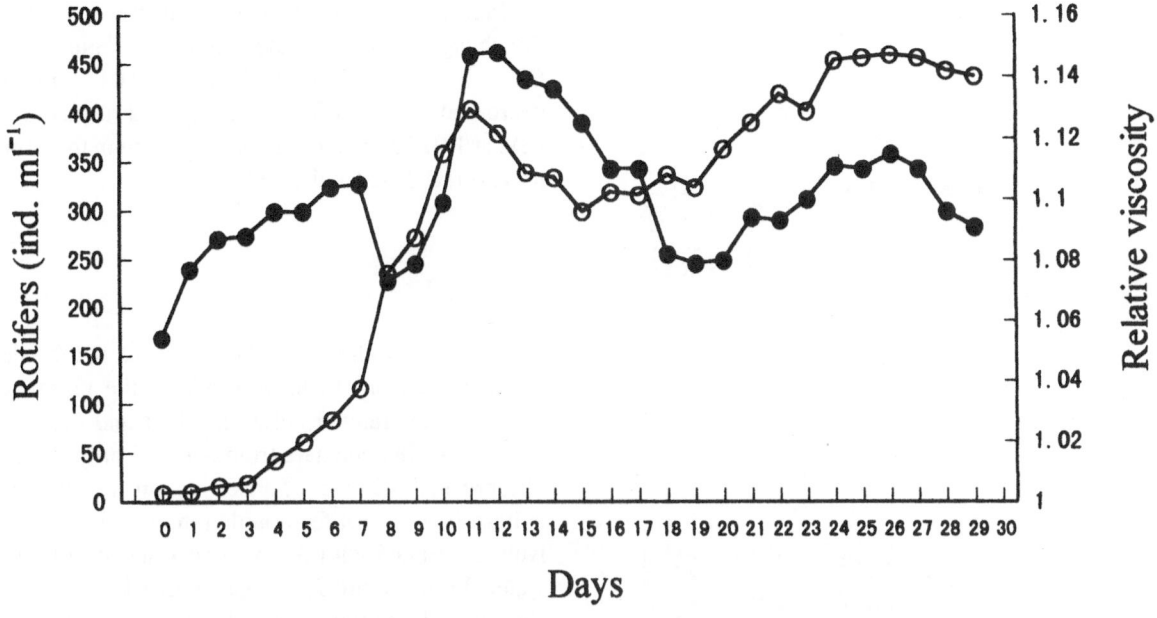

Figure 1. Population growth of *B. plicatilis* (open circles) and relative water viscosity (closed circles) during 29 day culture period.

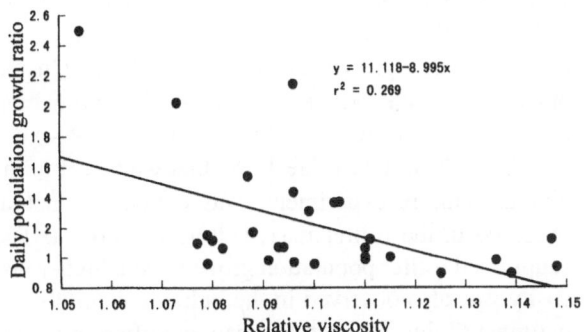

Figure 2. Correlation between viscosity of culture water and daily population growth ratio of *B. plicatilis*.

Table 1. Mean and SD of life span and number of offspring, and intrinsic rate of natural increase (r) of *B. plicatilis* reared in sea water (22 ppt in salinity) with different viscosities by dissolving methyl cellulose. Values that do not share a superscript are significantly different ($p < 0.05$, Tukey HSD test)

Relative vicosity	n	Life span	Number of offspring	r
1	10	8.0 ± 0.0^a	20.8 ± 3.0^a	0.853
1.022	10	7.7 ± 0.5^a	19.4 ± 2.0^a	0.917
1.031	10	7.7 ± 0.5^a	13.1 ± 2.2^b	0.748
1.078	10	8.0 ± 0.0^a	11.8 ± 2.9^b	0.784
1.169	10	6.3 ± 2.0^a	8.1 ± 3.7^c	0.640

Results

Figure 1 presents daily data of rotifer population size and water viscosity in a 500 ml culture. Through the inoculation of rotifers at 10 ind ml^{-1} and feeding of *N. oculata* at 7×10^7 cells ml^{-1} on day 0, culture viscosity increased 1.054 times. The R.V. value increased for the initial 7 days, and it decreased when rotifers grew exponentially on day 8. After day 8, viscosity changes corresponded to the population growth pattern. The highest viscosity was 1.148 on day 13 when rotifer density was 380 ind/ml. Negative correlation was detected between daily population growth rates and R.V. values (Figure 2, $y = 11.118 - 88.995x$, $r^2 = 0.269$, df = 28, $F = 10.25$, $P < 0.01$).

Life table characteristics of amictic females cultured in sea water at different viscosities are presented in Table 1. Viscosity of culture water significantly affected the life span and number of offsprings (ANOVA: life span – $F = 5.62$, df = 4, $P < 0.01$; number of offsprings – $F = 36.03$, df = 4, $P < 0.01$). There were no significant differences in life span of amictic females except when cultured at R.V. = 1.169. Compared to controls, the number of offspring was significantly reduced at viscosities higher than 1.031. The intrinsic rate of natural increase (r) ranged between 0.640 (R.V. = 1.169) and 0.917 (R.V. = 1.022).

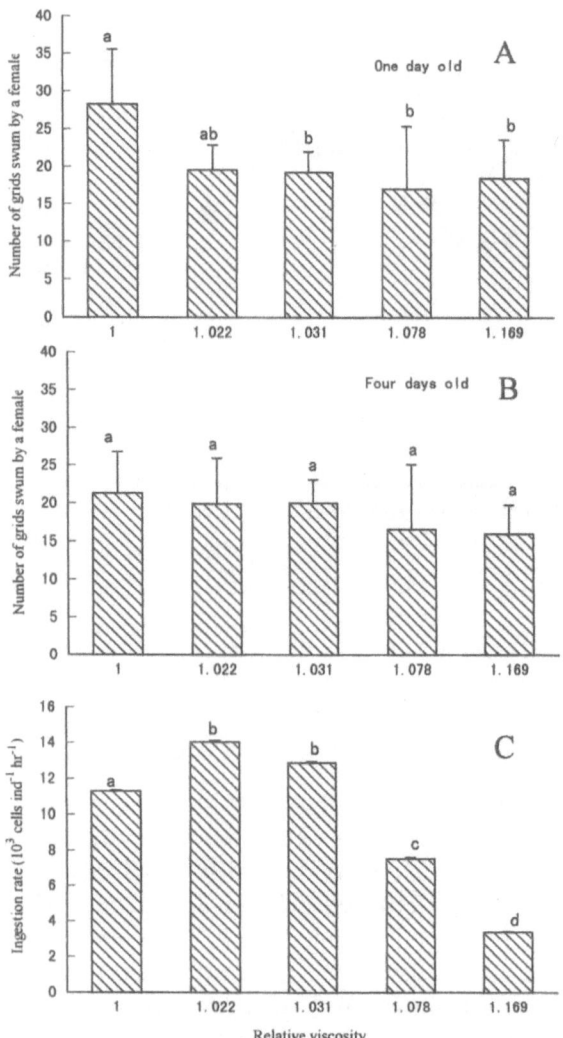

Figure 3. Number of 1 mm² grids swum by one day old (A) and four days old (B) *B. plicatilis* females at different viscosities, and ingestion rate (cells ind⁻¹ h⁻¹) at different viscosities (C). Vertical bars indicate standard deviation. Values that do not share a superscript are significantly different ($P < 0.05$, Tukey HSD test).

The number of 1 mm² grids swum by a rotifer is shown in Figure 3. Viscosity significantly influenced the swimming speed for 24-hour-old rotifers (ANOVA, $F = 4.21$, df = 4, $P < 0.01$). Such effect was detected only against the control (R.V. = 1). Rotifers swam slower when the relative water viscosity (R.V.) was higher than 1.031. The number of grids swum by a rotifer did not differ when the R.V. was higher than 1.022. For 96-hour-old rotifers it did not differ among treatments. The ratio of swimming to attached individuals was not affected by culture viscosity in either newborn or egg-bearing females.

Ingestion rates were also significantly affected by the change of culture water viscosity (Figure 3; ANOVA, $F = 165.32$, df = 4, $P < 0.01$). Ingestion decreased when rotifers were exposed to media with R.V. at 1.078 and 1.169. It was higher than the control, however, at 1.022 and 1.031.

Discussion

In the batch culture condition where rotifer density changed from 10 to 460 ind/ml, the viscosity of culture water ranged between 1.054 and 1.143. Although we obtained data from one trial only, we could observe the increase of viscosity during the rotifer culture period. Rotifer population growth was negatively correlated with the relative viscosity of culture water. From Figure 2, it appears that lower viscosity (less than 1.11) more strongly affected rotifer reproduction. Only small changes occur with increasing the viscosity from 1.11 to 1.15. Factors other than viscosity, such as free ammonia concentration (Yu & Hirayama, 1986) and bacterial flora (Yu et al., 1990), should be taken into account in explaining the low population growth in the batch culture. On the other hand, predominance of vitamin B₁₂ producing bacteria stabilizes rotifer cultures (Yu et al., 1994).

The highest R.V. value 1.169 that we tested in individual culture experiments was equivalent to that observed in the rotifer mass culture (Figure 1). The suppressed rotifer population growth at the higher viscosity was also observed in the individual culture experiment (Table 1). Both lifespan and offspring number of individually cultured amictic females decreased with an increase of viscosity.

In order to clarify the mechanism of the lower rotifer reproduction at the higher viscosity, we examined swimming speed and ingestion rate of rotifers. These are recognized as indicators to assess physiological stress on rotifers from a variety of environmental factors (Snell et al., 1987; Juchelka & Snell, 1994) and as important components in the energy budget of rotifers that determines reproductive potential. A decrease in swimming speed was observed for younger rotifers (Figure 3) having a lower Reynold's number which are more susceptible to the effect of viscosity increase. The ingestion rate decreased markedly when R.V. was more than 1.078 (Figure 3). These data indicate that the increase of culture viscosity functions as sublethal stress for rotifers that decreases their reproductive ability. In our experiment, the ingestion rate of rotifers was faster than that of the control (R.V. = 1) when R.V. was 1.022 and 1.031. The mechanism of this action is unknown, but similar results were reported by Juchelka & Snell (1994) who measured ingestion rates when rotifers were exposed to several types of toxicants. Former publications (Hirayama & Ogawa, 1972; Hirayama et al., 1973; Starkweather, 1988) also met with the difficulty of directly correlating *B. plicatilis* population growth with its ingestion rate at different food densities.

A problem encountered in this study was that the regulation of water viscosity can be adjusted only by dissolving chemicals which may have a toxic effect on animals. In addition to methyl cellulose, we also employed sodium alginate to regulate water viscosity. However, the addition of sodium alginate resulted in 50–60% mortality of rotifers during the initial three days. With methyl cellulose, there were no significant differences in life span of amictic females except when cultured at R.V. = 1.169. Methyl cellulose is generally a non-toxic compound which is widely used for human food processing and cosmetics. We suspect that dissolved peptides may cause an increase in viscosity in rotifer culture water. We are attempting to clarify the characteristics of these peptides and to reduce their effect on the R.V. of culture water by the addition of various enzymes.

Phytoplankton mucus may kill marine fish by clogging their gills (Matsusato & Kobayashi, 1974). The mucilage produced by diatoms restricted fish trawling (Boalch & Harbour, 1977). Jenkinson (1986) explained these mechanisms by investigating non-newtonian property of laboratory phytoplankton cultures. The relative viscosity of sea water collected from an area where abundant decomposing macroalgae existed was 1.015 (M. Hamada, pers. comm.). In this experiment, the inoculation of rotifers on day 0 at 10 ind ml^{-1} and feeding of *N. oculata* at 7×10^6 cells ml^{-1} immediately caused a 5.4% increase in culture viscosity. The viscosity coefficients of freshwater and seawater (36 ppt S) increase from 0.802 and 0.866 mPa s to 1.794 and 1.890 mPa s, respectively, when the water temperature decreases from 30 to 5 °C (Miyake & Koizumi, 1948; Ohgushi, 1981). We conducted the current experiments at 25 °C, since most of the rotifer mass cultures are maintained around this temperature. In order to understand the rotifer life cycle in nature, however, it is of interest to continue research at lower temperatures where viscous forces predominate. Such a research also provides information for live food culturing conducted at colder regions.

Rotifers are widely used as a live food source for feeding fish larvae. Production of larvae, therefore, depends on the productivity and stability of rotifer cultures. Factors that cause instability of the rotifer mesocosm have been reviewed by several authors (Hirayama, 1987; Maeda & Hino, 1991). Among them, an increase of free ammonia, a change of bacterial flora and contamination by protozoan species are critical. Several methods, such as measurements of egg ratio (eggs per female) and swimming activity, have been established for the rapid assessment of the rotifer mass culture status (Snell et al., 1987). Current study suggests that the measurement of water viscosity can also be a means of detecting stress on rotifers by environmental conditions, where viscous substances are comparatively abundant. Recent developments enable rotifer mass cultures to reach more than 10 000 ind ml^{-1} mainly because of the availability of ample amounts of a preserved phytoplankton diet (Yoshimura et al., 1996, 1998). Increase of viscosity could be more significant in such an extreme environment.

Acknowledgments

We would like to express our thanks to Professors M. Hamada and T. Muramatsu for the guidance in viscosity measurement, and reviewers for improving the manuscript. A part of this study was supported by a grant-in-aid from the Ministry of Education, Culture and Science, Japan (no. 08660235).

References

Birch, L. C., 1948. The intrinsic rate of natural increase of insect populations. J. anim. Ecol. 17: 15–26.

Boalch, G. T. & D. S. Harbour, 1977. Unusual diatom off the coast of south-west England and its effect on fishing. Nature 269: 687–688.

Fuller, J. L. & G. L. Clarke, 1936. Further experiments on the feeding of *Calanus finmarchicus*. Biol. Bull. 70: 308–320.

Hagiwara, A., K. Hamada, S. Hori & K. Hirayama, 1994. Increased sexual reproduction in *Brachionus plicatilis* (Rotifera) with the addition of bacteria and rotifer extracts. J. exp. mar. Biol. Ecol. 181: 1–8.

Hirayama, K., 1987. A consideration of why mass culture of the rotifer *Brachionus plicatilis* with baker's yeast is unstable. Hydrobiologia 147: 269–270.

Hirayama, K. & S. Ogawa, 1972. Fundamental studies on physiology of rotifer for its mass culture– I. Filter feeding of rotifer. Bull. Jap. Soc. Sci. Fish. 38: 1207–1214.

Hirayama, K., K. Watanabe & T. Kusano, 1973. Fundamental studies on physiology of rotifer for its mass culture – III. Influence of phytoplankton density on population growth. Bull. Jap. Soc. Sci. Fish. 39: 1123–1127.

Jenkinson, I. R., 1986. Oceanographic implications of non-newtonian properties found in phytoplankton cultures. Nature 323: 435–437.

Juchelka, C. M. & T. W. Snell, 1994. Rapid toxicity assessment using rotifer ingestion rate. Arch. envir. Contam. Toxicol. 26: 549–554.

Koehl, M. A. R. & J. R. Strickler, 1981. Copepod feeding currents: Food capture at low Reynolds number. Limnol. Oceanogr. 26: 1062–1073.

Maeda, M. & A. Hino, 1991. Environmental management for mass culture of rotifer, Brachionus plicatilis. In W. Fulks & K. L. Main (eds), Rotifer and Microalgae Culture Systems. Proceedings of a U.S.–Asia Workshop, The Oceanic Institute, Honolulu, 1991: 125–133.

Mann, K. H. & J. R. N. Lazier, 1991. Dynamics of marine ecosystems. Blackwell Scientific Publications, Cambridge, USA, 466 pp.

Matsusato, T. & H. Kobayashi, 1974. Studies on death of fish caused by red tide. Bull. Nansei Fish. Res. Lab. 7: 43–67.

Miyake, Y. & M. Koizumi, 1948. The measurement of the viscosity coefficient of sea water. J. mar. res. 7: 63–66.

Nimura, Y., 1963. Effects of different concentrations of Chlamydomonas on the grazing and growth of Artemia with reference to the salinity of its living medium. Bull. Jap. Soc. Sci. Fish. 29: 424–433.

Ohgushi, M., 1981. Theory of Ships, Vol. III. Kaibundo Book Company, Tokyo, Japan, 280 pp. (In Japanese).

Rubenstein, D. I. & M. A. R. Koehl, 1977. The mechanism of filter feeding: Some theoretical considerations. Am. Nat. 111: 981–994.

Satuito, C. G. & K. Hirayama, 1986. Fat soluble vitamin requirements of the rotifer Brachionus plicatilis. In J. L. Maclean, L. B. Dizon & L. V. Hosillos (eds), The First Asian Fisheries Forum. Asian Fisheries Society, Philippines: 619–622.

Snell, T. W., M. J. Childress & E. M. Boyer, 1987. Assessing the status of rotifer cultures. J. World Aquacult. Soc. 18: 270–277.

Starkweather, P. L., 1988. Reproductive and functional responses of the rotifer Brachionus plicatilis to changing food density. Verh. int. Ver. Limnol. 23: 2001–2005.

Starkweather, P. L., 1995. Near-coronal fluid flow patterns and food cell manipulation in the rotifer Brachionus calyciflorus. Hydrobiologia 313/314: 191–195.

Yoshimura, K., A. Hagiwara, T. Yoshimatsu & C. Kitajima, 1996. Culture technology of marine rotifers and the implications for intensive culture of marine fish in Japan. Mar. Freshwat. Res. 47: 217–222.

Yoshimura, K., K. Usuki, T. Yoshimatsu, C. Kitajima & A. Hagiwara, 1997. Recent development of high density mass culture system of the rotifer Brachionus rotundiformis. Hydrobiologia 358: 139–144.

Yu, J.-P. & K. Hirayama, 1986. The effect of un-ionized ammonia on the population growth of the rotifer in mass culture. Nippon Suisan Gakkaishi 52: 1509–1513.

Yu, J.-P., Hino, A., Noguchi, T. & H. Wakabayashi, 1990. Toxicity of Vibrio alginolyticus on the survival of the rotifer Brachionus plicatilis. Nippon Suisan Gakkaishi 56: 1455–1460.

Yu, J.-P., K, Hirayama & A. Hino, 1994. The role of bacteria in mass culture of the rotifer Brachionus plicatilis. Bull. Natl. Res. Inst. Aquaculture, Suppl. 1: 67–70.

Hydrobiologia **387/388**: 495–498, 1998.
E. Wurdak, R. Wallace & H. Segers (eds), Rotifera VIII: A Comparative Approach.
© *1998 Kluwer Academic Publishers.*

Probiotic culture of the rotifer *Brachionus plicatilis*

H. Hirata[1], O. Murata[2], S. Yamada[2], H. Ishitani[2] & M. Wachi[2]
[1]*Faculty of Agriculture, Kinki University, Nakamachi 3327-204, Nara 631-8505, Japan*
[2]*Institution of Aquaculture, Kinki University, Shirahama 3153, Wakayama 649-2211, Japan*

Key words: Rotifera, *Brachionus*, probiotic materials, ecosystem, Uchishiro organisms

Abstract

We adapted a probiotic culture system originally developed for domestic animals and fish for the intensive culture of rotifers. In this modification, a crude starter fluid was made by incubating a mixture of some 40 species of bacteria (so-called Uchishiro Microorganisms) with the by-products of other food-related materials. The probiotic culture medium (PCM) was then completed by incubation of 50 g of starter fluid (25 °C) for 3–4 days in 1 L of 50% sea water with strong aeration. After processing, the medium may be used as food for the rotifer *Brachionus plicatilis*. We determined the optimal levels for feeding *B. plicatilis* cultures by providing a starter culture of 100 rotifers different concentrations of PCM and monitoring their population growth over five days. The highest population density of *B. plicatilis* (351 ind ml^{-1}) was achieved from a mixture of 20 ml PCM per 1 L rotifer water. From these experiments, we conclude that PCM is useful as food for rotifer culture if it is supplied at a satisfactory concentration.

Introduction

Probiotic culture systems (PCS) have been developed for a variety of domestic animals and fish (Kosasa, 1990; Fukushima & Nakano, 1995; Hirata, 1996; Nakano, 1996). The research we discuss here was conducted in order to determine whether that methodology could be applied to the culture of rotifer species important in intensive aquaculture systems. The special features of this technology are that probiotic medium is produced from materials that have little commercial use and that the final product is fed to the rotifer culture in the form of a liquid. PCS may provide the means to develop more economic ways of producing rotifers as food for commercially important species.

In this paper we describe the methods we used to produce the probiotic culture medium (PCM) and the outcome of two preliminary studies: (1) observations on the population density of bacteria growing in the PCM (Experiment 1) and (2) experiments to determine the optimal level of PCM to be fed to the rotifer *Brachionus plicatilis* (Experiment 2).

Materials and methods

Production of PCM

The method is based on the use of recycled organic material from other probiotic systems used in fish aquaculture (Figure 1). PCM was made by mixing 10 kg of heads of cultured fish (by-products of cultured yellowtail *Seriola quinqueradiata*), 10 kg of cultured *Ulva pertusa* (by-product of sea water purification in the fish farms), and 10 kg of dried bean curd refuse (by-products of food products), with 3 g of dried 'Uchishiro microorganisms' (UM) in a fermentator (Gomi-luck 50) for 4 h at 70 °C. UM have been used as organic fertilizer in 'Uchishiro farms' (Nagano Prefecture, Japan) for about 40 years. It consist of about 40 species of aerobic bacteria such as *Bacillus subtilis*, *B. licheniformis*, *B. cereus* and *B. macerans*. The material is processed to a dried powder which is ground into particles of about 1 mm diameter. The probiotic medium is then made by dissolving 50 g of the dried powder in 1 L of brackish water (=17‰ salinity). The medium is then incubated with strong aeration for 4 days.

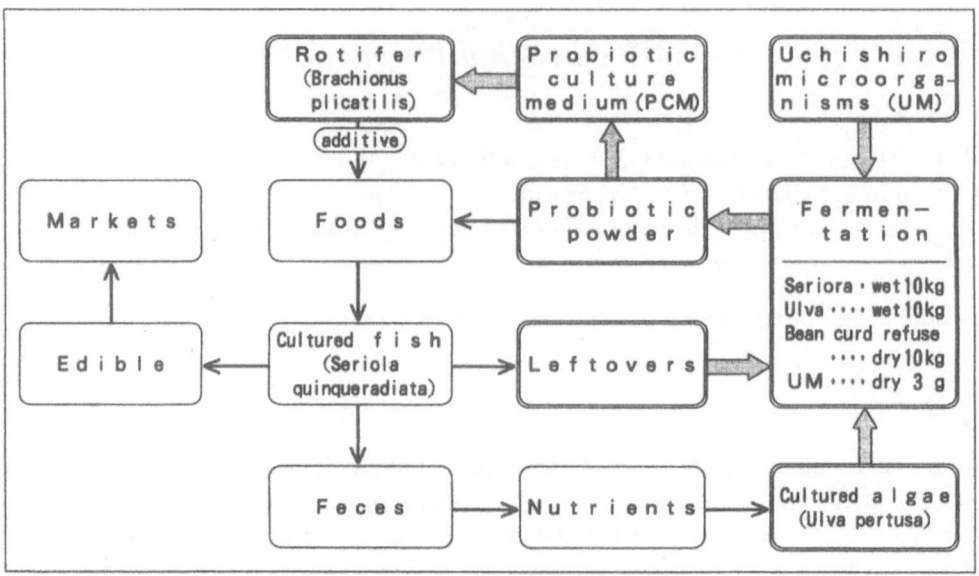

Figure 1. Flowchart for the probiotic culture of rotifers.

Determination of optimal feeding levels

In Experiment 1, bacterial population density during the production of the PCM was observed every other day by counting the number of colony forming units (CFUs) using standard microbiological methods for enumerating bacteria.

In Experiment 2 we prepared four tanks of the rotifer culture medium, each with a different amount of the PCM: Tank 1 – 10% (100 ml PCM for 1 L culture water); Tank 2 – 2% (20 ml PCM per 1 L); Tank 3 – 1% (10 ml PCM per 1 L); Tank 4 – no PCM. Two feeding frequencies were tested. Series A was fed on the first day only, and Series B was fed every other day. The initial rotifer population density in all tanks was about 100 individuals per ml. All rotifer cultures were in 3 units of 1 L culture water in 2 l transparent plastic bottles. Water temperature was held at 25 °C using a water bath. The light conditions of the culture room were not controlled.

Results

Experiment 1

The growth of bacteria (as CFU ml^{-1}) in the PCM in Experiment 1 are shown in Figure 2. The initial number of CFUs in each medium was about $10^{6.5}$ CFU ml^{-1}, and increased to $10^{11.5}$CFU ml^{-1} at day 4. Thus,

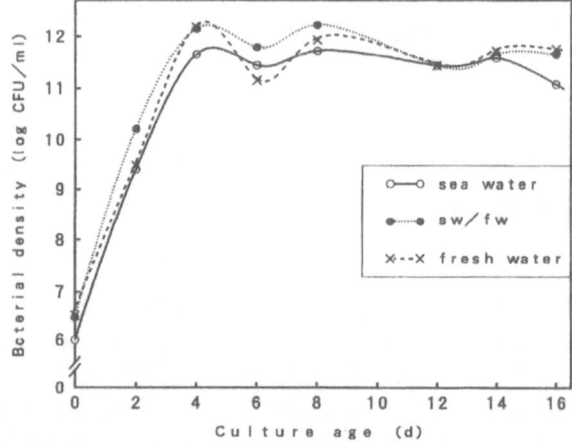

Figure 2. Bacterial density in probiotic fluid.

the generation time for bacterial growth in incubating PCM is about 5.8 h. The density of CFUs remained about the same thereafter.

From these results, we decided that the PCM cultures could be fed to *B. plicatilis* beginning on the 4th day. CFU showed no difference with respect to salinity, which varied from 0 to 34‰. This result suggests that UM consist of both marine and freshwater bacteria and that brackish water could be substituted as culture medium for the PCM.

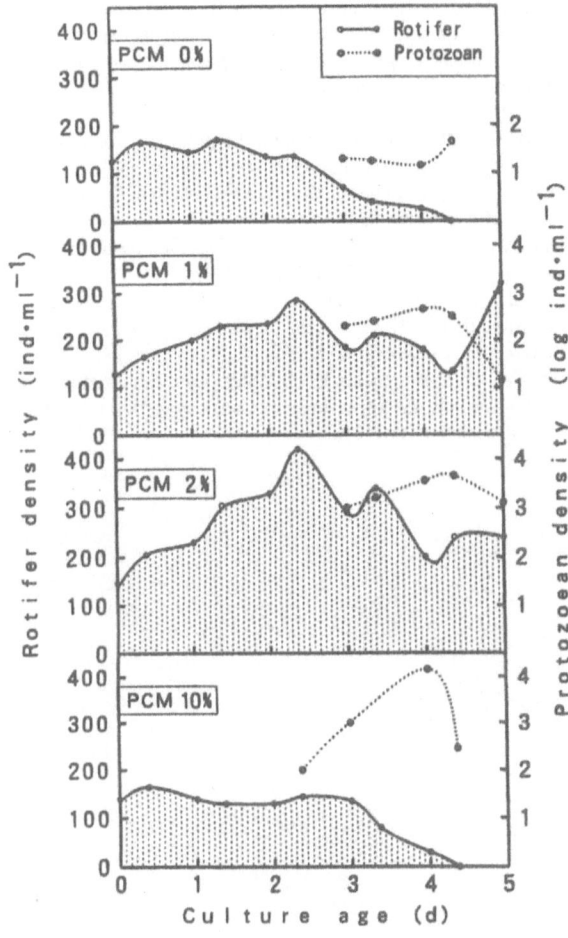

Figure 3. Daily variation of the population densities of rotifer and protozoan daily fed different amounts of PCM (Probiotic Culture Medium).

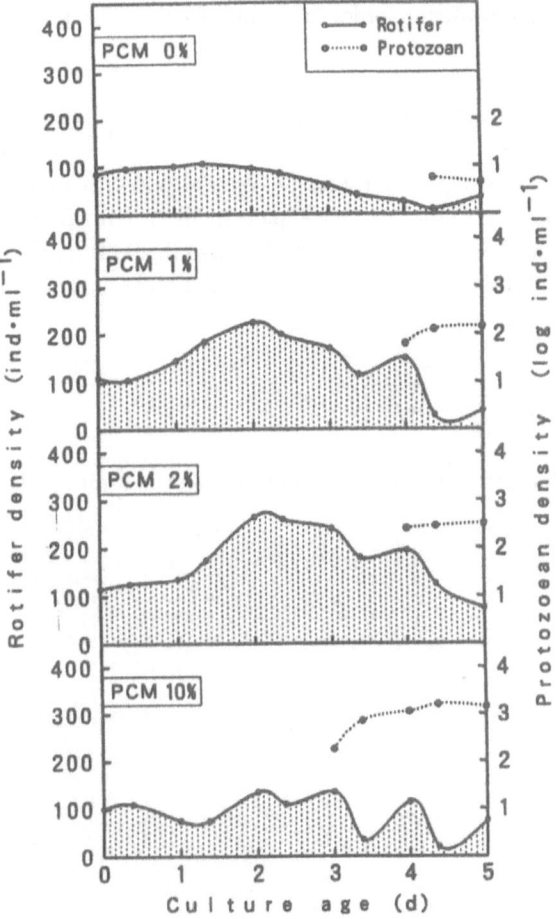

Figure 4. Daily variation of the population densities of rotifer and protozoan once (at initiation of culture) fed different amounts of PCM.

Experiment 2

Results obtained from both series A and B in Experiment 2 are presented in Figures 3 and 4, respectively. In series A, the rotifer densities in the 1 and 2% feeding groups showed rapid growth until the 2nd day. Thereafter, the densities decreased while the density of protozoans increased. These results suggest a negative correlation between rotifer and protozoan. This may be due to competition between rotifers and protozoans. In contrast, the rotifer density in the 0 and 10% feeding groups decreased and stood at 10 and 5 ind ml^{-1} and 245 ind ml^{-1} on the 2nd day, and further decreased gradually to zero and 10 ind ml^{-1}, respectively. These results are most likely due to either lack, or excess of food, respectively. Basing on the above-mentioned results, we judged that feeding with 1–2% PCF per day is the optimum feeding rate.

Discussion

It is well known that *B. plicatilis* can grow by feeding on microorganisms such as yeast and bacteria (Hirata & Mori, 1967; Hirata et al., 1982; Ushiro & Hino, 1989; Yamasaki & Hirata, 1990). Rao (in lit.) reported on the growth of rotifers in waste waters, and found that rotifers can grow on the microbial load therein. Our work confirms these reports. On the other hand, probiotic materials have been used recently as food supplement in fish culture. The fish fed such probiotics grew well without requiring drugs added (Kosasa, 1990; Nakano, 1996). The present experiments sug-

498

gest that recycled probiotic bacterial cultures can be
fed to rotifers in order to produce food for fish culture.

Acknowledgments

We acknowledge Elizabeth Wurdak for providing us
the opportunity to participate in the VIIIth International Rotifer Symposium. We also thank Terry Snell
and Robert Wallace who read and improved the manuscript.

References

Hirata, H., 1996. Probiotic culture of fish. Yoshoku 33(3): 142–145.

Hirata, H. & Y. Mori, 1967. Culture of rotifer *Brachionus plicatilis* fed on baker's yeast. Saibaigyogyo 5(2): 36–40.

Hirata, H., M. Ushiro & I. Hirata, 1982. Ecological succession of *Chlorella saccarophila*, *Brachionus plicatilis*, and autogenous bacteria in culture water. Mem. Fac. Fish., Kagoshima Univ. 31: 153–160.

Kosasa, M., 1990. Recent works on the probiotics for feeding in domestic animals. Feeding 30(7): 41–47.

Nakano, M., 1996. Studies on the probiotic culture of flounder. Proc. Symp. Ecol. Aquacul. 1–4.

Ushiro, M. & A. Hino, 1990. Microbial consumption by the rotifer *Brachionus plicatilis*. Kaiyo 22(1): 20–27.

Yamasaki, S. & H. Hirata, 1990. Relationship between food consumption and metabolism of rotifer *Brachionus plicatilis*. Nippon Suisan Gakkaishi 56: 591–594.

Yamasaki, S. & H. Hirata, 1993. Water quality maintenance and food recycling the waste matter of a rotifer ecocystem culture. Suisanzoshoku 41: 7–11.